GENERAL, ORGANIC, AND BIOLOGICAL CHEMISTRY

GENERAL, ORGANIC, AND BIOLOGICAL CHEMISTRY

CHEMISTRY FOR THE LIVING SYSTEM

M. LYNN JAMES

The University of Northern Colorado

JAMES O. SCHRECK

The University of Northern Colorado

JAMES N. BEMILLER

Southern Illinois University at Carbondale

D. C. HEATH AND COMPANY   Lexington, Massachusetts   Toronto

PREFACE

General, Organic, and Biological Chemistry is designed for use by students pursuing careers in the health sciences, including nursing, medical laboratory technology, health and physical education, home economics, dietetics, and environmental science. Thus, a major goal of the authors is to develop understanding of the chemical basis of life. The book surveys the basic principles and concepts of general, organic, and biological chemistry, and integrates applications of chemical principles and concepts through discussions of the molecular and physical bases of *physiological processes* and *clinical applications*. Only arithmetic and simple algebra are used in the necessary formula calculations.

Whenever possible we have discussed chemical principles within the context of health-related applications: caisson disease (diver's bends) illustrates the effect of changes in pressure on a gaseous solute; the use of embalming fluid to preserve biological specimens provides an interesting setting for determining the molarity of a solution; the effect of freezing on plant tissues demonstrates changes in state; distortion of the eardrum associated with inflammation of the Eustachian tube serves as an example of Boyle's Law. We discuss the nutritional aspects of the elements, the carcinogenic nature of hydrocarbons, blood substitutes, food additives, metabolic diseases, drugs, vitamins, hormones, and a variety of general applications throughout the text and in examples and problems.

Terms and their *definitions* are highlighted. Selected definitions are given in the margin of the page where the term (printed in boldface) is first used. Other definitions and terms (italicized for easy identification) are given within the body of the discussion. Important terms and concepts are defined and discussed the first time they are used, and reviewed later as needed.

Learning objectives are placed at the beginning of each major section to communicate in a consistent, orderly, and efficient manner what the authors expect students to learn from the section. The learning objectives have been derived from an analysis of knowledge and skills required in later sections and what others expect from students who have completed courses in chemistry for the health sciences. It is intended that, in addition to the specific behavior called for in a learning objective, students will also be able to demonstrate related behaviors. For example, although an objective might state "Define saturated fatty acid and unsaturated fatty acid" a student, in addition to writing definitions, could also be expected to select a correct definition from a list of possible definitions, determine if a statement is true or false, draw the structures of examples of saturated and unsaturated fatty acids, identify structures of saturated and unsaturated fatty acids, and when given structures of fatty acids, classify them as being saturated or unsaturated.

Problems (and questions) based on the learning objectives are provided at the end of almost every section. Answers follow these problems, giving students prompt and frequent feedback regarding their success in learning. Additional problems follow the summary of each chapter; answers to approximately one-half of these "Additional Problems" are given in the *Study Guide.*

Chapter summaries review important topics covered within the chapter, and introduce the next chapter. *Figures, tables,* and *key terms* are emphasized through the use of a *two-color display.*

Topics are presented in the following order: Chapters 1–9 cover the fundamental principles of general chemistry; Chapters 11–17 deal with organic chemistry; and Chapters 19–28 are concerned with biological chemistry. Chapter 10 provides a bridge from general to organic chemistry, and Chapter 18, from organic to biological chemistry.

This book is suitable for a full-year course. No background in chemistry is necessary. However, there is sufficient material to meet the needs of courses that have a high-school chemistry prerequisite. It is possible for instructors to delete certain sections or even complete chapters so that the book can be used in courses that are less than a full year in length.

Necessary mathematical skills are introduced in Chapter 1 and in Appendix A, and they are applied to specific topics through worked-out examples and exercises. The information necessary to write and manipulate numbers using scientific (exponential) notation is provided in Appendix A for those not already familiar with it. Only a rudimentary understanding of logarithms is needed for dealing with pH values, with the discussion concentrating on integer values of pH. (More exact noninteger values are referred to where appropriate for those desiring to use them.)

A *Study Guide* is available for use with this text. It includes additional questions and problems (with answers) to supplement those given in the text, answers (and sample solutions) to selected text problems, and self-tests with answers.

A *Laboratory Guide* is also available. It complements the text and provides practical laboratory experience in chemistry, with particular emphasis on common chemical techniques, especially those used in the health-related sciences.

We wish to express our appreciation to the reviewers whose ideas contributed to the improvement of the final manuscript: Robert Umans, Wellesley College; Neil Coley, Chabot College; Scott Mohr, Boston University; and Ethelreda Laughlin, Cuyahoga Community College. We also thank the staff of D. C. Heath and Company for their assistance. Most of all, however, we want to express our appreciation to our wives and children for their support and understanding during the many hours, days, and weeks involved in the preparation of this textbook. Without their support the project could never have been completed.

M. Lynn James
James O. Schreck
James N. BeMiller

CONTENTS

CHAPTER 1

SOME FUNDAMENTAL CONCEPTS AND TOOLS OF CHEMISTRY

1-1 CHEMISTRY AND YOU

An understanding of chemistry is important in many aspects of life and is especially significant when one is studying the living system. Applying chemical principles, a technician in a medical laboratory can determine the type of poison that a child has swallowed, a nurse can appreciate the meaning and significance of a factor known as blood pH, a doctor can prescribe drugs wisely, and a dietitian can design a proper diet.

Professionals in the health-related sciences are constantly called upon to make independent decisions concerning the solution of a variety of problems related to the care and treatment of others. In many cases they are expected not only to make wise decisions for themselves but also to guide auxiliary personnel whom they supervise. While performing these duties, you will not always find a policy, a "rule of thumb," or a person in authority for guidance when a problem arises. On such occasions, a sound understanding of the chemistry at work in the living system is essential. The basis for maintaining and restoring humans to good health lies in the chemical principles presented in this text.

Chemistry and chemicals are part of every aspect of life from the natural and synthetic materials that compose our environment to the processes within us that enable life to continue. Therefore, the study of chemistry not only has important implications in your future professional life but also can be an interesting and exciting experience in learning more about ourselves and the world in which we live.

We will begin our study of chemistry by considering a few basic scientific concepts. We will then discuss the concept of variables and the methods that a scientist uses to analyze the living system. Next we will look at two systems of measurement used in science and in our everyday life, and finally, we will consider some chemical calculations and the concepts of density and specific gravity.

1-2 BASIC CONCEPTS

LEARNING OBJECTIVES

1. Define chemistry, matter, the Law of Conservation of Energy, and the Law of Conservation of Mass.
2. Define and identify examples of chemical and physical properties, variables, dependent and independent variables, hypotheses, laws, chemical equations, theories, and models.

Chemistry is involved in the dyes that give colors to clothing, the foods that we eat, the burning of gasoline in our cars, the growth of plants, and the varied processes that take place in our bodies. An understanding of chemistry can, therefore, help us in many ways. We can remove stains from our clothing if we know what chemical to use to clean them and we can avoid being seriously burned if we know that a lighted cigarette can cause gasoline to explode. In this sense, we are all chemists, practical chemists, who daily apply chemical principles to our lives.

> **Chemistry** is the study of matter and the changes that matter undergoes.

1-2.1 CHEMISTRY, MATTER, AND ENERGY

The world around us is made up of matter. Matter exists in three forms called *solids, liquids,* and *gases.* In general, any substance that is not decomposed by heat can exist in any of these three forms. Water is such a substance. As ice it is classified a solid because it is hard and, for practical purposes, has a constant volume. The water we drink is, of course, a liquid having a more or less constant volume that assumes the shape of its container. When we boil water the steam that rises above the container is in the gaseous state of water. We sometimes refer to this as *water vapor.* Gases do not have a constant volume and they fill a container of any size.

> **Matter** is anything that occupies space and has mass.

When studying matter we consider its characteristics, or properties. The properties of matter fall into two general categories called chemical and physical properties. The tendencies of toast to burn and of bleach to remove stains are examples of chemical properties. Typical physical properties are color, taste, hardness, temperature, and odor.

> **Chemical properties** are those characteristics associated with the tendency of a substance to undergo a change through which a different substance is formed.

The changes that matter undergoes fall into the same categories as their properties, *chemical* and *physical.* An example of a chemical change is the reaction of glucose ($C_6H_{12}O_6$) with oxygen (O_2) in the body to produce carbon dioxide (CO_2) and water (H_2O). This reaction is shown by the following expression, called a chemical equation:*

> **Physical properties** are characteristics that do not involve the tendency of a substance to change to a different substance.

$$C_6H_{12}O_6 + 6\,O_2 \longrightarrow 6\,CO_2 + 6\,H_2O + \text{heat} \qquad \textbf{(1.1)}$$

The melting of ice (solid water) and the evaporation of perspiration are both illustrations of physical changes since different substances are not formed in either case.

> A **chemical equation** is a group of symbols and numbers that indicate the types and amounts of reacting substances and substances produced in a chemical change.

---

*As illustrated by equation 1.1, a chemical equation is a type of shorthand notation for a sentence. We can represent the statement *two plus two equals four* by $2 + 2 = 4$, where the symbols 2 and 4 stand for the words *two* and *four* respectively. Similarly, equation 1.1 reads *one molecule of glucose reacts with six molecules of oxygen to produce six molecules of carbon dioxide plus six molecules of water,* where the symbols $C_6H_{12}O_6$, O_2, CO_2, and H_2O stand for the appropriate chemical species. More will be said about chemical symbols, molecules, and equations in Chapters 3 and 4.

The combined effect of the "burning" of glucose (a chemical change), as illustrated by equation 1.1, and the cooling due to the evaporation of perspiration (a physical change) helps us to maintain the fairly constant body temperature of 98.6°F.

Most chemical and physical changes are accompanied by some type of energy change. For example, the melting of ice when it is exposed to sunlight involves a conversion of radiant energy from the sun into the kinetic energy of the moving particles of water. In this case the radiant energy is absorbed by surface particles of ice. This enables the particles to break away from the rest of the ice and become liquid. In the liquid state, the particles move about somewhat at random. Similarly, the warmth of the body is maintained when chemical energy in the reacting chemicals is converted into heat energy. For example, when glucose reacts with oxygen, as illustrated by equation 1.1, heat energy is produced with the CO_2 and the H_2O. In both the melting of ice and the burning of the glucose, no total energy is lost or gained when all forms of energy are considered. This fact is known as the Law of Conservation of Energy. In these and all such processes the form of the energy may change, but the total amount of energy remains the same.

> The **Law of Conservation of Energy** states that no energy is created or destroyed during chemical or physical changes.

A similar rule of conservation applies to the amount of matter involved in chemical and physical changes. When a log burns in a fireplace, the chemical reaction is comparable to the reaction of glucose with oxygen (equation 1.1), except that some of the wood does not react completely and is left in the form of ashes. According to the most accurate weighing instruments available, the total weight of the CO_2, H_2O, and ashes is known to be the same as the original weight of the wood plus the weight of the oxygen used to burn the log. The fact that the weight, or mass, remains the same is known as the Law of Conservation of Mass.*

> The **Law of Conservation of Mass** states that no matter is created or destroyed in a chemical or physical change.

1-2.2 SCIENTISTS AND THE SCIENTIFIC APPROACH

Scientists in general and chemists in particular have the ultimate goal of arriving at a correct understanding of the nature of things. A scientist should perform investigations with an open mind and accept the results of experiments in an unbiased manner. This is not always easy. Scientists sometimes read into their results what they hope to find rather than what the results actually suggest. An illustration of how this can happen is the various opinions that have been expressed concerning the use of marijuana as a hallucinogenic drug. Scientists and nonscientists alike have taken stands on both sides of the question of whether the use of marijuana is harmful to the body. Some claim that the use of marijuana is a perfectly harmless pastime. Others feel that it can bring about permanent damage to such things as the heredity-determining chromosomes within the body cells. It may take years to find the final answer to this question.

We should be careful to base our conclusions on acceptable scientific research

---

*Scientists now believe that matter is actually a special form of energy and that matter can be converted into other forms of energy and vice versa through nuclear reactions. It is believed that in every chemical process involving an energy change there is an associated change in the amount of matter. In such processes, however, the sum of "matter energy" plus other forms of energy remains constant. Thus, there is really only one law of conservation, that of energy. The amount of matter being produced or lost in these chemical processes is so small, however, that the change cannot be detected by available instruments and will be ignored in this book.

and then be willing to change when new evidence indicates that an earlier conclusion is wrong. This is especially important when controversial issues such as the use of marijuana are being considered.

Scientists organize their thinking in various ways to simplify and make the world more understandable. One way this is done is by identifying and controlling the variables that are being studied. Some variables commonly studied in the health-related sciences are body temperature, blood pressure, and weight. Variables fall into two main categories, dependent and independent variables. For example, your pulse rate changes as you exercise, when you are frightened, and as you age. Your pulse rate is considered the dependent variable, since it is altered by a change in the other variables. Your age is an independent variable since it changes by itself, while exercise and fright are independent variables since they cause a change in your pulse rate and, to some extent, can be controlled directly by you.*

As you study chemistry generally you will not be required to identify whether a specific quantity is acting as a dependent or an independent variable. However, it is important to be aware of the various relationships that exist between variables. Anyone who has been on a diet is well aware of the dependency of body weight on the type and amount of food that is eaten. The significance of these dependencies is also evident when a person is hospitalized. The monitoring of the various body functions (variables) such as body temperature, blood pressure, pulse rate, and composition of the urine and blood provides information about the person's state of health. Deviation of any of these functions from their normal values is brought about by variation in the state of health of the patient.

When studying living systems, scientists try to control the variables that are not of immediate interest. For example, the resting pulse rate is determined after the person has been resting for several minutes. The pulse rate will vary with age and other factors, including exercise. In determining the resting pulse rate, one factor, the level of activity, is controlled by keeping the person at rest. This way the doctor can detect health problems, unrelated to activity, that affect the pulse rate.

The conclusion reached when studying the dependency of one variable on the values of other variables is initially stated as a hypothesis. The hypothesis is usually proposed before all the facts have been collected and analyzed. Although it may be perfectly valid, the hypothesis is, initially, tentative and forms the basis for further investigations. For example, it was hypothesized at one time that mice and rats spontaneously generated from old rags, since they were frequently observed coming out of them. Research easily proved this hypothesis incorrect.

After many observations have been made and the accuracy of a hypothesis has been well established, we call it a law. The laws of conservation of mass and energy discussed in Section 1-2.1 are examples of this. Laws are assumed to hold true for all situations as long as the stipulated conditions are met. However, laws should not be accepted as absolute fact. For example, the Law of Gravity has been accepted for many years as applying to the force of attraction between any two objects in all circumstances. Recent research indicates that a slight variation in the

A **variable** is a quantity that may change in value.

A **dependent variable** changes as a result of a change in one or more other variables.

Independent variables either change by themselves or are controlled by us directly.

A **hypothesis** is a scientific generalization as originally presented.

A **law** is a concise summary of a number of facts that, so far as is known, is always valid under the given conditions.

---

*It should be noted that a given variable can be considered a dependent variable in some instances and an independent variable in others. For example, blood pressure can be controlled directly through a technique called biofeedback. In such cases it would be thought of as an independent variable. Blood pressure also is controlled by changes in other variables. Thus, it is also a dependent variable.

form of the Law of Gravity occurs when objects are close together.* Whenever a limitation such as this occurs, modifications must be made in the stipulated conditions, or in the form of wording of the law.

When our explanations are based upon a variety of observations and relate several hypotheses and laws they are called **theories**. The theory of evolution is probably the most famous theory ever proposed. In accordance with our definition, it deals with a broad and complex class of things, the vast variety of living species. Theories are typically concerned with areas of study that cannot be observed directly. For example, no one has actually observed lower forms of life evolve into man. Theories are only as valid as the hypotheses and laws on which they are based and the assumptions made in their development. We will consider several theories, including atomic theory and collision theory of chemical reactions as we proceed through our consideration of the principles of chemistry.

As scientists develop hypotheses, laws, and theories, they frequently use what are called **models**. Models are developed by following the *model-building process*. Part of this process is illustrated in the following poem.

A **theory** is an explanation of complex phenomena that involves many interrelated variables.

A **model** is a hypothetical description of the real world that is based upon, and thereby attempts to explain, the observations of the world that have been made.

The Blind Men and the Elephant

It was six men of Indostan
To learning much inclined,
Who went to see the elephant,
(Though all of them were blind)
That each by observation
Might satisfy his mind.

The first approached the elephant
And, happening to fall
Against his broad and sturdy side,
   At once began to bawl:
   "God bless me! but the elephant
   Is very like a wall!"

The second feeling of the tusk,
Cried "Ho!" what have we here,
So very round, and smooth, and sharp!
   To me 'tis very clear,
This wonder of an elephant is
   Very like a spear!"

The third approached the animal,
   And happening to take
The squirming trunk within his hands,
Thus boldly up he spake,
   "I see," quoth he, "the elephant
Is very like a snake!"

The fourth reached out his eager hand,
   And fell about the knee.
"What most this wondrous beast is like,
   Is very plain," quoth he!
" 'Tis clear enough, the elephant
Is very like a tree!"

The fifth, who chanced to touch the ear,
   Said: "E'en the blindest man
Can tell what this resembles most:
   Deny the fact who can,
This marvel of an elephant
   Is very like a fan!"

The sixth no sooner had begun
   About the beast to grope,
Than, seizing on the swinging tail
That fell within his scope,
   "I see," quoth he, "the elephant
Is very like a rope!"

And so these men of Indostan
Disputed loud and long
   Each in his own opinion
   Exceeding stiff and strong,
   Though each was partly in the right,
   And all were in the wrong.

Anonymous

---

*Newton's law of universal gravitation states that the force of gravity is directly proportional to the product of the two masses involved, m_1 and m_2, and inversely proportional to the square of the distance between their centers, d. Mathematically, this is expressed by the equation $F = G(m_1, m_2/d^2)$, where G is a proportionality constant. Evidence now suggests that when objects are separated by a few centimeters, G is no longer constant but varies slightly with the distance of separation.

In this fictitious story, each of the blind men constructed a mental model of what he "observed" with his hands. However, each completed only the first two steps in the process that scientists follow. This process is summarized in Table 1.1. The blind men made initial observations and constructed various models, but they stopped at that point. Scientists, like many of us, occasionally do the same thing and "blindly" argue for their ideas or models without sufficient observations. A good scientist always goes on to steps 3 and 4, and, in fact, repeats these two steps until little doubt remains concerning the validity of the model. Similarly, individuals in the health-related sciences always must be careful to obtain sufficient additional observations to support a diagnosis before the diagnosis is acted upon. The application of the model-building process will be seen in Chapter 2 as we consider the development of atomic theory.

Table 1.1 The Model-Building Process

General Procedure	Applied by a Health-Care Provider*
1. Make initial observations.	1. Make initial observations of signs and symptoms of the patient.
2. Construct mental model.	2. Make an initial diagnosis.
3. Make additional observations to test the validity of the model.	3. Complete laboratory tests, X-rays, a physical examination, and so forth to check the diagnosis.
4. Retain the model as is, revise it, or reject it, depending upon how well it agrees with the new observations.	4. Retain the diagnosis, revise it, or reject it, depending upon the results of the tests.
5. Use the model.	5. Treat the patient.

*Physician, nurse, and physician's assistant

PROBLEMS

ⓓ 1. When you "burn" gasoline in your car, as illustrated by the equation,

$$2\ C_8H_{18} + 25\ O_2 \longrightarrow 16\ CO_2 + 18\ H_2O + \text{heat}$$

is the weight of the gasoline (C_8H_{18}) and oxygen (O_2) that reacted greater than, equal to, or less than that of the carbon dioxide (CO_2) and water (H_2O) produced?

ⓓ 2. Is the fact that ice melts at 32°F (0°C) a chemical or physical property?

ⓓ 3. Would you expect the number of disease-fighting white blood corpuscles found per unit volume of blood to be a dependent or an independent variable?

ⓓ 4. Is the well-established fact that heart rate decreases as arterial blood pressure increases a hypothesis, a law, a theory, or a model?

ANSWERS TO PROBLEMS **1.** equal to **2.** physical property **3.** dependent variable **4.** law (Marey's Law)

ⓓ indicates that the problem is discussed in depth in *Study Guide with Solutions and Problems for General, Organic, and Biological Chemistry* by M. Lynn James and James O. Schreck. ⓢ (see p. 14) indicates that the solution to the problem appears in that book.

1-3 SCIENTIFIC MEASUREMENT

LEARNING OBJECTIVES

1. Define unit, system of units, standard unit, meter, degree Celsius, gram, liter, and calorie.
2. Give the decimal prefixes and their symbols for 10^3, 10^{-2}, 10^{-3}, and 10^{-6}.
3. Convert Fahrenheit temperature readings to the Celsius scale.
4. Give the basic metric units and symbols for length, temperature, mass, volume, and energy.

5. Make one-step unit conversions within the English and metric systems involving length, mass, volume, and energy measurements.
6. Make one-step unit conversions between the English and metric systems involving length, mass, and volume measurements.

In this section we will continue to familiarize ourselves with the language and tools of science. As chemists study the living system they use numbers and measurements. Without them, it would be impossible for scientists adequately to do their work and convey information to others. We will now learn about the numbers and units of measure used by scientists and how these quantities are related to one another. Simple conversions between various quantities also shall be made to prepare us for doing more complex conversions in Section 1-4.

1-3.1 MEASUREMENTS AND STANDARDS

Measurements play an important part in our daily lives. For example, it is necessary for us to know whether we are 5 or 500 miles from home and whether we weigh 115 or 175 pounds. In these and most measurements there are really two parts, the value, or number, and the unit.

Saying only that we live 5 away is not enough. We might mean 5 houses, 5 blocks, or 5 miles. We must include the proper unit, houses, blocks, or miles, to communicate clearly to others what we mean. A group of units such as miles, pounds, gallons, and minutes forms a system of units.

A system of units is a good system if it is easy to work with, accurate, and based upon well-defined universal standard units. Clothing sizes are measurements that are easy to work with. Most of us know the size of shoe or jacket or pants that we wear and we can readily locate, in any store, clothing that fits us. However, clothing sizes are not accurate and they are not the same throughout the world.

Systems of units that are accurate do, however, exist. We will deal briefly with two of these, the English system and the metric system. A third system, the apothecaries' system, is also used in some branches of the health-related sciences; however, chemists do not generally use the apothecaries' system so we will not consider it further here. A fourth system, the International System of Units (SI Units), was established in 1960 by the International Bureau of Weights and Measures. The basic SI units and metric units generally are the same or similar. Adoption of the complete set of SI units by the scientific world has been slow, and the health-related sciences, for the most part, have essentially ignored those SI units that differ from their metric counterparts. We will not study or use those units here.

A **unit** is a value, quantity, or magnitude in terms of which other values, quantities, or magnitudes are measured.

A **system of units** is a group of units that provides a basis on which quantities are compared and communicated.

Standard units have fixed values that are used in making measurements.

1-3.2 THE ENGLISH SYSTEM

The English system is the one familiar to most Americans. It was established before and during the Middle Ages. Its units were based upon common items of the age, changing from time to time and differing from place to place. For example, one definition of the inch was the length of three grains of barley taken from the middle of a head of barley, while another was the width of the thumb (see Figure 1.1). In addition to this, odd relationships arose between various units, such as 12 inches = 1 foot, 3 feet = 1 yard, and 16 ounces = 1 pound. As time passed, more accurate and consistent units than barley corn and thumb's width were established, but the odd relationships between some units remained. Table 1.2 is a list of the more important equalities relating various English system units. Remembering and working with these odd numbers makes using the English system fairly difficult.

As seen in Table 1.2 the English system does not use a single set of standard units, but often uses several different units for the same measurement. As an illustration, when we weigh something, we can express the amount in ounces, pounds, or tons. Convenience usually determines the choice of unit. The weight of small things such as a candy bar is given in ounces, we express our own weight in pounds, and farmers buy hay by the ton. Since the English system is not used in science, generally we will not use it in this text except in this chapter, where it is used to provide practice for you in learning to do chemical calculations.

Figure 1.1
Two different standards used to establish the inch in the English system.

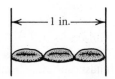

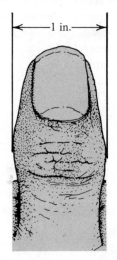

Table 1.2   Common English System Equalities

1 foot (ft) =	12 inches (in.)	1 T =	3 teaspoons (t)
1 yard (yd) =	3 ft	1 pound (lb) =	16 oz
1 mile (mi) =	5280 ft	1 ton =	2000 lb
1 gallon (gal) =	4 quarts (qt)		
1 qt =	2 pints (pt)		
1 pt =	2 cups		
1 cup =	8 ounces (oz)		
1 oz =	2 tablespoons (T)		

Conversion between units frequently is necessary in science and acquiring an ability to make these conversions is important if we are to understand and "do" chemistry. The following example is a simple illustration of how we can make what can be called one-step unit conversions.

EXAMPLE 1.1   If someone is 5 ft, 2 in. tall, how many inches tall is the person?
Since there are 12 inches in 1 foot, we may write a conversion factor, 12 in./1 ft. Multiplying this by 5 ft converts the unit from feet into inches.

$$5 \, ft \times \frac{12 \text{ in.}}{1 \, ft} = 60 \text{ in.}$$

Adding the extra 2 inches gives the person's total height as

$$60 \text{ in.} + 2 \text{ in.} = 62 \text{ in.}$$

In this example, the conversion factor, 12 in./1 ft, also could have been referred to as an equivalent ratio since it is the ratio of two equivalent quantities, 12 in. and 1 ft. On this basis, we can write 12 in./1 ft = 1. Thus, multiplication by the conversion factor is really multiplication by 1. The distance (5 ft) remains the same, only the units in which it is expressed are changed.

1-3.3 THE METRIC SYSTEM

The metric system is the system of measurement used in most countries.* In science it has been used almost exclusively for many years. The metric system is based upon a set of standard units with larger or smaller units related to the standard unit through a *decimal system.* In this system, quantities are related to one another by powers of ten. For example, the dollar, dime, and penny are related through a decimal system; there are 10 dimes and $100 = 10 \times 10 = 10^2$ pennies in 1 dollar.† Table 1.3 is a list of important decimal prefixes. The use of decimal numbers in the metric system avoids the confusion often caused by the odd conversion factors of the English system. Some of the important standard metric units, with the symbol commonly used to represent each unit, are given in Table 1.4.

The standard unit of length in the metric system is the meter, which is equal to the distance between two lines engraved on a bar of platinum-iridium alloy stored at 32°F (0°C) in a vault near Paris, France. This unit is equal to 39.37 in., slightly more than one yard. When the length, or distance, between two locations becomes so large that the meter is small by comparison, we refer to the distance in terms of thousands of meters or, using the decimal prefix *kilo-,* kilometers (km). For example, distances between cities in countries using the metric system are given in kilometers. Small distances, such as the size of one's waist, are usually measured in centimeters (1 cm = 0.01 m). Still smaller distances are measured in millimeters (1 mm = 0.001 m). Extremely small distances are sometimes reported as nanometers (1 nm = 10^{-9} m).

The standard unit for temperature in the metric system is the degree Celsius or degree centigrade. The Celsius scale is a decimal scale since there are $100 = 10^2$

The **meter (m)** is defined as the length equal to 1,650,763.73 wavelengths of the red-orange radiation from krypton-86.

The **degree Celsius (°C)** is $\frac{1}{100}$ of the temperature difference between the average sea-level boiling point (100°C) and freezing point (0°C) of water.

Table 1.3 Decimal Prefixes	
giga-, G, 10^9	*deci-*, d, 10^{-1}
mega-, M, 10^6	**centi-*, c, 10^{-2}
**kilo-*, k, 10^3	**milli-*, m, 10^{-3}
hecto-, h, 10^2	**micro-*, μ, 10^{-6}
deka-, da, 10^1	*nano-*, n, 10^{-9}

*Most frequently used prefixes.

---

*The metric system was devised in 1795. By 1979, it had been adopted or introduced in almost every country in the world. The United States is the only major nation in the world that does not use the metric system exclusively.

†If you are not familiar with scientific notations such as 10^2, you should refer to Appendix A for a discussion of the topic.

Table 1.4   Standard Metric Units

Quantity	Standard Units	Symbol
length	meter	m
temperature	degree Celsius	°C
mass	gram	g
volume	liter	L
time	second	sec
energy	calorie	cal

degrees between the freezing point and boiling point of water on this scale. It is easier to work with than the Fahrenheit scale, where there are 180 degrees between these two points. Conversion of Fahrenheit temperatures to Celsius temperatures can be made using the equation

$$°C = \tfrac{5}{9}(°F - 32.0°)$$

A comparison of the two scales is shown in Figure 1.2.

EXAMPLE 1.2   What is normal body temperature on the Celsius scale?
    Since normal body temperature is approximately 98.6°F, we have

$$°C = \tfrac{5}{9}(°F - 32.0°)$$
$$= \tfrac{5}{9}(98.6° - 32.0°)$$
$$= 37.0°C$$

The quantity of matter in a body is called its *mass*. For our purposes we will consider *mass* and *weight* to be synonymous.* The standard unit of mass in the metric system is the **gram**. A gram is much smaller than an ounce (28.35 g = 1 oz). Larger quantities of mass such as a person's weight are reported in kilograms (1 kg = 1000 g). For example, a 220-pound man weighs 100 kg. The smaller amounts of drugs in most medicines and in our daily vitamin requirement are frequently given in milligrams (1 mg = 0.001 g). The daily allowance of vitamin B_1 (thiamine) recommended for infants is 0.2 mg to 0.5 mg. Very small amounts are often measured in terms of micrograms (1 μg = 10^{-6} g).

One **gram (g)** is equal to the mass of one cubic centimeter of water at 4°C.

EXAMPLE 1.3   If 0.375 g of hydrogen cyanide (HCN) is enough to be a lethal dose for an adult (that is, enough to kill an adult), how many milligrams of hydrogen cyanide is a lethal dose?
    Table 1.5 lists the relationship between milligrams and grams. We can therefore write

$$1 \text{ mg} = 0.001 \text{ g}   \text{(or } 1000 \text{ mg} = 1 \text{ g)}$$

Mass and *weight* are terms that are frequently used interchangeably. Actually, the terms refer to separate physical concepts. The term *mass* refers to the amount of material in an object and the term *weight* refers to the gravitational force, or the attraction of the earth, acting on the object. As an example of this, an astronaut weighs less on the moon than on the earth due to the weaker gravitational attraction of the moon.

Figure 1.2   Temperature comparisons in the Fahrenheit and Celsius scales.

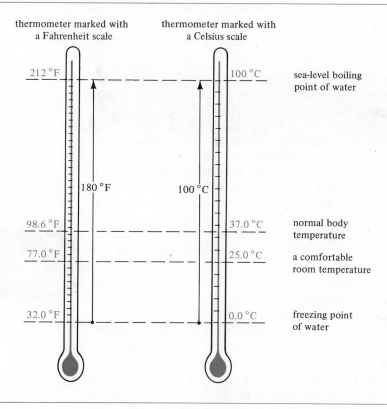

Thus, we have the conversion factor 1 mg/0.001 g = 1 (or 1000 mg/1 g = 1). Multiplying the amount of hydrogen cyanide by the second form of the conversion factor

$$0.375 \text{ g} \times \frac{1000 \text{ mg}}{1 \text{ g}} = 375 \text{ mg}$$

we arrive at the value we need.

The standard unit of volume in the metric system is the **liter**. It is slightly larger than a quart (1 L = 1.06 qt). Since the liter is a rather large volume for many laboratory uses, the milliliter (1 mL = 0.001 L) is usually used in chemical applications. From the definition of the liter it can be shown that 1 mL is equal to a cube with

A **liter (L)** is equal to the volume of 1000 cubic centimeters.

Table 1.5   Common Decimal Equalities for Any Unit
1 μ unit = 10^{-6} units or 10^6 μ unit = 1 unit
1 m unit = 0.001 unit or 1000 m unit = 1 unit
1 c unit = 0.01 unit or 100 c unit = 1 unit
1 k unit = 1000 units or 0.001 k unit = 1 unit

sides of 1 cm, that is, 1 cubic centimeter (1 cc) as shown in Figure 1.3. The unit cubic centimeter is frequently used in medicine. For example, a tetanus booster shot typically consists of 0.5 cc tetanus antitoxin.

Most chemical reactions either give off or absorb energy in the form of heat. We may eat because eating is a pleasurable experience, but more important, we eat so that we can have energy. The physical and chemical changes that take place in the body, called *metabolism,* give us the energy needed to do the things we do and to keep our bodies warm. Various units are used in the metric system to measure energy. We will use the **calorie** as the standard unit of energy. This unit is sometimes referred to as the *small calorie* or *gram calorie.* When dealing with human nutrition, such as when we count the calories in our diet, the large Calorie (Cal) is used. The large Calorie is equal to 1000 small calories; in other words, it is 1 kilocalorie (1 kcal). The large Calorie is always capitalized to distinguish it from the small calorie (cal) and in scientific literature it is abbreviated as Cal, or kcal. Table 1.6 gives caloric values of various foods in terms of large calories.

Although the metric system is used almost exclusively throughout this text, it is often helpful to be able to compare units of the metric system with units of measure in the English system. Table 1.7 gives the common equalities and conversion factors for converting from a measurement expressed in English units to the equivalent measurement expressed in metric units.

The process of converting from one unit to another is basically the same as that of changing money, a process with which you are familiar. If you were asked to change a five-dollar bill into quarters you would probably mentally calculate: "Since there are 4 quarters in 1 dollar, there must be $4 \times 5 = 20$ quarters in 5 dollars." For simple amounts like $5.00 the calculation seems quite easy, but most of us have to think twice if we are given a more complex amount such as $9.50. In such cases it is helpful to approach the problem from a more rigorous, mathe-

Figure 1.3
The milliliter is equal to the volume of a cube with sides of 1 cm each.

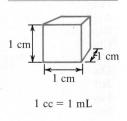

1 cc = 1 mL

A **calorie (c or cal)** is equal to the amount of energy required to raise the temperature of one gram of water by 1°C (from 14.5°C to 15.5°C).

Table 1.6	Caloric Values of Foods	
Food	Amount	Approximate Number of Calories
Egg	1 egg	75
Skimmed milk	6 oz	80
Malted milk	10–12 oz	450–500
Bacon	1 slice	30
Hamburger	2 cakes	85
Avocado	$\frac{1}{2}$ pear	265
Carrots	$\frac{3}{4}$ cup	40
Celery	3 stalks	15
Potato	$\frac{1}{4}$ cup, mashed	50
Tomato, fresh	1 medium	20
Potato chips	10 large	100
Potato salad with mayonnaise	$\frac{1}{2}$ cup	200
Popcorn	$1\frac{1}{2}$ cup	100
Apple	1 small	55
Banana	1 small	90
Grapefruit	$\frac{1}{2}$	40
Orange Juice	$\frac{1}{2}$ glass	50
Watermelon	$1\frac{1}{2}$-inch thick slice	190

Table 1.6 Caloric Values of Foods (*continued*)

Food	Amount	Approximate Number of Calories
Peanuts	10	50
Peanut butter	1 T	100
Apple pie	3-in. cut	200
Brownies	1 brownie	140
Cheese cake	2½-in. cut	275
Chocolate bar with nuts	1	250
Doughnuts	1	140
Jelly or jam	1 T	100
Layer cake, iced	1 medium portion	340
Sundae with nuts and whipped cream	Fountain size	400
Sugar	1 T	50
Soda	Fountain size	325
Carbonated soda	6 oz	80
Tea, plain	1 cup	0
Coffee, plain	1 cup	0
Coffee, with cream and sugar	1 cup	75–90
Beer	8 oz	110

matical point of view. To do this we can write the conversion factor 4 quarters/ $1.00 = 1 and then multiply the amount of money to be changed by this factor:

$$\$9.50 \times \frac{4 \text{ quarters}}{\$1.00} = 38 \text{ quarters}$$

This is basically the process we followed when we changed five dollars into quarters; however, we did not write it out quite as formally.

Table 1.7 English-Metric Equalities and Conversion Factors

Equalities	To Convert from English Unit	to Metric Unit	Multiply by
1 lb = 454 g	Pound	Gram	$\dfrac{454 \text{ g}}{1 \text{ lb}}$
1 kg = 2.20 lb	Pound	Kilogram	$\dfrac{1 \text{ kg}}{2.20 \text{ lb}}$
1 oz = 28.35 g	Ounce	Gram	$\dfrac{28.35 \text{ g}}{1 \text{ oz}}$
1 in. = 2.54 cm	Inch	Centimeters	$\dfrac{2.54 \text{ cm}}{1 \text{ in.}}$
1 m = 39.37 in.	Inch	Meter	$\dfrac{1 \text{ m}}{39.37 \text{ in.}}$
1 km = 0.621 mi	Mile	Kilometer	$\dfrac{1 \text{ km}}{0.621 \text{ mi}}$
1 L = 1.06 qt	Quart	Liter	$\dfrac{1 \text{ L}}{1.06 \text{ qt}}$

EXAMPLE 1.4   How many kilograms does a 143-pound person weigh?

Reference to Table 1.7 yields the conversion factor 1 kg/2.20 lb. Multiplying the person's weight by this factor gives the desired result:

$$143 \text{ lb} \times \frac{1 \text{ kg}}{2.20 \text{ lb}} = 65.0 \text{ kg}$$

You will find that hospital charts frequently give the patient's weight in kilograms.

Conversion from metric to English units can be done by inverting, or turning upside down, the conversion factors given in Table 1.7 and multiplying the resulting factor by the metric measurement. For example, 80 km is equal to 80 km $\times$ 0.621 mi/km = 50 mi.

PROBLEMS

5. What is the basic unit of weight or mass in the metric system?

S 6. Denver has been known for years as the "mile high city." Once the metric system has been totally adopted in the United States, how might we refer to Denver?

S 7. Physiological standards for substances in the blood are often expressed as the amount of each substance per 100. mL of blood. What part of a liter is 100. mL of blood?

S 8. For question 7, what part of a quart is 100 mL of blood?

S 9. A room temperature of 68°F is a bit cool. What is this temperature in °C?

S 10. How would you convert your weight in pounds into kilograms?

ANSWERS TO PROBLEMS   5. gram (sometimes the kilogram is used)   6. the "1.61-km-high city"   7. 0.100   8. 0.106 qt   9. 20°C   10. multiply your weight by 1 kg/2.20 lb

1-4 CHEMICAL CALCULATIONS – THE FACTOR UNIT METHOD

LEARNING OBJECTIVE

Perform multiple-step unit conversions involving the English and metric systems for length, mass, and volume.

In the last section the English and metric systems were introduced. Conversion factors within and between the two systems were given and one-step unit conversions were considered. Many of the chemical calculations that you shall be expected to do later in this course involve comparable conversions. In this section, we will deal with more complex calculations using a method referred to as the *unit cancellation,* or *factor unit, method.*

The unit conversions discussed in the previous section used the factor unit method. This method consists of four steps.

1. Determine the quantity given and the quantity desired.
2. Determine a series of conversion steps that go from what is given to what is desired.

3. Develop equivalent ratios (conversion factors) for each of the steps identified in step 2.
4. Multiply the quantity given by each of the conversion factors until the desired unit is reached.

The example that follows illustrates how these steps are performed.

EXAMPLE 1.5   A healthy adult may pass as much as $1\frac{1}{2}$ (1.5) qt of urine during a 24-hour period. How many cubic centimeters (cc) would this urine specimen be?

Step 1   Examination of the statement of the problem tells us that the quantity given is 1.5 qt and the quantity desired is the volume in cubic centimeters. This can be conveniently represented as

$$1.5 \text{ qt} \Longrightarrow ? \text{ cc}$$

where the arrow indicates the desired conversion.

Step 2   This step consists of identifying a series of units that take us step by step from the unit given to the unit desired. The only limitation imposed on this process is that we must know the necessary conversion factors for each step or be able to develop them.

For the above conversion, Table 1.7 gives us the conversion factor, 1 L/1.06 qt, for changing quarts to liters.

$$1.5 \text{ qt} \Longrightarrow ? \text{ L}$$

This conversion would seem to be a logical one since it takes us from the English unit quart to the metric unit liter and our final objective is to determine the volume in terms of another metric unit of volume, cubic centimeters.

The next step that would seem reasonable is to convert from liters to milliliters.

$$? \text{ L} \Longrightarrow ? \text{ mL}$$

The appropriate conversion factor for this step can be obtained from Table 1.5, where the equality 1 mL = 0.001 L can be obtained. This equality allows us to develop the correct conversion factor for this conversion. Finally, since 1 mL = 1 cc, as shown in Figure 1.3, we have the relationship necessary to make the final conversion.

$$? \text{ mL} \Longrightarrow ? \text{ cc}$$

The steps above can be completed one at a time or they can be completed together by the following series of conversions.

$$1.5 \text{ qt} \Longrightarrow ? \text{ L} \Longrightarrow ? \text{ mL} \Longrightarrow ? \text{ cc}$$

Step 2 of the factor unit method is the most difficult step. As illustrated above, it consists in identifying one or more equalities that gradually bring us closer to the desired unit. In a sense it is like planning a trip and deciding the route to be followed. For example, when traveling from New York City to Los Angeles it is helpful to know that Chicago and Denver are between the two cities. Once these cities have been identified, one can choose the correct highways that eventually will take us to our destination. In this case Chicago, Denver, and other appropriate cities can serve as overnight stops in our travels. Similarly, liters and milliliters were stopping places in our "travels" in the preceding problem.

Step 3   This step consists in obtaining the necessary conversion factors to complete the calculation. The appropriate conversion factor for the first unit change discussed in step 2 is listed in Table 1.7 as 1 L/1.06 qt. The conversion factor needed to make the second conversion can be obtained from the equality 1 mL = 0.001 L by dividing the left side of the equality by the right side. This gives us the conversion factors from equalities. When developing conversion factors, *the desired unit should always appear in the numerator while the old unit should be placed in the denominator.* Applying this to the conversion from milliliters to cubic centimeters, we have 1 cc/1 mL.

Step 4   The desired quantity can now be calculated by multiplying the quantity given by each of the above conversion factors.

$$1.5 \text{ qt} \times \frac{1 \text{ L}}{1.06 \text{ qt}} \times \frac{1 \text{ mL}}{0.001 \text{ L}} \times \frac{1 \text{ cc}}{1 \text{ mL}} = 1400 \text{ cc*}$$

The basic steps of the factor unit method outlined above will allow you to complete most of the calculations that you will be expected to do in this text. In the authors' opinion this method has several advantages over other methods that might be used. However, if you already feel comfortable in doing these calculations by some other method, or if your instructor prefers another method, there is no need to change to this one. The important thing is that you develop the skill needed to perform calculations.

PROBLEMS

⑤ 11. If you donate 1.00 pt of blood, how many milliliters of blood are you giving?

⑤ 12. Eighty milligrams of glucose per 100 cc of blood is a normal value. How many ounces of glucose does 80 mg represent?

13. Show how you would convert the measure of your height from feet and inches to centimeters.

⑤ 14. The toxicity of substances is often given by the LD_{50} value. This is calculated by determining the weight of the substance necessary to kill 50% of the test animals to which it is given, and then dividing that weight by the average weight of the animals. In this way the LD_{50} value is usually given as the milligrams or grams of the substance per gram or kilogram of body weight of the test animal. In applying this value to humans it is assumed that the substance will behave in a similar manner. The LD_{50} value of caffeine is 200 mg caffeine/kg body weight. A cup of strong coffee contains about 100 mg caffeine. How many cups of coffee could a 165-pound person drink before there would be a 50% risk of dying from caffeine poisoning? (Hint: First calculate the person's weight in kilograms, then determine how much caffeine would be necessary to have a 50% chance of killing someone of that weight, and, finally, convert this to the number of cups of coffee that are needed to yield this much caffeine.)

---

*Calculated by hand or with a calculator, the number 1415 cc is obtained, where four significant figures have been indicated in the answer. Since the original value (1.5 qt) contained only two significant figures it is not appropriate to report the answer to four significant figures. In general, the authors will follow the rule that an answer cannot contain more significant figures than the original value. For further discussion of significant figures, see Appendix A.

Ⓢ 15. What is a speed limit of 55 mph in terms of kilometers per hour?

Ⓢ 16. 100 g of wheat contain about 12 g of protein.

   (a) How many grams of protein does 1.00 lb of wheat contain?

   (b) How many ounces of protein does 1.00 lb of wheat contain? (Hint: Determine the conversion factor for converting from pounds of wheat to grams of wheat and use this to do part a. For part b, determine the conversion factor relating grams to ounces and convert the answer in part a to ounces.)

ANSWERS TO PROBLEMS **11.** 472 mL **12.** 0.0028 oz **13.** $(x \text{ ft} \times \frac{12 \text{ in.}}{1 \text{ ft}} + y \text{ in.}) \times \frac{2.54 \text{ cm}}{1 \text{ in.}}$

14. 50 cups **15.** 89 km/hr **16. (a)** 55 g **(b)** 1.9 oz.

1-5 DENSITY AND SPECIFIC GRAVITY

LEARNING OBJECTIVES

1. Define density and specific gravity.

2. Given the weight of a specific volume of liquid, calculate its density, and vice versa.

3. Given the density of a liquid, calculate its specific gravity, and vice versa.

In the preceding sections of this chapter we have learned a great deal about the language, tools, and techniques of science, including the health-related sciences. In Section 1-5 we will conclude this chapter by considering the related quantities, density and specific gravity. These have many applications in science in general and particularly in the health-related sciences.

1-5.1 DENSITY

Most of the quantities that were considered in the last two sections are derived quantities. A *derived quantity* is a quantity that is a combination of one or more other basic quantities. For example, volume is a derived quantity. Consider the volume of a box. This value is derived by multiplying the length of the box times its width times its height. If each of these distances is measured in centimeters, the resulting volume is expressed in the familiar unit cm^3, or cc. Scientists believe that there are only four basic quantities: length, mass, time, and temperature. Just as the volume of a box was obtained by multiplying three lengths (or distances) together, all quantities other than these four are derived by some combination of one or more of the basic quantities. The units of derived quantities may be written in terms of the units associated with the four basic quantities. For example, the speed of a car is usually given using a combination of two units, distance and time, such as miles per hour.

One important derived quantity that we have not looked at yet is called the

density. Density can be expressed mathematically as

$$D = \frac{m}{V} \qquad (1.2)$$

where m is the mass of the substance and V is its volume. Densities of liquids are usually expressed as g/mL, or g/cc.

The quantity density serves many useful purposes in science and in our everyday lives. For example, a service station attendant checks the acid in your car's battery and the antifreeze in your car's radiator by determining their density. The density of both fluids changes as their composition changes. The change in density is used to determine the composition and, therefore, the ability of the battery to start your car or the antifreeze to protect your car's radiator.

Similarly, the density of body fluids changes with our state of health. The density of urine is frequently measured to detect various health problems. Health technicians usually measure density with a special glass tube called a *hydrometer*. When it is used to determine the density of urine, this tube is frequently referred to as a *urinometer*. The hydrometer floats in the fluid being analyzed as a boat floats in water (see Figure 1.4). The greater the density of the fluid, the higher the hydrometer floats, and the lower the density, the more it sinks into the liquid.

EXAMPLE 1.5   If 250 mL of urine weighs 255 g, what is its density?

Using equation 1.2, we have

$$D = \frac{m}{V} = \frac{255 \text{ g}}{250 \text{ mL}} = 1.02 \text{ g/mL}$$

1-5.2 SPECIFIC GRAVITY

Specific gravity is another derived quantity of importance to us. Mathematically it is given by the equation

$$\text{sp gr} = \frac{m_1}{m_2} \qquad (1.3)$$

where m_1 is the mass of a given volume of the substance under consideration and m_2 is the mass of the same volume of some reference substance. For liquids and solids the usual reference substance is water.

Specific gravity is quite closely related to density as it is the ratio of the density of the substance being considered to the density of the reference substance; that is, $\text{sp gr} = D_1/D_{\text{ref}}$. When water is used as the reference substance, the specific gravity of the substance is essentially the same as its density since the density of water is essentially 1 g/mL. However, the specific gravity is unitless, whereas density typically has the unit g/mL.

The use of specific gravity is historically older than that of density and is gradually being replaced by density in most applications. Specific gravity, however, is still used at times in the health-related sciences, for example in urinalysis, and students in these fields should be familiar with the term and how to use it. As an example of the use of specific gravity in the health-related sciences, the specific gravity of urine, based upon water as the reference substance, can vary (1.005 sp gr to 1.025 sp gr),

Density (D) is equal to the mass of a substance per unit of volume.

Specific gravity (sp gr) denotes the ratio of the mass of a substance to the mass of an equal volume of a reference substance.

Figure 1.4 A urinometer floats at various levels in solutions of differing specific gravities. Shown increasing from left to right are (a) pure water, (b) urine from person with diabetes insipidus, (c) normal urine, and (d) urine from person with diabetes mellitus.

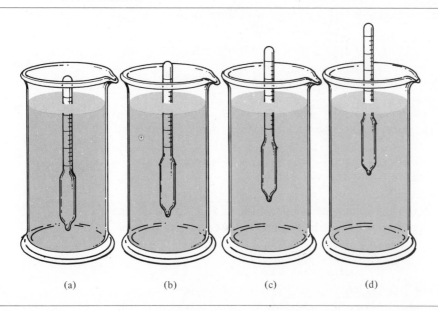

(a)   (b)   (c)   (d)

depending upon the amount of fluid intake by the individual and the amount of dissolved substances in the urine. In such cases the specific gravity increases with an increase in the amount of dissolved solids. Individuals suffering from the disorder diabetes insipidus have urine that has a very low specific gravity (close to that of water). On the other hand persons with a condition known as diabetes mellitus may excrete urine of high specific gravity caused by excessive quantities of glucose dissolved in the urine. Figure 1.4 depicts the specific gravity of samples of urine associated with these two conditions along with normal urine and water.

EXAMPLE 1.6   The specific gravity of blood is approximately 1.052.* Using this value, what is the weight in grams of 1.00 pt of blood?

If the volume associated with each of the two masses in equation 1.3 is taken as 1 mL, we may use the density form of the equation,

$$\text{sp gr} = \frac{D_{\text{blood}}}{D_{\text{H}_2\text{O}}}$$

As stated above, the density of water is essentially 1 g/mL, which allows us to write

$$\text{sp gr} = D_{\text{blood}} = 1.052 \text{ g/mL}$$

The unit pint is an English unit, while density (g/mL) and grams are metric

---

*Specific gravity values, such as that of blood, are frequently reported in the form 1.052 25/4, where the 25 indicates the temperature at which the density of the substance was determined and the 4 gives the temperature of the reference substance.

units. Thus, it is necessary to convert the volume into metric units. Using the factor unit method, we can make the following conversions:

$$1 \text{ pt} \Longrightarrow ? \text{ qt} \Longrightarrow ? \text{ L} \Longrightarrow ? \text{ mL}$$

Multiplying 1.00 pt by the conversion factors associated with these steps yields

$$1.00 \text{ pt} \times \frac{1 \text{ qt}}{2 \text{ pt}} \times \frac{1 \text{ L}}{1.06 \text{ qt}} \times \frac{1000 \text{ mL}}{1 \text{ L}} = 472 \text{ mL blood}$$

To obtain the desired weight we must make the conversion

$$472 \text{ mL} \Longrightarrow ? \text{ g}$$

The density provides the necessary conversion factor for this change. Therefore, we have

$$472 \text{ mL} \times \frac{1.052 \text{ g}}{1 \text{ mL}} = 497 \text{ g blood}$$

which is the weight called for in the statement of the problem. This final conversion is used frequently in chemistry and it would be a good idea to keep it in mind for later reference.

PROBLEMS

S 17. The density of cerebrospinal fluid (the fluid contained within the four ventricles or cavities of the brain and the central canal of the spinal cord) is about 1.006 g/mL. How much does 25.0 mL of the fluid weigh?

S 18. Carbon tetrachloride (CCl_4) is a rather toxic liquid that was used for many years as a household cleaner for clothing. 40.0 mL of CCl_4 weighs 63.56 g. What is the density of CCl_4?

S 19. As much as 1500.0 cc of saliva can be secreted in 24 hours. Its specific gravity is approximately 1.004. How much does 1500.0 cc of saliva weigh? (Hint: When solving this problem, assume that the density of the reference substance is 1.000 g/mL.)

ANSWERS TO PROBLEMS   **17.** 25.2 g   **18.** 1.59 g/mL   **19.** 1506 g

SUMMARY

The discussions in Chapter 1 covered some of the things that chemistry is concerned with and how scientists study and describe the world. We have seen the necessity of making measurements and the need for a system of measurements. We briefly looked at two systems of measurement: the English system and the metric system. Basic units and conversions within and between these two systems were covered. Last we looked at the terms *density* and *specific gravity* and performed calculations involving them. The knowledge and skills learned in this chapter will prove valuable to you throughout the course as we do other chemical calculations.

We are now ready to take a look at what the world is made of on the submicroscopic level. Chapter 2 discusses this briefly from a historical perspective and then considers the current model of the atom in some detail. Finally, the nature of radioactivity is discussed.

ADDITIONAL PROBLEMS

d 20. A tree grows through the process of photosynthesis by absorbing radiant energy from the sun. How does the amount of total energy present after the plant has grown compare with the total energy present before photosynthesis took place?

d 21. Indicate whether each of the following is a chemical or physical property. Justify your choice in each case.
   (a) the softness of a baby's skin
   (b) the redness of the hemoglobin in the blood
   (c) the ability of plants to convert carbon dioxide and water into cellulose through the process of photosynthesis
   (d) the aroma of food
   (e) the tendency of most people to convert excess food into fat in the body
   (f) the tendency of carbon dioxide gas to become solid CO_2 (dry ice) when its temperature is lowered to $-78°C$

d 22. Indicate whether each of the following is a constant or a variable. Justify each choice.
   (a) a person's weight
   (b) the mass of a level teaspoon of sugar
   (c) the temperature of the human body
   (d) the composition of urine
   (e) the composition of pure water
   (f) the composition of blood
   (g) the composition of a chocolate cake

d 23. Classify each of the following as a hypothesis, law, theory, or model.
   (a) To explain the color radiated by atoms when they are heated, it was proposed at one time that atoms are like the solar system with a tiny positively charged nucleus acting like the sun and small negative particles revolving about the nucleus like planets around the sun.
   (b) the well-established fact that the volume of the gas in a balloon decreases when the outside pressure of the balloon increases
   (c) the complex explanation of the behavior of gases including why the temperature affects their pressure, how gas particles collide with each other and with the walls of the container, and other characteristics of gases
   (d) the suggestion that a new drug may provide a cure for some dreaded disease

s 24. Convert the following temperatures to degrees C.
   (a) 32°F   (b) 212°F
   (c) 0°F   (d) 13°F
   (e) −40°F

s 25. Carry out the following conversions.
   (a) 75 in. $\Longrightarrow$? ft
   (b) 75 cm $\Longrightarrow$? m
   (c) 75 qt $\Longrightarrow$? gal
   (d) 75 mL $\Longrightarrow$? L
   (e) 75 mg $\Longrightarrow$? g
   (f) 75 kcal $\Longrightarrow$? cal

s 26. Carry out the following conversions.
   (a) 4.5×10^2 mg $\Longrightarrow$? kg
   (b) 350 cm $\Longrightarrow$? km
   (c) 21 msec $\Longrightarrow$? sec
   (d) 16.3 km $\Longrightarrow$? mm
   (e) 125 pt $\Longrightarrow$? gal
   (f) 5.0 mi $\Longrightarrow$? in.

s 27. Carry out the following conversions.
   (a) 75 mi $\Longrightarrow$? km
   (b) 75 lb $\Longrightarrow$? kg
   (c) 75 in. $\Longrightarrow$? mm
   (d) 75 oz $\Longrightarrow$? g
   (e) 75 pt $\Longrightarrow$? mL
   (f) 75 qt $\Longrightarrow$? mL

s 28. Carry out the following conversions.
   (a) 100 lb $\Longrightarrow$? mg
   (b) 100 mi $\Longrightarrow$? mm
   (c) 100 pt $\Longrightarrow$? L
   (d) 100 oz $\Longrightarrow$? kg

s 29. Calculate the specific gravity of ethyl alcohol (grain alcohol) at 36°C if its density is 0.7810 g/mL and the density of water is 0.9956 g/mL.

s 30. What is the density of cow's milk if one quart weighs 973 g?

NOTE: Answers to selected additional problems appear in *Study Guide with Solutions and Problems for General, Organic, and Biological Chemistry* by M. Lynn James and James O. Schreck.

CHAPTER 2

ATOMIC AND NUCLEAR THEORY—THE NATURE OF MATTER

Since the beginning of time, scientists and nonscientists alike have pondered the nature of their world. Through systematic study they have increased their understanding of how and why the world behaves the way it does. In this chapter we will briefly trace the major steps in the development of this understanding on the atomic level. After considering some of the major historical events in the development of atomic theory, we will examine the current model of the atom and briefly look at the phenomenon called radioactivity and some of its applications to the health-related sciences. With a knowledge of the atom and its parts, and how the atom behaves when it is not in contact with other atoms, you will be able to understand the chemistry that occurs when atoms combine.

2-1 EARLY ATOMIC THEORY

LEARNING OBJECTIVES

1. Give a brief review of the major steps that led to the development of Rutherford's model of the atom.

2. Define element, compound, atom, electron, and proton.

As we begin to study matter, it will be helpful if we have developed a firm understanding of scientists' beliefs concerning the nature of matter. This background will help us to appreciate the reasoning that scientists use and will give us a foundation on which to formulate our own conclusions regarding the nature of matter. Before we are finished with our consideration of the nature and behavior of matter, we will deal with some rather complex chemical species and learn about sophisticated chemical processes that occur in the living system. We will begin our study by considering some of the early historical events leading to scientists' current understanding of matter.

The origin of our present understanding of the atomic nature of matter was in the 17th and 18th centuries when it was recognized that matter occurs in essentially two distinct forms, **elements** and **compounds.** The oxygen (O_2) we breathe and the silver (Ag) found in jewelry and silverware are both elements since they are still only oxygen and silver when they are broken down into individual particles. The carbon dioxide (CO_2) and water (H_2O) we exhale as we breathe are compounds since they are composed of various elements that are chemically combined.

Early scientists found it necessary to assume that both elements and compounds are composed of tiny, discrete particles. They called these particles **atoms.** This early work was summarized in 1808 when John Dalton, an English chemist and physicist, postulated Dalton's atomic theory. This theory states that:

1. All matter is made up of individual particles called atoms that are indivisible and indestructible.
2. All atoms of a given element are identical.
3. Atoms of different elements have different masses and properties.
4. Atoms of different elements combine in simple ratios to form compounds.
5. When a compound decomposes, the original atoms are recovered unaltered.

The major points of Dalton's atomic theory are still considered to be true.

The composition of these atoms became apparent as a result of studies dealing with electrical charges. In 1883 an Englishman, Michael Faraday, studied the chemical changes that occur when electricity is passed through liquids, such as molten sodium chloride and sodium chloride dissolved in water. Faraday noted that elements, such as sodium (Na), chlorine (Cl_2), hydrogen (H_2), and oxygen (O_2), were produced in the process and that the passage of a given amount of electricity always resulted in the production of a specific amount of each substance. These observations suggest two things. First, matter must in some way involve electricity. Second, electricity, like matter, must exist in discrete particle-like amounts; that is, there must be a sort of "atom of electricity."

Other studies dealing with electrical charges such as static electricity and lightning also strongly suggested to the early scientists that matter must in some way contain some form of electrical charge. The nature of these charges was clarified in 1897, when the English physicist Sir J. J. Thomson studied the glowing discharge that appears in the cathode-ray tube pictured in Figure 2.1. This tube is similar to the familiar neon tube used in advertising signs, wherein a glow appears within an evacuated tube that contains metal electrodes connected to a source of electrical potential. Thomson found that the glow was caused by a stream of negative particles flowing between the two charged electrodes. From this he concluded that atoms must in some way contain the tiny negative particles that we know as **electrons.**

Since atoms of uncombined elements are normally electrically neutral, it was reasonable to assume that there must be a positive charge to balance the negative charge of the electrons. In recognition of this, Thomson, in 1898, proposed the "plum pudding" model of the atom. According to this model, electrons in the atom are embedded in a uniform sea of positive charge "like plums in pudding." Bulk

Elements are substances that cannot be broken down into simpler substances by ordinary chemical or physical means.

Compounds are homogeneous substances composed of two or more elements that have been united chemically.

An **atom** is the smallest portion of an element that can enter into a chemical combination.

Electrons are discrete particles having a negative charge of one and a mass $\frac{1}{1837}$ of that of a hydrogen atom.

Figure 2.1   Cathode-ray tube. Note how a stream of negative particles passes between the two electrodes.

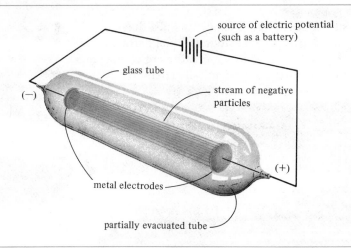

matter was assumed to consist of billions and billions of these extremely tiny ball-like atoms piled together like stacks of gum balls in a gum machine.

Proof of the existence of positively charged particles appeared in 1896 when Henri Bequerel, a French scientist, observed that uranium atoms gave off what appeared to be some form of invisible, penetrating ray. At the time, it was not clear what this radiation was; but in 1899, a New Zealander by the name of Ernest Rutherford discovered that the ray was actually positively charged particles to which he gave the name *alpha (α) particles.* Further reference to alpha particles shall be made in Section 2–5.

The next important step in the development of atomic theory was also contributed by Rutherford. In 1909 he was conducting experimental studies on the nature of matter. Rutherford bombarded thin sheets of gold foil with alpha particles and then observed what happened to the alpha particles after they hit the foil. Figure 2.2 illustrates the experimental apparatus. Most of the alpha particles passed straight through the gold foil. However, a significant fraction of the alpha particles were deflected at large angles (see Figure 2.2). The alpha particle is too heavy and moves too fast to be appreciably deflected by an atom composed of electrons embedded in a diffuse, positively charged fluid as proposed by Thomson. Rutherford, therefore, proposed a new model that stated that the atom consisted of a tiny, positively charged core, or atomic nucleus, with enough electrons outside the nucleus to produce an electrically neutral atom. Using this model, the deflection of the alpha particles is easily explained. As the particles came near the more massive nucleus, they were repelled by the positive charge of the nucleus and bounced off it at an angle. In this case, the angle at which the alpha particles were deflected would depend upon how close they came to a "head on" collision with the nucleus. Later evidence proved that the alpha particle is identical to the nucleus of a helium atom. Just as the negative charge of an atom was known to be associated with a particle, the electron, the positive charge of the atomic nucleus, was found to be associated with tiny, positively charged particles called protons.

The **atomic nucleus** is the inner core of the atom where the positive charge is found and which constitutes the main mass of the atom.

A **proton** is a discrete particle having a positive charge of one and a mass approximately equal to that of the hydrogen atom.

Figure 2.2   Rutherford's apparatus shows how a small but significant fraction of the α particles are deflected at large angles.

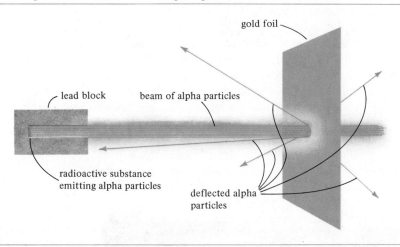

PROBLEMS

1. Of what does a cathode ray consist?
2. Define the term atom.

ANSWERS TO PROBLEMS   **1.** a stream of electrons   **2.** the smallest portion of an element that can enter into a chemical combination

2-2 THE PERIODIC BEHAVIOR OF THE ELEMENTS

LEARNING OBJECTIVES

1. Define atomic number, period, group, neutron, isotope, and atomic weight.
2. Discuss the periodic behavior of the elements as associated with an increase in atomic number.
3. Discuss the significance of the number of protons in the nucleus of an atom.

4. Discuss the basis upon which the periodic table is built.
5. Given an element from a list assigned by the instructor, give its name, symbol, atomic number, and atomic weight.

In Chapter 1 we developed a basic understanding of some of the methods and tools used by scientists in studying and reporting their measurements. This was continued in the last section where the existence of the atom and its fundamental nature was established. In Section 2-2 we will extend our understanding of the nature of the atom by considering the atom's characteristics, or properties, and how these properties vary from element to element. As a result, we will be able to organize the elements into various groups whose members have certain properties in common. By grouping elements in this manner, we will greatly simplify our later considerations of the behavior of the elements.

As more and more elements were discovered and their chemical behavior determined, a periodic picture of their properties began to emerge. Like the seasons of the year, various properties of the elements were found to go through a cyclic trend.

Scientists attempted to organize their knowledge of the properties of the elements so as to simplify their study. They originally arranged the elements in horizontal rows in order of increasing weight and in vertical columns by similarity of properties. This arrangement worked well for most elements, but anomalies appeared that could not be explained easily. Scientists gradually became aware of the presence of protons in the atoms of the elements and they recognized that each element differed from every other element by the number of protons in its nucleus. For example, they concluded that hydrogen (H) has one proton; helium (He), two; lithium (Li), three; and so forth. When the elements were arranged in horizontal rows in order of increasing numbers of protons and in vertical columns by similarity of properties, the anomalies disappeared. The modern periodic table given in Figure 2.3 and inside the front cover of this book uses this approach, listing the elements from left to right according to the number of protons (indicated by the number in the top left corner) and from the top to bottom in columns by similarity of properties.

As arranged in the modern periodic table, the elements exhibit a general variation in properties from left to right across each horizontal row. For example, potassium (K), which is located in the first column on the left, is a highly reactive metal; in fact, potassium is so reactive that it decomposes cold water violently to form hydrogen gas and potassium hydroxide and must be stored in a liquid such as kerosene to avoid reaction with moist air. In comparison, iron (Fe), which is found in the eighth column, will only react with pure water when the water is heated to form steam. Reaction with moist air occurs to form rust ($Fe_2O_3 \cdot x\ H_2O$), but much more slowly than in the case of potassium. Farther to the right, copper (Cu) is less reactive than iron, decomposing water only at high temperatures and reacting with air to a significant extent only when heated. Even farther to the right, the less metallic element germanium (Ge) is fairly unreactive and does not ordinarily react with water or with air.

Moving to the second to last column on the right, we see bromine (Br), an example of the *nonmetallic elements*. Bromine reacts violently with many metals. It slowly decomposes water when exposed to sunlight and attacks (reacts with) living tissue. On the other hand, krypton (Kr), which appears in the last column on the right, is so unreactive that only in recent years and through special techniques have scientists been able to cause it to undergo chemical reactions.

A similar variation occurs among the other elements in the periodic table. In each case, there is a left-to-right trend from very reactive metals to less reactive metals followed by less reactive nonmetals and on to quite reactive nonmetals, ending with an essentially unreactive, or inert, element. This pattern is referred to as *periodic behavior*. Thus, we call the chart with this arrangement of the elements the periodic table.

Returning to our consideration of protons, we cannot overemphasize the importance of the number of protons in an element. The number of electrons associated with an atom may vary as the atom reacts chemically with other atoms. We

Figure 2.3 Modern periodic table. Elements are listed from left to right by the number of protons and from top to bottom in columns by similarity of properties. Atomic numbers are in the top left corner. Atomic weights are shown below the symbols. Atomic weight values in parentheses are those of the isotopes of longest half-life.

PERIODIC TABLE OF ELEMENTS

IA	IIA		IIIB	IVB	VB	VIB	VIIB	VIIIB			IB	IIB	IIIA	IVA	VA	VIA	VIIA	0 VIIIA
1 H 1.008																		2 He 4.00
3 Li 6.94	4 Be 9.01												5 B 10.81	6 C 12.01	7 N 14.01	8 O 16.00	9 F 19.00	10 Ne 20.18
11 Na 22.99	12 Mg 24.31												13 Al 26.98	14 Si 28.09	15 P 30.97	16 S 32.06	17 Cl 35.45	18 Ar 39.95
19 K 39.10	20 Ca 40.08		21 Sc 44.96	22 Ti 47.90	23 V 50.94	24 Cr 52.00	25 Mn 54.94	26 Fe 55.85	27 Co 58.93	28 Ni 58.71	29 Cu 63.55	30 Zn 65.37	31 Ga 69.72	32 Ge 72.59	33 As 74.92	34 Se 78.96	35 Br 79.90	36 Kr 83.80
37 Rb 85.47	38 Sr 87.62		39 Y 88.91	40 Zr 91.22	41 Nb 92.91	42 Mo 95.94	43 Tc (99)	44 Ru 101.07	45 Rh 102.91	46 Pd 106.4	47 Ag 107.87	48 Cd 112.40	49 In 114.82	50 Sn 118.69	51 Sb 121.75	52 Te 127.60	53 I 126.90	54 Xe 131.30
55 Cs 132.91	56 Ba 137.34		57 La 138.91	72 Hf 178.49	73 Ta 180.95	74 W 183.85	75 Re 186.2	76 Os 190.2	77 Ir 192.2	78 Pt 195.09	79 Au 196.97	80 Hg 200.59	81 Tl 204.37	82 Pb 207.19	83 Bi 208.98	84 Po (210)	85 At (210)	86 Rn (222)
87 Fr (223)	88 Ra (226)		89 Ac (227)	104 Rf (261)	105 Ha (260)													

— *Transition Elements* —

Inner Transition Metals

Lanthanum series

58 Ce 140.12	59 Pr 140.91	60 Nd 144.24	61 Pm (147)	62 Sm 150.35	63 Eu 151.96	64 Gd 157.25	65 Tb (158.92)	66 Dy 162.50	67 Ho 164.93	68 Er 167.26	69 Tm 168.93	70 Yb 173.04	71 Lu 174.97

Actinium series

90 Th 232.04	91 Pa (231)	92 U 238.03	93 Np (237)	94 Pu (242)	95 Am (243)	96 Cm (247)	97 Bk (247)	98 Cf (251)	99 Es (254)	100 Fm (253)	101 Md (256)	102 No (254)	103 Lr (257)

will consider this in detail in Chapter 3. However, the number of protons in the nucleus of an atom of an element does not change during ordinary chemical reactions. In light of these factors, scientists identify each element by the number of protons in its nucleus. This number is called the element's atomic number. As mentioned previously, the number of protons, or atomic number, appears at the top of each square in the periodic table (see Figure 2.3).

The **atomic number** is the total charge of the nucleus of an atom as determined by the number of protons.

Another important feature of the periodic table is the symbols of the various elements. Writing out the name of each element is often a cumbersome task. To avoid this problem a system of abbreviations using one or two letters to represent each element has been developed. Frequently, the letter or letters come from the common name for the element. Thus, hydrogen is H, helium is He, and element 67, holmium, is Ho. Occasionally, the symbol for an element is taken from its Latin name. For example, sodium (*natrium*) is Na, potassium (*kalium*) is K, and iron (*ferrum*) is Fe.

Special names are also applied to various groupings of elements in the periodic table to aid in identifying each group. Elements to the left of the diagonal zig-zag line that begins at the left of element number 5, boron (B), are called *metals;* those to the right are known as *nonmetals;* and those bordering the line are sometimes referred to as *metalloids.* Horizontal rows of elements are termed periods in recognition of the periodic behavioral changes they undergo. Columns of elements are called groups, or *families,* because they consist of elements with similar chemical properties. Note that the columns in the periodic table are labeled A or B. The A groups are collectively called the *representative elements* while elements from IIIB to IB on the periodic table are referred to as the *transition elements.* A few of the groups are also given special names. Among these are the *alkali metals* (IA), the *halogens* (VIIA), and the *inert,* or *noble, gases* (0).

A **period** of elements on the periodic table is a horizontal row of elements.

A **group** of elements on the periodic table is a vertical row of elements.

In addition to protons and electrons, there is a third fundamental atomic particle, the neutron. This particle was discovered by an English scientist, James Chadwick, in the 1930's. Neutrons are located in the nucleus of atoms with the protons. Although an atom of a given element has a fixed number of protons, different types of the atom may contain different numbers of neutrons. For example, there are two important types of carbon atoms. Both types have six protons; however, one type has six neutrons and the other has eight. Each type of atom of an element is known as an isotope. Scientists frequently identify an isotope by the sum of the number of protons and neutrons. This value is referred to as the *mass number.* For example, the isotope of carbon that has 6 protons and 6 neutrons is carbon-12 and the isotope that has 6 protons and 8 neutrons is carbon-14. The identification of isotopes of the elements is a modern modification of Dalton's atomic theory, in which all atoms of a given element were thought to be identical.

A **neutron** is a discrete particle with no charge that has a mass approximately equal to that of the hydrogen atom.

Isotopes are atoms of a given element that differ in the number of neutrons.

A second number included below each symbol in the periodic table is the element's atomic weight (at. wt). The atomic weight is a measure of the relative heaviness of an average atom of an element. Two things are involved in determining this value: (1) atomic weight is a relative weight compared to the weight of an atom of carbon-12 and (2) the atomic weight of an element is the average weight of all the isotopes for that particular element based upon the relative amounts of each isotope found in nature.

The **atomic weight** of an element is the average weight of the naturally occurring isotopes of that element compared with the weight of the most abundant isotope of carbon (carbon-12) which has an assigned weight of exactly 12.

The use of carbon-12 as a reference standard for atomic weight is similar in some ways to the use of the thumb's width (the inch) as a standard for measuring small distances. For measurements of atomic weight a special unit, the *atomic mass unit* (amu)*, is used. The atomic weight of an element is, then, the weight of that element compared with the weight of carbon-12, which has an assigned weight of 12 amu. Hydrogen (H), which has three isotopes and an average weight of about 1/12 that of carbon-12, has an atomic weight of 1.008 amu. Helium (He), which is about 1/3 as heavy as carbon-12, has an atomic weight of 4.00 amu and magnesium (Mg), whose isotopes have an average weight of just over twice that of carbon-12, has an atomic weight of 24.31 amu.

Biochemists report atomic weights in a unit called the *dalton*, named in honor of John Dalton. The terms *dalton* and *amu* mean the same thing and can be used interchangeably. They are very nearly equal to the average weight (atomic weight) of hydrogen,

$$1 \text{ dalton} = 1 \text{ amu} \simeq \text{average weight of a hydrogen atom}$$

In agreement with the usual practice, the unit amu will be used in general chemistry and organic chemistry portions of this book, while the unit dalton will be used in the biochemistry portion.

When determining atomic weights, a representative sample of each element as it occurs in nature is collected and an average weight is calculated. We would determine the atomic weight of chlorine (Cl) in the following way. Chlorine consists of primarily two isotopes: chlorine-35 and chlorine-37. Approximately 75.5% of natural chlorine is chlorine-35 (34.97 amu) and 24.5% is chlorine-37 (36.97 amu). This results in an atomic weight for chlorine of about

$$0.755 \times 34.97 \text{ amu} + 0.245 \times 36.97 \text{ amu} = 35.45 \text{ amu}$$

With a knowledge of the atomic weight and atomic number of an element we can determine the number of neutrons in the most abundant isotope of nearly all of the naturally occurring elements. This is obtained by rounding off the atomic weight to the nearest whole number and subtracting the atomic number.

$$\text{number of neutrons} = \text{rounded off atomic weight} - \text{atomic number}$$

Applied to chlorine, we have

$$\text{number of neutrons} = 35 - 17 = 18$$

This agrees with the fact that 75.5% of naturally occurring chlorine has 18 neutrons. Exceptions to this rule do occur. For example, no natural isotope of silver (Ag) exists that has 61 neutrons, despite the fact that this is what is predicted by the rule. Similarly, the idea of a most abundant isotope for man-made elements, such as plutonium (Pu) does not apply and the rule cannot be used for these elements. We will not concern ourselves with these exceptions.

PROBLEMS

3. What does the atomic number represent?
4. What is significant about the behavior of the elements in any given column on the periodic chart?

---

*An alternative way of looking at the unit amu is to relate it to the more common unit, gram: 1 amu $= 1.67 \times 10^{-24}$ g. We can think of the amu as a convenient subunit of the more common unit gram, just as the ounce is a convenient subunit of the larger unit, pound.

5. Define the term isotope.
6. What is the basis for the order of the elements in the periodic chart?
7. Match the names and symbols for the following elements:

   (a) selenium   (i) Na
   (b) silicon   (ii) Si
   (c) silver   (iii) S
   (d) sodium   (iv) Se
   (e) strontium   (v) Sr
   (f) sulfur   (vi) Ag

ANSWERS TO PROBLEMS   **3.** the number of protons in the nucleus of an atom   **4.** they have somewhat similar properties   **5.** varieties of the same element, the atoms of which have the same number of protons but a different number of neutrons   **6.** their atomic number   **7.** a-iv; b-ii; c-vi; d-i; e-v; f-iii

2-3 BOHR'S ATOMIC THEORY

LEARNING OBJECTIVES

1. Give a brief description of Bohr's model of the atom.

2. Explain why Bohr's model was rejected.

In this section we will continue our consideration of the development of atomic theory by studying the Bohr model of the atom. Bohr's model played an important role in this development and will be valuable to us in visualizing the subatomic nature of atoms.

In 1913 the Danish physicist Niels Bohr refined Rutherford's model of the atom by postulating that the electrons move about the nucleus like planets in a miniature solar system (see Figure 2.4). Further, Bohr proposed that the electrons could only move about in certain paths, or orbits; that is, they are only found at

Figure 2.4   Bohr's model of the atom was like a miniature solar system, showing negative electrons moving about the positive nucleus.

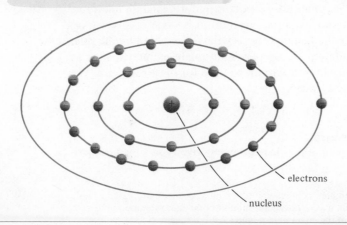

electrons

nucleus

Figure 2.5
A hydrogen atom radiates light of various colors depending on how far the electron "falls." Three possibilities for the electron fall are illustrated.

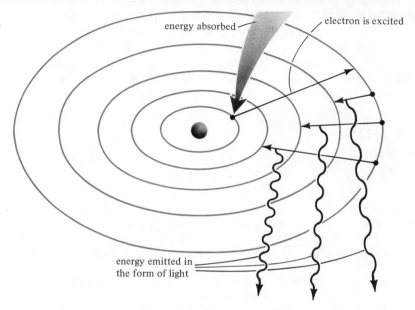

energy absorbed

electron is excited

energy emitted in the form of light

certain distances from the nucleus. This is comparable to a solar system where planets can only be located at certain distances from the sun; for example, one in which planets can only be located at 50 million miles, 90 million miles, 120 million miles, and other definite distances from the sun. Such a restriction on location would only be possible if the orbiting planets were limited to having only fixed amounts of energy associated with their motion. This is not the case for the planets in our solar system. As far as is known, planets can exist at any distance from our sun. Electrons in atoms, however, behave as if they are restricted to a certain distance from the nucleus. We frequently refer to the limitation of electronic energy to certain values by saying that the electrons exist in *"quantized" energy levels.*

As discussed in Chapter 1, models are only valid if they explain what we observe. In this regard, Bohr's model was both a success and a failure. One of its greatest successes was its ability to explain the origin of the colors seen in such things as fluorescent lights and flames in a fire. When the light from such sources is allowed to pass through a triangular piece of glass, known as a prism, a rainbow of colors appears. However in this case only certain colors of the rainbow are seen. The set of colors for a given element is known as the emission spectrum of the element and each color is associated with a specific amount of energy.

Bohr's model explained these colors by assuming that when an element is heated, its electrons absorb energy and jump to an orbit farther from the nucleus. An attractive force is always present between the nucleus and the electron, and sooner or later the electron falls back to an orbit that is closer to the nucleus. When the electron falls back, the atom radiates energy in the form of light. The color of this light depends upon how far the electron falls, that is, the difference in energy of the two energy levels between which the electron falls. Figure 2.5 illustrates this process for a hydrogen atom.

Bohr's model was particularly successful in explaining the colors emitted by hydrogen atoms. However, application of Bohr's model to other elements did not fare so well. Even major revisions of the model did not produce agreement between experimentally observed colors and those predicted theoretically. After several unsuccessful attempts by scientists to obtain agreement, Bohr's model was finally rejected and a radically new model developed. We will consider this new model in the next section.

PROBLEMS

8. What does it mean when one says that electrons exist in quantized energy levels?
9. Why was the Bohr model rejected?

ANSWERS TO PROBLEMS   **8.** Electrons in atoms appear to have only certain energy values. **9.** It failed to explain the colors observed for the spectra of elements other than hydrogen.

2-4 THE WAVE-MECHANICAL MODEL OF THE ATOM

LEARNING OBJECTIVES

1. Outline the main features of the wave-mechanical model of the atom.
2. Give the electron distribution for atoms by main energy level, sublevel, and orbital.

3. Draw the shapes of various s and p orbitals.

With the failure of Bohr's model, a German physicist by the name of Erwin Schrödinger, in 1926, proposed a new model for atoms. Although similar to Bohr's model in some respects, it is quite different in some fundamental aspects. The most important difference is that, rather than treating the electrons like orbiting planets as Bohr had done, Schrödinger used the mathematics that is used to describe the motion of a water wave or the wave associated with a violin string. In light of this, Schrödinger's model is frequently referred to as the wave-mechanical model. We will not be concerned with the mathematical details of this model, but it will be helpful for us to become acquainted with the results.

Its complex mathematical nature makes the wave-mechanical model difficult to visualize. A study of the rather unusual hotel shown in Figure 2.6 may clarify the way the model describes the distribution of electrons in atoms. Three distinct features of this or any other hotel are important to our consideration of atoms. First, when you stay in a hotel you occupy a room on a particular floor. Second, the room may be one of a set of rooms of a particular type such as a bridal suite or a presidential suite. Third, while staying in the hotel, you occupy a specific room and you do not move indiscriminately among the various rooms or floors.

The wave-mechanical model predicts a distribution of electrons in atoms similar to the distribution of people shown in the "atomic hotel" of Figure 2.6. The floors of the hotel are comparable to the *main energy levels* of the atom. For types of

Figure 2.6 "Atomic hotel" model for main energy levels and energy sublevels. Maximum number of people (electrons) for both main energy levels and sublevels are indicated in parentheses.

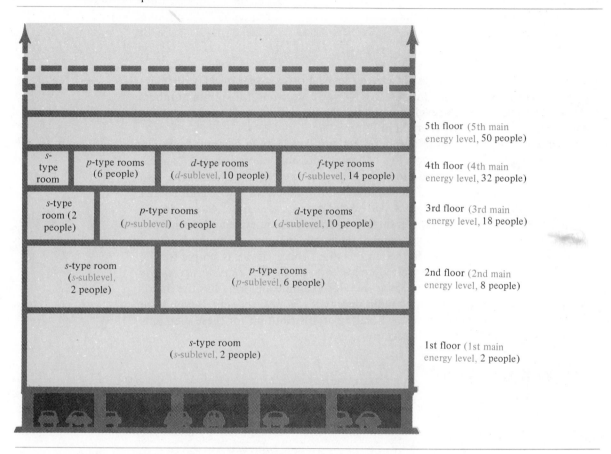

5th floor (5th main energy level, 50 people)

4th floor (4th main energy level, 32 people)

3rd floor (3rd main energy level, 18 people)

2nd floor (2nd main energy level, 8 people)

1st floor (1st main energy level, 2 people)

rooms the atom has *energy sublevels.* Instead of individual rooms, there are energy sub-sublevels or, as they are normally called, *orbitals.* We will consider each of these in some detail in this section.

2-4.1 MAIN ENERGY LEVELS

Electrons can be classified according to the main energy level they occupy in an atom in the same way that people might be classified according to the floor they occupy in a hotel (for example, a third-floor resident). Because of the attractive force that the nucleus has for electrons, electrons in a specific main energy level possess a fixed amount of energy that is associated with the average distance they are from the nucleus. This is comparable to the gravitational potential energy or the energy that people on various floors in a hotel have because of the force of gravity and their separation from the surface of the earth. As indicated on the right in Figure 2.6 each floor, or main energy level, can only accommodate a given number of people, or electrons. No atom has all the possible energy levels filled and, indeed, most atoms have one or more levels only partially filled, with many other levels having no electrons at all. It is useful, however, to be able to calculate the maximum number of electrons that a given main energy level can accommodate. This number can be calculated using the formula

$$\text{number of electrons} = 2n^2 \qquad \qquad \textbf{(2.1)}$$

where n is the number of the main energy level. For example, the fourth main energy level can contain a maximum of $2 \times 4^2 = 32$ electrons. Maximum numbers of electrons for the first five levels are, therefore, 2, 8, 18, 32, and 50, respectively. These numbers are given to the right in Figure 2.6.

In theory there is an infinite number of energy levels in each atom; however, electrons tend to occupy the position of lowest possible energy that is "open." When this occurs the atom is said to be in its *ground state.* Only when there is an input of energy from outside the atom do the electrons "elevate" to higher energy levels. When this occurs, the atom is said to be in an *excited state.* Atoms do not ordinarily exist in an excited state. In this book, discussions relative to the composition of atoms shall deal almost exclusively with atoms in their ground state.

The distribution of electrons in the various main energy levels is simple for the first 18 elements in the periodic table. *The electrons in the first 18 elements simply occupy the lowest unfilled main energy level.* For example, chlorine has the distribution 2, 8, and 7 for the first three main energy levels, respectively. Beyond the element argon the simplicity is lost. The electron distribution for these elements shall be discussed in the following section.

2-4.2 ENERGY SUBLEVELS

Further refinement of the wave-mechanical model requires that we consider the various types of energy sublevels or, in terms of the "atomic hotel," the various types of rooms on each floor. Figure 2.6 shows that the main energy levels are composed of sublevels. As seen in this figure the various sublevels are referred to as *s, p, d, f,* and so forth. Electrons in a specific energy sublevel behave quite differently than they would if they occupied a different type of sublevel. This behavior shall be considered when sub-sublevels (orbitals) and their shapes are discussed in Section 2-4.3

As in the case of the main energy levels, sublevels do not need to be completely full. It is, however, useful to know the maximum numbers of electrons each type of sublevel can accommodate. The maximum numbers for the *s, p, d,* and *f* sublevels are 2, 6, 10, and 14, respectively.

As may be seen in Figure 2.6, the number of sublevels in a given main energy level is equal to the number of that level. For example, there are four sublevels in the fourth main energy level. Therefore the number of sublevels increases by one for each level. A second feature of energy sublevels that is also illustrated in Figure 2.6 is that each successive main energy level includes the same sublevels as the previous one, plus an additional sublevel not found on the previous main energy level. For example, both the second and third main energy levels consist of an *s* and a *p* sublevel, but the third main energy level also has a *d* sublevel that does not exist in the second main level.

Figure 2.7 The higher the energy level is, the closer together are the relative energy values for energy sublevels. The arrows show the order of filling orbitals up to the 5s orbital.

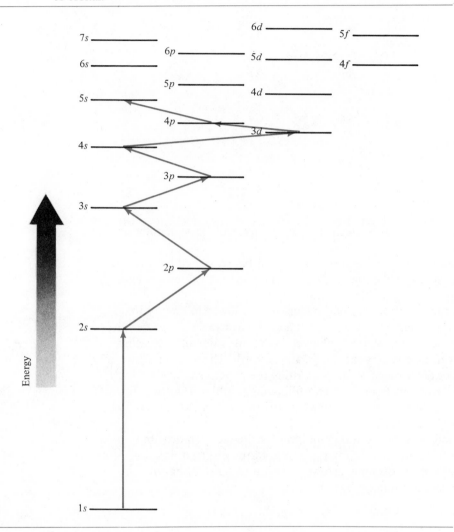

The distribution of electrons in the various energy sublevels may be conveniently represented by using the notation

$$(n\ell)^e$$

where n is the number of the main energy level, ℓ represents the sublevel involved, and e is the number of electrons in that sublevel. As indicated before, the atoms hydrogen through argon in the periodic table have a simple distribution. For these elements, electrons occupy the lowest unfilled main energy level and the s sublevel is filled before any electrons occupy the p sublevel on that energy level (see Figure 2.7). This results in the following order for the distribution of electrons: $1s$, $2s$, $2p$,

Figure 2.8   The order in which energy sublevels fill in can be determined by following each of the diagonal arrows from left to right.

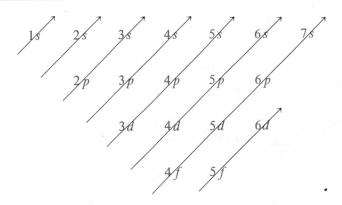

3s, 3p.* A sodium atom, with its eleven electrons, has the sublevel electron notation $(1s)^2(2s)^2(2p)^6(3s)^1$.

In all cases the ground-state electron distribution will be associated with electrons occupying sublevels having the lowest energies. The relative values for these energy sublevels are given in Figure 2.7. Careful examination of this figure shows that the s sublevel on the fourth main energy level (4s) is lower in energy than the d sublevel on the third main energy level (3d) and electrons will therefore fill the 4s sublevel before occupying the 3d sublevel. Next, in order of increasing energy, is the 3d sublevel, followed by the 4p, 5s, and so forth. The electron distribution of elements can be determined by simply writing the various sublevels in the order indicated by the arrows in Figure 2.7 and filling each sublevel in turn. For example, the notation for the sublevel distribution for potassium (which has 19 electrons) is $(1s)^2(2s)^2(2p)^6(3s)^2(3p)^6(4s)^1$.

An alternative method for arriving at the notation for the electron distribution of an element is given in Figure 2.8. Using this method, the electron notation is written by following each of the diagonal arrows from left to right. Either method leads to the following order for filling sublevels.

1s, 2s, 2p, 3s, 3p, 4s, 3d, 4p, 5s, 4d, 5p, 6s, 4f, 5d, 6p, 7s, 5f, 6d, and so forth.†

2-4.3 ORBITALS

Our model of the atom is almost complete now, but before we leave it we will consider how electrons are distributed within a given energy sublevel. Experimental evidence shows that each electron within a specific sublevel differs in its

---

*As given here, the order in which electrons fill sublevels is from left to right. The number of electrons occupying any specific sublevel will vary from atom to atom, and so the electrons are not included in this list.

†Some of the elements with atomic number greater than 20 have an electron distribution that differs slightly from this order and from the approach described for filling sublevels. This is particularly true of the transition elements. For our purposes, this variation is not significant and we will not discuss it.

behavior in some way from every other electron in that same sublevel. In fact no two electrons in any given atom behave identically. As before, it will be helpful to introduce this refinement of our model by looking at the "atomic hotel." Figure 2.9 extends this model to include the specific rooms, or **orbitals**.

As shown in this figure, each main energy level has various energy sublevels, with each sublevel consisting of one or more orbitals. Each orbital is capable of "housing" up to two electrons, as indicated by the people. Just as people in a real hotel tend to stay in their own rooms, electrons in atoms normally stay in a specific orbital and do not indiscriminately move about from orbital to orbital. Examination of Figure 2.9 also reveals that the number of orbitals in the various sublevels follows the order: one s orbital, three p orbitals, five d orbitals, and seven f orbitals.

When more than one orbital constitutes a given sublevel, the orbital is identified by a sort of "room number" that indicates its orientation about the nucleus. For example, the three p orbitals are identified as p_x, p_y, and p_z. The full significance of this notation will become more obvious when we consider the shapes of the orbitals later in this section.

An **orbital** is the region in space that an electron in an atom occupies.

Figure 2.9 "Atomic hotel" model for orbitals. Maximum number of people (electrons) in each orbital is 2. Sublevels are indicated by the heavy black lines.

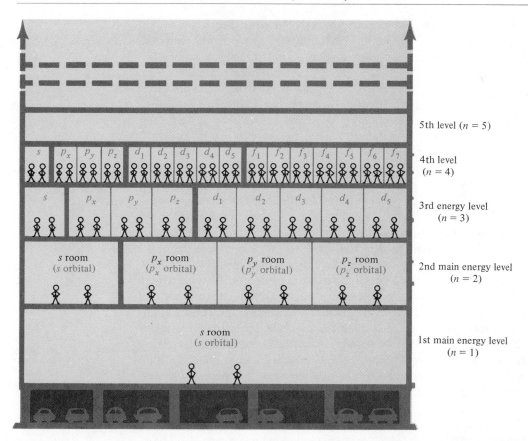

There are two main factors involved in the distribution of electrons among the various orbitals. First, electrons tend to occupy the lowest unfilled energy sublevel available, as discussed in Section 2-4.2. Second, electrons tend to occupy separate orbitals within a specific sublevel before they double up, forming an *electron pair* within a particular orbital. In other words, each orbital within a given sublevel is occupied by a single electron before any orbital in that sublevel has two electrons. This is due to a mutual repulsion between the negative electrons. By occupying separate orbitals the electrons can "maintain their distance" from each other. This will be more obvious when we consider the shapes of the various orbitals later in this section. The electrons tend to fill the lowest unfilled sublevel and so pair up whenever all the orbitals in a given sublevel have been singly occupied. To illustrate this behavior we will consider an atom of oxygen. The electron distribution of oxygen, by orbital, is represented by $(1s)^2(2s)^2(2p_x)^2(2p_y)^1(2p_z)^1$. As indicated here, the $2p_x$ orbital has two electrons, but the $2p_y$ and $2p_z$ orbitals each have only one electron rather than either one having two electrons.

For our purposes it will be sufficient to deal with the detailed electron distribution of the elements up through element 20 (Ca). That is, we will be concerned with just energy sublevels and orbitals up to $4s$.

Earlier in this chapter we mentioned that it is difficult to visualize the model of electronic motion predicted by wave-mechanics. However, it will be helpful in our later work if we have some idea of the shapes of the various orbitals. The wave-mechanical model does not allow us to predict precisely where an electron is at any given time. Rather, it allows us to predict the probability of an electron being in a certain region in space. Consider a comparable situation in life. A successful gambler depends on an ability to "play the odds;" that is, to determine the probability of a hand of cards taking the game or a particular horse winning the race. Just as the gambler cannot predict exactly when a specific horse will cross the finish line, the wave-mechanical model does not allow us to predict exactly when an electron will be at a specific position in the atom. It does, however, allow us to determine the probability of the electron being in a particular region in space, just as the gambler can determine the probability of a specific horse reaching the finish line first.

The following illustration will give you a feeling for the shape of atomic orbitals and thus the region in the space about the nucleus where the electron is likely to be found. Assume for the moment that you have been given the assignment of taking a time-exposure picture of a firefly as it moves about in the dark. Since the glow of the firefly goes off and on as it moves about, the film will only record the firefly's position when the firefly is giving off light. If the exposure time is sufficiently long, a cloud-like picture of white dots will appear on the film. Those regions that the firefly visited most frequently will have the greatest number of dots. In a similar manner, we can represent the probability of an electron being in a given region of space about the nucleus by a cloud-like picture, or orbital. Figure 2.10 illustrates the shape of the $1s$, $2s$, and $2p$ orbitals in this way. Those regions in the orbitals in this figure where there are many dots are the places where there is a high probability of finding the electron and those regions with few dots are the places with a low probability of finding the electron.

An alternative approach for representing the shapes of orbitals is shown in Figure 2.11 where lines that give the general shape of each orbital are drawn. These

Figure 2.10 Atomic orbitals pictured as a "probability cloud."

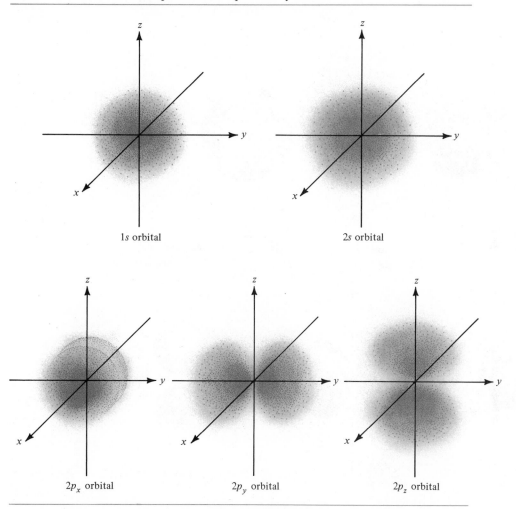

are contour lines. In drawing them, the intent is to enclose a region in space in which the electron can be found a certain percentage of the time. For example, the inner circle associated with the 1s orbital in Figure 2.11 indicates the region in which the electron is found 90% of the time, while the outer circle is associated with that region in which the electron is located 99% of the time. In light of this, these two lines are referred to as the 90% and 99% contour lines.

As seen in Figures 2.10 and 2.11, an electron in an s orbital occupies a spherical, or ball-like, region in space with the nucleus of the atom at the center. This is basically true for all s orbitals. As seen in the two figures, the major difference between the 1s and the 2s orbitals is their relative sizes and, therefore, the relative energy of the electrons that occupy them. Similarly, the 3s orbital will be even larger than either the 1s or 2s orbitals. The three p orbitals are considered to be concentrated around the x, y, and z axes, respectively. Because of this characteristic, they are identified as p_x, p_y, and p_z as mentioned earlier in this section. An electron

Figure 2.11   Atomic orbitals pictured by contour lines.

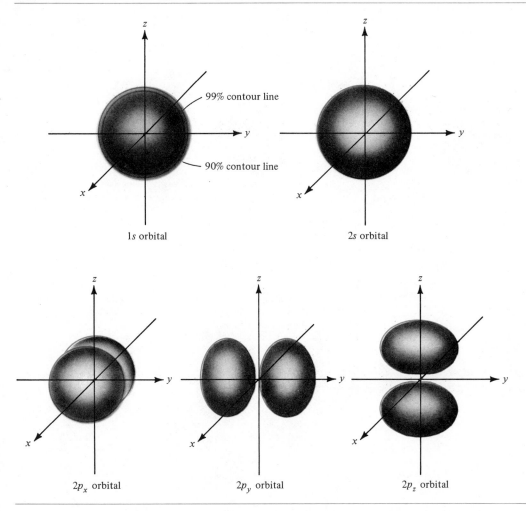

1s orbital

2s orbital

$2p_x$ orbital

$2p_y$ orbital

$2p_z$ orbital

in a *p* orbital occupies a "dumbbell," or three-dimensional, figure-eight-shaped, region in space. As with the *s* orbitals, the major difference between the orbitals 2*p*, 3*p*, 4*p*, and so forth, is their size and their energy.

In light of the fact that the p_x, p_y, and p_z orbitals are oriented at right angles to each other, we can now understand better why electrons tend to occupy separate orbitals within any given sublevel. By occupying separate orbitals, the *p* electrons are able to be as far apart as possible while in the same energy sublevel.

Early in this section it was stated that no two electrons in any given atom are identical. Thus far in our discussion we have distributed the electrons among the various orbitals where each orbital can contain a maximum of two electrons. However, no distinction has been made between the two electrons occupying the same orbital. Careful scientific examination of electrons indicates that they appear to spin like a top. The two electrons in any specific orbital spin in opposite directions however, and so behave differently from one another. Electrons in atoms, then, either are located in different main energy levels, sublevels, or orbitals or they possess opposite spin from one another.

PROBLEMS

10. What is the maximum number of electrons that can be found in the following?
    (a) the third main energy level
    (b) the s sublevel
    (c) the $3p_x$ orbital
11. Give the electron notation of sulfur by the following.
    (a) main energy levels
    (b) energy sublevels
    (c) individual orbitals

ANSWERS TO PROBLEMS   **10. (a)** 18   **(b)** 2   **(c)** 2   **11. (a)** 2, 8, 6
(b) $(1s)^2(2s)^2(2p)^6(3s)^2(3p)^4$ **(c)** $(1s)^2(2s)^2(2p_x)^2(2p_y)^2(2p_z)^2(3s)^2(3p_x)^2(3p_y)^1(3p_z)^1$

2-5 THE NUCLEUS AND RADIOACTIVITY

LEARNING OBJECTIVES

1. Describe alpha, beta, and gamma decay, and the effect that each has on living tissue.
2. Write the nuclear notation for any given isotope of an element.
3. Write equations for nuclear reactions for typical alpha, beta, and gamma decay processes.
4. Define half-life and do simple calculations based on half-life values.

The concepts of atomic theory presented in the earlier sections of this chapter will be useful as we consider the chemical behavior of the elements. However, before moving into this topic in the next chapter, we will briefly consider another aspect of atomic structure dealing with the atomic nucleus and the decay process that some nuclei undergo.

With the discovery that the nucleus of uranium undergoes a type of disintegration called radioactivity, Becquerel, in 1896, opened the door to the development of an important modern tool for the health-related sciences. In this section we will briefly consider the nature of radioactivity, its hazards, and some of its beneficial applications.

Since Becquerel's observation of the radioactivity of uranium, it has been found that almost all isotopes of the elements heavier than lead (atomic number 82) and many isotopes of the lighter elements undergo some form of radioactive decay. This characteristic seems to be associated with one or both of the following factors. First, when the nucleus of an atom is relatively large, as is the case of elements heavier than lead, it tends to be unstable. Second, when the ratio of neutrons to protons in a nucleus varies too much from a certain value associated with a stable

Table 2.1   Properties of Alpha, Beta, and Gamma Radiation

Type	Symbol	Nature	Mass	Penetration through Living Tissue
Alpha	α	He^{2+}	4.0026 amu	A few cells
Beta	β	Electron	0.000548 amu	Several millimeters
Gamma	γ	High energy X-ray	0	Deep into or through the body

nucleus, the nucleus tends to undergo spontaneous decay so as to achieve a more stable ratio.* An example of the latter process is the decay of carbon-14, which has 8 neutrons and 6 protons, into nitrogen-14, which has 7 of each. Three major types of radiation resulting from natural radioactivity have been observed. They are called *alpha* (α), *beta* (β), and *gamma* (γ) *radiation*. The important characteristics of these types of radiation are summarized in Table 2.1.

2-5.1 ALPHA DECAY

Alpha decay is the term applied to the emission of positively charged alpha parti-cles from the nucleus of atoms. An alpha particle is identical to the nucleus of a helium atom; that is, it consists of two protons and two neutrons. The alpha particle is spontaneously ejected by various radioactive elements. Alpha decay is particularly prevalent in the elements heavier than lead. When alpha decay occurs, the original atom loses two protons. Thus, it undergoes a *nuclear transmutation;* that is, the atom becomes a different element. For a nuclear transmutation involving an alpha decay, the element formed has an atomic number that is two less than the original element and a mass number that is decreased by four.

A convenient method for representing this and other types of radioactive decay uses nuclear notation. To represent a specific isotope of an element we use the form

$$_{P}^{P+N}X \quad \text{or} \quad {}_{\text{atomic number}}^{\text{mass number}}X$$

where X stands for the symbol of the element, P is the number of protons (the atomic number), and N is the number of neutrons. The sum of P and N is the mass number. To represent the most abundant isotope of radium, radium-226, we write $_{88}^{226}Ra$. This isotope undergoes alpha decay. The equation for the nuclear transmutation for radium-226 is given by

$$_{88}^{226}Ra \longrightarrow {}_{86}^{222}Rn + \alpha \tag{2.2}$$

or, written in an alternative form,

$$_{88}^{226}Ra \longrightarrow {}_{86}^{222}Rn + {}_{2}^{4}He^{2+} \tag{2.3}$$

Medical science has taken advantage of this decay process by using radium in radiation therapy for cancer.

---

*The stable ratio of neutrons to protons varies from element to element throughout the periodic table. For most light elements, the ratio is approximately unity. As one goes to heavier elements there is gradual increase in the ratio until, at the element lead, it has a value of about 1.5.

Radiation therapy depends on the fact that when living cells are exposed to radiation they can be destroyed. To understand how cell destruction can be brought about by alpha particles we must recall that an alpha particle has a positive charge. Associated with this charge is an electric field that surrounds the alpha particle. This electric field is like an invisible sphere of influence that reaches out from the alpha particle. If you have ever handled a magnet you have experienced the effect of a similar type of field, a magnetic field. If one magnet is brought near to another magnet or to something containing iron, you feel the push or pull of the magnet for the other object. Although less easily observed, a similar push or pull occurs between charged objects and is said to be associated with an electric field. As an alpha particle moves through matter, for example, living tissue, the electric field surrounding the particle acts on the electrons of some of the atoms of the material through which the particle passes, removing electrons from the atoms. When this happens, the chemically combined atoms, or molecules, that make up the cells are disrupted and cells are destroyed. Alpha particles travel at a relatively low speed and so are in the vicinity of each atom a relatively long time. They have ample opportunity to do a lot of biological damage to tissue through which they pass. On the other hand, alpha particles readily transfer their energy to the electrons they pull out of atoms. Thus, they soon slow down and so, do not penetrate far into materials. In the case of living tissue, alpha particles normally lose all their energy after traveling through a few cells. Thus, damage to the surface of the skin can occur. However, danger to cells deep in the body is minimal unless the substance emitting the alpha particles is taken internally or secondary gamma radiation occurs (see Section 2-5.3). In these cases, serious damage to living tissue can and does occur, unless the alpha emitter is used under medically controlled conditions.

2-5.2 BETA DECAY

A second form of radiation, *beta (β) decay*, occurs when an atom spontaneously ejects a beta particle from its nucleus. The beta particle is in reality an electron. The origin of the beta particle, however, is different. The beta particle is produced in the nucleus and is believed to be due to a neutron spontaneously changing into a proton and an electron*

$$_0^1 n \longrightarrow {}_1^1 p + \beta + \text{energy} \tag{2.4}$$

The beta particle is then thrown from the nucleus by the accompanying energy that is released in the process. When beta decay occurs, the mass number remains the same but the atomic number increases by one. An important example of beta decay is that of carbon-14.

$$_6^{14} C \longrightarrow {}_7^{14} N + \beta \tag{2.5}$$

or, written in an alternative form,

$$_6^{14} C \longrightarrow {}_7^{14} N + {}_{-1}^{0} e^{1-} \tag{2.6}$$

---

*Although the nuclear notation used here for neutrons and protons is not in strict agreement with the convention established earlier it is convenient in some cases to broaden the definition of the notation to include these. An electron also can be represented by $_{-1}^{0} e^{1-}$.

Carbon-14 is believed to be formed in the upper atmosphere by the bombardment of nitrogen-14 by neutrons, $_0^1n$, that have been produced by cosmic rays from outer space.

$$_7^{14}N + _0^1n \longrightarrow _6^{14}C + _1^1H \tag{2.7}$$

However, $_6^{14}C$ is continually decaying to $_7^{14}N$ and a relatively constant level of $_6^{14}C$ is thereby maintained in the atmosphere. The $_6^{14}C$ in the atmosphere does not exist as elemental carbon, but is usually combined with oxygen in the form of carbon dioxide (CO_2). Green plants incorporate the radioactive carbon-14 into their cellular structure along with the normal carbon-12 through a process of assimilation known as photosynthesis. As long as the plant is living, $_6^{14}C$ is continually added to the plant. However, as soon as the plant dies the intake of all carbon ceases and the amount of carbon-14 present begins to decrease because of beta decay. The decay of carbon-14 is a slow process and many years must pass before a significant change in its level is observed. Animals that eat the plants will have the radioactive isotope incorporated into their cells. As with plants, the intake of $_6^{14}C$ stops when the animal dies and the amount of carbon-14 begins to decrease. Archeologists and geologists have taken advantage of this fact, developing a process known as *carbon dating* for determining the age of carbon-containing substances. This process consists in determining the amount of carbon-14 remaining in the dead plant or animal and then calculating how long the specimen has been dead, based upon the known rate of decay of $_6^{14}C$.

Beta particles, like alpha particles, possess an electric field because of their charge and so beta particles tend to push electrons from atoms through which they pass. Disruption of molecules is, therefore, also caused by beta particles. The beta particle travels almost 100 times faster than the alpha particle and therefore spends much less time in the vicinity of each atom. The extent of damage to the living tissue through which beta particles have passed is generally less than that brought about by alpha particles. However, beta particles are much more penetrating than alpha particles, traveling through several millimeters of tissue before coming to a stop. The tissue damage that beta particles cause can, therefore, occur deeper inside the living system. The net effect is that exposure to beta emitters from outside the body is generally more dangerous than is exposure to alpha emitters.

2-5.3 GAMMA RADIATION

Gamma radiation is actually a form of energy called electromagnetic radiation. Another familiar form of electromagnetic radiation is the visible light by means of which we see things with our eyes. Other forms of electromagnetic radiation are the microwaves that are used to cook food in a microwave oven, radio waves that our radios convert into sound, and X-rays that allow the dentist to "take a picture" of our teeth. Each of these types of radiation differs from the others only in the amount of energy associated with them. The relative energies of these and other types of electromagnetic radiation are illustrated in Figure 2.12. Gamma rays are identical to high energy X-rays, but the source is different. Gamma rays originate from nuclei undergoing radioactive decay and X-rays are produced by a special form of cathode-ray tube called an X-ray tube.

Figure 2.12
The electromagnetic spectrum. All electromagnetic waves have the same fundamental character and the same speed in vacuum, but many aspects of their behavior depend on their energy.

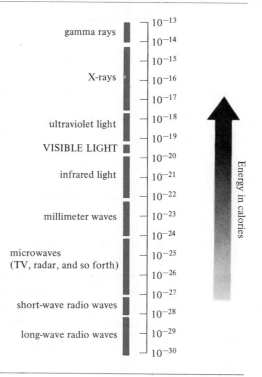

The production of gamma rays from the nucleus of an atom is actually what can be called a secondary effect. Alpha and beta decay are frequently accompanied by an emission of excess energy in the form of gamma rays. In the discussion concerning main energy levels in Section 2-4.1, it was pointed out that electrons in an atom can exist in either a ground state or an excited state, depending upon the amount of energy possessed by the electrons in the atom. A similar condition is believed to exist for the nucleus of an atom. The product nucleus that results from either alpha or beta decay initially may be in an excited state. Like atoms possessing electrons in excited states, the nucleus tends to lose energy and go to a more stable ground state. When this occurs, energy is given off in the form of gamma rays.

An example of a nuclear disintegration involving gamma rays is the beta decay of $^{60}_{27}\text{Co}$. Cobalt-60 decays according to the following equation,

$$^{60}_{27}\text{Co} \longrightarrow {}^{60}_{28}\text{Ni*} + \beta \qquad (2.8)$$

where the asterisk indicates that the resulting nickel atom is in an excited state. The excited nickel-60 atom then emits a high-energy gamma ray as it goes to the ground state. When gamma decay occurs, the mass number and the atomic number remain the same, so we have

$$^{60}_{28}\text{Ni*} \longrightarrow {}^{60}_{28}\text{Ni} + \gamma \qquad (2.9)$$

Just as electrons in atoms have several energy levels available to them, the nuclei of many elements can exist in several excited energy states, and so gamma radiation can take place in two or more steps. This is what actually occurs in the decay of $^{60}_{27}$Co into $^{60}_{28}$Ni, but for our purposes the simple reaction shown by the equation above will suffice. The fact that cobalt-60 decay produces high-energy gamma rays makes this disintegration process a very convenient source of radiation for the treatment of cancer.

In a similar manner, the alpha decay of radium, given by equation 2.2, involves subsequent gamma decay by the radon atoms produced. As the radon atoms are produced they are in an excited state and emit gamma rays,

$$^{222}_{86}\text{Rn}^* \longrightarrow {}^{222}_{86}\text{Rn} + \gamma \qquad \qquad \textbf{(2.10)}$$

The effectiveness of radium in treating cancer is primarily due to the gamma radiation rather than the alpha particles.

Gamma rays, unlike alpha and beta particles, lose all their energy in a single collision with an electron. However, gamma rays frequently will penetrate matter much more deeply than will the alpha and beta particles before this loss of energy occurs. Gamma rays are, therefore, much more dangerous to the living system than are alpha or beta particles. The process by which the gamma ray destroys cells is twofold. First, by transferring its energy into an orbital electron, it disrupts the molecule of which the electron is part. Second, as the gamma ray ejects the electron from the atom, the electron becomes equivalent to a beta particle. This high energy electron disrupts other atoms, bringing about additional molecular breakdown. Since all of this may occur well within the living system, damage to cells and tissue that are part of vital organs can result.

2-5.4 HALF-LIFE

Thus far in our discussion we have considered the major types of radiation resulting from radioactive decay. Nothing has been said, however, about how rapidly or how slowly the decay occurs. Some chemical and nuclear changes occur almost instantaneously while others are extremely slow. Determination of the exact time when all atoms of a given type have undergone a decay process is extremely difficult, if not impossible. In light of this, an alternative procedure is used for determining how rapidly a decay occurs. This procedure consists in determining the amount of time that would be required for half of the initial amount of a radioactive substance to undergo decay. This period of time is referred to as the **half-life,** $t_{1/2}$, of the substance. Half-life periods range from fractions of seconds to billions of years. For example, the half-life for the decay of $^{5}_{3}$Li is estimated to be about 10^{-21} sec while the decay of $^{238}_{92}$U has a half-life of about 4.51 billion years. The first reaction is so rapid that its $t_{1/2}$ value can only be estimated at best. The second decay process is so slow that it provides a valuable means for determining the age of samples of earth. For example, it is thought that approximately half of the uranium-238 on the earth has undergone decay. Therefore, the age of the earth is considered by most scientists to be approximately 4.5 billion years old.

In Section 2-5.2 we considered the use of $^{14}_{6}$C in carbon dating. With the introduction of the concept of half-life, we can examine this application in more detail and, at the same time, take a more complete look at the meaning of half-life. The half-life of $^{14}_{6}$C is 5730 years. In other words, regardless of how much carbon-14 there is in a sample there will be only half as much after 5730 years. After another

Half-life is the period of time required for the radioactivity of a substance to drop to half its original value.

Figure 2.13   Radioactive decay of carbon-14 over four half-lives. The color line gives actual amounts of carbon-14 at any specific time.

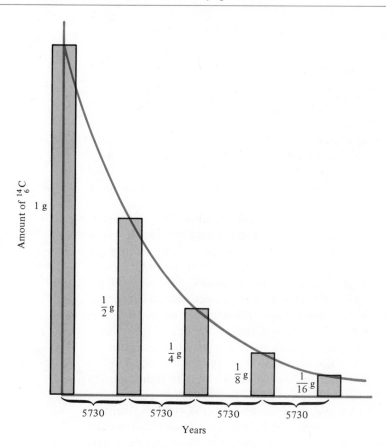

5730 years there will be only half of the first one-half portion left, or one fourth of the original amount. Figure 2.13 illustrates how this operates over several half-lives.

EXAMPLE 2.1   If a tree contained 20 g of $^{14}_{6}C$ when it died and 4 half-lives have passed since it died ($4 \times 5730 = 22,920$ years), how much $^{14}_{6}C$ will remain in the wood?

This may be calculated in the following steps.
After one half-life there will be

$$\tfrac{1}{2}(20) = 10 \text{ g}$$

After two half-lives there will be

$$\tfrac{1}{2}(10) = 5 \text{ g}$$

After three half-lives there will be

$$\tfrac{1}{2}(5) = 2.5 \text{ g}$$

After four half-lives there will be

$$\tfrac{1}{2}(2.5 \text{ g}) = 1.25 \text{ g}$$

Alternatively, the amount can be calculated in a single step by multiplying the original amount by $\frac{1}{2}$ four times.

$$(\tfrac{1}{2})(\tfrac{1}{2})(\tfrac{1}{2})(\tfrac{1}{2})(20 \text{ g}) = 1.25 \text{ g}$$

2-5.5 RADIATION AS A TOOL

In our study thus far we have seen that, in some ways, radioactivity is harmful to all forms of plant and animal life. However, when properly understood and wisely used, radioactivity provides us with a powerful tool in such fields as medicine, agriculture, and scientific research. Because of its importance in such areas, we will consider a few additional applications of radioactivity.

Nuclear medicine capitalizes on the property of various radioactive isotopes to undergo decay and the fact that the decay process can easily be detected and monitored by suitable instruments, such as the geiger counter. Using radioactive isotopes to learn more about the living system, in radiation therapy or as a diagnostic tool, is a common practice in today's modern hospitals. For example, by using the radioactive isotope of iron $^{59}_{26}\text{Fe}$, researchers have been able to determine that, unlike most other elements in the animal body, iron atoms in the hemoglobin of the red blood cells generally remain in the body unless the animal bleeds and new red blood cells must be produced to replace those lost. Feeding an animal a diet containing the $^{59}_{26}\text{Fe}$ isotope does not increase the level of this isotope in the body very much. This indicates that the iron has passed directly through the body and very little replacement of iron occurs.

Earlier, reference was made to the use of $^{226}_{88}\text{Ra}$ and $^{60}_{27}\text{Co}$ in the treatment of cancer. Another valuable radioactive isotope for both diagnosis and therapy for thyroid cancer is $^{131}_{53}\text{I}$. The thyroid glands are two large glands in the neck that produce hormones that affect the rate at which foods are transformed into waste products and energy. These glands tend selectively to concentrate iodine from the rest of the body, making it possible for the beta decay of $^{131}_{53}\text{I}$ to be used in the detection of thyroid malfunction and in treatment of thyroid cancer. Because most of the iodine taken into the body goes to the thyroid gland, there is very little danger to healthy tissue in other parts of the body from the radiation of iodine-131. Table 2.2 gives a sample list of some clinical uses of radiopharmaceuticals. Included are the applications, the isotope used, the half-life of the isotope, and the typical carriers used to get the isotope into the body.

One example of the application of radioisotopes to agricultural research is the use of fertilizers containing $^{30}_{15}\text{P}$. This isotope undergoes beta decay, and can, therefore, be used to determine phosphorus intake by the plant at various stages of growth. Through such studies increased crop yields have resulted and the rate of plant growth has been increased. In some areas of the United States this allows more than one crop of corn to be harvested during a growing season.

Another interesting application of radiation is in the area of food preservation and the sterilization of drugs and medical supplies. For example, it has been found that irradiation of foods destroys most of the bacteria present and makes it possible

Table 2.2 Clinical Uses of Radiopharmaceuticals

Use	Isotope	Half-Life	Carrier
Thyroid studies	^{131}I	8.08 days	Sodium iodide, thyroxine, and liothyronine
Kidney studies	^{131}I	8.08 days	Iodohippurate sodium
	^{197}Hg	65 hours	Chlormerodrin
	^{203}Hg	46 days	Chlormerodrin
Liver studies	^{131}I	8.08 days	Rose bengal sodium
	^{198}Au	64.8 hours	Gold colloid
Gastrointestinal tract studies	^{131}I	8.08 days	Triolein, oleic acid, and tolpovidone
	^{57}Co	270 days	Vitamin B_{12}
	^{51}Cr	27.8 days	Sodium chromate
Blood studies	^{131}I	8.08 days	Serum albumin
	^{51}Cr	27.8 days	Sodium chromate
	^{59}Fe	44.5 days	Ferric chloride, ferrous citrate, and ferrous sulfate
Brain studies	^{131}I	8.08 days	Serum albumin
	^{74}As	17.9 days	Sodium arsenate
	^{197}Hg	65 hours	Chlormerodrin
	^{99}Tc	6 hours	Pertechnetate
Eye studies	^{32}P	14.3 days	Sodium phosphate
Electrolyte studies	^{22}Na	2.58 years	Sodium chloride
	^{42}K	12.5 hours	Potassium chloride
	^{45}Ca	165 days	Calcium chloride
	^{47}Ca	4.56 days	Calcium chloride
Cancer therapy	^{131}I	8.08 days	Sodium iodide
	^{131}I	8.08 days	Ethiodized oil
	^{198}Au	64.8 hours	Gold colloid
	^{32}P	14.3 days	Chromic phosphate
Thyroid therapy	^{131}I	8.08 days	Sodium iodide
Blood disease therapy	^{32}P	14.3 days	Sodium phosphate

to store the food for long periods of time without refrigeration. Similar irradiation of drugs and medical supplies makes them surgically sterile.

Another application of interest is the use of radioisotopes in migration studies of insects and other pests. The pests are allowed to feed on substances containing radioactive isotopes, then traps are set at various locations so that the directions and distances traveled by the insects can be determined. Still another application is the use of radiation in nutrition studies. For example, cattle feed containing radioactive substances can be used to determine the type and quantity of minerals that cattle need in their diets to build bones and muscles.

One last application of radioactivity that may well have the greatest overall influence is the production of electrical energy through nuclear generating plants. As other sources of energy, such as oil and gas, are depleted, this source of energy may become crucial if we are to maintain the high standard of life enjoyed in various parts of the world.

As we have seen, radioactivity provides a valuable tool for the health-related sciences. Great potential good can be achieved through the wise use of nuclear chemistry.

PROBLEMS

12. Of what does an alpha particle consist?
13. What is the nuclear notation for barium-138?
14. One of the products of atomic bombs is $^{90}_{38}$Sr. Since strontium tends to be stored in the bones, the beta decay of $^{90}_{38}$Sr may present a serious health hazard to anyone exposed to fall-out from an atomic bomb. Write the equation for the decay of $^{90}_{38}$Sr.
15. Define half-life.

ANSWERS TO PROBLEMS   **12.** two protons and two neutrons   **13.** $^{138}_{56}$Ba
14. $^{90}_{38}$Sr $\longrightarrow$ $^{90}_{39}$Y $+\ \beta$ **15.** the period of time required for the radioactivity of a substance to drop to half its original value

---

SUMMARY

The discussion in this chapter has centered around the nature and characteristics of the matter that makes up the world. We have found that our world is made up of tiny, individual particles called atoms. These atoms, in turn, are composed of yet smaller particles called protons, neutrons, and electrons. The properties of atoms depend on how many of these smaller particles are present, especially the number of protons and electrons. The protons and neutrons are located in a tiny nucleus, while the electrons move about at relatively large distances outside the nucleus. Individual electrons within an atom differ from one another by the amount of energy that they possess. These differences in energy can be explained by assuming that the electrons are distributed in a set of main energy levels composed of various energy sublevels that consist of different orbitals. Finally, we concentrated our attention on the nucleus, especially the tendency of some isotopes to undergo radioactive decay. We examined three forms of radiation, alpha, beta, and gamma, and examples of the harmful and helpful aspects of such radiation were presented.

ADDITIONAL PROBLEMS

16. Define the following terms.
    (a) compound   (b) element
    (c) proton   (d) neutron
    (e) atomic weight
17. According to Dalton's theory, how do atoms of different elements differ?
18. What subatomic particle determines what type of element an atom is?
19. Using the periodic table given in Figure 2.3 and your knowledge of the behavior of the various groups of elements, predict the reactivity of the following.
    (a) sodium (Na)   (b) chlorine (Cl)
    (c) argon (Ar)

20. Give the corresponding name or symbol, atomic number, and atomic weight of the following elements.
    (a) boron   (b) K
    (c) aluminum   (d) Ca
    (e) magnesium   (f) Cu
    (g) lead   (h) Ni
    (i) mercury   (j) Fe
21. Why is Bohr's model of the atom frequently referred to as the solar system model?
22. How did Bohr explain the colors radiated by elements when they are heated?
23. How many sublevels and orbitals are present in the third main energy level?

24. Give the electron distribution by main energy level, sublevel, and orbital for the following elements.
    (a) Na (b) P
    (c) Ca (d) Al
    (e) C   (f) F
    (g) Ar

25. Using a single Cartesian coordinate system (x-, y-, and z-axis), draw the p_z and p_y orbitals, using contour lines.

26. What is the difference between the orbits described in the Bohr model and the orbitals used in the wave-mechanical model?

27. How far do beta particles typically travel through living tissue?

28. Write equations for the following nuclear reactions.
    (a) alpha decay of $^{239}_{94}Pu$
    (b) beta decay of $^{3}_{1}H$
    (c) gamma decay of $^{99}_{43}Tc*$
    (d) beta decay of $^{99}_{43}Tc$
    (e) alpha decay of $^{214}_{83}Bi$
    (f) gamma decay of $^{64}_{28}Ni*$

29. The half-life of carbon-14 is 5730 years. If a plant dies today and contains 50 mg of carbon 14, how much of this isotope will be present in 11,460 years?

CHAPTER 3

CHEMICAL BONDS—
BONDS OF LIFE

In Chapter 2 we learned that all matter is composed of atoms and that these atoms are composed of protons, neutrons, and electrons. Our study in Chapter 2 was primarily concerned with elements in their uncombined, or atomic, form. The helium used to fill "lighter-than-air" balloons and the neon used in signs are both examples of such uncombined elements. Uncombined elements do not, however, play a very significant role in our lives. Only when atoms chemically combine with one another do we obtain the variety of substances that make up our bodies and provide us with the food we eat, the medicine we take, the clothes we wear, and the almost endless list of substances that surround us.

In this chapter we will consider the main factors that are involved in forming these substances, that is, how atoms are bonded or held together to form compounds. We will discuss the two major types of bonds, ionic and covalent. With a knowledge of these factors we will be able to understand better why substances behave the way they do, why some compounds are necessary for life while others will destroy it, why some substances are helpful and others are harmful. It can be important in the health-related sciences to know the names of various compounds. We will, therefore, conclude the chapter with a consideration of how to recognize and write the names and formulas of some of the more common and important substances that surround us.

3-1 INERT GAS ELECTRON DISTRIBUTION

LEARNING OBJECTIVES

1. Describe the relationship of the outermost, occupied main energy level to the properties of the elements.
2. Write the characteristic electron distribution that the inert gases have in common.

3. Discuss the significance of the inert gas electron distribution to atoms combined with one another.

In Chapter 2 we learned that electrons in atoms are distributed among the various atomic orbitals in an orderly fashion. We saw that, in each case, electrons occupy those orbitals where they can have the lowest energy. We also saw that a cyclic, or periodic, recurrence of properties is evidenced by the elements in the periodic table. A careful look at the properties and the electron distribution of the inert gases can help our understanding of how and why atoms combine chemically. We will consider these factors in this section.

The chemical and physical properties of an element depend primarily upon the number of electrons present in each atom of the element and especially upon the number of electrons in the *outermost, occupied main energy level* (the last or highest energy level containing electrons). In general, each element in a given group has the same number of electrons in its outermost, occupied main energy level and these electrons have a similar distribution among the various energy sublevels and orbitals. For example, all of the group IA elements have a single electron occupying an outer *s* orbital, as seen below.

and

$$H: (1s)^1$$
$$Li: (1s)^2(2s)^1$$
$$Na: (1s)^2(2s)^2(2p)^6(3s)^1$$
$$K: (1s)^2(2s)^2(2p)^6(3s)^2(3p)^6(4s)^1$$

Each of these elements except hydrogen is a very reactive metal. Similarly, the elements in group VIIA have two electrons in the *s* sublevel and five electrons in the *p* sublevel in their outermost, occupied level. As an illustration of this we have

$$F: (1s)^2(2s)^2(2p)^5 \qquad \text{and} \qquad Cl: (1s)^2(2s)^2(2p)^6(3s)^2(3p)^5$$

These and the other elements of group VIIA are very reactive nonmetals. Likewise all group 0 elements except helium have eight electrons in their outermost occupied main energy level and are very unreactive gases. In each case, the electron distribution in the atoms is the principal factor that determines the chemical and physical properties of an element. The importance of this fact shall be illustrated in Section 3-2 when we consider how the properties of an element change when the number of electrons in its atoms changes.

Atoms combine with one another by modifying their electron distribution; that is, *chemical bonds* are formed when atoms either change the number of electrons associated with each atom or alter the arrangement of the electrons about the atoms involved in the chemical bond. Of particular significance in this regard is the electron distribution of the inert gases to which reference has just been made. All of these elements, except helium, possess filled *s* and *p* sublevels in the outermost, occupied main energy level. This may be represented by means of sublevel notation as

$$(ns)^2(np)^6$$

where *n* is the number that identifies the outermost, occupied level for the specific element. For example, neon has the electron distribution

$$Ne: (1s)^2(2s)^2(2p)^6$$

with $n = 2$. The electrons in argon are distributed according to

$$\text{Ar: } (1s)^2 (2s)^2 (2p)^6 (3s)^2 (3p)^6$$

with $n = 3$.

As suggested by the group name, inert or noble gases, these elements do not readily combine with other elements. This is due to a particularly stable, or "desirable," electron arrangement. In other words, the inert gases do not tend to undergo a change in their electron distribution by chemically combining with other atoms.

In general, atoms tend to change their electron distribution so as to achieve a more stable electron arrangement. In doing this, atoms of elements other than the inert gases tend to combine with one another to become like an inert gas in terms of electron distribution. To do this, atoms gain, lose, or share electrons with one another until they become like the inert gas whose atomic number is closest in value to that of the element undergoing change. When this occurs the element is said to be *isoelectronic* with the inert gas.

PROBLEM

1. Which electrons in an atom are the most important in determining the chemical properties of an element?

ANSWER TO PROBLEM   **1.** those in the outermost occupied main energy level

3-2 TYPES OF BONDS

LEARNING OBJECTIVES

1. Define the following terms: chemical reaction, ion, chemical formula, ionic bond, electronegativity, covalent bond, molecule, molecular orbital, sigma orbital, and pi orbital.
2. Write chemical equations incorporating the sublevel electron distribution for the reaction of metals with non-metals to form ionic compounds.
3. Predict whether a bond between atoms is ionic or covalent based upon a knowledge of their electronegativities.
4. Describe the rule of eight and the rule of two and identify which rule is followed by each atom in a given compound.
5. Write Lewis electron-dot formulas for elements and simple compounds.

6. Describe what constitutes a single, a double, and a triple bond and identify which type of bond is involved in each bond in a given compound.
7. Predict whether a covalent bond in a given compound will be polar or non-polar and indicate the partial charges on atoms involved in polar bonds.
8. Specify whether atomic or molecular orbitals are involved in specific bonds of compounds.
9. Predict whether a molecular orbital in a given compound will be a sigma or pi orbital, based upon the atomic orbitals involved in forming the bond. Draw the resulting molecular orbital.

Much of what we have learned thus far has been designed to prepare us to understand the nature of chemical bonds. In this section we will begin our consideration of chemical bonds as we discuss the fundamental aspects of bond formation.

3-2.1 IONIC BONDS

As mentioned in Section 3-1, atoms tend to gain, lose, or share electrons to become isoelectronic with the nearest inert gas. In this section we will be concerned with bond formation that is due, primarily, to the gain and loss of electrons in atoms, that is, bonds due to a transfer of electrons between atoms.

Sodium is a very reactive metal. In fact, sodium is so reactive that when it comes in contact with water it reacts by tearing the water apart to release hydrogen as illustrated by the equation

$$Na + 2\,H_2O \longrightarrow H_2 + 2\,NaOH + energy$$

The vigor with which this occurs becomes evident when the released hydrogen ignites with the air and the remaining sodium melts. Chlorine, on the other hand, is a very reactive gaseous nonmetal. Because of its reactivity, chlorine is used in small amounts to purify water and as a disinfectant. In larger amounts it becomes a poison and can lead to death if a sufficient amount is inhaled.

The reactivity of these two elements is also illustrated when they join to form sodium chloride (NaCl),

$$2\,Na + Cl_2 \longrightarrow 2\,NaCl + energy \qquad (3.1)$$

As indicated in this equation, energy is released. Under certain conditions the reaction can occur with explosive force. Energy factors, such as this, shall be considered in more detail in Chapter 7, but now, it is sufficient to know that, in most cases, the greater the reactivity, the greater the amount of energy that is released.

When sodium reacts with chlorine, a sodium atom gives, or donates, an electron to a chlorine atom. This may be represented by the following equations, using sublevel electron distribution.

$$Na[(1s)^2(2s)^2(2p)^6(3s)^1] \longrightarrow Na^+[(1s)^2(2s)^2(2p)^6] + e^- \qquad (3.2)$$

$$Cl[(1s)^2(2s)^2(2p)^6(3s)^2(3p)^5] + e^- \longrightarrow Cl^-[(1s)^2(2s)^2(2p)^6(3s)^2(3p)^6] \qquad (3.3)$$

In this chemical reaction the sodium has attained the same electron distribution as neon and chlorine has become like argon.* Each sodium atom still has eleven protons but by losing an electron it ends up with only ten electrons, and is *isoelectronic* with neon. As a result of this change, sodium has a net charge of $+1$ as indicated by the plus sign that appears as a superscript following the symbol for sodium in equation (3.2). Similarly, by gaining an electron, chlorine has a total of 18 electrons and only 17 protons, giving it a net charge of -1. This is indicated by the minus sign after Cl in equation (3.3). The charged particles Na^+ and Cl^- are

A **chemical reaction** is the process during which one or more components are transformed into new substances.

---

*The number 2, found in three places in equation 3.1, has been ignored in equations 3.2 and 3.3 in order to make the illustration more clear. The reason for the number 2 will become more obvious when we consider the covalent bond.

called ions. In the process of the reaction, chlorine has become isoelectronic with argon.

Equation (3.1) shows that sodium chloride is formed as represented by the chemical formula, NaCl. By undergoing the electron changes shown in equations (3.2) and (3.3), both elements have undergone significant changes in their properties. A very reactive metal and a poisonous gas are now chemically united to form a crystalline solid that is so unreactive that we use it to season our food daily. It is significant that sodium ions are a necessary chemical species in the living system where, among other things, they act to preserve a proper balance between calcium and potassium to maintain normal heart action.

Sodium chloride normally exists as a solid composed of sodium and chloride ions arranged in a definite repeating pattern. Figure 3.1 illustrates a small crystalline cube of NaCl. Alternate layers of ions exist in the crystal, each interior ion surrounded by six ions of opposite charge. In a compound such as sodium chloride, the ions are held together by the electrical attraction of the oppositely charged ions. The resulting bond is called an ionic bond, and the compound is referred to as an *ionic compound.*

Ionic bonds are typically associated with compounds formed by the reaction of metals with nonmetals, due to the difference in the tendency of various atoms to draw, or attract, electrons toward them. Reference was made in Chapter 2 to the periodic behavior of the elements. One of the more important properties that illustrates this periodic variation is the electron-attracting ability of the elements. This tendency to attract electrons is known as electronegativity and generally increases from left to right and from bottom to top on the periodic table. The net effect of these trends is a gradual increase in the electronegativity of the elements proceeding from the lower left hand portion of the periodic table to the upper right hand corner. Thus, chlorine has a higher electronegativity than sodium,

An **ion** is an electrically charged atom or group of chemically bonded atoms.

A **chemical formula** consists of a group of symbols and subscripts that indicate the number and kinds of atoms in a chemical species.

An **ionic bond** is the force of attraction that binds together unlike charged ions to form a distinct chemical species.

Electronegativity is a measure of the innate tendency of an atom of an element to draw electrons to itself in a chemical bond.

Figure 3.1
A tiny crystal of NaCl has alternate layers of ions. Each interior ion, like the central Na$^+$ shown here, is surrounded by six ions of opposite charge.

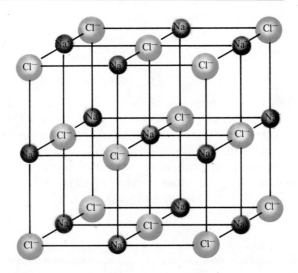

Figure 3.2 Electronegativity values of some of the elements. The arrows show general trend in increasing electronegativity.

H 2.1																	He . . .
Li 1.0	Be 1.5											B 2.0	C 2.5	N 3.0	O 3.5	F 4.0	Ne . . .
Na 0.9	Mg 1.2											Al 1.5	Si 1.8	P 2.1	S 2.5	Cl 3.0	Ar . . .
K 0.8	Ca 1.0	Sc 1.3	Ti 1.5	V 1.6	Cr 1.6	Mn 1.5	Fe 1.8	Co 1.8	Ni 1.8	Cu 1.9	Zn 1.6	Ga 1.6	Ge 1.8	As 2.0	Se 2.4	Br 2.8	Kr . . .
Rb 0.8	Sr 1.0	Y 1.2	Zr 1.4	Nb 1.6	Mo 1.8	Tc 1.9	Ru 2.2	Rh 2.2	Pd 2.2	Ag 1.9	Cd 1.7	In 1.7	Sn 1.8	Sb 1.9	Te 2.1	I 2.5	Xe . . .
Cs 0.7	Ba 0.9	La-Lu 1.1-1.2	Hf 1.3	Ta 1.5	W 1.7	Re 1.9	Os 2.2	Ir 2.2	Pt 2.2	Au 2.4	Hg 1.9	Tl 1.8	Pb 1.8	Bi 1.9	Po 2.0	At 2.2	Rn . . .
Fr 0.7	Ra 0.9	Ac-Lw 1.1-															

lithium than cesium, and fluorine than francium. These trends can be seen in Figure 3.2 where values of electronegativity are listed.

Values of electronegativities given in Figure 3.2 are relative values on what is known as the Pauling scale. A scale of this type is comparable to ranking how good a day has been. A typical scale for the quality of a day might be from 0 to 10, with a 10 indicating an unbelievably great day and 0, a day when everything went wrong. For the Pauling scale of electronegativities the "electron-appealing" element fluorine was assigned a value of 4.0 and then all other elements were ranked in comparison with fluorine. On the basis all the elements to the left of and below fluorine have lower electronegativity values because they have less tendency to attract electrons.

When atoms combine chemically, a sort of "electron tug of war" takes place. The two atoms involved in forming the bond compete for electrons in the outermost occupied main energy levels in each other. If one atom has sufficient "electron appeal" compared with the other, one or more electrons will be drawn away from the "less appealing" atom and an ionic bond will be formed. In this way, chlorine (EN = 3.0) pulled an electron from sodium (EN = 0.9), and an ionic bond was formed. This may be represented by the following equation.

$$\underbrace{Na + Cl}_{e^-} \longrightarrow Na^+ + Cl^-$$

The degree to which an electron is pulled or transferred to form compounds varies from one pair of atoms to another. In general, whenever the difference in electronegativities of the bonded atoms is at least 1.7, the bond is considered ionic.* Applied to NaCl, we have the difference in electronegativities, $\Delta(EN)$, given by

$$\begin{aligned}\Delta(EN) &= (EN)_{Cl} - (EN)_{Na}\\ &= 3.0 - 0.9\\ &= 2.1\end{aligned}$$

which is in agreement with the ionic nature assumed.

A second approach to predicting ionic bonds is the rough rule *elements that are far apart on the periodic table form ionic bonds.* Using this rule, rubidium fluoride (RbF) would be assumed to be ionic and carbon monoxide (CO) would not. In most cases of concern to us, this rule will suffice.

In the formation of NaCl a one-electron transfer takes place. A similar process occurs when a magnesium atom and an oxygen atom react, except that two electrons are transferred as illustrated by the equation

$$\begin{aligned}Mg[(1s)^2(2s)^2(2p)^6(3s)^2] &+ O[(1s)^2(2s)^2(2p)^4] \longrightarrow\\ &Mg^{2+}[(1s)^2(2s)^2(2p)^6] + O^{2-}[(1s)^2(2s)^2(2p)^6]\end{aligned} \qquad \textbf{(3.4)}$$

A slightly more complicated transfer takes place when sodium reacts with oxygen. In this case the oxygen atom tends to acquire two electrons and become like neon; however, if a sodium atom loses two electrons, it will "overshoot" and become isoelectronic with fluorine. This problem is easily solved by involving two sodium atoms and letting each atom donate an electron to the oxygen. This reaction may be represented by the equation

$$2\,Na + O \longrightarrow 2\,Na^+ + O^{2-} \qquad \textbf{(3.5)}$$

In this equation the electron notation has been omitted. In general, the electron notation for the various species involved in chemical reactions is not included and they shall not be given in the majority of the equations in the remainder of this book.

More complex reactions can also take place. For example, when aluminum reacts with oxygen, two aluminum atoms each lose three electrons, and three oxygen atoms each gain two electrons for a total of six electrons transferred.

$$\begin{aligned}2\,Al[(1s)^2(2s)^2(2p)^6(3s)^2(3p)^1] &+ 3\,O[(1s)^2(2s)^2(2p)^4] \longrightarrow\\ &2\,Al^{3+}[(1s)^2(2s)^2(2p)^6] + 3\,O^{2-}[(1s)^2(2s)^2(2p)^6]\end{aligned}$$

In all these reactions, each atom achieved an inert gas electron distribution. This is true for many reactions where ionic bonds are formed. Most atoms of the representative elements that are in ionic compounds have eight electrons in their outermost occupied main energy level. This fact is often referred to as *the rule of eight,* or *the octet rule.* Exceptions to this rule occur for the transition elements where electrons can also be located in *d* sublevels in the outermost occupied main energy level of the ions and for the very light elements [up to boron (B) on the periodic table], which tend to become isoelectronic with helium. The latter ele-

---

*Chemists frequently use a simpler approach than this, classifying all salts (compounds composed of a metal and a nonmetal) as ionic. In some cases the difference in electronegativities is less than 1.7, but the compounds, in general, have the same basic properties as compounds that fulfiill this criterion and so are referred to as ionic compounds.

ments tend to react with other elements in such a way that they have two electrons in the first main energy level. This tendency is known as *the rule of two.* For example in the reaction of lithium with fluorine.

$$Li[(1s)^2(2s)^1] + F[(1s)^2(2s)^2(2p)^5] \longrightarrow Li^+(1s)^2 + F^-[(1s)^2(2s)^2(2p)^6]$$

lithium follows the rule of two.

Recognition of these two rules led an American chemist, G. N. Lewis, to propose an alternative method for illustrating the electron distribution for elements. Using the *Lewis electron-dot formulas,* or *kernel notation,* we give the symbol of an element and then indicate the number of electrons in the outermost occupied main energy level by surrounding the symbol with a corresponding number of dots. The four elements referred to in equations (3.1) and (3.2) are given by

$$Na\cdot \qquad \cdot\ddot{\underset{..}{C}}l\!: \qquad Mg\cdot \qquad \cdot\ddot{O}\!:$$

As shown here, the dots are usually positioned on the sides of an invisible square surrounding the symbol, $\boxed{X}$.

Application of this to equation (3.1) gives

$$Na^{\times} + \cdot\ddot{\underset{..}{C}}l\!: \longrightarrow Na^+ + {}^{\times}\!\ddot{\underset{..}{C}}l\!:^- \tag{3.6}$$

Two things should be noted about this equation. First, no electron dots appear around the sodium ion, Na^+. This agrees with the common practice of always considering the same main energy level of a given element throughout the equation. For the sodium atom, this is the third level. Since the sodium ion has no electrons in that level, none are shown. The second feature is the use of different symbols ($\cdot$ and $\times$) to represent the electrons on different elements, as seen on the sodium atom in equation (3.6). This bookkeeping aid is used to assist in accounting for electrons throughout the equation. This will be especially helpful when Lewis electron-dot formulas are given for complex ions. In such cases it is very easy to lose, or misplace, electrons as they are carried from left to right through the equation. Using different symbols minimizes this error. However, one must be careful to not assume that the electron is different in some way because a different symbol is used. All electrons are basically identical, differing only when they occupy different orbitals. The electron that was transferred from the sodium atom to the chlorine atom in equation (3.6) cannot be identified in the chloride ion despite the bookkeeping trick.

3-2.2 COVALENT BONDS

As discussed in the previous section, atoms with electronegativity values that are very different combine chemically through a transfer of one or more electrons. Chemical bonds are not restricted to such a transfer, however. Many compounds are formed that involve elements with very similar electronegativities. This is particularly true for compounds involving nonmetals. Such compounds can even form between atoms with identical electronegativities. In all of these cases neither atom has a strong enough attraction for the electrons on the other atom to remove any electrons; therefore, a transfer of electrons does not occur. In such cases, a type of bond that involves a sharing of electrons is formed. This is called a **covalent bond.** In these bonds, the shared electrons count for each atom forming the bond; therefore, each atom can achieve the electron distribution of an inert gas.

Each covalent bond between atoms involves two, and only two, electrons. The simplest example of a covalent bond is the bond between two hydrogen atoms. Each hydrogen atom has a tendency to gain an electron and become more like helium. However, since the two atoms are identical, neither is capable of drawing an electron from the other atom. In this case the atoms essentially compromise and share the two electrons as illustrated by the equation

$$\text{H}^{\textbf{.}} + {}_{\text{x}}\text{H} \longrightarrow \text{H}{}_{\text{x}}^{\textbf{.}}\text{H} \tag{3.7}$$

The electrons in this pair of shared electrons are called *bonding electrons.* Such a shared pair of bonding electrons make up what is referred to as a *single bond.* In an alternative Lewis electron-dot formula, a single line is used to represent the bonding electron pair. Using this *line formula,* the bond between hydrogen atoms is written H—H. Both types of Lewis electron-dot formulas will be useful to us in emphasizing the bonds involved in various chemical species. In most cases, however, the ordinary chemical formulas referred to earlier in this section are used. Thus, hydrogen has the formula H_2.

The bonding of two fluorine atoms is similar to that of two hydrogen atoms. Each fluorine atom has seven electrons in the outermost occupied main energy level and each tends to gain an additional electron in order to have eight electrons. Again, a compromise is reached as shown by the formulas

$$:\overset{..}{\underset{..}{\text{F}}}{}_{\text{x}}^{\text{xx}}\overset{\text{xx}}{\underset{\text{xx}}{\text{F}}}{}_{\text{x}} \quad \text{or} \quad \text{F}—\text{F}$$

As illustrated on the right, electrons not involved in bond formation are generally omitted when line formulas are written.

A slightly more complicated process takes place when two nitrogen atoms combine. In this case, each atom is three electrons short of the electron distribution of the nearest inert gas, neon. In the compromise between the two nitrogen atoms that occurs, three pairs of electrons are shared. A total of three bonds is formed, as illustrated by the formulas

$$:\text{N}{}_{\text{x}}^{\text{xx}}{}_{\text{x}}\text{N}{}_{\text{x}} \quad \text{or} \quad \text{N}\equiv\text{N}$$

The combination of bonds given here is referred to as a *triple bond.* Similar multiple bonds involving the sharing of two pairs of electrons also exist in many compounds. There are referred to as *double bonds.*

The chemical species H_2, F_2, and N_2 are examples of **molecules.*** More specifically, these substances exist as *diatomic molecules.* This fact was ignored for simplicity in equations (3.3), (3.4), (3.5), and (3.6). In H_2, F_2, and N_2, the bonding electrons are equally shared between identical atoms. Covalent bonds of this type are known as **nonpolar bonds.** Nonpolar bonds are bonds between atoms that have the same electronegativity.

---

*In this text, we will restrict the term *molecule* to chemical species involving two or more atoms held together with covalent bonds and possessing no net charge. A more general label, *unit particle,* can be used in reference to any distinct chemical species, including atoms, ions, and molecules.

A **covalent bond** is the force of attraction that joins two atoms by sharing a pair of electrons to form a distinct chemical species.

A **molecule** is the smallest combination of atoms of an element or compound that has a stable, independent existence and possesses no net charge.

Nonpolar bonds are covalent bonds with positive and negative charges evenly distributed between two bonded atoms.

Thus far, we have looked at the two extremes of bonding: (1) bonds formed by a transfer of electrons between different atoms and (2) bonds formed by an equal sharing of electrons between identical atoms. Most compounds involve bonding that lies somewhere between these two extremes. These bonds still involve a sharing of electrons, but there is no longer an equal sharing. An important example of this type of bonding is provided by the compound HCl. Known to chemists as hydrogen chloride or, when in solution, hydrochloric acid, HCl is an essential constituent of the gastric juice contained in the stomach.

When hydrogen (EN = 2.1) and chlorine (EN = 3.0) combine, a covalent bond is formed, as illustrated by the formulas

$$H {\overset{\times}{\cdot}} \overset{..}{\underset{..}{Cl}}: \quad \text{or} \quad H—Cl$$

Since the difference in electronegativity is only 0.9, the chlorine atom does not have sufficient "electron appeal" to draw the electron completely away from the hydrogen atom. Thus, an ionic bond does not form. On the other hand, since the two atoms do not have an equal attraction for the electrons, they do not share the bonding electrons equally. The covalent bond that is formed involves an unequal sharing of electrons, where the bonding electron pair is drawn more toward the chlorine atom.

This unequal sharing can be viewed as a sort of partial ionic bond, with the two atoms having a partial charge. This arrangement can be represented by the following line formula,

$$H^{\delta+}—Cl^{\delta-}$$

The partial charges are indicated by means of the Greek letter delta, δ, and the appropriate sign.

The bond in HCl is an example of what we refer to as a **polar bond** in recognition of the charge separation suggested in the formula above. We often refer to such molecules as *polar molecules*. Other examples of diatomic molecules that are polar are carbon monoxide (CO), hydrogen bromide (HBr), and nitrogen monoxide (NO). Most compounds in the living system are composed of polar molecules.

So far we have dealt with rather simple diatomic molecules. The great variety of compounds that make up the world is possible because of the ability of most atoms to bond with two or more atoms at the same time. Nitrogen exhibits this ability. Nitrogen tends to form three bonds, as illustrated by the triply bonded N_2 molecule. These three bonds may be formed with the same atom, as in the case of N_2, or may involve three different atoms. An example of the latter is the molecule formed by nitrogen and hydrogen. Hydrogen can only form a single bond. Therefore, when an atom of hydrogen reacts with an atom of nitrogen it cannot satisfy the bonding needs of the nitrogen. As a result, nitrogen tends to form single bonds with three hydrogen atoms as shown in the formulas

$$H {\overset{\times}{\cdot}} \overset{..}{\underset{\overset{\times}{\underset{\cdot}{}}}{N}} {\overset{\times}{\cdot}} H \quad \text{or} \quad H—\overset{|}{\underset{|}{N}}—H$$
$$\overset{}{\underset{H}{}} \qquad\qquad\quad \overset{}{\underset{H}{}}$$

A **polar bond** is a covalent bond where there is an unequal charge distribution between the two bonded atoms.

In general the number of covalent bonds that an atom of an element is capable of forming is equal to the number of electrons it would need to gain or lose to become like the nearest inert gas. Thus carbon normally forms four bonds and

Figure 3.3
Covalent bond formation. The lines associated with each element represent potential covalent bonds.

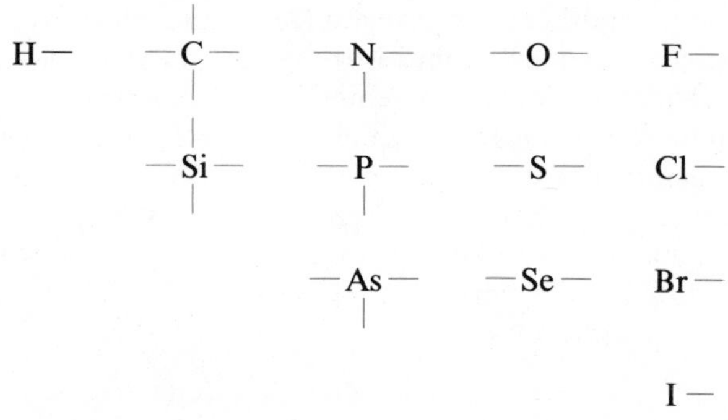

$$H- \qquad -C- \qquad -N- \qquad -O- \qquad F-$$
$$-Si- \qquad -P- \qquad -S- \qquad Cl-$$
$$-As- \qquad -Se- \qquad Br-$$
$$I-$$

oxygen forms two bonds. Figure 3.3 illustrates the usual number of covalent bonds formed by the more common elements in the periodic table. Each line represents a potential covalent bond for a given element.

The ability of atoms to form multiple bonds, such as in the case of ammonia, greatly increases the number of possible compounds that can be formed from the various elements. An extension of this tendency to form multiple bonds makes it possible for an almost limitless number of substances to form. Certain atoms, most notably carbon, are capable of bonding together in long chains and even chains with side, or branch, chains attached to the main chain. This characteristic of carbon is central to the chemistry of carbon, which shall be considered in the organic chemistry portion of the text.

A good example of the chain-forming ability of atoms is the compound ethyl alcohol.

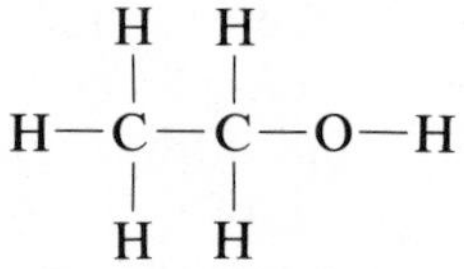

$$\begin{array}{ccc} & H & H \\ & | & | \\ H- & C- & C-O-H \\ & | & | \\ & H & H \end{array}$$
ethyl alcohol

This substance is commonly known as grain alcohol and is the principal component in alcoholic beverages. In medicine it is used as an antiseptic and as a preservative in certain medicines such as cough medicine. Comparison of the bonds formed by the carbon, hydrogen, and oxygen atoms in ethyl alcohol with those predicted in Figure 3.3 shows that each atom has formed the expected number of bonds. Another alcohol that illustrates a type of branch chain is isopropyl alcohol, also known as rubbing alcohol.

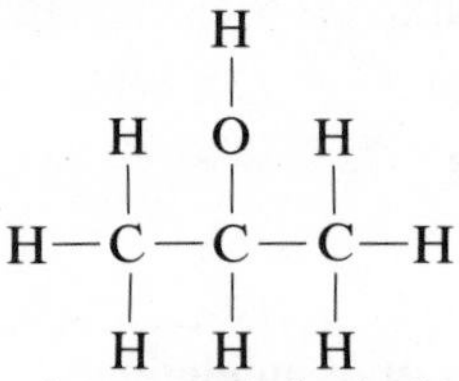

$$\begin{array}{cccc} & & H & \\ & & | & \\ & H & O & H \\ & | & | & | \\ H- & C- & C- & C-H \\ & | & | & | \\ & H & H & H \end{array}$$
isopropyl alcohol

As seen in this structure, the O—H (hydroxyl) group, typical of alcohols, is attached to the central atom and acts as a small branch chain. Comparison of the properties of the two compounds gives a good illustration of how properties of similar substances can vary. Although ethyl alcohol can be consumed in reasonable amounts without serious ill effects, as little as 100 mL of isopropyl alcohol can prove fatal. These and other alcohols shall be considered in depth in Chapter 14.

3-2.3 MOLECULAR ORBITALS

In the discussion of ionic and covalent bonds, we considered the transfer and sharing of electrons, but little reference was made to the orbitals that the bonding electrons occupy. When bonds form between atoms, all the electron orbitals are affected to a certain extent. However, the major change may be assumed to involve only the bonding electrons, so we will concentrate our attention on them.

In the case of ionic bonds, the electrons are assumed to undergo a transfer from atomic orbitals in the metal atom to atomic orbitals in the nonmetal atom. For sodium chloride this means that an electron in the $3s$ orbital of sodium transfers to the $3p_z$ orbital of chlorine to pair up with the lone electron already occupying that orbital.

$$Na[(1s)^2(2s)^2(2p)^6(3s)^1] + Cl[(1s)^2(2s)^2(2p)^6(3s)^2(3p_x)^2(3p_y)^2(3p_z)^1] \longrightarrow$$
$$Na^+[(1s)^2(2s)^2(2p)^6] + Cl^-[(1s)^2(2s)^2(2p)^6(3s)^2(3p_x)^2(3p_y)^2(3p_z)^2]$$

When covalent bonds are formed, we assume that a more complex process than that for ionic bonding occurs. In this case the bonding electrons are believed to occupy a different region in space than they did in the separated atom. This region is referred to as a **molecular orbital**. In the reaction between two hydrogen atoms, each atom has an electron in a $1s$ orbital. As the atoms combine, the region in space that they occupy is changed. This change is illustrated in Figure 3.4, where the reaction is shown, emphasizing the orbitals that are involved. As seen here, the molecular orbital differs significantly in shape from the original $1s$ atomic orbitals. The molecular orbital shown in Figure 3.4 is an example of a **sigma (σ) orbital**. The bond that is formed is known as a *sigma bond*. For our purposes, it may be assumed that all single covalent bonds are sigma bonds.

A **molecular orbital** is the region in space occupied by a bonding pair of electrons.

A **sigma orbital** is a molecular orbital that is symmetrical about the line joining the two bonded atoms.

Figure 3.4   Two hydrogen atoms with $1s$ electrons react to form a hydrogen molecule with a bonding pair of electrons in a σ bond.

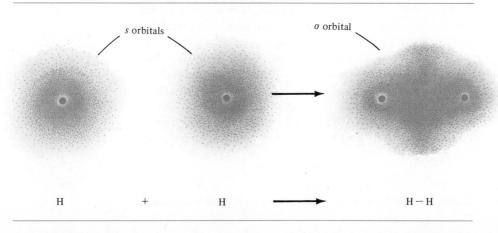

| | | s orbitals | | σ orbital |

| H | + | H | → | H $-$ H |

Figure 3.5   (a) Two fluorine atoms each with one $2p_z$ electron react to form a fluorine
molecule with a σ bond. (b) A hydrogen atom with a $1s$ electron reacts with
a chlorine atom with a single $3p_z$ electron to form a σ bond. (Only orbitals
involving bonding electrons are shown in order to improve the clarity of the
figures.)

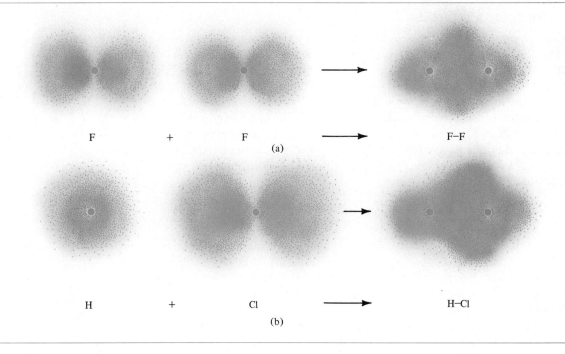

F   +   F   ⟶   F—F
(a)

H   +   Cl   ⟶   H—Cl
(b)

Sigma orbitals can be formed by two atoms involving s electrons, as in the case
of H_2, or two atoms involving p electrons, as in the case of F_2, or an atom with an
s electron bonding with an atom with a p electron, as involved in the formation
of HCl. The reactions associated with the latter two compounds are illustrated in
Figure 3.5. Sigma orbitals are symmetrical about a straight line passing through the
nuclei of the two atoms that are bonded and the region of greatest probability for
locating the bonding electrons occurs between the two nuclei. This distribution of
the bonding electrons contributes to the "electronic glue" that holds the molecule
together. An illustration of this type of symmetry is an apple. If a long needle is
passed through the stem and core, the apple is symmetrical about the needle; that
is, the apple is equally distributed about the needle at any distance from the needle.
Figures 3.4 and 3.5 illustrate this *line symmetry* as it applies to the sigma orbitals.

When multiple bonds are formed between two atoms, another type of molecular
orbital may be formed. This can be illustrated by considering again the triple bonds
in a molecule of N_2. The orbital electron distribution for an atom of nitrogen is

$$(1s)^2(2s)^2(2p_x)^1(2p_y)^1(2p_z)^1$$

One of the p electrons on each atom, for instance the one in the p_x orbital, forms
a sigma bond as pictured at the top of Figure 3.6. The other two p orbitals on each

atom, p_y and p_z, are oriented parallel to the comparable orbitals on the other atom. The two p_y orbitals are parallel, as are the two p_z orbitals. Neither orbital can form a sigma bond since they do not have the necessary end-to-end orientation illustrated for F_2 in Figure 3.5. This orientation is illustrated in the middle and bottom portions of Figure 3.6. We see that molecular orbitals are formed, but these two orbitals involve a region in space that has a different shape than does the sigma orbital. These orbitals are shaped somewhat like the two parts of a sliced hot dog bun, and are called **pi (π) orbitals.** A pi orbital occurs both above and below a plane, or sheet, passing through the nuclei of the atoms involved in the bond. A pi orbital has what is called *planar* symmetry. The bond associated with a pi orbital is known as a *pi bond.*

A **pi orbital** is a molecular orbital that lies above and below a plane passing through the nuclei of the two bonded atoms.

Figure 3.6 The reaction of two nitrogen atoms involves three molecular orbitals (shown separately here for clarity). One σ and two π molecular orbitals are formed.

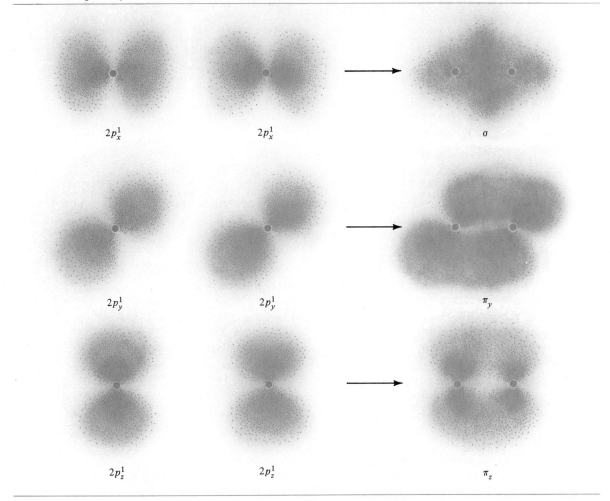

$2p_x^1$ $2p_x^1$ σ

$2p_y^1$ $2p_y^1$ π_y

$2p_z^1$ $2p_z^1$ π_z

The two types of molecular orbital, sigma and pi orbitals, can be assumed to account for covalent bonding. In each case a covalent bond is associated with a shared pair of electrons. In addition, atoms may have one, two, or three such bonds holding them together. As might be expected, the greater the number of bonds between two atoms, the greater the strength of the bond.

PROBLEMS

⚐ 2. Which of the following pairs is most likely to form an ionic bond: Br and Cl, K and F, or H and H?

⚐ 3. Would you expect Be to follow the rule of eight or the rule of two?

4. Write the Lewis electron-dot formula for the following: (a) BrCl (b) KF (c) CO_2

⚐ 5. Is there a single, double, or triple bond between the carbon and each oxygen in CO_2? Are the bonds polar or nonpolar?

6. Draw a rough sketch of a sigma orbital involved in the bond between hydrogen and fluorine in HF.

⚐ 7. Would you expect the molecular orbitals between nitrogen and the three hydrogens in NH_3 to be sigma or pi orbitals?

ANSWERS TO PROBLEMS **2.** K and F **3.** the rule of two **4. (a)** :Br×Cl× **(b)** K+ ×F:−

(c) O×C×O **5.** double bonds polar bonds **6.** H ⬭ F **7.** sigma

---

3-3 FORMULAS AND NAMES OF INORGANIC COMPOUNDS

LEARNING OBJECTIVES

1. Determine the oxidation numbers of each element in atoms, molecules, ions, and compounds.

2. Write chemical formulas of any of the elements and common binary and

ternary compounds as directed by the instructor.

3. Write names of common binary and ternary compounds as directed by the instructor.

In Section 3-2 we discussed the nature of the chemical bond. When bonds form, compounds are produced. By understanding how bonds are formed we have been able to determine the chemical formulas of some relatively simple compounds. However, the complexity of many chemical species makes it difficult, if not impossible, to derive their formulas using the procedure followed by Section 3-2. In Section 3-3 we will expand our ability to write chemical formulas, while developing an ability to name chemical compounds based upon their formulas. An ability to write chemical formulas and name compounds shall be helpful throughout the remainder of the course and in your professional career.

3-3.1 OXIDATION NUMBERS

In Section 3-2, the chemical formulas of several compounds were obtained by following the rule of two or the rule of eight. In this section we will consider a more workable approach to writing chemical formulas based upon an arbitrary set of numbers called oxidation numbers. These numbers give the combining tendency of an element. As an example, we will consider the substance calcium chloride, which is used to raise the calcium content of the blood in disorders resulting from a lack of adequate calcium. This substance has the formula $CaCl_2$, indicating that one atom of calcium combines with two atoms of chlorine. In other words, the combining abilities of the two atoms are in a ratio of 2 to 1. These two numbers are, basically, the oxidation numbers of the two elements, except we also include a sign, $+$ or $-$, when referring to the oxidation number of a given element. In general, the oxidation number of the more electronegative element is negative and the oxidation number of the less electronegative element is positive. The oxidation numbers, or *oxidation states,* of calcium and chlorine in $CaCl_2$ are, therefore, $+2$ and -1, respectively.

The **oxidation number** of an element is the charge on an ion of the element or the charge that an atom in a covalent compound would have if the bonding electrons were completely transferred to the more electronegative atom in the bond.

Applying oxidation numbers to writing chemical formulas is similar to working a crossword puzzle. In both cases you may need to establish a few key parts of the puzzle before you can complete the entire picture. To play the game of formula-writing we first must know the oxidation numbers of a core group of elements, then we must know how to apply them. Then, if we do not know the oxidation number of a specific element in some chemical species, usually we can determine it from the ones we do know. The following rules give us the necessary information to do this.*

1. *The oxidation number of elements not combined with other elements is 0.* This includes both monoatomic (one atom) and diatomic (two atom) species such as Na, Mg, H_2, and O_2.
2. *The oxidation number of hydrogen in compounds is $+1$.* Thus, the oxidation state of hydrogen in HCl, NH_3, and CH_4 is $+1$ in each case.
3. *The oxidation number of oxygen in compounds is -2.* Thus the oxidation state of oxygen in BaO, H_2O, $(CH_3COO)_2Ca$, and PO_4^{3-} is -2 in each case.
4. *Group IA elements have a $+1$ oxidation number in compounds.* For example, sodium has a $+1$ oxidation number in NaCl.
5. *Group IIA elements have a $+2$ oxidation number in compounds.* As discussed previously, calcium has a $+2$ oxidation number in $CaCl_2$.
6. *The sum of the oxidation numbers for all the atoms in a chemical species is equal to the charge of the species.* For neutral compounds such as NaCl and HCl, the sum is equal to 0. For polyatomic ions such as NH_4^+ and SO_4^{2-}, it is equal to their charge; that is, $+1$ and -2, respectively.
7. *The oxidation number of the more electronegative element in a binary (two element) compound is generally equal to the group number minus 8.* For example, the oxidation number of sulfur in PbS is $6 - 8 = -2$ and of bromine in $AlBr_3$ is $7 - 8 = -1$.

In addition to the rules above, it is helpful to memorize the oxidation numbers of a few metals, particularly some of the more common transition metals. Table 3.1 lists some of these values. In several cases the transition metals have two or more possible oxidation numbers. Similarly, most of the nonmetals can exist in several *oxidation states.* In order to determine which oxidation number applies to one of

---

*Exceptions to some of these rules occur; however, these exceptions do not concern us.

Table 3.1   Common Oxidation Numbers of Metal Ions

Ion	Oxidation Number	Ion	Oxidation Number
Al	+3	Ag*	+1
Fe*	+2, +3 **(ferr)**	Sn	+2, +4 **(stann)**
Ni*	+2	Hg	+1, +2 **(mercur)**
Cu*	+1, +2 **(cupr)**	Pb	+2, +4 **(plumb)**
Zn	+2		

*Metals marked with an asterisk have other oxidation numbers than the ones shown but those given are the more important ones.

these elements in a particular compound, we must know the oxidation numbers of the other elements in that compound.

EXAMPLE 3.1   What are the oxidation numbers (ON) of the elements in the chemical species (a) Fe_2O_3, (b) $Al_2(SO_4)_3$ (used in treating sewage, as an agricultural pesticide, and as an antiseptic and detergent in medicine), and (c) ClO_4^-?

(a) For Fe_2O_3, Rule 3 gives oxygen an oxidation number of -2. Because Fe has two possible oxidation numbers (see Table 3.1), we must determine which oxidation number applies in this compound. Using Rule 6, we can write

$$2(ON)_{Fe} + 3(ON)_O = 0$$

or

$$2(ON)_{Fe} + 3(-2) = 0$$

or

$$(ON)_{Fe} = +\frac{3 \times 2}{2} = +3$$

(b) For $Al_2(SO_4)_3$, Table 3.1 gives $+3$ for Al and rule 3 yields -2 for O. Sulfur may have one of several oxidation numbers. Applying rule 6 gives

$$2(ON)_{Al} + 3(ON)_S + 3 \times 4(ON)_O = 0$$

or

$$2(+3) + 3(ON)_S + 3 \times 4(-2) = 0$$

or

$$(ON)_S = \frac{(3 \times 4 \times 2) - (2 \times 3)}{3} = +6$$

(c) For ClO_4^- Rule 3 gives -2 for O, but Cl may exist in several different oxidation states. Using rule 6, we have

$$(ON)_{Cl} + 4(ON)_O = -1$$

or

$$(ON)_{Cl} + 4(-2) = -1$$

or

$$(ON)_{Cl} = -1 + (4 \times 2) = +7$$

3-3.2 WRITING INORGANIC FORMULAS

The chemical formula is a convenient means for representing chemical species. We have already found it useful in many instances in the text. In this section we will consider the basic rules for writing formulas for some of the more common chemicals. We will deal with three classes: (1) elements, (2) binary compounds, and (3) ternary compounds. For the latter two classes we will concentrate our attention on compounds formed by the reactions of metallic species with nonmetallic species (such compounds are called salts).

Elements

The formulas of elements generally fall into two classes, atoms and diatomic molecules. The formulas of all the elements except the *seven diatomic molecules* H_2, N_2, O_2, F_2, Cl_2, Br_2, and I_2 can be written using the symbol alone.* For example Na, Fe, He, Cu, and P.

Binary Compounds

These are two-element compounds whose formulas are arrived at by means of the following rules:

1. *Write the symbol of the more metallic element first.* As an example, for the compound aluminum oxide this is Al, Al.
2. *Write the symbol of the less metallic element next.* For aluminum oxide this is O, AlO.
3. *Assign oxidation numbers to both elements.* For aluminum and oxygen these are Al: $+3$ and O: -2, $Al^{3+}O^{2-}$.
4. *Interchange oxidation numbers as subscripts, ignoring the sign in each case.* For aluminum oxide this yields Al_2O_3.
5. *Divide through by the greatest common divisor.* In the above example, and in most cases, this step can be ignored. Occasionally formulas are obtained which require such division. An example is the compound lead(IV) oxide, a highly poisonous substance used in the manufacture of pigments. The number four in Roman numerals indicates that the oxidation number of lead is $+4$. Applying rules 1, 2, and 3 for writing formulas for two-element compounds to this compound gives us Pb: $+4$, O: -2. Rule 4 then gives us the formula Pb_2O_4, which contains just twice as many atoms of each type as what should be written in the formula. Applying Rule 5 gives PbO_2, which is the correct formula.

Ternary Compounds

These are compounds that contain three different elements. We will be concerned primarily with oxygen-containing ternary compounds, especially those containing what are known as *polyatomic ions.* Table 3.2 contains a list of the more common polyatomic ions with their names.

The following are rules for writing formulas for ternary compounds.

1. *Write the symbol of the metallic element first.* If we are writing the formula for barium phosphate the symbol of the metal is Ba, Ba.

---

*There are exceptions to this rule, such as sulfur, which ordinarily exists in the form S_8, and phosphorus, which has a formula of P_4. We will not deal with these and other exceptions in this text.

Table 3.2 Formulas of Common Polyatomic Ions

Positive ion	Carbon-containing	Subgroup VA	Subgroup VIA	Subgroup VIIA
NH_4^+ ammonium	CO_3^{2-} carbonate	NO_3^- nitrate		
		NO_2^- nitrite		
		PO_4^{3-} phosphate	SO_4^{2-} sulfate	ClO_3^- chlorate
		$H_2PO_4^-$ dihydrogen phosphate	HSO_4^- hydrogen sulfate	
		HPO_4^{2-} hydrogen phosphate		
	HCO_3^- bicarbonate or hydrogen carbonate	AsO_4^{3-} arsenate	SeO_4^{2-} selenate	BrO_3^- bromate
	$C_2H_3O_2^-$ or CH_3COO^- acetate			IO_3^- iodate
	$C_2O_4^{2-}$ or $(COO^-)_2$ oxalate			

2. *Write the formula of the polyatomic ion next.* From Table 3.2 this is PO_4^{3-}, $BaPO_4$.

3. *Assign oxidation numbers to the metal and the polyatomic ion.* For barium we have Ba: $+2$. The oxidation number of an ion is equal to the charge on the ion; therefore, we have for PO_4^{3-} : -3, $Ba^{2+}PO_4^{3-}$.

4. *Interchange oxidation numbers as subscripts.* If more than one polyatomic ion is to be indicated, parentheses should be drawn around the polyatomic ion before the subscript is added. For barium phosphate this yields $Ba_3(PO_4)_2$.

5. *Divide through by the greatest common divisor.* In $Ba_3(PO_4)_2$ this is not necessary because the greatest common divisor is the number 1. A compound in which this must be done is calcium sulfate, a substance used to make plaster casts. Applying rules 1, 2, and 3 for writing formulas of ternary compounds gives Ca: $+2$, SO_4^{-2}: -2. Rule 4 then gives $Ca_2(SO_4)_2$, which upon division by 2 yields the correct formula $CaSO_4$.

The rules above give the basic information for writing chemical formulas. They are based upon the fact that chemical compounds obey the *principle of electroneutrality*. This principle states that while individual ions can have a net positive or negative charge the overall compound will have no net charge. There are various classes of substances that do not at first appear to fit these rules. Some of these shall be covered in Section 3-3.3. As you become familiar with naming these substances your ability to write formulas will increase.

3-3.3 INORGANIC NOMENCLATURE

The term *nomenclature* refers, in general, to the system of names used in a particular branch of knowledge or art. Applied to chemistry it is concerned with the names of the various chemical species. In this section we will deal with the names of some of the more common compounds that are in the class of substances referred to as inorganic compounds. These are primarily the noncarbon-containing compounds. As in Section 3-3.2, we will consider compounds in the two general categories, binary and ternary.

Binary Compounds

We give the general rules first, then modifications of these rules. The rules are:

1. *State the name of the more metallic element first.* If we are naming $BaCl_2$ (a substance once used in treating complete heart block as a cardiac muscle stimulant), the name of the metal is barium.

2. *State the stem of the less metallic element.* A list of common stem names is given in Table 3.3. For chlorine this is *chlor-*

Table 3.3 Common Nonmetal Stem Names

C: *carb-* (salt) *carbon-* (acid)	N: *nitr-* P: *phosph-* (salt) *phosphor-* (acid) As: *arsen-*	O: *ox-* S: *sulf-* (salt) *sulfur-* (acid) Se: *selen-*	F: *fluor-* Cl: *chlor-* Br: *brom-* I: *iod-*

3. *Add the suffix -ide after the stem of the nonmetal.* The first three steps yield the correct name, *barium chloride.*

4. *If the metal has more than one oxidation state then either of the following approaches can be used.*

   (a) *The Stock system.* This approach consists of giving the oxidation number of the metal in Roman numerals, surrounded by parentheses, between the name of the metal and the stem of the nonmetal. Therefore, on applying rules 1 to 4 for naming binary compounds, SnF_2 becomes tin(II) fluoride and SnF_4 is named tin(IV) fluoride.

   (b) *The ous–ic method.* When there are only two important oxidation states, the higher oxidation state can be indicated by the suffix *-ic* following the stem name of the metal. Metallic stem names are given in parentheses in Table 3.1. The lower oxidation state is given by the suffix *-ous.* Using this method, in conjunction with Rules 1 to 4 for naming binary compounds, we see that SnF_2 becomes stannous fluoride, a name you may recognize as the active ingredient in some fluoride toothpastes. Using this method for naming SnF_4, we have stannic fluoride.

5. When the compound is composed of two nonmetals, the Greek prefixes given in Table 3.4 are used to indicate the number of atoms of each type that are in the compound. Thus, N_2O_3 is dinitrogen trioxide. The prefix *mono-* is usually omitted except in a few cases such as the very toxic gas carbon monoxide (CO).*

6. *Many hydrogen-containing compounds produce what are called acidic solutions when they are dissolved in water.* These can, in general, be identified by the fact

Table 3.4
Greek Prefixes

1: *mono-*
2: *di-*
3: *tri-*
4: *tetra-*
5: *penta-*

---

*Many exceptions to this rule are in common usage. For example, the general anesthestic N_2O is called nitrous oxide and laughing gas, rather than the proper names dinitrogen monoxide or dinitrogen oxide.

that hydrogen appears first in the formula. Exceptions to this occasionally occur, for example, H_2O, but this rule can be helpful in identifying acids while the student is gaining experience. These solutions are represented by putting the symbol aq in parentheses after the formula, for example, HCl(aq). The names for these solutions are derived differently than those for binary compounds. Their name is written by giving the prefix *hydro-* then the stem of the nonmetal followed by a suffix *-ic,* and then the word acid. All stems are the same as before except the acid form is used for phosphorus and sulfur. Thus HCl(aq) becomes hydrochloric acid and H_2S(aq) becomes hydrosulfuric acid.*

7. *Three important types of ternary compounds are named as binary compounds.* These are compounds containing ammonium (NH_4^+), hydroxide (OH^-), and cyanide (CN^-). Thus, NaOH is named sodium hydroxide, the deadly substance HCN(aq) is referred to as hydrocyanic acid, and NH_4Cl is called ammonium chloride.

Ternary Compounds

The names for ternary compounds are derived similarly to those for binary compounds.

1. *State the name of the metallic element first.* Metals with two or more oxidation states are named in the same way as binary compounds. Thus for the compound $CuNO_3$ we have, using the Stock system, copper(I), and using the ous–ic approach, cuprous for the name of the metal.

2. *State the name of the polyatomic ion.* Table 3.2 gives the names of the more common polyatomic ions. The ion NO_3^- is seen in this table to have the name nitrate. The name of the compound can therefore be written as either copper(I) nitrate or cuprous nitrate. In the example above the *-ate* given after the stem *nitr-* is a suffix associated with the common polyatomic ions.

Other polyatomic ions with different numbers of oxygen atoms than the common ion often exist. These are named similarly to the above except a different suffix may be used and, in some cases, a prefix is added. We will, in general, not be concerned with them; however, examples of them are

$NaClO$:   sodium hypochlorite
$NaClO_2$:   sodium chlorite
$NaClO_3$:   sodium chlorate
$NaClO_4$:   sodium perchlorate

3. *Certain hydrogen-containing polyatomic compounds are also acids.* They are named by replacing the suffix *-ate* in the name of the polyatomic ion with the suffix *-ic* and adding the word acid. Thus H_2SO_4 is named sulfuric acid and CH_3COOH is

---

*The formulas of these binary acids are frequently written without indicating whether or not the substance is in the pure form [HCl (g)] or in solution [HCl (aq)]. In such cases they can be named as the pure substance according to rules 1–3 (hydrogen chloride) or by rule 6 (hydrochloric acid). Proper representation of the formula should clearly indicate which form is being referred to, but this is not always done in actual practice.

called acetic acid. These acids can be identified by the appearance of hydrogen first in the formula or by the presence of the carboxyl group, COOH. A series of acids that are comparable to those given above for the sodium salts are

$HClO$:   hypochlorous acid
$HClO_2$:   chlorous acid
$HClO_3$:   chloric acid
$HClO_4$:   perchloric acid

PROBLEMS

8. Determine the oxidation number of bromine in each of the following chemical species.
   Ⓢ (a) HBr (b) Br_2 Ⓢ (c) HBrO (d) $HBrO_2$ (e) $HBrO_3$ (f) $HBrO_4$
9. Write the correct chemical formulas for the following chemical species.
   (a) carbon   (b) bromine   (c) hydrogen iodide
   (d) copper(II) bromide   (e) magnesium carbonate   (f) silver oxalate
10. Write the correct names for the following compounds.
   (a) SrO   (b) PbO   (c) CO_2 (d) HI(aq)
   (e) $Mg(OH)_2$ (f) K_3PO_4 (g) H_3PO_4

ANSWERS TO PROBLEMS   **8. (a)** -1 **(b)** 0 **(c)** $+1$ **(d)** $+3$ **(e)** $+5$ **(f)** $+7$
9. (a) C **(b)** Br_2 **(c)** HI(g) **(d)** $CuBr_2$ **(e)** $MgCO_3$ **(f)** $Ag_2C_2O_4$ **10. (a)** strontium oxide **(b)** lead(II) oxide or plumbous oxide **(c)** carbon dioxide **(d)** hydroiodic acid **(e)** magnesium hydroxide **(f)** potassium phosphate **(g)** phosphoric acid

---

SUMMARY

For the most part the world is composed of compounds. In this chapter we have considered the stability of the inert gases and the significance of this factor to how other elements combine with one another to form these compounds. We saw that two principal types of compounds formed are ionic and covalent compounds. We found that ionic compounds consist of charged atoms, or ions, formed by a transfer of electrons and bound together by their mutual electrical attraction. Covalent compounds, on the other hand, were found to consist of atoms held together through a sharing of one or more electron pairs. We learned that these shared electrons occupy a different region in space than they originally occupied and that this space is called a molecular orbital. Two types of molecular orbitals were identified as sigma and pi.

After covering the theories of bonding, we dealt with the rules for writing chemical formulas and naming compounds, including what we referred to as oxidation numbers. With this information we should have some understanding of how and why compounds form.

In the next chapter we will learn how a chemist measures these substances and we shall do various calculations involving them. We will consider how these substances react with each other and how a chemist determines the composition of an unknown substance.

ADDITIONAL PROBLEMS

11. What is the sublevel electron distribution for the electrons in the outermost occupied main energy level for the following?
    (a) He   (b) Ar   (c) Ne

12. Write the sublevel electron distribution that the following elements will tend to achieve when they chemically combine with another element. (Assume an ionic reaction occurs.)
    (a) K   (b) Li   (c) Be
    (d) N   (e) S   (f) F
    (g) Al

13. Write balanced equations for the reactions of the following pairs of elements. (Give the sublevel electron distribution for each species. Assume ionic bonds are formed.)
    (a) $K + F \longrightarrow$
    (b) $Mg + S \longrightarrow$
    (c) $Al + P \longrightarrow$
    (d) $Ca + Cl \longrightarrow$
    (e) $Li + O \longrightarrow$
    (f) $Be + P \longrightarrow$

14. Write Lewis electron-dot formulas for the following chemical species.
    (a) Al   (b) S   (c) Br_2
    (d) HI   (e) LiF   (f) MgO
    (g) Na_2O (h) CCl_4

15. Using values of electronegativities given in Figure 3.2 or the relative position of elements in the periodic table, predict whether the following chemical species will have polar or nonpolar bonds. Indicate the partial charge on each atom for the polar species identified.
    (a) Br_2 (b) H_2O (c) CF_4

    (d) SO_2 (e) CI_4 (f) OF_2
    (g) HI   (h) NCl_3

16. At elevated temperatures (greater than 750°C), sulfur exists primarily as S_2 molecules. Draw the Lewis electron-dot formula and indicate whether it is bonded with single, double, or triple bonds.

17. Determine the oxidation state of each element in the following chemical species.
    ⒟ (a) Ba   ⒮ (b) BaO
       (c) BaI_2 ⒮ (d) I_2
       (e) AlN   ⒮ (f) ZnF_2
       (g) $FeCl_2$ ⒮ (h) Hg_2Cl_2
       (i) $LiIO_3$ ⒮ (j) $MgSeO_4$
       (k) $CuSO_4$ (l) K_2SO_3
       (m) BrO_4^- (n) $HC_2H_3O_2$
       (o) $Hg_3(PO_4)_2$ (p) $Ni(ClO_3)_2$

18. Write the correct formulas for the following chemical species.
    (a) boron   (b) magnesium
    (c) nitrogen   (d) potassium iodide
    (e) barium bromide   (f) aluminum sulfide
    (g) iron(III) fluoride (h) nickel carbonate
    (i) nickel bicarbonate (j) aluminum iodate
    (k) calcium arsenate   (l) sodium acetate
    (m) lead(IV) oxalate   (n) mercury(II) nitrate

19. Write correct names for the following compounds.
    (a) $NiCl_2$ (b) ZnS
    (c) FeN   (d) CuSe
    (e) P_2O_5 (f) OF_2
    (g) HF(aq)   (h) $H_2Se(aq)$
    (i) $(NH_4)_2S$ (j) $Ca(CN)_2$
    (k) $HgCO_3$ (l) $HgHCO_3$
    (m) $HBrO_3$ (n) H_2SeO_4

CHAPTER 4

THE MOLE CONCEPT
AND CHEMICAL EQUATIONS

In Chapter 3 we found that atoms can combine in a variety of ways. The compounds that are produced by these combinations make up the world about us and are the subject of our study in chemistry. In this chapter we will consider a quantity of matter known as the mole and apply the factor-unit method to calculations involving the mole. We will also extend our consideration of chemical reactions to look at how some typical chemical compounds react with each other to produce other compounds and how the equations for these reactions are balanced. Finally we will apply the factor-unit method to these equations and relate the quantities of chemicals involved in a reaction to each other through the quantity the mole.

4-1 THE MOLE CONCEPT

LEARNING OBJECTIVES

1. Define molecular weight and mole.
2. Discuss the factors of importance in the choice of the quantity the mole.
3. Calculate the molecular weight of compounds.
4. Determine the weight of a mole for elements and compounds.

5. Calculate the number of unit particles and individual atoms in a mole of a compound.
6. Make conversions between mole, unit particles, and grams for a given chemical species.

In our considerations thus far, we have examined matter from simple atoms to some rather complex chemical species. We have taken a theoretical look at how electrons exist in the atom and we have arrived at a sophisticated model for electron behavior called the wave-mechanical model. We have talked about how these atoms combine to form compounds. In all of this you should remember that no one

Figure 4.1 Cluster of uranium atoms. This photograph was taken by means of a scanning transmission electron microscope. (Courtesy of M. Isaacson, M. Ohtsuki, and M. Utlaut, The Enrico Fermi Institute)

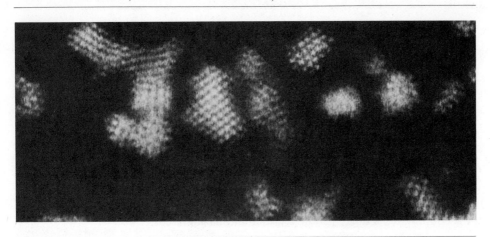

actually has seen a clear picture of an atom with its protons and neutrons in the nucleus and electrons moving about the nucleus. Certainly no one has seen an electron in either an atomic or molecular orbital even with the most powerful microscope available (see Figure 4.1). However, this inability to see individual atoms and molecules is not an insurmountable barrier to the scientist. The wave-mechanical theory discussed in Chapters 2 and 3 is accepted because it does a thorough job of explaining what can be observed and there are no reasons to question the theory seriously. However, if new theories are developed that explain what is observed better than does our current model, then these would be adopted.

Not only are we limited because we are not able clearly to see individual atoms and molecules, but we also cannot handle them, weigh them, observe them as they react, or do much at all with them individually. In order to do these things we need a large enough collection of atoms so that we can work with them easily. In short, we need a convenient number of atoms or molecules.

In this section we will consider a convenient number of atoms or molecules that we call the mole and relate it to the quantities with which we are already familiar, grams, atoms, and molecules. The quantity the mole will be useful to us throughout the text and will be important to you as you work with substances and solutions in your professional career.

4-1.1 THE MOLE

We frequently meet situations in everyday life where items are dealt with in convenient amounts. For example, when we buy eggs we usually buy them in the convenient size of a dozen. Likewise, carrots and radishes are usually sold in bunches of a convenient size. In each case, there are enough items so that we usually do not have to go back later that day or the next to buy more. On the other hand, we have not bought so much that we cannot carry the food out of the store or consume it in a reasonable amount of time. The chemist's unit for measuring a "convenient bunch" of atoms and molecules is based on the atomic weight of the elements.

In Chapter 2 atomic weight was introduced with the unit atomic mass unit (amu). The unit amu is useful for comparing the relative weights of the atoms of different elements, but it is not ordinarily used to measure the weight of a substance. The unit of weight commonly used by chemists is, as discussed in Chapter 1, the gram. It is important to remember that *the atomic weight of the element measured in grams is a convenient amount for chemists to use.* For example, you can hold 32 g of sulfur in your hands or put it in a 100-mL beaker. Similarly 4 g of gaseous helium have a volume of approximately 22 L, or about 6 gal. Because of its importance to us, the atomic weight of an element in grams is given the special name *gram-atomic weight* (g at. wt).

When we deal with compounds we have a quantity that is comparable to atomic weight, called **molecular weight.** As with atomic weight, the unit for molecular weight is amu. The molecular weight of a substance is determined by multiplying the atomic weight of each element in the compound by the number of atoms of that element in the formula of the substance, and then adding these numbers. This procedure is illustrated in the following example.

> The **molecular weight** is the average weight of a unit particle of a chemical species compared with the weight of an atom of carbon-12.

EXAMPLE 4.1   Common table sugar (sucrose) has the formula $C_{12}H_{22}O_{11}$. What is its molecular weight, measured in atomic mass units?

Applying the procedure discussed above we have

Element	Number of Atoms		Atomic Weight		
C	12	×	12 amu	=	144 amu
H	22	×	1 amu	=	22 amu
O	11	×	16 amu	=	176 amu
					342 amu

Compounds actually are weighed in terms of grams, not amu's. The molecular weight of most substances in grams (the gram-molecular weight) is also a convenient amount with which to work. For example, 342 g of sucrose is approximately a cup of sugar, the amount of sugar that you add to $\frac{1}{2}$ gal of water and a beverage mix, such as Kool-Aid. The terms *molecular weight* and *gram-molecular weight* are frequently applied to substances composed of atoms and ions in addition to those actually consisting of molecules. The authors will follow this practice in most cases in this text. Because of the importance of the gram-molecular weight, a sort of nickname, the term **mole,** has been coined by chemists to represent this quantity. This term is used almost universally instead of the more cumbersome name, gram-molecular weight.

> A **mole** is the amount of some chemical species corresponding to its atomic or molecular weight in grams.

During your course in chemistry you will be concerned with the quantity mole, again and again. The mole is used as the "convenient bunch" of atoms or molecules referred to earlier. A mole of iron is a piece of iron equal to the atomic weight in grams; that is, 55.85 g. A mole of sulfate ions, SO_4^{2-}, weighs

$$(g\ at.\ wt)_S + 4(g\ at.\ wt)_O = 32\ g + 4(16\ g) = 96\ g$$

Figure 4.2   A mole of each of the following: copper (63.5 g, in pennies), table sugar (342 g), water (18 g), table salt (58.5 g), and iron (56 g, in nails). One mole of each of these items is a convenient amount. (Courtesy of Warren R. Buss)

and a mole of water, H_2O, weighs

$$2(\text{g at. wt})_H + (\text{g at. wt})_O = 2(1 \text{ g}) + 16 \text{ g} = 18 \text{ g}$$

Figure 4.2 shows a mole of several common substances and illustrates the convenient size of a mole.

In addition to designating the molecular weight in grams, the term *mole* is also associated with a specific number of particles of each substance, just as the term *dozen* is associated with the number 12. The term *dozen* represents any 12 items: 12 eggs, or 12 doughnuts, or 12 people. Similarly, when we refer to a mole of some chemical species we are referring to a specific number of *unit particles,** or individual atoms, ions, or molecules of that substance. The number of particles in a mole is considerably larger than 12. A mole consists of 6.02×10^{23} unit particles.† A mole of sodium is therefore 6.02×10^{23} atoms of sodium and weighs 23.0 g. A mole of SO_4^{2-} is 6.02×10^{23} sulfate ions and weighs 96.1 g. In addition, a mole of SO_4^{2-} is composed of 6.02×10^{23} atoms of sulfur and $4 \times 6.02 \times 10^{23} = 24.1 \times 10^{23}$ atoms of oxygen, all bonded together in the form of SO_4^{2-} ions. The number 6.02×10^{23} is known as *Avogadro's number.*

---

*As discussed in Chapter 3, ionic compounds such as NaCl consist of many millions of individual ions positioned in an orderly crystalline arrangement. No specific ions are actually matched as a formula, such as NaCl, would suggest. In light of this, use of the term *molecule* really is not appropriate. In order to avoid this problem we can use the general term *unit particle,* which refers to a particle consisting of the number of atoms of each type that appear in the formula; the term then can be used to refer to atoms, ions, or molecules.

†More exactly, a mole consists of 6.023×10^{23} unit particles. The significance of the 0.003 portion of the number is small and it shall be ignored throughout the text. Note should also be made that the number 6.023×10^{23} is an experimentally determined number resulting from the choice of the gram-molecular weight as the basis of the mole and not a defined number such as 1 doz = 12 items.

EXAMPLE 4.2   (a) How many molecules of sucrose ($C_{12}H_{22}O_{11}$) are in a mole of sucrose? (b) How many total atoms are present in a mole of $C_{12}H_{22}O_{11}$?

(a) From the above discussion we can conclude that there are 6.02×10^{23} molecules of $C_{12}H_{22}O_{11}$ in a mole.

(b) To obtain the total number of atoms in a mole of sucrose, we must add the number of atoms of each type. These are calculated as follows.

Element	Atoms per Molecule		Avogadro's Number		Atoms per Mole
C	12	$\times$	6.02×10^{23}	$=$	72×10^{23}
H	22	$\times$	6.02×10^{23}	$=$	132×10^{23}
O	11	$\times$	6.02×10^{23}	$=$	66×10^{23}

$$270 \times 10^{23} \text{ total atoms}$$
$$= 2.70 \times 10^{25} \text{ atoms}$$

Relationships other than those between moles and grams and moles and unit particles shall be developed in later chapters. At this point, you should remember the following equalities that summarize these two relationships:

$$1 \text{ mole} = 1 \text{ gram-molecular weight*} \qquad (4.1)$$
$$1 \text{ mole} = 6.02 \times 10^{23} \text{ unit particles} \qquad (4.2)$$

A potentially troublesome situation arises whenever we refer to a mole of one of the seven diatomic elements (H_2, N_2, O_2, F_2, Cl_2, Br_2, and I_2). The term *mole* is ordinarily used to refer to Avogadro's number of unit particles. Since a unit particle for these seven elements is a diatomic molecule, a mole of one of these elements normally refers to 6.02×10^{23} molecules, or a total of $2 \times 6.02 \times 10^{23}$ atoms of the element. For example, 1 mol† of oxygen consists of 6.02×10^{23} O_2 molecules and weighs 32.0 g (twice the gram-atomic weight).

At times it is convenient to refer to a mole of atoms for these diatomic elements. In such cases, be sure to specify that atoms are involved. For example, if you are referring to a mole of hydrogen atoms, write 1 mol H atoms, not just 1 mol H.

4-1.2 MOLE CONVERSIONS

Many calculations in chemistry are basically problems involving a conversion of units. This is particularly true for problems involving moles. These calculations can be done using the factor-unit method discussed in Chapter 1. Before this is done, it is helpful to consider two examples of conversions involving the unit dozens.

EXAMPLE 4.3   Assume for a moment that we are buying eggs for a camping trip and we need eight dozen eggs. In addition, assume that the only store available is a country store that stocks only fresh farm eggs that have not been packaged by the dozen. To buy eight dozen eggs we can count the eggs out a dozen at a time or we can simply count out the total number of eggs needed. To use the latter method, we must know how many eggs are in eight dozen, that is, we need to convert from

---

*Do not interpret this equation as 1 mole = 1 gram, as is sometimes done. This is not correct. The equality indicates that 1 mole = the molecular weight of the substance in grams.

†The abbreviation mol is the accepted representation when referring to the amount of substance in moles. This term shall be used throughout the text.

eight dozen eggs to the total number of eggs needed. For this calculation we have the conversion factor

$$\frac{12 \text{ eggs}}{1 \text{ doz eggs}} = 1$$

which allows us to make the necessary conversion as follows.

$$8 \text{ doz eggs} \times \frac{12 \text{ eggs}}{1 \text{ doz eggs}} = 96 \text{ eggs}$$

EXAMPLE 4.4   As a second illustration we will consider that we have been asked to bake an enormous cake. The recipe calls for 5.0 lb of eggs, but we do not have a scale on which to weigh them. However, we do read in our recipe book that a dozen large eggs weigh 1.5 lb. With this relationship we can write the conversion factor

$$\frac{1 \text{ doz large eggs}}{1.5 \text{ lb large eggs}}$$

Using this and the previous conversion factor we can make the conversions

$$5.0 \text{ lb eggs} \Longrightarrow ? \text{ doz large eggs} \Longrightarrow ? \text{ large eggs}$$

through the following calculation

$$5.0 \text{ lb eggs} \times \frac{1 \text{ doz large eggs}}{1.5 \text{ lb eggs}} \times \frac{12 \text{ large eggs}}{1 \text{ doz large eggs}} = 40 \text{ large eggs}$$

which is the number needed to give us 5.0 lb of eggs.

The conversions involved in mole calculations are basically the same type as we used for calculations with dozens. To make these conversions we first must develop conversion factors comparable to those already developed, relating dozens to a number (12) and to the weight (1.5 lb for large eggs). Equalities 4.1 and 4.2 allow us to write the comparable conversion factors.

$$\frac{1 \text{ gram-molecular weight}}{1 \text{ mol}} \qquad\qquad \textbf{(4.3)}$$

$$\frac{6.02 \times 10^{23} \text{ unit particles}}{1 \text{ mol}} \qquad\qquad \textbf{(4.4)}$$

These give us the necessary informaton to convert from moles to grams and from moles to unit particles and vice versa. As illustrated in Examples 4.1 and 4.2 we can now write the equalities 342 g $C_{12}H_{22}O_{11}$/1 mol $C_{12}H_{22}O_{11}$ and 6.02×10^{23} molecules $C_{12}H_{22}O_{11}$/1 mol $C_{12}H_{22}O_{11}$.

We shall begin our consideration of mole calculations by making a conversion from moles to grams. This calculation is similar to the conversion of dozens of eggs to number of eggs given above.

EXAMPLE 4.5   Formaldehyde (CH_2O) is the principal source of the pungent, irritating odor noticeable around preserved biological specimens. The embalming fluid used in these cases consists of 1.33 mol of CH_2O in 100.0 mL of solution. How many grams does 1.33 mol of CH_2O represent?

The gram-molecular weight of CH_2O is (12.0 g + 2(1.0 g) + 16.0 g) = 30.0 g. Applying this to equation 4.3 yields

$$\frac{30.0 \text{ g } CH_2O}{1 \text{ mol } CH_2O}$$

Using this in conjunction with the factor-unit method gives

$$1.33 \text{ mol } CH_2O \times \frac{30.0 \text{ g } CH_2O}{1 \text{ mol } CH_2O} = 40.0 \text{ g } CH_2O$$

Mention should be made that the value 100.0 mL of solution did not directly enter into the calculation, but simply determined the amount of CH_2O under consideration.

EXAMPLE 4.6   A regular-size drinking glass holds approximately 12.0 oz, or about 340. g of water. If you drink a glass of water of this size, how many molecules of water are you drinking?

In this problem we must make the following conversions:

$$340. \text{ g } H_2O \Longrightarrow ? \text{ mol } H_2O \Longrightarrow ? \text{ molecules } H_2O$$

Both conversion factors 4.3 and 4.4 are necessary to do the calculation. For H_2O these become

$$\frac{18.0 \text{ g } H_2O}{1 \text{ mol } H_2O} \tag{4.5}$$

and

$$\frac{6.02 \times 10^{23} \text{ molecules } H_2O}{1 \text{ mol } H_2O} \tag{4.6}$$

Conversion factor 4.5 must be inverted in order to make the necessary conversion. That is,

$$\frac{1 \text{ mol } H_2O}{18.0 \text{ g } H_2O} \tag{4.7}$$

Using conversion factors 4.6 and 4.7 we can do the required calculation

$$340. \text{ g } H_2O \times \frac{1 \text{ mol } H_2O}{18.0 \text{ g } H_2O} \times \frac{6.02 \times 10^{23} \text{ molecules } H_2O}{1 \text{ mol } H_2O}$$

$$= 1.13 \times 10^{25} \text{ molecules } H_2O$$

To give you some idea as to how large this number is, consider a stack of 1.13×10^{25} dimes. A single dime is approximately 1 mm thick. Therefore, it would require a stack of dimes approximately 11 quintillion kilometers (11×10^{18} kilometers), or 7 quintillion miles high. This is equivalent to 75 billion stacks of dimes reaching from the earth to the sun. Quite a large number!

PROBLEMS

⑤ 1. Lysine ($C_6H_{14}N_2O_2$) is one of ten compounds known as essential amino acids.* Calculate its molecular weight.

⑤ 2. How many moles of NH_3 do 2.4×10^{24} molecules of NH_3 make up?

⑤ 3. How many moles of bromine atoms are contained in 2.0 mol of Br_2?

⑤ 4. When we take a breath we inhale approximately 1.5 L of air. This contains, among other gases, about 0.013 mol of oxygen. How many molecules of oxygen do we inhale in a single breath?

⑤ 5. A penny weighs approximately 3.1 g. Assuming that it is pure copper, how many atoms of copper are in one penny?

ANSWERS TO PROBLEMS   **1.** 146 amu   **2.** 4.0 mol NH_3 **3.** 4.0 mol Br atoms
4. 7.8×10^{21} molecules O_2.   **5.** 2.9×10^{22} atoms Cu

4-2 CHEMICAL EQUATIONS AND THE MOLE CONCEPT

LEARNING OBJECTIVES

1. Define acid, base, reactant, and product.
2. Given the reactants in a chemical reaction predict the products obtained in (1) direct combination reactions, (2) single replacement reactions, (3) double replacement reactions, and (4) reactions of oxygen with carbon-containing compounds.

3. Balance chemical equations of the type discussed in Objective 2.
4. Apply the mole concept to balanced chemical equations for the calculation of unit particles, moles, and grams of one substance from a given amount of another substance.

The previous section dealt with the conversion of units involving moles, molecules, atoms, and grams as applied to single substances. Similar conversions can be made between two or more substances when they are related to one another through a chemical equation. In Chapter 3 we saw how atoms react to form compounds. Many compounds can also react with one another to form different compounds.

In the living system a great variety of such reactions can and do occur. In general these reactions are beneficial; however, reactions occasionally occur that are harmful to the living system and, in some cases, prove fatal. For example particular care must be exercised when taking combinations of medicines. Many people have died as a result of taking, at the same time, two or more medicines that reacted chemically. Many of the reactions that occur in the living system shall be dealt with in the biochemistry portion of the text (Chapters 18–28). In this section we shall consider some common, less complex reactions. It is important to realize, however, that the principles that apply here are those that are involved in the life processes.

---

*An essential amino acid is one that is necessary for growth and maintenance of the body, but must be obtained from the food we eat because the human body is not able to produce it. More will be said about amino acids in Chapter 21.

When reactions occur, definite relationships exist between the number of unit particles of each type undergoing reaction and the number of unit particles, produced. In this section we shall discuss these relationships and how the mole concept applies to chemical reactions.

4-2.1 WRITING CHEMICAL EQUATIONS

In Chapter 3 we learned that atoms frequently combine to form more complex chemical species called compounds. These compounds can, in many cases, react with one another to form different compounds. In this section we shall consider four general types of reactions: (1) direct combination, (2) single replacement, (3) double replacement, and (4) the reaction of oxygen with carbon-containing (organic) compounds.

Direct Combination Reactions

Direct combination reactions are usually associated with the combination of either (1) two elements or (2) two compounds. The direct combination of elements was dealt with in Chapter 3 and will not be covered further here. The direct combination of compounds can be associated with a wide variety of reactions, but the discussion in this chapter will be limited to the reaction of metal and nonmetal oxides with water.

1. *Nonmetal oxides in water.* As certain nonmetal oxides combine with the water they produce compounds known as **acids.** An example of this is the compound sulfur trioxide (SO_3), a common air pollutant produced when sulfur-containing coal is burned. Sulfur trioxide reacts with water, such as in rain drops, in the following manner.

An **acid** is a compound that dissociates to produce hydrogen ions in an aqueous solution. (*Arrhenius* concept)

$$H_2O(l) + SO_3(g) \longrightarrow H_2SO_4(aq)^* \qquad \textbf{(4.8)}$$

The sulfuric acid produced is a corrosive substance that chemically attacks such things as the marble in outdoor statues, resulting in serious damage. In addition SO_3 is very irritating to the eyes and the respiratory tract. Many other nonmetal oxides react in a similar manner. A few of these nonmetal oxides with the acid produced are listed in Table 4.1.

Table 4.1 Nonmetallic Oxides and the Corresponding Acids Formed in Water

Nonmetal Oxides	Acids	Names
SO_3	H_2SO_4	Sulfuric
SO_2	H_2SO_3	Sulfurous
CO_2	H_2CO_3	Carbonic
N_2O_5	HNO_3	Nitric
P_4O_{10}	H_3PO_4	Phosphoric

---

*The letters in parentheses, such as (aq), represent the physical state of the associated chemical species. The letter g represents the gaseous state, l the liquid state, s a solid, and aq an aqueous, or water, solution. It is easier for us to understand some equations if we are aware of the physical state of each species. In such cases, as in equation 4.8, these symbols shall be included.

Table 4.2   Metallic Oxides and the Corresponding
                 Bases Formed in Water

Metallic Oxide	Bases	Names
Na_2O	NaOH	Sodium hydroxide
K_2O	KOH	Potassium hydroxide
Rb_2O	RbOH	Rubidium hydroxide
MgO	$Mg(OH)_2$	Magnesium hydroxide
CaO	$Ca(OH)_2$	Calcium hydroxide
SrO	$Sr(OH)_2$	Strontium hydroxide
BaO	$Ba(OH)_2$	Barium hydroxide

2. *Metallic oxides in water.* Oxides of many of the more reactive metals combine with water to produce compounds known as bases. A good example of this is the reaction that takes place when sodium oxide (Na_2O) is placed in water.

$$Na_2O(s) + H_2O(l) \longrightarrow 2\ NaOH(aq) \qquad (4.9)$$

A **base** is a compound that dissociates to produce hydroxide ions in an aqueous solution. (*Arrhenius* concept)

The sodium hydroxide that is produced commonly is known as lye. It is a very caustic substance used for cleaning sink traps and toilets and was used in pioneer days to make soap. Sodium hydroxide is particularly dangerous if it gets into the eyes because it can cause blindness. Other examples of metallic oxides with the associated bases are listed in Table 4.2. In general, the oxides of the metals in groups IA and IIA react in this manner. Care should be taken, however, not to assume that all metallic oxides react in the same way. For example, iron(III) oxide (Fe_2O_3) and aluminum oxide (Al_2O_3) have a very limited solubility in water and do not react with it to a measurable extent. The aluminum in cookware reacts with oxygen in the air to form a protective coating of Al_2O_3, keeping the relatively reactive aluminum from reacting with food that is cooked in the pan. If such a reaction occurred between aluminum and many substances, the aluminum compounds formed would be poisonous.

Single Replacement Reactions

Single replacement reactions are usually of two general types: (a) metal atoms replacing (taking the place of) other metal ions in solutions and (b) nonmetal atoms replacing nonmetal ions.

1. *Metal ions generally replace other metal ions in a solution* if the metal doing the replacing is more reactive than the metal whose ion it replaces. Table 4.3 gives a list of the relative reactivity of metal atoms in terms of their ability to become ions. Metal atoms having the greatest reactivity are positioned higher in the list. Hydrogen is included in the list because it shares various characteristics with metals, including the ability to combine with nonmetals. This list of metals is known as the *activity series.*

An illustration of this type of reaction is the replacement reaction between zinc and hydrochloric acid.

$$Zn(s) + 2\ HCl(aq) \longrightarrow ZnCl_2(aq) + H_2(g) \qquad (4.10)$$

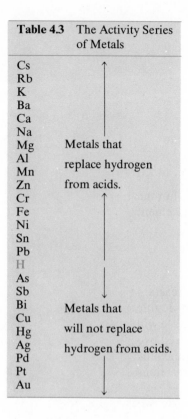

Table 4.3   The Activity Series of Metals

Cs
Rb
K
Ba
Ca
Na
Mg   Metals that
Al
Mn   replace hydrogen
Zn   from acids.
Cr
Fe
Ni
Sn
Pb
H
As
Sb
Bi   Metals that
Cu
Hg   will not replace
Ag   hydrogen from acids.
Pd
Pt
Au

This reaction shows the laboratory method for producing hydrogen gas. Any metal above hydrogen in the activity series could be used to produce the hydrogen from an acid; however, the higher the metal is in the series, the faster the reaction occurs.

In a similar way, one metal will replace another metal in a compound if the first metal is higher on the activity series. For example, zinc replaces copper in copper sulfate, in the same way that it replaces hydrogen in hydrochloric acid. This is given by the equation

$$Zn(s) + CuSO_4(aq) \longrightarrow ZnSO_4(aq) + Cu(s) \qquad \textbf{(4.11)}$$

These single replacement reactions typically occur in solution, as illustrated in equations 4.10 and 4.11. For example, solid zinc ordinarily will not react with solid copper sulfate because the two are not normally mixed together on the atomic level.

$$Zn(s) + CuSO_4(s) \longrightarrow \text{No reaction} \qquad \textbf{(4.12)}$$

Single replacement reactions are the basis upon which automobile batteries and flashlight dry cells operate. Batteries and dry cells are examples of electrochemical cells. We will consider this topic in Chapter 9 when we study electrochemistry.

2. *Nonmetal atoms replace other nonmetal ions* in certain cases when the atom doing the replacing is more reactive than the atoms whose ion is being replaced.

Table 4.4	Typical Nonmetal Elements and the Nonmetal Ions They Replace	
Replacing Elements	**Ions Replaced**
F_2	Cl^-, Br^-, I^-
Cl_2	Br^-, I^-
Br_2	I^-

An illustration of this is given by the equation

$$Cl_2(g) + 2\,NaBr(s) \longrightarrow 2\,NaCl(s) + Br_2(g) \qquad \textbf{(4.13)}$$

It is not easy to predict when one nonmetal can replace another nonmetal, but the elements listed in Table 4.4 provide some typical examples of such replacement reactions.

Double Replacement Reactions

Double replacement reactions are basically a "trading partners" type of reaction such as when two couples go to a dance and trade dates for dances. In double replacement reactions, however, the partners remain permanently traded. We will consider two classes of double replacement reactions: (1) the reaction between two different salts (compounds composed of a metallic species and a nonmetallic species) and (2) the reaction between an acid and a base. Both types of reactions typically take place in solution.

In considering these reactions it is important that we realize that just mixing two chemicals does not always lead to a reaction. The right chemistry must be present for reaction to occur. That is, there must be a sort of driving force that causes the reaction to take place. For double replacement reactions the driving force results from one of three conditions: one or both of the substances produced must be (1) more covalent (less polar) than the substances reacting, (2) an insoluble substance, or (3) a gas. These will be illustrated below.

1. *Two salts will undergo a double replacement reaction* if one of the substances produced is relatively insoluble. These reactions typically occur in solution. An example of this type is the reaction of sodium chloride with silver nitrate in aqueous solution.

$$NaCl(aq) + AgNO_3(aq) \longrightarrow NaNO_3(aq) + AgCl(s) \qquad \textbf{(4.14)}$$

The silver chloride is insoluble in water and falls to the bottom of the container as a *precipitate*. The precipitate provides the driving force for the reaction in that it is removed from the solution in which the reaction is taking place. In general, predictions concerning the solubility of substances are difficult to make and problems in this text will not require such predictions.

2. *Acids and bases react to produce a salt plus water.* Probably the most common example of this type is the reaction of hydrochloric acid with sodium hydroxide.

$$HCl(aq) + NaOH(aq) \longrightarrow NaCl(aq) + H_2O(l) \qquad \textbf{(4.15)}$$

In this reaction the driving force is in the formation of the covalent substance, water. All the other substances involved in the reaction are ionic compounds in solution and do not influence the reaction significantly. The remaining species are in the form of the ions H^+, Cl^-, Na^+ and OH^-, which can be envisioned moving about in the solution as individual ions. Because of the importance of this type of reaction to chemistry and the health-related sciences, we shall consider it in more detail in Chapter 6 when we discuss acid-base neutralization, and then again in Chapter 8 in conjunction with ionic equilibria.

Compounds Consisting of C, H, and O React with O_2 to Produce CO_2 and H_2O

Although a broad variety of organic compounds exist we shall consider only two general types at this point in the text: (a) those compounds composed of only carbon and hydrogen, known as hydrocarbons and (b) those compounds composed of carbon, hydrogen, and oxygen, known as sugars. These and other compounds will be considered in more detail in the organic chemistry portion of the text (Chapters 10–17).

1. *Hydrocarbons* form an important class of compounds. Included in this group are compounds that are contained in cooking gases, candles, gasoline, and motor oil. All of these compounds share the common characteristic that they react with oxygen, or burn, to produce carbon dioxide and water. This reaction, known as *combustion,* is the most important source of energy available in developed countries. Because there is a limit to the amount of naturally occurring hydrocarbon, such as petroleum, great concern is expressed in many parts of the world with respect to how long these resources will last and how efficiently they are used.

As an example of the reaction of hydrocarbons with oxygen, we will consider the combustion of natural gas. Although natural gas is a mixture of several gases, including methane (CH_4), ethane (C_2H_6), and propane (C_3H_8), it is, for the most part, methane and we will deal only with the combustion of that substance. When methane undergoes complete combustion (that is, when it reacts completely with oxygen), it does so according to the equation

$$CH_4(g) + 2\ O_2(g) \longrightarrow CO_2(g) + 2\ H_2O(g) + heat \qquad \textbf{(4.16)}$$

As indicated, heat is produced during the reaction and is normally the reason for burning the gas.

When insufficient oxygen is available, incomplete combustion of hydrocarbons occurs and either elemental carbon or carbon monoxide is produced. When burning gasoline in a car, a certain amount of carbon monoxide normally results. Gasoline, like natural gas, is also a mixture of hydrocarbons. In this case, the mixture is composed of heavier molecules that exist as a liquid at room temperature. One such molecule is octane (C_8H_{18}). When octane undergoes incomplete combustion in the engine of a car, a mixture of carbon monoxide and carbon dioxide usually results. For simplicity, the following equation is written with only the carbon monoxide considered.

$$2\ C_8H_{18}(l) + 17\ O_2(g) \longrightarrow 16\ CO(g) + 18\ H_2O(g) + energy \qquad \textbf{(4.17)}$$

The resulting carbon monoxide replaces the oxygen in red blood cells and in extreme cases can lead to death by asphyxiation; that is, death due to a lack of oxygen. Carbon monoxide can and does at times become a serious air pollutant, especially where there are many automobiles and atmospheric conditions such that the air containing carbon monoxide remains in a particular geographic area for several days.

2. *Sugars* are composed of carbon, hydrogen, and oxygen and on complete combustion produce carbon dioxide and water. One group of sugars, disaccharides, consists of two simpler sugars bound together chemically. More shall be said about these compounds in Chapter 19. For now we are interested, primarily, in their reaction with oxygen. Several disaccharides have the same chemical formula, $C_{12}H_{22}O_{11}$, but differ in their *structural formulas*. The structural formula is the arrangement and position of the atoms of which the compound is composed. Three common disaccharides of this type are sucrose (table sugar), lactose (the sugar in milk from mammals), and maltose (sugar derived from the breakdown of starch). The overall reaction of each of these sugars with oxygen can be represented by the equation

$$C_{12}H_{22}O_{11}(s) + 12\ O_2(g) \longrightarrow 12\ CO_2(g) + 11\ H_2O(l) + energy \qquad (4.18)$$

Keep in mind that the process by which this reaction takes place in the living systems consists of many steps and is, therefore, more complex than is shown in this simple equation. The energy produced from the burning of the many foods, such as the above example of sugar, helps to keep our bodies warm and gives us the energy to perform our daily tasks. The carbon dioxide leaves our body as we breathe and the water is expelled through our breath, in perspiration, and in urine.

4-2.2 BALANCING CHEMICAL EQUATIONS

When chemicals react with one another they do so in certain definite ratios. For example, water reacts with dinitrogen pentaoxide to produce nitric acid according to the equation

$$H_2O(l) + N_2O_5(g) \longrightarrow 2\ HNO_3 \qquad (4.19)$$

That is, one molecule of H_2O reacts with one molecule of N_2O_5. The complete combustion of butane (C_4H_{10}) in a butane lighter or heater, as shown in Figure 4.3, follows the equation

$$2\ C_4H_{10}(g) + 13\ O_2(g) \longrightarrow 8\ CO_2(g) + 10\ H_2O(g) \qquad (4.20)$$

where *13 molecules of oxygen* are required to react with every *2 molecules of butane*, and *8 molecules of carbon dioxide*, and *10 molecules of water* result from the reaction. In both reactions, specific amounts of each **reactant** combined and specific amounts of each **product** were formed. This characteristic is true in general for all chemical reactions. To represent this characteristic in equation form, coefficients, or numbers, are placed in front of those chemical formulas where more than one unit particle is involved. For example, the number 2 was placed in front of the formula HNO_3 in equation 4.19. When no number is indicated, such as before the H_2O and N_2O_5, it is understood that only one unit particle is involved. When all the coefficients necessary to account for the correct numbers of atoms and molecules have been included in an equation, we consider it *balanced*.

Writing balanced chemical equations involves three fundamental steps:

1. *Determine what reactants and products are involved.*

A **reactant** is any chemical species that undergoes a chemical change.

A **product** in a chemical reaction is a chemical species formed as a result of a chemical change.

Figure 4.3
In a butane lighter, the reaction between butane (C_4H_{10}) and oxygen (O_2) produces carbon dioxide (CO_2) and water (H_2O), plus energy in the form of fire.

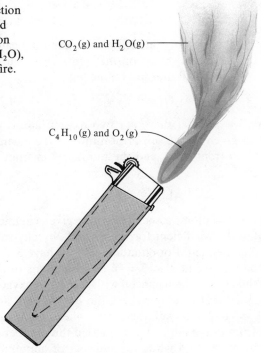

$CO_2(g)$ and $H_2O(g)$

$C_4H_{10}(g)$ and $O_2(g)$

2. *Write correct formulas of each chemical species with an arrow separating reactants and products and a plus sign between each substance.*
3. *Add numbers, as needed, in front of the formulas for the various chemical species to indicate how many of each species are involved in the reaction.*

The manner in which step 1 is performed depends upon the specific reaction under consideration. For this step the reactant species are usually known and formulas for them can be written. You can determine the products if you know the nature of the reaction taking place. Section 4-2.1 covered some of the more common types of reactions. Products for these reactions can be predicted with some success. Reactants and products for other reactions can be identified by referring to other portions of this book or through laboratory experiences. For step 2 the material in Chapter 3 enables us to write correct formulas for a relatively wide variety of both reactants and products as needed. Finally, the equation is balanced as discussed in the material that follows.

To illustrate this procedure for obtaining a balanced chemical equation, we will first consider the direct combination reaction between aluminum and oxygen. In this case the reactants are the elements aluminum (Al) and oxygen (O_2), with the formulas shown.

$$Al + O_2 \longrightarrow \qquad\qquad (4.21)$$

In a reaction such as this, between a metal and a nonmetal, the formula of the product is obtained following the rules given in Section 3-3.2 for binary compounds. Here, the compound formed is Al_2O_3. We can now write the more complete equation.

$$Al + O_2 \longrightarrow Al_2O_3 \qquad (4.22)$$

To balance this equation we note that there is only one aluminum atom indicated on the left side of the arrow, but there are two shown on the right. The number 2 is therefore placed in front of the symbol for aluminum on the left.

$$2\, Al + O_2 \longrightarrow Al_2O_3 \qquad (4.23)$$

Moving on to consider the oxygen we see that there are two oxygen atoms on the left side and three on the right as given by the subscript numbers. To balance this element the number 3 is placed in front of O_2 and the number 2 in front of Al_2O_3 yielding

$$2\, Al + 3\, O_2 \longrightarrow 2\, Al_2O_3 \qquad (4.24)$$

In chemical equations, the total number of atoms associated with a given chemical species is determined by multiplying the coefficient for the formula by any subscript that follows the symbol for that element. For equation 4.24 we have $3 \times 2 = 6$ oxygen atoms to the left of the arrow and $2 \times 3 = 6$ oxygen atoms to the right of the arrow. The equation is then said to be balanced with respect to oxygen.

Each time a new coefficient is added it is wise to recheck the earlier work and determine if the coefficients added previously are still the correct ones. In this case the addition of the coefficient 2 in front of the Al_2O_3 has increased the number of aluminum atoms to the right of the arrow to 4 while the number of aluminum atoms on the left remains at 2. This is easily remedied by changing the number 2 to 4 on the left hand side of the equation. With this change the equation is completely balanced,

$$4\, Al + 3\, O_2 \longrightarrow 2\, Al_2O_3 \qquad (4.25)$$

As a second, more complex example of obtaining a balanced chemical equation, we will consider the double replacement, or neutralization, reaction between phosphoric acid and barium hydroxide. Frequently it is helpful to obtain the formulas for reactants; that is, do step 2 before attempting to predict products. Applying the procedure discussed in Section 3-3.2, we have the formulas H_3PO_4 and $Ba(OH)_2$ for the two reactants in this reaction. The reactant portion of the reaction can now be written

$$H_3PO_4 + Ba(OH)_2 \longrightarrow \qquad (4.26)$$

Determination of the products in this reaction can now be accomplished in at least two different ways. The first method is to complete equation 4.26 by "trading partners" as follows.

$$H_3PO_4 + Ba(OH)_2 \longrightarrow BaPO_4 + HOH \qquad (4.27)$$

Alternatively, we can remember that the reaction of an acid with a base produces a salt plus water as pointed out in Section 4-2.1. Because a salt consists of a metallic species combined with a nonmetallic species, we can conclude that the salt, barium phosphate, is formed in addition to the water. Using either method the polyatomic ions, $PO_4{}^{3-}$ and OH^-, are carried over intact to the right side of the arrow.

When we write these initial formulas, the subscripts 3 and 2 that indicate how many hydrogen atoms and hydroxide ions, respectively, are involved in the reactant species to the left of the arrow are not carried over to the product side. *The number of each atom or polyatomic ion needed in the formula for each product species is determined independently of how many atoms or ions are present in the reactants.*

Accomplishing step 1 for other types of reactions differs in some of the details, but the same basic principles apply. The nature of the products for many reactions is predictable and in most instances at least an initial formula can be written.

Applying step 2 to the products in equation 4.14 leads to the formula $Ba_3(PO_4)_2$ for barium phosphate. The formula for water appearing in this equation has the correct number of atoms of hydrogen and oxygen, two and one, respectively. However, the elements are arranged in the rather unconventional form HOH so as to emphasize the "exchange of partners" that occurred in reaction 4.27. It will be convenient to leave the formula in this form until the equation is balanced. Modification of equation 4.27 to include the correct formula for barium phosphate yields the equation

$$H_3PO_4 + Ba(OH)_2 \longrightarrow Ba_3(PO_4)_2 + HOH \qquad \textbf{(4.28)}$$

As illustrated in the previous reaction between aluminum and oxygen, step 3 (balancing the equation) consists in placing appropriate coefficients in front of the formulas, so that there are equal numbers of atoms of each type on both sides of the arrow. For the types of reactions we have been considering in this chapter, we do this normally by inspection; that is, we look at the equation and intelligently try numbers until the correct combination has been obtained.

We use the same basic procedure to balance equation 4.28 as was used to obtain equation 4.25, except that when balancing equations that contain polyatomic ions, it is usually best to treat these ions as a unit rather than to consider the individual atoms of these ions. With this in mind, we retained the formula of water in equation 4.28 in the form HOH to keep the appearance of the OH group.

To balance equation 4.28, we find there are three hydrogen atoms on the left and only one on the right (remember the second hydrogen in HOH will be considered later as part of the OH group). The number 3 is then placed as a coefficient in front of the formula for water to equalize the number of hydrogen atoms on both sides.

$$H_3PO_4 + Ba(OH)_2 \longrightarrow Ba_3(PO_4)_2 + 3\ HOH \qquad \textbf{(4.29)}$$

There is only one PO_4^{3-} ion to the left of the arrow, while there are two PO_4^{3-} ions to the right. A 2 must therefore be added as a coefficient in front of the H_3PO_4 to adjust for this inequality.

$$2\ H_3PO_4 + Ba(OH)_2 \longrightarrow Ba_3(PO_4)_2 + 3\ HOH \qquad \textbf{(4.30)}$$

Checking the hydrogen atoms again, we see that there are 6 atoms on the left side but only 3 non-OH hydrogen atoms on the right. The number 3 in front of the water must therefore be doubled to balance the hydrogen atoms on the two sides of the equation.

$$2\ H_3PO_4 + Ba(OH)_2 \longrightarrow Ba_3(PO_4)_2 + 6\ HOH \qquad \textbf{(4.31)}$$

Consideration of the number of barium atoms on each side of the equation requires that the number 3 be added in front of $Ba(OH)_2$. After checking the number of atoms or ions of each species on both sides of the equation, we can see that we have arrived at a balanced chemical equation.

$$2\ H_3PO_4 + 3\ Ba(OH)_2 \longrightarrow Ba_3(PO_4)_2 + 6\ HOH \qquad (4.32)$$

The only modification that is usually made is to write the formula of water in the more traditional form, H_2O.

$$2\ H_3PO_4 + 3\ Ba(OH)_2 \longrightarrow Ba_3(PO_4)_2 + 6\ H_2O \qquad (4.33)$$

4-2.3 THE MOLE AND BALANCED EQUATIONS

As discussed in the previous section, a balanced chemical equation represents an equality of individual atoms as the reactants change to products. In this section we will extend this idea and show the relationships between the various reactants and products on a molecular, mole, and gram basis. For this we will consider the combustion of butane as given earlier by equation 4.20.

$$2\ C_4H_{10}(g) + 13\ O_2(g) \longrightarrow 8\ CO_2(g) + 10\ H_2O(g) + energy \qquad (4.20)$$

As discussed earlier, this equation can be read on a molecular basis as indicating that 2 molecules (unit particles) of C_4H_{10} and 13 molecules (unit particles) of O_2 react together to yield 8 molecules (unit particles) of CO_2 and 10 molecules (unit particles) of H_2O. This quantitative relationship between elements and compounds as measured by formulas and equations is referred to as *stoichiometry*. Balanced chemical equations are often referred to as *stoichiometric equations*.*

Stoichiometric equations allow us to do calculations relating the amounts of the various reactants and products involved in a given reaction.

EXAMPLE 4.7   Based upon equation 4.20, how many molecules of H_2O will be produced by the complete combustion of 20 molecules of C_4H_{10}?

To do this calculation we must be able to obtain a relationship between molecules of C_4H_{10} reacting and molecules of H_2O produced in the context of this reaction. The stoichiometric equation provides this information: for every 2 molecules of C_4H_{10} consumed in this reaction there are 10 molecules of H_2O formed, or

$$\frac{10\ molecules\ H_2O}{2\ molecules\ C_4H_{10}}$$

Applying the factor-unit method, we can arrive at the desired answer.

$$20\ molecules\ C_4H_{10} \times \frac{10\ molecules\ H_2O}{2\ molecules\ C_4H_{10}} = 100\ molecules\ H_2O$$

---

*Nonstoichiometric reactions can also occur when there is an excess of one or more reactants. For example, when butane (C_4H_{10}) burns in air there is normally much more oxygen present than the stoichiometric amount needed according to equation 4.20. Under these conditions the reaction still follows equation 4.20. The excess O_2 remains unaffected and simply is left after the reaction is complete. In this case the amount of products resulting is dependent upon the amount of butane available and the butane is referred to as the *limiting reagent*. If conditions are such that the oxygen is the limiting reagent, then the nature of the reaction changes and a mixture of CO and CO_2 is produced along with the H_2O. The stoichiometry would now be dependent on a new equation that reflects this reaction.

Comparable calculations could also be made involving the O_2 and CO_2 in equation 4.20 with the result that for the complete combustion of 20 molecules of C_4H_{10}, 130 molecules of O_2 would be required and 80 molecules of CO_2 would be produced. For each compound in the reaction, we have just 10 times the amount specified in equation 4.20. On this basis the combustion of butane can be written

$$20\ C_4H_{10} + 130\ O_2 \longrightarrow 80\ CO_2 + 100\ H_2O \qquad \textbf{(4.34)}$$

which is read *20 molecules of butane react with 130 molecules of oxygen to produce 80 molecules of carbon dioxide and 100 molecules of water.* This equation still represents a balanced chemical equation.

In a similar way we can write balanced chemical equations based upon 100 or 1,000,000 (10^6) times the amount given by equation 4.20 and obtain the equations

$$200\ C_4H_{10} + 1300\ O_2 \longrightarrow 800\ CO_2 + 1000\ H_2O \qquad \textbf{(4.35)}$$

and

$$2 \times 10^6\ C_4H_{10} + 13 \times 10^6\ O_2 \longrightarrow 8 \times 10^6\ CO_2 + 10 \times 10^6\ H_2O \quad \textbf{(4.36)}$$

In each case the relative number of molecules remains the same. Writing the equation with 6.02×10^{23} times the number of each kind given in equation 4.20 gives

$$2 \times 6.02 \times 10^{23}\ C_4H_{10} + 13 \times 6.02 \times 10^{23}\ O_2 \longrightarrow$$
$$8 \times 6.02 \times 10^{23}\ CO_2 + 10 \times 6.02 \times 10^{23}\ H_2O \qquad \textbf{(4.37)}$$

The equation is still balanced but we can now consider it in terms of moles, since one mole of a substance consists of 6.02×10^{23} unit particles of the substance. Equation 4.37 tells us that in this reaction 2 moles of C_4H_{10} combine with 13 moles of O_2 to produce 8 moles of CO_2 and 10 moles of water. This is usually represented in equation form by the following.

$$2\ C_4H_{10}(g) + 13\ O_2(g) \longrightarrow 8\ CO_2(g) + 10\ H_2O(l) + \text{energy} \qquad \textbf{(4.38)}$$

A comparison of this equation with equation 4.20 shows that the two are identical. Equation 4.20 may, then, be considered in terms of molecules or moles.

EXAMPLE 4.8   How many moles of oxygen are required for the complete combustion of 6.84 moles of butane?

To do this calculation we use the same basic reasoning as was used in Example 4.7. That is, we must relate moles of oxygen to moles of butane. Equation 4.38 provides the necessary relationship,

$$\frac{13\ \text{mol}\ O_2}{2\ \text{mol}\ C_4H_{10}}$$

Multiplying this conversion factor by the quantity given (6.84 mol C_4H_{10}) yields the desired quantity.

$$6.84\ \text{mol}\ C_4H_{10} \times \frac{13\ \text{mol}\ O_2}{2\ \text{mol}\ C_4H_{10}} = 44.5\ \text{mol}\ O_2$$

From a molecular point of view we cannot have fractional numbers of molecules since molecules must exist as whole molecules. However, from the mole perspective we may have any fraction or multiple of a mole of each substance as illustrated in Example 4.8.

The last application of stoichiometric equations we will investigate deals with the relative numbers of grams of the various substances related through some reaction. We will consider two such applications: (1) relating moles of one substance to grams of another substance and (2) using grams for each substance.

The simple relationships involving molecules and moles derived from the coefficients in the balanced chemical equations must be expanded when calculations are made with respect to grams of the various substances. This is due to the fact that the molecular weights of the various species in a balanced chemical equation can differ considerably. For example, the molecular weight of butane is 58 while that of water is just 18.

Conversion from moles of one substance to grams of a second substance involves two-step calculations; one step is a mole-to-mole conversion such as that given in Example 4.8 and the other step involves a mole-to-gram conversion such as that given in Example 4.5. The following example illustrates a calculation using these steps.

EXAMPLE 4.9   How many grams of H_2O will be produced with 7.15 mol of CO_2 during the complete combustion of an adequate amount of C_4H_{10} based upon equation 4.38?

For this calculation we must make the following conversions:

$$7.15 \text{ mol } CO_2 \Longrightarrow ? \text{ mol } H_2O \Longrightarrow ? \text{ g } H_2O$$

The conversion factor for the first step can be obtained from equation 4.38,

$$\frac{10 \text{ mol } H_2O}{8 \text{ mol } CO_2}$$

The conversion factor for the second step is given by the gram-molecular weight,

$$\frac{18.0 \text{ g } H_2O}{1 \text{ mol } H_2O}$$

Using these factors in conjunction with the factor-unit method, we have the answer

$$7.15 \text{ mol } CO_2 \times \frac{10 \text{ mol } H_2O}{8 \text{ mol } CO_2} \times \frac{18.0 \text{ g } H_2O}{1 \text{ mol } H_2O} = 161 \text{ g } H_2O$$

The second type of calculation dealing with grams involves the conversion of grams of one substance to grams of a second substance. This conversion is basically the same as that considered in Example 4.9, except an additional change from grams to moles is necessary. A calculation of this type is given in the following example.

EXAMPLE 4.10   How many grams of CO_2 will be produced upon the complete combustion of 100 g of C_4H_{10}?

The solution to this problem is obtained by making the following conversions.

$$100. \text{ g } C_4H_{10} \Longrightarrow ? \text{ mol } C_4H_{10} \Longrightarrow ? \text{ mol } CO_2 \Longrightarrow ? \text{ g } CO_2$$

and the appropriate conversion factors, in the order needed, are

$$\frac{1 \text{ mol } C_4H_{10}}{58.0 \text{ g } C_4H_{10}}, \quad \frac{8 \text{ mol } CO_2}{2 \text{ mol } C_4H_{10}}, \quad \text{and} \quad \frac{44.0 \text{ g } CO_2}{1 \text{ mol } CO_2}$$

where the first and last factors are simply forms of the gram-molecular weight and the middle factor is obtained from equation 4.38. The desired quantity is obtained by means of the calculation

$$100. \text{ g } C_4H_{10} \times \frac{1 \text{ mol } C_4H_{10}}{58.0 \text{ g } C_4H_{10}} \times \frac{8 \text{ mol } CO_2}{2 \text{ mol } C_4H_{10}} \times \frac{44.0 \text{ g } CO_2}{1 \text{ mol } CO_2} = 303 \text{ g } CO_2$$

Many variations of problems similar to those illustrated can be considered, including molecules to moles or grams, or vice versa. Space will not allow further consideration of such problems here; however, the basic principles used in these examples apply and comparable calculations give the desired answers.

PROBLEMS

6. A compound that produces hydroxide ions in an aqueous solution is known as what type of substance?
7. Complete and balance the following chemical reactions. If no reaction occurs, indicate this by placing the letters N.R. to the right of the arrow.
    (a) $SO_2(g) + H_2O(l) \longrightarrow$
    (b) $Rb_2O(s) + H_2O(l) \longrightarrow$
    (c) $Ag(s) + CuSO_4(aq) \longrightarrow$
    (d) $F_2 + KBr \longrightarrow$
    (e) $Na_2CO_3(aq) + CaCl_2(aq) \longrightarrow$
        (Assume that one of the potential products is relatively insoluble.)
    (f) $HNO_3(aq) + Al(OH)_3(aq) \longrightarrow$
    (g) $C_6H_{14}(l) + O_2(g) \longrightarrow$
    (h) $C_5H_{10}O_5(s) + O_2(g) \longrightarrow$
⑤ 8. The hot flame of an acetylene torch is caused by the combustion of acetylene (C_2H_2) according to the equation

$$2 C_2H_2(g) + 5 O_2(g) \longrightarrow 4 CO_2(g) + 2 H_2O(g) + \text{energy}$$

Using this equation, complete the following calculations.
    (a) How many molecules of O_2 are required for the complete combustion of 16 molecules of C_2H_2?
    (b) How many moles of CO_2 can be produced from 27.5 mol of O_2?
    (c) How many grams of water can be produced by the combustion of 11.6 mol of C_2H_2?
    (d) How many grams of CO_2 will be produced with 85.6 g of H_2O?

ANSWERS TO PROBLEMS   **6.** a base   **7. (a)** $SO_2(g) + H_2O(l) \longrightarrow H_2SO_3(aq)$
(b) $Rb_2O(s) + H_2O(l) \longrightarrow 2 RbOH(aq)$ **(c)** $Ag(s) + CuSO_4(aq) \longrightarrow$ N.R.
(d) $F_2(g) + 2 KBr(s) \longrightarrow 2 KF(s) + Br_2(g)$ **(e)** $Na_2CO_3(aq) + CaCl_2(aq) \longrightarrow$
$2 NaCl(aq) + CaCO_3(s)$ **(f)** $3 HNO_3(aq) + Al(OH)_3(aq) \longrightarrow Al(NO_3)_3(aq) + 3 H_2O(l)$
(g) $2 C_6H_{14}(l) + 19 O_2(g) \longrightarrow 12 CO_2(g) + 14 H_2O(g)$ **(h)** $C_5H_{10}O_5(s) + 5 O_2(g) \longrightarrow$
$5 CO_2(g) + 5 H_2O(g)$ **8. (a)** 40 molecules   **(b)** 22.0 mol   **(c)** 209 g   **(d)** 418 g

SUMMARY

Since individual atoms and even the most complex compounds are very, very small, chemists found it necessary to identify a collection of atoms or molecules of a convenient size, called the mole. Although a mole consists of an astronomical number of unit particles (6.02×10^{23}), it is usually just about the right size to hold in your hand, weigh, or put in a beaker. From its definition, the mole can be related to the number of atoms or molecules and to the weight of a substance. Thus conversions between various units can be made. In addition to conversions involving a single substance, the mole concept can be extended to conversions between different substances that are related to one another through a balanced chemical equation. We considered a few simple classes of chemical reactions where we could predict with some success the products that result. We then covered the rules necessary to balance these chemical equations. Finally the mole concept was applied to stoichiometric calculations involving these equations.

In the next chapter we will examine a new topic, the three states of matter (solids, liquids, and gases) and the forces and energies involved in each state. Our major concern, however, shall be with the behavior of gases as represented by what are known as the gas laws. Application of these characteristics to the living system shall also be made.

ADDITIONAL PROBLEMS

9. How many molecules are in a mole of the analgesic (pain killer) Darvon (dextropropoxyphene), $C_{22}H_{29}NO_2$?

10. List three factors that make the quantity mole an appropriate amount of matter.

11. Calculate the gram-molecular weight of the following chemicals.
    - ⑤ (a) Darvon analgesic ($C_{22}H_{29}NO_2$)
    - (b) the tranquilizer meprobamate ($C_9H_{18}N_2O_4$)
    - ⑤ (c) glucose ($C_6H_{12}O_6$)
    - (d) the local anesthetic Novocain (procaine hydrochloride, $C_{13}H_{20}N_2O_2HCl$)
    - (e) the drug cocaine hydrochloride ($C_{17}H_{21}NO_4HCl$).

12. The recommended daily dietary allowances of some vitamins for a nursing mother are listed below. Calculate the number of moles and molecules required in each case.

13. A typical prescription of the sleep-inducing drug, Valium muscle relaxant (diazepam), $C_6H_{13}ClN_2O$, consists of 5.0 mg per tablet. How many (a) moles and (b) molecules of Valium are contained in a 5.0-mg tablet?

14. Calculate the number of individual atoms of each type in one mole of each compound listed in problems ⑤ 9, ⑤ 11, and 13.

15. Make the following conversions.
    - ⑤ (a) 7.00 mol Na $\Longrightarrow$? atoms Na
    - ⑤ (b) 7 atoms Na $\Longrightarrow$? mol Na
    - ⑤ (c) 7.00 mol Na $\Longrightarrow$? g Na
    - ⑤ (d) 7.00 g Na $\Longrightarrow$? mol Na
    - (e) 7.00 mol H_2O $\Longrightarrow$? molecules H_2O
    - (f) 7 molecules H_2O $\Longrightarrow$? mol H_2O
    - (g) 7.00 mol H_2O $\Longrightarrow$? g H_2O
    - (h) 7.00 g H_2O $\Longrightarrow$? mol H_2O
    - (i) 7.00 mol CH_3COOH $\Longrightarrow$? molecules CH_3COOH

Vitamin	Formula	Amount
⑤ C (ascorbic acid)	$C_6H_8O_6$	60 mg
⑤ Niacin (nicotinic acid)	$C_6N_5NO_2$	20 mg
B$_2$ (riboflavin)	$C_{17}H_{20}N_4O_6$	2.0 mg
B$_6$ (pyridoxine hydrochloride)	$C_8H_{11}NO_3HCl$	2.5 mg
B$_{12}$ (cyano-cobalamin)	$C_{63}H_{88}CoN_{14}O_{14}P$	6.0 μg

(j) 7 molecules $CH_3COOH \Longrightarrow$
      $?\ mol\ CH_3COOH$
(k) 7.00 mol $CH_3COOH \Longrightarrow$
      $?\ g\ CH_3COOH$
(l) 7.00 g $CH_3COOH \Longrightarrow$
      $?\ mol\ CH_3COOH$

⑤ 16. Make the following conversions.
(a) 15.0 mol $O_2 \Longrightarrow ?\ mol\ O$ atoms
(b) 15.0 mol $O_2 \Longrightarrow ?$ molecules O_2
(c) 15.0 mol $O_2 \Longrightarrow ?$ atoms O
(d) 15.0 mol O atoms $\Longrightarrow ?$ atoms O
(e) 15.0 mol $O_2 \Longrightarrow ?\ g\ O_2$
(f) 15.0 mol O atoms $\Longrightarrow ?\ g\ O_2$

17. Identify the reactants and products for the following reaction.

$C_{12}H_{22}O_{11}(s) + heat \longrightarrow$
      $12\ CO_2(g) + 11\ H_2O(g)$

18. Complete and balance the following chemical reactions. If no reaction occurs, indicate this by placing N.R. to the right of the arrow.
(a) $CO_2(g) + H_2O(l) \longrightarrow$
(b) $P_4O_{10}(g) + H_2O(l) \longrightarrow$
(c) $SrO(s) + H_2O(l) \longrightarrow$
(d) $Fe_2O_3(s) + H_2O(l) \longrightarrow$
(e) $Ni(s) + AgNO_3(aq) \longrightarrow$
(f) $Mg(s) + H_2SO_4(aq) \longrightarrow$
(g) $H_2(g) + KCl(aq) \longrightarrow$
(h) $Br_2(g) + KCl(s) \longrightarrow$
(i) $Br_2(g) + KI(s) \longrightarrow$
(j) $NaCl(aq) + KBr(aq) \longrightarrow$
(Assume all potential products are soluble.)
(k) $NaCl(aq) + Pb(NO_3)_2 \longrightarrow$
(Assume one of the potential products is soluble.)
(l) $H_3PO_4(aq) + Ca(OH)_2(aq) \longrightarrow$
(m) $HI(aq) + KOH(aq) \longrightarrow$
(n) $C_9H_{20}(g) + O_2(g) \longrightarrow$
(o) $C_{18}H_{38}(g) + O_2(g) \longrightarrow$
(p) $C_7H_{14}O_7(s) + O_2(g) \longrightarrow$
(q) $C_3H_6O_3(s) + O_2(g) \longrightarrow$

⑤ 19. The complete metabolism of ethyl alcohol in the human body can be represented by the following equation.

$2\ C_2H_5OH + 7\ O_2 \longrightarrow 4\ CO_2 + 6\ H_2O$

Perform the following calculations based upon this equation.
(a) 70 molecules $C_2H_5OH \Longrightarrow$
      $?$ molecules H_2O
(b) 70 molecules $O_2 \Longrightarrow ?$ molecules H_2O
(c) 70.0 mol $C_2H_5OH \Longrightarrow ?\ mol\ H_2O$
(d) 70.0 mol $O_2 \Longrightarrow ?\ mol\ CO_2$
(e) 70.0 mol $C_2H_5OH \Longrightarrow$
      $?$ molecules H_2O
(f) 70.0 mol $O_2 \Longrightarrow ?$ molecules C_2H_5OH
(g) 70 molecules $C_2H_5OH \Longrightarrow ?\ mol\ H_2O$
(h) 70 molecules $O_2 \Longrightarrow ?\ mol\ H_2O$
(i) 70.0 mol $C_2H_5OH \Longrightarrow ?\ g\ H_2O$
(j) 70.0 mol $CO_2 \Longrightarrow ?\ g\ C_2H_5OH$
(k) 70.0 g $C_2H_5OH \Longrightarrow ?\ g\ H_2O$
(l) 70.0 g $CO_2 \Longrightarrow ?\ g\ H_2O$

⑤ 20. Common baking soda ($NaHCO_3$) has been used for years as a remedy for indigestion. It is safe to use this substance occasionally, but overuse can lead to an excessive amount of base in the blood, a condition called alkalosis. $NaHCO_3$ reacts with the acid found in the stomach, HCl, according to the following equation.

$NaHCO_3(aq) + HCl(aq) \longrightarrow$
      $NaCl(aq) + H_2O(l) + CO_2(g)$

The CO_2 leaves the stomach in the form of a gas and is the substance that causes us to burp. A typical amount of $NaHCO_3$ that would be used for this purpose is approximately one gram. For 1.00 g of $NaHCO_3$, calculate the following.
(a) weight of HCl consumed
(b) weight of NaCl produced
(c) moles of H_2O produced
(d) molecules of CO_2 produced

CHAPTER 5

STATES OF MATTER, CHANGES IN STATE, AND THE GAS LAWS

In Chapter 4 we learned about the quantity of matter called the mole and used it to relate unit particles and grams of matter to each other. We then considered several typical types of chemical reactions and learned how to balance chemical equations for these reactions. Finally we did calculations involving moles, unit particles, and grams for various substances related to one another through a balanced chemical equation.

In this chapter we will examine the three states of matter: solid, liquid, and gas. We will look at the characteristics of each state and the factors involved when a substance is converted from one state to another. After these factors have been considered we will concentrate our attention on the behavior of gases and describe their properties through various gas laws. Finally we will consider the significance of the gas laws to the living system.

5-1 STATES OF MATTER AND CHANGES IN STATE

LEARNING OBJECTIVES

1. Define change in state, melting point, vapor pressure, pressure, boiling point, specific heat, heat of fusion, heat of vaporization, and hydrogen bond.
2. Specify the two types of forces of attraction that hold unit particles close together in the solid and liquid states.
3. Describe the natures of solids, liquids, and gases, including the types of energy associated with each state.
4. Describe the changes in state associated with the melting and boiling process.
5. Draw and label the heating curve for water covering the temperature range from less than 0°C to greater than 100°C.
6. Using words and diagrams, discuss (1) the physical forces associated with polarity and (2) hydrogen bonding in water, and the effect that these forces have on the properties of water.

In Chapter 1, the three states of matter, solid, liquid, and gas, were introduced. Each state of matter plays an important role in life. We depend upon the rigidness of solids to give strength to the bones in our bodies. Blood must be a liquid in order to flow through our veins to bring the necessary nutrients and oxygen to each cell of our bodies. Similarly, we live in a sea of gases and depend upon this "sea" to obtain the oxygen we need in order to burn the food we eat.

In this section we will consider the states of matter in more depth and discuss the energy considerations that are involved as substances change from one state into another. We will also consider a special type of bond known as the hydrogen bond and learn how hydrogen bonds give water some unique and important characteristics. In addition we will briefly consider how hydrogen bonds play an important role in the structure of protein molecules in the living system.

5-1.1 ENERGY AND CHANGES IN STATE

In Chapter 3 we considered two types of chemical bonds, ionic and covalent, and found that these bonds are the basis upon which chemical compounds form. These bonds are strong forces of attraction that hold together the atoms within the unit particles of compounds.

In the case of ionic compounds, the individual ions are not paired on a one-to-one basis as suggested by the formula, NaCl; rather, each ion is surrounded by several ions of opposite charge. This is illustrated in Figure 3.1 where the individual Na^+ and Cl^- ions are generally surrounded by six ions of opposite charge. We can see that ionic bonds not only hold together the individual ions in a chemical bond, but also hold together the millions of ions present in ionic substances in the solid and liquid states.

For nonionic substances various physical forces provide the attraction that holds the unit particles close together in the solid and liquid states. For example, in a covalent compound such as butane (C_4H_{10}), covalent bonds act to hold the carbon and hydrogen atoms together in each molecule, but these bonds have little influence between molecules. For nonionic substances these physical forces, though weaker than ionic bonds, act as a sort of glue that keeps the unit particles in solids and liquids from separating. Changes in state, such as melting or boiling substances, consists in overcoming these ionic and physical forces by the addition of heat to the solid or liquid substance.

A **change in state** is the process by which a substance passes from one state (solid, liquid, or gaseous) to another.

From a molecular point of view, *solids* are composed of unit particles held together in a rigid, orderly, crystalline pattern.* Three typical arrangements are pictured in Figure 5.1. A solid is like a theater that has seats arranged in rows that are not only located on the floor of the building, but also stacked in layers on top

---

*Solids sometimes are defined as substances that are hard and have a definite shape and volume. Included under this definition are substances such as glass, where the unit particles do not occur in an orderly crystalline pattern. From the point of view of this text, glass is considered a very thick, or viscous, liquid, somewhat like honey, that pours very slowly. This is evidenced by the broadening of window panes that can be observed at the bottom of the pane as the glass flows downward over a period of years. This is particularly noticeable in older glass.

Figure 5.1 Typical crystalline patterns showing the orderly arrangement of unit particles within solids.

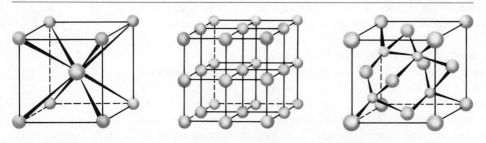

of each other. In theaters (solids), the people (unit particles) can move about in their seats (positions) and enjoy a good stretch. However, the unit particles cannot move about freely, but are restricted by the ionic, or physical, forces referred to earlier.

In chemistry the motion in place and stretching is referred to as *vibration* and the unit particles are said to have *vibrational energy*. The amount of vibrational energy that a substance has depends upon (a) the type of substance involved, (b) the strength of the forces holding the unit particles in place, and (c) the temperature of the substance. As heat is added and the temperature increases, the vibrational energy increases.

If the temperature of a solid becomes high enough, some of the unit particles on the surface of the solid can acquire enough energy so that they can actually break out of position in the crystalline solid and form a liquid. When this begins to occur we say that the solid is melting. The temperature at which this occurs is known as the **melting point**.

In the *liquid state,* the particles remain relatively close together due to the continued influence of the forces of attraction between the unit particles, but they can now move about more freely. Consider the theater illustration again. Converting a solid into a liquid is, in some ways, comparable to expending the energy to remove the seats and turn the theater into a very crowded, three-dimensional dance hall. The people (unit particles) are still fairly close together and still continue to move about (vibrate), but they can now work their way, or move about, throughout the crowded hall and turn around (rotate in position). The motion of the unit particles through the liquid is referred to as *translation* and the particles are said to possess *translational,* or *kinetic, energy.* In a similar way, the motion of the unit particles in position is spoken of as *rotation* and the accompanying energy is referred to as *rotational energy.* Unit particles of liquid possess translational, rotational, and vibrational energy in addition to the electronic and nuclear energy considered in Chapter 2.

Returning to the theater illustration, when most solids melt, the people (unit particles) can leave their seats and move into the aisles (expand) and occupy the entire theater. In a similar manner, when a substance changes from a solid to a liquid it typically expands approximately 20%. The individual unit particles in the liquid are believed to be essentially the same size as they were in the solid; however they are now, on the average, a bit farther apart.

As the liquid particles move about, some move quite rapidly while others are relatively slow and still others have speeds in between. As the particles move about

The **melting point** is that temperature at which a solid changes into a liquid.

they continually collide with one another and with the walls of the container. As a result of these collisions, the particles may change their direction and may speed up or slow down. As a result of this, some of the particles may be moving fast enough (that is, they may acquire sufficient translational energy) to overcome the remaining force of attraction between unit particles and break away from the rest of the bulk liquid. In this way the liquid can slowly evaporate. As the temperature of the liquid is raised, the average speed of the particles increases and more particles acquire enough energy to leave the liquid, causing it to evaporate more rapidly. As the temperature continues to rise, it reaches a point at which the vapor pressure of the particles leaving the liquid is equal to the pressure of the atmosphere pushing down on the liquid and the substance boils. The temperature at which this occurs is called the boiling point.*

As the unit particles enter the vapor, or gaseous, state they become separated from one another and move about independently. In the air we breathe the particles are, on the average, approximately 10 times farther apart than they would be in the liquid state. This results in a volume change of about 1000 times. When many liquids evaporate or boil, the volume they occupy increases approximately 100–1000 times.†

The particles in the gaseous state possess translational, rotational, vibrational, and electronic energy but move about at a higher speed than they did while in the liquid state. The motion of gases can be pictured as that of a swarm of rapidly moving bees. The gaseous particles are moving about at enormous speeds. For example, the average speed of a gaseous water molecule at room temperature is about 6×10^4 cm/sec, or 1300 mi/hr! Since the water molecules are on the average only about 3×10^{-7} cm apart, they do not travel far between collisions, but their average speed between collisions still would be 6×10^4 cm/sec.

All of us have seen the condensation of water vapor on the side of an ice-cold glass. In this case the process of evaporation has been reversed and gaseous water, or water vapor, present in the air condensed on the cold glass. When a gaseous water molecule strikes the cold glass, it transfers some of its energy to the glass and, as a result, does not have enough energy to break away from the physical force of attraction between the water molecules and the surface of the glass or between it and other water molecules that may have already collected on the glass. Similarly, when we put an ice cube tray filled with water into the freezer the water molecules lose energy to the colder surroundings until they can no longer move about at random and, therefore, are trapped in the crystalline pattern of ice. This process of freezing occurs at the temperature at which ice melts, 0°C. Since this behavior is true for all liquids, the melting point and freezing point of a substance are the same. The use of the terms *melting* or *freezing* simply suggests which process is taking place, rather than indicating different temperatures.

The ability of water to exist as a solid, liquid, or gas is typical of all substances that do not decompose when they are heated or cooled. Another example is table salt which melts when the temperature reaches 801°C and becomes a vapor above

Vapor pressure is the outward push, or force, per unit surface area, of the particles of a liquid as they tend to escape into the gaseous state.

Pressure is a force, per unit area, exerted over the surface of a body.

The **boiling point** is that temperature at which the vapor pressure of the liquid becomes equal to the pressure of the atmosphere above the liquid.

---

*Since the pressure of the atmosphere varies depending upon the altitude above sea level, we commonly use a reference boiling point known as the *normal boiling point.* This is simply the boiling point under average sea-level atmospheric pressure (referred to as 1 atmosphere pressure).

†The value 100–1000 given here assumes that the resulting vapor was collected when the temperature and pressure were at normal room conditions.

Figure 5.2   Heating curve for water.

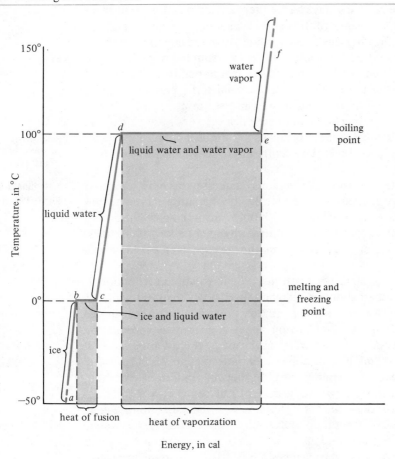

1465°C. The high melting and boiling points of table salt as compared with a substance such as water is due to the fact that the ions in the table salt are held together by the much stronger force of attraction that is associated with ionic bonds. In general, ionic compounds have relatively high melting and boiling points. On the other hand, a substance like oxygen does not become a liquid until it is cooled to −183°C. In this case, there are very weak physical forces between the oxygen molecules and thus, very low temperatures must be reached for these weak physical forces to hold the molecules together in the liquid or solid state.

Not all substances can exist in all three states. For example, table sugar, sucrose ($C_{12}H_{22}O_{11}$), decomposes at about 200°C by losing water from its structure to form a brown, syrupy mass called caramel. Therefore sugar ordinarily cannot exist as either a liquid or a vapor.

The process of going from a solid to a liquid and then to a gas can be represented graphically by what is known as a heating curve. Figure 5.2 gives the heating curve for water. The plot shows how the temperature of the substance (the dependent variable) changes with a change in the quantity of heat added (the independent variable). As can be seen from this diagram, the addition of heat to ice that is below 0°C causes it to warm up until it reaches the melting point. During the melting process no change in temperature occurs and the line b–c in Figure 5.2 is horizontal. This is a general characteristic of all pure substances. Once the ice is completely melted, the addition of more heat raises the temperature of the water, line c–d, until the boiling point is reached. At this point the temperature again remains constant, line d–e, until all the liquid has been changed into a gas, or vapor.

It is useful to remember that the temperature remains constant at the boiling point. When we are in a hurry while boiling food, we have a natural tendency to turn up the heat to cook the food faster. Heating the food more strongly just boils the liquid away at a higher rate. Once a liquid is boiling, it is as hot as it can be.

If we continue to heat the water vapor, the temperature of the vapor increases, line e–f. If the vapor is under a constant pressure, for example that of the atmosphere, the volume of the vapor will increase as the temperature increases due to the greater force of the gaseous molecules as they strike the sides of the container. This effect shall be dealt with in more detail in Section 5-2.

A quantitative measure of the increase of the temperature of a substance with heating is provided by the **specific heat**. The value of the specific heat for a substance depends upon the nature of the substance involved and the physical state it is in. For water in its three states we have the following specific heats: ice, 0.49 cal/g-°C; water, 1.00 cal/g-°C; and water vapor, 0.48 cal/g-°C. Liquid water has one of the highest specific heats of any substance. The values of specific heat for some typical substances listed in Table 5.1 illustrate this point.

Another factor that can be seen in the heating curve in Figure 5.2 is the amount of heat required to melt or boil a gram of the substance. The quantity of heat on the left is called the **heat of fusion** and the quantity on the right is known as the **heat of vaporization**.* The length of the solid horizontal line in Figure 5.2 in each

The **specific heat** is that amount of heat that is needed to increase the temperature of one gram of a substance by one degree centigrade.

The **heat of fusion** is the amount of heat required to convert one gram of a solid to a liquid at its melting point.

The **heat of vaporization** is the amount of heat required to convert one gram of a liquid to a vapor at its normal boiling point.

Table 5.1   Specific Heats of Elements and Compounds at 25°C

Substance	Specific Heat (cal/g-°C)
Al(s)	0.217
C(grap)	0.167
Cu(s)	0.092
Ethyl alcohol	0.581
Au(s)	0.031
Fe(s)	0.113
Pb(s)	0.0305
Na(s)	0.29
NaCl(s)	0.21
H_2SO_4(l)	0.339
H_2O(s)	0.49
H_2O(l)	1.00
H_2O(g)	0.48

*Although the heat of vaporization is usually considered to be associated with the boiling of a substance, heat is also associated with the evaporation of a liquid. In the case of the evaporation of perspiration, heat is withdrawn from our bodies producing a cooling effect.

case is related to the amount of heat required to change one gram of the substance from one state to another. In the case of water, the heat of fusion, line b–c, is equal to 80 cal/g. Eighty calories is enough heat to warm one gram of water from 0°C to 80°C, since each calorie will raise the temperature of water by 1°C. A similar relationship exists between the length of the line d–e in Figure 5.2 and the heat of vaporization. For water the heat of vaporization is 540 cal/g. This is a relatively high value as is evident when we realize that this is enough heat to raise the temperature of 5.4 g of water from 0°C to 100°C.

As discussed in Chapter 1, energy is conserved in all physical and chemical processes. Therefore, the same amount of heat is given off when a substance condenses as was absorbed when it evaporated. Similarly, the same amount of heat is evolved when the substance is cooled as is absorbed when it is heated to the original temperature. Because of the large amount of heat given off when steam condenses (540 cal/g) and then cools from the boiling point of 100°C to a body temperature of 37°C (63 cal/g), there is a real danger associated with contact with steam. A steam burn can be very serious, particularly if a large portion of the body is burned.

Before we conclude our consideration of changes in state we will examine the effect of pressure on the boiling point of a substance. Since the boiling point of a liquid is reached when its vapor pressure becomes equal to the atmospheric, or outside pressure, the boiling point of a substance changes as this outside pressure changes. For example, if you have ever camped in high, mountainous areas you may have noticed that it takes longer to cook soup or to boil an egg there than it does at lower elevations. In this case the atmospheric pressure in the mountains is lower because we are above some of the atmosphere and, as a result, the boiling point is lower and the food cooks at a lower temperature. In general the boiling point of a liquid increases with an increase in pressure and decreases with decrease in pressure.

Several applications of this behavior are made to take advantage of both higher and lower boiling points. Chemists frequently do what is called a vacuum distillation, a technique that depends upon a reduction in the outside pressure to cause a substance to boil at a lower temperature. This is particularly valuable when the substance tends to decompose at a higher temperature.

Another common application of this effect is in the pressure cooker. A higher pressure develops inside the sealed container by means of a valve, called a petcock, that only allows gases to escape when a higher-than-atmospheric pressure is reached. The higher pressure raises the boiling point and the food cooks faster since it is being cooked at a higher temperature. Another advantage of the pressure cooker is that more bacteria are killed and in a shorter time than at the regular boiling point. When foods such as string beans are canned, they are processed under pressure so that the botulism bacteria* are destroyed.

---

*Botulism is a severe form of food poisoning caused by botulin, the toxin produced by the *Clostridium botulinum* bacteria. Botulism is associated with the growth of the bacterium in improperly canned or preserved meats and vegetables. The bacteria cannot grow in foods that are relatively acidic such as tomatoes, but they thrive in what are known as low-acid foods such as string beans. Such foods contain too few hydrogen ions to restrict the growth of the bacterium. Low-acid foods should always be boiled a minimum of ten minutes to make certain that any botulinus toxin that may be present in the food is destroyed.

5-1.2 WATER, THE HYDROGEN BOND, AND CHANGES IN STATE

In the previous section reference was made to the unique properties of water. Included in these is the fact that water expands on freezing. Since water plays an important role in the living system we will consider this behavior and some of its ramifications in various aspects in life. The water molecule is angular in shape, with an angle of 104.5° between the bonds. Each bond is a polar bond due to the difference in electronegativities of the hydrogen (EN = 2.1) and oxygen (EN = 3.5) atoms. The polar nature of water provides part of the physical force of attraction between hydrogen atoms in one molecule and oxygen atoms in different molecules. When ice is formed from liquid water, this attraction leads to a long-range superstructure as illustrated in Figure 5.3. As seen in this figure, holes of empty space develop as ice crystals form. The tendency of ice to form this superstructure leads to the expansion of water when it freezes, since holes are formed in the ice that were not present in the liquid water.

$$\overset{\delta-}{O}$$
$$\overset{\delta+}{H}\underset{104.5°}{}\overset{\delta+}{H}$$

Water

   The property of expansion on freezing has both good and bad aspects. A beneficial application of this behavior is in the removal of warts by a physician by freezing them with liquid nitrogen. Liquid nitrogen exists at a temperature of −196°C or less. Contact of living tissue with liquid nitrogen freezes the water in the tissue. As the water in the tissue freezes it expands and breaks the membranes of the cells, killing them in the process. Frostbite acts in a similar way except that

Figure 5.3   Ice forms in a superstructure made up of oxygen and hydrogen atoms arranged in a hexagonal pattern of six-membered oxygen rings.

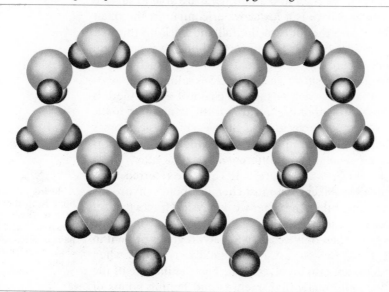

Figure 5.4
Hydrogen bonds in water, as illustrated by the dashed lines, allow each water molecule to be bridged to up to four other water molecules in ice.

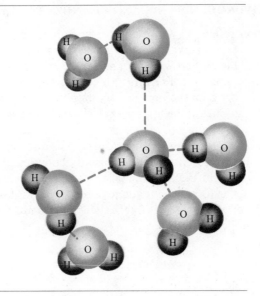

desirable tissue is killed. In severe cases of frostbite so much tissue is killed that amputation of the affected part of the body may be necessary. The damage to tissue by freezing is similar in some respects to that which occurs when we are burned by a hot object. The steam burn referred to previously is an example of the latter. Damage to tissue by freezing is, therefore, frequently referred to as a "freeze burn."

Similar destruction can occur in plant tissues. The first good freeze in the fall, considered to be 28°F (−2.2°C) or lower, kills most unprotected annual plants such as tomatoes and petunias. Many trees, on the other hand, are able to survive the freezing temperatures of winter. There is a lower water content in the plant tissue because the sap flows down into the roots in the fall and back up through the tree in the spring. Companies that process frozen foods limit the destruction of tissue and the resulting loss in texture by freezing the food so quickly that build up of the long range superstructure of ice shown in Figure 5.3 is minimized.

Reference to Figure 5.3 shows that in each hexagonal chain of atoms there is an alternating pattern of oxygen-hydrogen-oxygen, and so forth. In addition to the regular covalent bonds within each water molecule and the weaker physical forces due to the polarity of water that act between molecules, a second type of bond is formed between oxygen and hydrogen atoms on different water molecules. These bonds are illustrated by the dashed lines in Figure 5.4. They are referred to as hydrogen bonds. Although generally weaker than most chemical bonds, hydrogen bonds are, in some cases, much stronger than the other weak physical forces that act between molecules such as those associated with the polarity of water. As the name *hydrogen bond* suggests, this type of bond only occurs when hydrogen is present in the molecule and the bond is, for the most part, only significantly strong when nitrogen, oxygen, or fluorine atoms are also involved in it. The significance of the hydrogen bond is evident in the abnormally high melting and boiling points of

A **hydrogen bond** is a bond formed in which a proton (hydrogen nucleus) is shared by two pairs of electrons associated with two other atoms.

water compared with those of the heavier H_2S molecule ($-82.9°C$ and $-61.8°C$, respectively) where little hydrogen bonding occurs. Hydrogen bonds serve an extremely important function in many compounds and are vital to the living system. An important example of hydrogen bonding is found in a class of compounds called *proteins*. Proteins are complex compounds that occur naturally in plants and animals and are essential for the growth and repair of animal tissue. Proteins exist as long-chain molecules consisting of thousands of atoms joined in various structures by hydrogen bonding such as the coiled or helical structure pictured in Figure 5.5. In this figure we can see that hydrogen bonds, represented by the dashed line between oxygen and hydrogen atoms, act to keep the molecule in this coiled shape.

Heating protein tends to break hydrogen bonds, as is evident when an egg is cooked. The white of the egg is composed of the protein albumin which, on being heated, undergoes a change in its chemical and physical properties. This is due to, among other things, the breaking of the hydrogen bonds and resulting alteration of a type of hydrogen-bonded structure referred to as a tertiary structure. This change is noted as the egg turns white and becomes firm. Further reference shall be made to proteins and hydrogen bonding in Chapters 20 and 21.

Figure 5.5
The coiled, or helical, structure of proteins is maintained by the hydrogen bonds illustrated by the dashed lines. (Only a portion of the protein and associated hydrogen bonds is shown for clarity.)

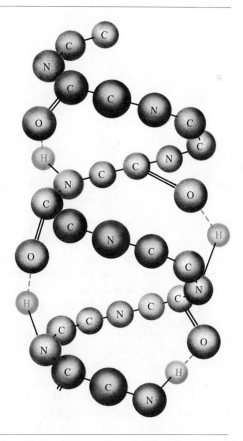

Although hydrogen bonds are weaker than chemical bonds, they still have considerable strength, as illustrated by large values for the heat of vaporization and the melting and boiling points of water. If water freezes in the engine block of a car, it can crack the block due to the enormous pressure associated with the expansion of water as it freezes. This same pressure has been used to crack large rocks to obtain large stone blocks for constructing buildings. To do this, holes are drilled in the rock, filled with water, stoppered in some way, and then allowed to freeze in the middle of winter. After the spring thaw, the rocks are found cleaved along the line of the holes.

PROBLEMS

1. Define the phrase change in state.
2. Which force of attractions in solids is stronger, the ionic or the physical force?
3. What physical state is associated with the unit particles moving about at random at relatively large distances from one another?
4. What do we call the amount of heat necessary to melt one gram of a substance?
5. What is a unique characteristic of water when it freezes?

ANSWERS TO PROBLEMS   1. It is the process by which a substance passes from one state to another.   2. the ionic force   3. the gaseous state   4. the heat of fusion   5. it expands

5-2 THE GAS LAWS

LEARNING OBJECTIVES

1. Define standard pressure and standard temperature.
2. Make conversions between degrees Celsius and degrees Kelvin.
3. State Boyle's law, Charles' law, the temperature - pressure relationship for gases, the combined gas law, and Dalton's law of partial pressure. Perform calculations using each of these laws.
4. Perform conversions from moles to volume and volume to moles at STP.

In Section 5-1 the three states of matter were examined in some detail. We will now turn our attention to the behavior of gases as observed experimentally. Experience with actual gases led scientists to formulate what are called the gas laws. Historically the nature of gases was among the earliest areas of science studied. Since air is a mixture of gases and easily can be put in a container, early scientists had a ready source for study. The following sections in this chapter deal with the results of these studies.

5-2.1 BOYLE'S LAW

In 1662 the English scientist Robert Boyle observed that the volume of a trapped gas varies inversely with the total pressure on the gas. In other words, when the pressure exerted against a gas is increased, the volume of the gas decreases. This is illustrated in Figure 5.6. Simply stated, this is analogous to squeezing a sponge.

Figure 5.6
Boyle's Law. As the pressure due to the bricks increases, the volume decreases proportionally.

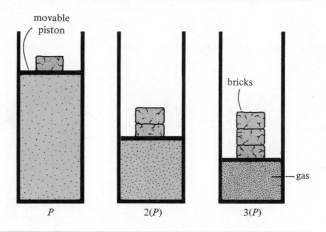

When we put pressure on the sponge by squeezing it, the sponge becomes smaller. The same thing happens to the trapped gas in Figure 5.6 when we "squeeze" it by increasing the number of bricks on the piston. Since the amount of gas and the temperature of the gas also affect its volume these were held constant when Boyle did his work. Briefly stated, Boyle's law says: *If a given mass of gas is considered and if its temperature is held constant, the volume of the gas is inversely proportional to the pressure.*

The various gas laws that shall be covered in this chapter provide excellent examples of the concept of independent and dependent variables and the control of certain variables so that the relationship between other variables can be more easily observed. Boyle maintained the gas at a constant temperature and also kept the amount, or number of moles, of gas constant. By doing this, he was able easily to determine the effect that a change in pressure has on the volume. In this case, the pressure acts as the independent variable and the volume is the dependent variable.

Mathematically Boyle's law is given by the expression

$$V_{T,n} \propto 1/P$$

which is read *when the temperature and number of moles are held constant the volume is inversely proportional to pressure.* Written as an equality, it is

$$V_{T,n} = k'/P$$

where k' is the appropriate proportionality constant. For most applications the proportionality constant is eliminated by considering the volume of the gas under two different pressures. Rearranging the above equation, we can write, in general,

$$P \times V_{T,n} = k' \tag{5.1}$$

Applying equation 5.1 to two sets of values for the volume and pressure and omitting the subscripts T and n as is common, we can write

$$P_1 \times V_1 = k' = P_2 \times V_2 \tag{5.2}$$

Figure 5.7
A Torricelli barometer consists of a glass tube sealed on one end, filled with mercury, and inverted in a pool of mercury.

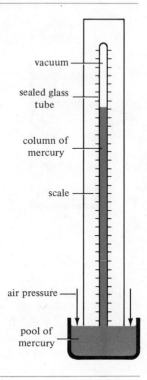

vacuum

sealed glass tube

column of mercury

scale

air pressure

pool of mercury

or, eliminating k', we obtain

$$P_1 V_1 = P_2 V_2 \qquad\qquad (5.3)$$

This equation allows us to do calculations related to the pressure and volume of a contained gas. Before we do these calculations however it will be helpful for us to consider how pressure is measured and the units that are associated with it.

Pressure is normally measured by means of an apparatus known as a barometer. A simplified drawing of one type, a Torricelli barometer, is given in Figure 5.7. This type of barometer is basically a glass tube that has been sealed on one end and then filled with mercury and inverted, with the open end placed in a pool of mercury. When this is done the mercury will fall until the weight of the mercury per unit of area is just equal to the weight of air per unit of area on top of the pool of mercury. The length of the column of mercury gives us a measure of the air pressure. At sea level the length of this column of mercury is, on the average, 760 mm. The unit mmHg, taken from this method of measuring air pressure, is frequently used as a unit of air pressure. The pressure at sea level is referred to as the **standard pressure.** A second unit of pressure, the atmosphere (atm), is also commonly used. Note that 1 atm = 760 mmHg.* With this information we are now prepared to do calculations using Boyle's Law.

Standard pressure is 1 atm, or 760 mmHg.

---

*Two additional units of pressure, torr and psi (pounds per square inch), also are occasionally used. The torr is equivalent to mmHg and was chosen to honor Evangelista Torricelli, who developed the type of barometer pictured in Figure 5.7. The unit psi is related to atm by 1 atm = 14.7 psi and is frequently used as the unit of pressure on tanks of gases such as the oxygen tank used in hospitals.

EXAMPLE 5.1   If 10 mL of air are drawn into a hypodermic syringe when the atmospheric pressure is 640 mmHg and the tip of the needle is then sealed, what volume will the gas occupy if the plunger is pushed with sufficient force to change the pressure to 2.5 atm?

   To solve a problem of this type we can use equation 5.3. We must begin by identifying the different variables in the equation. Since the initial pressure and volume are 640 mmHg and 10 mL, respectively, we have $P_1 = 640$ mmHg and $V_1 = 10$ mL. Thus we have $P_2 = 2.5$ atm and V_2 is the unknown quantity. Solving for V_2 in equation 5.3 we have

$$V_2 = \frac{P_1 V_1}{P_2} \tag{5.4}$$

Substituting the appropriate values we have

$$V_2 = \frac{640 \text{ mmHg} \times 10 \text{ mL}}{2.5 \text{ atm}}$$

Inspection of the units indicates that the two units of pressure are not consistent. Unless one of them is changed, a unit mmHg mL/atm will be obtained. This is not a unit of volume. This problem can be resolved by converting one of the pressure units into the other. To convert the unit atm to mmHg, we can use the relationship given above (1 atm = 760 mmHg). Therefore we have

$$2.5 \text{ atm} \times \frac{760 \text{ mmHg}}{1 \text{ atm}} = 1900 \text{ mmHg}$$

Substituting this into the expression for V_2 we have

$$V_2 = \frac{640 \text{ mmHg} \times 10 \text{ mL}}{1900 \text{ mmHg}} = 3.4 \text{ mL}$$

   In the living system there are generally no closed regions that contain gases, as did the syringe in the example above or a balloon. The closest thing to this condition is when we have a cold and the Eustachian tube becomes inflamed and blocked so that air connot freely enter and leave the region of the inner ear (see Figure 5.8). In this case, there may be a difference between the atmospheric pressure outside the ear and the pressure in the inner ear. Because of this pressure difference, the eardrum tends to move as the plunger in Example 5.1 does. As a result, the eardrum becomes distorted, resulting in a distortion in the way we hear things. This distortion is particularly noticeable if we experience a change in altitude, such as when we travel in the mountains or fly in an airplane that does not have a pressurized cabin. Chewing gum or cracking one's jaw frequently will relieve the problem temporarily by opening the Eustachian tube to allow air to flow into or out of the inner ear and thus equalize the pressure.

   Another application of the principles dealt with in Boyle's law is the process of breathing. We lower the pressure in our lungs by contracting the diaphragm, causing it to flatten out downward. Since the pressure in the lungs is now lower than the atmospheric pressure, air flows into the lungs. As the diaphragm is relaxed it rises, increasing the pressure in the lungs and forcing air to flow out. Since the temperature of the air and the amount of air that we breathe are not constant, a

Figure 5.8
When the Eustachian tube becomes inflamed, it can block the flow of air between the inner ear and the pharynx.

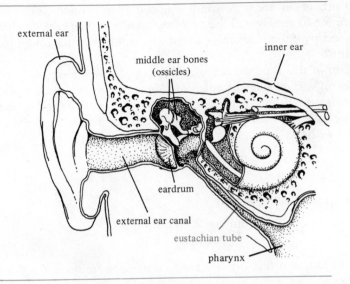

more general relationship than Boyle's law actually applies (see Section 5-2.4). However, the major effect is one of pressure and to that extent can be understood through Boyle's law.

5-2.2 CHARLES' LAW

Approximately one hundred years after Robert Boyle formulated his law, two individuals, Jacques Charles (1746–1823) and Joseph Louis Gay-Lussac (1779–1850), independently discovered that *for a fixed mass of a gas under constant pressure the volume of the gas is directly proportional to the absolute temperature of the gas*. Known generally as Charles' Law, this can be written

$$V_{P,n} \propto T \tag{5.5}$$

where the temperature is considered to be the independent variable and the volume is acting as the dependent variable. Written as an equality, it is

$$V_{P,n} = k''T \tag{5.6}$$

where k'' is the proportionality constant. This expression applies to a gas at any temperature as long as there is a fixed amount of gas and a constant pressure. Therefore the equation can be applied to two sets of conditions to eliminate k'' in the same way k' was removed from Boyle's law to obtain equation 5.3. On doing this we obtain

$$\frac{V_1}{T_1} = \frac{V_2}{T_2} \tag{5.7}$$

EXAMPLE 5.2   If an air-tight plastic bag that is partially filled with air has a volume 1.50 L when it is in the freezer compartment of a refrigerator at −20°C, what ume will it have if it is placed in a hot room at 100°F?

In this problem it is important to convert both temperatures to what is known as the *absolute, or Kelvin, temperature scale.* Failure to do so will lead to an incorrect value for the volume. The importance of this is illustrated in this example, since a negative value for the volume would be obtained if a temperature of $-20°C$ were used. A negative volume is like a negative weight: neither is possible.

The absolute temperature scale has a zero point 273.2° below the zero point in the Celsius scale. Basically it can be considered the temperature at which all molecular motion ceases. Calculation of the Kelvin temperature is made by adding 273.2° to the Celsius value

$$\text{deg K} = °C + 273.2° \tag{5.8}$$

Using this equation, $-20°C$ on the Kelvin scale is

$$\text{deg K} = -20°C + 273.2° = 253.2°K$$

Conversion of Fahrenheit temperatures to the Kelvin scale is begun by converting to degrees Celsius as discussed in Chapter 1 and then solving for the value in degrees Kelvin. Applying this to the temperature 100°F gives

$$\text{deg C} = \tfrac{5}{9}(°F - 32°) = \tfrac{5}{9}(100° - 32) = 37.8°C$$

Converting to Kelvin we have

$$\text{deg K} = 37.8°C + 273.2° = 311.0°K$$

We are now ready to calculate the final volume, V_2, of the air in the plastic bag at the higher temperature. Identification of the matching values of the temperature and volume gives $T_1 = 253.2°K$, $V_1 = 1.50$ L, $T_2 = 311.0°K$, and V_2 is the unknown quantity. Solving for V_2 in equation 5.7 gives

$$V_2 = \frac{V_1 \times T_2}{T_1} = \frac{1.50 \text{ L} \times 311.0°K}{253.2°K} = 1.84 \text{ L}$$

Since the living system is essentially a constant-temperature system, Charles' law is not of prime importance to the health-related sciences. One application, however, is the effect of temperature on the air we breathe. Except on quite hot days when the air temperature exceeds body temperature, there is a general warming of the air as we inhale. As the air is warmed it will undergo an expansion according to Charles' law and will occupy a slightly larger volume in the lungs. Although the expansion normally is of little consequence, it does illustrate the relationship between volume and temperature that exists for gaseous substances.

5-2.3 PRESSURE-TEMPERATURE RELATIONSHIP FOR GASES

The pressure of a gas in a constant volume container, such as an automobile tire, increases with an increase in temperature. This relationship is comparable to Charles' law and may be stated as follows: *For a fixed mass of a gas that is maintained at a constant volume the pressure of the gas is directly proportional to the absolute temperature of the gas.* Mathematically this relationship can be written as

$$P^{V,n} \propto T \tag{5.9}$$

with the temperature arbitrarily chosen as the independent variable and the pressure functioning as the dependent variable. Written as an equality this relationship becomes

$$P_{V,n} = k'''T \tag{5.10}$$

or, for two different temperatures and the accompanying pressures,

$$\frac{P_1}{T_1} = \frac{P_2}{T_2} \tag{5.11}$$

Known by various names such as Gay-Lussac's law and Amontons' law, this relationship becomes important in many applications in the world about us.

If we understand Gay-Lussac's law, we can appreciate the warning given on aerosol spray cans that they should not be stored above a certain temperature, usually 120°F (49°C). At higher temperatures the pressure inside the can may become so great that the can will burst. Two factors contribute to this effect: the increase in pressure due to an increase in temperature as discussed in this section and the tendency of the liquid in the can to vaporize, or boil, as the temperature increases. Both processes contribute to the overall pressure within the can. Calculations based upon Gay-Lussac's law are done in the same manner as those involving Charles' law except pressure is considered rather than volume.

5-2.4 THE COMBINED GAS LAW

From the discussion in Section 5-2.1 we learned that the volume of a gas is inversely proportional to the pressure [$V_{T,n} \propto (1/P)$] when the amount of gas and its temperature are held constant. In Section 5-2.2 we found that the volume of a gas is directly proportional to the absolute temperature ($V_{P,n} \propto T$) when the amount of gas and the applied pressure are constant. We know that the size, or volume, of a balloon increases when we blow air into it. Stated scientifically, this is: *When the temperature and pressure of a gas are held constant the volume of the gas is directly proportional to the number of moles of gas (n) present.* In mathematical terms,

$$V_{P,T} \propto n \tag{5.12}$$

In real-life situations we seldom experience conditions where most of the variables are held constant. For example, most of us probably have experienced the disappointment as a child of receiving a carnival balloon filled with helium and waking up the next morning to find the balloon much smaller and perhaps floating only part way toward our bedroom ceiling or even lying on the floor. Several factors probably combined to make the balloon shrink. First, and foremost, helium atoms tend to pass through the walls of the balloon, thus decreasing the amount of gas (n) in the balloon. Second, the bedroom may have been colder in the morning than it was the previous day. Finally, the pressure of the atmosphere varies from day to day and a high could have passed into our geographical area during the night. Since nothing was held constant in this case all three independent variables, moles of gas, temperature, and pressure, must be included in an expression of the volume. This is accomplished by combining them into what is appropriately called the *combined,* or *general gas law.* This law takes the form

$$V \propto \frac{nT}{P} \tag{5.13}$$

We read this: *The volume of a gas is directly proportional to the number of moles of gas and to the absolute temperature and is inversely proportional to the applied pressure.* Written as an equality the law becomes

$$V = \frac{nRT}{P} \qquad (5.14)$$

or

$$PV = nRT \qquad (5.15)$$

where R is a constant of proportionality known as the gas constant. The value of R depends upon the units used for the volume, temperature, and pressure, but it is independent of their magnitude. When the volume is expressed in liters, the temperature in degrees Kelvin and the pressure in atmosphere, R has a value of 0.0821 L-atm/°K-mol.

EXAMPLE 5.3   The lung capacity of an average adult male is approximately 6500 mL. How many moles of air will the lungs hold at normal body temperature (37°C) and a pressure of 700 mmHg?

For this calculation we are asked to solve for the number of moles (n), so equation 5.15 must be rearranged to the form

$$n = \frac{PV}{RT} \qquad (5.16)$$

Since the value of R is based upon pressure in atmospheres, temperature in degrees Kelvin, and volume in liters all of these quantities must be converted to the appropriate units. To make these changes we have

$$6500 \text{ mL} \times \frac{1 \text{ L}}{1000 \text{ mL}} = 6.5 \text{ L}$$

$$700 \text{ mmHg} \times \frac{1 \text{ atm}}{760 \text{ mmHg}} = 0.92 \text{ atm}$$

and

$$°K = 37°C + 273.2° = 310.2°K$$

Therefore, using equation 5.16 we can calculate the number of moles,

$$n = \frac{PV}{RT} = \frac{0.92 \text{ atm} \times 6.5 \text{ L}}{0.0821 \dfrac{\text{L-atm}}{°K\text{-mol}} \times 310.2°K} = 0.23 \text{ mol}$$

Gases that follow the equation $PV = nRT$ are known as *ideal gases.* This equation is, therefore, also referred to as the *ideal gas law* or *the universal gas law.* Note that no reference is made to the specific gas involved when using this equation. Thus, one mole of any gas will occupy the same volume at any specific temperature and pressure. For convenience, a reference, or *standard condition* of temperature and pressure has been chosen. The standard pressure is 1 atm, or 760 mmHg as indicated previously. Standard temperature is 0°C or 273.2°K. *One mole of any gas*

Standard temperature is 0°C, or 273.2°K.

occupies 22.4 L at standard temperature and pressure (STP). Stated as an equality, *for STP conditions,*

$$1.00 \text{ mol gas} = 22.4 \text{ L} \qquad (5.17)$$

On this basis conversions between moles of a gas and the volume of a gas at STP can be made using the factor-unit method.

EXAMPLE 5.4   A common size for a room is 12 ft by 12 ft by 8 ft high. In terms of volume, this is a room having a volume of approximately 3.26×10^4 L. How many moles of air are in such a room at STP?

For this calculation we can use the relationship given by equation 5.17 and solve for the number of moles directly.

$$3.26 \times 10^4 \text{ L} \times \frac{1 \text{ mol}}{22.4 \text{ L}} = 1.46 \times 10^3 \text{ mol}$$

Comparison of the calculations in Examples 5.3 and 5.4 shows that both involve a conversion from volume to moles. However, the first calculation is applicable to any conditions of temperature and pressure while the latter calculation deals only with STP conditions.

Since equation 5.15 describes the behavior of gases in general it may be rearranged and applied to two sets of temperature and pressure. Solving for R in both sets of conditions we have

$$\frac{P_1 V_1}{n_1 T_1} = R \quad \text{and} \quad \frac{P_2 V_2}{n_2 T_2} = R$$

Since both expressions are equal to R they must be equal to each other, and we can write

$$\frac{P_1 V_1}{n_1 T_1} = \frac{P_2 V_2}{n_2 T_2} \qquad (5.18a)$$

When the number of moles is constant the equation takes the form

$$\frac{P_1 V_1}{n T_1} = \frac{P_2 V_2}{n T_2} \qquad (5.18b)$$

When the number of moles has been cancelled, the equation becomes

$$\frac{P_1 V_1}{T_1} = \frac{P_2 V_2}{T_2} \qquad (5.19)$$

Whereas equation 5.15 applies in general, equation 5.19 is restricted to those cases where the number of moles is constant. All but one of the six variables appearing in equation 5.19 must be known before calculations can be performed using it.

EXAMPLE 5.5   If a weather balloon having a volume of 950 L is released at sea level (1 atm pressure) and at a temperature of 77°F (25°C), what volume will it have at 35,000 ft where the pressure is about 0.250 atm and the temperature is −40°C?

As in the previous cases, we must use absolute temperature here. Making this change we have

$$T_1 = 25°C + 273.2° = 298.2°K$$

and

$$T_2 = -40°C + 273.2° = 233.2°C$$

The corresponding pressures and volumes are $P_1 = 1$ atm, $V_1 = 950$ L, $P_2 = 0.250$ atm, and V_2 is the quantity to be determined. Rearranging equation 5.19 to solve for V_2 gives

$$V_2 = \frac{P_1 V_1 T_2}{T_1 P_2} = \frac{1 \text{ atm} \times 950 \text{ L} \times 233.2°K}{298.2°K \times 0.250 \text{ atm}} = 2970 \text{ L}$$

Although the lower temperature tends to decrease the volume, the relative change in the pressure is more important and causes a net increase in the volume. The net effect of the changes in the two variables is an overall increase of $\frac{2970}{950} = 3.13$ times. Care must be exercised by meteorologists in loading such a balloon so that it does not expand to the point of bursting as it gets to high altitudes.

5-2.5 DALTON'S LAW OF PARTIAL PRESSURE

In addition to giving us his atomic theory, John Dalton, in 1803, formulated *Dalton's law of partial pressure*. This law states that *the total pressure of a mixture of gases is equal to the sum of the partial pressures of the individual gases.* Written in mathematical form, it is

$$P_T = P_1 + P_2 + P_3 + \cdots \qquad \textbf{(5.20)}$$

where P_T is the total pressure of the gaseous mixture and P_1, P_2, P_3, and so forth are the partial pressures, or portions of the total pressure that each gas in the mixture exerts. Analyzed from a slightly different point of view, Dalton's law is equivalent to saying that the pressure exerted by an individual gas (P_i) in a mixture of gases is equal to the product of its percentage, or pressure fraction (*PF*) in the mixture and the total pressure. Written as a mathematical equality, it is

$$P_i = (PF)_i P_t \qquad \textbf{(5.21)}$$

The most common mixture of gases that we experience is the air we breathe. Percentage composition of air is given in Table 5.2. We can see that the four major gases near the earth's surface are nitrogen, oxygen, argon, and water vapor. The water vapor is listed with a range of compositions since the total amount present depends upon the temperature of the air and the relative humidity. The latter factor, relative humidity, is the amount of water vapor present in the air compared with the total amount of water vapor that the air can hold. Appendix B lists values of the vapor pressure of water at various temperatures. These values correspond to the vapor pressure associated with 100% relative humidity.

EXAMPLE 5.6   What are the partial pressures of the three major components in air at STP?

For STP conditions the temperature is 273°K and $P_T = 760$ mmHg. Although the temperature has been specified, it will not affect the partial pressures or percent

Table 5.2   Composition of Air Near the Earth's Surface

Nonvariable Constituent	Content by Volume*	Variable Constituent	Normal Content by Volume*
Nitrogen	78.084%	Water vapor	0.1–1%
Oxygen	20.946%	Carbon dioxide	0.033%
Argon	0.934%	Ozone	0.02–0.07 ppm
Neon	18.18 ppm	Ammonia	0.01 ppm
Helium	5.24 ppm	Sulfur dioxide	0.0002 ppm
Methane	2.00 ppm	Carbon monoxide	0.1 ppm
Krypton	1.14 ppm	Nitrogen dioxide	0.001 ppm
Hydrogen	0.50 ppm	Radon	trace
Nitrous oxide	0.50 ppm	Dust	trace
Xenon	0.087 ppm		

*1 ppm = 0.0001 vol %.

composition and thus will not enter into the calculation. Applying equation 5.21 to calculate the partial pressure of the three components we have,

$$P_{N_2} = 0.781 \times 760 \text{ mmHg} = 594 \text{ mmHg}$$

$$P_{O_2} = 0.209 \times 760 \text{ mmHg} = 159 \text{ mmHg}$$

and $$P_{Ar} = 0.00934 \times 760 \text{ mmHg} = 7.1 \text{ mmHg}$$

We find application of Dalton's law to the living system when we study the percent composition, or partial pressure, of oxygen in various parts of the respiration system as listed in Table 5.3. The inspired air is simply the air we breathe in,* the alveolar air is the air in the air cells (sacs of the lungs) of the pulmonary artery and the alveolar capillaries (blood vessels that bring oxygen-deficient blood to the lungs to pick up oxygen), and the expired air is the air we breathe out. Since the partial pressure in alveolar air is greater than that in the alveolar capillaries, oxygen diffuses from the alveolar air sacs into the capillaries. Thus the oxygen content of the pulmonary blood leaving the lungs is increased. This oxygen is then transported to the cells of the body where it is involved in providing us with energy.

Table 5.3   Percentages of Dried Gases in Breathed Air

Gas	Inspired Air	Alveolar Air	Expired Air
Oxygen	20.96	14.2	16.3
Nitrogen	79.00	80.3	79.7
Carbon dioxide	0.04	5.5	4.0

PROBLEMS

6. What is the standard pressure?
7. The fact that the volume of a gas is directly proportional to its absolute temperature is known as what?

*Differences in the percentage composition of N_2, O_2, and CO_2 in Tables 5.2 and 5.3 are due to the fact that the values given in Table 5.3 are for dry air, which increases the percentage composition of each remaining gas.

⑤ 8. A balloon having an initial volume of 5.5 L in Death Valley (elevation −280 ft) under a total pressure of 1.1 atm is carried to the summit of Mt. Whitney (elevation 14,496 ft) where the total pressure is 0.66 atm. What is the new volume of the balloon? (Assume the temperature remained constant.)

⑤ 9. If an automobile tire is inflated to a pressure of 32.0 psi at a temperature of 60°F, what will be the pressure of the tire after driving on a hot day when the tire temperature reaches 150°F? (Assume that the volume of the tire remains the same.)

⑤ 10. A given amount of gas has a volume of 1.00 L when the temperature and pressure are at STP. What will be the volume if the conditions are changed to 819°C and 0.50 atm?

⑤ 11. What volume will 0.63 mol of hydrogen occupy at STP?

ANSWERS TO PROBLEMS   **6.** 1 atm or 760 mmHg   **7.** Charles' Law   **8.** 9.2 L   **9.** 37.5 psi
10. 8.0 L   **11.** 14 L

SUMMARY

In this chapter we have considered the three states of matter and the factors involved when a change in state takes place. We found that solid and liquid particles are held closely together by ionic and physical forces and melting and boiling of a substance involves the weakening or breaking of these forces. One particular force of attraction, the hydrogen bond, was found to be especially important in giving water and various biochemical compounds such as proteins unique and vital chemical and physical properties. After considering these changes in state we turned our attention to the gaseous state and discussed various gas laws and how they describe the behavior of gases and allow us to calculate the properties of gases and we studied a few of their applications to the living system.

In the next chapter we will look at a special type of mixture called a solution. We will consider the factors involved in the solution process, the units used to express the concentration of solutions, and some of the properties of solutions.

ADDITIONAL PROBLEMS

12. The force, per unit area, exerted over the surface of a body is known as the _____.

13. What are the meanings of the terms *heat of fusion* and the *heat of vaporization*?

14. Physical forces of attraction are important in holding unit particles of what types of substance together in the solid state and liquid state?

15. What are the major differences in the behavior of the unit particles in the solid and in the liquid states?

⑭ 16. Describe the process that takes place when a liquid boils.

17. What happens to the temperature as a solid melts?

18. Draw a diagram of the structure of ice.

19. What is the standard temperature in degrees Kelvin?

20. If a gas occupies
   ⑤ (a) 5.0 L when the pressure is 1.0 atm, what will the volume be if the pressure changes to 2.0 atm?
   ⑤ (b) 3.0 L when the pressure is 14.7 atm, what will the volume be if the pressure changes to 26.7 atm?
   (c) 105 mL when the pressure is 508 torr, what will the volume be if the pressure changes to 470 torr?
   (d) 27.3 cc when the pressure is 652 mmHg,

what will the volume be if the pressure changes to 740 mmHg?

(Assume a constant temperature.)

21. If the pressure on a quantity of
    - ⓢ (a) air is 0.70 atm when the volume is 200 L, what must the pressure be changed to so that the volume of the gas will be 100 L?
    - ⓢ (b) Ne is 10.6 atm when the volume is 9.6 L, what must the pressure be changed to so that the volume of the gas will be 28.8 L?
    - (c) CH_4 is 392 torr when the volume is 0.280 mL, what must the pressure be changed to so that the volume of the gas will be 4.68 mL?
    - (d) CO is 1022 mmHg when the volume is 333 cc, what must the pressure be changed to so that the volume of the gas will be 2333 cc?

(Assume a constant temperature.)

22. If a gas occupies a volume of
    - ⓢ (a) 5.0 L when the temperature is 300°K, what will the new volume be when the temperature is changed to 150°K?
    - ⓢ (b) 3.0 L when the temperature is 273°K, what will the new volume be when the temperature is changed to 400°K?
    - (c) 105 mL when the temperature is 25°C, what will the new volume be when the temperature is changed to 250°C?
    - (d) 27.3 cc when the temperature is −14°C, what will the new volume be when the temperature is changed to 28°C?

(Assume a constant pressure.)

23. If the temperature of a gas is
    - ⓢ (a) 250°K when the volume is 200. L, what must the temperature be changed to so that the volume of the gas will be 100. L?
    - ⓢ (b) 390°K when the volume is 9.6 L, what must the temperature be changed to so that the volume of the gas will be 3.2 L?
    - (c) 98°C when the volume is 0.280 mL, what must the temperature be changed to so that the volume of the gas will be 2.80 mL?
    - (d) −40°C when the volume is 333 cc, what must the temperature be changed to so that the volume of the gas will be 330 cc?

(Assume a constant pressure.)

24. If a gas is under a pressure of
    - ⓢ (a) 2.00 atm when the temperature is 273°K, what will the new pressure be when the temperature is changed to 273°C?
    - ⓢ (b) 800 mmHg when the temperature is 273°C, what will the new pressure be when the temperature is changed to 0°C?
    - (c) 608 torr when the temperature is 25°C, what will the new pressure be when the temperature is changed to 125°C?
    - (d) 22.4 atm when the temperature is −40°C, what will the new pressure be when the temperature is changed to 320°K?

(Report answers using the pressure units given and assume a constant volume.)

25. If the temperature of a fixed volume of gas is
    - ⓢ (a) 27°C when the pressure of the gas is 1.00 atm, what must the temperature be changed to so that the pressure is 3.00 atm?
    - ⓢ (b) 244°K when the pressure of the gas is 625 torr, what must the temperature be changed to so that the pressure is 375 torr?
    - (c) 410°C when the pressure of the gas is 1440 mmHg, what must the temperature be changed to so that the pressure is 760 mmHg?
    - (d) −62°C when the pressure of the gas is 218 torr, what must the temperature be changed to so that the pressure is 2.18 atm?

(Report temperature in degrees Celsius.)

26. A gas has a volume of
    - ⓢ (a) 1.00 L when the temperature and pressure are STP. What will the volume be if the conditions are changed to 546°C and 2.00 atm?
    - ⓢ (b) 6.57 L when the temperature and pressure are 388°K and 800 torr. What will the volume be if the conditions are changed to 600°K and 400 torr?
    - (c) 210 mL when the temperature and pressure are 20°C and 0.800 atm. What will the volume be if the conditions are changed to 40°C and 1.20 atm?
    - (d) 71.3 mL when the temperature and pressure are −14°C and 615 mmHg.

What will the volume be if the conditions are changed to 333°K and 1.03 atm?

(Assume that the number of moles of gas remains constant.)

27. A gas has a pressure of
    ⑤ (a) 1.00 atm when it has a volume and temperature of 22.4 L and 273°K. What will the pressure of the gas be when the volume and temperature are 44.8 L and 273°C?
    ⑤ (b) 685 torr when it has a volume and a temperature of 7.17 L and 288°C. What will the pressure of the gas be when the volume and the temperature are 58.3 L and 388°C?
    (c) 1.90 atm when it has a volume and a temperature of 25.8 mL and 68°C. What will the pressure of the gas be when the volume and the temperature are 5.11 mL and −68°C?
    (d) 859 mmHg when it has a volume and temperature of 0.930 L and 515°K. What will the pressure of the gas be when the volume and the temperature are 3.34 L and 90.3°C?

(Assume that the number of moles of gas remains constant.)

28. A gas has a temperature of
    ⑤ (a) 273°K when the volume and pressure are 22.4 L and 1.00 atm. What temperature will be associated with a volume and pressure of 11.2 L and 2.00 atm?
    ⑤ (b) 267°K when the volume and pressure are 4.83 L and 752 torr. What temperature will be associated with a volume and pressure of 27.0 L and 577 torr?
    (c) 66.1°C when the volume and pressure are 569 mL and 368 torr. What temperature will be associated with a volume and pressure of 351 mL and 541 torr?
    (d) 116°C when the volume and pressure are 18.3 L and 888 mmHg. What temperature will be associated with a volume and pressure of 12.0 L and 1850 mmHg?

(Report all answers in degrees Celsius and assume that the number of moles of gas remains constant.)

29. A quantity of gas has a volume, pressure, and temperature of
    ⑤ (a) 22.4 L, 1 atm, and 0°C
    ⑤ (b) 1.00 L, 2.00 atm, and 273°C.
    (c) 488 mL, 576 torr, and −91.0°C.
    (d) 748 mL, 76.0 mmHg, and 436°C.
    How many moles of gas are present in each case?

30. What pressure will be exerted by
    ⑤ (a) 1.00 mol of gas when the temperature and volume are 273°K and 22,400 mL?
    ⑤ (b) 28.0 mol of a gas when the temperature and volume are 182°C and 90.8 L?
    (c) 3.28 mol of a gas when the temperature and volume are 462°K and 619 mL?
    (d) 0.179 mol of a gas when the temperature and volume are 569°C and 136 mL?

31. What volume will be occupied by
    ⑤ (a) 0.100 mol of a gas when the temperature and pressure are 400°K and 5.00 atm?
    ⑤ (b) 20.6 mol of a gas when the temperature and pressure are 400°C and 262 mmHg?
    (c) 4.61 mol of a gas when the temperature and pressure are 80°C and 621 torr?
    (d) 0.0654 mol of a gas when the temperature and pressure are −78°C and 0.268 atm?

32. What temperature will be associated with
    ⑤ (a) 10.0 mol of a gas when the pressure and volume are 2.50 atm and 820 L?
    ⑤ (b) 0.461 mol of a gas when the pressure and volume are 511 torr and 62.0 L?
    (c) 67.0 mol of a gas when the pressure and volume are 1350 mmHg and 974 L?
    (d) 0.832 mol of a gas when the pressure and volume are 81.8 atm and 308 mL?

⑤ 33. What volume does each of the following gases occupy at STP?
    (a) 0.500 mol H_2 (b) 2.5 mol O_2
    (c) 18.0 g He   (d) 18×10^{23} molecules of N_2

34. How many moles of each of the following gases are there at STP?
    ⑤ (a) 33.6 L Ne   ⑤ (b) 5.6 L F_2
    (c) 1792 mL Ar   (d) 100 mL H_2

35. Using the data given in Table 5.3, calculate the the partial pressures (in mmHg) of O_2, N_2, and CO_2 in
    ⑤ (a) alveolar air   (b) expired air when the total pressure is 1.00 atm

CHAPTER 6

SOLUTIONS – THEIR BEHAVIOR AND CONCENTRATION

In Chapter 5 we considered the three states of matter: solids, liquids, and gases. Each of these states can be associated with pure substances and with mixtures of substances, including what are called solutions. The gold jewelry we wear and the stainless steel tableware we use are examples of solid solutions. The soft drinks we consume and the blood plasma that flows through our veins are both liquid solutions. The air we breathe is a gaseous solution.

In this chapter we will center our attention on liquid solutions. We will learn about the various factors involved in the process of forming a solution and how the amount of a substance that can be dissolved changes with a change in temperature and pressure. We will then consider some of the units of solution concentration that are important in chemistry and in the health-related sciences. Finally, we will discuss the effect that substances have upon the properties of liquids in which they are dissolved, including the osmosis that is vital to living things.

6-1 SOLUTIONS AND SOLUTION THEORY

LEARNING OBJECTIVES

1. Define solution, solvent, solute, electrolyte, nonelectrolyte, strong electrolyte, weak electrolyte, solvation, and solubility.
2. Describe the characteristics of solutions of electrolytes and nonelectrolytes.
3. Discuss the significance of energy and disorder in the solution process.

4. Be able to predict qualitatively the solubility of a given solute in a specific solvent based upon a knowledge of the nature of the substances involved.
5. Be able to predict qualitatively the effect of a change in temperature and pressure on the solubility of a substance.

In this section we will consider various types of liquid solutions with particular emphasis on water, or aqueous, solutions such as those found in the various body fluids. These body solutions contain the ions and molecules that are important for the maintenance of life. Two factors are important in forming solutions: whether energy is gained or lost when the solution forms and the tendency of substances to mix. We will briefly consider each of these factors and their effect on whether or not and to what extent solutions are formed. In addition we will consider the effect that temperature and pressure have on how much substance will dissolve in a liquid to form a solution.

A **solution** is a homogeneous mixture of two or more substances that are intimately mixed together on the molecular level.

6-1.1 CLASSES OF COMPOUNDS IN SOLUTIONS

In Chapter 3 two classes of compounds were introduced: ionic and covalent. Basically, ionic compounds are compounds in which one or more chemical bonds have been formed by an electron transfer between two atoms. In general, ionic compounds are solids at room temperature and consist of a large crystalline arrangement like that shown in Figure 3.1. When a solution is prepared by dissolving an ionic solid in a solvent, it dissociates. Individual ions of the solute break away from the bulk solid and float about somewhat independently in the solution. For example, when NaCl is dissolved in water, ions of Na^+ and Cl^- move about freely in the solution. When these ionic substances dissolve they produce a solution that is capable of conducting an electric current.* In light of this behavior the substance is referred to as an electrolyte.

A **solvent** is the component in a solution that is present in the greater amount.

A **solute** is a substance dissolved in a solvent to produce a solution.

Electrolytes form an important class of substances in the living system, serving three general functions in the body.

An **electrolyte** is a substance that, when in the pure molten state or in solution, exists as ions and is capable of conducting an electric current.

1. Many electrolytes are essential minerals such as the calcium ion (Ca^{2+}) which is important in the formation of bones and teeth, in blood clotting, and in normal muscle and nerve activity.
2. Electrolytes control the passage of ions and molecules between various parts of the body. More shall be said about this characteristic in Section 6-3.
3. Electrolytes help maintain the acid-base balance required for normal cellular activities.

The most abundant electrolyte ions in the human body are Na^+, K^+, Cl^-, and HPO_4^{2-}. Figure 6.1 gives a comparison of relative amounts of the more important ions in the human body based upon the concentration unit milliequivalents per liter. We will discuss this unit in Section 6-2. Chapter 10 presents a more complete discussion of the importance of electrolytes to the living system.

Not all substances conduct an electric current in the pure state or in solution. For example many covalent substances such as water, isopropyl alcohol (C_3H_7OH),

---

*An electric current, or electricity, consists of a flow of electrons through some material. Electrons can easily flow through a copper wire and therefore copper is considered a good conductor. Other materials such as the plastic insulation surrounding the wire do not allow the electrons to flow easily and we speak of such materials as nonconductors, or insulators. In solutions the electric current consists of a flow of charged ions.

Figure 6.1   Relative amounts of electrolytes, based on the concentration unit mEq/L in the liquid part of the blood (plasma), the fluids surrounding the cells (interstitial fluid), and the fluid within the cells (intracellular fluid). The left half of each column represents the positive ions and the right half represents the negative ions. (From *Principles of Anatomy and Physiology, 2nd Edition* by Gerald J. Tortora and Nicholas P. Anagnostakos. Copyright © 1975, 1978, by Gerald J. Tortora and Nicholas P. Anagnostakos. Reprinted by permission of Harper & Row, Publishers, Inc.)

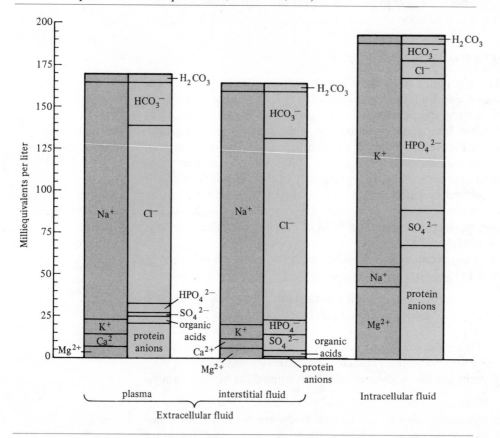

and gasoline are very poor conductors both in the pure liquid state and when mixed to form a solution.* This is due to the fact that neither of these substances exists as an ion in either state. Such substances are referred to as nonelectrolytes.

The failure to conduct electricity is not limited to solutions of two liquids. For example, when the covalent compound table sugar (sucrose, $C_{12}H_{22}O_{11}$) is dis-

A **nonelectrolyte** is a substance that does not dissociate into ions in solution and does not conduct an electric current.

---

*You may have been warned about touching electrical appliances or turning lights off and on when bathing because of the danger of being electrocuted. It isn't the water itself but the electrolytes that are dissolved in the water that will conduct an electric current. For example, perspiration contains, among other things, salts such as NaCl. As we bathe, these salts are washed off our skin and dissolve in the bath water to produce a solution that is capable of conducting electricity quite readily. If there is an electrical short of any kind, a strong current can flow through the water and through the body, disrupting the physiological processes that depend on normal "current flow" in the body.

solved in water the resulting solution does not conduct an electric current. The process of dissolving sugar is, in some ways, similar to that which occurs when ionic substances, such as sodium chloride, dissolve. However, when sugar dissolves, the covalently bonded molecules remain intact and no ions are available to carry the electric current. In light of this behavior, such substances are also considered to be nonelectrolytes.

Not all covalent substances remain intact in solution. When some polar covalent substances are dissolved in water they are torn apart into ions or ionized by the attractive forces they have for the water. The resulting solution is capable of conducting an electric current. Probably the most common example of such a compound is hydrogen chloride. In the pure state hydrogen chloride is a polar covalent compound that does not conduct electricity to a great extent. However, when it is dissolved in water it ionizes completely into H^+ and Cl^- ions and, therefore, behaves in the same manner as an ionic compound. Such compounds are also considered electrolytes.

Electrolytes such as sodium chloride and hydrogen chloride that exist as ions in aqueous solutions are referred to as **strong electrolytes**. This can be represented for HCl by the equation

$$HCl(g) \xrightarrow{H_2O} H^+(aq) + Cl^-(aq) \qquad (6.1)$$

A **strong electrolyte** is an electrolyte that is completely dissociated into ions in solution.

The arrow pointing to the right indicates that the ionization reaction goes to completion; that is, none of the HCl remains intact. The appearance of H_2O above the arrow indicates that water is acting as the solvent. Because of the presence of the $H^+(aq)$ and $Cl^-(aq)$ ions, an aqueous solution of HCl is a very good conductor of electricity.

Some electrolytes have less ability to conduct electricity than others because they are only partially ionized in aqueous solutions. As an example of this, acetic acid (CH_3COOH) ionizes according to the equation

$$CH_3COOH(aq) \xrightleftharpoons{H_2O} H^+(aq) + CH_3COO^-(aq) \qquad (6.2)$$

where the arrows pointing in opposite directions indicate that the reaction does not go to completion but reactants and products are both present in the solution. Substances that behave in this manner are known as **weak electrolytes**. More shall be said about strong and weak electrolytes in Chapter 8 when we discuss ionic equilibria.

A **weak electrolyte** is an electrolyte that is only partially dissociated into ions in solution and is therefore a relatively poor conductor of electricity.

6-1.2 FACTORS INVOLVED IN THE SOLUTION PROCESS

Two fundamental factors are important in the process of forming a solution: energy factors and disorder factors. *Natural, or spontaneous, processes, such as the formation of solutions, tend to occur in such a way that (1) energy is given up and/or (2) the various substances become more disordered (mixed up).* An example of the first factor is a bowl of hot soup. It spontaneously cools, because it gives up energy in the process. An input of heat, or energy from an even hotter source such as a hot burner on a stove, is required to warm the soup. An example of the second factor is that weeds in a garden naturally sprout in a disordered arrangement, but effort is required to get straight, orderly rows of corn and beans.

Applying the second factor to the solution process, two or more different substances spontaneously tend to mix together, since they can become disordered in the process of mixing. This is illustrated when food coloring is added to water. The two immediately begin to mix as evidenced by the spread of the color throughout the water. This continues until the two substances are totally "mixed up." Similar mixing occurs whenever we make any solution. The substances mix together on the molecular level and thus experience an increase in the state of disorder.*

Energy factors often counteract the natural tendency of substances to mix and therefore many substances do not dissolve to a great extent in each other. For example, your skin does not dissolve very much when you wash your hands in water.

From the point of view of energy, three major processes must take place for a solid or liquid to disolve in a liquid solvent: (1) *the solute particle must break away from the bulk solute,* (2) *a hole must be created in the solvent* to make room for the solute particle by breaking the intermolecular forces that hold solvent molecules together in the liquid state, and (3) *solute-solvent bonds must form* (that is, solvation must occur). These processes are illustrated separately in Figure 6.2. In reality, all three processes will be taking place simultaneously.

Solvation is the interaction of solvent molecules with solute particles to form aggregates whose particles are loosely bonded together.

The first and second processes both require energy since it takes energy to break bonds. Thus these two processes are said to be *endothermic* (heat-absorbing). The third process gives off energy as new interactions between solute and solvent occur.

Figure 6.2   The three major processes taking place in the solution process. (a) Solute particles break away from the bulk solute. (b) Holes are created in the solvent to make room for the solute particle. (c) Solvation takes place.

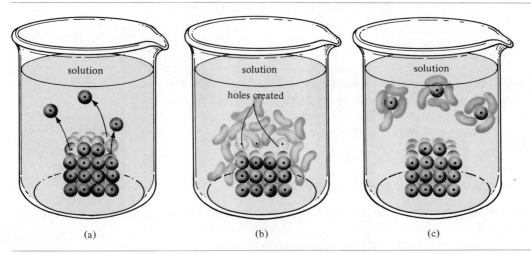

(a)   (b)   (c)

---

*Chemists use a special term, *entropy,* to indicate the degree of disorder of a system. The tendency of things to become disordered can be written as "All spontaneous processes in the universe tend to occur so as to have an increase in entropy."

Thus, the solvation process is referred to as *exothermic* (heat-producing). Since spontaneous processes tend to occur so that energy is given off or lost, a solute, in general, will only dissolve to a great extent if the energy given off as a result of solvation is greater than the energy required to break solute-solute bonds and solvent-solvent bonds in the first two processes.

As a result of the factors above, the solubility of solutes in various solvents differs. In some cases, such as in solutions of ethyl alcohol in water, the two substances are totally *soluble* in each other, dissolving in one another in any proportion. Other substances may be essentially *insoluble,* that is, do not dissolve in each other to a significant degree. An example of this is vegetable oil and water, which are essentially insoluble in each other.

The **solubility** of a substance is the extent to which the substance will dissolve in a given solvent.

In many cases, the two substances are partially soluble. For example sodium chloride has a solubility of 35.7 g in 100 g of water at 0°C. Similarly, table sugar has a solubility of 179 g in 100 g of water at 0°C and 487 g in 100 g of water at 100°C. A solution with the maximum amount of solute dissolved in the solvent is referred to as a *saturated solution.* One that has less than this amount dissolved is considered an *unsaturated solution.*

These examples illustrate the general rule, *like dissolves like,* or stated more explicitly: *polar solvents tend to dissolve polar and ionic solutes, and nonpolar solvents tend to dissolve nonpolar solutes.* The complete solubility of ethyl alcohol and water is an example of the solubility of polar substances since both substances are quite polar. The solubility of NaCl in water illustrates the solubility of ionic substances in polar solvents. An example of the solubility of nonpolar substances in one another is the use of the nonpolar substance carbon tetrachloride (CCl_4) as a dry-cleaning solvent. Since nonpolar solutes such as skin oils and grease readily dissolve in carbon tetrachloride, they are easily removed from clothing by that compound.*

The function of water as a solvent has important implications to the human organism in a variety of ways. About 60% of red blood cells, 75% of muscle tissue, and 92% of blood plasma consist of water. Water is a vital solvent that dissolves and suspends the many substances found in the fluids of the body. Because of its polar nature water dissolves the necessary electrolytes and nutrients that go into our body cells and the waste products that go out of the cells. Its tendency to form hydrogen bonds, as discussed in Chapter 5, is also important in dissolving such substances as sugars. Water in the blood also forms a dilute solution with some of the oxygen we inhale, as it carries the oxygen to body cells. Similarly, a large part of the carbon dioxide that is carried from the cells to the lungs to be exhaled is dissolved in the water in the blood.

The human body does not tend to cool down or heat up rapidly when exposed to cold or heat because of (1) the high specific heat of water (see Chapter 5), (2) the large amount of water in the human body which accounts for 65–75% of body weight, and (3) the fact that water does not conduct heat readily. The water in the body, then, helps to maintain a fairly constant body temperature despite variation in environmental conditions.

---

*The use of CCl_4 as a dry-cleaning solvent has, for the most part, been discontinued due to evidence that it causes liver and kidney damage.

6-1.3 EFFECT OF TEMPERATURE AND PRESSURE ON SOLUBILITY

Although the energy and disorder factors (as evidenced by the nature of the solvent and solute) basically determine whether and to what extent substances dissolve in one another, the solubility of the substances is also influenced by temperature and pressure. In most cases an increase in the temperature of the solution increases the solubility of solutes in solvents. For example, if you have ever made rock candy by dissolving sugar in water, you probably warmed the water in order to increase the amount of sugar that dissolved. This effect is illustrated by the solubility values of sugar at 0°C and 100°C given earlier in Section 6-1.2.

An important exception to the tendency of substances to increase in their solubility with an increase in temperature is the effect of temperature on the solubility of gases in water. In this case the solubility decreases as the temperature increases. This can be seen when a glass of cold water is left sitting on the table for some time. As the temperature of the water increases to room temperature bubbles of air frequently can be seen collecting on the inside of the glass. In this case some of the air that was dissolved at the lower temperature comes out of solution and forms the bubbles.

The characteristic of gases to decrease in their solubility in water is an important factor in what is known as thermal water pollution. Many industrial firms, such as electrical generating plants, use water as a coolant and then discharge this water into lakes and streams. The discharged water typically has a temperature as much as 10°C higher than the water into which it is discharged. This additional warm water raises the overall temperature of the lake or stream and therefore decreases the limited amount of dissolved oxygen already present in the water. In addition, the increased temperature speeds up chemical and biological processes, including those involved in various reactions with oxygen such as an increased rate of respiration by fish. Since the level of oxygen has already been lowered due to the higher temperature, fish and microorganisms die and pollute the water.

The effect of temperature on the solubility of a solute depends upon whether the solution process is exothermic or endothermic. *When heat is evolved (produced) as the solute dissolves, an increase in temperature decreases the solubility of the solute.* This is represented by the arrows above the following equation

$$\text{solute} \xrightarrow[\underset{T\downarrow}{\longrightarrow}]{\overset{T\uparrow}{\longleftarrow}\ \text{solvent}} \text{solution} + \text{heat} \qquad (6.3)$$

As indicated by the arrows below the equation, *a decrease in temperature increases the solubility. When heat is absorbed during the solution process, an increase in temperature increases the solubility and a decrease in temperature decreases the solubility* as follows:

$$\text{heat} + \text{solute} \xrightarrow[\underset{T\downarrow}{\longleftarrow}]{\overset{T\uparrow}{\longrightarrow}\ \text{solvent}} \text{solution} \qquad (6.4)$$

A change in pressure has little effect on the solubility of solids and liquids in a liquid solvent. However, *an increase in pressure on a gaseous solute dissolving in a liquid solvent increases the solubility of the gas significantly.* An interesting example of this is the bends, or caisson disease. When deep sea divers descend to great depths in the ocean, the increased pressure of the air brings about an increased solubility of the air in the blood. As the diver begins to ascend to the surface of the water, the pressure inside the diving suit decreases with the accompanying decrease in the solubility of the dissolved gases. If the ascension is made too rapidly, some of the dissolved nitrogen from the air comes out of the solution in the blood and forms bubbles inside the blood vessels similar to the air bubbles that appear on the inside of a warm glass of water. These bubbles can both stimulate and damage nerves, causing great pain and, in some cases, paralysis or death. This can be avoided if the oxygen necessary for respiration is mixed with another gas, such as argon, rather than the nitrogen normally found in air. The argon has such a limited solubility that there is not enough argon dissolved in the blood to create a problem as the diver returns to the surface of the water.

PROBLEMS

1. Define the term strong electrolyte.
2. What are some of the characteristics of nonelectrolyte solutions?
3. What two fundamental factors are important in the process of forming a solution?
4. A benzene molecule (C_6H_6) is nonpolar. How soluble will benzene be in water?
5. Ammonium nitrate [$NH_4NO_3(s)$] has been used as a urinary acidifier and diuretic. When $NH_4NO_3(s)$ dissolves in water, heat is absorbed according to the following equation

$$\text{heat} + NH_4NO_3(s) + H_2O(l) \longrightarrow NH_4NO_3(aq)$$

What effect will a decrease in temperature have upon the solubility of $NH_4NO_3(s)$?

ANSWERS TO PROBLEMS **1.** an electrolyte that is totally dissociated into ions in solution **2.** The uncharged solute molecules move about individually in the solution. As a result, the solution does not conduct an electric current to any great extent. **3.** the tendency to lose energy and the tendency to become disordered **4.** essentially insoluble **5.** the solubility will decrease

6-2 CONCENTRATION UNITS

LEARNING OBJECTIVES

1. Define concentration, percent by weight, percent by volume, molarity, gram-equivalent weight, mEq/L, and normality.

2. Perform calculations using the units listed in objective 1.

In our work in previous chapters we have dealt with various measures of matter including atoms, grams, and moles. In most cases these measures have been what might be called single-unit quantities. For example, when we refer to a gram of a substance we are simply dealing with a weight of the substance, and a single unit, the gram, is sufficient to describe the amount of substance involved. When the substance is dissolved in a solvent these simple units are not adequate to convey all the information necessary. For example, 1 g of a solute might be dissolved in 1 g, 100 g, or some other amount of solvent. In order to specify the composition of the solution more clearly, we must not only specify the quantity of solute present but also the amount of solvent or solution involved. When we do this we are specifying the concentration of the solution. Several concentration units are important in chemistry and in the health-related sciences. Among these we will consider percent by weight and volume, weight/volume percent, molarity, milliequivalents/L, and normality. Each of these shall be discussed in the subsections that follow. Typical calculations involving these concentration units shall be demonstrated.

> The **concentration** of a solution is the amount of solute present in a specified amount of the solvent or solution.

6-2.1 PERCENT BY WEIGHT AND VOLUME

One of the simplest concentration units is percent by weight, or simply percent.* Percent by weight can be represented by the equation

$$x \, \text{wt}\% = \frac{x \text{ g solute}}{100 \text{ g solution}} \tag{6.5}$$

> **Percent by weight** is the number of grams of solute per 100 grams of solution.

where x is the numerical value of the percent by weight.

An example of the use of this unit is a physiological saline solution, used to replenish body water in cases of dehydration, that consists of a 0.89 percent by weight solution of NaCl in water. Such a solution also is referred to as *isotonic*, since the osmotic pressures of the solution and of blood and body cells are equal. More shall be said about osmotic pressure in Section 6-3.

EXAMPLE 6.1   How much sodium chloride is needed to prepare 250 g of physiological saline solution?

From equation 6.5 and the concentration of physiological saline solutions given in the preceding discussion, we can write

$$0.89 \, \text{wt}\% = \frac{0.89 \text{ g NaCl}}{100 \text{ g solution}}$$

Multiplying this by the quantity of the solution needed gives the weight of NaCl

$$250 \text{ g solution} \times \frac{0.89 \text{ g NaCl}}{100 \text{ g solution}} = 2.2 \text{ g NaCl}$$

A parallel concentration unit, percent by volume, is occasionally used, particularly when both solute and solvent are liquids. Percent by volume can be expressed mathematically as

$$y \, \text{vol} \% = \frac{y \text{ mL solute}}{100 \text{ mL solution}} \tag{6.6}$$

> **Percent by volume** is the number of milliliters of solute per 100 mL of solution.

---

*Although the term *percent* is frequently used to represent percent by weight, it also may represent percent by volume and weight/volume percent, as discussed later in this section.

Table 6.1 Physiological Standards, Average Normal for Blood

Substance	Amount*
Ammonia	40–80 μg
Ascorbic acid, or vitamin C (whole blood)	0.4–1.5 mg
Calcium, total (serum)	9–11 mg
Chloride (serum)	10–10.6 mEq
Cholesterol, total (serum)	120–240 mg
Glucose, fasting level (whole blood)	60–100 mg
Hemoglobin (whole blood)	14–18 g (males)
	12–16 g (females)
Iron, inorganic (serum)	75–175 mg
Lactic acid (whole blood)	6–16 mg
Magnesium (serum)	0.15–0.25 mEq
Phosphate, inorganic (serum)	3.0–4.5 mg
Potassium (serum)	0.35–0.45 mEq
Protein (serum)	6.0–8.0 g
Sodium (serum)	13.6–14.5 mEq
Sulfates, as sulfur inorganic (serum)	0.8–1.2 mg
Triglycerides (serum)	1.0–15.0 mg
Urea (whole blood)	20–40 mg

*Values are expressed per 100 mL (per deciliter).

where y is the value of the concentration in volume percent. Calculations involving percent by volume are similar to those involving percent by weight except a volume unit such as milliliters is used rather than the weight unit, grams.

A third percent concentration unit used in the health-related sciences involves a mixture of weight of solute and volume of solution. This unit is the *weight/volume percent,* given by the equation

$$z \text{ wt/vol } \% = \frac{z \text{ wt solute}}{100 \text{ mL solution}} \tag{6.7}$$

where z is the value of the weight/volume percent of the solution and the weight of the solute is expressed in grams or milligrams. For example, the composition of blood is frequently expressed as grams or milligrams of solute per 100 mL, or per deciliter, of blood. Table 6.1 lists representative values of some substances found in blood with their concentration based upon this concentration unit. Calculations using weight/volume percent are also similar to those for percent by weight, but one must be careful to use the proper unit of weight, grams or milligrams, for the solute and milliliters for the solution.

6-2.2 MOLARITY

One of the most important concentration units for chemists is the unit **molarity**, M. Expressed mathematically, molarity is given by

$$w M = \frac{w \text{ mol solute}}{1000 \text{ mL solution}} \tag{6.8}$$

The **molarity** (M) of a solution is the number of moles of solute per 1000 mL, or 1 liter of solution.

where w is the value of the molarity. For example, concentrated hydrochloric acid has a concentration of 12 molar, represented as 12 M. This is also frequently written in the form [HCl(aq)] = 12 M, where the brackets inform the reader that *the concentration of the substance within the brackets is being given in terms of molarity.* Various calculations involving molarity can be performed. We shall consider four different types. To begin with we shall look at the calculations of grams of solute in a given volume of solution.

EXAMPLE 6.2   How many grams of HCl are present in 150 mL of concentrated HCl?

Using equation 6.8 and the molarity of concentrated HCl given above, we can write

$$12\ M\ (\text{HCl}) = \frac{12\ \text{mol HCl}}{1000\ \text{mL solution}}$$

To do this calculation we must perform two conversions. First we must convert from moles of HCl to grams of HCl. Then we must determine how many grams of HCl are present in 150 mL of solution as stated in the problem, rather than the 1000 mL of solution given in the concentration unit.

To accomplish the first step we multiply the concentration by the gram-molecular weight.

$$\frac{12\ \text{mol HCl}}{1000\ \text{mL solution}} \times \frac{36.5\ \text{g HCl}}{1\ \text{mol HCl}} = \frac{440\ \text{g HCl}}{1000\ \text{mL solution}}$$

Now we multiply this result by the volume of the solution being considered, 150 mL, and obtain the weight of HCl.

$$150\ \text{mL solution} \times \frac{440\ \text{g HCl}}{1000\ \text{mL solution}} = 66\ \text{g HCl}$$

Another type of calculation involves the determination of the molarity of a solution when we know the amount of solute present in a given amount of solution. For this calculation we shall determine the molarity of the embalming fluid referred to in Example 4.5.

EXAMPLE 6.3   Fifty milliliters of the embalming fluid used to preserve biological specimens contains 20 g of CH_2O. What is the molarity of such a solution?

From the statement of the problem it is clear that two conversions must be performed:

$$\frac{20\ \text{g } CH_2O}{50\ \text{mL solution}} \overset{\Longrightarrow}{\Longrightarrow} \frac{\text{No. mol } CH_2O}{1000\ \text{mL solution}}$$

We must convert from grams of CH_2O to moles of CH_2O. Then we must determine how many moles of CH_2O are in 1000 mL of solution, rather than the 50 mL of solution referred to in the problem. The order in which these two conversions are completed is not important, but for convenience let us first convert from 20 g of CH_2O to moles of CH_2O. As before we use the g-molecular weight as follows,

$$\frac{20 \text{ g } CH_2O}{50 \text{ mL solution}} \times \frac{1 \text{ mol } CH_2O}{30 \text{ g } CH_2O} = \frac{0.67 \text{ mol } CH_2O}{50 \text{ mL solution}}$$

To convert this to the amount of CH_2O in 1000 mL of solution, we have

$$1000 \text{ mL solution} \times \frac{0.67 \text{ mol } CH_2O}{50 \text{ mL solution}} = 13 \text{ mol } CH_2O$$

Since the 13 mol of CH_2O are present in 1000 mL of solution we can write the following to obtain the molarity:

$$\frac{13 \text{ mol } CH_2O}{1000 \text{ mL solution}} = 13 \text{ } M$$

Most solutions in chemistry and in the living system are not nearly as concentrated as the concentrated HCl(aq) referred to in Example 6.2 and the embalming fluid just considered. As an example of a more common concentration value and a third type of calculation involving molarity, we shall consider the molarity of NaCl in blood serum.

EXAMPLE 6.4   Normal blood serum (the liquid portion of the blood) has, among other solutes, a concentration of NaCl of approximately 0.14 M. What volume of blood serum would contain 1.0 g of NaCl?
    For this solution we can use equation 6.8 and write

$$0.14 \text{ } M \text{ NaCl} = \frac{0.14 \text{ mol NaCl}}{1000 \text{ mL solution}}$$

To do this calculation we must convert 1.0 g of NaCl to the equivalent amount in moles of NaCl. For this we have

$$1.0 \text{ g NaCl} \times \frac{1 \text{ mol NaCl}}{58.5 \text{ g NaCl}} = 0.017 \text{ mol NaCl}$$

To calculate the volume of blood serum containing 1.0 g of NaCl we can invert the above concentration expression to obtain the necessary conversion factor as follows:

$$0.017 \text{ mol NaCl} = \frac{1000 \text{ mL solution}}{0.14 \text{ mol NaCl}} = 1.2 \times 10^2 \text{ mL solution}$$

which is the volume of blood serum that contains 1.0 g of NaCl.

Solutions are generally prepared in one of two ways as seen in Figure 6.3. The first method is to mix the appropriate amount of pure solute and solvent to obtain the desired solution. For example, physiological saline solutions easily can be prepared in this manner by mixing NaCl and H_2O (see Example 6.1). The second approach is to mix an appropriate amount of solvent and concentrated solution of solute to obtain the desired concentration. For example, pure HCl is a rather

Figure 6.3   Preparation of solutions by mixing (a) pure solute and pure solvent and (b) a more concentrated solution and pure solvent.

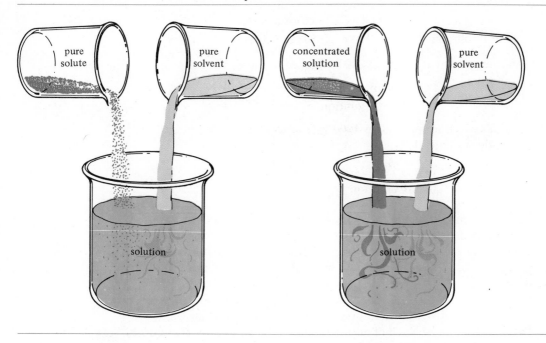

toxic gaseous compound. To prepare a solution of HCl by the first method would involve bubbling HCl(g) through water until the proper amount of HCl(aq) is present in the solution. Doing this would prove to be quite difficult and dangerous. In such cases the desired solution is prepared by taking a more concentrated solution, such as 12 M HCl, and diluting it by adding water.

When such dilutions are performed, the total number of moles of solute in the two solutions remains the same. This allows us to relate the initial molarity, M_i, and volume, V_i, and the final molarity, M_f, and volume, V_f, to one another through the equation

$$M_i \times V_i = M_f \times V_f \tag{6.9}$$

EXAMPLE 6.5   How many milliliters of concentrated HCl (12 M HCl) solution must be used in order to prepare 500 mL of a 0.10 M solution?

We must first identify the given quantities: $M_i = 12\ M$, $V_f = 500$ mL, $M_f = 0.10\ M$. We can then solve for V_i in equation 6.9 and substitute the given quantities for the variables and obtain the volume of the concentrated HCl needed.

$$V_i = \frac{M_f \times V_f}{M_i} = \frac{0.10\,M \times 500\ \text{mL}}{12\,M} = 4.2\ \text{mL}$$

To prepare the final solution, we therefore dilute 4.2 mL of concentrated HCl with enough water to yield a total volume of 500 mL of solution.

6-2.3 EQUIVALENTS, mEq/L, AND NORMALITY

The concentration units introduced so far are very convenient and useful in many respects. However, none of them provide an adequate measure of the reacting capacity of the solution. As an example of this, consider acids and their ability to produce hydrogen ions in solution and thus, to react with bases to produce a salt plus water. The amount of an acid required to produce 1 mole of hydrogen ions in solution varies from acid to acid because: (1) the gram-molecular weights of the various acids differ and (2) the number of ionizable hydrogen ions per unit particle differs. An example of the first effect above is that a greater weight of HBr (81 g) is required to produce one mole (1 g) of H^+ than would be needed if HCl (36.5 g) were used. This can be seen if we consider the ionization of one mole of each acid as given by the equations

$$HBr \xrightarrow{H_2O} H^+(aq) + Br^-(aq)$$
$$(81 \text{ g}) \qquad (1 \text{ g}) \qquad (80 \text{ g})$$

and

$$HCl \longrightarrow H^+(aq) + Cl^-(aq)$$
$$(36.5 \text{ g}) \qquad (1 \text{ g}) \qquad (35.5 \text{ g})$$

As seen in these equations, the greater weight of HBr is needed because of the greater weight of bromine as compared with chlorine. On this basis we can see that percent by weight is not an adequate concentration unit for comparing acids since solutions of different acids with the same percent by weight will produce different amounts of hydrogen ion depending upon their gram-molecular weight. Considering the second factor above, acids can be classified as monoprotic, diprotic, or triprotic depending upon how many hydrogen ions they produce in solution; that is, one, two, or three, respectively. Table 6.2 lists some typical acids under each classification. On this basis, 1 liter of a 1 *M* solution of an acid can produce one, two, or three moles of hydrogen ions depending upon the type of acid involved.

It is important to be able to relate acids in terms of their ability to produce hydrogen ions in solution. A unit of comparison is provided by what is known as the **gram-equivalent weight**, or as it is frequently termed, the **equivalent weight** of the substance. *The equivalent weight of an acid can be calculated by dividing the gram-molecular weight of the acid by the number of ionizable hydrogen atoms in the acid.* On this basis HCl has an equivalent weight of (36.5 g/1) = 36.5 g and H_2SO_4 has an equivalent weight of (98 g/2) = 49 g.

The **gram - equivalent weight (or equivalent weight)** of a substance is the amount in grams of a substance that supplies or will react with 1 mole of H^+ or will gain or lose 1 mole of electrons to produce a neutral atom.

Table 6.2	Classification of Acids According to Their Ionizable Hydrogens
Class	Acids
Monoprotic	HCl, HBr, HI, HNO_3, CHOOH, CH_3COOH
Diprotic	H_2SO_4, H_2CO_3 $H_2C_2O_4$, $[(COOH)_2]$
Triprotic	H_3PO_4

The term *equivalent weight* is applied by chemists to a variety of compounds. For our purposes, however, we will limit our consideration to acids, bases, and electrolyte ions. *For bases, the equivalent weight can be determined by dividing the gram-molecular weight by the number of ionizable hydroxide ions in the base.* NaOH therefore has an equivalent weight of $(40 \text{ g}/1) = 40$ g and the equivalent weight of $Al(OH)_3$ is $(78 \text{ g}/3) = 26$ g. *For electrolyte ions, the equivalent weight can be determined by dividing the gram-atomic weight or gram-molecular weight of the ion by the charge on the ion.* Therefore, the equivalent weight of Na^+ is $(23 \text{ g}/1) = 23$ g, that of Ca^{2+} is $(40 \text{ g}/2) = 20$ g, and that of SO_4^{2-} is $(96 \text{ g}/2) = 48$ g.

The equivalent weight of an acid is the weight of the acid in grams necessary to produce a mole of hydrogen ions in solution. For example, $\frac{1}{3}$ mole of H_3PO_4 (32.67 g) will produce a mole of H^+ ions as given by the equation

$$\tfrac{1}{3} H_3PO_4(aq) \longrightarrow H^+(aq) + \tfrac{1}{3} PO_4^{3-}(aq) \qquad \textbf{(6.10)}$$

The equivalent weight of the base is the weight of the base in grams needed to produce a mole of hydroxide ions, or, more correctly, the amount of base in grams necessary to react with 1 mole of hydrogen ions. As an illustration of this, $\frac{1}{2}$ mole of $Mg(OH)_2$ (29 g) can react with 1 mole of H^+ according to the equation

$$\tfrac{1}{2} Mg(OH)_2(aq) + H^+(aq) \longrightarrow \tfrac{1}{2} Mg^{2+}(aq) + H_2O(l) \qquad \textbf{(6.11)}$$

The equivalent weight of electrolyte ions is the weight in grams of the ion that will either give up or accept one mole of electrons in forming a neutral atom. For example, $\frac{1}{2}$ mole of Ca^{2+} (20 g) will accept 1 mole of electrons according to the equation

$$\tfrac{1}{2} Ca^{2+} + e^- \longrightarrow \tfrac{1}{2} Ca \qquad \textbf{(6.12)}$$

When the quantity gram-molecular weight was introduced, a special term, the *mole,* was also introduced where one mole refers to an amount of a substance equal to the gram-molecular weight of the substance. In a similar manner a new term, *equivalent,* is used with the quantity gram-equivalent weight. *One equivalent refers to an amount of the substance equal to its gram-equivalent weight.*

1 equivalent of HCl = 36.5 g HCl

1 equivalent of $Al(OH)_3$ = 26 g $Al(OH)_3$

and   1 equivalent of Ca^{2+} = 20 g Ca^{2+}

In addition to the use of the term *equivalents,* a subunit *milliequivalents* (0.001 equivalent) is often used in the health-related sciences. For example, as seen in Figure 6.1, concentration of substances in various body fluids is frequently reported in milliequivalents of solute per liter of solution, or as it is usually written, *mEq/L.* * Determination of concentrations in terms of mEq/L is illustrated by the following example.

---

*As is illustrated by the concentration unit mEq/L, the nature of the substances referred to in the numerator and denominator is not always specified. As a general rule, the student can assume that the solute is the substance referred to in the numerator and the solution is indicated in the denominator. Exceptions to one or both of these assignments exist; however, we shall not concern ourselves with those concentration units.

EXAMPLE 6.6   Blood serum contains approximately 1.6×10^{-4} g of K^+ per mL of serum. What is the concentration of potassium ions in milliequivalents per liter?

To perform the calculation we must make the following conversions:

$$\frac{1.6 \times 10^{-4} \text{ g } K^+}{1.0 \text{ mL solution}} \Longrightarrow \frac{\text{No. of mEq } K^+}{1 \text{ L solution}}$$

To convert from 1.6×10^{-4} g K^+ to mEq K^+ we begin by converting to mg K^+ by using the conversion factor 1000 mg K^+/1 g K^+.

$$\frac{1.6 \times 10^{-4} \text{ g } K^+}{1.0 \text{ mL solution}} \times \frac{1000 \text{ mg } K^+}{1 \text{ g } K^+} = \frac{0.16 \text{ mg } K^+}{1.0 \text{ mL solution}}$$

Conversion from 0.16 mg K^+ to mEq K^+ involves determining the equivalent weight of K^+ (39 g K^+/1 = 39 g K^+), writing a conversion factor involving grams and equivalents (1 Eq K^+/39 g K^+), and dividing the numerator and denominator of the conversion factor by 1000:

$$\frac{1 \text{ Eq } K^+/1000}{39 \text{ g } K^+/1000} = \frac{1 \text{ mEq } K^+}{39 \text{ mg } K^+} \tag{6.13}$$

We can write the right-hand side in the equation as above because the prefix *milli-* means *one one-thousandth*. Using this form of the conversion factor we can make the conversion from mg K^+ to mEq K^+:

$$\frac{0.16 \text{ mg } K^+}{1.0 \text{ mL solution}} \times \frac{1 \text{ mEq } K^+}{39 \text{ mg } K^+} = \frac{4.0 \times 10^{-3} \text{ mEq } K^+}{1.0 \text{ mL solution}}$$

Conversion of the number of milliequivalents of K^+ per milliliter of solution to the amount per liter of solution can be made using the conversion factor 1000 mL/L,

$$\frac{4.0 \times 10^{-3} \text{ mEq } K^+}{1.0 \text{ mL solution}} \times \frac{1000 \text{ mL solution}}{1 \text{ L solution}} = \frac{4.0 \text{ mEq } K^+}{1 \text{ L solution}}$$

which is the desired concentration.

As mentioned in Section 6-2.1 the unit mg solute/100 mL solution is also used to give the concentration of substances in body fluids. Conversions from mg solute/100 mL solution to mEq/liter are illustrated in the following example.

EXAMPLE 6.7   Normal blood serum has a magnesium ion concentration of approximately 2.3 mg Mg^{2+}/100 mL solution. What is the concentration of Mg^{2+} in milliequivalents per liter?

To determine the concentration in mEq/L we must make the following conversions:

$$\frac{2.3 \text{ mg } Mg^{2+}}{100 \text{ mL solution}} \Longrightarrow \frac{\text{No. of mEq } Mg^{2+}}{1 \text{ L solution}}$$

To convert from mg Mg^{2+} to mEq Mg^{2+}, we first determine the equivalent weight of Mg^{2+}. Since Mg^{2+} has a +2 charge, its gram-equivalent weight is (24 g/2) =

12 g. Following the reasoning used in Example 6.6, we can write the conversion factor 1 mEq Mg^{2+}/12 mg Mg^{2+}. Using this relationship we have

$$\frac{2.3 \text{ mg Mg}^{2+}}{100 \text{ mL solution}} \times \frac{1 \text{ mEq Mg}^{2+}}{12 \text{ mg Mg}^{2+}} = \frac{0.19 \text{ mEq Mg}^{2+}}{100 \text{ mL solution}}$$

To convert from 0.19 mEq Mg^{2+} per 100 mL solution to the number of mEq Mg^{2+} per L, we can again use the conversion factor 1000 mL/L:

$$\frac{0.19 \text{ mEq Mg}^{2+}}{100 \text{ mL solution}} \times \frac{1000 \text{ mL solution}}{1 \text{ L solution}} = \frac{1.9 \text{ mEq Mg}^{2+}}{\text{L solution}}$$

Chemists use a concentration unit very similar to mEq/L that is called **normality** and represented by N. In equation form the normality is given by

$$x\,\text{N} = \frac{x \text{ Eq of solute}}{1000 \text{ mL of solution}} = \frac{x \text{ Eq of solute}}{1 \text{ L of solution}} \qquad \textbf{(6.14)}$$

The **normality** (N) of a solution is the number of equivalents of a solute per 1000 mL, or 1 liter of solution.

where x is the value of the normality of the solute.

To relate normality to milliequivalents per liter, we can use the conversion factor 1 Eq/1000 mEq. For a solution having a concentration of q mEq/L this becomes,

$$N = \frac{q \text{ mEq}}{L} \times \frac{1 \text{ Eq}}{1000 \text{ mEq}} = \frac{q \text{ Eq}}{1000 \text{ L}} = \left(\frac{q}{1000}\right)N \qquad \textbf{(6.15)}$$

In simple terms, the normality is just one one-thousandth as large as the concentration in milliequivalents per liter.

As illustration of this relationship we can calculate the normality of Mg^{2+} in blood serum as discussed in Example 6.7. Using equation 6.15 and a concentration of 1.9 mEq Mg^{2+}/L solution we have

$$N(\text{Mg}^{2+}) = \frac{1.9 \text{ mEq Mg}^{2+}}{\text{L solution}} \times \frac{1 \text{ Eq Mg}^{2+}}{1000 \text{ mEq Mg}^{2+}} = 1.9 \times 10^{-3} \frac{\text{Eq Mg}^{2+}}{\text{L}}$$

$$= 1.9 \times 10^{-3} \, N \text{ Mg}^{2+}$$

and we see that the normality is one one-thousandth of the concentration in mEq/L. Normality is similar to molarity except the unit used for the solute is the equivalent weight. A 1 N solution of H_2SO_4 consists of 1 equivalent of H_2SO_4 per 1000 mL of solution, or 49 g of H_2SO_4 per 1000 mL of solution. In light of the discussion of equivalent weight given earlier in this section, *solutions of similar substances having the same normality have the same reacting capacity.* Therefore 1 liter of a 1 N solution of NaOH and 1 liter of a 1 N solution of $Mg(OH)_2$ are both capable of reacting with 1 mole of hydrogen ions. We will use this fact again when we consider what is called acid-base neutralization in Chapter 8.

Calculations involving normality can be done in a manner comparable to those for molarity except the solute is dealt with in terms of equivalents rather than moles. As an alternative method, we can convert from normality to molarity and then perform the major part of the calculation in terms of molarity. To make the normality-molarity conversion, the following relationship is used:

$N = M \times$ the number of ionizable H's (or OH's, or the ionic charge) per unit particle   **(6.16)**

Table 6.3	Relative Molarity and Normality of Common Acids and Bases		
Acids and bases		M	N
HCl		1	1
HNO_3		1	1
CH_3COOH		1	1
H_2SO_4		1	2
$H_2C_2O_4, [(COOH_2)]$		1	2
H_3PO_4		1	3
NaOH		1	1
$NH_4OH, [(NH_3 + H_2O)]$		1	1
$Mg(OH)_2$		1	2
$Ca(OH)_2$		1	2
$Al(OH)_3$		1	3

On this basis a 1 M H_2SO_4 solution is 2 N, since H_2SO_4 has 2 ionizable hydrogens. Table 6.3 illustrates the relationship between molarity and normality for a few common acids and bases.

EXAMPLE 6.8   What volume of 0.250 N H_3PO_4 can be prepared from 15.0 g of H_3PO_4?

Using the alternative method referred to in the previous discussion we can determine the molarity of the solution by rearranging equation 6.16 to the form

$$M = \frac{N}{\text{No. } H^+\text{'s in } H_3PO_4} = \frac{0.250}{3} = 0.0833 \ M$$

Now we can complete the calculation using molarity. First, using the definition of molarity we can write

$$0.0833 \ M = \frac{0.0833 \ \text{mol } H_3PO_4}{1000 \ \text{mL solution}}$$

To convert this to a form with grams we multiply by the gram-molecular weight.

$$\frac{0.0833 \ \text{mol } H_3PO_4}{1000 \ \text{mL solution}} \times \frac{98.0 \ \text{g } H_3PO_4}{1 \ \text{mol } H_3PO_4} = \frac{8.16 \ \text{g } H_3PO_4}{1000 \ \text{ml solution}}$$

To obtain the volume of solution that can be prepared from 15.0 g of H_3PO_4, we can invert the last relationship to obtain the necessary conversion factor and multiply by 15.0 g of H_3PO_4,

$$15.0 \ \text{g } H_3PO_4 \times \frac{1000 \ \text{mL solution}}{8.16 \ \text{g } H_3PO_4} = 1840 \ \text{mL solution}$$

To illustrate normality calculations using the concept of equivalents directly we will consider the following example.

EXAMPLE 6.9   A saturated solution of milk of magnesia, $Mg(OH)_2(aq)$, consists of 1.18×10^{-3} g of $Mg(OH)_2$ per 100 mL of solution. What is the normality of the solution?

To determine the normality of the solution we must make two conversions indicated by the following,

$$\frac{1.18 \times 10^{-3} \text{ g Mg(OH)}_2}{100 \text{ mL solution}} \Longrightarrow \frac{\text{No. of Eq Mg(OH)}_2}{1000 \text{ mL solution}}$$

To convert from 1.18×10^{-3} g of $Mg(OH)_2$ to the number of equivalents of $Mg(OH)_2$ we need to know the equivalent weight of $Mg(OH)_2$. Following the procedure discussed earlier in this section we have

$$\text{equivalent wt} = \frac{\text{gram-molecular weight}}{\text{No. of OH}^- \text{ per unit particle}} = \frac{58.3 \text{ g Mg(OH)}_2}{2}$$

$$= 29.2 \text{ g Mg(OH)}_2$$

or

$$\frac{29.2 \text{ g Mg(OH)}_2}{1 \text{ Eq Mg(OH)}_2}$$

which, on inverting, provides the necessary conversion factor to convert from grams of $Mg(OH)_2$ to equivalents of $Mg(OH)_2$.

$$\frac{1.18 \times 10^{-3} \text{ g Mg(OH)}_2}{100 \text{ mL solution}} \times \frac{1 \text{ Eq Mg(OH)}_2}{29.2 \text{ g Mg(OH)}_2} = \frac{4.04 \times 10^{-5} \text{ Eq Mg(OH)}_2}{100 \text{ mL solution}}$$

Conversion of the volume of the solution to 1000 mL, or 1 L, involves the conversion factor 1000 mL/L,

$$\frac{4.04 \times 10^{-5} \text{ Eq Mg(OH)}_2}{100 \text{ mL solution}} \times \frac{1000 \text{ mL solution}}{1 \text{ L solution}} = \frac{4.04 \times 10^{-4} \text{ Eq Mg(OH)}_2}{1 \text{ L solution}}$$

$$= 4.04 \times 10^{-4} N$$

Dilution problems involving normality are calculated in the same manner as in Example 6.5 using the relationship,

$$V_i \times N_i = V_f \times N_f^* \tag{6.17}$$

Do not use normality and molarity units in the same calculation. Calculations with mixed units frequently are incorrect because the normality and molarity of a given solution are not always the same, as illustrated in Table 6.3.

PROBLEMS

6. The concentration unit number of moles of solute per 1000 mL of solution is known as what?

---

*The number of total equivalents of solute does not change upon dilution. Since the product of the volume in liters and normality gives the number of equivalents

$$L \times \frac{\text{Eq}}{L} = \text{Eq}$$

then the products $V_i \times N_i$ and $V_f \times N_f$ can be equated. In addition, volumes in milliliters can also be used since this is the same as multiplying both sides of the equation by 1000.

7. Ringer's solution is an aqueous solution consisting of 0.86 wt/vol % NaCl, 0.030 wt/vol % KCl, and 0.033 wt/vol % $CaCl_2$ that is used for intravenous infusions. The amount of the solution that is administered varies, but as much as 2000 mL or more can be given during a 24-hour period. How many grams of each of the substances above are contained in 2000 mL of Ringer's solution? (The percentage in each case is given in terms of grams of solute per 100 milliliters of solution.)

8. Since the various percentages of the Ringer's solution referred to in problem 7 are given in grams of solute per 100 milliliters of solution, what is the molarity of each substance in the solution?

9. Calculate the gram-equivalent weight of each of the acids in Table 6.2.

10. Ringer's solution, referred to in problem 7, has the following concentrations of electrolytes in milligrams per 100 milliliters: Na^+, 338; K^+, 15.5; Ca^{2+}, 11.9; Cl^-, 558. What is the concentration of each electrolyte ion in milliequivalents per liter?

Ⓢ 11. What volume of 1.5×10^{-3} N $Ca(OH)_2$ can be prepared from 1.0 gram of $Ca(OH)_2$?

ANSWERS TO PROBLEMS   **6.** molarity   **7.** Ⓢ NaCl, 17 g; Ⓢ KCl, 0.60 g; and $CaCl_2$, 0.66 g **8.** Ⓢ 0.15 M, Ⓢ 4.0×10^{-3} M, and 3.0×10^{-3} M **9.** Ⓢ HCl, 36.5 g; HBr, 81 g; HI, 128 g; HNO_3, 63 g; HCOOH, 46 g; CH_3COOH, 60 g; Ⓢ H_2SO_4, 49 g; H_2CO_3, 31 g; $(COOH)_2$, 45 g; Ⓢ H_3PO_4, 32.7 g   **10.** Ⓢ Na, 147; Ⓢ K^+, 3.96; Ca^{2+}, 5.95; Cl^-, 157 mEq/L   **11.** 18 L

6-3 COLLIGATIVE PROPERTIES OF SOLUTIONS

LEARNING OBJECTIVES

1. Define colligative properties, osmosis, and osmotic pressure.
2. Discuss the effect of a solute on the vapor pressure, boiling point, and freezing point of a solvent.
3. Discuss osmosis, and osmotic pressure and its applications to the living system.

In this section we will consider how the properties of pure liquids are modified when a solute is added to a liquid to form a solution. Among these changes are a lowering of the vapor pressure of the solvent, an increase in the boiling point of the solution, and the lowering of the freezing point. Of greatest importance in the health-related sciences is the effect known as osmosis. Osmosis is what causes the wrinkling of the skin when one takes a long bath. Osmosis has an important function in the living system in the passage of water and electrolytes into and out of cells.

6-3.1 VAPOR PRESSURE EFFECT

In Chapter 5 we learned that some of the molecules in a liquid acquire sufficient energy to break away from the bulk liquid and evaporate. Solids behave in this way when they *sublime.* In this case surface molecules of the solid gain enough energy so that they can leave the bulk solid. The effect due to sublimation is much slower than that due to evaporation but, given enough time, an ice cube can sublime just

Figure 6.4   A phase diagram for water (the dashed lines) and an aqueous solution (the solid lines). The normal boiling point is that temperature at which the vapor pressure is 1 atm.

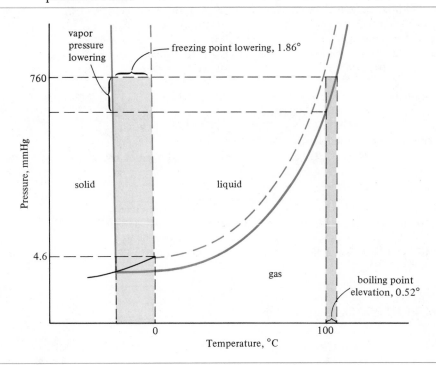

as a glass of water can evaporate. For example, some people hang their washing out in the middle of the winter when the temperature never rises above the freezing point, yet within a day or so the clothes dry because of sublimation.

Evaporation and sublimation are due to the vapor pressure of the liquid and solid, respectively. Experience shows that when a *nonvolatile* solute (a solute with a negligible vapor pressure) is dissolved in a liquid the vapor pressure of the liquid is lowered. This can be seen in Figure 6.4 which gives what is called the phase diagram of water. In this figure the dashed curve extending to the upper right-hand corner represents the vapor pressure of pure water. The lower solid curve represents the vapor pressure of an aqueous solution of a nonvolatile nonelectrolyte. As can be seen in this figure, the vapor pressure of the solution is lower than that of pure water at all temperatures. A similar effect applies to the solid-liquid lines that extend almost vertically upward on the left side of the plot. This lowering of the vapor pressure brings about an increase in the boiling point and lowering of the freezing point as discussed in the next section.

6-3.2 BOILING-POINT ELEVATION AND FREEZING-POINT LOWERING

As indicated in Chapter 5, a substance boils when its vapor pressure is equal to the pressure of the atmosphere that is exerted upon it. When the pressure of the atmosphere is 1 atm (760 mmHg), the boiling point of a substance is referred to as the *normal boiling point*. Reference to Figure 6.4 shows that the normal boiling point of pure water is 100°C. However as also can be seen in this figure, the vapor pressure of the aqueous solution has dropped below 760 mmHg, so the solution no longer will boil at 100°C when the atmospheric pressure is 760 mmHg. In order to get the solution to boil, the temperature must be raised until the vapor pressure is equal to 760 mmHg. To boil aqueous solutions containing 1 mole of nonvolatile nonelectrolytes per 1000 grams of water, the temperature must be raised by 0.52°C, or to a temperature of 100.52°C. This value, *0.52°C,* is referred to as the *boiling point elevation constant.*

Colligative properties, such as vapor-pressure lowering and boiling-point elevation are, for the most part, related to the concentration of particles in the solution. For dilute solutions, the colligative properties are essentially independent of the nature of the particles. Therefore the greater the concentration of the solution, the greater the effect on the property of the solution. When a solute is added to a solvent, the freezing point is affected in a manner similar to the way the boiling point is affected, except the freezing point is lowered, not raised. For an aqueous solution containing 1 mole of a nonvolatile nonelectrolyte per 1000 grams of water the freezing point is lowered by *1.86°C*. This value is known as the *freezing point lowering constant.* Figure 6.4 also illustrates the relationship between vapor pressure lowering and freezing point lowering.

> **Colligative properties** of solutions are properties which are dependent primarily upon the concentration of the solution and to a lesser extent upon the nature of the solute present.

Reference was made previously to the fact that colligative property values are related to the number of particles in the solution. This is important in two respects. First, the value of the colligative property increases with an increase in the concentration of the solute, as already mentioned. Second, when the solute is a strong electrolyte, the number of solute particles is increased by a factor equal to the number of ions per unit particle of electrolyte. As a result, the colligative property is increased by the same factor. For example, sodium chloride is essentially twice as effective as $C_6H_{12}O_6$ (on a mole basis) in lowering the freezing point of water because sodium chloride dissociates into two ions (Na^+ and Cl^-) in aqueous solutions while $C_6H_{12}O_6$ remains intact as individual molecules.

6-3.3 OSMOSIS AND OSMOTIC PRESSURE

Whenever pure water is separated from an aqueous solution by a *semipermeable, or selectively permeable, membrane* (a membrane through which only certain substances including water can pass), it tends to move from the region of pure water to the aqueous solution. Similarly, water moves from a solution of low solute concentration (high solvent concentration) to one of higher solute concentration (lower solvent concentration). As this movement occurs the volume of the less concentrated solution (less solute) decreases, while the volume of the more concentrated solution (more solute) increases as shown in Figure 6.5. Basically, the movement of solvent is associated with a difference in vapor pressure, with the solvent moving from the region of high vapor pressure (low solute concentration) to the region of low vapor pressure (high solute concentration).

> **Osmosis** is the tendency of solvent to flow from a solution of lower solute concentration to a solution of higher solute concentration through a semi-permeable membrane.

Known as osmosis, this colligative property is of prime importance in the living system. For example, the cell membrane of red blood cells is a semipermeable membrane. If pure water were intravenously fed into the blood stream it would dilute the electrolytes and nonelectrolytes in the blood plasma. Water would then

Figure 6.5   Osmosis causes a movement of solvent from the dilute solution (Figure a—on the right) to the concentrated solution (Figure a—on the left), bringing about changes in solution volumes (Figure b). Application of pressure reverses the movement (Figure c).

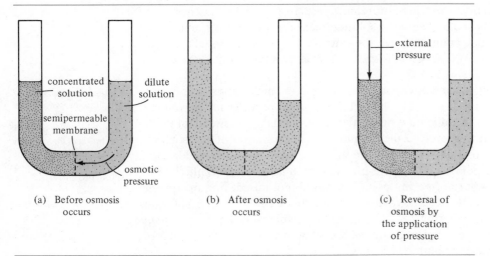

(a) Before osmosis occurs

(b) After osmosis occurs

(c) Reversal of osmosis by the application of pressure

tend to flow into the red blood cells, increasing the size of the cell until the cell membrane would eventually burst. The rupturing of the cell membranes in such cases is due to **osmotic pressure.** Figure 6.5 illustrates osmotic pressure and how it functions in solutions. The average osmotic pressure of blood in contact with pure water is about 7.7 atm. To avoid the problem of rupturing the cell membranes, *isotonic solutions* are used. An isotonic solution is a solution that has a concentration of electrolytes and/or nonelectrolytes that will exert an osmotic pressure equivalent to that of the concentration of the solution to which it is added. When an isotonic solution enters the blood stream, the movement of water into and out of the cell remains the same and the cells are unaffected.

When the living system suffers dehydration or blood loss that is not so great that a whole-blood transfusion is necessary, physiological saline solution can be injected safely to replace body fluids. Similarly, an isotonic solution of 5% by weight glucose solution can be used in intravenous feeding when oral feeding is difficult or impossible.

Passage of water and other substances through cell membranes is not restricted to those cases in which a solution enters the living system during intravenous infusion, but is taking place continually in the body and is vital to the maintenance of the life of the cell. Water is exchanged between the interior of the cell (protoplasm) and the solution surrounding each cell. In addition, many dissolved substances, such as nutrients and body wastes, are able to penetrate the cell membrane. In this way the exchange of water between the cell and its surroundings is accompanied by the exchange of these other substances. An example of this is the passage of oxygen and carbon dioxide into and out of the cell, respectively.

The ability of substances to pass through the cell membrane depends upon several factors and varies from substance to substance. Certain generalizations can

Osmotic pressure is the pressure exerted by a solution on the semipermeable membrane that separates it from a solution of higher concentration.

be made, however. Water and gases such as carbon dioxide, oxygen, and nitrogen pass readily and rapidly through the cell membrane. Many organic substances such as alcohol and ether also easily penetrate the cell membrane. Other more complex organic substances such as glucose and amino acids are somewhat slower in their penetration. Even slower in their passage are simple inorganic salts, acids, and sugars such as sucrose. There is little, if any, penetration of the membrane by most complex solutes in the protoplasm, including proteins, most sugars, and many salts.

In addition to the isotonic solutions referred to earlier in this section there are two other classes of solutions. The first, referred to as *hypotonic,* is solutions that have a lower osmotic pressure than that of the cells. Such a solution is a 0.66% by weight solution of NaCl. Animal cells placed in such a solution tend to take in water and swell. Water in contact with blood, as noted earlier in the discussion, is acting hypotonically. The second class of solutions, known as *hypertonic,* has a higher osmotic pressure than do cells and therefore tends to shrink living cells placed in them. These two classes of solutions are destructive to cells and are normally avoided.

In the living system the passage of water and dissolved substances through the cell membranes does not depend solely upon the basic laws of osmosis and diffusion. The living cell and cell membrane play an active part in the exchange process. In many cases specific solutes are concentrated in certain cells at a concentration many times greater than that of the solutes in the surrounding solutions. For example, many marine algae have a concentration of iodine within their cells that is more than a million times greater than is present in the sea water in which they live. Cells have a means of forcing unit particles of particular substances to move in a direction opposite to that dictated by the laws of diffusion and osmosis. The process involved in such cases is called active transport. The details of this process are beyond the scope and purpose of this text, but the process appears to be associated with specific proteins and other molecules in the cell membrane that influence the movement of substances across the cell membrane. The transport machinery is called *metabolic pumps* and scientists speak of *sodium pumps* and *potassium pumps.* More shall be said about these in Chapter 9 when we consider the passage of nerve impulses along nerve cells.

Under normal conditions, active transport enables the body cells to regulate the entry of nutrients into the cells at a rate that satisfies the food requirements of the cells. For example, when the cells are surrounded by a solution that is very low in an essential nutrient, the cells are able to draw the nutrient into them through the process of active transport. Active transport, therefore, is vital to the survival of the living organism.

PROBLEMS

12. Define osmosis.
13. What does the addition of a solute to a solvent do to the freezing point of the solvent?

ANSWERS TO PROBLEMS   **12.** the tendency of a solvent to flow from a solution of lower concentration to a solution of higher concentration through a semipermeable membrane **13.** It lowers the freezing point of the solvent.

SUMMARY

In this chapter we have investigated the nature of solutions, including the various factors involved in forming solutions and in determining the solubilities of substances in solvents. We found that certain concentration units were useful in describing solutions quantitatively. Solutions vary from pure substances in their colligative properties and we considered some of these properties as applied to aqueous solutions. Of particular importance to the health-related sciences is osmosis. In the next chapter we will turn our attention to other important topics, chemical kinetics and chemical equilibrium.

ADDITIONAL PROBLEMS

14. What is the term for the interaction of solvent molecules with solute particules when they are loosely bonded together?

15. Of what does conduction of an electric current in solution consist?

16. For any natural or spontaneous process, what does the system tend to do in terms of its degree of disorder?

⒟ 17. Will KI be more soluble in the polar solvent $H_2O(l)$ or in the nonpolar solvent $C_6H_{14}(l)$?

18. What effect will an increase in pressure have upon the solubility of ethyl ether ($C_2H_5OC_2H_5(l)$) in water?

19. How many grams of
    ⒮ (a) $AlCl_3$ are in 850 g of a 1.20 percent by weight solution?
    ⒮ (b) HCl are in 300 g of a 5.0 percent by weight solution?
    (c) NaOH are in 16 g of a 0.5 percent by weight solution?
    (d) Na_2SO_4 are in 1550 g of a 15.0 percent by weight solution?

20. How many milliliters of
    ⒮ (a) ethyl alcohol (C_2H_5OH) are in 90.0 mL of a 5.0 percent by volume solution?
    ⒮ (b) CCl_4 are in 2.5 L of a 1.5 percent by volume solution?
    (c) acetone (CH_3COCH_3) are in 5.00 mL of a 35.0 percent by volume solution?

21. Using the values given in Table 6.1, calculate the number of milligrams of the following in 5000 mL of whole blood. [This volume of blood is a typical value for all of the blood in a person weighing approximately 70 kg (150 lb).]
    ⒮ (a) ascorbic acid   ⒮ (b) glucose (fasting level)   (c) lactic acid   (d) iron

22. What is the molarity of a solution that contains
    ⒮ (a) 10.4 g of Na_2CO_3 in 400 mL of solution?
    ⒮ (b) 4.62 g of $(COOH)_2$ in 1200 mL of solution?
    (c) 78.4 g of H_3AsO_4 in 208 mL of solution?

23. How many milliliters of a
    ⒮ (a) 0.500 M solution of $C_{12}H_{22}O_{11}$ are needed to obtain 25.0 g of solute?
    ⒮ (b) 2.00 M solution of NaOH are needed to obtain 100 g of solute?
    (c) 5.4 M solution of KBr are needed to obtain 23 g of solute?

24. How many grams of
    ⒮ (a) KOH would be required to prepare a O.40 M solution having a total volume of 750 mL?
    ⒮ (b) Na_2CO_3 would be required to prepare a 1.3 M solution having a total volume of 15 mL?
    (c) $NaClO_4$ would be required to prepare a 2.00 M solution having a total volume of 550 mL?

25. What will the molarity of a solution of
    ⒮ (a) HCl if 1200 mL of a 1.5 M solution is diluted to 2000 mL?
    ⒮ (b) NaOH if 15 mL of a 0.10 M solution is diluted to 25 mL?
    (c) NaCl if 100 mL of a 1.0 M solution is diluted to 1000 mL?

26. Determine the gram-equivalent weight of the following acids and bases.
    ⒮ (a) H_3AsO_4 ⒮ (b) H_2TeO_4
    (c) $Ba(OH)_2$ ⒮ (d) $Pb(OH)_4$
    (e) $HC_2H_3O_2$, [(CH_3COOH)]

27. Normal blood-serum electrolyte levels measured as milligrams of electrolyte ion per 100 mL of solution are:

⑤ (a) Na^+, 314 to 326 mg   ⑤ (b) K^+, 16 to 22 mg   (c) Ca^{2+}, 8.5 to 10.5 mg   (d) Cl^-, 345 to 380 mg

What is the concentration of each ion in milliequivalents per liter?

28. What is the normality of a solution that contains
    ⑤ (a) 24.3 g of $Ca(OH)_2$ in 1200 mL of solution?
    ⑤ (b) 30.2 g of H_2SO_3 in 900 mL of solution?
    (c) 3.86 g of $Ba(OH)_2$ in 1420 mL of solution?

29. How many grams of
    ⑤ (a) $Al(OH)_3$ are in 1400 mL of a 0.20 N solution?
    ⑤ (b) HNO_3 are in 900 mL of a 1.1 N solution?
    (c) $Ba(OH)_2$ are in 1.00 mL of a 1.00 N solution?

30. What will be the normality of a solution of
    ⑤ (a) HCl if 1200 mL of a 1.5 N solution of it is diluted to 2000 mL?

⑤ (b) NaOH if 15 mL of a 0.10 N solution of it is diluted to 25 mL?
(c) H_2S of 175 mL of a 0.500 N solution of it is diluted to 500 mL?

31. How many milliliters of H_2O must be added to
    ⑤ (a) 950 mL of a 0.423 N solution of H_2SO_4 to give a 0.250 N solution?
    ⑤ (b) 600 mL of a 0.85 N solution of H_3PO_4 to give a 0.50 N solution?
    (c) 100 mL of a 1.0 N solution of KOH to give a 1.0×10^{-4} N solution?
    (d) 100 mL of a 0.50 N solution of $Ba(OH)_2$ to give a 0.75 N solution?

32. What are the four colligative properties discussed in this chapter?

33. When solutions having different concentrations are separated by a semipermeable membrane in which direction does the solvent flow?

CHAPTER 7

CHEMICAL KINETICS AND CHEMICAL EQUILIBRIUM

In the earlier chapters of this text we studied various aspects of chemical reactions. In Chapter 3 we found that reactions between atoms are basically associated with a rearrangement of the electrons in the atoms that consists of either a transfer or a sharing of electrons. In Chapter 4 we looked at several basic types of reactions and practiced predicting whether or not a reaction will occur and, if it does, what products are formed. In all of this material, little has been said about the step-by-step process by which a reaction occurs, how fast it occurs, and to what extent it occurs. In Chapter 7 we will look at these topics.

Of particular importance is a type of reaction known as a reversible reaction. This type of reaction is associated with chemical equilibrium. Some time will be spent on this important topic and we will build a foundation that will enable us to understand better the nature of the weak electrolytes that were introduced in Chapter 6.

7-1 REACTION RATES AND COLLISION THEORY

LEARNING OBJECTIVES

1. Define chemical kinetics, reaction mechanism, rate of reaction, rate equation, and catalyst.
2. Specify the order of a chemical reaction based upon the rate equation.
3. Predict the effect of a change in temperature and concentration on the rate of a chemical reaction.
4. Discuss the general characteristics of the way catalysts operate and their effect on the rate of reaction.
5. Discuss the basic principles of collision theory in terms of molecular orientations and energy.
6. Draw, label, and discuss a diagram of potential energy versus progress of reaction for endothermic and exothermic reactions.

The world we live in is continually undergoing change. Many of these are physical and chemical changes. In this section we will take a more detailed look at chemical changes. Life processes are basically chemical reactions and much of what follows in later chapters is a detailed consideration of various classes of reactions that take place in the living system. In this discussion we will extend our knowledge of chemical reactions and build a foundation that will help us to understand better the concept of chemical equilibrium presented in the next section.

7-1.1 BASIC CONCEPTS

In our earlier contact with chemical reactions we were primarily concerned with what chemical species are reacting and what substances we can expect to form; that is, we learned to identify the reactants and products. In that discussion, little was said about how a reaction actually takes place and how fast a reaction occurs. The study of such processes is termed chemical kinetics.

> **Chemical kinetics** is the branch of chemistry that deals with the speed of reactions and the stepwise process they follow.

Chemical reactions are, in some ways, similar to taking a trip. Some trips are what we might call one-step trips; that is, we go directly from our point of origin to our destination without any stops or visits in between. Other trips are more complex, including several visits along the way. The reaction path for a chemical reaction is similar. Some reaction paths are simple one-step reactions, while others occur by a much more complex process and involve many individual steps, or *elementary reactions*. The process by which the overall reaction occurs is known as the reaction mechanism.

An example of a simple one-step mechanism is the reaction of nitrogen monoxide (NO) with ozone (O_3). This reaction is believed to occur through a simple bimolecular (two-molecule) collision as shown in Figure 7.1 and illustrated by the equation,*

> A **reaction mechanism** consists of the stepwise changes, or elementary chemical reactions, that reacting molecules undergo as they are converted from reactants to products.

$$NO(g) + O_3(g) \longrightarrow NO_2(g) + O_2(g) \qquad (7.1)$$

In this case the balanced chemical equation represents the mechanism by which the reaction takes place. In general, reaction mechanisms are much more complex than the overall chemical equation would suggest. For example, the complete oxidation of glucose in the human body as represented by the overall chemical equation,

$$C_6H_{12}O_6(aq) + 6\,O_2(aq) \longrightarrow 6\,CO_2(aq) + 6\,H_2O(l) \qquad (7.2)$$

actually occurs by a complex mechanism (see Chapters 24 and 25).

A second important aspect of our study of chemical kinetics deals with the rate of reaction. The rate of reaction can be considered the speed of reaction. As applied to a car or train, the term *speed* refers to a change in distance during a

> The **rate of reaction** is the change in the concentration of reactants or products per unit of time.

---

*This reaction occurs in the upper atmosphere. Exhaust from jet airplanes includes, among other things NO which reacts with the ozone already in the atmosphere. Depletion of the ozone is of some concern since the gas acts as a screen to shield us from harmful ultraviolet radiation from the sun.

Figure 7.1 The bimolecular collision mechanism for the reaction of NO(g) with O_3(g) occurs by a simple collision.

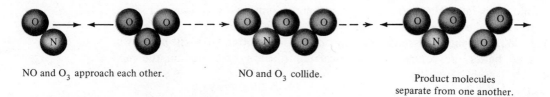

NO and O_3 approach each other.   NO and O_3 collide.   Product molecules separate from one another.

given period of time. In the case of chemical reactions the speed, or rate of reaction, indicates how much the concentration of a reactant or product changes during a given amount of time.

The rate of reaction is dependent upon several factors including the concentration of reactants, temperature, the nature of the substances reacting, and the presence of what are known as catalysts. Most of these factors will be studied in some detail in Section 7-1.2; however, at this point in our discussion it will be helpful to consider how the rate of reaction is related to the concentration of reactants. For a reaction following a mechanism such as the one given in the equation

$$A + B \longrightarrow AB \tag{7.3}$$

the rate of reaction (the dependent variable) is directly proportional to the concentration of both A and B (the independent variables). Mathematically, we can say

$$\text{Rate} \propto [A] \tag{7.4}$$

and

$$\text{Rate} \propto [B] \tag{7.5}$$

In this case the rate of reaction will increase if either [A] or [B] or both are increased, and will decrease if either or both of them are decreased.

Equations 7.4 and 7.5 are usually combined into the single expression,

$$\text{Rate} \propto [A][B] \tag{7.6}$$

and the expression is converted into an equality by the addition of the proportionality constant, k.

$$\text{Rate} = k\,[A][B] \tag{7.7}$$

The constant k is referred to as the *rate constant* and equation 7.7 is called the **rate equation**. Chemists frequently report the rate of a chemical reaction by giving the value of the rate constant k. Fast reactions have large values of k and slow reactions have small values of k. The rate equation allows us to calculate the rate of reaction if we know the value of k and the concentrations of A and B in the solution being studied. An alternative method for reporting how fast a reaction is occurring is to report the half-life of the reaction. This is essentially what was done in Chapter 2 when the half-lives for the various radioactive decay processes were reported. In this case, a shorter half-life indicates a faster reaction.

The **rate equation** is the equation relating the rate of reaction to the concentrations of chemical species that determine the rate of reaction.

The rate of reaction for many reactions depends upon the concentration of one or more chemical species squared or raised to some other power. For example, the reaction between NO and H_2

$$2\,NO + 2\,H_2 \longrightarrow N_2 + 2\,H_2O \qquad (7.8)$$

follows the rate equation,

$$Rate = k[NO]^2[H_2]^1 \qquad (7.9)$$

It is significant that the coefficient for H_2 in equation 7.8 is not the same as the exponent for $[H_2]$ in equation 7.9. This is because the mechanism for the reaction is more complex than equation 7.8 suggests. This situation occurs very frequently, so the research chemist must be careful to avoid assuming they are the same.

The exponents 2 and 1 associated with [NO] and $[H_2]$ in equation 7.9 give us what is called the *order of reaction* for NO and H_2, respectively. The order of reaction with respect to a particular chemical species is simply the power, or exponent, to which the concentration of that species must be raised in order to explain how the rate of reaction varies with variation in the concentration of that species. Reaction 7.8, therefore, is considered to be second order in NO and first order in H_2 as given by the exponents in equation 7.9. In a similar manner, the overall order of the reaction is given by the sum of the exponents, or individual orders. Thus, the overall order of equation 7.9 is third order.

A second illustration of the order of a reaction is the radioactive decay process discussed in Chapter 2. This is a simple first-order reaction that depends only on the amount of the radioactive isotope being considered. Based on this fact, the rate of alpha decay of $^{226}_{88}Ra$ by the reaction

$$^{226}_{88}Ra \longrightarrow\; ^{222}_{86}Rn + \alpha \qquad (7.10)$$

is given by the rate equation*

$$Rate = k[Ra] \qquad (7.11)$$

To further illustrate the relationship between the order of a reaction and the rate of reaction, we will consider how the rate of reaction changes as the concentration of a chemical species changes. For example, if we consider equation 7.11, we can see that doubling the concentration of Ra will double the rate ($[2]^1 = 2$). Similarly, if we examine equation 7.9, we can determine that doubling the concentration of NO will increase the rate of that reaction by a factor of four ($[2]^2 = 4$). Likewise, for a reaction that follows the rate equation $Rate = k[A]^3$, doubling the concentration of A will increase the rate of reaction by a factor of eight ($[2]^3 = 8$).

7-1.2 FACTORS AFFECTING THE RATE OF REACTION

The nature of the substances reacting is the most important factor involved in the rate of reaction. Mixing chemicals together does not guarantee that they will react.

---

*When reactions are first-order with respect to some chemical species, the usual practice is to omit the superscript 1 that appears in equation 7.9. The superscript 1 is used in equation 7.9 for emphasis. Ordinarily, we will not include it throughout the rest of the text.

As seen in Chapter 4, zinc will react vigorously with hydrochloric acid according to the equation

$$Zn(s) + 2\,HCl(aq) \longrightarrow ZnCl_2(aq) + H_2(g) \qquad (7.12)$$

while silver and hydrochloric acid will not react at all.

$$Ag(s) + 2\,HCl(aq) \longrightarrow N.R. \qquad (7.13)$$

Another illustration of this idea is that sodium will react violently with hydrochloric acid,

$$2\,Na(s) + 2\,HCl(aq) \longrightarrow 2\,NaCl(aq) + H_2(g) \qquad (7.14)$$

while lead reacts very slowly, according to the equation

$$Pb(s) + 2\,HCl(aq) \longrightarrow PbCl_2(s) + H_2(g) \qquad (7.15)$$

The same basic reaction is taking place in each of these equations: a metal is replacing hydrogen from hydrochloric acid. However, the rate for each reaction varies drastically depending upon the nature of the metal involved. The rate of reaction, then, is primarily dependent upon the nature of the reactants involved.

In general, *an increase in temperature increases the rate of reaction.* This is similar to the effect that an increase in temperature has upon the evaporation of a liquid. In the case of evaporation, a higher temperature increases the number of liquid molecules that have enough energy to escape from the bulk liquid; therefore, the rate of evaporation is greater. In the case of a chemical reaction, a higher temperature increases the number of molecules that have enough energy to react; therefore, the rate of reaction is greater. For many reactions, *the rate of reaction approximately doubles for every 10°C rise in temperature and is decreased by half for a 10°C drop in temperature.* More shall be said about the relationship between temperature and rates of reaction when we discuss collision theory in Section 7-1.3.

Rates of chemical reactions in the living system are also affected by temperature. This is evidenced by the increased rates of movement of insects, reptiles, and other cold-blooded animals with an increase in temperature. Similarly, the chirping rate of crickets has been found to be temperature-dependent. Counting the chirps can be used as a means of obtaining temperature readings.

Reaction processes in warm-blooded animals have a similar temperature dependence. For example, as the body temperature is lowered, metabolic processes slow down as evidenced by the hibernation of animals in the winter when little food is required. Doctors take advantage of this effect and slow down metabolic processes by cooling the blood when heart surgery is performed.

When dealing with living organisms the simple relationship between increased temperature and increased rate of reaction does not always apply. This is illustrated by the reaction that takes place during the souring of milk due to bacteria. Milk undergoes the reaction of souring much more rapidly when it is at room temperature than when it is cooled in the refrigerator, due to the increased activity and growth in the number of bacteria at the higher temperature. This agrees in general with the relationship between increased temperature and increased rate of reaction. However, as the temperature of the milk is raised above room temperature a point is reached where the rate of the souring reaction is drastically decreased, even if the temperature is returned to room temperature. This is because, for the most part, the bacteria that cause the milk to sour are killed at the higher temperature. This is the basis for the pasteurization process.

A similar inverse dependence on temperature also occurs in the human body. An increase in body temperature above normal also has a negative effect: body reactions slow down and one can suffer heat stroke if the body temperature goes above 105°F (40.6°C). For humans the optimal temperature is approximately 98.6°F (37.0°C).

Catalysts can also play a significant role in determining the rate of reaction in many chemical reactions. As an example of this, the decomposition of $KClO_3$ by heating according to the following equation

$$2 \, KClO_3(s) \xrightarrow{\text{heat}} 2 \, KCl + 3 \, O_2 \qquad (7.16)$$

A **catalyst** is a substance that increases the rate of a chemical reaction without being consumed by the reaction and can be recovered, chemically unchanged, when the reaction is completed.

occurs fairly slowly. However adding a small amount of manganese(IV) oxide greatly increases the rate of reaction. In this instance, MnO_2 is acting as a catalyst since it can be reclaimed unaffected when the reaction is over. *Catalysts increase the rate of reaction by providing an alternative mechanism by which the reaction can take place more rapidly.* More shall be said about this when we consider collision theory in Section 7-1.3.

Catalysts are of prime importance in the living system. Enzymes are complex proteins produced by living cells that act as catalysts in biological systems. For example, pepsin is an enzyme in gastric juice that catalyzes the digestion of proteins. Similarly, salivary α-amylase is the enzyme found in saliva that catalyzes the breakdown of starch. Enzymes also play a key role in all metabolic processes such as the oxidation of glucose referred to previously.

7-1.3 COLLISION THEORY

As suggested in Section 7-1.1, the mechanisms by which various reactions occur are often complex. It is not our purpose to discuss these in detail; however, the knowledge of one general type of mechanism, collision theory, is particularly valuable in understanding chemical kinetics. In order for two substances, A_2 and B_2, to react, they must come together, or collide. Reaction between A_2 and B_2 cannot, of course, occur any faster than the rate at which they collide. This approach to explaining chemical reactions is referred to as collision theory.

If we assume that the rate of reaction for the equation

$$A_2(g) + B_2(g) \longrightarrow 2 \, AB(g) \qquad (7.17)$$

depends only upon a simple bimolecular collision such as the reaction of NO and O_3 discussed in Section 7-1.1, certain factors become important. One of these is the fact that most reactions proceed at a much lower rate than would be predicted based upon how often they collide. In other words, on the average it takes more than one collision in order for the reactants to undergo the change to products. In part, this can be explained by assuming that only certain orientations of the reactant molecules with respect to one another allow the molecules to react. For

Figure 7.2   (a) A possible acceptable collision orientation and (b) a possible unacceptable collision orientation.

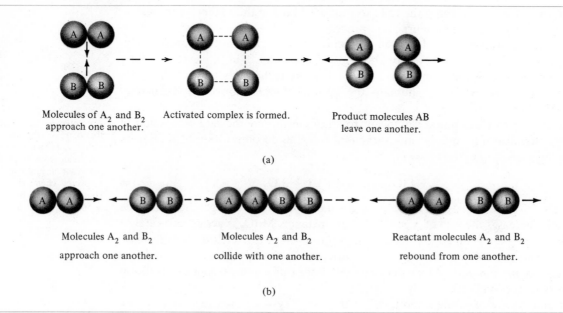

Molecules of A_2 and B_2 approach one another.   Activated complex is formed.   Product molecules AB leave one another.

(a)

Molecules A_2 and B_2 approach one another.   Molecules A_2 and B_2 collide with one another.   Reactant molecules A_2 and B_2 rebound from one another.

(b)

example, it might be that the reaction occurs only when the reactant molecules come together with some acceptable relative positions, or orientation, such as is illustrated in Figure 7.2(a). Molecules colliding with the unacceptable orientation shown in Figure 7.2(b) will simply rebound, like colliding billiard balls, without undergoing reaction.

A second factor that determines the rate of reaction deals with the energy aspects of the reaction. Figure 7.3 gives the relative energies of the various chemical species involved in a reaction such as that given by equation 7.17. The potential energy of the various chemical species is plotted versus the progress of the reaction. The latter quantity is simply a measure of how far through the reaction any pair of reactant molecules has progressed.

To begin with, molecules A_2 and B_2 have a certain amount of potential energy. This energy is represented by the horizontal straight line $x-y$ passing through the curve. As the two molecules come together, there will be an increase in their potential energy. This is represented by the hill as we go from left to right across the plot in Figure 7.3, indicating that the electron clouds surrounding A_2 and B_2 start to interact. Think of what happens when you squeeze two sponges together. A repulsive force due to the elasticity of the sponges causes them to separate when you let go of them. Similarly, the reacting molecules also will experience a repulsive force due to the negatively charged electrons in each molecule. In order for the reaction of A_2 and B_2 to occur, then, the two molecules must not only collide with the proper orientation but they must be moving fast enough so that they can be sufficiently close to allow the necessary rearrangements of atoms and electrons to take place to form products. The point at which this occurs is referred to as the *transition state* and the chemical species associated with this state

$$
\begin{array}{c}
B\text{---}B \\
\vert \quad\ \ \vert \\
A\text{---}B
\end{array}
$$

activated complex

is known as the *activated complex*. After reaction has occurred the two product molecules begin to separate and the potential energy of the molecular system decreases until the product molecules are at relatively large distances from one another. This is represented by the decline in energy on the right side of the hill in Figure 7.3.

Basically, the two reactant molecules will react if they have the proper orientation and sufficient energy to form the activated complex. To do this, they must

Figure 7.3 Potential energy diagram for an exothermic reaction.

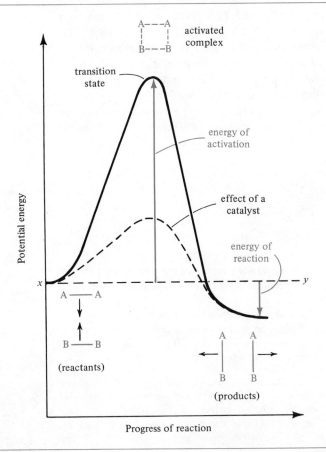

have energy at least equal to the *energy of activation.* The energy of activation is equal to the difference in energy of the activated complex and the reactants, as indicated in Figure 7.3. Reacting molecules that do not have enough energy to go over this energy barrier can be pictured as going part way up the hill on the left in Figure 7.3, then coming back down, or rebounding, without reacting. An increase in temperature increases the energy of the molecules; therefore, a higher temperature increases the number of molecules that have sufficient energy to react and increases the rate of reaction.

The concepts above also help to explain how a catalyst speeds up the reaction. Basically, the catalyst provides an alternative path that has a lower energy of activation. In a simple way a catalyst acts like a mountain tunnel that allows traffic to pass from the reactant side to the product side without going over the high mountain pass. This is illustrated by the dotted curve in Figure 7.3. Since the catalyzed path has a lower energy of activation, a large number of reactants have sufficient energy to react via this path, increasing the rate of reaction.

One other aspect of the potential energy diagram, Figure 7.3, is important to us. In this figure we can see that the potential energy of the products is lower than that of the reactants. The difference in energy between these two values is equal to the energy of reaction. When the potential energy of the products is lower than that of the reactants, as is the case in Figure 7.3, energy is given off and the reaction is exothermic. If the potential energy of the products is higher than that of the reactants, heat is absorbed during the reaction and the process is endothermic.

In this section we have considered some aspects of chemical kinetics. We have looked at some of the factors that influence the rate of reaction and how these rates are described. Little has been said, however, about those cases in which a reaction only goes partially to completion. Such reactions form a very important part of the chemistry that surrounds us. This topic shall be covered in Section 7-2 as chemical equilibrium is introduced.

PROBLEMS

1. Define the term rate of reaction.
2. The reaction

$$CO(g) + Cl_2(g) \longrightarrow COCl_2(g)$$

   follows the rate equation   Rate $= k[CO][Cl_2]^{3/2}$

   What is the order with respect to (a) CO (b) Cl_2 (c) the overall order?
3. What effect does a decrease in temperature normally have upon the rate of reaction?
4. What effect does the addition of an appropriate catalyst to a reaction vessel have upon the rate of reaction?
5. What two factors influence the rate of reaction based upon collision theory?
6. The activation energy is equal to the difference in energy of what two chemical species?
7. The energy of reaction is equal to the difference in energy of what two chemical species?
8. What is the state associated with the activated complex?

ANSWERS TO PROBLEMS   **1.** the change in the concentration of reactants or products per unit of time   **2. (a)** first   **(b)** three halves   **(c)** five halves   **3.** it decreases the rate   **4.** it increases the rate   **5.** the molecular orientation of the colliding molecules and the energy of activation for the reaction   **6.** the activated complex and the reactants   **7.** the reactants and the products   **8.** the transition state

7-2 CHEMICAL EQUILIBRIUM

LEARNING OBJECTIVES

1. Define equilibrium, equilibrium constant, and LeChatelier's principle.
2. Write equilibrium constant expressions for chemical reactions at equilibrium.

3. Describe the effect on the point of equilibrium of changes in the concentration of reactants and products, the temperature, the pressure, and the effect of the addition of a catalyst.

In Section 7-1 we considered chemical kinetics. We found that most reactions occur by rather complex mechanisms. We also learned that various factors influence the rate of reaction, including the concentration and nature of the reactants, and the temperature and presence of appropriate catalysts. In this section we will look at a special type of reaction called a reversible reaction. This type of reaction results in the condition called chemical equilibrium. Chemical equilibrium has many important applications in the living system, as we will discover in Chapter 8 and throughout the remainder of the text. For the present discussion we will consider the basic principles concerning chemical equilibrium and the factors that affect it.

7-2.1 REVERSIBLE REACTIONS AND EQUILIBRIUM

Thus far in our discussion we have dealt primarily with reactions that go totally from reactants to products, with essentially no unit particles of reactants remaining when the reaction is complete. For example, at a temperature of 25°C, hydrogen and iodine react to form hydrogen iodide according to the equation,

$$H_2(g) + I_2(g) \longrightarrow 2\ HI(g) \tag{7.18}$$

and essentially no $H_2(g)$ or $I_2(g)$ remains when the reaction is over. In such cases the reaction is considered to have *gone to completion*. However, as the temperature is raised, the $HI(g)$ tends to break apart, or decompose, and re-form hydrogen and iodine according to the equation,

$$2\ HI(g) \longrightarrow H_2(g) + I_2(g) \tag{7.19}$$

At 425°C, for example, the amount of HI formed by the reaction of equal amounts of H_2 and I_2 is less than 90% complete.

The fact that the $HI(g)$ begins to decompose to form $H_2(g)$ and $I_2(g)$ does not mean that the reaction given by equation 7.18 is no longer occurring. This reaction does continue to occur and, as indicated in the previous section, it will probably occur at an even greater rate because of the elevated temperature. What has happened, however, is that the reaction given by the equation 7.19 tends to counterbalance the first reaction, so we can no longer say that the reaction represented by equation 7.18 has gone to completion. Reactions that can go in either direction are considered to be *reversible reactions*. In the above illustration equation 7.18 is *the forward reaction* and equation 7.19 is *the reverse reaction*. If some elevated temperature is maintained, a point is reached where the forward rate of reaction is equal to

Figure 7.4
Water vapor in equilibrium with
liquid water.

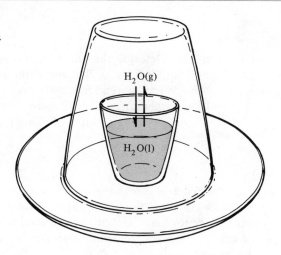

the reverse rate of reaction. When this happens the system is said to have achieved **chemical equilibrium.**

Chemical equilibrium can perhaps be best understood by considering the physical equilibrium associated with the evaporation of a liquid such as water as discussed in Chapter 5. Such evaporation can be stopped simply by inverting a large glass over a smaller glass containing water, as illustrated in Figure 7.4. Evaporation will continue for a short time, but now the gaseous water molecules that leave the liquid water are trapped. As the water molecules move about in the gas phase, they occasionally strike the surface of the water, are caught (condense), and once again become part of the liquid water. When the glass of water is first covered, the rate of evaporation is generally much greater than the rate of condensation since there are usually relatively few water molecules in the gas phase. As the number of water molecules increases, the rate of condensation increases until the rate of evaporation and the rate of condensation are equal. At this point the level of the water no longer drops and we say that the water vapor and liquid water are in equilibrium. It is important to realize that in this case the evaporation and condensation have not stopped or even slowed down, but are simply occurring at the same rate.

Applying this reasoning to the chemical equilibrium between H_2, I_2, and HI we have a similar situation. If, initially, we add H_2 and I_2 to a reaction vessel at 425°C they will immediately begin to react to form HI. To start with there is no HI in the vessel and so there cannot be any decomposition of it. However, as soon as some HI molecules are formed, they can begin to decompose to form H_2 and I_2. At first there are so few HI molecules the rate of decomposition of HI will be much less than its rate of formation. As these reactions proceed the amount increases and, therefore, so does its rate of decomposition. The amount of H_2 and I_2 decreases as a result of the forward reaction, resulting in some decrease in the rate of formation of HI. Eventually the two rates become equal and we say that equilibrium has been reached. As with the evaporation process the two reactions, 7.18 and 7.19, have not

Chemical equilibrium is that condition in a chemical reaction when the forward and reverse reactions occur at the same rate.

stopped but are simply occurring in opposite directions. This is generally indicated by combining equations 7.18 and 7.19 as

$$H_2(g) + I_2(g) \rightleftharpoons 2\,HI(g) \tag{7.20}$$

where the two arrows indicate that both reactions are still taking place.

You must be careful when dealing with equilibrium processes to avoid assuming that the concentrations of reactants and products are the same. A common error is to assume that since the forward and reverse rates are equal, the concentrations of the various chemical species are also equal. In general, this is not true. For equilibrium reactions such as equation 7.20, the concentrations of the various reactants and products, *the point of equilibrium,* is dependent upon several factors including initial concentrations, temperature, and pressure. These shall be discussed in Section 7-2.3.

7-2.2 EQUILIBRIUM CONSTANTS

When a condition of chemical equilibrium has been achieved, the concentrations of reactants and products stop changing, remaining constant unless some outside disturbance disrupts the point of equilibrium. In addition, it has also been found experimentally that the ratio of the product (or multiplication) of product concentrations to the product (or multiplication) of reactant concentrations each raised to a power equal to the stoichiometric coefficient in the balanced chemical equation is constant. As applied to equation 7.20, this is written

$$\frac{[HI]^2}{[H_2][I_2]} = K \tag{7.21}$$

where the portion to the left of the equal sign is referred to as the *equilibrium constant expression* and K is the equilibrium constant.

For the more general equation

$$m\,M + n\,N \rightleftharpoons p\,P + q\,Q$$

the equilibrium constant expression takes the form

$$K = \frac{[P]^p[Q]^q}{[M]^m[N]^n} \tag{7.22}$$

An **equilibrium constant** is the ratio of the product of product concentrations to the product of reactant concentrations at equilibrium, where each concentration is raised to a power equal to the stoichiometric coefficient in the balanced chemical equation.

Equilibrium constant expressions are normally written with the product concentrations in the numerator and reactant concentrations in the denominator, as shown in equations 7.21 and 7.22. Such expressions allow us to make certain calculations involving chemical reactions at equilibrium and are valuable in determining the point of equilibrium. It is important to note that equilibrium constants maintain a fixed value at each temperature regardless of the starting concentrations of each component.

One other aspect is important for us to consider when writing equilibrium constant expressions. Equilibrium processes can involve substances in two or more phases. For example the equilibrium

$$2\,C(s) + O_2(g) \rightleftharpoons 2\,CO(g) \tag{7.23}$$

involves solid carbon and a mixture of gaseous oxygen and carbon monoxide. For this reaction the concentration of the pure component, carbon, is constant. This may at first seem incorrect, but a little thought should help us to understand this fact.

The constant nature of the concentration of pure components can be appreciated by considering pure water. Since pure water has a density of 1.00 g/mL, 1 L of water (1000 mL) will weigh 1000 g. On this basis we can calculate the molar concentration of pure water as follows,

$$\frac{1000 \text{ g } H_2O}{1 \text{ L } H_2O} \times \frac{1 \text{ mol } H_2O}{18.0 \text{ g } H_2O} = \frac{55.5 \text{ mol } H_2O}{1 \text{ L } H_2O} = 55.5 \ M \qquad (7.24)$$

When reference is made to the molarity of some substance, it is always stated on the basis of how much of the substance is present in 1 L of the substance or 1 L of a solution containing the substance. In the case of pure water, calculating the molarity involves determining how much water is in each liter, regardless of how much total water is present. From this point of view the molarity of water in a bucket, a bathtub, or a lake full of pure water is always 55.5 M. The same reasoning applies to quantities less than a liter, such as a glass of water, except we now consider the number of moles of water *that would be present if* there were 1 L. The answer is still 55.5 M.

Returning to our consideration of equation 7.23, the equilibrium constant expression

$$K = \frac{[CO]^2}{[C]^2[O_2]} \qquad (7.25)$$

can be rearranged to the form

$$K[C]^2 = \frac{[CO]^2}{[O_2]} \qquad (7.26)$$

Since both K and $[C]$ are constant, the product $K[C]^2$, must also be a constant K'. The expression given in equation 7.26 can be written as

$$K' = \frac{[CO_2]^2}{[O_2]} \qquad (7.27)$$

which is a simpler equilibrium constant expression. This procedure applies whenever the equilibrium process involves a pure solid or liquid or when one of the components, such as the solvent in a solution, remains essentially constant. We will take advantage of this method in Chapter 8 when ionization constant expressions are developed.

7-2.3 FACTORS THAT INFLUENCE CHEMICAL EQUILIBRIUM

Since chemical equilibrium can be considered the point at which two chemical reactions are occurring in opposite directions at the same rate, then the same factors that influence the rate of reaction can influence the point of chemical equilibrium. *The most important factor in determining the point of chemical equilibrium is the nature of the chemical species involved.* As an example of this we can consider the equilibrium that HCl(g) and HF(g) each establish in aqueous solutions. Both of these substances dissolve in water to produce acidic solutions. HCl totally ionizes in aqueous solutions according to the following equation.

$$HCl(aq) \xrightarrow{H_2O} H^+(aq) + Cl^-(aq) \qquad (7.28)$$

In cases such as this where essentially all of the HCl(aq) is in the form of the product species, $H^+(aq)$ and $Cl^-(aq)$, we can say that the equilibrium is shifted totally toward products. The single arrow pointing toward products also suggests that the reaction has gone to completion.

Despite the apparent close similarity of HF(g) and HCl(g), HF(g) only partially dissociates in aqueous solutions according to the equation,

$$HF(aq) \underset{\longleftarrow}{\overset{H_2O}{\rightleftharpoons}} H^+(aq) + F^-(aq) \qquad (7.29)$$

We say that the equilibrium is shifted toward reactants in equilibrium processes such as this, where very little product is present. The lengths of the two arrows are, at times, used to indicate in which direction the equilibrium is shifted.

It is difficult to predict the direction in which an equilibrium reaction is shifted based only upon the chemical formulae. The examples of HCl and HF illustrate this quite well. However if we know the value of the equilibrium constant for the reaction we can easily determine in which direction the equilibrium is shifted. *A large value for the equilibrium constant indicates that the equilibrium is shifted to a great extent toward the product side.* For example, the equilibrium constant for equation 7.28 is so large that it is considered to be infinity. *Small values (values much less than unity) are associated with equations that are shifted significantly toward reactants.* As an example of this, the equilibrium constant for the reaction

$$2\,NO_2(g) \rightleftharpoons 2\,NO(g) + O_2(g) \qquad (7.30)$$

is equal to 1.8×10^{-14} at 25°C. In this case the predominant species is NO_2.

Moderate values for K *(values of* K *that are near unity) indicate that the equilibrium is not shifted appreciably toward reactants or products.* For example, the value of *K* for the reaction,

$$H_2(g) + CO_2(g) \rightleftharpoons H_2O(g) + CO(g) \qquad (7.31)$$

is 1.6 when the temperature is 986°C. In this case we can conclude that the value obtained by the multiplication of H_2 and CO_2 is approximately equal to the value obtained by the multiplication of H_2O and CO since their ratio is close to unity. More shall be said about the use of equilibrium constants in Chapter 8.

Note that the concentrations of the various chemical species added to a reaction vessel do not alter the value of the equilibrium constant. For example, in reference to the equilibrium between H_2, I_2, and HI as represented by equation 7.20, the equilibrium constant *K* applies at each specific temperature regardless of whether we start with only H_2 and I_2, with HI, with any other combination of two, or with all three. For this equilibrium, *K* has a value of 54.7 at 425°C regardless of the initial concentrations of the various species added.

Be careful not to interpret the constancy of the equilibrium constant as indicating that the concentration of each species is the same regardless of the initial amounts of each substance added to the reaction vessel. For example, let us assume that equal amounts of H_2 and I_2 are added to one vessel and twice as much H_2 as I_2 is added to the other. The final concentrations of H_2 and I_2 in the two vessels will not be the same. In the first vessel the concentrations of H_2 and I_2 will always be equal since they combine on a one-to-one basis as indicated by equation 7.20. In the second vessel, two concentrations will never be the same and in fact the ratio of H_2 to I_2 will become increasingly large as the equilibrium point is approached. In general, the concentration of both H_2 and I_2 will be different than the amounts in the first vessel. In each vessel, however, the amounts of H_2, I_2, and HI will be such that the calculation of K using equation 7.21 gives a value of 54.7 when the vessels are at a temperature of 425°C. More will be said about such calculations in Chapter 8.

The differing amounts of H_2, I_2, and HI in the two vessels just referred to can be considered to be associated with two different points of equilibrium. In general, the amounts of reactants and/or products added to a reaction vessel affect the point of equilibrium because the concentrations of all species involved at equilibrium depend upon the initial concentrations present. In addition, the point of equilibrium of a reaction in equilibrium can be shifted to some new point by the addition or removal of one or more of the various chemical species involved. For example, if more H_2 is added to either of the two reaction vessels after equilibrium has been reached some of the remaining I_2 molecules will combine with some of the added H_2 molecules to produce additional HI until a new point of equilibrium is achieved. In this case, the value of K has not changed but the concentration of each species has shifted. We refer to such a change in concentration by saying that the equilibrium has shifted toward products since the concentration of the product, HI, has increased while the concentration of the reactant, I_2, has decreased.

From the point of view of chemical kinetics the added H_2 increases the forward rate of reaction (the rate of reaction of H_2 with I_2 to form HI). Therefore, the reaction is shifted toward products. As this shifting occurs there is a resulting increase in the reverse rate as the concentration of HI increases. At some point the two rates once again become equal and equilibrium is once again established. The new point of equilibrium will be associated with the new concentrations of reactants and products but will be due to the fact that the forward and reverse rates of reaction are equal.

In general, *the addition of one or more reactant species to a chemical system in equilibrium will shift the equilibrium toward products while the addition of one or more product species will shift the equilibrium toward reactants.* Conversely, *removal of some reactant particles will shift the equilibrium toward the reactant side and removal of product particles shifts the equilibrium toward the product side.*

These shifts in equilibrium can also be explained through LeChatelier's principle. With the addition of more H_2 to the equilibrium represented by equation 7.20, there was an excess amount of H_2 present, placing stress on the equilibrium. By shifting the reaction to the product side, some of the excess H_2 was used up by combining with some of the I_2 and a new point of equilibrium was established. Removal of a substance creates stress because there is a deficiency of that species and the equilibrium shifts to the side of the equilibrium that has lost some substance to make up part of that deficiency. Therefore, removal of I_2 in equation 7.20 will cause a shift in the equilibrium toward reactants.

LeChatelier's principle states that whenever stress is brought to bear on a system at equilibrium, the system will adjust itself to a new point of equilibrium so as to minimize the stress.

The temperature has an effect on most equilibrium processes and also changes the value of the equilibrium constant. In general, *an increase in temperature shifts the equilibrium toward the product side if the reaction is endothermic and toward the reactants if the reaction is exothermic. Lowering the temperature* has the opposite effect, *shifting endothermic reactions toward the reactants and exothermic reactions toward the products.* This is essentially the same as the effect of temperature on the solubility of substances (see Chapter 6).

As an illustration of this effect, equation 7.31 is an endothermic reaction and can be written in the form,

$$\text{heat} + H_2(g) + CO_2(g) \rightleftharpoons CO(g) + H_2O(g) \tag{7.32}$$

An increase in temperature would be expected to shift the equilibrium toward products. In support of this prediction, the equilibrium constant is 0.10 at 417°C and 1.6 at 986°C. The higher value of K at the higher temperature indicates a shift toward products as the temperature is increased.

We can also use LeChatelier's principle to explain the effect of temperature on the equilibrium. An increase in temperature favors a shift that uses up heat and a decrease in temperature favors a shift that yields heat. For endothermic reactions this means a shift toward products, while exothermic reactions will shift toward reactants.

Pressure changes have very little effect on equilibrium processes involving liquids and solids. However, for equilibrium processes occurring in the gas phase a pressure change can affect the equilibrium point significantly. In general, *an increase in pressure shifts a gaseous equilibrium to the side of the equation that has the fewest unit particles and a decrease in pressure shifts the equilibrium to the side with the greatest number of particles.* Basically, the pressure effect is an effect associated with the higher or lower concentrations brought about by the change in volume accompanying the change in pressure. Applying this reasoning, we know that the equilibrium

$$N_2(g) + 3 H_2(g) \rightleftharpoons 2 NH_3(g) \tag{7.33}$$

will shift toward the product side if the pressure is increased since there are 2 unit particles on the product side and $1 + 3 = 4$ unit particles on the reactant side. This effect can be explained by LeChatelier's principle. When stress is applied, due to an increase in pressure, it can be relieved by reducing the number of unit particles exerting a pressure within the reaction vessel. Therefore an increase in pressure shifts the equilibrium to the side with the fewer unit particles. Decreasing the pressure will have the opposite effect. For an equilibrium such as

$$H_2(g) + CO_2(g) \rightleftharpoons H_2O(g) + CO(g) \tag{7.34}$$

the number of unit particles for the product side is equal to that on the reactants side. In cases such as this an increase in total applied pressure has no effect. This makes sense considering LeChatelier's principle, because no pressure stress is relieved by a shift in the equilibrium.

The last factor to be considered is the effect of a catalyst. We know that catalysts increase the rate of reaction. This applies equally to both the forward rate and reverse rate. The net effect is that there is no change in the point of equilibrium by the addition of a catalyst, but equilibrium simply is reached more rapidly.

PROBLEMS

9. State LeChatelier's principle.
10. What is the equilibrium constant expression for the equilibrium

$$2 NO_2(g) \rightleftharpoons N_2O_4(g)$$

11. How does the equilibrium constant expression for the equilibrium*

$$CO(g) + H_2O(g) \rightleftharpoons CO_2(g) + H_2(g)$$

compare with that associated with equation 7.34?
12. The reaction

$$N_2(g) + 3 H_2(g) \longrightarrow 2 NH_3(g)$$

is an exothermic process. What effect will the following have upon the system when it is at equilibrium?
(a) addition of $H_2(g)$ (b) removal of $N_2(g)$ (c) an increase in temperature (d) a decrease in pressure (e) addition of a catalyst

---

ANSWERS TO PROBLEMS **9.** Whenever a stress is brought to bear on a system at equilibrium, the system will adjust itself to a new point of equilibrium so as to minimize the stress.

10. $K = \dfrac{[N_2O_4]}{[NO_2]^2}$ **11.** it is the inverse of that for equation 7.34 **12. (a)** shift the equilibrium to the right **(b)** shift the equilibrium to the left **(c)** shift the equilibrium to the left **(d)** shift the equilibrium to the left **(e)** none other than speeding up the attainment of equilibrium

---

SUMMARY

In this chapter we have considered two of the most important topics in chemistry, chemical kinetics and chemical equilibrium. The latter topic shall be of particular importance to us and we will consider it in more detail in Chapter 8 as we discuss various equilibrium processes involving ions in aqueous solutions. Many processes in the human body fall into this category and we will consider a few of the more important examples in Chapter 8.

ADDITIONAL PROBLEMS

13. What is the branch of chemistry that deals with the rate of reactions and the stepwise process they follow?
14. What is the overall order for the reactions given below as represented by the rate equations that follow each equation?

(a) $2 NO(g) + Cl_2(g) \longrightarrow 2 NOCl(g)$,
$$Rate = k[NO]^2[Cl_2]$$
(b) $CH_3CH_2NO_2(aq) + OH^-(aq) \longrightarrow H_2O(l) + CH_3CHNO_2^-(aq)$,
$$Rate = k[CH_3CH_2NO_2][OH^-]$$

---

*Comparison of this equilibrium with that given by equation 7.34 shows that reactants and products in the two equations have reversed sides. Always be careful to write the equilibrium constant expression based upon the equation currently under consideration.

(c) $SO_2Cl_2(g) \longrightarrow SO_2(g) + Cl_2(g)$,
$$Rate = k[SO_2Cl_2]$$
(d) $CO(g) + Cl_2(g) \longrightarrow COCl_2(g)$,
$$Rate = k[CO][Cl_2]^{1/2}$$

15. What effect will raising the temperature by 20°C (first by 10°C, and then by 10°C more) have on the rate of reaction of many reactions?

16. What overall effect does a chemical reaction have on a catalyst?

17. Collision theory states that most reactions occur only when reacting molecules have a certain acceptable orientation. What effect does this have on the actual rate of reaction compared with the anticipated rate of reaction that is based upon the number of collisions of reacting molecules?

18. Draw a potential-energy versus progress-of-reaction diagram for an endothermic reaction and label the following positions.
    (a) reactants   (b) activated complex
    (c) products

19. In the diagram in Problem 18, show the effect of a catalyst by drawing a dashed curve in the plot.

20. From the point of view of collision theory, why does an increase in temperature increase the rate of reaction?

21. What is the equilibrium constant expression for the following equilibrium?
$$N_2(g) + 3\ H_2(g) \longrightarrow 2\ NH_3(g)$$

☑ 22. Write the equilibrium constant expression for the following equilibrium.
$$N_2H_4(l) + O_2(g) \longrightarrow$$
$$N_2(g) + 2\ H_2O(g) + heat$$

23. What effect will each of the following changes have upon the equilibrium in Problem 22?
    (a) removal of N_2 (b) addition of H_2O
    (c) increase in temperature   (d) increase in pressure   (e) addition of a catalyst

CHAPTER 8

ACIDS, BASES, AND IONIC EQUILIBRIA

The living system is composed of thousands of chemical substances, each affecting or being affected by the processes taking place in the body. Two of the more important types of substances are the acids and bases that determine the acidity of various body fluids such as blood and urine. The maintenance of correct levels of acidity in the living system is crucial to life. Equilibrium processes similar to those discussed in the last chapter act to control these levels. In this chapter we will extend our consideration of acids and bases, the equilibrium that exists between them, and the action of substances called buffers in controlling the proper levels of acidity in the living system.

8-1 CONCEPTS OF ACIDS AND BASES

LEARNING OBJECTIVES

1. Define and identify acids and bases according to the Arrhenius concept and the Brønsted-Lowry concept, including conjugate acids and conjugate bases.	2. Define neutralization and titration and solve for the normalities and volumes of solutions of acids and bases involved in a titration.

In this section we will broaden our understanding of acids and bases as we extend our inquiry from the limited definitions of Arrhenius first introduced in Chapter 4 to those of Brønsted and Lowry. We will also consider in greater depth the neutralization reaction between acids and bases introduced in Chapter 4 and discuss what is known as titration. We will find that solvation of hydrogen ions by water occurs to produce ions such as the hydronium ion (H_3O^+). The concepts discussed in this section provide a foundation upon which we will build in later sections in this chapter and throughout the remainder of the book. The applications of these basic principles in the health-related sciences are numerous and the knowledge you develop of these basic principles now will greatly expand your knowledge and appreciation of this later material.

8-1.1 ARRHENIUS CONCEPT OF ACIDS AND BASES

Acids and bases were introduced in Chapter 3 and their reaction with one another was considered as one type of double replacement reaction in Chapter 4. These discussions were based upon the Arrhenius concept of acids and bases. *An Arrhenius acid is a hydrogen-containing compound that produces hydrogen ions (H^+) in solution. An Arrhenius base is an OH-containing compound that produces hydroxide ions (OH^-) in solution.*

Hydrochloric acid, HCl(aq), is a typical Arrhenius acid. HCl(aq) is produced in the human body by cells in the lining of the stomach. This acid serves several functions in the body, including acting as an acid in the digestion process and inhibiting the multiplication of bacteria. In excess, HCl(aq) is the source of stomach upset (acid stomach). As discussed in Chapter 7, HCl(aq) totally ionizes into aqueous solutions according to the equation

$$HCl(aq) \xrightarrow{H_2O} H^+(aq) + Cl^-(aq) \qquad (8.1)$$

Arrhenius acids can be placed in two general categories: inorganic acids such as HCl, H_2SO_4, HNO_3, and H_2CO_3, and organic acids such as formic acid (HCOOH), the poisonous substance in the sting of an ant, and acetic acid (CH_3COOH), the principal component of vinegar (see Section 16-1). As noted in Section 3-3.3 the formula of an inorganic acid is written with the hydrogen appearing first in the formula. Organic acids are carbon-containing acids identifiable by the presence of the carboxyl group, COOH, which appears last in the formula. The formula for an organic acid occasionally is written following the inorganic method, giving the carboxyl hydrogen first. On this basis we have $HCHO_2$ for formic acid and $HC_2H_3O_2$ for acetic acid. The authors will, in general, write the formula of an organic acid using the first method.

Arrhenius bases are generally metallic hydroxides. In this text the consideration of those bases shall be limited to this class of compounds. A common Arrhenius base is magnesium hydroxide, $Mg(OH)_2$. Although $Mg(OH)_2$ is only very slightly soluble in water, the part that does dissolve exists as ions of Mg^{2+} and OH^-.

$$Mg(OH)_2(s) \rightleftharpoons Mg^{2+}(aq) + 2\ OH^-(aq) \qquad (8.2)$$

8-1.2 NEUTRALIZATION REACTIONS

The complete reaction of an acid with a base is known as neutralization. This may be illustrated by the equation for the reaction of $Mg(OH)_2(aq)$ with HCl(aq)

$$2\ HCl(aq) + Mg(OH)_2(aq) \longrightarrow 2\ H_2O(l) + MgCl_2(aq) \qquad (8.3)$$

Neutralization is the chemical reaction of an acid with a base in which equivalent amounts of each combine to form a salt and water.

If neutralization has occurred, only $H_2O(l)$ and $MgCl_2(aq)$ will remain. Equation (8.3) forms the basis upon which milk of magnesia (a suspension of $Mg(OH)_2$) acts as an antacid as it neutralizes the excess HCl in the stomach. In practice, a person does not ingest an amount of $Mg(OH)_2$ equivalent to the total $HCl(aq)$ present in the stomach. Doing this would prove fatal. Only a fraction of this amount is normally taken and, therefore, only the excess acid is neutralized.

The neutralization process is used in the laboratory to determine the amount of acid or base present in a solution. The acid or base of unknown concentration is neutralized by the addition of the other species from a solution of known concentration. This process is called titration. Acid-base titration calculations are similar to the dilution problems considered in Chapter 6.

Titration is the term applied to the determination of the concentration of a solution of an acid, base, oxidizing agent, or reducing agent by the measured addition of an appropriate reagent of known concentration until the equivalence point is reached.

EXAMPLE 8.1   Gastric juice, the digestive fluid produced in the stomach, contains hydrochloric acid. If a 20.0 mL sample of gastric juice having a concentration of 0.063 M is titrated with 3.0×10^{-6} M $Mg(OH)_2$, how many milliliters of $Mg(OH)_2$ solution will be required?

For acid-base titration problems we usually use concentrations in terms of normality since we are interested in the reacting capacities of the species. Therefore the first step is to convert the concentration given as molarity to normality as discussed in Chapter 6. In this earlier discussion we learned that for acids and bases the normality is related to the molarity through the equation

$$N = M \times \text{Number of ionizable H's (or OH's)}$$

Applying this to the current problem we obtain

$$N(\text{HCl}) = 0.0630 \ M \times 1 \ \text{H}^+ = 0.0630 \ N$$

and $$N(\text{Mg(OH)}_2) = 3.0 \times 10^{-6} \ M \times 2 \ \text{OH}^- = 6.0 \times 10^{-6} \ N$$

The equation applicable to titration problems is comparable to equation 6.17 used for dilution problems since the number of equivalents of acid is equal to the number of equivalents of base when the titration is complete. Therefore we have

$$V_{\text{acid}} \times N_{\text{acid}} = V_{\text{base}} \times N_{\text{base}} \qquad \text{(8.4)}$$

where V is the volume of acid or base and N is the normality of each. Solving for the unknown volume of base using equation 8.4 gives

$$V_{\text{base}} = \frac{V_{\text{acid}} \times N_{\text{acid}}}{N_{\text{base}}} = \frac{20.0 \ \text{mL} \times 0.0630 \ N}{6.0 \times 10^{-6} \ N} = 2.1 \times 10^5 \ \text{mL} = 210 \ \text{L*}$$

8-1.3 THE HYDRATION OF HYDROGEN IONS

Let us now return to the statement given previously that acids produce hydrogen ions in solution. This is not strictly correct because individual hydrogen ions do not exist as separate particles in the solution. The hydrogen ions (the protons)

---

*We calculated a large volume of base and used an extremely diluted solution of $Mg(OH)_2$ in this example because of the very limited solubility of $Mg(OH)_2$. Milk of magnesia products sold commercially consist of solid $Mg(OH)_2$ suspended in an aqueous solution. In an actual laboratory titration, a more soluble base such as NaOH is commonly used.

formed when the acid molecule dissociates were solvated as they combined with solvent molecules to form the solvated species, HSol$^+$, where Sol represents solvent (see Chapter 6). In the case of an aqueous solution the special term *hydration* is used in place of the term *solvation* and the hydronium ion (HH_2O^+ or H_3O^+) is formed.* On this basis, equation 8.1 is written

$$HCl(aq) + H_2O(l) \longrightarrow H_3O^+(aq) + Cl^-(aq) \qquad \textbf{(8.5)}$$

Both methods (equations 8.1 or 8.5) are frequently used in writing the reaction of an acid with water; however, *it should always be understood that the hydrogen ion actually exists in the form of solvated ions in the form of solvated ions in aqueous solutions.*

8-1.4 THE BRØNSTED-LOWRY CONCEPTS OF ACIDS AND BASES

In 1923 Johannes Brønsted of Denmark and J. M. Lowry of England independently proposed a concept of acids and bases that is more general than that of Arrhenius. Their definition of an acid, a **Brønsted-Lowry acid,** is based upon the tendency of one substance to donate protons to another substance. Similarly, the tendency of substances to accept protons provides the definition of a **Brønsted-Lowry base.** An example of these two definitions is when ammonia (NH_3) in the presence of cyanide ions (CN^-) donates protons to the cyanide ions according to the equation

A **Brønsted-Lowry acid** is a substance that donates a proton in a chemical reaction.

A **Brønsted-Lowry base** is a substance that accepts a proton in a chemical reaction.

$$NH_3 + CN^- \rightleftharpoons NH_2^- + HCN \qquad \textbf{(8.6)}$$

In this reaction NH_3 is acting as the Brønsted-Lowry acid since it donates a proton and CN^- is acting as the Brønsted-Lowry base since it accepts a proton. In the reaction given by equation 8.6 no hydrogen ions (or H_3O^+ ions) were produced, therefore NH_3 was not acting as an Arrhenius acid. No OH^- ions were produced and so CN^- did not act as an Arrhenius base. The Brønsted-Lowry concept, therefore, broadens the definitions of acids and bases to include substances that are not included within the more narrow Arrhenius concept. The Brønsted-Lowry concept is like a larger umbrella that not only covers those substances that qualify as Arrhenius acids and bases but also includes additional substances not covered by the "Arrhenius umbrella." Figure 8.1 illustrates the greater range of the Brønsted-Lowry definition.

Arrhenius acids, like HCl, *are also acids from the Brønsted-Lowry point of view,* since they can donate protons as illustrated in equations 8.3 and 8.5. *Arrhenius bases,* such as $Mg(OH)_2$, *are also Brønsted-Lowry bases* since the OH^- ion readily accepts protons as illustrated by equation 8.3 and the equation

$$CH_3COOH + OH^- \rightleftharpoons CH_3COO^- + H_2O \qquad \textbf{(8.7)}$$

One special class of compounds, amines, acts as Brønsted-Lowry bases but not as Arrhenius bases. Amines have the general formula RNH_2, where R usually consists

---

*In reality a hydrogen ion, or proton, exists as a mixture of more complex species represented by $H(H_2O)_x{}^+$, where x may be one or more. For simplicity, however, the simpler formulas H^+ or H_3O^+ shall be used throughout the text.

Figure 8.1
The Brønsted-Lowry concept of acids and bases acts like a "larger umbrella" to include some substances not covered by the "Arrhenius acid-base umbrella."

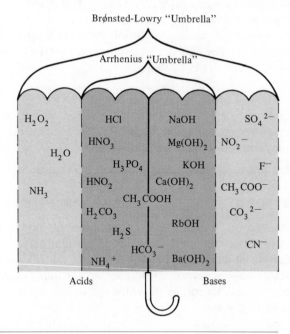

of a carbon-containing group such as CH_3. For example, methyl amine (CH_3NH_2) behaves as a Brønsted-Lowry base as illustrated by the equation

$$HCl + CH_3NH_2 \longrightarrow CH_3NH_3{}^+ + Cl^- \qquad \textbf{(8.8)}$$

Amines are very important in the living system and more shall be said about their tendency to act as bases in Section 17-3.

Special terms are applied to the products formed when Brønsted-Lowry acids and bases combine. The substance formed when a Brønsted-Lowry base gains a proton is referred to as a **conjugate acid.** In equation 8.8, $CH_3NH_3{}^+$ is the conjugate acid. In a similar manner the substance formed when a Brønsted-Lowry acid loses a proton is called a **conjugate base.** In equation 8.8, Cl^- is the conjugate base. *The formula of the conjugate base of a Brønsted-Lowry acid is written by subtracting a proton from the formula of the acid.* For example, the conjugate base of $HCO_3{}^-$ is $CO_3{}^{2-}$. *The conjugate acid of a Brønsted-Lowry base is obtained by adding a proton to the formula of the base.* For CN^- the conjugate acid is, therefore, HCN. Table 8.1 gives a list of Brønsted-Lowry acids and their conjugate bases.

It is important to note that *whether a substance is acting as an acid or base from the Brønsted-Lowry point of view depends upon the other chemical species that are present.* In the presence of HCl, water behaves as a Brønsted-Lowry base as shown in equation 8.5. In the presence of NH_3, water behaves as a Brønsted-Lowry acid and reacts according to the equation

$$NH_3(aq) \quad + \quad H_2O \quad \rightleftharpoons \quad NH_4{}^+(aq) \quad + \quad OH^- \quad \textbf{(8.9)}$$

| Brønsted-Lowry base | Brønsted-Lowry acid | conjugate acid | conjugate base |

More shall be said about this in the next section.

A **conjugate acid** is the substance formed when a Brønsted-Lowry base accepts a proton.

A **conjugate base** is the substance formed when a Brønsted-Lowry acid loses a proton.

Table 8.1 Relative Strengths of Acids and Bases

Acid		$-\text{H}^+$ →	Conjugate Base
Name	Formula	Formula	Name
Perchloric acid	$HClO_4$	ClO_4^-	Perchlorate ion
Sulfuric acid	H_2SO_4	HSO_4^-	Hydrogen sulfate ion
Hydroiodic acid	HI	I^-	Iodide ion
Hydrobromic acid	HBr	Br^-	Bromide ion
Hydrochloric acid	HCl	Cl^-	Chloride ion
Nitric acid	HNO_3	NO_3^-	Nitrate ion
Hydronium ion	H_3O^+	H_2O	Water
Hydrogen sulfate ion	HSO_4^-	SO_4^{2-}	Sulfate ion
Phosphoric acid	H_3PO_4	$H_2PO_4^-$	Dihydrogen phosphate ion
Nitrous acid	HNO_2	NO_2^-	Nitrite ion
Hydrofluoric acid	HF	F^-	Fluoride ion
Acetic acid	CH_3COOH	CH_3COO^-	Acetate ion
Carbonic acid	H_2CO_3	HCO_3^-	Hydrogen carbonate ion
Hydrogen sulfide	H_2S	HS^-	Hydrosulfide ion
Hydrocyanic acid	HCN	CN^-	Cyanide ion
Ammonium ion	NH_4^+	NH_3	Ammonia
Hydrogen carbonate ion	HCO_3^-	CO_3^{2-}	Carbonate ion
Hydrogen peroxide	H_2O_2	HO_2^-	Hydroperoxide ion
Water	H_2O	OH^-	Hydroxide ion
Ammonia	NH_3	NH_2^-	Amide ion
Hydroxide	OH^-	O^{2-}	Oxide ion

Increasing Acid Strength (left). Increasing Base Strength (right). Conjugate Acid ⟶ Base, $+\text{H}^+$

*Acids enclosed within this brace are the strong acids. They exist only as ions in dilute aqueous solutions.

PROBLEMS

1. Define the following terms.
   (a) Arrhenius acid   (b) neutralization   (c) Brønsted-Lowry base
2. Classify the following substances as Arrhenius acids, Arrhenius bases, or neither.
   (a) $NaOH$ (b) H_3PO_4 (c) $C_2H_5NH_2$ (d) $Ba(OH)_2$
   (e) C_3H_7OH (f) $HClO_4$ (g) $Al(OH)_3$ (h) $LiOH$
   (i) KOH (j) HIO (k) $C_6H_5NH_2$ (l) $NaCl$
   (m) H_2SO_3 (n) HCO_3^- (o) $C_5H_{11}OH$ (p) KNO_3
3. What volume of a 0.100 M solution of NaOH is needed to titrate 400 mL of a solution containing citric acid,

$$HOOC-CH_2-\underset{\underset{COOH}{|}}{\overset{\overset{OH}{|}}{C}}-CH_2-COOH$$

   with a concentration of 0.250 M? (Note: Citric acid has three acidic carboxyl groups.)
4. Identify and label the Brønsted-Lowry **and** conjugate acids and bases in each of the following equations:
   (a) $HNO_3(aq) + H_2O(l) \longrightarrow H_3O^+(aq) + NO_3^-(aq)$
   (b) $NH_3(aq) + HCl(aq) \longrightarrow NH_4^+(aq) + Cl^-(aq)$
   (c) $HCl(aq) + F^-(aq) \longrightarrow Cl^-(aq) + HF(aq)$
   (d) $HCO_3^-(aq) + H_2O(l) \longrightarrow CO_3^{2-}(aq) + H_3O^+(aq)$
   (e) $HCO_3^-(aq) + HI(aq) \longrightarrow H_2CO_3(aq) + I^-(aq)$

ANSWERS TO PROBLEMS   **1. (a)** a compound that produces hydrogen ions in solution **(b)** the chemical reaction of an acid with a base in which equivalent amounts of each substance combine to leave only a salt and water   **(c)** a substance that accepts a proton in a chemical reaction   **2.** Arrhenius acids, b, f, j, m, n; Arrhenius bases, a, d, g, h, i; neither, c, e, k, l, o, p   **3.** 3000 mL   **4.** Brønsted-Lowry acids: **(a)** HNO_3 **(b)** HCl **(c)** HCl   **(d)** HCO_3^- **(e)** HI   Brønsted-Lowry bases: **(a)** H_2O **(b)** NH_3 **(c)** F^- **(d)** H_2O **(e)** HCO_3^- Conjugate acids: **(a)** H_3O^+ **(b)** NH_4^+ **(c)** HF   **(d)** H_3O^+ **(e)** H_2CO_3 Conjugate bases: **(a)** NO_3^- **(b)** Cl^- **(c)** Cl^- **(d)** CO_3^{2-} **(e)** I^-

8-2 STRENGTHS OF ACIDS AND BASES

LEARNING OBJECTIVES

1. Write ionization-constant expressions for acids and bases based upon the corresponding chemical equations.	2. Compare the relative strengths of acids and bases based upon the value of the corresponding ionization constant.

In this section we will build upon the foundation established in Chapter 7 and in Section 8-1. We will develop a deeper understanding of the concepts of strong and weak electrolytes as applied to acids and bases and of the use of equilibrium-constant expressions and the ionization constants to obtain some idea of the relative strengths of acids and bases. This material will add to our general knowledge of acids and bases, enabling us to understand better such important topics as blood pH and buffers that are presented in later sections of this chapter.

8-2.1 RELATIVE STRENGTHS OF ACIDS AND BASES

As indicated in equation 8.5, HCl totally ionizes in water to form H_3O^+ and Cl^- ions. Similarly, the base sodium hydroxide (NaOH), known commercially as lye, totally dissociates in water to produce Na^+ and OH^- ions. Many other acids and bases behave in a similar manner. As discussed in Chapter 6 such substances are examples of strong electrolytes. When the substance is an acid or a base, the terms strong acid or strong base frequently are used. *Defining the strength of an acid or base in terms of its degree of ionization, or dissociation into ions is, basically, an Arrhenius concept.* From this point of view, a weak acid, such as lactic acid, is an

> A **strong acid** is an acid that is totally dissociated in an aqueous solution to produce H_3O^+ ions (Arrhenius concept).
>
> A **strong base** is a base that is totally dissociated in aqueous solution to produce OH^- ions (Arrhenius concept).
>
> A **weak acid** is an acid that is only partially dissociated in an aqueous solution to establish an equilibrium between the undissociated species and hydronium ions (Arrhenius concept).

$$\underset{\text{lactic acid}}{\underset{|}{CH_3CHCOOH}}$$
$$OH$$

acid that only partially ionizes in aqueous solution to establish the equilibrium

$$CH_3CHCOOH(aq) + H_2O \rightleftharpoons H_3O^+(aq) + CH_3CHCOO^-(aq) \qquad \textbf{(8.10)}$$
$$\quad\;| \qquad\qquad\qquad\qquad\qquad\qquad\qquad\qquad\qquad\quad |$$
$$\quad OH \qquad\qquad\qquad\qquad\qquad\qquad\qquad\qquad\qquad OH$$

This partial ionization is a common feature of organic acids. Similarly, some bases are weak bases since they only partially ionize in an aqueous solution. For example, silver hydroxide (AgOH) establishes the equilibrium

$$AgOH(aq) \overset{H_2O}{\rightleftharpoons} Ag^+(aq) + OH^-(aq) \qquad \textbf{(8.11)}$$

A **weak base** is a base that is only partially dissociated in an aqueous solution to establish an equilibrium between the undissociated species and hydroxide ions (Arrhenius concept).

in aqueous solutions with most of the compound existing in the un-ionized form.

From the Brønsted-Lowry point of view, a strong acid is an acid that readily donates protons to other substances. A weak acid is an acid that has a limited tendency to donate protons. Comparison of equations 8.5 and 8.10 indicates that HCl is a much stronger acid than lactic acid because all HCl molecules donate their protons to the solvent (water), while only a fraction of the lactic acid molecules do so.

The strength of a base, from the Brønsted-Lowry point of view, depends upon the relative affinity of the substance for protons. For example, the oxide ion (O^{2-}) is a very strong base. When a salt such as Na_2O is dissolved in water, the oxide ion reacts with the water according to the equation

$$O^{2-} + H_2O \rightleftharpoons 2OH^- \qquad \textbf{(8.12)}$$

This is the basis upon which the statement, *oxides of the more reactive metals combine with water to produce bases,* was made in Section 4-2. The chloride ion is an example of a very weak Brønsted-Lowry base. When a salt such as NaCl is added to water, there is no measurable tendency for the Cl^- ion to take, or accept, protons from the water.

$$Cl^-(aq) + H_2O(l) \longrightarrow N.R. \qquad \textbf{(8.13)}$$

As we have seen in the previous discussion, acids and bases vary in their relative strengths with respect to donating and accepting protons. Table 8.1 lists some of the more common Brønsted-Lowry acids and bases in order of relative strength. The higher the acid is on the list the greater is its tendency to donate protons. The lower the base is on the list, the greater is its tendency to accept protons. As we read Table 8.1 from left to right, a proton is removed from the acid to yield the conjugate base. For example, to obtain the conjugate base of perchloric acid we have, $HClO_4$ minus H^+ gives ClO_4^-. Reading Table 8.1 from right to left, a proton is added to the base to yield the conjugate acid. The conjugate acid of the hydroxide ion is, therefore, OH^- plus H^+, giving H_2O. We can see that *the stronger the acid, the weaker is its conjugate base* and *the weaker the acid, the stronger is its conjugate base. The stronger the base, the weaker is its conjugate acid* and *the weaker the base, the stronger is its conjugate acid.* For example, the strong base OH^- has an extremely weak conjugate acid, H_2O; the weak acid HCN has a relatively strong conjugate base, CN^-; and the strong acid HNO_3 has a very weak conjugate base, NO_3^-. The characteristic of weak acids and bases to have strong conjugate bases and acids is important in what is known as hydrolysis (see Section 8-4).

In Section 8-1.3 it was mentioned that water has the capability of acting as either an acid or a base, depending upon the nature of the other species in the solution. Many hydrogen-containing substances exhibit this characteristic. The relative strengths of the two substances involved determine whether the hydrogen-containing substance will act as an acid or a base in the reaction. For example, in the presence of HNO_2, HCO_3^- tends to react according to the equation

$$HCO_3^- + HNO_2 \rightleftharpoons H_2CO_3 + NO_2^-$$

since HNO_2 is a stronger acid (proton donor) than HCO_3^- (see Table 8.1). On the other hand, in the presence of NH_3, HCO_3^- tends to react according to the equation

$$HCO_3^- + NH_3 \rightleftharpoons CO_3^{2-} + NH_4^+$$

since HCO_3^- is a stronger acid than NH_3 (see Table 8.1). *In each case an equilibrium is established, with the point of equilibrium depending upon the relative strengths of the substances involved.*

8-2.2 IONIZATION CONSTANTS

In Chapter 7 we found that equilibrium constants provide a convenient measure of the relative amounts of reactants and products involved in an equilibrium process. The same principles apply to ionic equilibria. In Section 7-2.2 it was pointed out that when an equilibrium constant expression includes concentration terms for solids, pure liquids, or the solvent in a solution, the concentration of these species can be incorporated into the equilibrium constant. Equation 7.27 was derived in this manner. This same procedure becomes important when we write equilibrium-constant expressions for ionization processes. This can be illustrated for the ionization of acetic acid.

$$CH_3COOH(aq) + H_2O(l) \rightleftharpoons H_3O^+(aq) + CH_3COO^-(aq) \qquad \textbf{(8.14)}$$

Writing an equilibrium constant expression for this process, we have

$$K = \frac{[H_3O^+][CH_3COO^-]}{[H_2O][CH_3COOH]} \qquad \textbf{(8.15)}$$

This expression is a perfectly valid equilibrium constant expression that provides a measure of the relative amounts of reactants and products. However, chemists commonly use a simpler form of the equilibrium expression that is based upon the fact that the concentration of water (55.5 M) varies only negligibly during the reaction. This form is obtained by moving the term representing the concentration of water in equation 8.15 to the left side of the equal sign. This results in the expression

$$K[H_2O] = K_i = \frac{[H_3O]^+[CH_3COO^-]}{[CH_3COOH]} \qquad \textbf{(8.16)}$$

which is known as the *ionization constant expression*. Since the product $K[H_2O]$ is essentially constant, it can be expressed in terms of K_i, the *ionization constant*. Values of K_i for some typical acids are listed in Table 8.2. These values sometimes are called *acid* ionization constants and are represented by the symbol K_a. Ionization constant expressions can also be written for bases. Applying the principles above to the equilibrium,

$$NH_3(aq) + H_2O(l) \rightleftharpoons NH_4^+(aq) + OH^-(aq) \qquad \textbf{(8.17)}$$

Table 8.2 Ionization Constant Values of Some Common Acids at 25°C

Acid	Reaction	K_a
Perchloric	$HClO_4(aq) + H_2O \longrightarrow H_3O^+(aq) + ClO_4^-(aq)$	∞*
Hydrochloric	$HCl(g) + H_2O \longrightarrow H_3O^+(aq) + Cl^-(aq)$	∞
Sulfuric	$H_2SO_4(aq) + H_2O \longrightarrow H_3O^+(aq) + HSO_4^-(aq)$	∞
Nitric	$HNO_3(aq) + H_2O \longrightarrow H_3O^+(aq) + NO_3^-(aq)$	∞
Hydrogen sulfate ion	$HSO_4^-(aq) + H_2O \rightleftharpoons H_3O^+(aq) + SO_4^{2-}(aq)$	1.2×10^{-2}
Acetylsalicylic	$CH_3COOC_6H_4COOH(aq) + H_2O \rightleftharpoons H_3O^+(aq) + CH_3COOC_6H_4COO^-(aq)$	3.27×10^{-4}
Formic	$HCOOH(aq) + H_2O \rightleftharpoons H_3O^+(aq) + HCOO^-(aq)$	1.77×10^{-4}
Carbonic	$H_2CO_3(aq) + H_2O \rightleftharpoons H_3O^+(aq) + HCO_3^-(aq)$	1.70×10^{-4}
Lactic	$CH_3\underset{\underset{OH}{\mid}}{C}HCOOH(aq) + H_2O \rightleftharpoons H_3O^+(aq) + CH_3\underset{\underset{OH}{\mid}}{C}HCOO^-(aq)$	1.38×10^{-4}
Acetic	$CH_3COOH(aq) + H_2O \rightleftharpoons H_3O^+(aq) + CH_3COO^-(aq)$	1.79×10^{-5}
Boric	$H_3BO_3(aq) + H_2O \rightleftharpoons H_3O^+(aq) + H_2BO_3^-(aq)$	6.4×10^{-10}
Hydrocyanic	$HCN(aq) + H_2O \rightleftharpoons H_3O^+(aq) + CN^-(aq)$	4.90×10^{-10}
Hydrogen carbonate ion	$HCO_3^-(aq) + H_2O \rightleftharpoons H_3O^+(aq) + CO_3^{2-}(aq)$	6.31×10^{-11}

*The symbol ∞ represents the quantity infinity.

we have

$$K_i = K_b = \frac{[NH_4^+][OH^-]}{[NH_3]} \qquad \textbf{(8.18)}$$

where K_i, or K_b, is the *base ionization constant*. Values of K_b for some of the more common bases are given in Table 8.3.

Ionization constants are useful in providing a measure of the degree of dissociation of a substance; that is, they help us to measure the degree to which the substance has reacted with water. In agreement with what we learned in Chapter 7, a lower value of K_i indicates that the equilibrium is shifted less toward products; that is, the acid or base is weaker. For example, an acid such as HCl, with a K_a value of infinity, is totally ionized as illustrated by the following equation for a 0.100 M solution.

initial concentrations: 0.100 M 55.5 M 0.00 M 0.00 M

$$HCl(aq) + H_2O(l) \longrightarrow H_3O^+(aq) + Cl^-(aq) \quad \textbf{(8.19)}$$

*final concentrations
at equilibrium:*   0 M 55.4 M 0.100 M 0.100 M

For a comparable solution of CH_3COOH with a value of K_a equal to 1.79 $\times 10^{-5}$, we have

*initial
concentrations:* 0.100 M 55.5 M 0.00 M 0.00 M

$$CH_3COOH(aq) + H_2O(l) \rightleftharpoons H_3O^+(aq) + CH_3COO^-(aq) \quad \textbf{(8.20)}$$

*final
concentrations
at equilibrium:* 0.09867 M 55.5 M 0.00133 M 0.00133 M

As we can see in equation 8.20, CH_3COOH is only partially broken up into ions.

Table 8.3   Ionization Constant Values of Some Common Bases at 25°C

Base	Reaction	K_b
Sodium hydroxide	$NaOH(aq) \xrightarrow{H_2O} Na^+(aq) + OH^-(aq)$	∞
Magnesium hydroxide	$Mg(OH)_2(aq) \xrightarrow{H_2O} Mg^{2+}(aq) + 2\,OH^-(aq)$	∞
Ethylamine	$C_2H_5NH_2(aq) + H_2O \rightleftharpoons C_2H_5NH_3^+(aq) + OH^-(aq)$	5.6×10^{-4}
Methylamine	$CH_3NH_2(aq) + H_2O \rightleftharpoons CH_3NH_3^+(aq) + OH^-(aq)$	4.42×10^{-4}
Silver hydroxide	$AgOH(aq) \xrightarrow{H_2O} Ag^+(aq) + OH^-(aq)$	1.1×10^{-4}
Ammonia	$NH_3(aq) + H_2O \rightleftharpoons NH_4^+(aq) + OH^-(aq)$	1.76×10^{-5}
Bicarbonate	$HCO_3^-(aq) + H_2O \rightleftharpoons H_2CO_3(aq) + OH^-(aq)$	2.34×10^{-8}
Pyridine	$C_5H_5N(aq) + H_2O \rightleftharpoons C_5H_5NH^+(aq) + OH^-(aq)$	2.0×10^{-9}
Pyrimidine	$C_4H_4N_2(aq) + H_2O \rightleftharpoons C_4H_4N_2H^+(aq) + OH^-(aq)$	1.56×10^{-14}
Urea	$H_2NCONH_2(aq) + H_2O \rightleftharpoons H_2NCONH_3^+(aq) + OH^-(aq)$	1.5×10^{-14}

Substitution of the final equilibrium concentrations in equation 8.16 yields the appropriate value of K_a.

$$K_a = \frac{(0.00133\ M)(0.00133\ M)}{(0.09867\ M)} = 1.79 \times 10^{-5}$$

Equations 8.19 and 8.20 illustrate the relationship between the degree of ionization and the ionization constant. *A value of infinity for the ionization constant is typical of all strong acids and bases.* On the other hand, *moderately weak acids,* such as carbonic acid and acetic acid, and *moderately weak bases,* such as ammonia, *have K_i values in the range of 10^{-4}–10^{-6}. Extremely weak acids,* such as boric acids, *and bases,* such as urea, *can have K_i values as low as 10^{-10} or less.* Figure 8.2 illustrates the variation of acid and base strength with values of K_i.

The strength of an acid has physiological significance in the living system. For example, if HCl were a weak acid it could not function as well in aiding digestion and inhibiting bacterial growth. Carbonic acid (H_2CO_3) is able to act as a buffer in the blood because it is a relatively weak acid, and boric acid can be used as an antiseptic wash in the eyes since it is extremely weak, while use of a stronger acid would damage the eyes.

PROBLEM

5. Write ionization constant expressions for (a) HCN and (b) $C_2H_5NH_2$ and indicate whether they are strong, moderately weak, or extremely weak. (Equilibrium reactions and ionization constants for these species are given in Tables 8.2 and 8.3.)

ANSWERS TO PROBLEM   **5. (a)** $K_i = \dfrac{[H_3O^+][CN^-]}{[HCN]} = 4.90 \times 10^{-10}$, extremely weak

(b) $K_i = \dfrac{[C_2H_5NH_3^+][OH^-]}{[C_2H_5NH_2]} = 5.6 \times 10^{-4}$, moderately weak

Figure 8.2
The strength of an acid or base is measured
by the value of the ionization constant.

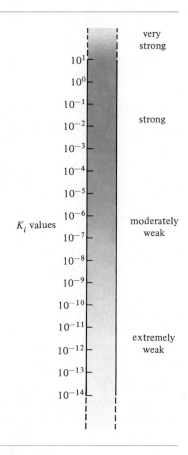

K_i values

10^1	very strong
10^0	
10^{-1}	strong
10^{-2}	
10^{-3}	
10^{-4}	
10^{-5}	
10^{-6}	moderately weak
10^{-7}	
10^{-8}	
10^{-9}	
10^{-10}	
10^{-11}	extremely weak
10^{-12}	
10^{-13}	
10^{-14}	

8-3 IONIZATION OF WATER

LEARNING OBJECTIVES

1. Calculate the concentration of the hydronium ion or the hydroxide ion given the concentration of (a) the other species, (b) the acid, or (c) the base in the reaction.

2. Calculate pH values based upon the hydronium ion or hydroxide ion concentrations.
3. Calculate the hydronium ion concentration associated with a given pH.

In this section we will refer to the concepts of acids, bases, and ionic equilibria developed in the previous sections as we discuss the ionization of water. We will consider the use of pH as a measure of the hydrogen ion (hydronium ion) concentration of solutions. Many of the quantities and concepts learned previously will begin to take on new meaning and relevance in this material. In addition, you will learn about some of the chemistry of the health-related sciences such as blood pH.

8-3.1 THE ION-PRODUCT CONSTANT FOR WATER (K_w)

Water is a major component of living systems. We learned in Section 8-1.4 that water has the important characteristic of being able to act as either a Brønsted-Lowry acid or base. This tendency allows two water molecules to react according to the following equation

$$H_2O + H_2O \rightleftharpoons H_3O^+(aq) + OH^-(aq) \tag{8.21}$$

However, the extent of this reaction is so slight that the degree of ionization is approximately 1 ionization per 500 million molecules of water. This is comparable to only 2 separations, or divorces, in 1 billion marriages. Quite a small fraction!

Despite the very slight degree of ionization of water, *the equilibrium given by equation 8.21 is extremely important and is involved in every aqueous solution regardless of what other equilibria may be present.* Since an equilibrium is involved in equation 8.21, an equilibrium constant expression can be written for the process.

$$K = \frac{[H_3O^+][OH^-]}{[H_2O][H_2O]} \tag{8.22}$$

As in the case of the ionization processes discussed in Section 8-2.2, a simpler form of equilibrium expression can be written by removing terms for the concentration of water, which is essentially constant, from the right side of equation 8.22. On doing this we obtain

$$K[H_2O][H_2O] = K_w = [H_3O^+][OH^-] \tag{8.23}$$

where the resulting constant, K_w, known as the *ion-product constant for water,* remains constant at any given temperature unless the concentrations of the various solutes that may be present become quite large and thus affect the concentration of H_2O. The value of K_w can be determined by considering pure water. Analysis of pure water at 25°C shows that the hydronium ions and hydroxide ions are both 1.0×10^{-7} M. Substituting these values into equation 8.23 gives

$$K_w = [H_3O^+][OH^-] = (1.0 \times 10^{-7})(1.0 \times 10^{-7}) = 1.0 \times 10^{-14}$$

Although the value of 1.0×10^{-14} was obtained for the case with pure water, it applies to all aqueous solutions. The significance of this can be appreciated by considering the effect of adding acetic acid to water. In this case two interrelated equilibria are set up simultaneously, one represented by the horizontal equilibrium, the other by the vertical equilibrium.

$$
\begin{array}{c}
CH_3COOH(aq) \\
+ \\
H_2O \\
\updownarrow \\
H_2O + H_2O \rightleftharpoons H_3O^+(aq) + OH^-(aq) \\
+ \\
CH_3COO^-(aq)
\end{array}
\tag{8.24}*
$$

---

*Equations such as 8.24 are written to emphasize the interrelationship of various equilibria and do not attempt to represent balanced chemical equations. For example, in equation 8.24 there is an extra water on the reactant sides compared with the product sides.

As would be expected, the addition of an acid to water will increase the acidity, or $H_3O^+(aq)$ concentration. As a result of this increase, the horizontal equilibrium, involving water, shifts toward reactants as some of the additional $H_3O^+(aq)$ from the vertical reaction reacts with the part of the $OH^-(aq)$ present in the solution. A decrease in the $OH^-(aq)$ concentration results, but not all traces of $OH^-(aq)$ are removed. The concentration of $OH^-(aq)$ simply decreases until new equilibrium concentrations of each species are reached. At this point, the product of $[H_3O^+]$ and $[OH^-]$ has once again become equal to 1.0×10^{-14}.

EXAMPLE 8.2 What is the equilibrium concentration of OH^- in a solution when enough CH_3COOH has been added to increase the equilibrium concentration of H_3O^+ to 1.0×10^{-3} M?

Rearranging equation 8.23 we have

$$[OH^-] = \frac{K_w}{[H_3O^+]} = \frac{1.0 \times 10^{-14}}{1.0 \times 10^{-3}} = 1.0 \times 10^{-11}\ M$$

As discussed earlier in this section the hydronium-ion and hydroxide-ion concentrations in pure water are both 1.0×10^{-7} M. This concentration is applicable to many other aqueous solutions. For example, the addition of NaCl to pure water does not affect the value of $[H_3O^+]$ and $[OH^-]$ and they remain at 1.0×10^{-7} M after the solution is prepared. *Solutions that have a concentration of $[H_3O^+]$ and $[OH^-]$ equal to 1.0×10^{-7} M are considered to be neutral solutions.* On the other hand, *solutions having a hydronium ion concentration greater than 1.0×10^{-7} M are referred to as acidic solutions,* and solutions in which the $[OH^-]$ concentration is greater than 1.0×10^{-7} M, ($[H_3O^+]$ less than 1.0×10^{-7}) M are considered to be basic solutions.

8-3.2 THE pH AND pK_i SCALES

The concentration of H_3O^+ in solution frequently is quite small. These values can be easily represented by scientific notation such as 1.0×10^{-7} M. However, an alternative method has been developed and is in more general use in science and medicine. This approach is known as the pH scale. This scale defines the pH in terms of the reciprocal, or inverse concentration of H_3O^+ through what is known as a logarithmic relationship. The pH is defined by the equation

$$pH = \log \frac{1}{[H_3O^+]} = -\log [H_3O^+] \qquad \textbf{(8.25)}$$

or, written in the alternative or antilogarithmic form,

$$[H_3O^+] = 10^{-pH} \qquad \textbf{(8.26)}$$

When the pre-exponential portion associated with $[H_3O^+]$ is unity, for example 1×10^{-2}, then equation 8.26 can be used as is and the pH can be obtained by direct comparison. This is illustrated in the following example.

EXAMPLE 8.3 Gastric juice has a hydronium ion concentration range from approximately 1.0×10^{-1} M to 1.0×10^{-3} M. What is the pH range associated with these values?

Since $[H_3O^+] = 10^{-pH}$, comparison of this relationship with the value $[H_3O^+] = 1.0 \times 10^{-1}$ M yields a pH of 1. Comparison with the value of $[H_3O^+] = 1.0 \times 10^{-3}$ M gives a value of 3 for the pH value of the lower concentration. Therefore, the approximate pH range of gastric juice is 1–3.

Most solutions that you will encounter do not have simple hydronium concentrations such as 1×10^{-1} M or 1×10^{-3} M. For example, the hydronium ion concentration in normal blood plasma is approximately 4×10^{-8} M. Determination of pH values for these cases is more complex and we will not consider them in detail. However, we can readily determine the pH range within which such values fall. In the case of blood plasma, the value 4×10^{-8} M lies between 1×10^{-7} M and 1×10^{-8} M; therefore, the pH of blood plasma is between 7 and 8. (The actual pH value is 7.40.)

The opposite calculation, the calculation of the hydronium ion concentration from pH values, is a relatively simple process when the pH is a whole number. In this case the hydronium ion concentration is obtained by substituting the pH value into the equation

$$[H_3O^+] = 10^{-pH}$$

For example, a solution of NaOH having a pH of 12 will have a hydronium ion concentration of 1×10^{-12} M.

For nonwhole number pH values, the same procedure can be used; however, the result is a bit complex. For example the pH of the intercellular fluid in muscles is 6.1. Substituting this value into equation 8.30 gives $[H_3O^+] = 10^{-6.1}$ M. Although this number is a valid exponential number that can be converted into an ordinary exponential number through logarithmic calculations, we will not concern ourselves with the detailed calculation. Suffice it to say that 6.1 is between 6 and 7, and, therefore, the hydronium ion concentration is between 1×10^{-6} M and 1×10^{-7} M. Since 6.1 is close to 6, it would be safe to assume that the H_3O^+ concentration is nearer to 1×10^{-6} M than it is to 1×10^{-7} M. (The actual value is 7.9×10^{-7} M, or 0.79×10^{-6} M, which is almost equal to 1×10^{-6} M.)*

pH values provide a convenient measure of the acidity of a solution. *Acidic solutions have a pH less than 7, basic solutions have a pH greater than 7, and neutral solutions have a pH of 7.* A common error in using pH values is to assume that a higher pH value indicates greater acidity. Actually the reverse is true. Consider the relationship between pH and the hydronium ion concentration, $[H_3O^+] = 10^{-pH}$. As you can see in this expression as the pH value increases, the value of $[H_3O^+]$ decreases. Figure 8.3 illustrates the inverse relationship between $[H_3O^+]$ and the pH.

Table 8.4 lists pH values of some common substances. Comparison of equal concentrations of hydrochloric acid and acetic acid illustrates that weak acids have higher pH values (are less acidic) than strong acids of the same overall concentration. A similar comparison of the weak base NH_3 with NaOH indicates an

---

*Conversion of noninteger pH values into hydronium ion concentrations and vice versa can be done relatively easily using electronic calculators with logarithmic and antilogarithmic function keys or through the use of tables of logarithms.

Figure 8.3
The hydrogen ion concentration and the pH vary in inverse order.

$[H_3O^+]$	pH
10^{-1}	1
10^{-2}	2
10^{-3}	3
10^{-4}	4
10^{-5}	5
10^{-6}	6
10^{-7}	7
10^{-8}	8
10^{-9}	9
10^{-10}	10
10^{-11}	11
10^{-12}	12
10^{-13}	13
10^{-14}	14

Table 8.4 Approximate pH Values of Some Common Substances

Substance	pH Value	Substance	pH Value
		Acids	
1 N hydrochloric	0.1	1 N acetic	2.4
0.1 N hydrochloric	1.1	0.1 N acetic	2.9
0.01 N hydrochloric	2.0	0.01 N acetic	3.4
		Bases	
1 N sodium hydroxide	14.0	1 N ammonia	11.6
0.1 N sodium hydroxide	13.0	0.1 N ammonia	11.1
0.01 N sodium hydroxide	12.0	0.01 N ammonia	10.6
Saturated magnesium hydroxide (milk of magnesia)	10.5		
		Biological Materials	
Human feces	4.6–8.4	Human duodenal (small intestines) contents	4.8–8.2
Human interstitial fluid and blood	7.4	Human liver intracellular fluids	6.9
Gastric juice	1.2–3.0		
		Foods	
Cabbage	5.2–5.4	Pears	3.6–4.0
Carrots	4.9–5.3	Peas	5.8–6.4
Cherries	3.2–4.0	Potatoes	5.6–6.0
Dill pickles	3.2–3.6	Rhubarb	3.1–3.2
Egg white	7.6–8.0	Strawberries	3.0–3.5
Grapes	3.5–4.5	Vinegar	2.4–3.4
Lemons	2.2–2.4	Wines	2.8–3.8
Peaches	3.4–3.6		

opposite effect; that is, weak bases have lower pH values (are more acidic) than strong bases having the same concentration. Most biological fluids, with the exception of gastric juice, are near a neutral pH of 7. Most foods, on the other hand, are somewhat acidic, having pH values less than 7.

A similar approach to that of pH is frequently used to represent values of ionization constants. As evident in Tables 8.2 and 8.3, ionization constants often have very large or very small values. In light of this, pK_i values defined by the equation,

$$pK_i = -\log K_i \tag{8.27}$$

are commonly used in science because of the simpler numbers involved. They are determined and operate in the same manner as pH values. pK_i values shall be used later in the text.

PROBLEMS

6. Calculate the concentration of H_3O^+ in an aqueous solution when the concentration of OH^- is the following:
   - ⑤ (a) $1.0 \times 10^{-4} M$
   - ⑤ (b) $3.5 \times 10^{-9} M$
   - (c) $4.6 \times 10^{-10} M$
   - (d) $0.0029 M$

   Indicate whether each solution is acidic, basic, or neutral.

7. Calculate or estimate the approximate pH or pH range of the following substances based on the concentration of hydronium ions given.
   - ⑤ (a) $[HCl] = 10^{-5} M$
   - (b) $[HNO_3] = 10^{-1} M$
   - (c) Oranges, $10^{-3} M$ to $10^{-4} M$
   - (d) Soft drinks, $10^{-2} M$ to $10^{-4} M$
   - ⑤ (e) $[H_3O^+] = 10 M$
   - (f) $[H_3O^+] = 1.0 \times 10^{-9} M$
   - (g) $[H_3O^+] = 1.0 \times 10^{-14} M$
   - ⑥ (h) Milk, $5.0 \times 10^{-7} M$ to $2.5 \times 10^{-7} M$
   - ⑥ (i) Apples, $1.26 \times 10^{-3} M$ to $5.0 \times 10^{-4} M$
   - (j) Human saliva, $3.2 \times 10^{-7} M$ to $3.2 \times 10^{-8} M$

8. Calculate or estimate the hydronium ion concentrations for the following solutions with the given pH values.
   - ⑤ (a) 2.0
   - ⑤ (b) 6.0
   - (c) beer, 4.0–5.0
   - (d) 0.1 N KCN, 11.0
   - (e) 0.01 N KOH, 12.0
   - ⑤ (f) Human spinal fluid, 7.3–7.5
   - ⑤ (g) Human urine, 4.8–8.4
   - (h) Tomatoes, 4.0–4.4

ANSWERS TO PROBLEMS **6. (a)** $1.0 \times 10^{-10} M$, basic **(b)** $2.9 \times 10^{-6} M$, slightly acidic **(c)** 2.2×10^{-5}, acidic **(d)** 3.4×10^{-12}, very basic **7.** (approximate pH ranges given in parentheses) **(a)** 5 **(b)** 1 **(c)** 3–4 **(d)** 2–4 **(e)** −1 **(f)** 9 **(g)** 14 **(h)** $(7 > pH > 6) 6.3$–6.6 **(i)** $(4 > pH > 2) 2.9$–3.3 **(j)** $(8 > pH > 6)$ 6.5–7.5 **8.** (approximate ranges given in parentheses) **(a)** $10^{-2} M$ **(b)** $10^{-6} M$ **(c)** 10^{-4}–$10^{-5} M$ **(d)** $10^{-11} M$ **(e)** $10^{-12} M$ **(f)** $(10^{-7}$–$10^{-8} M) 5.0 \times 10^{-8}$–$3.2 \times 10^{-8} M$ **(g)** $(10^{-4}$–$10^{-9} M) 1.6 \times 10^{-5}$–$4.0 \times 10^{-9} M$ **(h)** $(10^{-4}$–$10^{-5} M) 1 \times 10^{-4}$–$4.0 \times 10^{-5} M$

8-4 CHANGES IN IONIC EQUILIBRIA

LEARNING OBJECTIVES

1. Define hydrolysis, common-ion effect, and buffer.

2. Write balanced chemical equations for hydrolysis, common-ion, and buffer reactions.

3. Predict the effect on the pH when a salt is added to an aqueous solution.

4. When given the name or formula of a specific acid or base, give the formula of a salt that contains the common ion of that acid or base.

5. When given the names or formulas of an acid-salt (conjugate base) pair or base-salt (conjugate acid) pair, write two-dimensional equations illustrating the common-ion effect.

6. Identify buffer pairs from a list of acid-salt (conjugate base) or base-salt (conjugate acid) pairs.

7. Discuss the action of a given buffer pair in minimizing changes in pH when small amounts of acids or bases are added to a buffered solution.

In this section we will use the background material developed in the three previous sections of this chapter as we discuss hydrolysis, the common ion effect, and buffers. We will study the function of these important processes in the living system. The operation of buffers is particularly important and we will consider in some depth how buffers function to maintain an almost constant pH in the blood.

8-4.1 HYDROLYSIS

In Section 4-2 and again in Section 8-2.1 reference was made to the fact that the addition of a metal oxide to water results in the formation of a base. This is illustrated by the equation,

$$Na_2O(aq) + H_2O(l) \longrightarrow 2\ NaOH(aq) \qquad \textbf{(8.28)}$$

As a result of this reaction, the solution is basic. This is an example of hydrolysis. Hydrolysis is the reaction of a metal ion and/or a nonmetal ion with water to change the H^+ and OH^- concentration of an aqueous solution.* We will consider the reaction of nonmetal ions first.

> **Hydrolysis** is the reaction of a salt with water, resulting in a change in the acidity of the solution.

The reaction of nonmetal ions with water is essentially the reaction of the bases listed in Table 8.1 with water. We will limit our consideration to bases that do not contain ionizable hydrogen.† These bases fall into three groups depending upon their relative position on the right in Table 8.1: (1) Those above H_2O, (2) those between H_2O and OH^-, and (3) those below OH^-. If one knows the position of the base in Table 8.1, predictions concerning the hydrolysis of these bases can be made easily as follows.

1. *Bases above H_2O do not react with water and, therefore, do not affect the acidity of aqueous solutions.* This lack of reaction is due to the fact that these ions are weaker bases than water and therefore do not remove protons from water. This behavior is illustrated for Cl^- by equation 8.13 and for NO_3^- by the equation

$$NO_3^-(aq) + H_2O(l) \longrightarrow N.R. \qquad \textbf{(8.29)}$$

2. *Bases between H_2O and OH^- react to some degree with water to produce a basic solution.* These ions establish an equilibrium of the type

$$A^-(aq) + H_2O(l) \rightleftharpoons HA(aq) + OH^-(aq) \qquad \textbf{(8.30)}$$

where A^- is the nonmetal ion of the added salt that acts as a Brønsted-Lowry base and HA is the conjugate acid of A^- that is formed when A^- combines with H^+. As an example of this, the addition of a salt containing acetate ions (CH_3COO^-) to water results in the following equation

$$CH_3COO^-(aq) + H_2O(l) \rightleftharpoons CH_3COOH(aq) + OH^-(aq) \qquad \textbf{(8.31)}$$

The point of equilibrium for reactions of this type depends upon the relative position of the base in Table 8.1. The lower the ion is, the more the equilibrium is shifted to the right and the more basic is the resulting solution.

3. *Bases below OH^- react completely with H_2O to produce a basic solution.* This is illustrated by equations 8.12 and 8.28. Ions below OH^- are such strong bases that the reaction goes to completion.

---

*In a more general sense, the term *hydrolysis* refers to any reaction with water during which water is added to a compound causing it to split into two parts. One part of the compound then reacts chemically with the H in water while the other part reacts with the OH. In addition to the use of the term in the context given, certain reactions of organic compounds are also called nydrolysis reactions. These shall be considered later in the text.

†Conjugate bases such as $H_2PO_4^-$, HPO_4^{2-}, and HCO_3^- may react as either an acid or a base, depending upon the pH of the solution. We will not concern ourselves with this characteristic.

We will now turn our attention to the positive ion in salts. Some positive ions react with water to produce acidic solutions. These ions fall into two major classes: (1) nitrogen-containing ions such as NH_4^+ and $CH_3NH_3^+$ and (2) metal ions that react with water.

1. *Nitrogen-containing ions such as* NH_4^+ *and amines react with water to produce an acidic solution.* The most common example of this is the reaction of the ammonium ion

$$NH_4^+(aq) + H_2O(l) \rightleftharpoons NH_3(aq) + H_3O^+(aq) \qquad \textbf{(8.32)}$$

The degree to which this type of hydrolysis occurs depends upon the relative acid strength of the positive ion.

2. *Certain metal ions react with water to produce an acidic solution* according to the following equation

$$M^{n+}(aq) + H_2O(l) \rightleftharpoons MOH^{(n+)-1}(aq) + H^+(aq) \qquad \textbf{(8.33)}$$

where M^{n+} represents a metal ion with a charge of $n+$. Metals that react with water fall into the following groups: (a) all $1+$ metal ions except the Group IA metal ions below Li^+; (b) all $2+$ metal ions other than the Group IIA metal ions below Be^{2+}, and (c) all $3+$ metal ions. On this basis, Al^{3+} reacts with water as follows:

$$Al^{3+}(aq) + H_2O(l) \rightleftharpoons AlOH^{2+}(aq) + H^+(aq) \qquad \textbf{(8.34)}$$

to produce an acidic solution. The acidity of the solution depends upon the point of equilibrium in equation 8.33 for each metal ion. However, we will not concern ourselves with the degree to which each reaction takes place.

In the preceding discussion we concentrated on the reactions of positive and negative ions with water. In reality these positive and negative ions are always found together in the form of salts. Depending upon which positive and negative ions constitute the salt, hydrolysis by one or both ions may occur, as discussed in the preceding material. The acidity or basicity of the resulting solution depends upon the nature of the positive and negative ions in the salt. A few examples of specific salts and their behavior will be helpful in seeing how hydrolysis functions.

As indicated previously, a 0.89 wt% solution of NaCl (saline solution) is frequently given intravenously to maintain body fluids of hospitalized patients. Any significant change in the pH of the blood on mixing with the saline solution would have adverse affects on the living system. Consider the salt, NaCl. The Na^+ ion is one of the metal ions that does not react with water; therefore, it will not tend to change the pH of the solution. Similarly, the Cl^- ion does not hydrolyze since it is a weaker conjugate base than water. Therefore NaCl does not hydrolyze and NaCl solutions may be mixed safely with blood.

The addition of a salt such as ammonium nitrate (NH_4NO_3) to water, on the other hand, does result in a change in pH. The NH_4^+ ion reacts with water according to equation 8.32, but the NO_3^- is a weak conjugate base that does not react with water, as indicated by equation 8.29. The net effect in this case is that the acidity of the solution is increased and the pH is decreased.

The opposite effect is noted on adding a salt such as barium cyanide ($Ba(CN)_2$) to water. The Ba^{2+} does not hydrolyze, but the CN^- reacts with water according to the equation

$$CN^-(aq) + H_2O(l) \rightleftharpoons HCN(aq) + OH^-(aq) \qquad \textbf{(8.35)}$$

In this case, the overall effect is to produce a basic solution.

In some cases predicting the net effect that a salt has when it is added to water is more complicated than these examples suggest. This is the case when both the positive and negative ions tend to react with water. An example of such a salt is ammonium cyanide (NH_4CN). As illustrated in the last two examples, both NH_4^+ and CN^- tend to react with water, but they have opposite effects on the pH. In such cases the relative strengths of the substances to act as an acid or base become important. For NH_4CN, the net effect is to produce a slightly basic solution. In general, it will not be important for us to be able to make such determinations.

8-4.2 COMMON ION EFFECT

Another type of change in ionic equilibria is the **common ion effect**. As an example of this we can consider an aqueous solution of NH_3 that has established the horizontal equilibrium shown in equation 8.36. If NH_4Cl is added to this solution it will dissociate into NH_4^+ and Cl^- as given by the vertical equation*

The **common ion effect** is the shifting of an ionic equilibrium by the addition of a chemical species that contains an ion that is common to the substance involved in the equilibrium.

$$NH_4Cl(aq)$$
$$\downarrow$$
$$NH_3(aq) + H_2O(l) \rightleftharpoons NH_4^+(aq) + OH^-(aq) \qquad \textbf{(8.36)}$$
$$+$$
$$Cl^-(aq)$$

As illustrated in this equation, the $NH_4^+(aq)$ ion is common to both reactions. The additional NH_4^+ coming from the NH_4Cl causes the horizontal reaction to shift to the left as some of the extra NH_4^+ reacts with the hydroxide ions to establish a new equilibrium position. As a result, the basicity of the solution decreases. The common ion effect also occurs in solutions of weak acids, and is the basis upon which a buffer solution operates. Both of these aspects shall be considered in some detail in the next section.

8-4.3 BUFFERS

The maintenance of a nearly constant pH is important in many aspects of chemistry and is of significance in almost all parts of living systems. Various portions of the body are maintained at different, but essentially constant, pH values. Among these, blood plasma (the liquid part of the blood) and interstitial fluid are both kept at a

---

*The equilibrium $2 H_2O \rightleftharpoons H_3O^+ + OH^-$ is still occurring in the solution, but it is not included in the equation to avoid making the equation too complex. The student should keep in mind that this equilibrium is present in all aqueous solutions.

pH of 7.4, intercellular fluids in the muscles have a pH of 6.1, and the liver has a pH of 6.9.

The pH of blood must remain between 7.35 and 7.45 for the body to function optimally. If the pH falls into the range of 6.8 to 7.34, a condition called *acidosis* exists. Similarly an increase in the pH to between 7.46 and 7.8 results in what is termed *alkalosis*. If the pH is not returned to the normal range of 7.35 to 7.45 within a reasonable time, irreparable damage can occur. In the case of acidosis, the victim suffers a depression of the central nervous system. If the pH of the blood falls below 7, the depression of the nervous system is so acute that the individual becomes disoriented and may go into a coma. On the other hand, the major physiological effect of alkalosis is overexcitability of the nervous system resulting in extreme nervousness, muscle spasms, and, in severe cases, convulsions. If the pH falls below 6.8 or rises above 7.8, death occurs. You can better appreciate how small the pH range that is acceptable in normal blood is when you realize that a range of 7.35 to 7.45 represents a difference in the concentration of H_3O^+ of approximately 1.0×10^{-8} M. Death can result when the variation in $[H_3O^+]$ is as small as 2.0×10^{-8} M.

Maintenance of the nearly constant pH values in blood is accomplished through three primary control systems: (1) the buffer mechanism, (2) the respiratory mechanism, and (3) the renal, or urinary, mechanism. We will consider only the first of these three. **Buffers** generally consist of a compound pair made up of a weak acid, HA, and its conjugate base, A^-; or a weak base, B, and its conjugate acid HB^+. Examples of the first type are the buffer systems H_2CO_3/HCO_3^-, which maintains a constant pH in blood and interstitial fluids, and $H_2PO_4^-/HPO_4^{2-}$, which is the principal buffer in intracellular fluids. An example of a weak base—conjugate acid pair is the combination NH_3/NH_4^+.

A **buffer** is a substance that maintains a nearly constant pH in a solution when relatively small amounts of an acid or base are added to the solution.

In a dynamic system, such as the human body, processes are taking place continually that either contribute or take away H_3O^+ ions. Such changes would have adverse effects if they were not controlled. To understand how the change in H_3O^+ is minimized, we will consider how the H_2CO_3/HCO_3^- pair operates in the blood to maintain a nearly constant pH.

Under normal conditions the buffer pair H_2CO_3/HCO_3^- functions quite adequately and quickly to minimize the change in the concentration of H_3O^+ in the blood. In this case an equilibrium exists between the weak acid, H_2CO_3, and its conjugate base, HCO_3^-, as represented by the horizontal portion of the following equation.

$$HA(aq) + H_2O$$
$$H_2CO_3(aq) + H_2O \rightleftharpoons H_3O^+(aq) + HCO_3^-(aq) \qquad (8.37)$$
$$A^-(aq)$$

Any additional acid, HA, that enters the blood produces $H_3O^+(aq)$ ions, as illustrated by the vertical equation. The resulting H_3O^+ is essentially a common ion to the horizontal equilibrium. As discussed in the preceding section most of the additional H_3O^+ will combine with the $HCO_3^-(aq)$ to shift the horizontal equilibrium to the left as indicated by the arrow above the horizontal equilibrium in equation 8.37. By this means, the change (increase) in the H_3O^+ concentration is minimized and the solution has been buffered.

The loss of H_3O^+ through various processes will shift the equilibrium to the right. This is illustrated in equation 8.38 where the addition of hydroxide ions is considered.

$$\begin{array}{c} OH^-(aq) \\ + \\ \xrightarrow{} \\ H_2CO_3(aq) + H_2O \rightleftharpoons H_3O^+(aq) + HCO_3^- \\ \Updownarrow \\ 2\,H_2O \end{array} \qquad (8.38)$$

In this case the added base (OH^-) tends to decrease the acidity of the solution by reacting with the H_3O^+ and thus decreasing its concentration in the solution. In the presence of the H_2CO_3/HCO_3^- buffer pair this effect is minimized as the horizontal equilibrium in equation 8.38 shifts to the right, as indicated by the arrow above the equation, due to the removal of the product species H_3O^+. This shift minimizes the change (decrease) in the H_3O^+ concentration and the solution is buffered.

The buffering processes given by equations 8.37 and 8.38 are only valid as long as there are adequate amounts of HCO_3^- and H_2CO_3 present to result in the necessary shift in the horizontal equilibrium. As long as this is true the stress on the buffer system will not be too great for it to handle and the pH will remain nearly constant. Aiding in this process are the respiratory and renal mechanisms spoken of earlier in this section.

If equimolar amounts of H_2CO_3 and HCO_3^- were present in the blood, a pH of approximately 6.4 would be maintained. In order to maintain the desired pH of 7.4, the ratio of concentrations of HCO_3^- to H_2CO_3 must be maintained at a level of 20 to 1. Under normal conditions the pH of the blood is kept at this optimum condition. However, stresses occasionally come into play that are too strong for the body to adjust to and medical help is required. Two main sources of such stresses are related to metabolic and respiratory processes. Both can increase or decrease the pH of the blood. Metabolic acidosis is due to a deficiency in the HCO_3^- level in the blood and can be caused by such traumas as severe diarrhea and prolonged fasting or starvation. Metabolic alkalosis, on the other hand, is caused by persistent vomiting or excessive intake of HCO_3^- such as when a person takes too much baking soda ($NaHCO_3$) to relieve an acid stomach. Respiratory acidosis can occur when there is a prolonged condition of shallow breathing or an obstruction in the windpipe. For example, pneumonia, emphysema, and drug overdose can result in respiratory acidosis. The opposite condition, respiratory alkalosis, is associated with hyperventilation (rapid and deep breathing leading to an excessive loss of CO_2). Respiratory alkalosis can, therefore, occur in cases of anxiety, rapid breathing at high altitude, and prolonged crying.

Other buffer systems, such as the $H_2PO_4^-/HPO_4^{2-}$ and NH_3/NH_4^+ pairs, operate on the same fundamental basis as the H_2CO_3/HCO_3^- buffer. The weak-base, conjugate-acid pair, NH_3/NH_4^+ establishes the equilibrium

$$NH_3(aq) + H_2O \rightleftharpoons NH_4^+(aq) + OH^-(aq) \qquad (8.39)$$

Changes in the pH of a solution are minimized by shifts in this equilibrium. In general, any compound pair, such as those discussed above, will act as a buffer system if sufficient concentrations of the two species are present to adjust for the acid-base stresses placed upon the system.

PROBLEMS

9. Write balanced chemical reactions for any hydrolysis reactions that occur when the following substances are added to water.
   (a) CH_3COONa (b) NH_4ClO_4 (c) K_2CO_3
   (d) $C_2H_5NH_3Cl$ (e) NaCN
   What effect will the addition of water have on the (1) acidity of the resulting solution and (2) pH?
10. Indicate a salt that contains a common ion for each of the following acids and bases.
    (a) HCN   (b) NH_3 (c) H_2CO_3
    (d) CH_3NH_2 (e) HCO_3^-
11. Which of the following compound pairs can act as a buffer system?
    (a) HCl/NaCl   (b) CH_3COOH/CH_3COONa
    (c) CH_3NH_2/CH_3NH_3Cl (d) $NaOH/H_2O$
    (e) $CH_3NH_2/NaCl$ (f) $CH_3CHCOOH/CH_3CHCOO^-$
                                          | |
                                         OH   OH

ANSWERS TO PROBLEMS   **9. (a)** $CH_3COO^- + H_2O \rightleftharpoons CH_3COOH + OH^-$; decrease acidity, increase pH   **(b)** $NH_4^+ + H_2O \rightleftharpoons NH_3 + H_3O^+$; increase acidity, decrease pH   **(c)** $CO_3^{2-} + H_2O \rightleftharpoons HCO_3^- + OH^-$; decrease acidity, increase pH   **(d)** $C_2H_4NH_3^+ + H_2O \rightleftharpoons C_2H_5NH_2 + H_3O^+$; increase acidity, decrease pH   **(e)** $CN^- + H_2O \rightleftharpoons$ HCN + OH^-; decrease acidity, increase pH   **10. (a)** NaCN   **(b)** NH_4Cl **(c)** Na_2CO_3 **(d)** CH_3NH_3Cl **(e)** $NaHCO_3$ **11.** b, c, f

---

SUMMARY

Among the many chemical species present in the human body, two of the more important are acids and bases. In this chapter we have broadened our understanding of acids and bases to include not only the Arrhenius concept but also the Brønsted-Lowry approach. Neutralization reactions between acids and bases were also considered with the titration calculations associated with these reactions. We also learned that weak acids and weak bases only partly ionize in solution and establish an equilibrium involving the undissociated and ionic species. Special equilibrium constant expressions called ionization constant expressions were developed and used. Using values for the associated ionization constants, we were able to compare the relative strengths of weak acids and bases. One of the most important ionic equilibrium processes is the one associated with the ionization of water. This process was considered in some depth and calculations of hydronium ion and hydroxide ion were made. The concept pH was then introduced and the method used to calculate pH from H_3O^+ and OH^- concentrations and the reverse calculations were discussed. Finally, the ideas concerning relative acid-base strength were used in developing the principles at work in hydrolysis, the common ion effect, and buffers. Each of these aspects was discussed in some depth.

In the next chapter we will conclude our consideration of general chemistry by studying another important aspect of chemistry, oxidation-reduction reactions and electrochemistry.

ADDITIONAL PROBLEMS

12. Define or describe the following terms.
    (a) titration
    (b) Brønsted-Lowry acid
    (c) hydrolysis
13. (a) Write correct formulas for the conjugate acid for each of the following substances when they act as Brønsted-Lowry bases.
    (a) H_2O (b) CO_3^{2-} (c) HPO_4^{2-}
    (d) NH_3 (e) $C_2H_5COO^-$ (f) CH_3NH_2
    (g) HSO_4^-
    (b) Write correct formulas for the conjugate bases for those substances in part (a) that can act as Brønsted-Lowry acids.
    [Answers given to parts (a), (d), and (g).]
14. Calculate the concentration of OH^- present in aqueous solutions of the following.
    ⑤ (a) 0.010 M NaOH
    ⑤ (b) 0.010 M HCl
    ⑤ (c) 1.4×10^{-3} M HNO_3
    (d) 6.4×10^{-4} M HI ($K_i = \infty$)
    ⑤ (e) 0.010 M NH_3 at equilibrium
    ⑤ (f) 0.010 M CH_3COOH at equilibrium
15. Calculate the pH values of the following substances based on the concentration of hydronium ions given. (Give approximate values where necessary.)
    ⑤ (a) Human gastric contents: $10^{-1}/10^{-3}$ M
    ⑤ (b) 0.01 M NaOH: 10^{-12} M
    ⑤ (c) 0.01 M NH_3: 2.51×10^{-11} M
    (d) Pancreatic juice:
        1.58×10^{-8} $M/1.00 \times 10^{-8}$
    (e) Cow's milk: 2.51×10^{-7} M
16. Calculate the hydronium ion concentration for the following solutions, given the pH values listed. (Give approximate values where necessary.)
    ⑤ (a) 0.1 N trisodium phosphate: 12.0
    (b) beans: 5.0/6.0
    ⑤ (c) drinking water: 6.5/8.0
    ⑤ (d) human milk: 6.6/7.6

    (e) human bile: 6.8/7.0
    (f) lemon juice: 2.3
17. Given the following salts and reactions with water, indicate which reactions actually occur and the effect on the pH of the solution that is observed.
    ⓓ (a) $KClO_4$: $K^+ + 2 H_2O \longrightarrow KOH + H_3O^+$
    $ClO_4^- + H_2O \longrightarrow HClO_4 + OH^-$
    ⓓ (b) NH_4I: $NH_4^+ + H_2O \longrightarrow NH_3 + H_3O^+$
    $I^- + H_2O \longrightarrow HI + OH^-$
    ⓓ (c) $CH_3CHCOORb$: $Rb^+ + 2 H_2O \longrightarrow$
       | $RbOH + H_3O^+$
       OH
    $CH_3CHCOO^- + H_2O \longrightarrow CH_3CHCOOH + OH^-$
       | |
       OH   OH
    ⓓ (d) Na_2SO_4: $Na^+ + 2 H_2O \longrightarrow$
                                        $NaOH + H_3O^+$
    $SO_4^{2-} + H_2O \longrightarrow$
                                        $HSO_4^- + OH^-$
    (e) $CH_3NH_3NO_3$: $CH_3NH_3^+ + H_2O \longrightarrow$
                                        $CH_3NH_2 + H_3O^+$
    $NO_3^- + H_2O \longrightarrow$
                                        $HNO_3 + OH^-$
18. Write two-dimensional reactions for the acid or base given in Problem 10, parts (a), ⓓ (b), and (d), and the associated common-ion-containing substance that you listed. Indicate what happens to the equilibrium upon addition of the common ion and what happens to the pH of the solution. (Answers are given to parts (a) and (d).)
19. By means of words and equations, discuss how the following buffer systems minimize the change in pH when a relatively small amount of (1) acid and (2) base is added to the solution.
    ⑤ (a) $H_2PO_4^-/HPO_4^{2-}$
    ⑤ (b) NH_3/NH_4^+
    (c) CH_3COOH/CH_3COO^-
    (d) $C_5H_5N/C_5H_5NH^+$

CHAPTER 9

OXIDATION-REDUCTION REACTIONS AND ELECTROCHEMISTRY

In Chapter 8 we studied ionic equilibria in solution. We found that various ions and the equilibria they maintain are important in the living system. For example, we learned that maintenance of proper levels of hydronium ions in the body is crucial to life and that the body controls these levels of acidity by means of buffer systems. In this chapter we will continue our consideration of ions in solution, but we will be concerned with the current that the ions carry and the changes in charge, or oxidation number, that they experience. Such changes consist of a transfer of electrons between atoms, or change in oxidation state of the atom and are referred to as oxidation-reduction reactions. We will learn that atoms and ions transfer these electrons directly from one atom to another or indirectly through a conductor by means of what is called an electrochemical cell. Two general types of electrochemical cells, voltaic and electrolytic, shall be considered. Finally we will briefly examine how oxidation-reduction reactions are involved in the living system and how potential differences between ions function in the passage of nerve impulses in the body.

9-1 BASIC CONCEPTS

LEARNING OBJECTIVE

Define and give examples of oxidation-reduction reactions, oxidation, reduction, reducing agent, oxidizing agent, and electrochemical cells.

In our earlier studies we considered various types of reactions that involve the transfer of electrons. In Chapter 3, for example, we learned that ionic bonds are formed when metal atoms give electrons to nonmetal atoms. In Chapter 4 we studied various types of reactions including single-replacement reactions in which metal atoms give electrons to other metal ions and nonmetal atoms take electrons from other nonmetal ions. These reactions are examples of oxidation-reduction reactions. Such reactions play an important role in the world about us and in the living system.

In order to understand oxidation-reduction reactions better, let us consider the single-replacement reaction between solid zinc and a solution of copper(II) sulfate that was introduced in Section 4-2.1.

An **oxidation-reduction reaction** involves an electron transfer or a change in the oxidation states of two or more chemical species.

$$Zn(s) + CuSO_4(aq) \longrightarrow ZnSO_4(aq) + Cu(s) \qquad (9.1)$$

Since the $CuSO_4$ and $ZnSO_4$ are both strong electrolytes, they exist in aqueous solutions as ions. This allows us to write equation 9.1 as a total ionic equation.

$$Zn(s) + Cu^{2+}(aq) + SO_4^{2-}(aq) \longrightarrow Zn^{2+}(aq) + SO_4^{2-}(aq) + Cu(s) \qquad (9.2)$$

The $SO_4^{2-}(aq)$ ions appear on both sides of the arrow in this equation and, therefore, are essentially unaffected by the reaction of Zn with $Cu^{2+}(aq)$. The $SO_4^{2-}(aq)$ can therefore be omitted from the equation without any loss to our understanding of what is taking place in the reaction. Doing this we can rewrite the equation as a net ionic equation

$$Zn(s) + Cu^{2+}(aq) \longrightarrow Zn^{2+}(aq) + Cu(s) \qquad (9.3)$$

In this equation the electron transfer is indicated through the change in oxidation number of the two species. We can clarify this reaction further by breaking it down into two steps: (1) the loss of electrons by the Zn(s)

$$Zn(s) \longrightarrow Zn^{2+}(aq) + 2 e^- \qquad (9.4)$$

and (2) the gain of electrons by the $Cu^{2+}(aq)$

$$Cu^{2+}(aq) + 2 e^- \longrightarrow Cu(s) \qquad (9.5)$$

When a chemical species has lost electrons, as in equation 9.4, it is considered to have undergone oxidation. Equations such as 9.4 are frequently referred to as oxidation half-reactions. When a chemical species has gained electrons, as in equation 9.5, it is considered to have undergone reduction. Equation 9.5 is an example of a reduction half-reaction.

Oxidation is the process in which a chemical species loses electrons, becoming more positive in its oxidation state.

Reduction is the process in which a chemical species gains electrons, becoming less positive in its oxidation state.

As the term *oxidation-reduction* (or, as it is sometimes called, *redox*) suggests, the two processes always occur simultaneously. Whenever one chemical species is *oxidized* (gives up electrons), some other species must be *reduced* (take electrons) because electrons must be conserved. In equation 9.3, the zinc was oxidized. However, zinc was also acting as the agent to bring about the reduction of copper ions. We can therefore refer to the zinc as the reducing agent, or reductant. Similarly, as the copper ions were being reduced they oxidized the zinc. Therefore, we refer to the copper as the oxidizing agent, or oxidant. In all such electron-transfer processes the electron goes from the chemical species with the lesser attraction for electrons to the one with greater attraction.

A **reducing agent** is a chemical species that brings about the reduction of another chemical species.

An **oxidizing agent** is a chemical species that brings about the oxidation of another chemical species.

For the reaction given by equation 9.3, think of copper ions in the solution colliding with zinc atoms in the solid zinc and receiving electrons from the zinc as shown in Figure 9.1. As this takes place, the resulting zinc ions move off into the

Figure 9.1
Zinc replaces copper in the
solution.

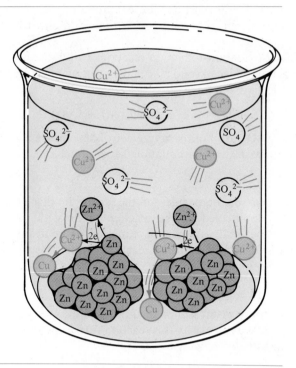

solution while the Cu atoms "fall" out of the solution as solid copper. In this case there is direct physical contact between the Cu^{2+} ions and the solid zinc atoms. The major process of interest to us in this reaction is the transfer of electrons from the zinc atoms to the copper ions.

As illustrated by this reaction zinc is more reactive, or active, than copper (see Chapter 4). This observation is confirmed if the roles of the two elements are reversed and copper metal is placed in a solution containing zinc ions. When this is done no reaction takes place between solid copper and zinc ions.

Returning to equation 9.3, an alternative approach for this "passing of electrons" can be arranged if we have electrons pass from the zinc to the copper ions through a conducting wire, rather than allow them to come in direct contact with each other. Such an apparatus is a type of electrochemical cell. We will consider these cells in some detail in the next section.

PROBLEMS

1. The process in which a chemical species gains electrons is known as what?
2. For the oxidation-reduction reaction

$$3 \text{ Mg(s)} + \text{Fe}_2(\text{SO}_4)_3(\text{aq}) \longrightarrow 3 \text{ MgSO}_4(\text{aq}) + 2 \text{ Fe}$$

identify the following.
(a) substance oxidized   (b) substance reduced
(c) oxidizing agent   (d) reducing agent

ANSWERS TO PROBLEMS   **1.** reduction   **2. (a)** Mg   **(b)** Fe^{3+} $(Fe_2(SO_4)_3)$ **(c)** Fe^{3+} $(Fe_2(SO_4)_3)$ **(d)** Mg

9-2 ELECTROCHEMICAL CELLS

LEARNING OBJECTIVES

1. Define voltaic cell, electrode, anode, cathode, cell potential, standard cell potential, electrolytic cell, and decomposition potential.
2. Diagram and label voltaic and electrolytic cells based upon a given oxidation-reduction reaction.
3. Write half-cell reactions for electro-chemical cells based upon the overall oxidation-reduction reaction. Calculate values of standard cell potentials based upon tabular values of standard reduction potentials.
4. Classify an electrochemical cell as either voltaic or electrolytic based upon the sign of the standard cell potential.

In the previous section we discussed some of the basic principles concerning oxidation-reduction reactions. In this section we will apply these basic principles to electrochemical cells. To do this we will consider two types of electrochemical cells, voltaic and electrolytic, and discuss what is known as their cell potential. The basic difference between voltaic cells and electrolytic cells will be covered and an ability to differentiate between them will be acquired.

An **electrochemical cell** is a combination of two electrodes and an electrolyte.

9-2.1 VOLTAIC CELLS

When an electrochemical cell involving Cu and $ZnSO_4$ is constructed it is an example of a galvanic, or voltaic cell. The basic features of a voltaic cell are shown in Figure 9.2. In a voltaic cell the reacting species are placed in separate containers to keep them from coming into direct contact with each other and transferring the electrons directly. As seen in this figure, a zinc bar is placed in a solution of $ZnSO_4$ and a solid copper bar is placed in the solution of $CuSO_4(aq)$.

A **voltaic cell** is an electrochemical cell in which chemical reactions produce an electric current.

The metal bars consisting of copper and zinc are called electrodes and provide electrical contact between the solutions and the electron-carrying wire. The reaction taking place at the zinc electrode is the oxidation and is represented by the equation

$$Zn(s) \longrightarrow Zn^{2+}(aq) + 2\ e^- \qquad (9.6)$$

An **electrode** in an electrochemical cell is the junction between the external circuit (a wire) and the electrolyte.

The zinc electrode is called the anode. At the copper electrode the reduction of copper takes place according to the equation

$$Cu^{2+}(aq) + 2\ e^- \longrightarrow Cu(s) \qquad (9.7)$$

The **anode** in an electrochemical cell is the electrode at which oxidation occurs.

The copper electrode is referred to as the cathode.

As these two reactions occur, electrons pass through the wire from the anode to the cathode as shown in Figure 9.2. This flow of electrons is an electrical current and can be used to operate light bulbs, heat objects, run motors, and so forth. A comparable flow of electrons takes place within the solution, but these electrons

The **cathode** in an electrochemical cell is the electrode at which reduction occurs.

Figure 9.2 A typical voltaic cell consisting of a zinc half-cell and a copper half-cell.

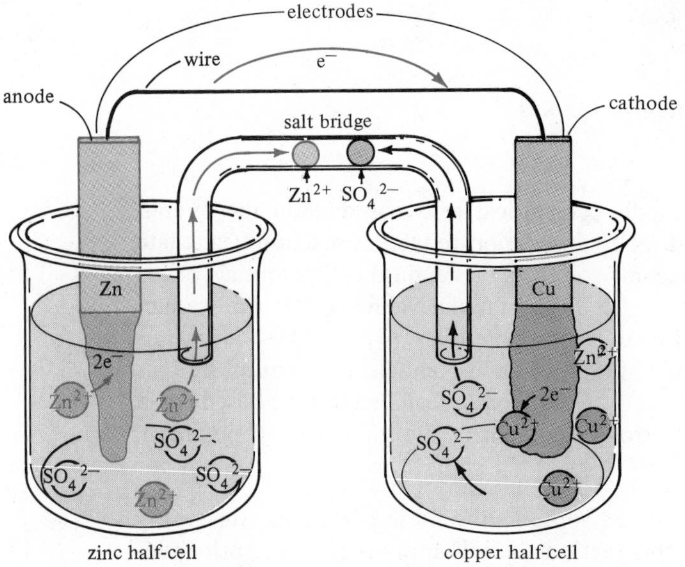

are associated with the ions in the solution. Negative ions flow from the right compartment in Figure 9.2 to the left and positive ions flow in the reverse direction. This ionic current flow can occur through a permeable membrane or through the U-shaped tube known as a *salt bridge* that is pictured in Figure 9.2.

9-2.2 STANDARD ELECTRODE POTENTIALS

In our consideration of single-replacement reactions in Chapter 4 we learned that some metal atoms have the potential for replacing other metal ions in solution and others do not. We used the relative reactivity of metals given in Table 4.3 to decide whether or not a particular pair of metals will undergo such a reaction. Elemental metal atoms high in the series are able to replace metal ions in compounds lower in the series. In our discussion of electrochemical cells it is useful to be able to make a more quantitative comparison between two metals and their tendency to undergo such reactions. This comparison is provided by the **cell potential**.

The cell potential is equal to the electrical potential, or voltage, measured between the two electrodes in an electrochemical cell. For example, when there is a concentration of $Zn^{2+}(aq)$ and $Cu^{2+}(aq)$ of 1 mol of metal ion per 1000 g of water, a pressure of 1 atm, and a temperature of 25°C, the cell diagrammed in Figure 9.2 is found to have a cell potential of 1.100 volts (v). Such cell potentials are referred to as **standard cell potentials**.*

The **cell potential** of an electrochemical cell is a measure of the force with which electrons are pushed or pulled through the external wire.

The **standard cell potential** is the potential of an electrochemical cell when the concentration of the ions in contact with the elemental electrode is 1 mol per 1000 g of solvent, the temperature is 25°C, and the pressure is 1 atm.

---

*It is more accurate to say that the standard cell potential is associated with a cell in which the behavior of the ion, as represented by what is known as its activity, is equal to one. An understanding of the concept of activity requires a knowledge of an area of chemistry known as thermodynamics. We shall not concern ourselves with the details of thermodynamics.

Although cell potentials can be measured using actual cells in the laboratory, frequently it is convenient to be able to calculate these potentials from what are known as *standard half-cell potentials*. To understand the nature of such calculations, it is helpful to consider the voltaic cell pictured in Figure 9.2 in more detail. Electrochemical cells basically consist of two parts (half-cells), an *oxidation half-cell* and a *reduction half-cell*. Since oxidation always occurs at the anode and reduction always occurs at the cathode, we can refer to the reaction

$$Zn(s) \longrightarrow Zn^{2+}(aq) + 2\,e^- \tag{9.8}$$

as the *anode half-cell reaction* and the reaction

$$Cu^{2+}(aq) + 2\,e^- \longrightarrow Cu(s) \tag{9.9}$$

as the *cathode half-cell reaction*. The combination of the two half-cell reactions results in the overall cell reaction

$$Zn(s) + Cu^{2+}(aq) \longrightarrow Zn^{2+}(aq) + Cu(s) \tag{9.10}$$

Different but comparable reactions can take place by replacement of one or both half-cell reactions with some other half-cell. For example, replacing Zn(s) and $ZnSO_4$(aq) in Figure 9.2 with Mg(s) and $MgSO_4$(aq) will result in the anode half-cell reaction

$$Mg(s) \longrightarrow Mg^{2+}(aq) + 2\,e^- \tag{9.11}$$

The overall cell reaction will be

$$Mg(s) + Cu^{2+}(aq) \longrightarrow Mg^{2+}(aq) + Cu(s) \tag{9.12}$$

In the context of our current discussion, the only significant difference between this cell reaction given by equation 9.12 and that given by equation 9.10 is that the reaction given in equation 9.12 occurs more vigorously and thus the cell potential is larger (2.7 v). Both reactions involve a flow of electrons through the external wire and a flow of ions within the solutions.

Several anode half-cell reactions could be used in place of the Zn(s), $ZnSO_4$(aq) half-cell and several cathode half-cell reactions could be substituted for the Cu(s), $CuSO_4$(aq) half-cell and still provide a voltaic cell. One important half-cell that is frequently substituted in this manner is the hydrogen half-cell,

$$H_2(g) \longrightarrow 2\,H^+(aq) + 2\,e^- \tag{9.13}$$

pictured on the left in Figure 9.3. *For reference purposes the hydrogen half-cell is arbitrarily assigned a potential 0.000 v.* When the standard cell potential of the cell pictured in Figure 9.3 is measured it is found to have a voltage of 0.337 v. This can be represented by the equation

$$H_2(g) + Cu^{2+}(aq) \longrightarrow 2\,H^+(aq) + Cu(s), \quad \mathscr{E}^\circ = 0.337\ v \tag{9.14}$$

where the symbol $\mathscr{E}^\circ$ represents the standard cell potential. Since the hydrogen half-cell has been assigned a value of zero we can write

$$H_2(g) \longrightarrow 2\,H^+(aq) + 2\,e^-, \quad \mathscr{E}^\circ = 0.00\ v \tag{9.15}$$

Subtracting equation 9.15 from 9.14 leaves us with

$$Cu^{2+}(aq) \longrightarrow Cu(s) - 2\,e^-, \quad \mathscr{E}^\circ = +0.337\ v \tag{9.16}$$

Figure 9.3   A standard voltaic cell consisting of a hydrogen half-cell and a copper half-cell.

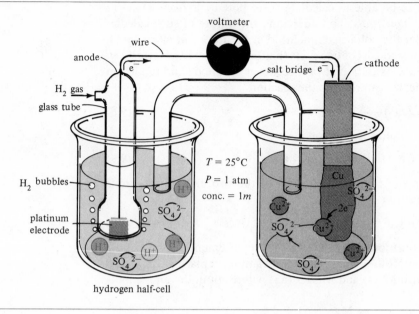

or, moving the electrons to the left of the arrow, we have

$$Cu^{2+}(aq) + 2\ e^- \longrightarrow Cu(s), \quad \mathscr{E}^\circ = +0.337\ v \qquad (9.17)$$

A similar procedure can be used to determine the half-cell potential for zinc. We pair a zinc half-cell with a hydrogen electrode and observe a potential of $+0.763$ v. In this cell the hydrogen is undergoing a reaction that is the reverse of that given by equation 9.15. However, the hydrogen half-cell potential is still assigned a value of zero. For this cell we can follow a similar procedure to that above and obtain

$$Zn(s) + 2\ H^+(aq) \longrightarrow Zn^{2+}(aq) + H_2(g), \quad \mathscr{E}^\circ = +0.763\ v \qquad (9.18)$$

$$\underline{-\ (2\ H^+(aq) + 2\ e^- \longrightarrow H_2(g)), \qquad\qquad\quad \mathscr{E}^\circ = +0.00\ v \qquad (9.19)}$$

$$Zn(s) - 2\ e^- \longrightarrow Zn^{2+}(aq), \qquad\qquad\qquad \mathscr{E}^\circ = +0.763\ v \qquad (9.20)$$

or, moving the electrons to the right of the arrow,

$$Zn(s) \longrightarrow Zn^{2+}(aq) + 2\ e^-, \quad \mathscr{E}^\circ = +0.763\ v \qquad (9.21)$$

Now that we have obtained the half-cell potentials for the oxidation of zinc and the reduction of copper, we can combine the two reactions, equations 9.17 and 9.21, to determine the overall cell potential for equation 9.10. From equation 9.21 we have

$$Zn(s) \longrightarrow Zn^{2+}(aq) + 2\ e^-, \quad \mathscr{E}^\circ = +0.763\ v$$

and from equation 9.17

$$Cu^{2+}(aq) + 2\ e^- \longrightarrow Cu(s), \quad \mathscr{E}^\circ = +0.337\ v$$

Table 9.1 Standard Reduction Half-Cell Potentials*

Element	Electrode Reaction			Standard Electrode (Reduction) Potential $\mathscr{E}°$ (volts)
	Oxidized Form		Reduced Form	
Potassium	$K^+ + e^-$	$\rightleftharpoons$	K	−2.925
Calcium	$Ca^{2+} + 2\,e^-$	$\rightleftharpoons$	Ca	−2.87
Sodium	$Na^+ + e^-$	$\rightleftharpoons$	Na	−2.714
Magnesium	$Mg^{2+} + 2\,e^-$	$\rightleftharpoons$	Mg	−2.37
Aluminum	$Al^{3+} + 3\,e^-$	$\rightleftharpoons$	Al	−1.66
Manganese	$Mn^{2+} + 2\,e^-$	$\rightleftharpoons$	Mn	−1.18
Zinc	$Zn^{2+} + 2\,e^-$	$\rightleftharpoons$	Zn	−0.763
Chromium	$Cr^{3+} + 3e^-$	$\rightleftharpoons$	Cr	−0.74
Iron	$Fe^{2+} + 2\,e^-$	$\rightleftharpoons$	Fe	−0.440
Cadmium	$Cd^{2+} + 2\,e^-$	$\rightleftharpoons$	Cd	−0.40
Cobalt	$Co^{2+} + 2\,e^-$	$\rightleftharpoons$	Co	−0.277
Nickel	$Ni^{2+} + 2\,e^-$	$\rightleftharpoons$	Ni	−0.250
Tin	$Sn^{2+} + 2\,e^-$	$\rightleftharpoons$	Sn	−0.136
Lead	$Pb^{2+} + 2\,e^-$	$\rightleftharpoons$	Pb	−0.126
Hydrogen	$2\,H^+ + 2\,e^-$	$\rightleftharpoons$	H_2	0.00
Copper	$Cu^{2+} + 2\,e^-$	$\rightleftharpoons$	Cu	+0.337
Iodine	$I_2 + 2\,e^-$	$\rightleftharpoons$	$2\,I^-$	+0.5355
Silver	$Ag^+ + e^-$	$\rightleftharpoons$	Ag	+0.7991
Mercury	$Hg^{2+} + 2\,e^-$	$\rightleftharpoons$	Hg	+0.854
Bromine	$Br_2(l) + 2\,e^-$	$\rightleftharpoons$	$2\,Br^-$	+1.0652
Platinum	$Pt^{2+} + 2\,e^-$	$\rightleftharpoons$	Pt	+1.2
Oxygen	$O_2 + 4\,H^+ + 4\,e^-$	$\rightleftharpoons$	$2\,H_2O$	+1.23
Chlorine	$Cl_2 + 2\,e^-$	$\rightleftharpoons$	$2\,Cl^-$	+1.3595
Gold	$Au^+ + e^-$	$\rightleftharpoons$	Au	+1.68
Fluorine	$F_2 + 2\,e^-$	$\rightleftharpoons$	$2\,F^-$	+2.87

*nonbio-organic values

Adding the two equations and their associated half-cell potentials we have

$$Zn(s) + Cu^{2+}(aq) \longrightarrow Zn^{2+}(aq) + Cu(s), \quad \mathscr{E}° = +1.100 \text{ v}$$

This is the potential that was reported previously, based upon measurements using an actual Zn, Cu^{2+} cell. Following this procedure, we can calculate the value for the standard cell potential for any combination of half-cells if we know the corresponding standard half-cell potentials.

Table 9.1 lists half-cell potentials. The standard half-cell potentials given are reduction potentials, as indicated by the list of electrode reactants. It is significant that the reaction of zinc given in this table is opposite to that of equation 9.21.

$$Zn^{2+} + 2\,e^- \longrightarrow Zn, \quad \mathscr{E}° = -0.763 \text{ v} \qquad \textbf{(9.22)}$$

The potential has the same value, but it has the opposite sign. For all oxidation half-cell reactions, *the standard oxidation half-cell potential is equal in magnitude but opposite in sign to the standard reduction half-cell potential.* This fact allows us to calculate the overall cell potential for any combination of half-cell reactions as illustrated in the following example.

EXAMPLE 9.1   What is the standard cell potential of a voltaic cell based upon the following overall equation?

$$2 \, Al(s) + 3 \, Pb^{2+}(aq) \longrightarrow 2 \, Al^{3+}(aq) + 3 \, Pb(s) \qquad (9.23)$$

We begin by separating the overall cell reaction into an oxidation half-cell reaction

$$2 \, Al(s) \longrightarrow 2 \, Al^{3+}(aq) + 6 \, e^- \qquad (9.24)$$

and a reduction half-cell reaction

$$3 \, Pb^{2+}(aq) + 6 \, e^- \longrightarrow 3 \, Pb(s) \qquad (9.25)$$

In these reactions it is necessary to include more than one atom of each type in order to conserve electrons. That is, we want to be sure that the same number of electrons gained are lost. This does not change the value of the standard half-cell potential for a given reaction. Values from Table 9.1 may be used as given and, in general, the value of the standard half-cell potential is independent of the number of electrons (in this case 6) involved. In equation 9.24, aluminum is undergoing oxidation and the sign of the standard reduction potential for Al, given in Table 9.1, must be changed.

$$2 \, Al(s) \longrightarrow 2 \, Al^{3+}(aq) + 6 \, e^-, \quad \mathscr{E}^\circ = +1.66 \, v \qquad (9.26)$$

In equation 9.25, Pb is undergoing reduction, and the standard reduction potential for Pb, given in Table 9.1, can be used without change in sign.

$$3 \, Pb^{2+}(aq) + 6 \, e^- \longrightarrow 3 \, Pb(s), \quad \mathscr{E}^\circ = -0.126 \, v \qquad (9.27)$$

Adding equation 9.26 and 9.27 and their associated half-cell potentials gives the desired overall equation and standard cell potential.

$$2 \, Al(s) + 3 \, Pb^{2+}(aq) \longrightarrow 2 \, Al^{3+}(aq) + 3 \, Pb(s), \quad \mathscr{E}^\circ = 1.53 \, v \qquad (9.28)$$

Redox reactions similar to those discussed above are the basis on which a car battery or a dry cell operates. The flow of electrons from the anode to the cathode provides the electricity needed to start a car, to operate a flashlight, and to power a portable radio. The basic principles associated with the redox reactions above come into play in many of the life-giving reactions taking place in the living system. Some of these shall be considered briefly in Section 9-3, and in more detail in the biochemistry portion of the text.

9-2.3 ELECTROLYTIC CELLS

When we studied voltaic cells in Section 9-2.1, we learned that a voltaic cell spontaneously produces an electric current by means of a chemical reaction. The reverse process is also possible; that is, an electric current can be used to bring about a chemical reaction. This is the underlying concept for the **electrolytic cell.** To illustrate this we will consider an electrochemical cell similar to the one given in Figure 9.1, except Ag(s) and $Ag^+(aq)$ are substituted for Zn(s) and $Zn^{2+}(aq)$. Figure 9.4 illustrates such a cell. Using values for the standard reduction potentials

An **electrolytic cell** is an electrochemical cell in which an electrical current is used to bring about chemical changes.

Figure 9.4 An electrolytic cell consisting of a silver half-cell and a copper half-cell connected by a battery.

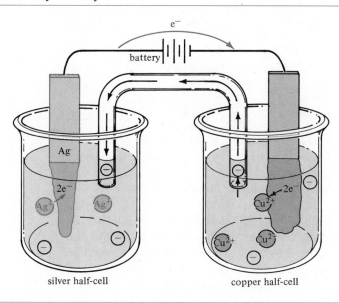

given in Table 9.1, we can calculate the standard overall cell potential for this process as follows.

$$2 \text{ Ag(s)} \longrightarrow 2 \text{ Ag}^+(aq) + 2 \text{ e}^-, \quad \mathscr{E}° = -0.799 \text{ v} \qquad \textbf{(9.29)}$$

$$\underline{\text{Cu}^{2+}(aq) + 2 \text{ e}^- \longrightarrow \text{Cu(s)}, \qquad\qquad\quad \mathscr{E}° = +0.037 \text{ v}} \qquad \textbf{(9.30)}$$

$$2 \text{ Ag(s)} + \text{Cu}^{2+}(aq) \longrightarrow 2 \text{ Ag}^+ + \text{Cu(s)}, \qquad \mathscr{E}° = -0.462 \text{ v} \qquad \textbf{(9.31)}$$

The negative value for this standard cell potential indicates that, if allowed to operate spontaneously, the reactions would take place in the opposite direction. In other words, the reaction indicated by the equation

$$2 \text{ Ag}^+(aq) + \text{Cu(s)} \longrightarrow 2 \text{ Ag(s)} + \text{Cu}^{2+}(aq) \qquad\qquad \textbf{(9.32)}$$

represents a possible voltaic cell reaction.

Placing a source of direct current, such as a battery (see Figure 9.4), between the wires joining the two electrodes can cause reaction 9.31 to occur if the potential of the battery is greater than 0.462 v. This potential is known as the **decomposition potential.**

> The **decomposition potential** is the minimum potential, or voltage required to bring about electrolysis.

EXAMPLE 9.2 Calculate the decomposition potential associated with the equation

$$3 \text{ Ni(s)} + \text{Al}_2(\text{SO}_4)_3(aq) \longrightarrow 3 \text{ NiSO}_4(aq) + 2 \text{ Al(s)} \qquad \textbf{(9.33)}$$

In the net ionic form, this equation becomes

$$3 \text{ Ni(s)} + 2 \text{ Al}^{3+}(aq) \longrightarrow 3 \text{ Ni}^{2+}(aq) + 2 \text{ Al(s)} \qquad \textbf{(9.34)}$$

For this equation we have the two half-cell reactions, with the value of their standard half-cell potentials

$$3 \text{ Ni(s)} \longrightarrow 3 \text{ Ni}^{2+}\text{(aq)} + 6 \text{ e}^-, \quad \mathscr{E}° = +0.250 \text{ v} \qquad \textbf{(9.35)}$$

and

$$2 \text{ Al}^{3+}\text{(aq)} + 6 \text{ e}^- \longrightarrow 2 \text{ Al(s)}, \quad \mathscr{E}° = -1.66 \text{ v} \qquad \textbf{(9.36)}$$

Adding these two equations gives

$$3 \text{ Ni(s)} + 2 \text{ Al}^{3+}\text{(aq)} \longrightarrow 3 \text{ Ni}^{2+}\text{(aq)} + 2 \text{ Al(s)}, \quad \mathscr{E}° = -1.41 \text{ v} \qquad \textbf{(9.37)}$$

Remember, the negative sign indicates that the reaction will not take place spontaneously. However, applying a potential greater than 1.41 v in opposition to the normal tendency for the reverse reaction will cause the Ni(s) to be oxidized and the Al^{3+} to be reduced. Therefore the decomposition potential is 1.41 v.

Although the cells pictured in Figure 9.4 and discussed in Example 9.2 are electrolytic cells, they are not the most common type. A more frequently used electrolytic cell is one that involves a single salt either in solution or in the molten (melted) state. For simplicity we will consider the case of a molten salt and examine the electrolysis of NaCl. Figure 9.5 illustrates a cell of this type.*

For this electrolysis we have

> **Electrolysis** is the process of causing nonspontaneous chemical reactions to occur through the use of an electric current.

$$2 \text{ NaCl} \xrightarrow{\text{electricity}} 2 \text{ Na} + \text{Cl}_2 \qquad \textbf{(9.38)}$$

or, written in the ionic form,

$$2 \text{ Na}^+ + 2 \text{ Cl}^- \xrightarrow{\text{electricity}} 2 \text{ Na} + \text{Cl}_2 \qquad \textbf{(9.39)}$$

As with previous cell reactions, we can separate the reaction into two half-cell reactions and assign standard electrode potentials to each reaction as follows.†

$$\text{Anode half-cell: } 2 \text{ Cl}^- \longrightarrow \text{Cl}_2 + 2 \text{ e}^-, \quad \mathscr{E}° = -1.36 \text{ v} \qquad \textbf{(9.40)}$$

and

$$\text{Cathode half-cell: } 2 \text{ Na}^+ + 2 \text{ e}^- \longrightarrow 2 \text{ Na}, \quad \mathscr{E}° = -2.71 \text{ v} \qquad \textbf{(9.41)}$$

Adding the two half-cell reactions together gives the overall cell reaction with a standard cell potential of -1.36 v $+ (-2.71$ v$) = -4.07$ v. The decomposition potential is, therefore, $+4.07$ v.

With reference to the process above there are two additional terms that are often used in chemistry. As can be seen in Figure 9.5, the negative chloride ions are attracted to the anode where they are oxidized. Negative ions are, therefore, frequently referred to as **anions**. The positive sodium ion is attracted to the cathode where it is reduced and is, therefore, referred to as a **cation**.

> An **anion** is a negative ion.

> A **cation** is a positive ion.

---

*Special equipment is necessary for this cell because of the high temperature (801°C) necessary to melt NaCl. This equipment was omitted in Figure 9.5 so that emphasis could be placed upon the basic process involved.

†For the conditions associated with pure molten NaCl, the values of the standard reduction potentials listed in Table 9.1 will not apply. The same basic principles, however, do apply, so this deviation has been ignored in these calculations.

Figure 9.5   A simplified electrolytic cell for the electrolysis of molten NaCl.

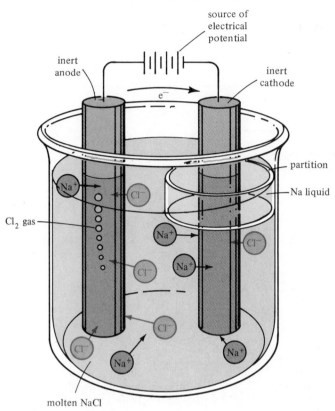

Electrolysis processes such as that given by equation 9.38 can be used to obtain pure metals. For example, one method for obtaining metallic sodium is based upon the electrolysis reaction given by equation 9.38. Similar electrolysis processes are involved in obtaining aluminum and in refining (purifying) copper metal.

In the previous sections we have considered the fundamental principles of electrochemistry. Little reference was made to the living system despite the integral part that redox reactions play in the functions of the body. In part this is because those reactions are quite complex. We will briefly cover some of the electrochemistry that takes place in the living system in the next section.

PROBLEMS

3. What is an electrochemical cell in which electrical current is used to bring about chemical changes?

4. Diagram a voltaic cell based upon the equation

$$2 \, Cr(s) + 3 \, NiSO_4(aq) \longrightarrow Cr_2(SO_4)_3(aq) + Ni(s)$$

and label all parts.

5. Write half-cell reactions for the reaction given in Problem 4 and calculate the standard cell potential with this voltaic cell using values for standard reduction half-cell potentials found in Table 9.1.

☐ ⓢ 6. Is the following equation associated with a voltaic or an electrolytic cell?

$$Cu(s) + Cd^{2+}(aq) \longrightarrow Cu^{2+}(aq) + Cd(s)$$

ANSWERS TO PROBLEMS   **3.** an electrolytic cell

4.

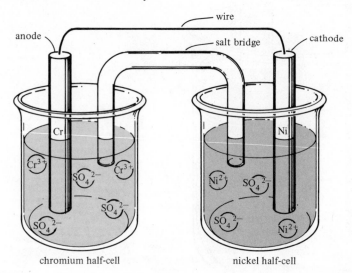

chromium half-cell   nickel half-cell

5. anode half-cell: $2\ Cr(s) \longrightarrow 2Cr^{3+}(aq) + 6\ e^-$, $\mathscr{E}° = 0.74$ v; cathode half-cell: $3\ Ni^{2+}(aq) + 6\ e^- \longrightarrow 3\ Ni(s)$, $\mathscr{E}° = -0.25$ v; $\mathscr{E}°$ cell $= 0.49$ v   **6.** electrolytic cell

---

9-3 ELECTROCHEMISTRY IN LIFE

LEARNING OBJECTIVES

1. Give examples of redox reactions involved in everyday processes.
2. Discuss the three general types of redox reactions that take place in the living system.

3. Briefly discuss the mechanism involved in the propagation of nerve impulses.

In the previous sections of this chapter we studied the basic principles involved in oxidation-reduction reactions and in electrochemical cells. In this section we will briefly consider how redox reactions and electrical potentials play an important role in life and in the living system. For the present discussion, we will limit ourselves to a brief consideration of the major factors involved.

9-3.1 REDOX REACTIONS IN LIFE

The term *oxidation* was originally associated with the reaction of oxygen with other substances. For example when coal is burned the carbon in the coal is oxidized according to the equation

$$C(s) + O_2(g) \longrightarrow CO_2(g) \qquad\qquad \textbf{(9.42)}$$

When iron rusts it is being oxidized according to the overall equation*

$$4\ Fe(s) + 3\ O_2(g) \longrightarrow 2\ Fe_2O_3(s) \qquad\qquad \textbf{(9.43)}$$

An even more impressive example of oxidation is associated with the flash produced in a flashbulb when the aluminum wire inside the bulb burns in the atmosphere of pure oxygen there

$$2\ Al + 3\ O_2 \longrightarrow 2\ Al_2O_3 \qquad\qquad \textbf{(9.44)}$$

Oxidation-reduction reactions involving oxygen enter into almost every aspect of life from the cleansing of wounds and bleaching of hair with hydrogen peroxide (H_2O_2) to the burning of gasoline in our cars, a candle as we dine, and wood in a fireplace as all of these react with oxygen in the air. Indeed, life itself ceases if the body is denied oxygen for a matter of minutes. Our urgent need for oxygen is illustrated when we hold our breath. Few of us can do so for more than a minute.

In the living system, oxidation can occur in three ways: (1) addition of oxygen through direct reactions, (2) removal of hydrogen from a substance, and (3) the loss, or direct transfer, of electrons from a chemical species. We will briefly consider examples of each of these here and then in more detail in later sections of the book.

Oxidation by means of the addition of oxygen through a direct reaction does not occur often in the living system. It is illustrated by the oxidation of the female hormone progesterone ($C_{21}H_{30}O_2$) to 11-hydroxyprogesterone ($C_{21}H_{30}O_3$) as the body transforms progesterone into other hormones. More shall be said about progesterone and its reactions in Section 14-2.3 and in the biochemistry portion of this text.

Oxidation through the removal of hydrogen (dehydrogenation) is illustrated by the oxidation of succinic acid ($HOOC(CH_2)_2COOH$) to fumaric acid ($HOOC(CH)_2COOH$). Once again an enzyme, succinate dehydrogenase, plays an important role in the reaction. This oxidation process is represented by the equation,

$$HOOC(CH_2)_2COOH + \left\{ \begin{array}{c} \text{oxidized form of} \\ \text{flavin adenine} \\ \text{dinucleotide} \end{array} \right\} \xrightarrow{\text{succinate dehydrogenase}}$$

$$HOOC(CH)_2COOH + \left\{ \begin{array}{c} \text{reduced form of} \\ \text{flavin adenine} \\ \text{dinucleotide} \end{array} \right\} \qquad \textbf{(9.45)}$$

Here, the removal of the two hydrogen atoms from $HOOC(CH_2)_2COOH$ is associated with the oxidation of succinic acid just as the addition of oxygen

---

*The exact mechanism for the rusting of iron is not known for a certainty. Equation 9.43 does, however, summarize the overall redox reaction that takes place.

Figure 9.6 Coupled redox reactions occurring in the respiratory chain. $\mathscr{E}^\circ$ values are reduction potentials for each half-cell reaction. The details of this mechanism shall be dealt with later, in the biochemistry portion of the text.

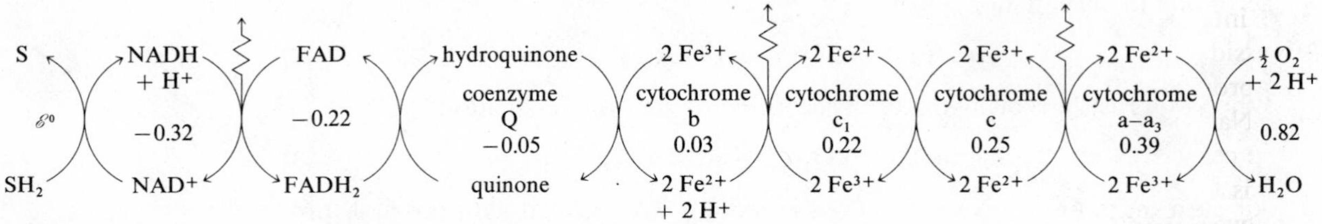

S	NADH + H$^+$	FAD	hydroquinone coenzyme Q	2 Fe^{3+} cytochrome b	2 Fe^{2+} cytochrome c$_1$	2 Fe^{3+} cytochrome c	2 Fe^{2+} cytochrome a–a$_3$	$\frac{1}{2}$O$_2$ + 2 H$^+$
$\mathscr{E}^\circ$	−0.32	−0.22	−0.05	0.03	0.22	0.25	0.39	0.82
SH$_2$	NAD$^+$	FADH$_2$	quinone	2 Fe^{2+} + 2 H$^+$	2 Fe^{3+}	2 Fe^{2+}	2 Fe^{3+}	H$_2$O

resulted in the oxidation of progesterone. The oxidation of succinic acid occurs in the mitochondria of cells and is an integral part of a mechanism known as the tricarboxylic acid cycle. The details of this cycle will be discussed in Chapter 25.

Oxidation by the loss, or direct transfer, of electrons is illustrated by the oxidation of iron, found in a group of complex enzymes called cytochromes, from Fe^{2+} to Fe^{3+}. This reaction can be represented by the equation

$$\left\{\begin{array}{c} 2\ Fe^{2+} \\ \text{in reduced} \\ \text{cytochromes} \end{array}\right\} \longrightarrow \left\{\begin{array}{c} 2\ Fe^{3+} \\ \text{in oxidized} \\ \text{cytochromes} \end{array}\right\} + 2\ e^- \qquad (9.46)$$

This process consists of the series of reactions shown in Figure 9.6. Known as the respiratory chain, or the cytochrome electron transport system, this process also takes place in the mitochondria. Note the reaction on the far right in Figure 9.6. Here, the electrons lost by the iron in equation 9.46 are shown being picked up by molecular oxygen according to the equation

$$2\ e^- + \tfrac{1}{2}O_2 \longrightarrow O^{2-} \qquad (9.47)$$

The oxygen then unites with hydrogen to form water. Associated with each redox reaction in the living system is a cell potential comparable to the cell potentials referred to earlier. This is illustrated in Figure 9.6 where the standard reduction potentials for each half-cell are given by the numbers within each loop. In the final reaction of the respiratory chain, substances are oxidized and water is produced. By means of coupled reactions such as those shown in Figure 9.6 the human body is able to use approximately 70% of the energy released while the other 30% is given off as heat.

9-3.2 PROPAGATION OF NERVE IMPULSES

Another aspect of electrical activity in living cells is the conduction of nerve impulses and muscle contraction. This conduction is not like the flow of electrons along a conductor or the movement of ions through a solution. Rather, impulses occur in nerves and muscles due to the movement of ions in a path that is perpendicular to the nerve, as illustrated in Figure 9.7. Thus, it is more appropriate to say that the impulse is propagated along the nerve, rather than conducted. The complex mechanism by which this propagation occurs is not fully understood. However, the following simplified explanation is sufficient for our purposes.

One of the characteristics of the cell membrane of nerve cells is its ability to maintain an unequal distribution of ions between the space inside and the space outside the cell. This unequal distribution, or charge separation, has a potential of approximately -0.070 v. The propagation of the nerve impulse involves a layer of Na^+ ions that are on the outside of the cell. When the nerve receives an impulse, the cell wall becomes permeable to the Na^+ ions so that they can enter the cell. This change in concentration causes a local change in the potential difference between the inside and outside of the cell. As a result, the permeability of the cell is propagated along the nerve wall in a wave-like manner, discharging the ionic layer along the nerve. The Na^+ ions are immediately pumped out, readying the

Figure 9.7
The propagation of a nerve impulse involves a layer of Na^+ ions that pass through the cell wall as it becomes permeable to them.

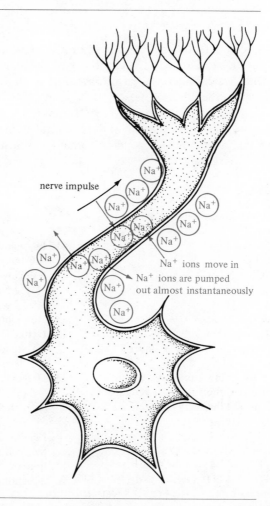

cell for any new impulse that may be received. The velocity at which the impulse travels along the nerve depends on such variables as nerve diameter, but, typically, has values from less than 1 to as much as 100 m/sec. Muscle cells act in a similar manner when Ca^{2+} ions enter the cells. The result is a contraction in the muscle.

PROBLEMS

6. List three examples of redox reactions in the world about us.
7. The oxidation of progesterone to give 11-hydroxyprogesterone can occur by what mechanism?
8. What ion plays a significant role in the propagation of nerve impulses?

ANSWERS TO PROBLEMS **6.** bleaching hair, rusting iron, burning wood, and so forth **7.** oxidation by means of the addition of oxygen through a direct reaction **8.** Na^+

SUMMARY

In this chapter we examined oxidation-reduction reactions. First, we discussed the basic principles and definitions involved in such processes and applied these to electrochemical cells. Then, we considered the role played by oxidation-reduction reactions and cell potentials in the life processes.

Here ends our consideration of what is commonly referred to as general chemistry. In the next chapter, we will consider some of the chemical elements essential to life and introduce another major area of chemistry, organic chemistry.

ADDITIONAL PROBLEMS

9. Which of the following reactions are oxidation-reduction reactions?
   (a) $2\,Mg + O_2 \longrightarrow 2\,MgO$
   (b) $S + O_2 \longrightarrow SO_2$
   (c) $NaOH + HCl \longrightarrow NaCl + H_2O$
   (d) $C_6H_{12}O_6 + 6\,O_2 \longrightarrow 6\,CO_2 + 6\,H_2O$
   (e) $H_2O + SO_2 \longrightarrow H_2SO_3$
   (f) $2\,Na + ZnSO_4 \longrightarrow Na_2SO_4 + Zn$
   (g) $2\,Fe + 3\,Cl_2 \longrightarrow 2\,FeCl_3$
   (h) $AgNO_3 + NaCl \longrightarrow NaNO_3 + AgCl$
   (i) $CH_4 + 2\,O_2 \longrightarrow CO_2 + 2\,H_2O$
10. For each of the following equations, identify the substance oxidized, the substance reduced, the oxidizing agent, and the reducing agent.
   (a) $4\,Ag + O_2 \longrightarrow 2\,Ag_2O$
   (b) $2\,Na + 2\,H_2O \longrightarrow 2\,NaOH + H_2$
   (c) $Cu + 2\,AgNO_3 \longrightarrow Cu(NO_3)_2 + 2\,Ag$
   (d) $4\,Al + 3\,Pb^{4+} \longrightarrow 4\,Al^{3+} + 3\,Pb$
   (e) $3\,Sb + 5\,HNO_3 + 2\,H_2O \longrightarrow$
   $$3\,H_3SbO_4 + 5\,NO$$

   (f) $Cr_2O_3 + 2\,Na_2CO_3 + 3\,KNO_3 \longrightarrow$
   $$2\,Na_2CrO_4 + 2\,CO_2 + 3\,KNO_2$$
11. In each of the following reactions identify the substance reduced, the substance oxidized, the reducing agent, and the oxidizing agent.
   (a) $2\,Na(s) + Mg^{2+} \longrightarrow 2\,Na^+ + Mg(s)$
   (b) $3\,K(s) + Al^{3+} \longrightarrow 3\,K^+ + Al(s)$
   (c) $Co(s) + Pb^{2+} \longrightarrow Co^{2+} + Pb(s)$
12. Define
   (a) standard cell potential
   (b) decomposition potential
13. Diagram a simple voltaic cell composed of
   (a) Ni and Mg electrodes
   ⑤ (b) Sn and Al electrodes
   (c) Co and Pb electrodes
Label all parts of the diagram and indicate the half-cell reactions for each electrode, the direction of flow in the external circuit, and the direction of flow of ions.

14. What is the voltage of cells that have the following cell reactions:
    § (a) $Fe^{2+} + Ni(s) \longrightarrow Fe(s) + Ni^{2+}$
    § (b) $Pb(s) + Fe^{2+} \longrightarrow Pb^{2+} + Fe(s)$
        (c) $Sn(s) + Cu^{2+} \longrightarrow Sn^{2+} Cu(s)$
    Will the reactions occur spontaneously?

15. Diagram an electrolytic cell for the electrolysis of molten
    § (a) $AlCl_3$ § (b) $HgCl_2$
        (c) AgBr
    Calculate the decomposition potential for each compound.

16. Diagram and label electrochemical cells in which the following cell reactions can take place.
        (a) $Cu + 2\,AgNO_3 \longrightarrow Cu(NO_3)_2 + 2\,Ag$
    § (b) $Co + Hg^{2+} \longrightarrow Co^{2+} + Hg$
        (c) $2\,KBr \longrightarrow 2\,K + Br_2$
    § (d) $Ni + ZnSO_4 \longrightarrow NiSO_4 + Zn$
    Write half-cell reactions for each reaction and calculate the standard cell potential for the overall reaction and indicate whether the cell is voltaic or electrolytic. Where appropriate, determine the decomposition potential.

17. Write reactions for the following oxidation processes.
    (a) the combustion of the fatty acid $C_{16}H_{32}O_2$
    (b) the burning of acetylene (C_2H_2) in an acetylene torch

18. What are the three general types of redox reactions that take place in the living system?

ⓓ 19. Briefly describe the mechanism involved in the propagation of nerve impulses.

CHAPTER 10

ELEMENTS ESSENTIAL TO LIFE: INTRODUCTION TO ORGANIC CHEMISTRY

Present evidence indicates the existence of more than 100 elements. Only about 24 of the elements are known to be essential for a healthy human life. From Table 10.1 we can see that the human body is composed of (based on total number of atoms): 99% hydrogen, oxygen, carbon, and nitrogen; less than 1% calcium (Ca), phosphorus (P), chlorine (Cl), potassium (K), sulfur (S), sodium (Na), and magnesium (Mg); and trace amounts (less than 0.01% total) of chromium (Cr), cobalt (Co), copper (Cu), fluorine (F), iodine (I), iron (Fe), manganese (Mn), molybdenum (Mo), selenium (Se), silicon (Si), tin (Sn), vanadium (V), and zinc (Zn). The need for these 24 elements for human life has been established beyond doubt. In Figure 10.1 the periodic table shows the positions of these elements. For purposes of discussion in this text, the elements essential to life fall into three broad classes: the trace elements, the intermediate elements, and the major or most abundant elements. The average elemental composition of a 70-kg (154-lb) human and the function of these elements is summarized in Table 10.2.

10-1 THE TRACE ELEMENTS

LEARNING OBJECTIVES

1. Define trace element.
2. List the thirteen trace elements essential to human life.
3. Associate each of the following health conditions with the appropriate trace element: anemia, hemosiderosis, lack of pigmentation, sexual immaturity, skeletal abnormalities, pernicious anemia, white muscle disease, impaired glucose metabolism, inhibition of cholesterol synthesis, goiter, dental caries, mottled enamel, and Wilson's disease.

Table 10.1 The Chemical Elements of Life

Element	Percent Composition in the Human Body*	
H	63	
O	25.4	~99%
C	9.4	
N	1.4	
Ca	0.31	
P	0.22	
Cl	0.03	
K	0.06	<1%
S	0.05	
Na	0.03	
Mg	0.01	
trace elements	0.01	

*Data expressed as percent of the total number of atoms in the human body.

It has been recognized for some time that very low concentrations of iron are essential to human life. Iron is a **trace element**. Only recently have we begun to appreciate the full extent and the complexity of the interactions between other trace elements in the environment and human health.

A **trace element** is an element that is found in very small amounts in the living system.

Currently, scientists are expanding the number of metabolic systems known to be affected by trace elements. Furthermore, they are uncovering previously unsuspected relationships between trace-element concentrations and abnormal states of health. From our current knowledge of trace elements, it seems likely that the evaluation of trace-element concentrations soon will play a basic role in the diagnosis of illness. Hopefully we will also develop the ability to manipulate trace-element concentrations to prevent illness.

Many trace elements are beneficial to the living system. At present at least thirteen trace elements have been identified as essential to human health. Table 10.3 summarizes these elements and their functions in living systems. In the following sections, we will study each of the thirteen trace elements in detail.

Iron

Iron is one of the trace metals found in greatest abundance in biological systems. Out of the total of about 7 g of iron in a 70-kg adult, 70% is present in the form of the proteins hemoglobin and myoglobin. Hemoglobin and myoglobin are proteins that contain ferrous (Fe^{2+}) iron. In this oxidation state, iron is red and imparts a red color to flesh. Hemoglobin transports oxygen from the lungs to all parts of the body. Myoglobin, which is structurally similar to hemoglobin, functions to store oxygen in muscle cells. About 15% of the iron in biological systems is stored in the liver, spleen, marrow, and kidneys, and the rest is involved in the formation of proteins and in oxidation-reduction reactions.

Figure 10.1 The chemical elements of life. The most abundant elements in the living systems are shown as ▪. The seven intermediate elements are shown as ▪. Thirteen elements are called the trace elements and are shown as ▪.

	I	II												III	IV	V	VI	VII	VIII
1	1 H																		2 He
2	3 Li	4 Be												5 B	6 C	7 N	8 O	9 F	10 Ne
3	11 Na	12 Mg												13 Al	14 Si	15 P	16 S	17 Cl	18 Ar
4	19 K	20 Ca	21 Sc	22 Ti	23 V	24 Cr	25 Mn	26 Fe	27 Co	28 Ni	29 Cu	30 Zn		31 Ga	32 Ge	33 As	34 Se	35 Br	36 Kr
5	37 Rb	38 Sr	39 Y	40 Zr	41 Nb	42 Mo	43 Tc	44 Ru	45 Rh	46 Pd	47 Ag	48 Cd		49 In	50 Sn	51 Sb	52 Te	53 I	54 Xe
6	55 Cs	56 Ba	57 La	72 Hf	73 Ta	74 W	75 Re	76 Os	77 Ir	78 Pt	79 Au	80 Hg		81 Tl	82 Pb	83 Bi	84 Po	85 At	86 Rn
7	87 Fr	88 Ra	89 Ac	104 Rf	105 Ha	106													

58 Ce	59 Pr	60 Nd	61 Pm	62 Sm	63 Eu	64 Gd	65 Tb	66 Dy	67 Ho	68 Er	69 Tm	70 Yb	71 Lu
90 Th	91 Pa	92 U	93 Np	94 Pu	95 Am	96 Cm	97 Bk	98 Cf	99 Es	100 Fm	101 Md	102 No	103 Lw

Table 10.2 Essential Elements and Their General Functions

Element	Grams/70-kg Person	*in vivo* Function*
Major		
O	43,550	
C	12,590	Structural components
H	6,580	
N	1,815	
Intermediate		
Ca	1,700	Structure of bone and teeth
P	680	Basic metabolism, structure of bone and teeth
K	250	Balance osmotic pressure
Cl	115	As the anions, Cl^- and SO_4^{2-}, maintain
S	100	electroneutrality, osmotic pressure, and cell volume
Na	70	Balance osmotic pressure
Mg	42	Structure of bones, regulation of metabolism
Trace		
Fe	7	Transportation and storage of oxygen
Mn, Co, Cu, Zn, Mo, Si, V, Cr, Se, I, Sn, F	trace	Primarily enzyme catalysis

In vivo refers to processes in living systems.

Table 10.3 The Trace Elements

Element	Ionic Form	Biochemical Function
Co	Co^{3+}	Component of Vitamin B_{12}
Cr	?	Lowers blood glucose level, increasing effectiveness of insulin
Cu	Cu^+, Cu^{2+}	Necessary for pigment formation, bone strength, strong blood vessels, and hemoglobin synthesis
F	F^-	Reduces dental caries
Fe	Fe^{2+}	Transports oxygen in hemoglobin and stores oxygen in myoglobin
I	I^-	Used in the formation of the iodine-containing hormone of the thyroid gland
Mn	Mn^{2+}	Affects the activity of some enzymes
Mo	MoO_4^{2-}	A component of the enzyme nitrogenase which catalyzes the conversion of N_2 to NH_3 in nitrogen-fixing bacteria
Se	?	Functions in some cellular redox reactions, prevents white muscle disease in livestock
Si	SiO_3^{2-}	Structural component in skin, cartilage, and ligaments
Sn	?	Not yet classified
V	VO_4^{3-}	Involved in redox reaction in lipid metabolism; possibly inhibits cholesterol synthesis
Zn	Zn^{2+}	Promotes growth rate and sexual development

An average daily human diet contains about 15 mg of iron, of which about 1 mg is absorbed daily. This is ordinarily enough to compensate for the small losses from the body, such as bleeding. The iron content of the body is regulated by the rate of intake. For example, this rate is increased during pregnancy and when necessary to compensate for iron lost in menstrual bleeding. If lost iron is not replenished, a condition known as *anemia* results. The symptoms of this condition are bodily weariness, fatigue, and apathy. Anemia is remedied by the administration of ferrous salts, such as ferrous fumarate and ferrous gluconate, that are easily absorbed in the intestine. The human body appears to have no means for excretion of excessive amounts of iron. An excess of ingested iron causes *hemosiderosis,* which is associated with insoluble granules of iron-containing substances (hemosiderin deposits) in the liver and a characteristic discoloration of the skin. This condition can result from drinking water containing too much iron, cooking food in iron pots, drinking too much red wine, and ingesting too much Geritol vitamin supplement with iron.

Copper

An adult human ingests approximately 2–5 mg of copper per day, about 30% of which is absorbed. The human body contains about 100 mg of copper. Both the intake and excretion of copper is highly regulated. This regulation is essential because an excess of copper is toxic. The accumulation of copper in the brain and liver causes *Wilson's disease.* The individual suffering from this disease can experience neurological problems and liver and kidney failure. If untreated, the disease is fatal, because red blood cells, bursting with copper, rupture.

Copper is so well distributed in the foods that we eat (shellfish and animal livers) that a deficiency never has been observed in humans. Occasionally a copper deficiency does occur in animals other than humans, sometimes because absorption of Cu^{2+} is opposed by Zn^{2+}. These copper-deficient animals exhibit bone defects, lack hair color, have weak arteries, and experience an impaired hemoglobin synthesis. Studies indicate that a copper-containing enzyme is either lacking or inactive in albinos. This enzyme, tyrosinase, is necessary for the formation of melanin, the black pigment of skin and hair. Since albinos are deficient in pigment, they are extremely sensitive to light and may die early from the harmful effects of the ultraviolet rays of the sun.

An **enzyme** is a protein that exhibits catalytic activity.

Copper is a necessary element in the development of the embryo. Copper is also required in early infancy, for the formation of two essential proteins: elastin, a chief constituent of the walls of the large blood vessels, and collagen, an integral part of tendons and bones. If a pregnant woman's diet is deficient in copper, her child will develop weak and fragile bones. In the extreme case, the fetus may die due to rupture of its aorta.

The formation of the hemoglobin molecule depends on copper as well as iron. There is a copper-containing enzyme in the blood that ensures that iron is moved from storage sites in the intestine, liver, and spleen to the bone marrow where hemoglobin is produced.

Zinc

The average human ingests 10–15 mg of zinc per day and the total zinc content of a 70-kg person is 1.4–2.3 g. Zinc is poorly absorbed, but tissue concentrations, particularly in the prostate gland, are relatively high. The metal plays an essential role in a multitude of enzymes.

The importance of zinc in human nutrition was first recognized in a group of dwarfed, sexually immature patients whose diets had consisted mainly of wheat flour and clay. The initial diagnosis was that these patients were suffering, in part, from an iron deficiency. However, iron supplements improved symptoms that were not characteristic of iron depletion. In fact, the patients started growing and maturing. It was then suspected that some other nutrient had been added, inadvertently, to the supplemented diets. Zinc was suggested when clinical tests showed that the patients had an increased level of a zinc enzyme in the blood. When the patients were given a pure zinc supplement, their rates of growth and sexual development increased markedly.

Manganese

Manganese can exist in eight different oxidation states. Biologically, the oxidation states $+2$ and $+3$ are the most important. Mn^{2+} is the most stable state and so, will be discussed here. Manganese is found in tissues and, in higher concentrations, in bone. A deficiency leads to degeneration of sexual characteristics and *skeletal abnormalities* such as bone fragility and bowing of legs.

A large number of enzymes require manganese for activation and some enzymes that require magnesium can utilize manganese instead. Manganese and magnesium are powerful central-nervous-system depressants and can cause general anesthesia (loss of sensation). Higher concentrations produce headaches, psychotic behavior, and drowsiness.

Cobalt

Cobalt, as Co^{3+}, is essential for the formation of **vitamin B_{12}**. Vitamin B_{12} controls a condition known as *pernicious anemia,* a disease that produces symptoms of fatigue and general weakness. These symptoms are the result of a lack of the red blood cells that carry the hemoglobin molecule, and thus oxygen, to the cells. Pernicious anemia results, not from lack of B_{12} in the diet, but from poor absorption of the vitamin. Vitamin B_{12} is not made by humans but is formed, principally, by bacteria. Since most plants contain little or no bacteria, pernicious anemia symptoms are sometimes observed among strict vegetarians.

A **vitamin** is a compound that is necessary for normal metabolic functioning of the body, but is not synthesized in the body and, therefore, must be provided in the diet.

Selenium

Although selenium is an extremely toxic element, it has been shown to be essential for the maintenance of liver cells in rats. Trace amounts have also been shown to prevent *white muscle disease* (a type of muscular dystrophy) in livestock.

When available in the correct amounts, selenium is incorporated in a highly specific fashion into certain functional proteins of the cell. Because selenium and sulfur are in the same group (group VI), they are chemically similar. When organisms receive more than trace amounts of selenium, selenium is substituted indiscriminately for sulfur in many cellular constituents. In general, these selenium compounds are more reactive than the corresponding sulfur compounds. Consequently, the cell may encounter problems that eventually lead to the death of the organism.

Chromium

In chromium-deficient animals, glucose is not removed from the blood effectively. This is the type of response that is observed in an animal deficient in the **hormone** insulin. The passage of glucose from the blood into the cells of the body has long been known to be regulated by insulin. The abnormal metabolism of glucose due to a deficiency of insulin results in **diabetes**. These observations have led scientists to investigate the relationship between chromium and diabetes. It appears that the role chromium plays in lowering the glucose level in the blood is to increase the effectiveness of insulin. This was demonstrated in a recent study in which the rate of glucose-removal was measured in 15 malnourished children before and after chromium supplementation. All the children showed impaired glucose removal prior to chromium supplementation. Nine showed an immediate and marked improvement following supplementation. Furthermore, all the children started to gain weight when given a regular dietary supplement containing chromium.

A **hormone** is a regulatory substance produced by one organ that exerts its effect on other tissues of the body.

Diabetes mellitus is an impairment of glucose metabolism, usually caused by the lack of insulin or ineffective insulin, that is characterized by high sugar levels in the blood and urine.

Molybdenum

Molybdenum has long been recognized as an element essential for the growth of plants. Only recently has it been proven to be a necessary animal nutrient as well. Molybdenum appears to play the role, in the operation of enzymes, of controlling a variety of processes, such as **nitrogen fixation**.

Molybdenum acts as a catalyst for nitrogenase, which itself catalyzes the conversion of atmospheric nitrogen to ammonia. Molybdenum not only activates nitrogenase, but also is involved in regulating the synthesis of nitrogenase. Studies show that when cultures of certain bacteria are grown in the absence of molybdenum, the microorganisms do not synthesize nitrogenase. Today, commercially produced molybdenum supplements are used to increase crop productivity in soils lacking this essential element.

A deficiency of molybdenum in the diet of animals also leads to a decrease in the activity of certain enzymes. For example, it has been shown that children deficient in the enzyme containing this metal secrete no urinary sulfate and suffer neurological defects. Urinary sulfate may be the major route by which sulfur is removed from organic compounds in the animal body.

Nitrogen fixation is a natural process by which nitrogen in the air is converted into compounds than can be used by plants.

Silicon

Silicon, in the form SiO_2, makes up the structural unit of diatoms. Diatoms are the principal component of cells that float near the surface of inland and oceanic waters and are the primary source of food for all water-dwelling animals. Some higher plants and sponges also accumulate silicon. In the chick, silicon is found in active calcification sites of young bone. It is present in low amounts in the internal organs of mammals but makes up approximately 0.01% of the skin, cartilage, and ligaments. Some studies indicate that silicon may function as a biological crosslinking agent in connective tissue.

Vanadium

Vanadium is present in marine organisms, for example, tunicates. These animals contain vanadocytes, green blood cells that are 4% V^{3+}. Vanadium also has been shown to be essential in the diet of rats. The adult human contains a total of about 30 mg of vanadium and it is suggested that vanadium may function in lipid **metabolism**. Because vanadium can assume a variety of oxidation states (from +2 to +5), a possible oxidation-reduction function is suggested. In high doses vanadium *inhibits cholesterol synthesis*. Evidence seems to indicate that cholesterol is associated with cardiovascular problems. Vanadium it also reported to inhibit development of dental caries (tooth decay) by stimulating mineralization of teeth.

Metabolism is the sum of all the chemical reactions of the living organism.

Tin

Little is known about the function of tin in humans, but it seems to be essential to the health of both plants and animals. Tin compounds have been detected in the central nervous system; however, there is no known involvement with specific enzyme systems.

Iodine

A small gland in the neck, the thyroid gland, contains most of the iodine in the human body. The iodine is incorporated chiefly in two hormones that stimulate the chemical activity in certain tissues. For example, a deficiency of the thyroid hormone, thyroxine, is reflected in an overall lower metabolic rate during fasting or resting; that is, a lower energy production to maintain vital cellular activity. A deficiency of this hormone, due to an insufficient amount of iodine, manifests itself in an enlargement of the thyroid. This condition is called *goiter*. The gland enlarges because the body tries to compensate for the reduced level of iodine by increasing the production of thyroid hormone through an increase in the number of cells in the thyroid. Fish is a good source of iodine, but most people increase their iodine intake by using iodized salt.

Fluorine

Fluorine, as fluoride ion (F^-), is present in humans in highly variable amounts, depending upon the diet and water supply. Fluoride has been shown to be essential to the diet of animals and it is most likely required by humans. The beneficial effect of fluoride in the prevention of *dental caries* is well documented. In areas where small amounts of fluoride (1 ppm) are either present or were added to the drinking water, the incidence of dental caries is remarkably low. On the other hand, in those areas where the fluoride concentration is high (8 ppm), the teeth are strong but unattractive. Teeth subjected to continuous high concentrations of fluoride are covered with white-to-gray spots. This condition is known as *mottled enamel.* The mechanism by which fluoride protects the integrity of tooth structure is not well understood. Studies show that fluoride is deposited in the tooth enamel and the resulting enamel is more resistant to acid-producing bacteria.

PROBLEM

1. Identify the trace element that fits each of the following descriptions.
    (a) increases blood glucose level by increasing the effectiveness of insulin
    (b) a deficiency of this element is caused by bleeding
    (c) essential for the formation of vitamin B_{12}
    (d) associated with an enzyme that causes pigmentation
    (e) increases sexual development and growth

ANSWERS TO PROBLEM   **1. (a)** chromium   **(b)** iron   **(c)** cobalt   **(d)** copper   **(e)** zinc

10-2 THE INTERMEDIATE ELEMENTS

LEARNING OBJECTIVES

1. List the six intermediate elements essential to the living system.
2. Write the ionic form for each intermediate element in the living system.
3. Describe *in vivo* functions for each intermediate element.

The elements present in the living system in intermediate amounts are calcium, phosphorus, chlorine, sulfur, sodium, and magnesium.

Calcium

Among the minerals of the earth's crust that contain calcium, the most important are fluorspar (CaF_2), calcium phosphate ($Ca_3(PO_4)_2$), and calcium carbonate ($CaCO_3$). Fluorspar provides the natural fluoridation of our water supplies. Calcium phosphate is thought to arise from decayed teeth, bones, and sea shells. Calcium carbonate, or limestone, is a source of a large quantity of calcium in our drinking water. Milk and many vegetables are also rich sources of calcium.

Calcium is the fifth most abundant element in the body, making up about 2.5% of an adult's body weight. Calcium ions are excluded from cells and most of the calcium is found in bones and teeth. Humans require calcium, phosphorus, and vitamin D for the formation of bones and teeth. Bone consists of a dense matrix of a protein and the crystalline calcium phosphate mineral, hydroxyapatite ($Ca_{10}(PO_4)_6(OH)_2$). A small amount of Mg^{2+} is present in place of Ca^{2+} in bones and a very small fraction of the hydroxide ion is replaced by fluoride ion which has a bone-strengthening effect. The principal function of vitamin D is in the control of calcium and phosphorus metabolism. This metabolism occurs in the liver and the kidneys. Persons with damaged kidneys often suffer severe demineralization of their bones.

Calcium is important in the process of *blood clotting*. In fact, calcium salts are sometimes administered to hasten blood clotting. In the presence of sodium or potassium citrates or oxalates, the calcium ion will complex with the citrate.

$$3 \ Ca^{2+} + 2 \ (C_6H_5O_7)^{3-} \longrightarrow Ca_3(C_6H_5O_7)_2$$

$$\text{citrate anion} \qquad\qquad \text{calcium citrate}$$

The formation of calcium citrate effectively removes Ca^{2+} thus lengthening blood-clotting time. Blood donations are usually taken into potassium citrate for this reason.

If the level of calcium in the blood falls, the calcium is replenished by drainage from the bones. This leads to *osteomalacia*, a condition in which the bones bend. Blood calcium concentration is regulated by parathyroid glands and probably prevents *osteoporosis* (brittleness of bone) in the aged. In addition, a calcium deficiency in the blood causes muscular twitchings and eventually convulsions. On the other hand, too much calcium results in a deadening of nerve impulses and muscle reponse.

Phosphorus

Ninety percent of the phosphorus in the body occurs in the bones and teeth as the phosphate ion in the inorganic calcium salt, $Ca_3(PO_4)_2$. In addition, the phosphate ion is part of three important anions in the body. Phosphate anions play an important role in maintaining **electroneutrality** of body fluids. Only dihydrogen phosphate and monohydrogen phosphate anions occur in significant concentrations in body fluids.

Electroneutrality is the balance of positively charged cations with negatively charged anions in body fluids.

$$H_3PO_4 \ \overset{-H^+}{\rightleftharpoons} \ H_2PO_4^- \ \overset{-H^+}{\rightleftharpoons} \ HPO_4^{2-} \ \overset{-H^+}{\rightleftharpoons} \ PO_4^{3-}$$

| phosphoric acid | dihydrogen phosphate anion | monohydrogen phosphate anion | phosphate ion |

The phosphate ion is also involved in supplying the immediate energy needs of the cell, and it is a component of many molecules that are essential to normal cellular functions. For example, phosphate is an important part of DNA and RNA molecules that control heredity and protein synthesis. These features will be considered in detail in Chapters 20 and 22.

Magnesium

Many enzymes are dependent upon magnesium. A special function of magnesium is participation in **photosynthesis** as a component of chlorophyll. In the presence of light energy, chlorophyll produces the green color associated with plants. The magnesium in our diets, then, comes from a variety of green vegetables as well as nuts, seafoods, and cereals. About 60% of the magnesium in the body is found in bones. Magnesium is necessary for nerve impulse transmissions, muscle contrac-

Photosynthesis is the formation of carbohydrates in green plants from carbon dioxide and water in the presence of sunlight and chlorophyll.

tion, and the metabolism of carbohydrates. As indicated earlier, Mg^{2+} can often be replaced by Mn^{2+} with full activity for enzymes that require Mg^{2+}. High concentrations of Ca^{2+} are often antagonistic to Mg^{2+}. This antagonism is clearly seen in the effect of the two ions on nerve and muscle action. A deficiency of Mg^{2+} or an excess of Ca^{2+} in the surrounding medium leads to increased nerve and muscle action. On the other hand, excess magnesium leads to anesthesia. It is of interest that there is a larger amount of Mg^{2+} in hibernating animals than is normally present.

Magnesium salts (such as $MgSO_4$ in Epsom salts and milk of magnesia) are purgatives, used to empty the bowels. Mg^{2+} cannot pass through the intestinal wall and so, pulls water through the wall into the intestine to hydrate itself. A hydrate is a compound formed when a substance, such as $MgSO_4$, combines with water.

$$MgSO_4 + 7\,H_2O \longrightarrow MgSO_4 \cdot 7\,H_2O$$

Magnesium also is necessary for all phosphate transferring systems. In certain tissues, a large percentage of the magnesium present is complexed with the energy-storing molecule ATP (adenosine triphosphate).

Potassium, Sodium, and Chlorine

Sodium and potassium salts often exist as their chloride salts, NaCl and KCl. We obtain some of the NaCl and KCl we need from the plants we eat. However, plants have about ten times as much potassium as sodium. Consequently, it is necessary to supplement our diets by adding NaCl.

The primary function of K^+, Na^+, and Cl^- ions is to control the electrical balance in cells, tissue fluids, and blood. These ions also are involved in the transport of oxygen and carbon dioxide in the blood. Moreover, they help maintain the normal level of *nerve and muscle response*. For example, sodium ions depress the activity of muscle enzymes and are required for proper muscle contractions, and potassium ions permit the heart muscle to relax between beats.

Sodium chloride is the source of hydrochloric acid of the gastric juices and sodium bicarbonate is a buffer used in the maintenance of the acid/base balance and the transport of CO_2 in body fluids. Sodium chloride is wasted in large quantities through perspiration; as a result a salty taste is often associated with sweat. Both Na^+ and K^+ ions exist in the body as salts of phosphoric, sulfuric, or organic acids such as uric or lactic acid. The sodium salt of uric acid is fairly insoluble and, when deposited in the cartilage, produces gout. Decreases in the concentration of the Na^+ and K^+ ions through perspiration or by precipitation, as in gout, affects muscle and nerve response. This leads to nausea, muscle cramps, and exhaustion. Athletes often replenish the salt lost through perspiration by taking salt tablets or drinking beverages with high salt content.

Sulfur

Most of the sulfur present in the living system exists as sulfide, disulfide, thiol, and sulfate (see Figure 10.2). These groups are utilized in a number of physiologically

Figure 10.2 Common forms of sulfur in living systems.

S^{2-}	$-SH$	$-S-S-$	SO_4^{2-}
sulfide	thiol or sulfhydryl group	disulfide group	sulfate

active proteins. One example is insulin. It is the disulfide groups in the insulin molecule, among other factors, that gives this protein its unique physiological activity to control the blood glucose level. Much of the sulfur metabolized in the living system is either excreted in the urine or incorporated into a variety of physiologically active compounds.

PROBLEM

2. Identify the intermediate element that fits each of the following descriptions.
   (a) functions in the blood-clotting mechanism
   (b) certain salts of this element are used as purgatives
   (c) excreted as its chloride in perspiration
   (d) occurs in bones and teeth in conjunction with calcium and magnesium
   (e) component of chlorophyll
   (f) insulin contains groups that have this element in them

ANSWERS TO PROBLEM **2. (a)** calcium **(b)** magnesium **(c)** sodium **(d)** phosphorus **(e)** magnesium **(f)** sulfur

10-3 THE MAJOR ELEMENTS

LEARNING OBJECTIVES

1. List the four major elements that make up a human being's body weight.
2. Summarize the features of the carbon atom that make it a unique, integral part of the molecules of living matter.

The four elements hydrogen, carbon, nitrogen, and oxygen make up more than 95% of the human adult's body weight. Table 10.4 shows some chemical forms of these elements in the human body. As shown, these elements are incorporated in

Table 10.4 Chemical Forms of Major Elements in the Human Body

Element	Chemical Form
Hydrogen	H^+, H_2O, organic molecules, biomolecules
Oxygen	H_2O, various ions such as SO_4^{2-}, $H_2PO_4^-$, organic molecules, biomolecules
Carbon	HCO_3^-, organic molecules, biomolecules
Nitrogen	NO_2^-, NO_3^-, organic molecules, biomolecules

a variety of covalent compounds (see Chapter 3) called *organic molecules*. These organic molecules are composed of carbon with some combination of hydrogen, oxygen, and nitrogen. Ultimately, the organic molecules form the fundamental components, or building blocks, of living matter. These component molecules are called biomolecules. Such molecules include amino acids, proteins, nucleic acids, lipids, carbohydrates, enzymes, and genes. Evolution of these molecules leads to cells and, ultimately, life. The chemical evolution of the four major elements into living systems is outlined in Figure 10.3.

Biomolecules are the organic molecules that are synthesized by human beings or are obligatory components of the diet.

Eight elements account for 98% of the atoms in the earth's crust. These elements include (in order of abundance): oxygen, silicon, aluminum, iron, calcium, sodium, potassium, and magnesium. We have already seen from Table 10.1 that oxygen, calcium, sodium, potassium, and magnesium are more abundant in the human body than are silicon, iron, and aluminum. Hydrogen and oxygen account for 88.4% of the atoms in the human body (see Table 10.1), carbon accounts for another 9.4%, and nitrogen, 1.4%. The remaining twenty elements (seven intermediate and thirteen trace elements) account for about 0.7% of the body's atoms.

Three major factors appear to have directed the evolution and chemistry of living organisms. The first is the universal presence of water. Water is a unique compound because it is a stable liquid at physiological temperatures (about 98.6°F). Reactions of living organisms take place in an aqueous medium because many ions essential to the living organism (see Table 10.4) can only exist in an aqueous medium.

The second factor influencing the selection of elements essential for life is related to the size of the atom and the charge per unit volume. Elements of atomic numbers 43, 61, and 92–105 are man-made (see Figure 10.1) and are unavailable

Figure 10.3 Evolution of major elements to organic molecules, biomolecules, and ultimately, life. Major elements are incorporated in more complex biomolecules that are not indicated in this figure.

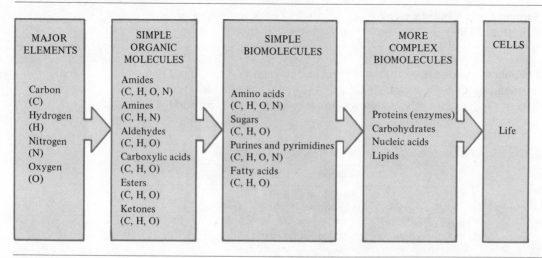

Figure 10.4   Some possible arrangements for carbon atoms.

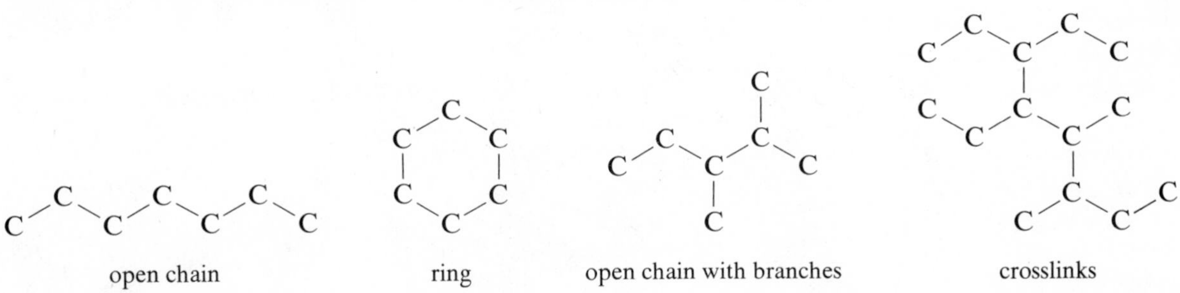

open chain   ring   open chain with branches   crosslinks

to living systems. The elements with atomic numbers 84, 85, and 87–92 are too radioactive to be useful to living organisms (see Chapter 2). The inert gases (Group VIII) are virtually inert. Other elements can be excluded on the basis of their relative unavailability [lanthanides (57–71) and actinides (89–105)] and their high toxicity (mercury, lead, and arsenic). The four most abundant atoms in the human body (hydrogen, carbon, oxygen, and nitrogen) are the smallest and the lightest elements that can achieve stable electronic configurations by sharing electrons with other atoms. This is an important step in forming chemical bonds leading to stable molecules (see Chapter 3). The seven next most abundant elements in living organisms all have atomic numbers below 21. The remaining thirteen elements essential to life are needed only in trace concentrations.

The third factor involves the chemical properties of carbon. Carbon is in the same periodic group as silicon, the second most abundant element in the earth's crust (146 times more abundant than carbon). Like carbon, silicon has the capacity to share four electrons and form four covalent bonds. Why, then, is carbon used as the central building block in forming biomolecules, rather than silicon? The unique features of carbon, but not silicon, are:

1. Carbon can bond with itself to form chains of hundreds of atoms and rings of varying sizes. Furthermore, these chains and rings may have branches and cross links. Some possibilities are illustrated in Figure 10.4.

2. Carbon can form both single and multiple bonds with atoms of many elements, including the halogens, oxygen, nitrogen, sulfur, and phosphorus, and with other carbon atoms. Carbon forms bonds to itself and to other atoms by sharing four electrons and thus can form four covalent bonds (see Chapter 3). These two features of the versatility of carbon are reflected in the fact that millions of organic compounds are found on the earth.

3. Carbon dioxide plays an important role in life processes due, for the most part, to its existence as a gas at physiologic temperatures. Carbon dioxide is soluble in water and always exists as a single molecule. Discrete molecules of silicon dioxide (SiO_2) do not exist; rather, SiO_2 forms a network of silica (SiO_4) tetrahedra. In this network the tetrahedra form hard, dense quartz, which does not melt under a temperature of 1600°. A portion of the silica (SiO_4) tetrahedra is shown in Figure 10.5.

Figure 10.5   A portion of the silica (SiO$_4$) tetrahedra.

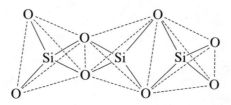

10-4 APPROACH TO THE STUDY OF ORGANIC CHEMISTRY

LEARNING OBJECTIVES

1. Define organic chemistry, covalent bond, functional group, substitution reaction, addition reaction, and elimination reaction.

2. Recognize the common functional groups in organic molecules: carbon–carbon single, double, or triple bonds; halogen; hydroxyl; carbonyl; carboxyl; and amino.

3. Identify the functional group associated with the following class names: alkanes, alkenes, alkynes, aromatics, halides, alcohols, aldehydes and ketones, carboxylic acids, esters, amides, amines, and ethers.

Organic chemistry is of interest to the student of the health-related sciences because of the close relationship between the compounds of carbon and life processes. One of the features of carbon described above is its ability to form a covalent bond with itself or another atom. Consequently a large number of compounds of carbon (*organic compounds*) is possible. Scientists have characterized millions of organic compounds and have divided these compounds into various classes. Each of the classes is characterized by a functional group. The functional groups that we will study, with their names and the name given to the class of organic compounds containing each type of functional group, are tabulated in Table 10.5. In our study of organic compounds we will learn a method for communicating about functional groups called *nomenclature*. Then, we will consider the kinds of chemical reactions that are characteristic of the functional group and relate them to one another. To further illustrate the interrelationship of functional groups, we will integrate some of the applicable reactions of life processes in our study of organic chemistry.

Organic chemistry is the study of the chemistry of the compounds of carbon.

A **functional group** is a collection of chemically bonded atoms with a characteristic set of properties.

10-4.1 THE COVALENT BOND IN ORGANIC COMPOUNDS

The characteristic bond of organic compounds is the covalent bond (see Section 3-2). In organic compounds carbon can form one or more covalent bonds to another carbon atom or to other atoms. If one carbon atom shares two electrons with a second carbon atom or a different type of atom, a single bond is formed.

Table 10.5 Organic Compounds by Functional Group and Class

Functional Group Name	Functional Group	Class Name
Carbon–carbon single bond	$-\overset{\mid}{C}-\overset{\mid}{C}-$	Alkane
Carbon–carbon double bond	$-C=C-$	Alkene
Carbon–carbon triple bond	$-C\equiv C-$	Alkyne
Aromatic*	(benzene ring)	Aromatic, or compounds derived from benzene
Halide	$-F, -Cl, -Br,$ or $-I$	Organic halide
Hydroxyl	$-OH$	Alcohol or phenol
Thiol	$-SH$	Thiol (mercaptan)
Ether	$-\overset{\mid}{C}-O-\overset{\mid}{C}-$	Ether
Sulfide	$-\overset{\mid}{C}-S-\overset{\mid}{C}-$	Sulfide
Disulfide	$-\overset{\mid}{C}-S-S-\overset{\mid}{C}-$	Disulfide
Carbonyl	$-\overset{O}{\overset{\parallel}{C}}-$	Aldehyde or ketone
Carboxyl	$-\overset{O}{\overset{\parallel}{C}}-OH$	Carboxylic acid
Ester	$-\overset{O}{\overset{\parallel}{C}}-O-\overset{\mid}{C}-$	Ester
Amide	$-\overset{O}{\overset{\parallel}{C}}-N-$	Amide
Amino	$-N-H, -N-H,$ or $-N-$	Amine
Cyano	$-C\equiv N$	Nitrile

*A cyclic system of six carbons with a unique overlap of *p* orbitals (indicated by circle).

Double and triple bonds are also possible. A double bond involves sharing four electrons between two atoms, and a triple bond involves sharing six electrons between two atoms.

Single covalent bonds: C—C   C—Cl   C—O   C—N   *each atom shares two electrons*

Double covalent bonds: C=C   C=O   C=N   *each atom shares four electrons*

Triple covalent bonds: C≡C   C≡N   *each atom shares six electrons*

Each kind of covalent bond has a characteristic length and bond strength. Also, the covalent bond imparts a particular shape to each molecule in which it exists. In Chapter 11 we will consider the shapes imparted to several molecules as a result of the covalent bonds in the molecule. We also will see how the shape of a molecule can influence our sense of smell.

10-4.2 CHEMICAL REACTIONS

The chemical reactions of the various classes of organic compounds are, for the most part, determined by the functional group. The chemical reactions that we will be studying are the interconversions of the organic functional groups themselves. The two predominant reactions are **substitution** and **addition**. A third reaction sometimes encountered is **elimination**. With the exception of carbon–carbon single bonds, which are chemically unreactive, each of the functional groups in Table 10.5 undergoes at least one of these reactions. For example, the addition reaction is characteristic of carbon–carbon double bonds (C=C), carbon–carbon triple bonds (C≡C), and carbon–oxygen double bonds (C=O). For example, the reaction between ethene, an alkene, and bromine as shown is an addition reaction.

A **substitution reaction** is a reaction in which one atom or a group of atoms is replaced by another atom or group of atoms.

An **addition reaction** is a reaction in which two molecules combine to form a single molecule.

An **elimination reaction** is a reaction in which an organic molecule gives up a group of atoms.

$$
\begin{array}{c}
\text{Br} \ \ \text{H} \\
| \ \ \ | \\
\text{H}-\text{C}=\text{C}=\text{H} + \text{Br}_2 \longrightarrow \text{H}-\text{C}-\text{C}-\text{H} \\
| \ \ | \ \ \ \ \ \ \ \ \ \ \ \ \ \ \ \ | \ \ \ | \\
\text{H} \ \ \text{H} \ \ \ \ \ \ \ \ \ \ \ \ \ \ \text{H} \ \ \text{Br}
\end{array}
$$

ethene   bromine
an alkene

Substitution is characteristic of carbon–chlorine single bonds (C—Cl) and carbon–oxygen single bonds (C—OH). For example, the reaction between methyl chloride, an organic halide, and sodium hydroxide as shown is a substitution reaction.

$$
\begin{array}{c}
\text{H} \text{H} \\
| \ | \\
\text{H}-\text{C}-\text{Cl} + \text{NaOH} \longrightarrow \text{H}-\text{C}-\text{OH} + \text{NaCl} \\
| \ | \\
\text{H} \text{H}
\end{array}
$$

methyl chloride
an organic halide

Elimination is also characteristic of carbon–chlorine single bonds (C—Cl) and carbon–oxygen single bonds (C—OH). For example, ethyl alcohol, an alcohol, can eliminate a molecule of water as shown in the reaction below.

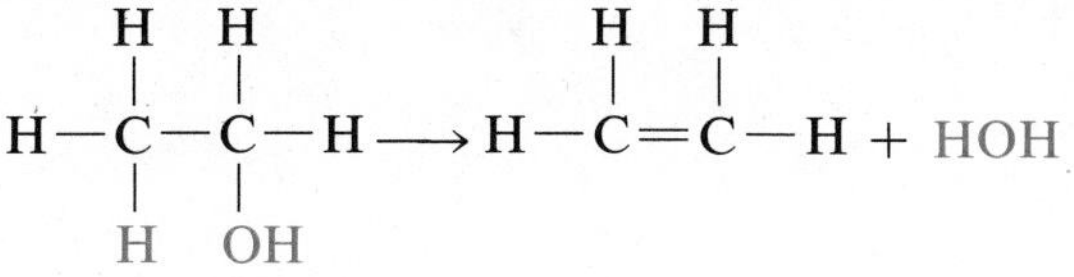

$$H-\overset{\overset{\displaystyle H}{|}}{\underset{\underset{\displaystyle H}{|}}{C}}-\overset{\overset{\displaystyle H}{|}}{\underset{\underset{\displaystyle OH}{|}}{C}}-H \longrightarrow H-\overset{\overset{\displaystyle H}{|}}{C}=\overset{\overset{\displaystyle H}{|}}{C}-H + HOH$$

ethyl alcohol
an alcohol

In this text, we will learn to recognize and name functional groups and consider the interconversions of the functional groups. Furthermore, we will study many examples from biochemistry, the third part of this text, to illustrate the chemistry of organic molecules.

PROBLEM

3. Identify the functional group (shaded) in each of the following.
   (a) a compound used in some sun tan lotions as a sunscreen

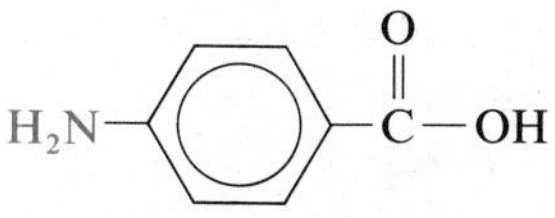

   $$H_2N-\underset{}{\bigcirc}-\overset{\overset{\displaystyle O}{||}}{C}-OH$$

   p-aminobenzoic acid

   (b) a compound with analgesic properties

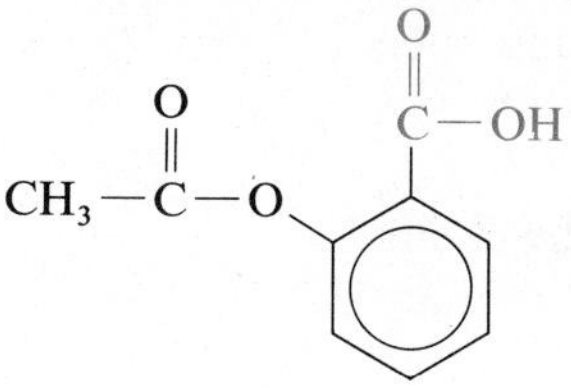

   $$CH_3-\overset{\overset{\displaystyle O}{||}}{C}-O-\underset{}{\bigcirc}\overset{\overset{\displaystyle O}{||}}{\underset{}{C}}-OH$$

   aspirin

   (c) a compound used as an antihypertensive

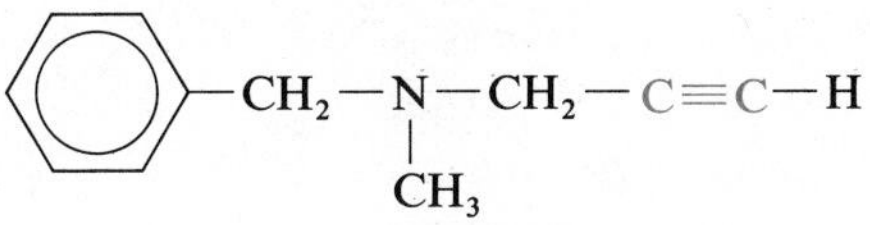

   $$\bigcirc-CH_2-\underset{\underset{\displaystyle CH_3}{|}}{N}-CH_2-C\equiv C-H$$

   pargyline

   (d) a nerve impulse transmitter

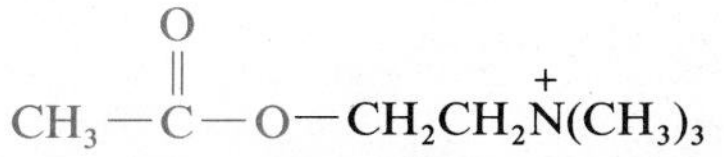

   $$CH_3-\overset{\overset{\displaystyle O}{||}}{C}-O-CH_2CH_2\overset{+}{N}(CH_3)_3$$

   acetyl choline

(e) a component of Neo-synephrine, a nasal decongestant

phenylephrine

ANSWERS TO PROBLEM **3 (a)** amino group **(b)** carboxyl group **(c)** carbon–carbon triple bond **(d)** ester **(e)** hydroxyl

---

10-5 ROLE OF ORGANIC COMPOUNDS IN OUR LIVES

Considering the unique features of carbon compounds (see Section 10-3), it should not surprise you to learn that organic compounds are the primary components of all that surrounds us. The essential needs of our daily life are food, fuel, and clothing. The principal components of food are carbon compounds. These carbon compounds include carbohydrates (starches and sugars), animal fats, vegetable oils, and proteins (milk, fish, and meat). The fuel we use for heating or in our cars is, primarily, petroleum which is chiefly natural gas and a mixture of many different organic compounds. Clothing is made from natural fibers, such as cotton and silk, and from synthetic fibers, such as Dacron polyester fiber and nylon. Both the natural and synthetic fibers are organic compounds. Organic compounds make up dyes for clothes and color film, perfumes, synthetic detergents, flavorings for food, oral contraceptives, pesticides, rubber tires, plastics, and many drugs, such as antihistamines for hay fever, sulfa drugs and antibiotics for infections, analgesics for pain, and psychochemicals for inducing behavioral changes.

Organic compounds also play an integral role in the chemistry of the living system. The digestion and metabolism of the foods are organic reactions that are catalyzed by enzymes that also are organic compounds. The chemistry of the living system is a relevant application of the chemistry of organic compounds, since most processes in the living organism involve organic substances. However, inorganic ions also play an important role in the chemistry of living systems. As we discussed in Section 10-1, Fe^{2+} is crucial for the transport of oxygen in the blood because it is an integral part of the hemoglobin molecule. It is also vital to the series of reactions making up the respiratory chain (see Sections 9-3.1 and 25-2). We learned in Chapter 10, and will study further in Chapter 23, that certain enzymes require Mg^{2+}, Zn^{2+}, Mn^{2+}, Cl^-, and other ions for their activity. Calcium is necessary for the formation of teeth and bone tissue which also contains considerable amounts of protein. Phosphate is essential for metabolism, and the bicarbonate ion is a component of the most important buffering system in the human body (see Section 8-4.3). As these few examples indicate, an important marriage of inorganic and organic chemistry takes place in all living organisms.

Thus far we have concentrated primarily on the chemistry of elements other than carbon. In Chapters 11–21 and 26, we will examine the chemistry of organic compounds; that is, compounds containing carbon and hydrogen and, sometimes, oxygen, nitrogen, and sulfur atoms. In Chapters 19–21 and 26, we will concentrate on the chemistry of the most abundant organic compounds of living cells. Then, in Chapters 20, 22, 24, 25, 27, and 28, we will describe some of the chemical reactions that are essential to human life. It is the study of these reactions that take place in living organisms that makes up the field of science known as *biochemistry*.

SUMMARY

About 24 elements are known to be essential for a healthy human life. These elements can be grouped in three broad categories: (1) trace elements, (2) intermediate elements, and (3) major, or most abundant, elements. The elements in each category and their general roles are summarized in Table 10.2.

Trace elements are found in very small amounts in the living system and function in many enzyme-catalyzed reactions. The function of the intermediate elements depends on the element. Among other processes, these elements are involved in electroneutrality maintenance, bone formation, and enzymatic processes. The major elements are hydrogen, carbon, nitrogen, and oxygen. They combine to form the fundamental structural components of living organisms, organic molecules (amides, amines, esters, and so forth). The systematic study of organic molecules is fundamental to organic chemistry.

The major elements in organic molecules tend to form covalent bonds with one another; thus, organic chemistry sometimes is called the chemistry of the covalent bond. The covalent bond is present in a unique collection of atoms called the functional group. Organic chemistry deals with the nomenclature and interrelationships of the functional groups. The latter is deduced from studying the chemical reactions of the various functional groups.

A good understanding of organic and inorganic chemistry is necessary for learning biochemistry. In the biochemistry section of this text we will study the organic and inorganic chemistry, as well as the biological aspects, of the cellular activities of the living cell.

ADDITIONAL PROBLEMS

4. What are the three broad classes of elements essential to life? Which elements are in each class?
5. Summarize the *in vivo* function of each of the broad classes of elements essential to life.
6. Define trace element.
7. Identify the trace element that fits each of the following descriptions:
   (a) found in greatest abundance in biological systems
   (b) incorporated in two hormonal regulators, and a deficiency causes enlargement of the thyroid
   (c) absence causes skeletal abnormalities such as bowing of legs
   (d) minute amounts prevent white muscle disease in animals
   (e) accumulation in the brain and liver causes mental illness and can lead to death
   (f) may function as a biological cross-linking agent in connective tissue
   (g) plays an important catalytic role in the conversion of atmospheric nitrogen to ammonia
   (h) absence can cause weak arteries
   (i) inhibits cholesterol synthesis
8. How do calcium ions affect blood clotting time?
9. How do phosphorus and calcium appear together in the body?

10. Why are green vegetables a good source of magnesium?

11. How does an excess of magnesium affect cells?

12. Why is Epsom salts an effective purgative?

13. What is the main function of potassium, sodium, and chloride ions?

14. What are the four most important forms of sulfur in the living system?

15. How are the four major elements incorporated in biomolecules?

16. Summarize the unique features of carbon that indicate why it is part of the structural components of living matter.

17. What type of bond is characteristic of organic molecules? Describe the bond.

18. How many electrons are shared between the two atoms of each of the following bonds.

   (a) $C=S$ (b) $C\equiv N$ (c) $C\equiv C$
   (d) $C-Cl$ (e) $C-O$

19. What is a functional group?

20. Name the functional group (in color) in each of the following structures.

   (a) 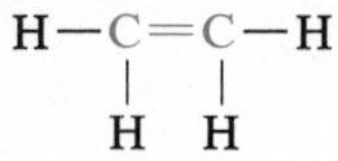 $H-C=C-H$
        $\quad\;\; | \quad\;\; |$
        $\quad\;\; H \quad\; H$

(b) $H-C=O$
     $\quad\;\; |$
     $\quad\;\; H$

(c) $H-C-C-OH$ with H, O (double bond)
     $\quad\;\; |$
     $\quad\;\; H$

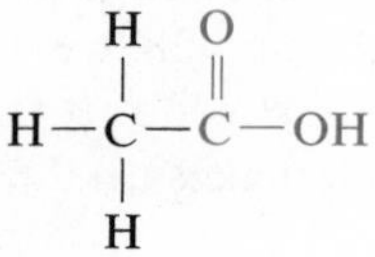

(d) 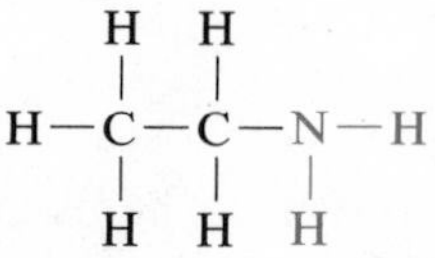 $H-C-C-N-H$
     with H, H / H, H, H

(e) 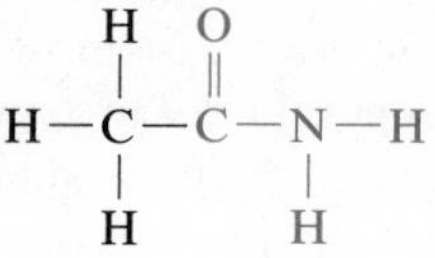 $H-C-C-N-H$
     with H, O (double bond) / H, H

(f) 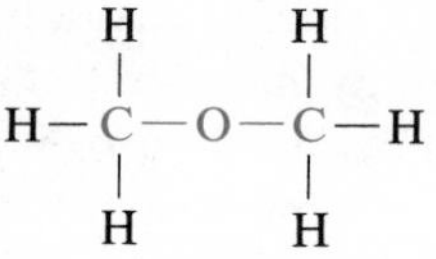 $H-C-O-C-H$
     with H, H / H, H

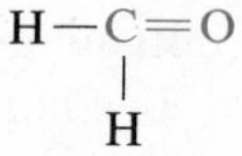

21. Define addition, substitution, and elimination reactions.

CHAPTER 11

HYDROCARBONS: ALKANES, ALKENES, AND ALKYNES

There are many organic compounds, called **hydrocarbons**, that are composed of only carbon and hydrogen. Many of these compounds are obtained from natural gas, petroleum, coal, and green plants. Hydrocarbons are very important in our lives. They can be burned to provide heat to keep us warm and to power our automobiles. Hydrocarbons also are used to make a variety of practical compounds that have found uses in plastic products and nonstick cookware, and as substitutes for rubber.

A **hydrocarbon** is an organic molecule that contains only carbon and hydrogen.

There are several kinds of hydrocarbons. In this chapter we will study three kinds: alkanes, alkenes, and alkynes. (We shall discuss another kind of hydrocarbon, aromatic hydrocarbon, in Chapter 12.) We will study the shapes associated with simple hydrocarbons and learn how our sense of smell depends on the shapes of molecules. The alkenes constitute a particularly important class of hydrocarbons because they undergo a variety of addition reactions that are useful to organic chemists.

11-1 STRUCTURE AND SHAPES

LEARNING OBJECTIVES

1. Define, recognize structures of, or give examples (with structures) of hydrocarbon, unsaturated hydrocarbon, and saturated hydrocarbon.
2. Determine the bonding differences between alkanes, alkenes, and alkynes.
3. Distinguish and recognize notations for sigma and pi bonds.

4. Describe the geometries or shapes of the simple organic molecules: methane, ethane, ethene, and ethyne.
5. Explain the importance of shape in organic molecules.

11-1.1 HYDROCARBON STRUCTURES

All alkanes, except methane, are characterized by the fact that they contain only carbon–carbon single bonds. Alkenes and alkynes contain at least one carbon–carbon double bond and one carbon–carbon triple bond, respectively. Alkanes are often referred to as saturated hydrocarbons. Alkenes and alkynes are called unsaturated hydrocarbons. Examples of simple saturated hydrocarbons, methane and ethane, and unsaturated hydrocarbons, ethene and ethyne, are shown.

A **saturated hydrocarbon** is a hydrocarbon that contains the maximum number of four atoms or groups bonded to each carbon.

An **unsaturated hydrocarbon** is a hydrocarbon that contains fewer than the maximum number of atoms or groups. *Unsaturated* implies that a multiple bond is present.

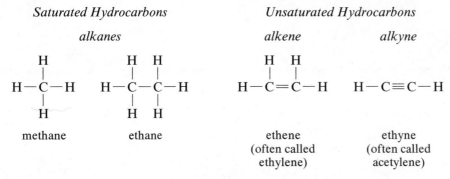

Saturated Hydrocarbons

alkanes

methane   ethane

Unsaturated Hydrocarbons

alkene   alkyne

ethene   ethyne
(often called   (often called
ethylene)   acetylene)

Figure 11.1 The tetrahedral carbon atom in methane and ethane. (a) A tetrahedral figure. (b) Methane in its tetrahedral shape. (c) Ball-and-stick model of methane. The black balls represent hydrogen and the color balls represent carbon. (d) The ethane molecule showing the two tetrahedra. (e) Ball-and-stick model of ethane.

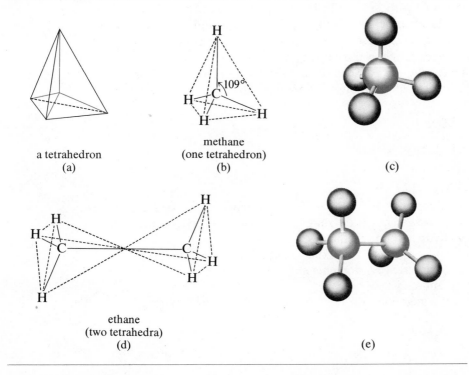

a tetrahedron
(a)

methane
(one tetrahedron)
(b)

(c)

ethane
(two tetrahedra)
(d)

(e)

The geometries, or shapes, associated with the molecules of these four simple hydrocarbons differ. Methane and ethane are tetrahedral, ethene is planar, and ethyne is a linear molecule.

In saturated hydrocarbons, carbon is bonded to four other atoms. In such a situation, the four atoms are found at the corners of a tetrahedron with the carbon atom at the center. Note the tetrahedral shapes of the methane and ethane molecules shown in Figure 11.1. The angle between any pair of bonds in either methane or ethane is the tetrahedral angle, 109°. All the bonds in methane and ethane are cylindrically symmetrical about an axis, or line, joining either the carbon–hydrogen or the carbon–carbon nuclei. Bonds that are symmetrical about a line joining two nuclei are called **sigma (σ) bonds.** Sigma bonds were described in Chapter 3.

> A **tetrahedron** is a geometric figure bounded by four plane sides.

> A **sigma bond** is a bond that is symmetrical about a line joining two nuclei.

Experimental evidence indicates that the ethene molecule is planar and the ethyne molecule, linear. The atoms of a molecule that lie in a plane form bond angles of 120°. Thus, each bond angle in ethene is 120°. The atoms of a molecule that lie along a continuous line form bond angles of 180°. Thus each bond angle in ethyne is 180°. The structures of ethene and ethyne are shown in Figures 11.2 and 11.3. In ethene there are two carbon atoms joined by two electron pairs (a carbon–carbon double bond). Only one of these pairs occupies the space along the bond axis. This one pair of electrons forms a sigma bond. The second pair of electrons occupies the regions of space above and below the plane of the molecule.

Figure 11.2   (a) The ethene molecule. (b) Ball-and-stick model of ethene.

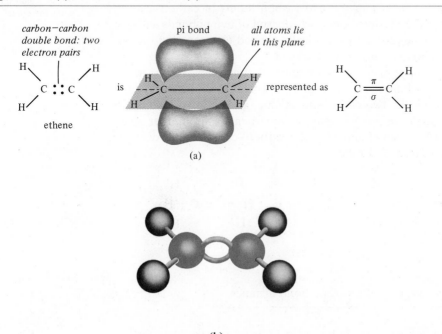

(a)

(b)

Figure 11.3   (a) The ethyne molecule. (b) Ball-and-stick-model of ethyne.

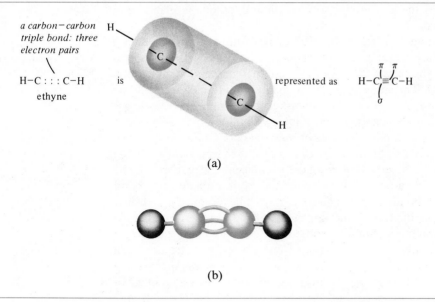

a carbon–carbon triple bond: three electron pairs

H−C ∶ ∶ ∶ C−H   is   represented as $H-C\overset{\pi\ \pi}{\underset{\sigma}{\equiv}}C-H$

ethyne

(a)

(b)

This kind of bond is a **pi (π) bond**. A pi bond is not symmetrical along the bond axis. Pi bonds were discussed in Chapter 3.

In ethyne, three electron pairs join the two carbon atoms (a carbon–carbon triple bond). The carbon–carbon triple bond is composed of two pi bonds and one sigma bond. The two pi bonds form a cylindrically shaped electron cloud about the carbon–carbon sigma bond as shown in Figure 11.3.

An increase in the number of pi bonds between two carbon atoms results in the shortening of the carbon–carbon bond distance. The two carbons in ethyne, then, are closer together than are the two carbons in ethene, and the two carbons in ethane are farther apart than are those in ethyne and ethene. The number given in parentheses under each structure shown is the carbon–carbon bond distance in angstrom units (one angstrom = 10^{-8} cm).

> A **pi bond** is a bond in which a pair of electrons occupies the region above and below the nuclei.

$$H-C\equiv C-H$$

ethyne
2 pi bonds
(1.21 Å)

$$\underset{H}{\overset{H}{\diagdown}}C=C\underset{H}{\overset{H}{\diagup}}$$

ethene
1 pi bond
(1.34 Å)

$$H-\underset{H}{\overset{H}{C}}-\underset{H}{\overset{H}{C}}-H$$

ethane
no pi bonds
(1.54 Å)

Increasing carbon–carbon bond distance
⟶

In summary, carbon forms only single bonds to hydrogen or to other carbon atoms in an alkane. An alkene always has at least one carbon–carbon double bond,

Figure 11.4   (a) The propene molecule showing the tetrahedral and planar arrangements. (b) The ball-and-stick model of propene.

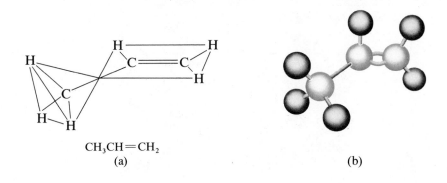

$$CH_3CH{=}CH_2$$
(a)   (b)

and an alkyne always has at least one carbon–carbon triple bond. The molecule of the alkanes methane and ethane exhibits a tetrahedral arrangement of atoms. In the molecule of the alkene ethene, the carbon–carbon double bond shows a planar arrangement of atoms. In the molecule of the alkyne ethyne, the carbon–carbon triple bond shows a linear arrangement of atoms. However, most hydrocarbon molecules have a combination of tetrahedral, planar, and linear arrangements of atoms. For example, some of the atoms in propene have a tetrahedral arrangement and the others have a planar arrangement as shown in Figure 11.4.

11-1.2 SENSE OF ODOR

We must have a good understanding of the shapes, or geometry, associated with molecules in order to discuss the sense of odor. According to the theory of odor, different olfactory nerve endings send signals that are characteristic of the odor to the brain centers where the signals are integrated and interpreted in terms of the character and intensity of the odor. It is now thought that there are seven primary odors and seven corresponding receptor sites of varying sizes, shapes, and chemical affinities. Thus, different odor sensations of a substance are determined by the shapes and sizes of the molecules concerned. The seven different odors are camphoric, musk, floral, ether, peppermint, pungent, and putrid. These seven odors and the hypothetical shapes of the olfactory receptor sites in the nasal cavity are illustrated in Table 11.1.

Molecules that have simple odors fit a given odor-class receptor site. If a molecule fits more than one site, it will have a complex odor. To give a particular odor, it is necessary that only a portion of the molecule fit the receptor site. However, the strength of the odor increases with an increase in the degree to which the molecule fits the site. For example, molecules can be chemically unrelated and yet fit the bowl-shaped receptor site for camphoric substances, thus possessing the camphoric odor.

Table 11.1   Shapes of Molecules and Associated Odors

Primary Odor	Familiar Example	Receptor Site*	Molecular Geometry
Camphoric	Moth balls		Spherical (bowl-shaped)
Ether	Dry-cleaning fluids		Rod-shaped
Floral	Roses		Disk with tail (kite-shaped)
Peppermint	Mint candy		Wedge-shaped
Musk	Damp basement		Disk-shaped
Pungent	Vinegar		Strong affinity for electrons
Putrid	Rotten eggs		Strong affinity for + center

*After an illustration from "The Stereochemical Theory of Odor" by John E. Amoore, James W. Johnston, Jr., and Martin Rubin, *Sci. Amer.* (February 1964), pp. 44–45.

As shown in Table 11.1, the molecules of the musk odor take the shape of a disk. The etheral odor is produced by rod-shaped molecules. The pleasant floral odor is due to molecules that have a "kite" shape (a disk with a tail). Molecules that have a cool, peppermint odor have the shape of a wedge. Note, however, that pungent and putrid molecules fit a receptor site because of electrical charge, not shape. The pungent molecules have a deficiency of electrons and, therefore, have a positive charge and a strong affinity for electrons. Putrid odors are due to molecules with an excess of electrons and are strongly attracted to the nuclei of adjacent atoms.

This concept that shape is related to function can be extended to include physiological actions of chemicals. Drug activity, for example, has been related to both shape and the charge distribution of the molecules of the drug. It is thought that there are receptor sites of definite shape in the brain and that the analgesic (pain-relieving) property of morphine and related compounds is due to their structural similarity to naturally occurring pain modulators (see Section 17-5.1).

11-2 NATURALLY OCCURRING HYDROCARBONS

LEARNING OBJECTIVE

Describe the kinds of hydrocarbons occurring in petroleum and green plants.

11-2.1 PETROLEUM AND COAL

Petroleum and coal are abundant natural sources of hydrocarbons. These sources were formed from the decomposition over millions of years of previously living plant and animal matter.

Natural gas, which is chiefly methane and is used as a heating fuel, is often found with petroleum. *Petroleum,* or crude oil, is a complex mixture of alkanes, alkenes, and aromatic hydrocarbons. Aromatic hydrocarbons contain a unique cyclic structure of six carbon atoms (see Chapter 12). Petroleum is usually dark-colored and viscous, and contains gases, liquids, and solids. The hydrocarbon molecules present in crude oil vary in size from those containing a single carbon atom to those containing 30 or more carbon atoms. The hydrocarbon molecules containing up to 4 carbon atoms are usually gases. Those hydrocarbon molecules containing 5 to 16 carbon atoms are liquids, and those with 17 or more are usually solids. To separate the hydrocarbons, petroleum is refined. Refining is accomplished by distillation in stages at different temperatures to a maximum of about 500°. Each fraction that is distilled differs in boiling range and in hydrocarbon composition from other fractions. The major hydrocarbon fractions consist mainly of alkanes and their uses are given in Table 11.2.

Many of the lighter liquid alkanes that result from the refinement of petroleum (C_5 and C_6, for example) can dissolve body oils on contact with the skin and cause dermatitis. On the other hand, heavier liquid alkanes act as softeners when applied to the skin. Mineral oil, a mixture of the heavier alkanes, is used on the skin to replace natural oils that have been washed away by frequent bathing and swimming. Vaseline petroleum jelly is a semisolid mixture of hydrocarbons that can be applied as a skin softener or as a protective film. For example, diaper rash is controlled because an infant's urine (a water solution) will not dissolve the protective film obtained when Vaseline is applied to an infant's skin.

Table 11.2 Composition and Uses of Petroleum

Fraction (bp < 20°–bp > 500°)	Composition	Use
Gas and liquefied gas	C_1—C_4 methane, ethane, propane, butane	Heating fuel
Naphtha	C_5—C_7 hydrocarbons	Solvents
Gasoline	C_6—C_{10} hydrocarbons	Motor fuels
Kerosene	C_{11}—C_{12} hydrocarbons	Jet engine fuel
Gas oil	C_{13}—C_{25} hydrocarbons	Diesel engine fuel, oil-burning furnace fuel
Lubricants	C_{26}—C_{38} hydrocarbons	Paraffin wax, petroleum jelly (Vaseline)
Residue		Asphalt

Coal, like petroleum, is an abundant source of hydrocarbons. However, petroleum technology has surpassed coal technology and we rely on petroleum as the major source of fuels and other chemicals. This situation may change in the near future, however, because dwindling crude oil reserves may induce us to turn to coal as a major source of fuel. In Section 11-5.1 we will see why petroleum and coal are useful heating fuels.

11-2.2 GREEN PLANTS

Isoprene is an unsaturated hydrocarbon that contains two double bonds (a diene). The five-carbon isoprene unit occurs in a wide variety of compounds found in nature. Compounds that contain the isoprene unit are called *terpenes*. Terpenes are the volatile, odoriferous constituents of many plants such as pine, basil, bayberry, rose, and eucalyptus. Examples of terpenes are shown in Figure 11.5.

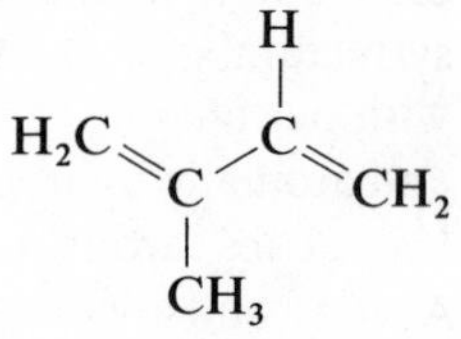

Some green plants make hydrocarbons that contain the carbon–carbon double and triple bonds. For example, *Coreopsis,* common garden plants with a daisy-like head, make the following hydrocarbons.

$$CH_2 = CH - C \equiv C - C \equiv C - C \equiv C - C \equiv C - CH = CH_2$$

$$CH_3 - CH = CH - C \equiv C - C \equiv C - CH = CH - CH = CH - CH = CH_2$$

Figure 11.5   Some natural terpenes and their sources. The isoprene units in each terpene are indicated.

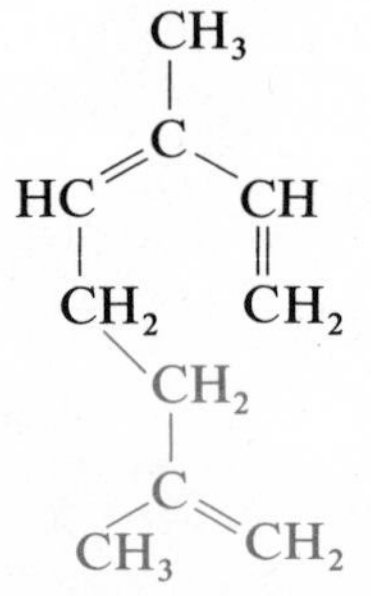

ocimene
(from basil)

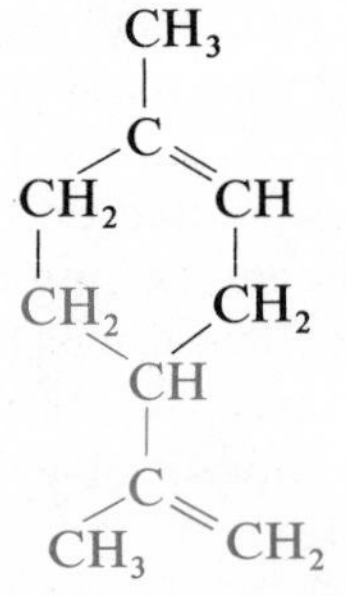

limonene
(from oil of lemon and
oil of orange)

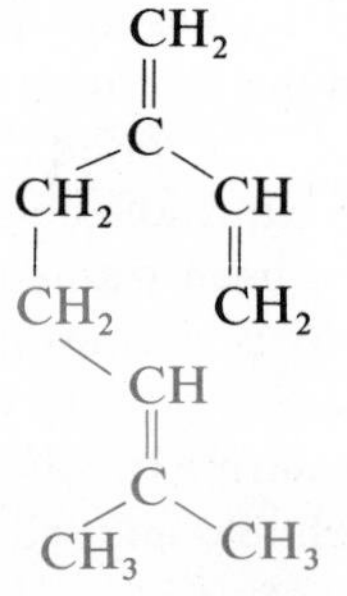

myrcene
(from bayberry)

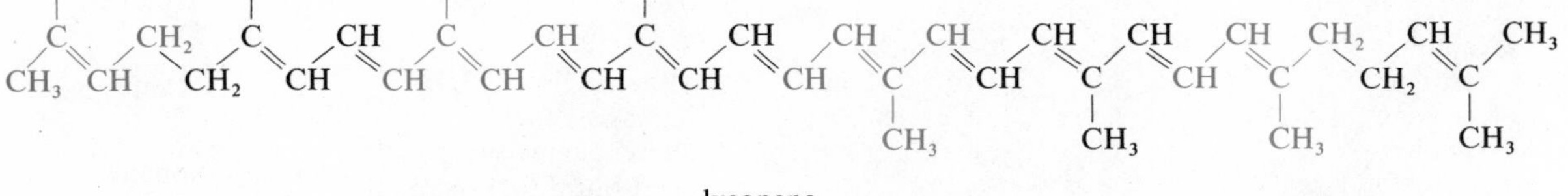

lycopene
(red pigment in ripe fruit, particularly tomatoes)

11-3 NOMENCLATURE OF HYDROCARBONS

LEARNING OBJECTIVES

1. Define or identify examples of cyclo-hydrocarbons, alkanes, alkenes, alkynes, homologs, and an alkyl group.
2. Given the structural formula of an alkane, alkene, alkyne, or cyclohydro-carbon, write its name using the IUPAC system.

3. Given the IUPAC name of an al-kane, alkene, alkyne, or cyclohydro-carbon, write its condensed structural formula.

11-3.1 ALKANES

A conventional method is used to name organic compounds. The method was devised by a group of scientists and is referred to as the IUPAC rules (International Union of Pure and Applied Chemistry). For all organic compounds, the IUPAC name consists of a prefix and a suffix. The prefix indicates the number of carbon atoms in the longest, continuous chain of carbon atoms and the suffix is an ending characteristic of that class of compounds. For **alkanes,** the characteristic ending is *-ane*. Note that this is the suffix found in the general name of this class of compounds: alk*ane*. We will discover later that a similar situation exists with alkenes and alkynes as well as other classes of organic compounds. The IUPAC names for ten alkanes and their **structural formulas** are listed in Table 11.3.

An **alkane** is a saturated hydrocarbon with the general formula C_nH_{2n+2}.

A **structural formula** is a notation that tells the exact number of atoms in a molecule and indicates which atoms are bonded to each other in the molecule.

Table 11.3   Names and Formulas of Alkanes

Molecular Formula	Structural Formula	Prefix	+ Suffix =	IUPAC Name
CH_4	CH_4	*Meth-*	*-ane*	Methane
C_2H_6	CH_3CH_3	*Eth-*	*-ane*	Ethane
C_3H_8	$CH_3CH_2CH_3$	*Prop-*	*-ane*	Propane
C_4H_{10}	$CH_3CH_2CH_2CH_3$ or $CH_3(CH_2)_2CH_3$	*But-*	*-ane*	Butane
C_5H_{12}	$CH_3CH_2CH_2CH_2CH_3$ or $CH_3(CH_2)_3CH_3$	*Pent-*	*-ane*	Pentane
C_6H_{14}	$CH_3CH_2CH_2CH_2CH_2CH_3$ or $CH_3(CH_2)_4CH_3$	*Hex-*	*-ane*	Hexane
C_7H_{16}	$CH_3CH_2CH_2CH_2CH_2CH_2CH_3$ or $CH_3(CH_2)_5CH_3$	*Hept-*	*-ane*	Heptane
C_8H_{18}	$CH_3CH_2CH_2CH_2CH_2CH_2CH_2CH_3$ or $CH_3(CH_2)_6CH_3$	*Oct-*	*-ane*	Octane
C_9H_{20}	$CH_3CH_2CH_2CH_2CH_2CH_2CH_2CH_2CH_3$ or $CH_3(CH_2)_7CH_3$	*Non-*	*-ane*	Nonane
$C_{10}H_{22}$	$CH_3CH_2CH_2CH_2CH_2CH_2CH_2CH_2CH_2CH_3$ or $CH_3(CH_2)_8CH_3$	*Dec-*	*-ane*	Decane

Each member of the alkane series listed in Table 11.3 differs from the next lower or the next higher member by a constant amount: one carbon and two hydrogen atoms. Such a series is called a *homologous series* and each member of the series is called a homolog. Propane is a homolog of ethane and butane.

Most alkanes are not simple, continuous chains of carbon atoms as shown in Table 11.3. Many alkanes are derived from those shown in Table 11.3 by replacement of hydrogen by alkyl groups. Such an alkane is referred to as a branched-chain alkane. An example is shown below.

$$CH_3 \longleftarrow \text{ an alkyl group has replaced a hydrogen}$$
$$|$$
$$CH_3CHCH_3 \longleftarrow \text{ continuous chain of carbon atoms as in propane}$$

a branched-chain alkane

There are many possible alkyl groups, but we will be concerned only with those listed in Table 11.4. An alkyl group is named by dropping the suffix -*ane* from the corresponding alkane and replacing it with the suffix -*yl* as shown in Table 11.4. For example, a methyl group is derived from methane by removal of any one of the four hydrogen atoms. The compound shown above contains a methyl group.

To complete the name of the compound shown above, we use the following IUPAC rules:

1. Choose the longest, continuous chain of carbon atoms. The IUPAC name of this hydrocarbon is the *parent name* of the compound (see Table 11.3).

Homologs are the members of a series of compounds in which each member differs from the next member by a CH_2 grouping.

An **alkyl group** is an assembly of covalently bonded atoms derived from the corresponding alkane molecule by the removal of a hydrogen atom.

Table 11.4   Some Common Alkyl Groups

Alkane*	Alkyl Group	Name
CH_4	CH_3-	Methyl
CH_3CH_3	CH_3CH_2-	Ethyl
$CH_3CH_2CH_3$	$CH_3CH_2CH_2-$	Propyl
$CH_3CH_2CH_3$	CH_3CH- $\quad\quad\|$ $\quad\; CH_3$	Isopropyl
$CH_3CH_2CH_2CH_3$	$CH_3CH_2CH_2CH_2-$	Butyl
$CH_3CH_2CH_2CH_3$	CH_3CH_2CH- $\quad\quad\quad\|$ $\quad\quad\; CH_3$	Secondary butyl or *sec*-butyl
$\quad\; CH_3$ $\quad\quad\|$ CH_3CHCH_3	$\quad\; CH_3$ $\quad\quad\|$ CH_3CHCH_2-	Isobutyl
$\quad\; CH_3$ $\quad\quad\|$ CH_3CHCH_3	$\quad\; CH_3$ $\quad\quad\|$ CH_3C- $\quad\quad\|$ $\quad\; CH_3$	Tertiary butyl or *tert*-butyl

*Hydrogen removed to give alkyl group is indicated in color.

2. Locate each alkyl group by numbering the longest, continuous chain of carbon atoms starting at the end of the chain nearest the alkyl group.

3. Name the alkyl group(s). Sometimes an alkyl group appears more than once. In these cases, the prefixes *di-, tri-,* and *tetra-* are used to indicate two, three, and four alkyl groups of the same kind. The position or location of each alkyl group must be indicated.

Using these rules the name of the alkane above is 2-methylpropane.

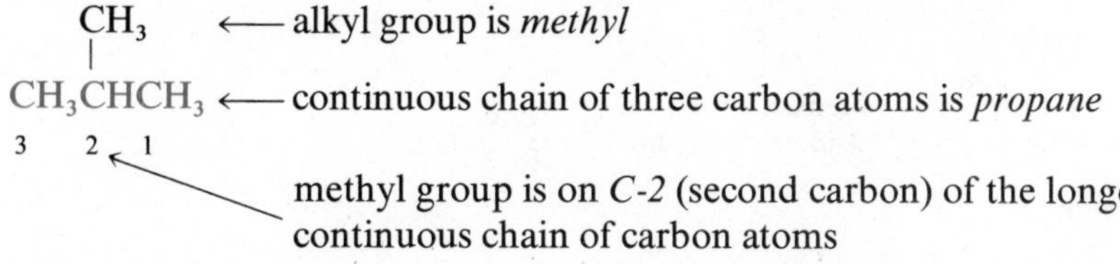

CH_3 ⟵ alkyl group is *methyl*

CH_3CHCH_3 ⟵ continuous chain of three carbon atoms is *propane*

   3   2 1

methyl group is on *C-2* (second carbon) of the longest continuous chain of carbon atoms

The structural formulas shown for 2-methylpropane and the alkanes in Table 11.3 are called condensed structural formulas because they do not show all the covalent bonds to each carbon atom. In the more complete or full structural formula (or simply, *structure*) of a hydrocarbon, all four covalent bonds to a carbon atom are shown as well as the one covalent bond between carbon and hydrogen. The complete and condensed structures for pentane are as follows.

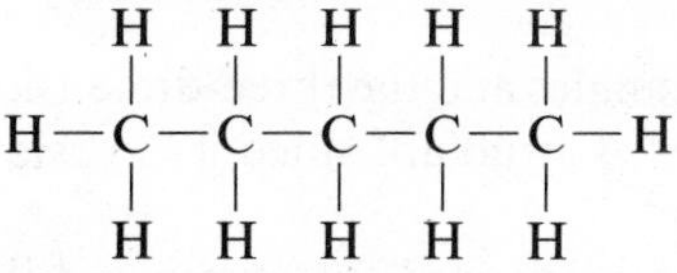

$$H-\overset{\overset{H}{|}}{\underset{\underset{H}{|}}{C}}-\overset{\overset{H}{|}}{\underset{\underset{H}{|}}{C}}-\overset{\overset{H}{|}}{\underset{\underset{H}{|}}{C}}-\overset{\overset{H}{|}}{\underset{\underset{H}{|}}{C}}-\overset{\overset{H}{|}}{\underset{\underset{H}{|}}{C}}-H$$

complete structure for pentane

$$CH_3CH_2CH_2CH_2CH_3 \quad \text{or} \quad CH_3(CH_2)_3CH_3$$

condensed structures for pentane

A comparison of the complete and condensed structures shows us that

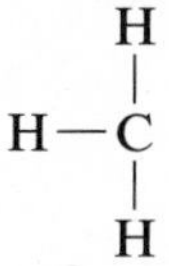

$$H-\overset{\overset{H}{|}}{\underset{\underset{H}{|}}{C}} \quad \text{condenses to} \quad CH_3$$

and

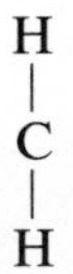

$$\overset{\overset{H}{|}}{\underset{\underset{H}{|}}{C}} \quad \text{condenses to} \quad CH_2$$

and

$$CH_2CH_2CH_2 \quad \text{condenses to} \quad (CH_2)_3$$

The complete and condensed structures for 2-methylpropane are

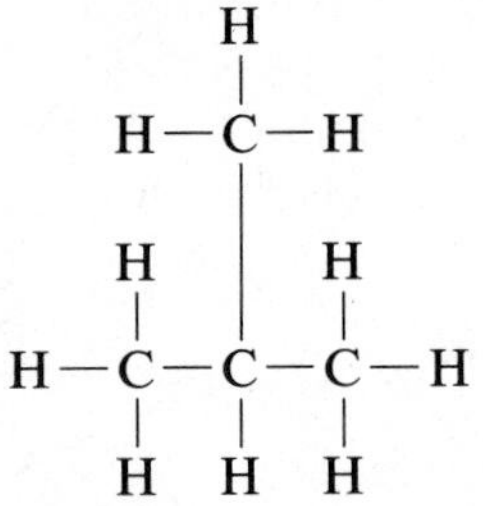

$$
\begin{array}{c}
H \\
| \\
H-C-H \\
\\
H \quad | \quad H \\
| \quad | \quad | \\
H-C-C-C-H \\
| \quad | \quad | \\
H \quad H \quad H
\end{array}
\qquad
\begin{array}{c}
CH_3 \\
| \\
CH_3CHCH_3
\end{array}
$$

complete structure for   condensed structure for
2-methylpropane   2-methylpropane

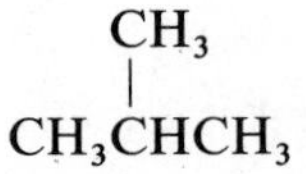

Again, a comparison shows us that

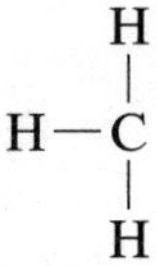

$$
\begin{array}{c}
H \\
| \\
H-C \quad \text{condenses to} \quad CH_3 \\
| \\
H
\end{array}
$$

and

$$
\begin{array}{c}
C \quad \text{condenses to} \quad CH \\
| \\
H
\end{array}
$$

Although none of these formulas depict the bond angles and the three-dimensional nature of the molecules, we will use the condensed structure since it is easier to write.

The following formulas are other examples of condensed structures and illustrate the nomenclature of alkanes just discussed. In these structures, the color areas represent the parent name.

$$
\begin{array}{c}
CH_3 \\
| \\
CH_3CHCH_2CH_3
\end{array}
\qquad
\begin{array}{c}
CH_3 \\
| \\
CH_3CCH_2CH_3 \\
| \\
CH_3
\end{array}
\qquad
\begin{array}{c}
CH_2CH_3 \\
| \\
CH_3CH_2CHCH_2CH_3
\end{array}
$$

2-methylbutane   2,2-dimethylbutane   3-ethylpentane
(*not* 3-methylbutane)   (*not* 2-dimethylbutane
                          or 2,2-methylbutane)

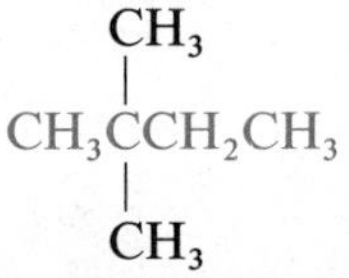

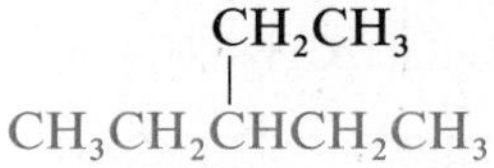

$$
\begin{array}{c}
CH_3 \quad CH_3 \\
| \quad\quad | \\
CH_3CHCH_2CHCH_3
\end{array}
\qquad
\begin{array}{c}
CH_3 \quad CH_3 \\
| \quad\quad | \\
CH_3CH_2CCH_2CHCH_2CH_3 \\
| \\
CH_3
\end{array}
$$

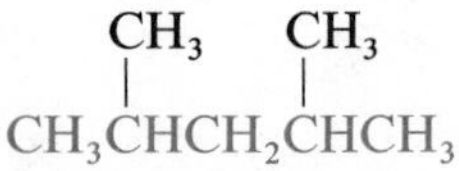

2,4-dimethylpentane 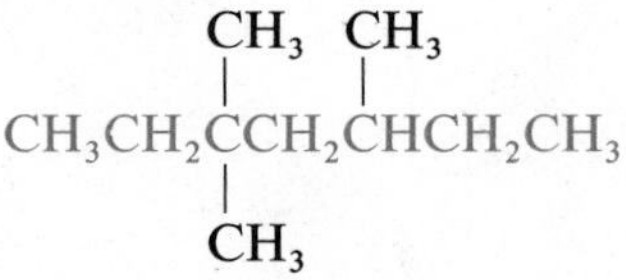   3,3,5-trimethylheptane

PROBLEMS

1. Identify the following alkyl groups.

(a) CH_3CH_2-

(b) 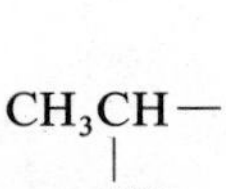 CH_3CH-
$\quad\quad\quad | $
$\quad\quad\quad CH_3$

(c) CH_3CH_2CH-
   $\quad\quad\quad\quad | $
   $\quad\quad\quad\quad CH_3$

(d) CH_3CHCH_2-
   $\quad\quad\quad\quad | $
   $\quad\quad\quad\quad CH_3$

(e) CH_3CHCH_3
   $\quad\quad\quad | $

2. Name the following alkanes by the IUPAC system.

(a) $CH_3CH_2CH_2CHCH_3$
   $\quad\quad\quad\quad\quad\quad | $
   $\quad\quad\quad\quad\quad\quad CH_3$

(b) $(CH_3)_4C$

(c) $CH_3(CH_2)_7CH_3$

(d) $CH_3CHCH_2CHCH_3$
   $\quad\quad\quad\quad | \quad\quad\quad | $
   $\quad\quad\quad\quad CH_3 \quad\quad CH_3$

(e)
   $\quad\quad\quad\quad\quad\quad CH_3$
   $\quad\quad\quad\quad\quad\quad | $
   $CH_3CHCH_2CH_2CHCH_3$
   $\quad\quad\quad | $
   $\quad\quad\quad CH_2$
   $\quad\quad\quad | $
   $\quad\quad\quad CH_3$

3. Write condensed structures for each of the following.

(a) 3-ethylpentane
(c) 2,3-dimethyl-4-ethyloctane
(e) 3-isopropylpentane

(b) 2,2-dimethylhexane
(d) 2,2,4-trimethylpentane

ANSWERS TO PROBLEMS   **1. (a)** ethyl group   **(b)** isopropyl group   **(c)** *sec*-butyl group

(d) isobutyl group   **(e)** isopropyl group   **2. (a)** 2-methylpentane   **(b)** $CH_3-\underset{\underset{\displaystyle CH_3}{|}}{\overset{\overset{\displaystyle CH_3}{|}}{C}}-CH_3$,

2,2-dimethylpropane   **(c)** $CH_3CH_2CH_2CH_2CH_2CH_2CH_2CH_2CH_3$, nonane
(d) 2,4-dimethylpentane   **(e)** 2,5-dimethylheptane   **3. (a)** $CH_3CH_2CHCH_2CH_3$
   $\quad\quad\quad\quad\quad\quad\quad\quad\quad\quad\quad\quad\quad\quad\quad | $
   $\quad\quad\quad\quad\quad\quad\quad\quad\quad\quad\quad\quad\quad\quad\quad CH_2$
   $\quad\quad\quad\quad\quad\quad\quad\quad\quad\quad\quad\quad\quad\quad\quad CH_3$

(b)
   $\quad\quad CH_3$
   $\quad\quad | $
   $CH_3CCH_2CH_2CH_2CH_3$
   $\quad\quad | $
   $\quad\quad CH_3$

(c)
   $\quad\quad CH_3 \quad CH_2CH_3$
   $\quad\quad | \quad\quad\quad | $
   $CH_3CHCHCHCH_2CH_2CH_2CH_3$
   $\quad\quad\quad\quad\quad | $
   $\quad\quad\quad\quad\quad CH_3$

(d)
   $\quad\quad CH_3 \quad CH_3$
   $\quad\quad | \quad\quad | $
   $CH_3CCH_2CHCH_3$
   $\quad\quad | $
   $\quad\quad CH_3$

(e) $CH_3CH_2CHCH_2CH_3$
   $\quad\quad\quad\quad\quad | $
   $\quad\quad\quad\quad\quad CHCH_3$
   $\quad\quad\quad\quad\quad | $
   $\quad\quad\quad\quad\quad CH_3$

11-3.2 ALKENES AND ALKYNES

Alkenes and alkynes are unsaturated hydrocarbons. Some common alkenes and alkynes are given in Table 11.5. Names for these hydrocarbons are derived by a method similar to that used for naming alkanes. The characteristic ending for an alkene is *-ene* and that for an alkyne is *-yne*. However, an extension of the IUPAC rules is needed in naming these two kinds of unsaturated hydrocarbons: (1) the $C=C$ or $C\equiv C$ must be part of the parent name and (2) the position of the $C=C$ or $C\equiv C$ in the parent name must be specified using the lowest possible number. In the following examples, the area in color indicates the carbon atoms in the parent name.

An **alkene** is a hydrocarbon that contains at least one carbon–carbon double bond and has the general formula C_nH_{2n}.

An **alkyne** is a hydrocarbon that contains at least one carbon–carbon triple bond and has the general formula C_nH_{2n-2}.

$CH_3CH_2CH=CH_2$

1-butene
(*not* 3-butene)

$CH_3C\equiv CCH_2CH_2CH_3$

2-hexyne
(*not* 4-hexyne)

$CH_3CH_2CH_2CH=CHCH_3$

2-hexene)
(*not* 4-hexene)

$$CH_3C=CHCH_3 \quad \overset{\displaystyle CH_3}{|}$$

2-methyl-2-butene
(*not* 3-methyl-2-butene)

$$CH_3CH_2CHCH_2C\equiv CH \quad \overset{\displaystyle CH_3}{|}$$

4-methyl-1-hexyne
(*not* 3-methyl-5-hexyne)

$$CH_3CH_2CHCH_2CH_2CH_3 \quad \overset{\displaystyle }{|} \quad CH_2C\equiv CH$$

4-ethyl-1-heptyne

For **polyunsaturated hydrocarbons,** the lowest numbers possible are used to locate the positions of each unsaturated linkage. For example,

A **polyunsaturated hydrocarbon** is a hydrocarbon that contains two or more double or triple bonds.

$$CH_2=C-CH=CH_2 \quad \overset{\displaystyle CH_3}{|}$$

2-methyl-1,3-butadiene
(also called isoprene)

$$\overset{1}{H}C\equiv CCH_2\overset{4}{C}\equiv CCH_2CH_2\overset{8}{C}\equiv CCH_3$$

1,4,8-decatriyne

Table 11.5   Alkenes and Alkynes

Alkene		Alkyne	
Molecular Formula	IUPAC Name	Molecular Formula	IUPAC Name
C_2H_4	Ethene	C_2H_2	Ethyne
C_3H_6	Propene	C_3H_4	Propyne
C_4H_8	Butene	C_4H_6	Butyne
C_5H_{10}	Pentene	C_5H_8	Pentyne
C_6H_{12}	Hexene	C_6H_{10}	Hexyne
C_7H_{14}	Heptene	C_7H_{12}	Heptyne
C_8H_{16}	Octene	C_8H_{14}	Octyne
C_9H_{18}	Nonene	C_9H_{16}	Nonyne
$C_{10}H_{20}$	Decene	$C_{10}H_{18}$	Decyne

11-3.3 CYCLOHYDROCARBONS

The prefixes used to indicate the number of carbons in the parent name for alkanes are also used for **cyclohydrocarbons.** Cyclopropane, cyclobutane, and cyclopentene are examples of cyclohydrocarbons. By convention, cyclohydrocarbons are represented by the appropriate geometrical figure and the hydrogen atoms attached to the ring carbons usually are not shown.

Cyclohydrocarbons are compounds that contain carbon atoms in a cyclic or ring structure.

cyclopropane   is used for   or

cyclobutane   is used for   or

cyclopentene   is used for   or

If a cyclohydrocarbon contains one or more alkyl groups bonded to a carbon atom of the ring, the smallest combination of numbers is used to locate their positions. In cycloalkenes, however, the double bond is always numbered $C_{\#1}$–$C_{\#2}$. Examples of compounds are 1-methylcyclohexene, 4-methylcyclohexene, cyclooctyne, 1,2-dimethylcyclopentane, and 1,1,3-trimethylcyclohexane.

1-methylcyclohexene   for   4-methylcyclohexene
(not 5-methylcyclohexene)   cyclooctyne

1,2-dimethylcyclopentane   1,1,3-trimethylcyclohexane
(not 1,3,3-trimethylcyclohexane)

PROBLEMS

4. Give the correct name for each of the following hydrocarbons.

(a) $CH_3CH=CHCH_2CH_2CH_3$

(b)

(c) $HC\equiv CCH_2CH_2\overset{\displaystyle CH_3}{\underset{\displaystyle |}{C}}HCH_3$

(d)

(e) $CH_3CH_2C\equiv CCH_2CH_3$

5. Write a condensed structural formula for each of the following hydrocarbons.

(a) 2-pentyne   (b) 1-hexene

(c) 1,4-dimethylcyclohexane   (d) 2-ethyl-1-butene

(e) 4,4-dimethyl-1-heptyne

ANSWERS TO PROBLEMS   **4. (a)** 2-hexene   **(b)** 1,1-dimethylcyclopentane   **(c)** 5-methyl-1-hexyne   **(d)** 3-methylcyclohexene   **(e)** 3-hexyne   **5. (a)** $CH_3C\equiv CCH_2CH_3$

(b) $CH_2=CHCH_2CH_2CH_2CH_3$ **(c)** $CH_3-\!\!\bigcirc\!\!-CH_3$ **(d)** $CH_2=C\overset{\displaystyle CH_2CH_3}{\underset{\displaystyle |}{C}}CH_2CH_3$

(e) $HC\equiv CCH_2\overset{\displaystyle CH_3}{\underset{\displaystyle \underset{\displaystyle CH_3}{|}}{\overset{\displaystyle |}{C}}}CH_2CH_2CH_3$

11-4 ISOMERS

LEARNING OBJECTIVES

1. Define, recognize structures of, or give examples (with structures) of isomers, structural isomers, geometric isomers, and conformations.
2. Given the structure of a hydrocarbon, draw its structural isomer.
3. Given the structure of a hydrocarbon, identify its primary, secondary, and tertiary carbon atoms.
4. Given an alkene or cycloalkane, draw its geometric isomers.
5. Given an alkane or cycloalkane, draw its conformations.

In this section we will consider **isomers** of alkanes, alkenes, and cycloalkanes. The types of isomerism we will discuss are structural isomerism, geometric isomerism, and conformation isomerism.

Isomers are compounds that have the same molecular formula but different structural formulas, physical properties, and chemical properties.

11-4.1 STRUCTURAL ISOMERISM

There are two **structural isomers** corresponding to the molecular formula C_4H_{10}: butane, a continuous chain alkane, and 2-methylpropane, a branched-chain alkane. One difference between these two isomers is that one of the carbon atoms in 2-methylpropane is bonded to three other carbons. No such carbon atom exists in butane. This difference may also be expressed in another way. **Carbon atoms** are designated as **primary**, **secondary**, and **tertiary** based on the number of other carbon atoms bonded to a given carbon atom.

Structural isomers are compounds that have the same molecular formula but differ from each other in the way that the atoms are bonded together.

A **primary carbon atom** is a carbon atom that is bonded to one other carbon atom, a **secondary carbon atom** is one that is bonded to two other carbon atoms, and a **tertiary carbon atom** is one that is bonded to three other carbon atoms.

$$CH_3CH_2CH_2CH_3 \qquad \overset{\displaystyle CH_3}{\underset{\displaystyle \text{butane}}{\vert}} CH_3CHCH_3$$

butane   2-methylpropane

Butane contains only primary and secondary carbon atoms and 2-methylpropane has only primary and tertiary carbon atoms.

Since structural isomers are compounds that correspond to the same molecular formula but have different structures, structural isomers have different properties. We can illustrate this difference by comparing octane and 2,2,4-trimethylpentane, two of the isomers of molecular formula C_8H_{18}. Both compounds are liquids commonly found in gasoline. Octane boils at 126° and 2,2,4-trimethylpentane boils at 99°. Each compound burns in oxygen (see Section 11-5.1) to produce carbon dioxide, water, and heat energy. This heat energy provides the force that moves the piston of an automobile. These two isomers produce essentially the same amount of energy when they burn. However, the branched-chain isomer (2,2,4-trimethylpentane) burns slower and less explosively than the continuous-chain isomer (octane). Rapid burning of the continuous-chain isomer causes the heat energy that is released to heat the engine block. This results in the common "knock" of an engine.

$$CH_3CH_2CH_2CH_2CH_2CH_2CH_2CH_3 \qquad CH_3CCH_2CHCH_3$$

octane   2,2,4-trimethylpentane

Structural isomerism also exists among alkenes and cyclohydrocarbons. For example, some isomers corresponding to the molecular formula C_4H_8 are 1-butene,

2-butene, and 2-methylpropene. Some isomers corresponding to C_5H_{10} are cyclopentane, methylcyclobutane, and 1,2-dimethylcyclopropane.

C_4H_8: $CH_3CH_2CH{=}CH_2$ $CH_3CH{=}CHCH_3$

$$CH_3\overset{\underset{\displaystyle |}{CH_3}}{C}{=}CH_2$$

1-butene   2-butene   2-methylpropene

C_5H_{10}:

cyclopentane   methylcyclobutane   1,2-dimethylcyclopropane

11-4.2 GEOMETRIC ISOMERISM

In alkenes the four atoms attached to the carbon–carbon double bond lie in the same plane. It takes considerable energy to break the pi bond of a carbon–carbon double bond and so, rotation about the carbon–carbon double bond is restricted. Thus a type of isomerism called geometric isomerism is possible. The difference between isomers is not where groups are located, but how they are oriented. The isomer that has like groups on the same side of the molecule is called the cis isomer. *cis*-2-Butene is an example of a cis isomer. The isomer with like groups on opposite sides is the trans isomer. *trans*-2-Butene is an example of a trans isomer. If one of the carbon atoms of the carbon–carbon double bond has like groups on the same carbon atom, no geometric isomer is possible.

cis-2-butene   *trans*-2-butene   2-methylpropene
(no geometric isomers)

{ two like groups on at least one carbon of C=C

The lack of rotation about a carbon–carbon single bond also makes geometric isomers possible for cyclic molecules. Cyclic molecules are usually represented as the appropriate geometric figure lying in the plane of the page. Atoms or groups attached to the ring carbon atoms are directed either upward or downward. In the cis isomer, the two groups lie on the same side of the plane. In the trans isomer, the two groups lie on the opposite sides of the plane. This is apparent in the representations of *cis*-1,2-dimethylcyclohexane and *trans*-1,2-dimethylcyclohexane.

cis-1,2-dimethylcyclohexane   *trans*-1,2-dimethylcyclohexane

The concept of geometric isomerism is involved in the processes of communication among insects. Lower animals communicate with one another just as man communicates with other members of his species. Animals pass on information to other animals for finding food, avoiding enemies, and selecting mates. To do this, animals rely on a system based on the exchange of chemical messages. Animals emit chemical substances that can attract the opposite sex, determine odor trails to be followed, mark territory, identify friends and nestmates, raise alarm in case of danger, or cause assembly in great numbers for migration or hibernation. The chemical message can trigger a sequence of physiological changes in the animal leading to moulting, new growth, and physical and sexual development of the recipient. This type of message is employed mainly among social animals to regulate and coordinate development of a colony. Such chemical messages are called *pheromones* when used among members of the same species for the purpose of communicating.

The most potent and spectacular of the pheromones are the insect sex attractants. The remarkable potency of these compounds is demonstrated by the minute amounts required to attract large numbers of insects. For example, male gypsy moths are called by 10^{-9} g of attractant and become sexually excited by as little as 10^{-12} g. Furthermore, the male gypsy moths have been attracted to females from one-fourth to one-half mile away. In moths, females produce the odor-emitting substance that is evaporated and dispersed by air currents produced by rapid vibration of the wings. Males detect the pheromone through olfactory receptors on their antennae (see Section 11-1). The males try to follow the female scent and are guided by an increased concentration of attractant to the precise resting place of the beckoning females. The pheromones identified for several insects include alkenes and cycloalkanes. These pheromones contain one or more functional groups (see Table 10.5), and are geometric isomers. Only one of the geometric isomers is biologically active. Some known insect attractants are given in Table 11.6.

What is the value of this information to chemists and entomologists? The increasing world population makes growing demands on food and fiber production. An awareness of our ecology necessitates the development of environmentally acceptable alternatives to synthetic insecticides, such as DDT or malathion, as a means of controlling pest insects. The sex pheromones offer a promising alternative because they could be used in insect control without detrimental side effects on other organisms and the ecosystem. One approach is to bait traps with pheromones. Optimum trapping efficiency is based on such factors as trap design, the type of pheromone dispenser, the amount and release rate of the pheromone, and the location of the trap. The advantage of this method is that one of the species is lured into the trap and destroyed and thereby removed from the reproductive cycle. The disadvantage is that sex pheromones are not absolutely species specific and for some species this approach is impractical.

An alternative approach is the use of pheromones to disrupt communication between males and females. Basically, this procedure prevents the males from finding the females. This is a promising approach, but the mechanism of this disruption is not clearly understood and the proper formulation of pheromone as disruptant is not known.

Table 11.6 Insect Sex Attractants

Insect	Biologically Active Geometric Isomer(s)
Boll weevil	
Female housefly	cis isomer
Honeybee	trans isomer
Silkmoth	

Pheromones are a promising and environmentally acceptable alternative to insecticides. However, much research still must be done in the area of formulation, the use of multiple pheromones, and the fate of the pheromones in the environment before we can put pheromones to practical use.

11-4.3 CONFORMATIONAL ISOMERISM

Many **conformations** are possible for alkanes. For example, some of the conformational isomers of pentane are shown. In each of these examples only the carbon skeleton is shown. For convenience, all the carbon atoms are assumed to be in the plane of the paper. Conformational isomer I is changed to II by rotation about the C_3—C_4 bond and II goes to III by rotation about the C_2—C_3 bond. Many other conformational isomers are possible.

Conformations are the many different isomers that arise by rotation about carbon – carbon single bonds.

$CH_3CH_2CH_2CH_2CH_3$

pentane

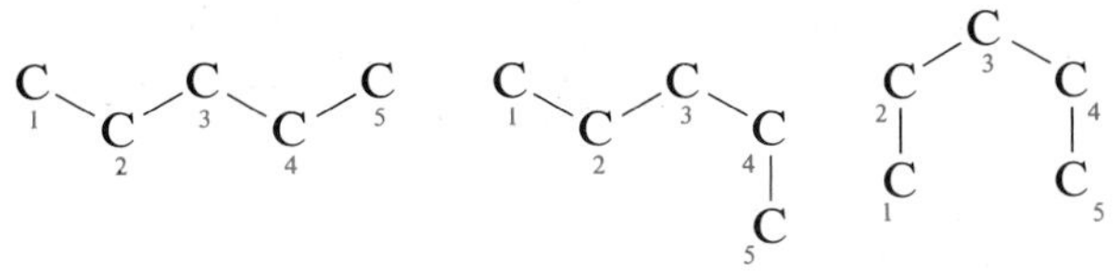

I   II   III

conformational isomers of pentane

For cyclohydrocarbons, conformational isomerism plays an important biological role. This is apparent in a comparison of the molecular models of cyclopentane and cyclohexane.

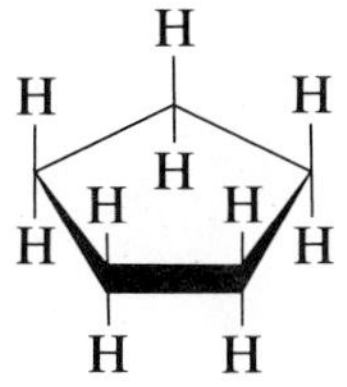

Molecular models show that all of the carbon atoms of cyclopentane are very near the same plane, with hydrogen atoms above and below the plane. Such a model is called a *Haworth formulation.* Cyclopentane does not show conformational isomerism.

Cyclohexane, unlike cyclopentane, exists in a nonplanar configuration and shows conformational isomerism. The two most important conformations are referred to as *boat* and *chair conformations.* The boat form is less stable because

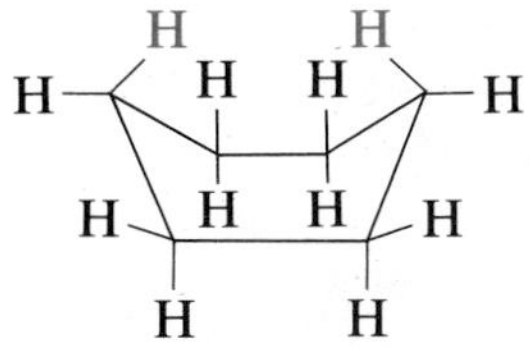

cyclopentane

cyclohexane
(boat)

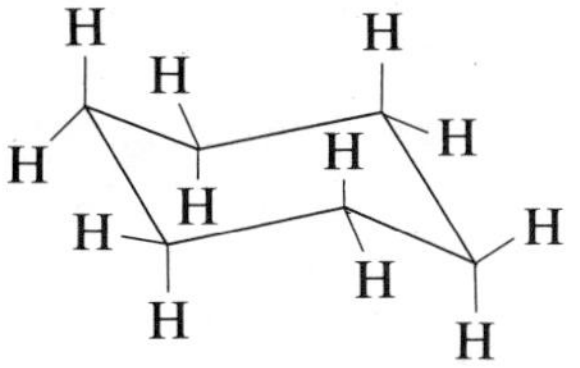

cyclohexane
(chair)

the two hydrogens (see structure) are close together and their electron clouds act to repel each other. We will not consider the boat form further. The hydrogen atoms in the chair form can be divided into two groups: axial (*a*) and equatorial (*e*) hydrogens. Axial hydrogens are easily distinguished from equatorial hydrogens because they alternate on carbons above and below the ring. The chair conformation of cyclohexane can exist in two indistinguishable conformations. The axial hydrogens in I are converted to equatorial hydrogens in II by rotations about carbon–carbon single bonds.

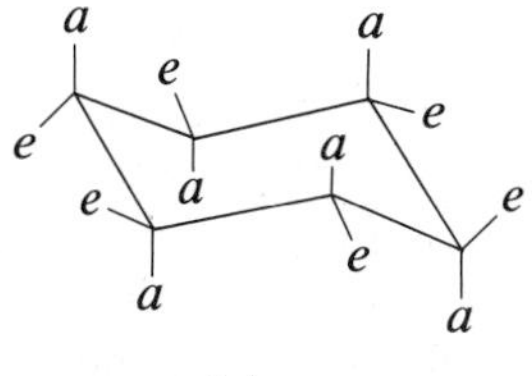

cyclohexane

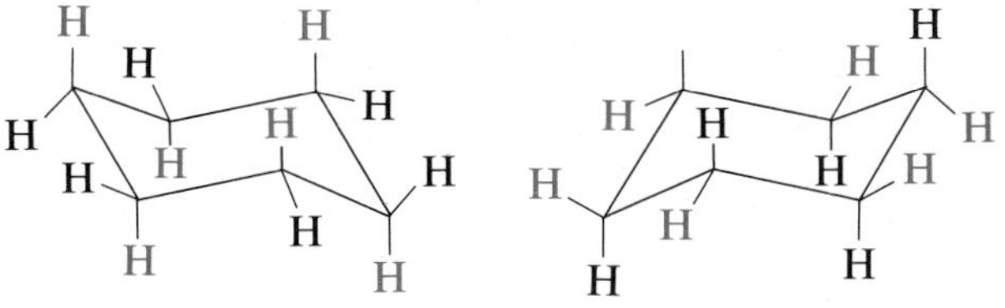

I   II

cyclohexane

For methylcyclohexane, two conformations are possible. In III the methyl group occupies an axial position and in IV it occupies an equatorial position. The more stable of these two conformations is the one in which the methyl group occupies an equatorial position. When the methyl group is axial, there are non-bonded interactions of the methyl group with the other two axial hydrogens. Such an interaction does not occur when the methyl group is in the equatorial position.

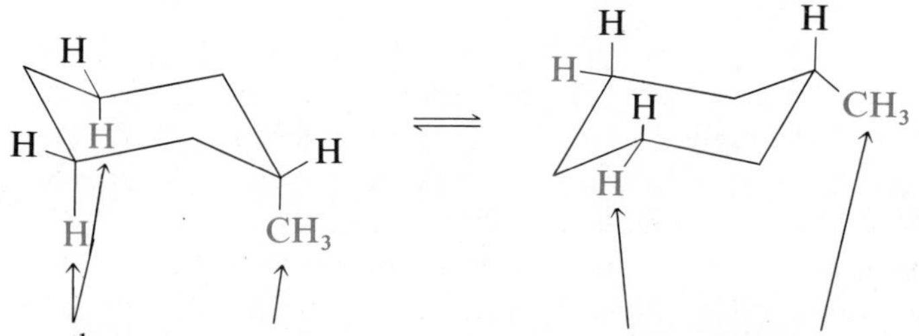

these groups are   these groups are
very close to one   separated from one
another   another

III   IV

methylcyclohexane

Conformational isomerism is very important in the living systems. Glucose, a source of energy for the living system, exists in a conformational form in which all the hydrogens occupy axial positions and the larger hydroxyl groups occupy equatorial positions.

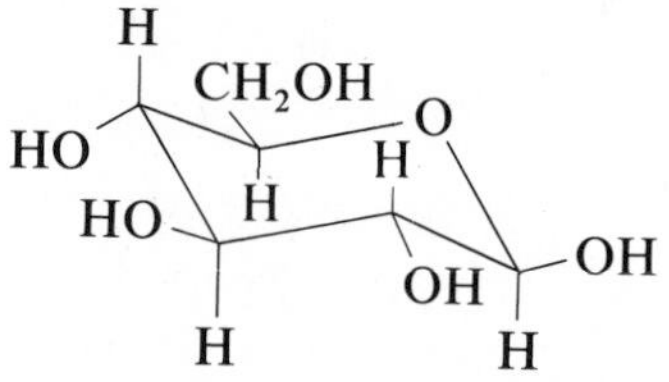

glucose

Although the chair conformation is the best representation of the six-membered ring as shown for methylcyclohexane and glucose, we will not use such representations in this text. Instead we will represent a six-membered ring as a hexagon. In the representation using the hexagon, the atoms or groups attached to the carbons of the ring are directed either upward or downward. Hydrogens attached to the carbons become part of the hexagon and are not normally shown. Using this representation, methylcyclohexane and glucose would appear as shown. This technique of representing structures with a hexagonal ring shall be discussed further in Chapter 19.

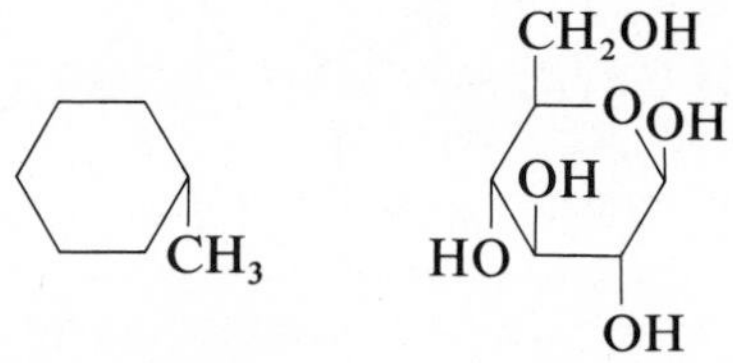

methylcyclohexane   glucose

PROBLEMS

6. Which of the following structures represent isomers? Which are identical?

(a) $CH_3CH_2CH_2CH_3$ and CH_3CHCH_3
$\qquad\qquad\qquad\qquad\qquad\quad |$
$\qquad\qquad\qquad\qquad\qquad CH_3$

(b) 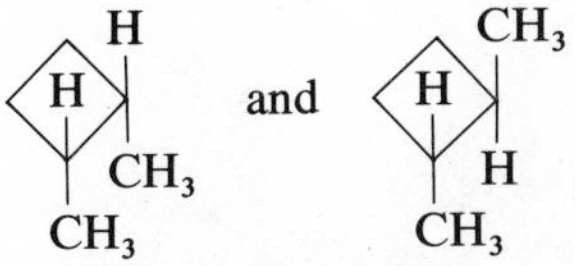 H, H, CH$_3$, CH$_3$ and CH$_3$, H, H, CH$_3$

(c) $CH_3CH_2CH_2OH$ and $HOCH_2CH_2CH_3$

(d) 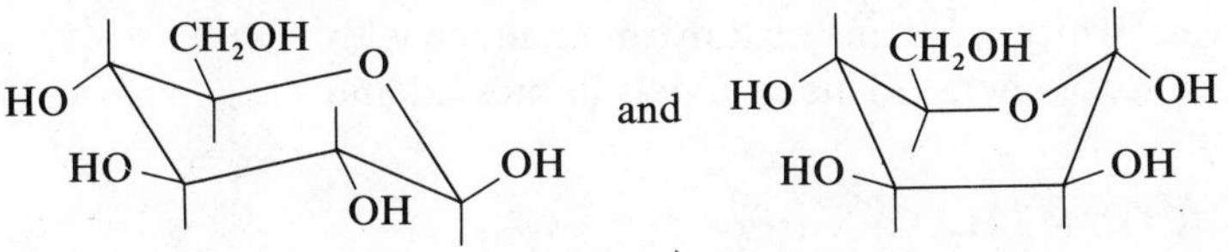 HO, CH$_2$OH, O, HO, OH, OH and HO, CH$_2$OH, O, OH, HO, OH

(e)

$\qquad$ Cl$\qquad$Cl$\qquad\qquad$Br$\qquad$Cl
$\qquad\quad$C=C$\qquad$and$\qquad$C=C
$\qquad$Br$\qquad$Br$\qquad\qquad$Cl$\qquad$Br

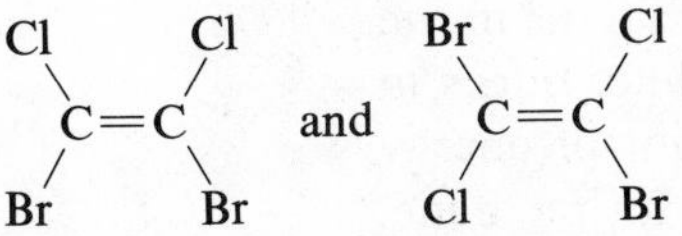

7. Write condensed structures for the three isomers of C_5H_{12}. Give correct IUPAC names.

ANSWERS TO PROBLEMS **6. (a)** structural isomers **(b)** geometric isomers. The first is the cis isomer; the other is the trans isomer **(c)** identical **(d)** conformational isomers **(e)** geometric isomers. The first is cis; the other is trans **7. (a)** $CH_3CH_2CH_2CH_2CH_3$,

$\qquad\qquad\qquad\qquad\qquad\qquad\qquad\qquad\qquad\qquad CH_3$
$\qquad\qquad\qquad\qquad\qquad\qquad\qquad\qquad\qquad\qquad\ |$

pentane **(b)** $CH_3CH_2CHCH_3$, 2-methylbutane **(c)** CH_3CCH_3, 2,2-dimethylpropane
$\qquad\qquad\qquad\qquad\ |$$\qquad\qquad\qquad\qquad\qquad\qquad\qquad\qquad\qquad |$
$\qquad\qquad\qquad\qquad CH_3$$\qquad\qquad\qquad\qquad\qquad\qquad\qquad\qquad CH_3$

11-5 REACTIONS OF HYDROCARBONS

LEARNING OBJECTIVES

1. Given the structure of an alkane, write the formulas for the products of its complete combustion.
2. Define and recognize examples of electrophilic addition reaction.
3. Write a statement of the Markovnikov rule.
4. Given the structure of an alkene or alkyne, write the structure for the addition product with a halogen, hydrogen, a hydrogen halide, water, or potassium permanganate.
5. Given the name of a carbonium ion of no more than four carbon atoms, write its structure and vice versa.
6. Given the structural segment of a polymer, write the structure of its monomer and vice versa.

As a class of organic compounds, alkanes are relatively unreactive. The presence of the pi bond in alkenes and alkynes increases the reactivity in these classes of compounds. In this section we will be concerned primarily with the reactions of alkenes and alkynes, but first we will consider one important reaction of an alkane, combustion. In Section 13-3.3 we will consider another reaction of an alkane, halogenation.

11-5.1 COMBUSTION OF ALKANES

Complete oxidation (or *combustion,* as it is sometimes called) of an alkane by oxygen produces carbon dioxide, water, and energy (see Section 4-2).

$$\text{alkane} + O_2 \longrightarrow CO_2 + H_2O + \text{energy}$$

Complete combustion of natural gas, which contains mainly methane, yields carbon dioxide, water, and large amounts of energy. This energy is harnessed and used to heat buildings.

$$CH_4 + 2\,O_2 \longrightarrow CO_2 + 2\,H_2O + \text{energy}$$

The amount of energy released increases with the number of carbon atoms in the alkane. For example, 2,2,4-trimethylpentane, a component of gasoline, burns in ample oxygen to form carbon dioxide, water, and heat. The amount of energy released in the form of heat for this alkane is much greater than that for methane. The heat released from the combustion of 2,2,4-trimethylpentane is harnessed and used to move the piston in an automobile engine.

$$
\begin{array}{c}
\quad\quad CH_3 \\
\quad\quad | \\
2\ CH_3CCH_2CHCH_3 + 25\,O_2 \longrightarrow 16\,CO_2 + 18\,H_2O + \text{energy} \\
\quad\quad | \quad\quad\ | \\
\quad\quad CH_3\ \ CH_3
\end{array}
$$

An important combustion reaction occurs in the living system. The metabolism or combustion of glucose ($C_6H_{12}O_6$) ultimately produces CO_2 and H_2O and is an important source of energy. The energy released in the combustion is stored and used by the body to perform mechanical work or other metabolic reactions (Chapter 24).

$$C_6H_{12}O_6 + 6\,O_2 \xrightarrow{\text{biological oxidation}} 6\,CO_2 + 6\,H_2O + \text{energy}$$

11-5.2 REACTIONS OF ALKENES

Alkenes, unlike alkanes, contain a very reactive functional group, the carbon–carbon double bond. The carbon–carbon double bond consists of a strong sigma bond and a weak pi bond. The reactions of alkenes occur at the carbon–carbon double bond and involve the breaking of the weaker pi bond. This results in the formation of two strong sigma bonds. The overall reaction of an alkene is addition and so it is called an *addition reaction.* In an addition reaction two molecules combine to form a single molecule. For example, the addition reaction between ethylene and hydrogen chloride forms a single molecule. The pi bond is broken and two new sigma bonds are formed. One is a carbon–hydrogen bond and the other is a carbon–chlorine bond.

$$\underset{\text{ethylene}}{\overset{\displaystyle H}{\underset{\displaystyle H}{\diagdown}}C\overset{\sigma}{\underset{\pi}{=}}C\overset{\displaystyle H}{\underset{\displaystyle H}{\diagup}}} + HCl \longrightarrow \underset{\text{ethyl chloride}}{H-\overset{\displaystyle H}{\underset{\displaystyle H}{\overset{|}{\underset{|}{C}}}}\overset{\sigma}{-}\overset{\displaystyle H}{\underset{\displaystyle Cl}{\overset{|}{\underset{|\sigma}{C}}}}-H}$$

Let us consider the addition reaction in more detail. The pi bond consists of two electrons and acts as a source of electrons. The carbon–carbon double bond

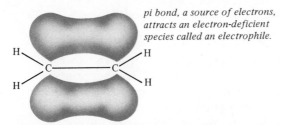

pi bond, a source of electrons, attracts an electron-deficient species called an electrophile.

reacts with **electrophilic reagents.** The reaction of an alkene with such reagents is an *electrophilic addition reaction.* Three common electrophilic reagents are the hydrogen halides, halogens, and water.

An **electrophilic reagent** is a species that is electron-deficient.

Addition Reaction Involving a Hydrogen Halide

The addition of hydrogen chloride (HCl), hydrogen bromide (HBr), or hydrogen iodide (HI) to an alkene forms an *alkyl halide.*

$$CH_2 = CH_2 + HCl \longrightarrow \underset{\substack{\text{ethyl chloride}\\ \textit{an alkyl halide}}}{\overset{\displaystyle CH_2 - CH_2}{\underset{\displaystyle H \quad\;\; Cl}{\big|\qquad\big|}}}$$

ethylene

How does this addition reaction take place? As we said before, the pi bond serves as a source of electrons and reacts with the electrophilic portion of HCl, which is H⁺. A cation is formed. We call this cation a **carbonium ion.**

A **carbonium ion** is a species containing a carbon atom that is electron-deficient and bears a positive charge.

$$\underset{\displaystyle H}{\overset{\displaystyle H}{\diagdown}}C=C\underset{\displaystyle H}{\overset{\displaystyle H}{\diagup}} + H^+ \longrightarrow H-\overset{\displaystyle H}{\underset{\displaystyle H}{\overset{|}{\underset{|}{C}}}}-\overset{\displaystyle H}{\underset{\displaystyle +}{\overset{|}{C}}}-H$$

a carbonium ion

The cation then reacts with the negative portion of HCl, which is the chloride ion Cl⁻.

$$H-\overset{\displaystyle H}{\underset{\displaystyle H}{\overset{|}{\underset{|}{C}}}}-\overset{\displaystyle H}{\underset{+}{\overset{|}{C}}}-H + Cl^- \longrightarrow H-\overset{\displaystyle H}{\underset{\displaystyle H}{\overset{|}{\underset{|}{C}}}}-\overset{\displaystyle H}{\underset{\displaystyle Cl}{\overset{|}{\underset{|}{C}}}}-H$$

ethyl chloride

Since the reaction with the electrophile (H^+) occurs first, the reaction is called an electrophilic addition reaction.

In another example, propene reacts with HBr as shown. In a reaction such as

$$CH_3-\overset{\overset{\displaystyle H}{|}}{C}=\overset{\overset{\displaystyle H}{|}}{C}-H + HBr \longrightarrow CH_3-\overset{\overset{\displaystyle H}{|}}{\underset{\underset{\displaystyle H}{|}}{C}}-\overset{\overset{\displaystyle H}{|}}{\underset{\underset{\displaystyle Br}{|}}{C}}-H + CH_3-\overset{\overset{\displaystyle H}{|}}{\underset{\underset{\displaystyle Br}{|}}{C}}-\overset{\overset{\displaystyle H}{|}}{\underset{\underset{\displaystyle H}{|}}{C}}-H$$

      propene   propyl bromide   isopropyl bromide

this where more than one product is formed, the major product can be determined by a comparison of the stability of the carbonium ions that give rise to each product. The carbon that bears the positive charge is designated as either primary, secondary, or tertiary depending on the number of carbon atoms to which it is directly bonded. If the carbon is bonded to one other carbon, the *carbonium ion* is said to be *primary*. In a *secondary carbonium ion* the carbon bearing the positive charge is bonded to two other carbons. A *tertiary carbonium ion* contains a positively charged carbon bonded to three other carbon atoms. The two carbonium ions formed from propene are shown.

a primary carbonium ion

a secondary carbonium ion

Experimental evidence tells us the following about the stability of carbonium ions: Tertiary carbonium ions are more stable than secondary carbonium ions and these are more stable than primary carbonium ions.

    order of stability of carbonium ions: tertiary > secondary > primary

For the two carbonium ions under discussion,

$$CH_3-\overset{\overset{\displaystyle H}{|}}{\underset{\underset{\displaystyle +}{}}{C}}-CH_3 \quad\text{is more stable than}\quad CH_3CH_2-\overset{\overset{\displaystyle H}{|}}{\underset{\underset{\displaystyle +}{}}{C}}-H$$

        secondary   primary
   carbonium ion   carbonium ion

The product of the reaction in question, then, is isopropyl bromide because it arises from the more stable isopropyl carbonium ion.

$$CH_3-\overset{\displaystyle +}{\underset{\displaystyle |}{\underset{\displaystyle H}{C}}}-CH_3 + Br^- \longrightarrow CH_3\underset{\displaystyle |}{\underset{\displaystyle Br}{C}}HCH_3$$

isopropyl carbonium ion   isopropyl bromide

What is responsible for this order of stability of carbonium ions? The alkyl groups (R) bonded to the positively charged carbon can increase the electron density at the positively charged carbon such that the R groups themselves become slightly positive. This increase in electron density at the positively charged carbon helps to make the ion more stable. The greater the number of alkyl groups bonded to the positively charged carbon, the more stable the carbonium ion.

$$R-\overset{\displaystyle R}{\underset{\displaystyle R}{C^+}} \quad \text{is more stable than} \quad R-\overset{\displaystyle R}{\underset{\displaystyle H}{C^+}} \quad \text{is more stable than} \quad R-\overset{\displaystyle H}{\underset{\displaystyle H}{C^+}}$$

tertiary   *secondary*   *primary*
carbonium ion   *carbonium ion*   *carbonium ion*

decreasing number of alkyl groups at positive site
⟶

The concept of electrophilic addition is summarized in a general rule known (after the scientist who formulated it) as the Markovnikov rule: In electrophilic addition to a carbon–carbon double bond, the more stable carbonium ion is formed. Stated in a simpler way, the hydrogen of the hydrogen halide adds to that carbon of the carbon–carbon double bond with the greatest number of hydrogens. For example, the addition of hydrogen iodide (HI) to 2-methylpropene gives *tert*-butyl iodide, not isobutyl iodide, as the product.

$$\underset{\text{no hydrogens}\nearrow}{CH_3\overset{\displaystyle CH_3}{\underset{\displaystyle |}{C}}=CH_2}\overset{\nwarrow 2\ hydrogens}{} + HI \longrightarrow CH_3\overset{\displaystyle CH_3}{\underset{\displaystyle |}{\underset{\displaystyle I}{C}}}CH_3 \qquad CH_3\overset{\displaystyle CH_3}{\underset{\displaystyle |}{C}}HCH_2I$$

2-methyl propene   *tert*-butyl iodide   isobutyl iodide is not formed

Addition Reaction Involving a Halogen

The product of the addition of a halogen (bromine or chlorine) to an alkene is called a *vicinal dihalide. Vicinal* indicates that the halogens are added on adjacent carbon atoms.

$$CH_2{=}CH_2 + Br_2 \longrightarrow \underset{\displaystyle Br}{\underset{\displaystyle |}{CH_2}}-\underset{\displaystyle Br}{\underset{\displaystyle |}{CH_2}}$$

ethene   1,2-dibromoethane
          a vicinal dihalide

$$CH_3CH{=}CH_2 + Cl_2 \longrightarrow CH_3CH{-}CH_2$$
$$\hspace{6cm}\underset{\displaystyle Cl}{|}\quad\underset{\displaystyle Cl}{|}$$

propene   1,2-dichloropropane
                              a vicinal dihalide

The addition of bromine is used as a test for the carbon–carbon double bond. The alkene is colorless, bromine is reddish-brown, and the product is colorless. As bromine is added to the colorless liquid alkene, you will observe that the reddish-brown color of bromine disappears. Once the addition of bromine to the double bond is complete, the excess bromine will give the solution a reddish-brown color. The color change is a positive test for the double bond, and also a test for unsaturation.

As well as being used to detect unsaturation in organic molecules as in the case with bromine, the addition of a halogen is also used to determine quantitatively the degree of unsaturation in lipids. Lipids are a class of water-insoluble esters (see Table 10.5) found in the living system (see Chapter 26). The lipids in which we are interested are the triglycerides which contain large amounts of polyunsaturation. The degree of unsaturation is defined as the amount of halogen absorbed by 100 g of triglyceride. The more halogen that is required, the higher is the degree of unsaturation. The halogen used in this test, iodine monobromide (IBr), is a combination of two halogens. The reaction of a typical triglyceride with IBr is shown.

$$CH_3(CH_2)_{10}CO_2CH_2$$
$$|$$
$$CH_3(CH_2)_{10}CO_2CH \;\; + \; 3\ IBr \longrightarrow$$
$$|$$
$$CH_3CH_2CH{=}CHCH_2CH{=}CHCH_2CH{=}CH(CH_2)_7CO_2CH_2$$

a triglyceride

$$CH_3(CH_2)_{10}CO_2CH_2$$
$$|$$
$$CH_3(CH_2)_{10}CO_2CH$$
$$|$$
$$CH_3CH_2CH{-}CHCH_2CH{-}CHCH_2CH{-}CH(CH_2)_7CO_2CH_2$$
$$\hspace{2cm}\underset{I}{|}\quad\underset{Br}{|}\quad\underset{I}{|}\quad\underset{Br}{|}\quad\underset{I}{|}\quad\underset{Br}{|}$$

a possible addition product

Nutritionists are interested in the amount of polyunsaturated **fatty acid** in our diets because of the concern about blood-cholesterol level. Evidence seems to indicate that cholesterol is associated with cardiovascular problems. The amount of saturated fatty acid in the diet may be related to the amount of cholesterol in the blood. Nutritionists recommend that we use cooking oils and margarines that have high polyunsaturated-to-saturated fatty acid ratios. Table 11.7 compares the fatty acid content in some common oils. It can be seen from Table 11.7 that safflower oil, soybean oil, and sunflower oil can help achieve a high polyunsaturated-to-saturated fatty acid ratio in our diets.

A **fatty acid** is an organic acid (see Table 10.5) with a hydrocarbon residue.

Table 11.7 Percent of Saturated and Unsaturated Fatty Acid in Common Oils*

Substance	Saturated	Monounsaturated[†]	Polyunsaturated[‡]
Coconut oil	91	8	0
Palm oil	47	43	10
Olive oil	9	84	5
Peanut oil	13	56	30
Cottonseed oil	27	25	48
Corn oil	15	51	34
Soybean oil	13	29	57
Sunflower oil	9	25	66
Safflower oil	7	19	74

*Figures have been rounded off to the nearest 1%. As a result, they may not total 100%.

[†]*Monounsaturated:* one carbon–carbon double bond

[‡]*Polyunsaturated:* two or more carbon–carbon double bonds

Addition Reaction Involving Water

The addition of water to an alkene forms an *alcohol.* Such an addition is difficult to achieve and is usually carried out in the presence of a catalyst. (The effect of a catalyst on the rate of a reaction was discussed in Section 7-1.3.) Furthermore, the yield of the product alcohol is poor. Like the addition of hydrogen halides, the addition of water involves a carbonium ion intermediate. The carbonium ion intermediate is not given in the equations showing the addition reactions involving water and ethylene and water and 2,3-dimethyl-2-butene.

$$CH_2{=}CH_2 + HOH \xrightarrow{\text{H}^+} \underset{\underset{\text{H}}{|}}{CH_2}{-}\underset{\underset{\text{OH}}{|}}{CH_2}$$

ethylene   ethyl alcohol
                            an alcohol

$$\underset{\underset{\text{CH}_3}{|}}{CH_3C}{=}CCH_3 + HOH \xrightarrow{\text{H}^+} \underset{\underset{\text{H}}{|}}{CH_3C}{-}\underset{\underset{\text{OH}}{|}}{CCH_3}$$

2,3-dimethyl-2-butene   2,3-dimethyl-2-butanol
                            an alcohol

The tricarboxylic acid cycle (TCA) plays an important role in the metabolism of carbohydrates, lipids, and proteins. It functions to produce energy in the form of a molecule called adenosine triphosphate (ATP) and provides metabolic intermediates for the formation of needed compounds. There are ten reactions in the TCA cycle. One of these reactions involves the addition of a molecule of water to the carbon–carbon double bond of fumaric acid to form malic acid. The reaction is catalyzed by the enzyme fumarase.

$$\underset{\text{fumaric acid}}{\overset{\displaystyle \text{HOOC}\diagdown\,\diagup\text{H}}{\underset{\displaystyle \text{H}\diagup\,\diagdown\text{COOH}}{\overset{\text{C}}{\underset{\text{C}}{\parallel}}}}} + \text{HOH} \xrightarrow{\text{fumarase}} \underset{\text{malic acid}}{\begin{array}{c}\text{COOH}\\|\\\text{HO}-\text{C}-\text{H}\\|\\\text{H}-\text{C}-\text{H}\\|\\\text{COOH}\end{array}}$$

There are several addition reactions that do not involve a carbonium ion intermediate. These include the addition of hydrogen, oxidation with potassium permanganate, and polymerization.

Addition Reaction Involving Hydrogen

The addition of hydrogen to an alkene is called *hydrogenation*. The product formed is an *alkane*. Like the addition of water to an alkene, the addition of hydrogen to an alkene is slow and proceeds more rapidly in the presence of a catalyst (see Section 7-1.3) such as nickel or platinum. The addition of a molecule of hydrogen to the pi bond occurs on the surface of the catalyst. The conversion of ethene and of 2-methylpropene to alkanes is shown.

$$\text{CH}_2{=}\text{CH}_2 + \text{H}_2 \xrightarrow{\text{Ni}} \underset{\qquad\;\;\text{H}\quad\;\text{H}}{\text{CH}_2-\text{CH}_2}$$

$$\underset{\text{ethene}}{} \qquad\qquad \underset{\substack{\text{ethane}\\ \textit{an alkane}}}{}$$

$$\underset{\text{2-methylpropene}}{\overset{\displaystyle \text{CH}_3}{\underset{|}{\text{CH}_3\text{C}{=}\text{CH}_2}}} + \text{H}_2 \xrightarrow{\text{Pt}} \underset{\substack{\text{2-methylpropane}\\ \textit{an alkane}}}{\overset{\displaystyle \text{CH}_3}{\underset{|}{\text{CH}_3\text{CHCH}_3}}}$$

Commercially, industry has developed large-scale hydrogenation processes for transforming liquid vegetable oils into solid fats. Margarine and cooking shortenings are prepared by the hydrogenation of vegetable oils. In one process hydrogen gas under pressure is bubbled into a tank of oil at about 200° that contains the finely divided nickel catalyst. The conversion of trioleylglycerol, a liquid vegetable oil, into tristearylglycerol, a solid fat, is shown.

$$\begin{array}{c}\text{CH}_3(\text{CH}_2)_7\text{CH}{=}\text{CH}(\text{CH}_2)_7\text{CO}_2\text{CH}_2\\|\\\text{CH}_3(\text{CH}_2)_7\text{CH}{=}\text{CH}(\text{CH}_2)_7\text{CO}_2\text{CH}\\|\\\text{CH}_3(\text{CH}_2)_7\text{CH}{=}\text{CH}(\text{CH}_2)_7\text{CO}_2\text{CH}_2\end{array} + 3\,\text{H}_2 \xrightarrow{\text{Ni}} \begin{array}{c}\text{CH}_3(\text{CH}_2)_7\text{CH}_2\text{CH}_2(\text{CH}_2)_7\text{CO}_2\text{CH}_2\\|\\\text{CH}_3(\text{CH}_2)_7\text{CH}_2\text{CH}_2(\text{CH}_2)_7\text{CO}_2\text{CH}\\|\\\text{CH}_3(\text{CH}_2)_7\text{CH}_2\text{CH}_2(\text{CH}_2)_7\text{CO}_2\text{CH}_2\end{array}$$

$$\underset{\substack{\text{trioleylglycerol}\\ \textit{a liquid vegetable oil}}}{} \qquad\qquad\qquad \underset{\substack{\text{tristearylglycerol}\\ \textit{a solid fat}}}{}$$

Oxidation of Alkenes

Cold, aqueous potassium permanganate ($KMnO_4$) oxidizes an alkene to a *glycol*. A glycol is a dihydroxy compound. The purple $KMnO_4$ is reduced to manganese dioxide (MnO_2), a brown solid. Unlike combustion (see Section 11-5.1), in which the carbon-containing compound is completely broken down to CO_2 and H_2O, this reaction with cold, aqueous $KMnO_4$ is a mild oxidation that results in the addition of two hydroxyl groups to the carbon–carbon double bond. The oxidation of ethylene and of propene is shown.

$$CH_2{=}CH_2 + KMnO_4 \xrightarrow{H_2O} \underset{\displaystyle\substack{| \quad |\\ OH \quad OH}}{CH_2{-}CH_2} + MnO_2 + KOH$$

ethylene   ethylene glycol
a glycol

$$CH_3CH{=}CH_2 + KMnO_4 \xrightarrow{H_2O} \underset{\displaystyle\substack{| \quad |\\ OH \quad OH}}{CH_3CH{-}CH_2} + MnO_2 + KOH$$

propene   propylene glycol
a glycol

Polymerization

Polymerization is a process by which small molecules are chemically converted to larger molecules called *polymers*. Polymers consist of repeating units of smaller molecules called *monomers*. Polymers can be formed under a variety of conditions: the presence of initiators, such as oxygen or peroxides; heat; and pressure. Polyethylene, made from the polymerization of ethylene, is used for electrical insulation, squeeze bottles, and some plastic cups. The polymerization process can be illustrated, as shown by ethylene. The wavy line in this reaction equation indicates

ethylene
a monomer   polyethylene
a polymer

that the unit in brackets repeats itself *n* times. For example, polyethylene may be written in an expanded form as shown and the value of *n* is between 100 and 1000 monomer units.

the portion of the polymer
shown as a product above

Polypropylene, like polyethylene, is made by polymerizing propylene. Polypropylene, however, is more flexible than polyethylene and is used to manufacture many of the items that are also made from polyethylene. Polypropylene is most useful for making such household items as pails, pans, and drinking glasses. It is also used in indoor–outdoor carpeting and in medical articles including syringe barrels and plungers in disposable syringes, ear drains, catheters, sutures, tubing, surgical mesh, blood filters, and hip joint repairs.

propylene   polypropylene

The general equation shown represents the polymerization of a monomer. Y and Z are usually groups other than hydrogen. Table 11.8 lists some commercially available polymers and a natural polymer, and describes their uses.

this carbon–carbon double bond   becomes   a carbon–carbon single bond

Table 11.8 Some Common Synthetic and Natural Polymers

Monomer	Polymer	Name	Uses
Synthetic			
vinyl chloride		Poly(vinyl chloride) (PVC)	Phonograph records, raincoats, shower curtains
1,1-dichloroethene		Poly(vinylidene chloride)	Packing film, plastic food wrap such as Saran
tetrafluoroethene		Poly(tetrafluoro-ethylene)	Chemically resistant, nonstick coating, such as Teflon, used in cookware

Table 11.8 Some Common Synthetic and Natural Polymers (*cont'd*)

Monomer	Polymer	Name	Uses
acrylonitrile		Poly (acrylonitrile)	Fibers and fabrics such as Orlon and Acrilan
styrene		Poly(styrene)	Foams, molded articles, insulation for refrigerators and air conditioners
methyl methacrylate		Poly(methyl methacrylate)	Safety glass, such as Lucite and Plexiglas, contact eye lenses and dentures
N-vinylpyrrolidone		Poly(N-vinyl pyrrolidone)	Blood plasma substitute, hair sprays
2-methylpropene		Butyl rubber	Tire inner tubes
Natural isoprene		Natural rubber	Tires after being vulcanized (heated with elemental sulfur) to provide extra strength

11-5.3 REACTIONS OF ALKYNES

The addition reactions of alkynes are similar to those of alkenes. However, alkynes contain twice as many pi bonds as do alkenes and, therefore, consume twice as much halogen, hydrogen halide, or hydrogen. For example,

$$H-C\equiv C-H \xrightarrow{Br_2} \underset{\substack{|\quad\ \ | \\ Br\ \ \ Br}}{CH=CH} \xrightarrow{Br_2} \underset{\substack{Br\ \ Br \\ |\quad\ \ | \\ |\quad\ \ | \\ Br\ \ Br}}{H-C-C-H}$$

ethyne   1,2-dibromoethene   1,1,2,2-tetrabromoethane

$$CH_3C\equiv CH \xrightarrow[Ni]{H_2} CH_3CH=CH_2 \xrightarrow[Ni]{H_2} CH_3CH_2CH_3$$
　　　propyne　　　　　　propene　　　　　　propane

$$CH_3C\equiv CCH_3 \xrightarrow{HCl} \underset{\substack{|\ \ | \\ H\ \ Cl}}{CH_3C=CCH_3} \xrightarrow{HCl} \underset{\substack{Cl \\ | \\ CH_3CH_2CCH_3 \\ | \\ Cl}}{}$$

　　2-butyne　　　　2-chloro-2-butene　　　2,2-dichlorobutane
　　　　　　　　　　　　　　　　　　　　(the Markovnikov rule applies)

In the case of water, however, alkynes react with only one mole:

$$CH_3C\equiv CCH_3 + HOH \xrightarrow{H^+} \underset{\substack{|\ \ | \\ H\ \ OH}}{CH_3C=CCH_3} \longrightarrow \underset{\substack{|| \\ O}}{CH_3CH_2CCH_3}$$

　　　　　　　　　　　　an unsaturated alcohol　　*a ketone*

The initial addition product, an unsaturated alcohol (often called an *enol*), is very unstable and isomerizes to a ketone (see Table 10.5). The ketone predominates in the product. This isomerization results in the movement of the double bond and a hydrogen atom and occurs frequently in biochemical systems. For example, in *glycolysis,* a series of steps in which glucose is oxidized in the living system to provide energy, one step involves the transformation of phosphoenol pyruvic acid to pyruvic acid.*

$$\underset{\substack{| \\ CH_2}}{\overset{\textstyle CO_2H}{\underset{\textstyle |}{C-OPO_3H_2}}} \xrightarrow{enzyme} \underset{\substack{|| \\ CH_2}}{\overset{\textstyle CO_2H}{\underset{\textstyle |}{C-OH}}} \longrightarrow \underset{\substack{| \\ CH_3}}{\overset{\textstyle CO_2H}{\underset{\textstyle |}{C=O}}}$$

　　phosphoenol　　　　pyruvic acid　　　pyruvic acid
　　pyruvic acid　　　　(enol form)　　　(keto form)

---

*Although these molecules are shown in the reaction equation in the un-ionized form, in actuality they exist in ionic forms at physiological pH. In this section of the text, the un-ionized forms are used for clarity in presenting the concepts.

PROBLEM

8. Describe the reaction and write the structure of the product of each equation.

(a) $CH_3CH_2CH_3$ + excess $O_2 \longrightarrow$

(b) $CH_3CH{=}CH_2$ + $Br_2 \longrightarrow$

(c) $CH_3CH{=}CHCH_3$ + HCl $\longrightarrow$

(d) $CH_3\underset{\underset{CH_3}{|}}{C}{=}CH_2$ + HI $\longrightarrow$

(e) 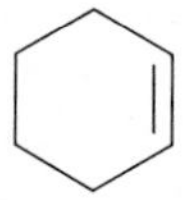 ⬡ + $H_2 \xrightarrow{Pt}$

(f) $CH_3CH{=}CHCH_3$ + $KMnO_4 \xrightarrow{H_2O}$

(g) $CH_3C{\equiv}CH$ + excess $Br_2 \longrightarrow$

(h) $CH_3C{\equiv}CCH_3$ + HOH $\xrightarrow{H^+}$

ANSWERS TO PROBLEM **8. (a)** complete combustion of an alkane (propane), products are CO_2 and H_2O **(b)** addition of a halogen to an alkene, product is $CH_3\underset{\underset{Br}{|}}{CH}{-}\underset{\underset{Br}{|}}{CH_2}$ **(c)** addition reaction of hydrogen chloride to an alkene, the only product is $CH_3\underset{\underset{Cl}{|}}{CH}{-}\underset{\underset{H}{|}}{CHCH_3}$

(d) addition reaction illustrating Markovnikov's rule, major product $CH_3{-}\underset{\underset{I}{|}}{\overset{\overset{CH_3}{|}}{C}}{-}CH_3$ **(e)** hydrogenation reaction, addition of hydrogen to a cycloalkene gives a cycloalkane, product is cyclohexane **(f)** oxidation reaction of an alkene by cold dilute potassium permanganate, product is a glycol $CH_3\underset{\underset{OH}{|}}{CH}{-}\underset{\underset{OH}{|}}{CH}{-}CH_3$ **(g)** addition of excess bromine to an alkyne, both pi bonds react, product is $CH_3{-}\underset{\underset{Br}{|}}{\overset{\overset{Br}{|}}{C}}{-}CHBr$ **(h)** addition of water to an alkyne, only one mole of water is consumed, product is an unsaturated alcohol and a ketone with the latter predominating, $CH_3\underset{\underset{OH}{|}}{C}{=}CHCH_3 \rightleftharpoons CH_3\underset{\underset{O}{\|}}{C}CH_2CH_3$

SUMMARY

Hydrocarbons, of which alkanes, alkenes, and alkynes are examples, contain only carbon and hydrogen. Alkanes (except methane) contain only carbon–carbon single bonds. Alkenes and alkynes contain at least one carbon–carbon double bond and one carbon–carbon triple bond, respectively. The carbon–carbon single bond, the carbon–carbon double bond, and the carbon–carbon triple bond impart definite shapes to organic molecules. The sense of odor appears to be related to the shapes of molecules. Hydrocarbons are found in a variety of sources in nature. The most important are petroleum, coal, and green plants.

Alkanes, alkenes, and alkynes are systematically named by the IUPAC system. Definite rules are followed in naming these compounds. In general, all alkyl groups are located and named, followed by the name of the longest continuous chain of carbon atoms (parent name).

Isomerism is possible for hydrocarbons. Three kinds of isomerism include structural isomerism, geometric isomerism, and conformational isomerism.

Combustion of alkanes (and other hydrocarbons) give carbon dioxide, water, and large amount of energy.

The typical reactions of alkenes and alkynes is addition. In general, the additions follow the Markovnikov rule. The pi bond of alkenes and alkynes serves as a source of electrons to electrophilic reagents. The typical reactions of these molecules include addition of hydrogen, halogen, hydrogen halide, and water.

Alkenes can be polymerized to a variety of marketable products.

ADDITIONAL PROBLEMS

9. Write structures or use words and equations to distinguish between each of the following.
   (a) alkene and alkyne
   (b) planar molecule and tetrahedral molecule
   (c) isoprene and terpene
   (d) homologs and isomers
   (e) alkane and alkyl groups
   (f) propyl and isopropyl groups
   (g) cis and trans isomers
   (h) structural isomer and conformational isomer
   (i) axial and equatorial hydrogens
   (j) combustion and hydrogenation
   (k) sigma and pi bonds
   (l) primary and secondary carbonium ions

10. Write structural formulas for each of the following.
    (a) 3,4-dimethyl-2-pentene
    (b) 3,3-diethylheptane
    (c) 1-hexene
    (d) 3-methyl-1-butyne
    (e) *cis*-2-pentene
    (f) cyclopentene
    (g) 3-octyne
    (h) 2,2-dimethyl-4-ethyldecane
    (i) *trans*-1,2-dimethylcyclobutane
    (j) 4-methyl-5-ethyl-1-heptene

11. Write the correct name for each of the following.

    (a) $CH_2{=}CHCHCH_2CHCH_3$
                    CH_3 CH_3

    (b)

    (c) $CH_3C{\equiv}CCHCH_2CH_3$
                    CH_3

    (d)

    (e) $CH_3CHCH_2CH_2CH_3$
               CH_2
               CH_3

    (f) $CH_3CHCH_2CHCH_3$
            CH_3 CH_3

    (g) $CH_2{=}C{-}CH{=}CH_2$ with CH_3

    (h) $HC{\equiv}CCCH_3$ with CH_3 and CH_3

    (i) $CH_3CH_2CH_2C{\equiv}CCH_2CH_3$

    (j) $CH_3CH_2CH{=}CH_2$
               CH_2
               CH_2
               CH_3

12. For each of the following molecular formulas, write the structure(s) requested.
    (a) C_4H_{10}, an alkane
    (b) C_5H_{10}, a cyclic hydrocarbon
    (c) C_6H_{10}, an alkyne
    (d) $C_2H_6O_2$, a glycol
    (e) C_5H_{12}, a pair of isomers
    (f) C_4H_8, geometric isomers
    (g) C_6H_{14}, the next higher homolog
    (h) C_4H_9, an isobutyl group
    (i) C_4H_{10}, a pair of conformations
    (j) C_4H_8, an alkene that has no geometric isomers

13. Distinguish between primary, secondary, and tertiary carbonium ions and give the structural formula of each. Write structures for the four butyl carbonium ions. Which is the most stable of the four butyl carbonium ions?

14. Write the structures and give the IUPAC names for the nine isomers of C_7H_{16}.

15. Name naturally occurring sources of hydrocarbons and indicate the general kind of hydrocarbon found in each.

16. Inositol is one of nine hexahydroxycyclohexanes. The hydroxyl group on C-1 is equatorial; all the rest are axial. Draw the chair conformational structure of inositol.

17. The reaction between acetylene and HCN gives acrylonitrile. Polymerization of acrylonitrile gives poly (acrylonitrile), which is known commercially as

Orlon acrylic fiber. Write equations to illustrate these reactions.

18. Write the structure of the organic product of each of the following.

(a) $CH_3CH=CH_2 + Cl_2 \longrightarrow$

(b) $+ KMnO_4 \xrightarrow{H_2O}$

(c) $CH_3CHCH_2CH_3 + \text{excess } O_2 \xrightarrow{\text{heat}}$
$\quad\quad |$
$\quad\quad CH_3$

(d) $HC\equiv CH + HOH \xrightarrow{H^+}$

(e) $CH_3C\equiv CCH_3 + 1 \text{ mol HBr} \longrightarrow$

(f) $\overset{H}{\underset{Cl}{\diagdown}}C=C\overset{H}{\underset{H}{\diagup}} \xrightarrow{\text{polymerize}}$

(g) $CH_3C=CH_2 + HOH \xrightarrow{H^+}$
$\quad\quad |$
$\quad\quad CH_3$

(h) $+ HCl \longrightarrow$

(i) $\overset{CH_3}{\underset{H}{CH_3\overset{|}{\underset{|}{C}}CH_2CH=CH_2}} + H_2 \xrightarrow{Ni}$

(j) $CH_3CH_2C\equiv CH + 2 \text{ mol Br}_2 \longrightarrow$

(k) $CH_3CH=CHCH_3 + HBr \longrightarrow$

(l) $+ HOH \xrightarrow{\text{fumarase}}$

CHAPTER 12

BENZENE AND RELATED AROMATIC COMPOUNDS

Many organic compounds with pleasant odors are obtained from a variety of natural sources. Typical examples are the compounds obtained from cinnamon tree bark and vanilla beans. The early organic chemists referred to these compounds as aromatic. Organic chemists soon recognized that not all aromatic compounds have pleasant odors. For example, some aromatic compounds were discovered and found to be foul-smelling and even toxic. Today the name *aromatic compound* refers to a class of organic compounds that possess a common structural feature: a planar cyclic structure of atoms with a unique overlap of *p* orbitals. We will begin our study of aromatic compounds with benzene, the simplest aromatic compound, and we will see why its structure is unique.

12-1 STRUCTURE OF BENZENE

LEARNING OBJECTIVES

1. Explain the difference between benzene and unsaturated compounds such as alkenes and alkynes.
2. Describe the resonance concept.
3. Explain why benzene is a resonance hybrid.
4. Describe the modern model of benzene.
5. Explain why benzene undergoes substitution reactions.
6. Define resonance hybrid and delocalized pi electrons.

Benzene is a hydrocarbon with the molecular formula C_6H_6. It has a cyclic structure and each carbon has one hydrogen atom bonded to it, as shown in the incomplete structure. This structure is incomplete, however, because each carbon atom has only three bonds (or six electrons). Since a carbon atom must have four bonds, the structure can be completed simply by adding extra bonds, as shown in the complete structure. This structure implies that benzene contains alternating

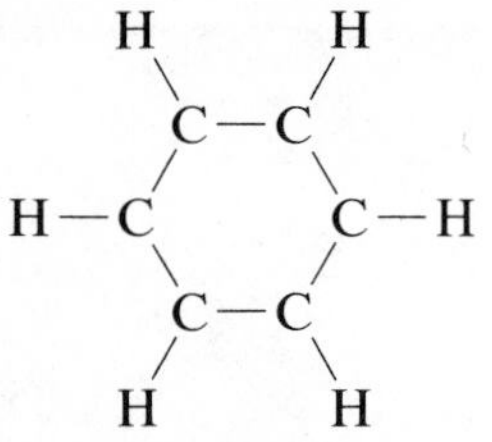

benzene
(incomplete structure)

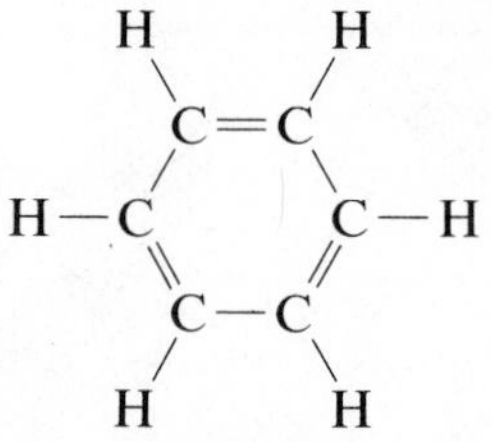

benzene
(complete structure)

double and single bonds. However, we know that benzene does not react with bromine in the same way as alkenes do (see Section 11-5.2). Therefore the double bonds shown in the complete structure of benzene are not the same as those in an alkene and benzene is not unsaturated in the sense that an alkene is unsaturated.

$$CH_3CH_2CH_2CH_2CH=CH_2 + Br_2 \longrightarrow CH_3CH_2CH_2CH_2CHCH_2Br$$
$$\overset{|}{Br}$$

colorless   *red*   *colorless*

benzene $+ Br_2 \longrightarrow$ no reaction

colorless   *red*   *red*

In 1865 in Germany, August Kekulé proposed a solution for this dilemma. He described benzene in terms of two structures, I and II, between which the benzene

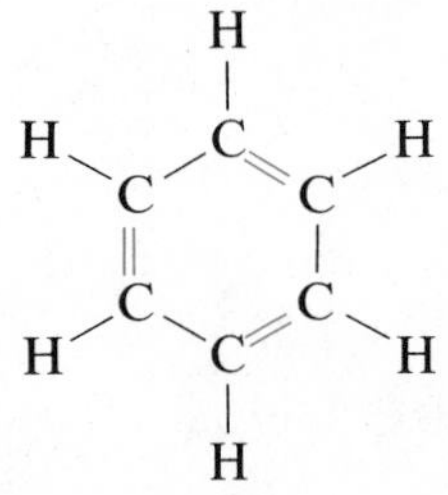

I

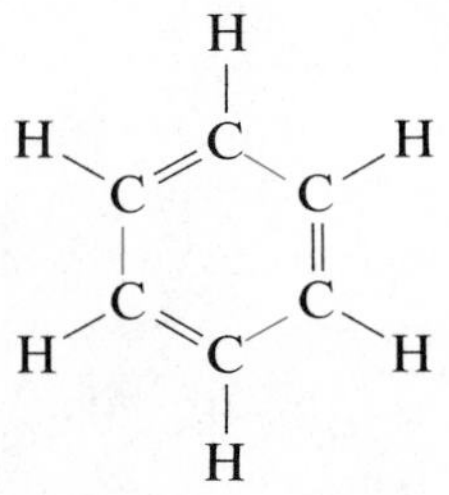

II

double bonds in I
are single bonds in II

molecule alternates. That is, the single and double bonds are rapidly alternating positions so that each carbon–carbon bond is neither single nor double but something intermediate. It can be seen that these two structures differ only in the position of the electrons. Structures that are similar to one another but differ from one another in the position of electrons are said to be in *resonance* and are called *resonance structures*. Therefore, benzene, which is neither structure I nor II, is a **resonance hybrid** of the two structures. In a resonance concept, only the hybrid

A **resonance hybrid** is a real molecule whose structure is described in terms of imaginary contributing structures.

has reality; the contributors are fictitious. Chemists have proposed interesting analogies for understanding the concept of resonance. One of these is to compare a resonance hybrid with the description of a rhinoceros as a cross between a unicorn and a dragon. The unicorn and the dragon are fictitious animals. Only the rhinoceros, the hybrid, has reality. So it is with benzene. The real structure of benzene is a hybrid of the two fictitious structures I and II above.

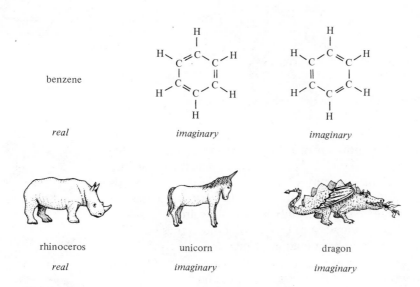

benzene

real   *imaginary*   *imaginary*

rhinoceros   unicorn   dragon

real   *imaginary*   *imaginary*

Today we know that in benzene all the carbon–carbon bond lengths are identical. The bond distance of 1.40 Å is the intermediate between the carbon–carbon single bond distance (1.54 Å) and the carbon–carbon double bond distance (1.34 Å) as suggested by structures I and II. In modern chemical terminology the structure of benzene is described as follows. Each carbon atom is bonded to three other atoms, two carbon atoms and one hydrogen atom. The six carbon atoms lie in a plane at the corners of a six-sided figure, a hexagon, and are joined by single bonds. These single bonds are sigma (σ) bonds. Each carbon atom of benzene has one p orbital occupied by a single electron. The p orbital of any one carbon overlaps equally the p orbitals of *both* carbon atoms to which it is bonded. The result is two continuous doughnut-shaped electron clouds, one lying above and the other lying below the plane of the six carbon atoms. In benzene there are twelve sigma bonds (six carbon–carbon bonds and six carbon–hydrogen bonds) and one pi cloud containing six electrons. All bond angles in benzene are 120°. Figure 12.1 illustrates the modern model of benzene.

In benzene the pi cloud electrons are associated with more than two carbon atoms. They are associated with all six carbon atoms and the electrons are said to be **delocalized**. When a pair of electrons associates with two carbon atoms, such as in ethene, a stable localized bond is formed. On the other hand, association with more than two carbon atoms leads to delocalized bonds.

Delocalized means that two pi electrons are associated with more than two carbon atoms.

Figure 12.1 Modern model of benzene. (a) Carbon–carbon and carbon–hydrogen sigma bonds. All bond angles are 120°. (b) Cyclic arrangement of six carbon atoms at the corners of a hexagon, each with a *p* orbital containing one electron. (c) Electron distribution in doughnut-shaped clouds.

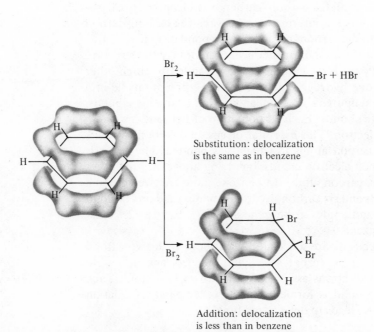

(a)   (b)   (c)

The result of delocalization is the formation of a stronger bond and a more stable molecule. As a result of delocalization, benzene is more stable than either structure I or structure II proposed by Kekulé. Structures I and II contain localized bonds in which a given pair of pi electrons bind together only two carbon atoms. In benzene the pi electrons participate in several bonds; that is, they are delocalized.

Because benzene has a delocalized cloud of pi electrons, it undergoes substitution rather than addition reactions. The bromination of benzene is a typical substitution reaction. The stability of the ring is preserved in substitution. Addition destroys the complete delocalization of the pi electrons of the ring.

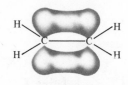

π electrons associated with two carbons: localized

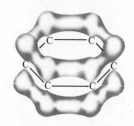

π electrons associated with six carbons: delocalized

Substitution: delocalization is the same as in benzene

Addition: delocalization is less than in benzene

We can summarize the characteristics of benzene, and aromatic compounds in general, as follows:

1. Possess a planar, cyclic structure.
2. Are stable because of delocalization of pi electrons.
3. Undergo substitution rather than addition reactions.

The Kekulé formulas (I and II) are seldom used for the structural formula of benzene. The Kekulé formulas are usually written as a regular hexagon and a hydrogen is understood to be attached to each angle of the hexagon (unless another group or atom is indicated). These are generally convenient to use when we need to account for the electrons. Benzene is a resonance hybrid of these two Kekulé formulas and is represented as a hexagon containing a circle. The circle stands for the cloud of six delocalized pi electrons.

12-2 NAMING AROMATIC COMPOUNDS DERIVED FROM BENZENE

LEARNING OBJECTIVES

1. Given the structural formula of a compound derived from benzene, write its name.

2. Given the name of an aromatic compound derived from benzene, write its structural formula.

3. Explain the distinction between the prefixes *ortho-, meta-,* and *para-.*

4. Given the structural formula of a compound containing a phenyl group, write its name.

5. Given the name of a compound containing a phenyl group, write its structural formula.

Many aromatic compounds are named as derivatives of benzene. In this section we will learn how to name **monosubstituted benzenes, disubstituted benzenes,** and **polysubstituted benzenes.** In general, monosubstituted benzenes are named by prefixing the name of the group or substituent to the word *benzene.* In some cases the monosubstituted benzene has a special name. For example, methylbenzene is called toluene.

A **monosubstituted benzene** is a benzene in which one hydrogen atom has been replaced by another atom or group. A **disubstituted benzene** is a benzene in which two hydrogen atoms have been replaced by other atoms or groups. A **polysubstituted benzene** is a benzene in which three or more hydrogen atoms have been replaced by other atoms or groups.

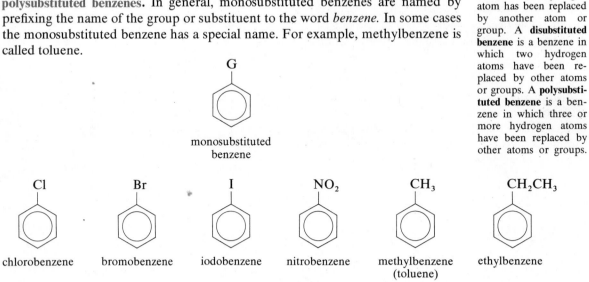

monosubstituted
benzene

chlorobenzene   bromobenzene   iodobenzene   nitrobenzene   methylbenzene
(toluene)   ethylbenzene

There are three kinds of disubstituted benzenes. If the two groups are located on adjacent carbons they are **ortho.** **Meta** substituted groups are separated by one carbon. If the groups are separated by two carbons, they are **para.**

Ortho- (*o*), *meta-* (*m*), and *para-* (*p*) are prefixes designating that two groups in a disubstituted benzene are separated by zero-, one-, and two-ring carbon atoms, respectively.

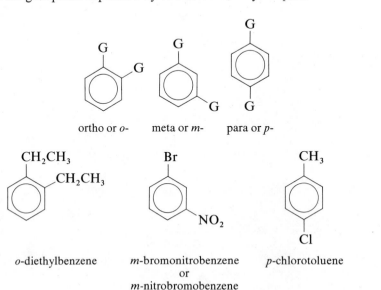

ortho or *o-*   meta or *m-*   para or *p-*

o-diethylbenzene   *m*-bromonitrobenzene
or
m-nitrobromobenzene   *p*-chlorotoluene

The prefixes *o-, m-,* and *p-* are used only when the ring is disubstituted. Derivatives of benzene that are polysubstituted are named using numbers to locate the substituents as shown. When the name toluene is used, the carbon bearing the methyl group is C-1.

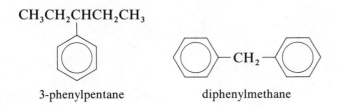

2,4-dinitrotoluene   2,6-dichloropropylbenzene   1-iodo-2,4-dichloro-5-nitrobenzene

The removal of a hydrogen atom from benzene gives rise to a phenyl group in much the same way that the methyl group is derived from methane.

benzene   phenyl group

Examples of nomenclature utilizing the phenyl group are 3-phenylpentane and diphenylmethane.

$CH_3CH_2CHCH_2CH_3$

3-phenylpentane

$—CH_2—$

diphenylmethane

PROBLEMS

1. Name the following derivatives of benzene.

(a)

I

(b)

Br

CH_3

(c)

CH_3

$CH—CH_3$

(d)

NO_2

CH_3

(e) $CH_3CH=CHCH_2CH_2—$

2. Write the structure for the following benzene derivatives.
   (a) butylbenzene
   (b) *m*-bromochlorobenzene
   (c) *p*-dimethylbenzene
   (d) 1,1-diphenylethane
   (e) 1-nitro-2,3-dibromobenzene

ANSWERS TO PROBLEMS   **1.** **(a)** iodobenzene or phenyliodide   **(b)** *o*-bromotoluene
(c) isopropylbenzene or 2-phenylpropane   **(d)** *p*-nitrotoluene   **(e)** 5-phenyl-2-pentene,
not 1-phenyl-3-pentene ($C=C$ gets the lowest number)

2.

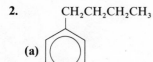

(a)

(b)

(c)

(d)

(e)

12-3 REACTIONS OF BENZENE

LEARNING OBJECTIVES

1. Given the reactants in a substitution reaction of benzene, write the structure(s) of the product(s).
2. Write the formula of the electrophile in substitution reactions of benzene.

3. Explain the importance of the resonance concept in substitution reactions of benzene.

The typical reaction of benzene and its derivatives is substitution. Benzene undergoes several substitution reactions. These include:

1. *Chlorination:* Substitution of chlorine for hydrogen.

$$\text{(benzene)} + Cl_2 \xrightarrow{FeCl_3} \text{(chlorobenzene)} + HCl$$

chlorine   chlorobenzene

2. *Bromination:* Substitution of bromine for hydrogen.

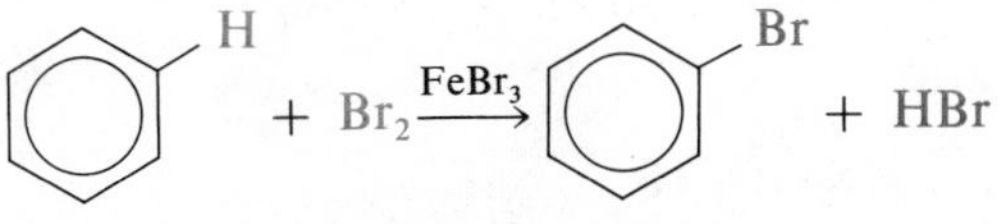

bromine   bromobenzene

3. *Nitration:* Substitution of a nitro group for hydrogen.

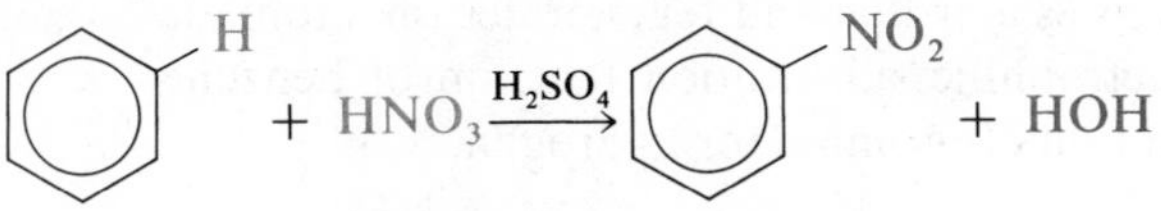

nitric acid   nitrobenzene

4. *Alkylation:* Substitution of an alkyl group (R) for hydrogen.

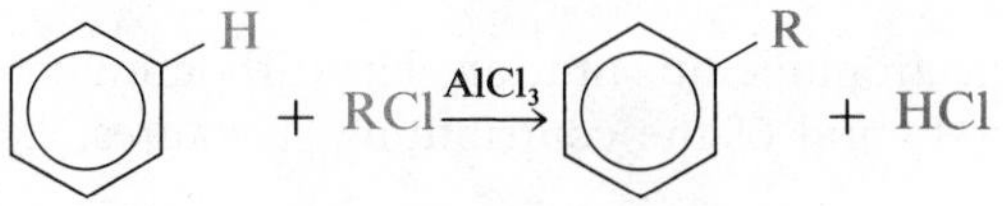

alkyl halide   alkylbenzene

How do these substitution reactions take place? In Chapter 11 it was explained that the pi cloud is a source of electrons that is available to a reagent that is seeking electrons. The pi cloud of the benzene ring also serves as a source of electrons and reacts with compounds that are deficient in electrons, namely, electrophilic reagents. The characteristic reaction of benzene is *electrophilic substitution.* The kinds of electrophilic substitutions occurring in benzene are those listed above: halogenation (such as chlorination or bromination), nitration, and alkylation. All these substitution reactions proceed by the same mechanism regardless of the reagent involved. Each of these reactions involves the initial reaction of an electrophile with benzene. The electrophile formed in each of the reactions shown appears in Table 12.1. We will use the bromination of benzene to illustrate electrophilic substitution in benzene.

Table 12.1	Electrophiles in Substitution Reactions of Benzene	
Substitution Reaction	**Electrophile**
Chlorination	Cl^+, chloronium ion
Bromination	Br^+, bromonium ion
Nitration	NO_2^+, nitronium ion
Alkylation	R^+, an alkyl carbonium ion

In the first step, the electrophile (Br^+) forms a bond to a carbon atom of benzene through the use of two of the pi electrons of benzene. The charged species that is formed is a familiar one: a carbonium ion (see Section 11-5.2).

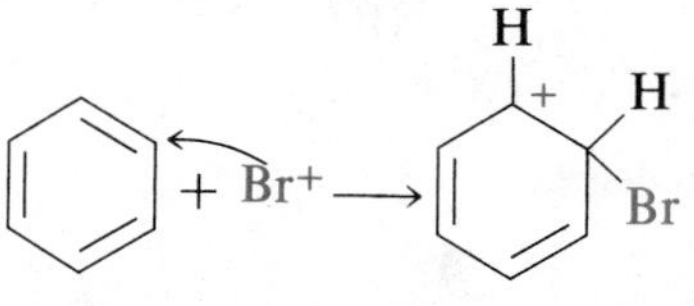

a carbonium ion

In the second step, the carbonium ion loses a proton and regenerates the aromatic system of six pi electrons. In this electrophilic substitution reaction of benzene then, the overall reaction is the substitution of bromine for hydrogen.

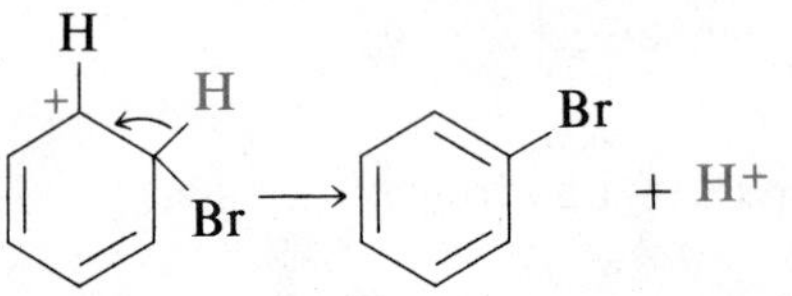

Resonance plays an important role in electrophilic substitution. The carbonium ion shown in step 1 above is a resonance hybrid of the contributing structures, III, IV, and V.

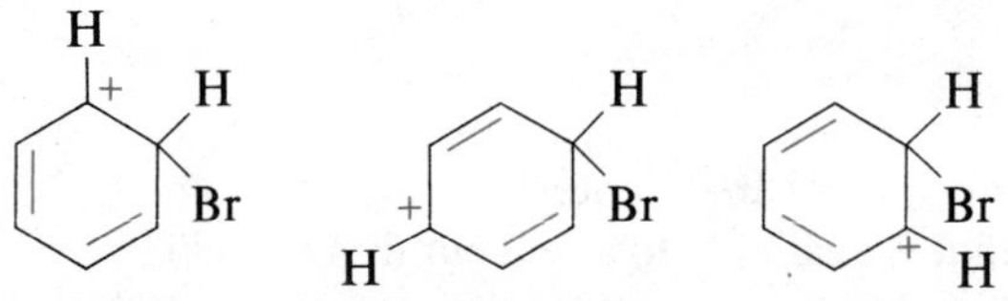

These structures are alike in all respects except in the positions of the pi electrons and of the positive charge. The positive charge is not localized on a single carbon atom, but is spread, or delocalized, over several carbon atoms. To represent this delocalization we often use a structure such as VI.

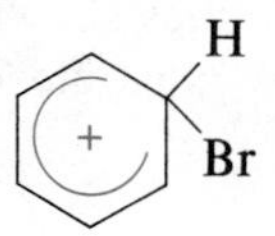

VI

As a consequence of delocalization of the positive charge, the actual carbonium ion, represented by VI, is more stable than any of the three contributing structures (III, IV, and V) in which the positive charge is localized. We have illustrated the mechanism of electrophilic substitution with the bromination of benzene, but you should remember that chlorination, nitration, and alkylation proceed by similar mechanisms.

12-4 SUBSTITUTION IN BENZENE DERIVATIVES

LEARNING OBJECTIVES

1. Given a substituent group on a benzene ring, classify it as an ortho/para director or a meta director.
2. Given the substitution reaction of a monosubstituted benzene, write the structure of the product.
3. Write equations for the two-step synthesis reactions using benzene or benzene derivatives as the starting compound.

12-4.1 ORTHO, META, AND PARA DIRECTORS

A monosubstituted benzene would be expected to form three disubstituted benzenes in electrophilic substitution reactions.

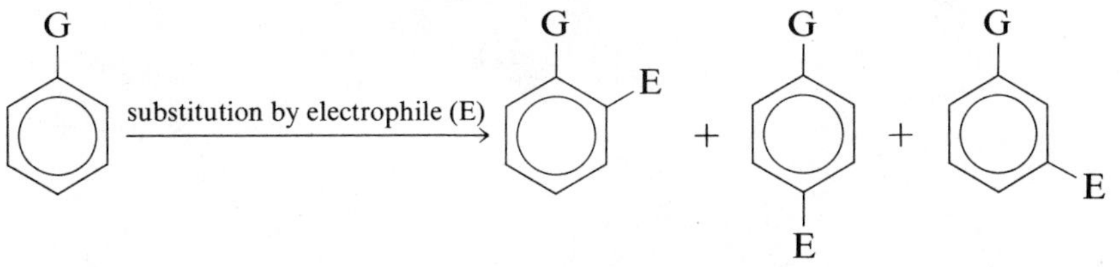

ortho para meta

Experimentally, one or two of the isomers are preferentially formed. The substituent (G) on the ring plays a large role in determining the position of the substitution.

Substituents generally fall into one of two categories: **ortho/para directors** and **meta directors.** Table 12.2 lists some common examples. Some typical examples are:

An **ortho/para director** is a substituent that directs an incoming group to the ortho and para positions. A **meta director** is a substituent that directs an incoming group to the meta position.

1. The nitration of bromobenzene produces *o*-bromonitrobenzene and *p*-bromonitrobenzene because the bromo group on the ring is an ortho/para director.

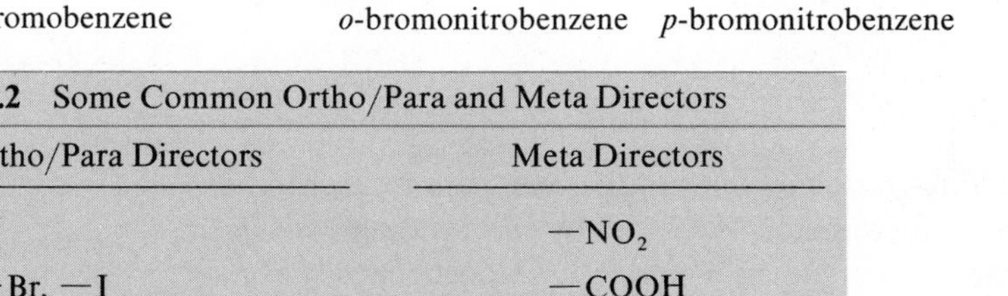

bromobenzene *o*-bromonitrobenzene *p*-bromonitrobenzene

Table 12.2 Some Common Ortho/Para and Meta Directors

Ortho/Para Directors	Meta Directors
$-OH$	$-NO_2$
$-Cl$, $-Br$, $-I$	$-COOH$
$-CH_3$, $-CH_2CH_3$, and so forth	$-SO_3H$

2. The bromination of nitrobenzene produces *m*-nitrobromobenzene because the nitro group on the ring is a meta director.

nitrobenzene   *m*-nitrobromobenzene

3. The chlorination of toluene produces a mixture of *o*-chlorotoluene and *p*-chlorotoluene because the methyl group of the ring is an ortho/para director.

    toluene   *o*-chlorotoluene   *p*-chlorotoluene

12-4.2 LABORATORY SYNTHESIS AND BIOSYNTHESIS

The ability of a group already on the benzene ring to direct an incoming group to a particular position on the ring plays an important role in chemical **synthesis** involving organic compounds. The following are examples of chemical synthesis.

> **Synthesis** is a series of reactions summarizing the reagents and reaction conditions for the preparation of a compound from a known source.

EXAMPLE 12.1   How can *p*-dichlorobenzene, a constituent of mothballs, be obtained (synthesized) from benzene?

    benzene   *p*-dichlorobenzene

In this example, the two chloro groups in the product are para to each other and the chloro group is an ortho/para director. We begin by chlorinating benzene, followed by a second chlorination of the initial product, chlorobenzene. Some *o*-dichlorobenzene is also formed but in minor amounts.

benzene   chlorobenzene   *p*-dichlorobenzene   *o*-dichlorobenzene
(major product)   (minor product)

EXAMPLE 12.2   How can *o*-chloronitrobenzene, a constituent of pigments for paints and inks, be obtained (synthesized) from benzene?

benzene   *o*-chloronitrobenzene

In this example, synthesis requires the introduction of chloro and nitro groups into the benzene ring. Furthermore the two groups are to be ortho to one another. We know that we must carry out both a chlorination and a nitration reaction. However, which do we decide to do first? The two substituents are ortho to each other. Of the two groups only chloro is an ortho director. Therefore we should chlorinate benzene and then nitrate the compound formed in chlorination of benzene.

benzene   chlorobenzene   *o*-chloronitrobenzene

The directive ability of a group already present on a benzene ring also plays an important role in the biosynthesis of several biologically active compounds such as L-dopa and thryoxine. L-Dopa has been used to aid individuals suffering from Parkinson's disease. The symptoms of Parkinson's disease, tremor and lack of muscle control, are reduced by L-dopa but exactly how the drug does this is not known. In the biosynthesis of L-dopa, under the control of the enzyme tyrosinase, a hydroxyl (OH) group is introduced into the aromatic ring of the amino acid tyrosine. The hydroxyl group takes the position on the aromatic ring ortho to the original hydroxyl group in tyrosine. This is exactly what we would predict, considering the directive influence of the hydroxyl group in tyrosine.

Similarly, iodination of tyrosine occurs at the two ortho positions just as we would predict, considering the directive influence of the hydroxyl group in tyrosine. Iodination of tyrosine leads to diiodotyrosine, an intermediate compound formed in the biosynthesis of the hormone thyroxine (see Section 13-2).

A **biosynthesis** is an *in vivo* reaction in which a molecule of the living system is made from another naturally occurring molecule.

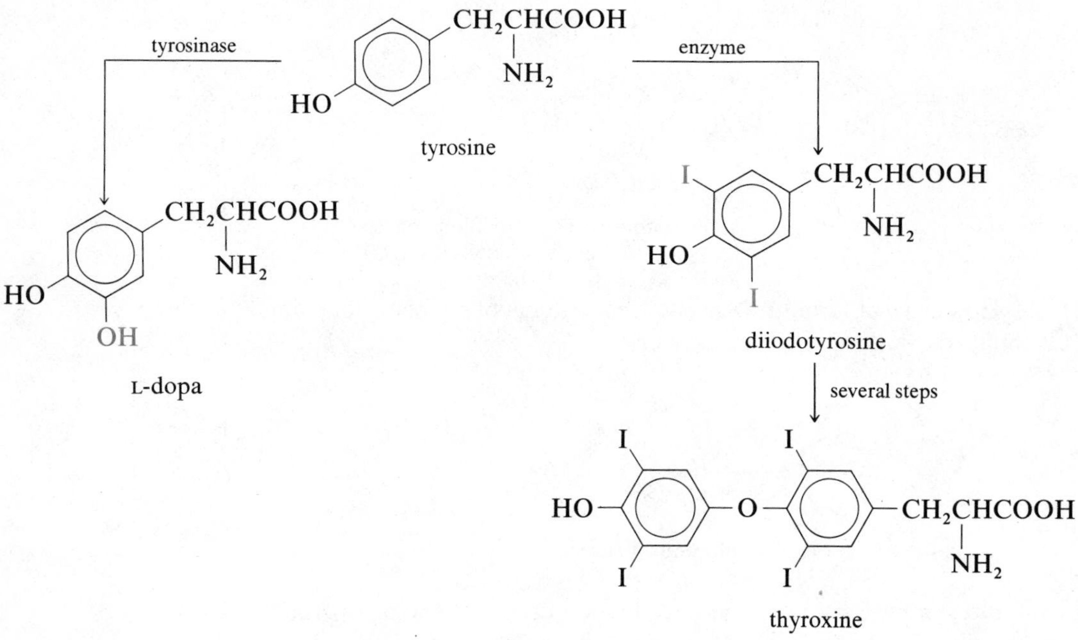

tyrosine

L-dopa

diiodotyrosine

thyroxine

PROBLEMS

3. Write the structure of the product of each substitution reaction.

(a) 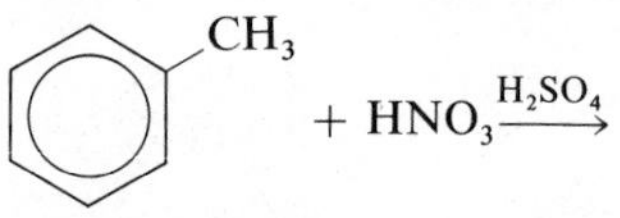 + HNO₃ $\xrightarrow{H_2SO_4}$

(b) 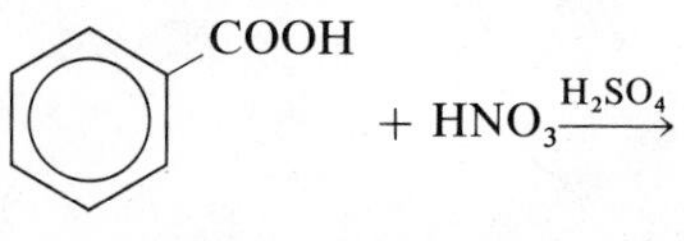 + HNO₃ $\xrightarrow{H_2SO_4}$

(c) 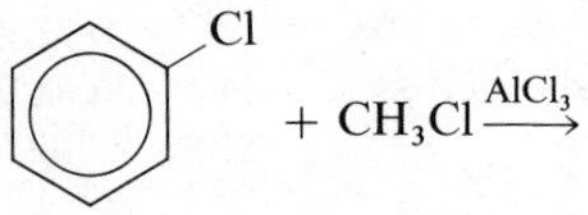 + CH₃Cl $\xrightarrow{AlCl_3}$

(d) 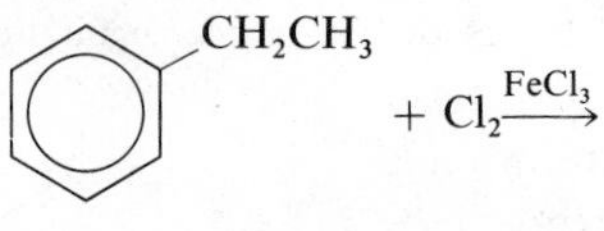 + Cl₂ $\xrightarrow{FeCl_3}$

4. Each of the following syntheses requires two steps. Write the correct sequence of steps necessary for each synthesis and give the necessary reagents.

(a) 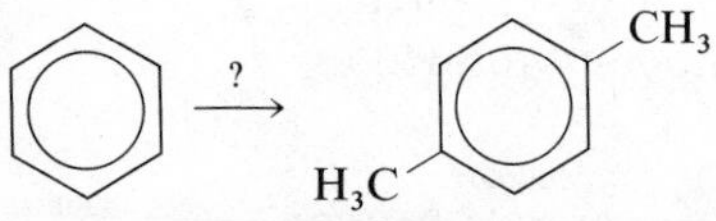 $\xrightarrow{?}$

(b) 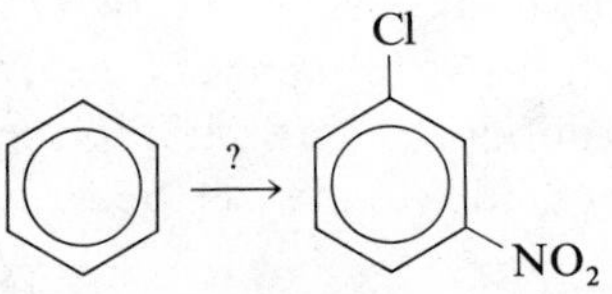 $\xrightarrow{?}$

(c) 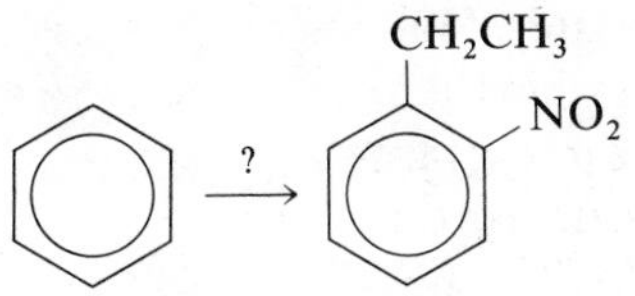 $\begin{array}{c}\text{benzene} \xrightarrow{\ ?\ } \end{array}$

CH$_2$CH$_3$, NO$_2$ product

ANSWERS TO PROBLEMS

3. (a)

CH$_3$, NO$_2$ / CH$_3$, NO$_2$

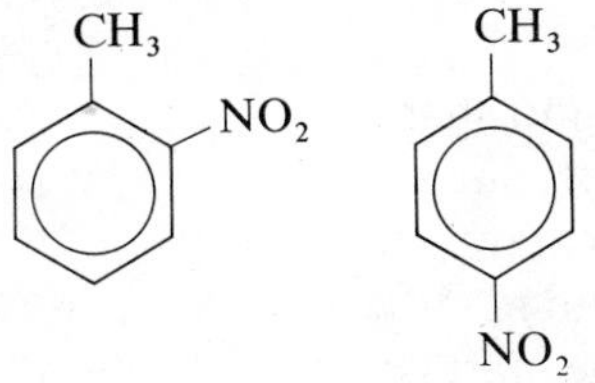

The CH$_3$ group is an ortho/para director.

(b)

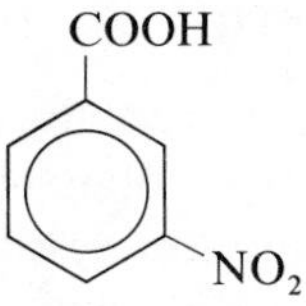

COOH, NO$_2$

The COOH group is a meta director

(c)

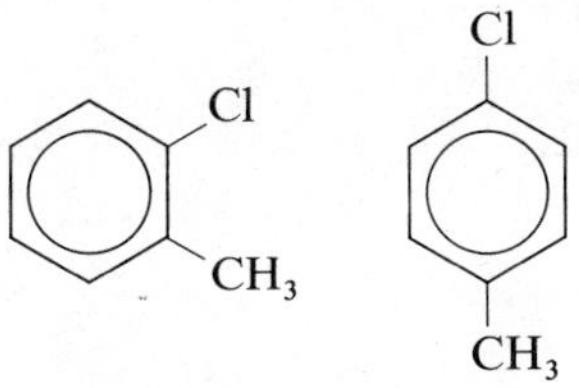

Cl, CH$_3$ / Cl, CH$_3$

The Cl group is an ortho/para director.

(d)

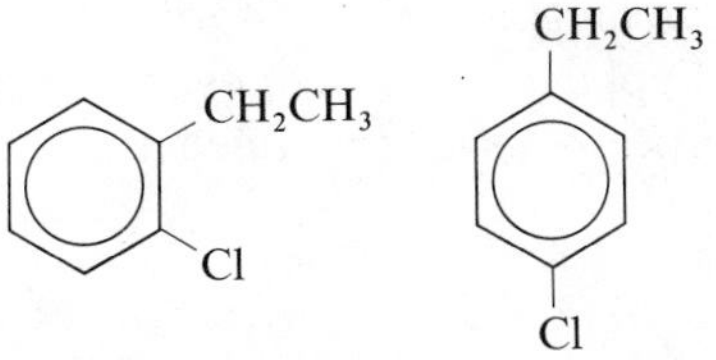

CH$_2$CH$_3$, Cl / CH$_2$CH$_3$, Cl

The CH$_3$CH$_2$ group is an ortho/para director.

4. (a) 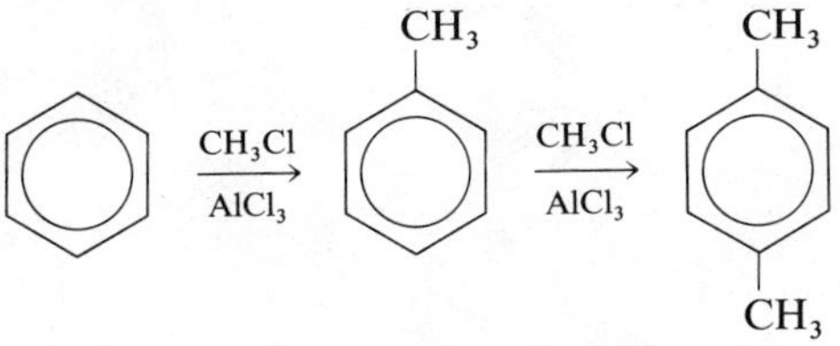 benzene $\xrightarrow[\text{AlCl}_3]{\text{CH}_3\text{Cl}}$ toluene $\xrightarrow[\text{AlCl}_3]{\text{CH}_3\text{Cl}}$ CH$_3$/CH$_3$

Some ortho product is also formed.

(b) 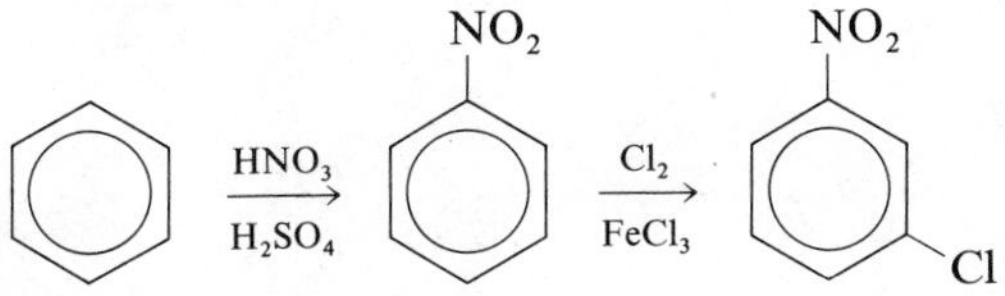 benzene $\xrightarrow[\text{H}_2\text{SO}_4]{\text{HNO}_3}$ NO$_2$ $\xrightarrow[\text{FeCl}_3]{\text{Cl}_2}$ NO$_2$, Cl

(c) 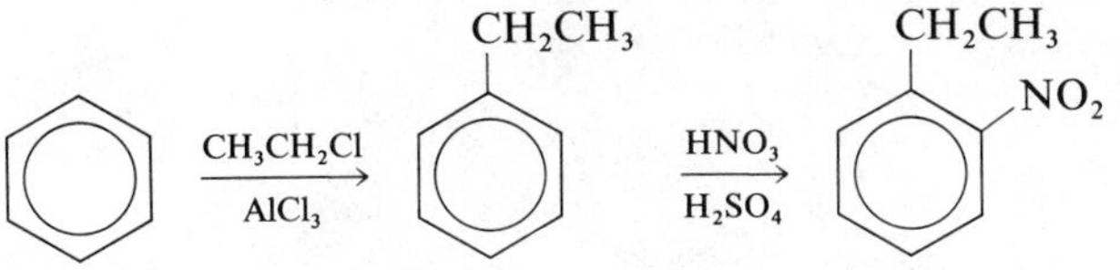 benzene $\xrightarrow[\text{AlCl}_3]{\text{CH}_3\text{CH}_2\text{Cl}}$ CH$_2$CH$_3$ $\xrightarrow[\text{H}_2\text{SO}_4]{\text{HNO}_3}$ CH$_2$CH$_3$, NO$_2$

Some para product is also formed.

12-5 OTHER TYPES OF AROMATIC COMPOUNDS

LEARNING OBJECTIVES

1. Define polynuclear aromatic compound and heterocyclic compound.
2. Draw or recognize structures of naphthalene, anthracene, phenanthrene, pyrrole, pyridine, purine, and pyrimidine.
3. Explain the need for foods, in our diets, which supply aromatic rings.

Aromatic compounds generally possess a planar cyclic structure, are stable because of delocalization of pi electrons, and undergo substitution rather than addition reactions. In addition to benzene and its derivatives, two other important aromatic compounds are polynuclear aromatic hydrocarbons and heterocyclic aromatic compounds.

12-5.1 POLYNUCLEAR AROMATIC HYDROCARBONS

Polynuclear aromatic hydrocarbons contain two or more aromatic rings and share a pair of carbon atoms. Examples of polynuclear aromatic hydrocarbons are shown in Table 12.3. Naphthalene is the simplest polynuclear aromatic hydrocarbon. It has been used in mothballs because it is toxic to some insects and larvae, and readily sublimes. Anthracene, another polynuclear aromatic hydrocarbon, is the starting compound for a number of dyes. Although many polynuclear aromatic hydrocarbons are carcinogenic (cancer-forming), naphthalene, anthracene, and phenanthrene do not appear to be carcinogenic. Note that in Table 12.3, the polynuclear aromatic hydrocarbons are identified as carcinogenic and noncarcinogenic.

Table 12.3   Some Typical Polynuclear Aromatic Hydrocarbons

Carcinogenic	Noncarcinogenic
benzanthracene	naphthalene
dibenzanthracene	anthracene
benzpyrene	phenanthrene

The polynuclear aromatic hydrocarbons may act as carcinogens in more than one way. They may either act directly or have to be activated. Regardless of the mode of action, the target of carcinogens is most likely the nucleic acids (see Chapter 20) of the cell: DNA, the genetic material, and RNA, which assists in protein synthesis. It is generally believed, but not proven, that these polynuclear aromatic hydrocarbon carcinogens interfere with the normal biological activity of DNA and RNA.

12-5.2 HETEROCYCLIC AROMATIC COMPOUNDS

Benzene and the polynuclear aromatic hydrocarbons have ring systems consisting only of carbon atoms. Some aromatic compounds have an element other than carbon present in the ring. Such compounds are called **heterocyclic aromatic compounds**. In many cases nitrogen is the atom that replaces carbon in this type of compound. Some examples of heterocyclic aromatic compounds are pyrrole, pyridine, purine, and pyrimidine.

Heterocyclic aromatic compounds are aromatic compounds which have an element other than carbon present in the ring.

pyrrole   pyridine   purine   pyrimidine

The four compounds shown as examples of heterocyclic aromatic compounds are important compounds in the living organism that, in general, do not occur free in nature but as components of some other important molecule. Pyrrole, for example, is found in both chlorophyll, the compound responsible for the green color in plants, and hemoglobin, the compound that transports oxygen from the lungs to other parts of the body. The structures of chlorophyll and hemoglobin are shown in Figure 12.2.

Plants can synthesize aromatic systems. Apparently, man is less able to do this. Certain amino acids containing aromatic rings are essential to our diet (see Section 28-3.3). This means that these amino acids cannot be synthesized by us from the compounds ordinarily present in the diet. Such amino acids are called *essential amino acids*. Essential amino acids, therefore, must be supplied in the diet. Some of the aromatic amino acids essential to man are phenylalanine, tyrosine, and tryptophan. Phenylalanine and tyrosine contain the normal benzene ring system. Tryptophan contains a benzene ring fused to a pyrrole ring. Such a combination is called an indole ring system.

phenylalanine   tyrosine   tryptophan

Figure 12.2   The structures of the heme portion of hemoglobin and chlorophyll contain the pyrrole rings as indicated in color. (a) When complexed with oxygen, heme enables hemoglobin to carry oxygen from the lungs to the tissue. (b) Chlorophyll is responsible for green-plant pigment.

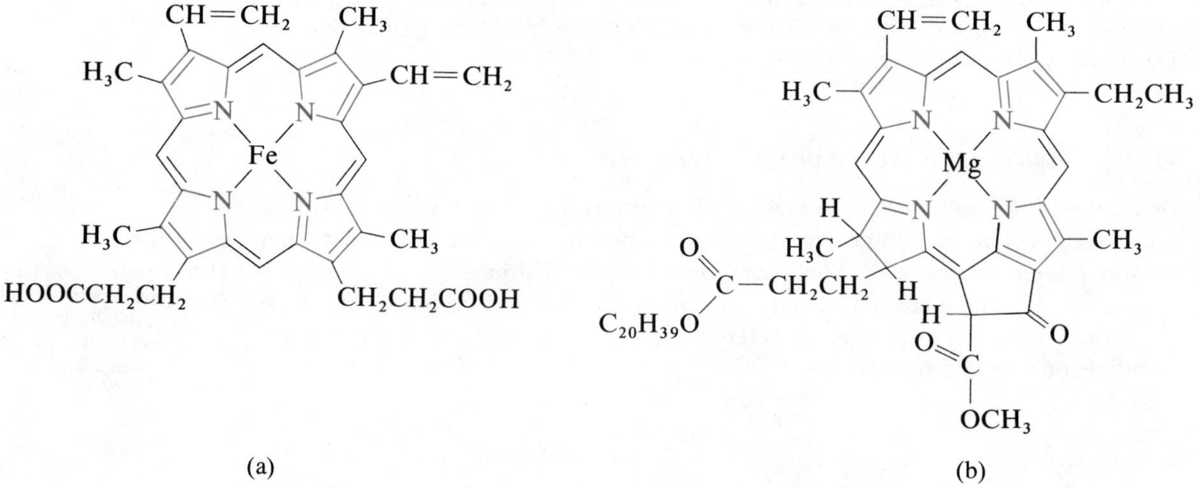

(a)

(b)

The essential components of the nucleic acids DNA and RNA are derivatives of purine and pyrimidine. Both DNA and RNA are important in the synthesis of proteins in the living system (see Chapter 20).

An important compound involved in oxidation-reduction reactions in biochemical systems is nicotinamide adenine dinucleotide (NAD$^+$). NAD$^+$, as shown in Figure 12.3, contains both pyridine and purine rings.

Figure 12.3   Nicotinamide adenine dinucleotide (NAD$^+$). The pyridine and purine rings are indicated in color.

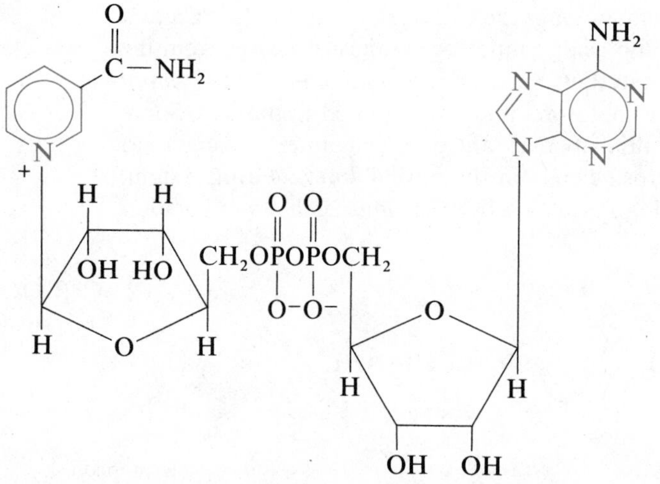

SUMMARY

Benzene has the molecular formula C_6H_6. Although benzene appears to be unsaturated, the unsaturation in benzene is different than that in alkenes and alkynes. Benzene is an aromatic compound. Aromatic compounds have a number of common properties. They have a planar cyclic structure, contain a pi cloud of delocalized electrons, and undergo substitution reactions. Examples are benzene, naphthalene, pyrrole, and pyridine.

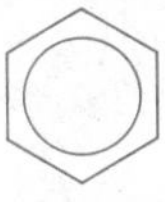

benzene 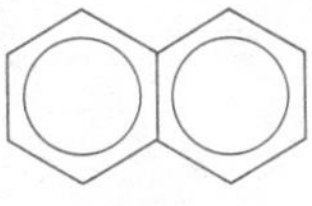   naphthalene

pyrrole    pyridine

Benzene is probably the most common aromatic compound. It is a resonance hybrid of two contributing structures and is more stable than either contributing structure.  and  are contributing structures to the structure of benzene, which is represented by  . Benzene and other aromatic compounds undergo a variety of electrophilic substitution reactions.

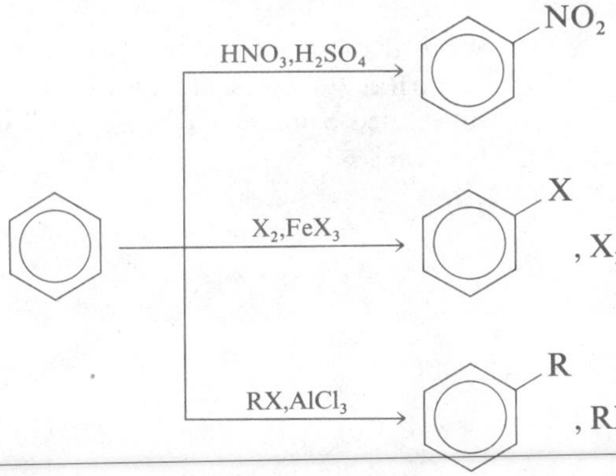

HNO_3, H_2SO_4 → NO_2

X_2, FeX_3 → X , X_2 is Cl_2 or Br_2

$RX, AlCl_3$ → R , RX is an alkyl halide

Monosubstituted aromatic compounds derived from benzene are named as derivatives of benzenes. Disubstituted benzenes are named using the prefixes *o*-, *m*-, and *p*-. Numbers are used to locate substituents in polysubstituted benzenes.

All substituents are classified as either ortho/para directors or meta directors. The positions taken up on the ring by an incoming group depends on the substituents already present on the ring. The directive ability of these substituents is important in the laboratory synthesis of aromatic compounds as well as aromatic compounds formed in the living system.

ADDITIONAL PROBLEMS

5. Write a structure for each of the following.
   (a) toluene
   (b) *o*-dimethylbenzene
   (c) propylbenzene
   (d) *m*-bromoethylbenzene
   (e) *p*-nitroethylbenzene
   (f) 2-phenyloctane
   (g) 2,5-dichloro-4-nitrotoluene
   (h) 3-phenylhexane
   (i) isopropylbenzene
   (j) pyrimidine

6. Name each of the following structures.

(a)

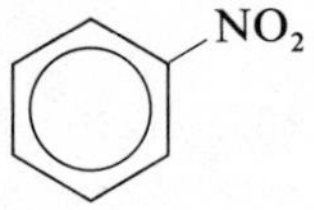

(b)

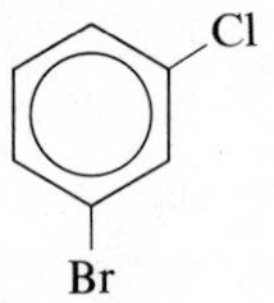

(c)

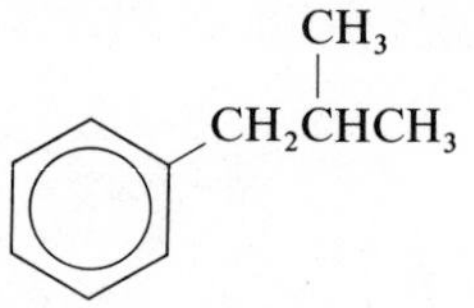

(d)

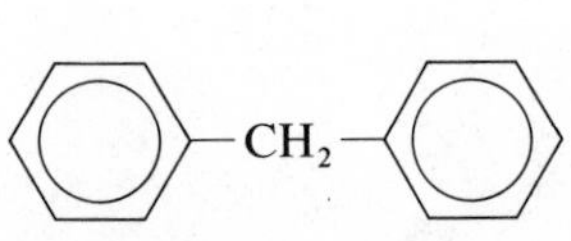

(e)

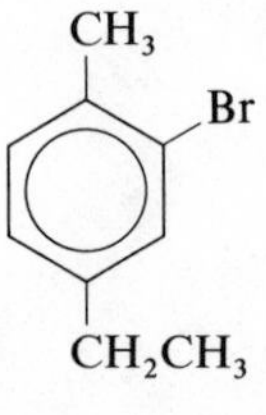

(f)

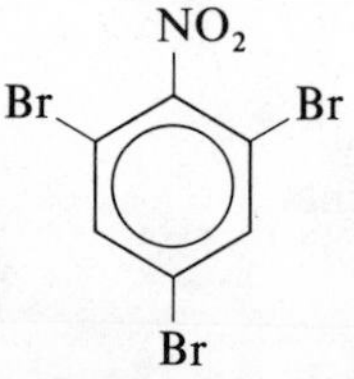

(g)

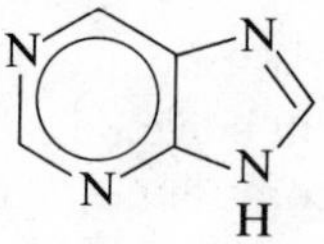

(h)

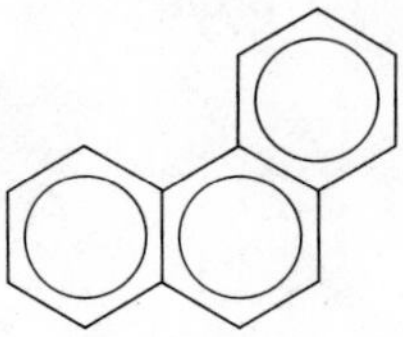

(i) 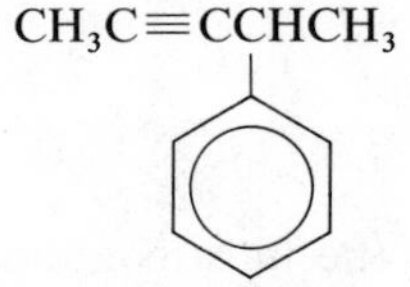 $CH_3C{\equiv}CCHCH_3$

(j)

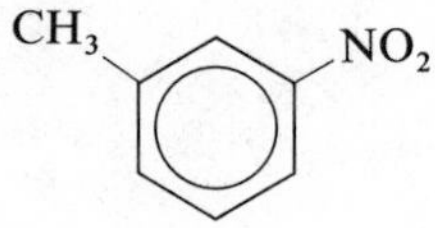

7. Write the structure of the product of each reaction [(a)–(c) represent the formation of a monosubstituted benzene and (d)–(j), a disubstituted benzene].

(a)  $+ CH_3CH_2Cl \xrightarrow{AlCl_3}$

(b) 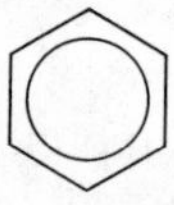 $+ Cl_2 \xrightarrow{FeCl_3}$

(c) 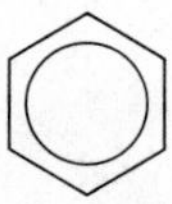 $+ HNO_3 \xrightarrow{H_2SO_4}$

(d) 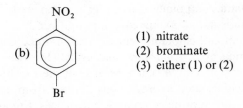 NO$_2$ + Br$_2$ $\xrightarrow{\text{FeBr}_3}$

(e) CH$_3$ + CH$_3$Cl $\xrightarrow{\text{AlCl}_3}$

(f) NO$_2$ + HNO$_3$ $\xrightarrow{\text{H}_2\text{SO}_4}$

(g) SO$_3$H + Cl$_2$ $\xrightarrow{\text{FeCl}_3}$

(h) Br + Br$_2$ $\xrightarrow{\text{FeBr}_3}$

(i) OH + HNO$_3$ $\xrightarrow{\text{H}_2\text{SO}_4}$

(j) CH$_2$CH$_3$ + Br$_2$ $\xrightarrow{\text{FeBr}_3}$

8. There are three bromonitrobenzene derivatives. Draw their structures and name each one.

9. There are four benzene derivatives corresponding to the molecular formula C$_8$H$_{10}$. Draw their structures and name each one.

10. Which of the following statements regarding benzene are false?
   (a) Benzene has a molecular formula C$_6$H$_6$.
   (b) Benzene is a hydrocarbon.
   (c) Benzene is an alkene.
   (d) Benzene is a resonance hybrid.
   (e) Benzene is a cyclic, nonplanar molecule.

(f) Benzene has a pi cloud containing six electrons.
(g) Benzene primarily undergoes addition reactions.
(h) Benzene contains three double bonds.
(i) Benzene contains a delocalized pi electron system.
(j) Benzene gives three monosubstituted derivatives.
(k) Benzene gives three disubstituted derivatives.

11. If you wanted to synthesize each of the following compounds from benzene, which of the two reactions would you perform first?

(a) CH$_3$, NO$_2$
    (1) methylate
    (2) nitrate
    (3) either (1) or (2)

(b) NO$_2$, Br
    (1) nitrate
    (2) brominate
    (3) either (1) or (2)

(c) Cl, NO$_2$
    (1) chlorinate
    (2) nitrate
    (3) either (1) or (2)

(d) CH$_3$, Cl
    (1) methylate
    (2) chlorinate
    (3) either (1) or (2)

12. Humans are less capable of synthesizing aromatic compounds *in vivo* than are plants, yet we need them to perform certain functions or make other aromatic compounds. How does the human body get its supply of aromatic compounds?

CHAPTER 13

ORGANIC HALIDES

Organic compounds containing halogen are rare in nature. Examples of such compounds are an iodinated compound produced by the thyroid gland in the living system and chloramphenicol, a chlorinated compound with antibiotic properties, formed by *Streptomyces venezuelae*. However, large quantities of organic halides are manufactured yearly because of their technological importance. For example, the chlorinated hydrocarbons derived from methane are widely used as solvents and the fluorine-containing substances Freons are used as refrigerants. Other fluorinated hydrocarbons are being studied by medical scientists as possible substitutes for blood. Still other halogenated compounds are used routinely as antiseptics or disinfectants, in X-ray examinations of soft tissue, and as insecticides such as DDT and lindane. To the organic chemist, the organic halides are very versatile because they can be used as precursors to many useful substances. In order to gain an appreciation for the versatility of the organic halides we will study their chemistry in this chapter.

13-1 STRUCTURE AND NOMENCLATURE

LEARNING OBJECTIVES

1. Define and recognize examples of alkyl halides and aryl halides.
2. Classify organic halides as primary alkyl, secondary alkyl, tertiary alkyl, or aryl.
3. Given the structure of an organic halide, write its IUPAC name and/or its common name.
4. Given the IUPAC name or common name of an organic halide, write its structural formula.

Organic halides are compounds in which a hydrogen atom of a hydrocarbon has been replaced by a halogen atom. The two common organic halides are the **alkyl halides** and the **aryl halides**. In the IUPAC system of nomenclature, the

An **alkyl halide** is one in which the halogen is attached to an alkyl group. An **aryl halide** is one in which the halogen is directly attached to an aromatic ring.

R—X

an alkyl halide

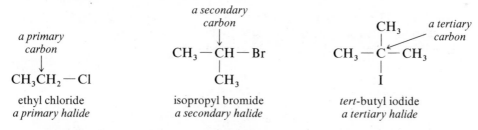—X

an aryl halide

halogen is considered a substituent on the hydrocarbon. The halogen group is referred to as chloro-, bromo-, or iodo-, and the position number is combined with the parent name of the hydrocarbon. Using the IUPAC system, we have 2-bromopentane, 2-methyl-4-chlorohexane, and 1,3-diiodocyclohexane.

$CH_3CHCH_2CH_2CH_3$
|
Br

2-bromopentane

$CH_3CHCH_2CHCH_2CH_3$
| |
CH_3 Cl

2-methyl-4-chlorohexane

1,3-diiodocyclohexane

A **primary alkyl halide** is one in which the halogen is attached to a primary carbon.

The halides containing simple alkyl groups (see Table 11.4) are named by specifying the name of the alkyl group followed by *chloride, bromide,* or *iodide.* When discussing alkyl halides, we often use the classification **primary, secondary,** or **tertiary alkyl halides.** Aryl halides were named in Section 12-2.

A **secondary alkyl halide** is one in which the halogen is attached to a secondary carbon. In a **tertiary alkyl halide** the halogen is attached to a tertiary carbon.

a primary carbon
↓
CH_3CH_2—Cl

ethyl chloride
a primary halide

a secondary carbon
↓
CH_3—CH—Br
|
CH_3

isopropyl bromide
a secondary halide

CH_3 *a tertiary carbon*
|
CH_3—C—CH_3
|
I

tert-butyl iodide
a tertiary halide

Compared to alkyl halides, aryl halides are very unreactive. In this chapter we will be concerned primarily with the chemistry of alkyl halides, but we will consider one aspect of the chemistry of aryl halides: reaction with an amino group of an amino acid.

PROBLEMS

1. Name the following organic halides.

(a) CH_3CHCH_2Cl
       |
      CH_3

(b) $BrCH_2CH_2CH_2CH_2CH_3$

(c) Cl ... NO_2

(d) $ClCH_2CH_2CH_2Cl$

(e) 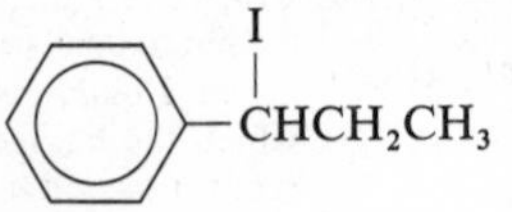 ⬡—$\overset{\overset{\displaystyle I}{\displaystyle |}}{C}HCH_2CH_3$

2. Which halide in question 1 fits the following description?
   (a) primary halide
   (b) aryl halide
   (c) secondary halide

ANSWERS TO PROBLEMS   **1.** **(a)** Common; isobutyl chloride, IUPAC,
1-chloro-2-methylpropane   **(b)** 1-bromopentane   **(c)** *m*-nitrochlorobenzene or
m-chloronitrobenzene   **(d)** 1,3-dichloropropane   **(e)** 1-iodo-1-phenylpropane
2. **(a)** a, b, d   **(b)** c   **(c)** e

13-2 THE USES OF ORGANIC HALIDES

LEARNING OBJECTIVE
List the uses of organic halides.

We have already seen from Table 11.8 that some polymeric organic halides are used to make materials that are commonly used in our daily lives. In this section we will consider some organic halides that are used as antiseptics and disinfectants, diagnostic agents, and blood substitutes. Finally, we will look briefly at two hormones secreted by the thyroid gland that are organic halides.

13-2.1 ANTISEPTICS AND DISINFECTANTS

Several organic halides are used as antiseptics and disinfectants. *Antiseptic* refers to action on living tissue. In some cases the antiseptic is used to kill microorganisms when only brief contact is available, as in mouth washes. On the other hand, the antiseptic also may be used during continuous, long exposure to prevent an increase in the number of microorganisms. *Disinfectants* are agents that act on inanimate objects. In addition to halogens and halogen-containing compounds, other classes of organic compounds possess antiseptic properties. These include phenolic compounds (see Table 14.4) and quaternary ammonium compounds (see Table 16.4).

Some of the typical halogenated organic compounds useful as antiseptics are povidone-iodine (marketed as Betadine), tetraiodopyrrole (marketed as Iodol), merbromin (marketed as Mercurochrome), and halazone (marketed as Pantocid). Povidone-iodine is a water-soluble complex of polyvinylpyrrolidone (see Table

11.8) and iodine. It is recommended for use on the skin and mucous membrane, and has been used successfully against some infections. It has also been used in pre- and post-operative and therapeutic antisepsis on skin, scalp, and genital tract. Tetraiodopyrrole is relatively insoluble in water and is used externally in a variety of ointments and suspensions. Merbromin is an organic mercurial compound in which mercury forms a direct bond with carbon. Aqueous solutions of merbromin (Mercurochrome), as well as formulations in a mixture of alcohol, acetone, and water, have antimicrobial activity on skin, mucous membranes, and wounds. Halazone is a well-known water purifier.

tetraiodopyrrole   merbromin   halazone

The mechanisms by which antiseptics and disinfectants exert their germicidal action are varied and complex. The most common mechanism appears to be the inactivation of certain enzymes. Iodinated compounds are transformed into iodine or hypoiodous acid (HOI). Iodine may become attached to a protein segment such as a tyrosine residue (see Section 12-4). Alternatively, iodine may form HOI which breaks down to hydroiodic acid (HI) and oxygen, the latter acting as an oxidant.

On the other hand, some antiseptics and disinfectants alter a protein's structure. For example, merbromin releases an ion, perhaps Hg^{2+}, which reacts with the protein's sulfhydryl groups ($-SH$) causing the protein to lose its biological activity.

13-2.2 DIAGNOSTIC AGENTS

Organic halides are also used as *diagnostic agents*. These serve the physician by permitting examination of the body and diagnosis of abnormalities and impairment of organ functions. Radiopaques are the most important type of diagnostic agent. These substances cause soft tissues such as the stomach, gallbladder, and urinary bladder to become visible during X-ray examination by causing absorption of X-rays and producing a shadow of positive contrast. A positive contrast is one in which the area to be visualized is denser than the surrounding tissue structure.

Many radiopaques are iodine-containing compounds. There appears to be no evidence that appreciable deiodination occurs with any iodine-containing radiopaque. However, elimination of a small amount of iodine from a radiopaque compound as iodide ion would not be a serious matter since iodide ion is a normal trace constituent of the body (see Section 10-1). The structures of several iodine-containing radiopaques and the organ or area involved follow.

$$CO_2H$$

3,5-diacetamido-2,4,6-
triiodobenzoic acid

(gallbladder,
urinary tract)

$$CH_3CH(CH_2)_8CO_2C_2H_5$$

ethyl 10-*p*-iodophenyl
undecanoate

(spinal canal,
lymphatic ducts)

$$CH_2CO_2(CH_2)_2CH_3$$

propyl 3,5-diiodo-4-oxo-
1-pyridine acetate

(bronchial tree)

13-2.3 BLOOD SUBSTITUTES

Recent experiments have provided evidence that some totally artificial media can temporarily supply oxygen to an organism without any necessarily lethal consequences. These experiments have opened the way for the development of artificial blood substitutes. The substances under investigation are perfluorocarbon derivatives, organic compounds in which all or most of the hydrogen atoms are replaced by fluorine atoms. These synthetic compounds are chemically and biologically stable fluids. Because they mimic the most important function of blood which is to transport oxygen to the tissues, these synthetic fluids are capable of standing in for blood in emergencies during an operation or in case of blood poisoning when the blood needs to be purified. They might also be used in treating diseases such as anemia when long term transfusion treatments are impractical. Because some oxygenated fluorocarbons actually carry about twice as much oxygen as does blood, there is a possibility that a blood-starved limb threatened with gangrene might be saved by treatment with oxygen saturated fluorocarbons. Perfluoroctyl bromide and perfluorodecalin are fluorocarbons that have been investigated as blood substitutes.

$$CF_3CF_2CF_2CF_2CF_2CF_2CF_2CF_2Br$$

perfluoroctyl bromide

perfluorodecalin

13-2.4 THYROID HORMONES

The two characteristic hormones secreted by the thyroid gland are thyroxine and triiodothyronine. Structurally they are aryl iodides and they are derived from tyrosine, an aromatic amino acid (see Section 12-4). Thyroxine is produced in much larger quantities than is triiodothyronine, but the latter is 5 to 10 times more active. These two hormones influence the oxygen consumption of nearly all the organs in the body except the brain. If the rate of oxygen consumption in an individual is below normal, it may be necessary to administer these thyroid hormones. On the other hand, if the rate of oxygen consumption is elevated above normal, it may be necessary to remove part of the thyroid gland to diminish the total thyroid hormone released.

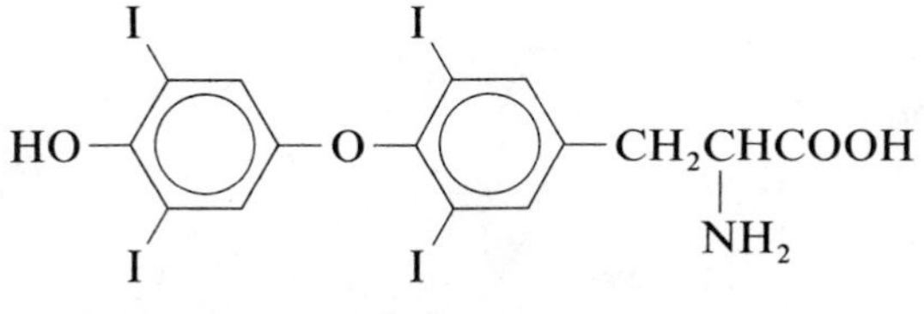

thyroxine

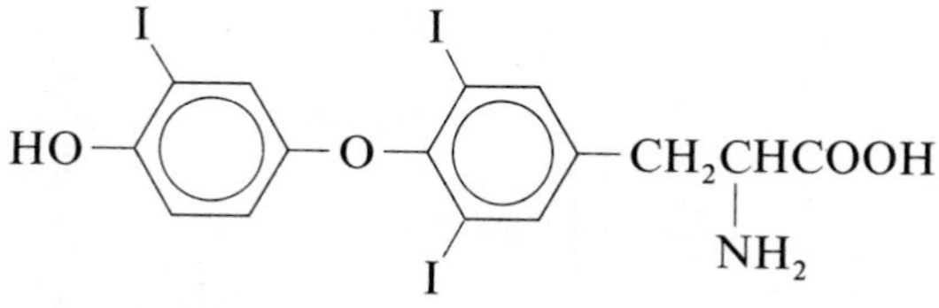

triiodothyronine

13-3 PREPARATIONS OF ORGANIC HALIDES

LEARNING OBJECTIVES

1. Given the reactants for the preparation of organic halides from alkanes, alkenes, alkynes, and aromatic hydrocarbons, complete the equation by writing the structures of the products.

2. Define and recognize examples of free radicals.
3. Explain how Freons have caused environmental changes.

13-3.1 REVIEW

We have learned how to prepare a number of different halides in previous chapters and it is useful to summarize these methods here. The sources of organic halides mentioned in the previous chapters were aromatic hydrocarbons and the unsaturated hydrocarbons: alkenes and alkynes.

Preparing Halides from Alkenes and Alkynes by an Addition Reaction (see Section 11-5.2)

1. Addition of bromine or chlorine.

$$CH_3CH{=}CH_2 + Br_2 \longrightarrow CH_3CH{-}CH_2$$
$$\phantom{CH_3CH{=}CH_2 + Br_2 \longrightarrow CH_3CH}\underset{Br}{|}\underset{Br}{|}$$

propene

1,2-dibromopropane

$$CH_3{-}C{\equiv}C{-}CH_3 + 2\,Cl_2 \longrightarrow CH_3{-}\underset{Cl}{\overset{Cl}{C}}{-}\underset{Cl}{\overset{Cl}{C}}{-}CH_3$$

2-butyne

2,2,3,3-tetrachlorobutane

2. Addition of hydrogen chloride, hydrogen bromide, or hydrogen iodide according to the Markovnikov rule.

$$CH_3CH{=}CH_2 + HCl \longrightarrow CH_3CH{-}CH_2$$

$$\underset{\text{Cl}}{|} \quad \underset{\text{H}}{|}$$

propene   2-chloropropane

$$CH_3C{\equiv}C{-}H \xrightarrow{\text{HI}} CH_3C{=}C{-}H \xrightarrow{\text{HBr}} CH_3C{-}C{-}H$$

propyne   2-iodo-1-propene   2-bromo-2-iodopropane

**Preparing Halides from Aromatic Hydrocarbons by a Substitution Reaction
(see Section 12-3)**

benzene   chlorobenzene

toluene   o-bromotoluene   p-bromotoluene

In summary, alkyl halides are generally prepared from alkenes and alkynes. Aryl
halides are formed from the halogenation of aromatic hydrocarbons in the presence
of the catalyst ferric bromide ($FeBr_3$) or ferric chloride ($FeCl_3$).

13-3.2 PREPARATION OF HALIDES FROM ALKANES

In addition to the available procedures described above, alkyl halides can be
formed from alkanes through the substitution of a hydrogen atom by a halogen
atom. Substitution of a halogen atom for a hydrogen atom occurs in the presence
of ultraviolet light (for example, sunlight). Chlorine and bromine are the halogens
most commonly used. Methane forms chloromethane (methyl chloride) when it is
chlorinated in the presence of ultraviolet light.

$$CH_4 + Cl_2 \xrightarrow{\text{ultraviolet light}} CH_3Cl + HCl$$

methane   methyl chloride

Halogenation, such as that for methane, proceeds in the following manner.

1. The ultraviolet light causes cleavage of chlorine molecules to chlorine atoms.

$$Cl_2 \xrightarrow[\text{light}]{\text{ultraviolet}} 2\ Cl\cdot$$

chlorine   chlorine
molecules   atoms

2. The chlorine atom abstracts (removes) a hydrogen atom from methane. The products are hydrogen chloride and an organic intermediate, a methyl **free radical**.

> A **free radical** is an atom or a group of atoms having an unpaired electron.

$$
\underset{\text{methane}}{H-\overset{\displaystyle H}{\underset{\displaystyle H}{C}}-H}
\ +\ \underset{\substack{\text{chlorine}\\\text{atom}}}{Cl\cdot}
\ \longrightarrow\
\underset{\substack{\text{methyl free}\\\text{radical}}}{H-\overset{\displaystyle H}{\underset{\displaystyle H}{C}}\cdot}
\ +\ \underset{\substack{\text{hydrogen}\\\text{chloride}}}{HCl}
$$

3. The free radical is very reactive and abstracts a chlorine atom from a chlorine molecule to form a molecule of the product, methyl chloride, and another chlorine atom.

$$
H-\overset{\displaystyle H}{\underset{\displaystyle H}{C}}\cdot
\ +\ Cl_2
\ \longrightarrow\
\underset{\substack{\text{methyl}\\\text{chloride}}}{H-\overset{\displaystyle H}{\underset{\displaystyle H}{C}}-Cl}
\ +\ \underset{\substack{\text{chlorine}\\\text{atom}}}{Cl\cdot}
$$

4. The newly formed chlorine atom can react with more methane. The halogenation of methane (or any alkane) is a substitution reaction involving a free radical. In the methyl free radical the carbon atom has seven electrons, one of which is not paired. The unpaired electron of the free radical occupies a p orbital. The methyl free radical is a planar species and the p orbital lies above and below the plane of the carbon and three hydrogens.

(120° methyl radical structure) is $H-\overset{\displaystyle H}{\underset{\displaystyle H}{C}}\cdot$ or simply $CH_3\cdot$

In the presence of excess methane the product is methyl chloride. However, in the presence of excess chlorine a variety of chlorinated methanes is formed. The number and kinds of chlorinated methanes depend on whether or not all the hydrogens of methane are substituted by chlorine.

$$
\underset{\text{methane}}{CH_4}
\ \xrightarrow{Cl_2}\
\underset{\substack{\text{methyl}\\\text{chloride}\\+\ HCl}}{CH_3Cl}
\ \xrightarrow{Cl_2}\
\underset{\substack{\text{methylene}\\\text{chloride}\\+\ HCl}}{CH_2Cl_2}
\ \xrightarrow{Cl_2}\
\underset{\substack{\text{chloroform}\\+\ HCl}}{CHCl_3}
\ \xrightarrow{Cl_2}\
\underset{\substack{\text{carbon}\\\text{tetrachloride}\\+\ HCl}}{CCl_4}
$$

Methyl chloride (chloromethane) has been used as a refrigerant. Methylene chloride (dichloromethane), chloroform (trichloromethane), and carbon tetrachloride (tetrachloromethane) have been used in degreasing and cleaning fluids. Chloroform was once used as an inhalation anesthetic.

Other alkanes react with halogen in the presence of ultraviolet light as illustrated by the bromination of ethane, which forms ethyl bromide, and toluene, which forms **benzyl** bromide.

A **benzyl group** is an alkyl group derived from toluene by removal of a methyl hydrogen.

$$CH_3CH_3 + Br_2 \xrightarrow[\text{light}]{\text{ultraviolet}} CH_3CH_2Br + HBr$$

ethane   ethyl bromide

toluene   benzyl bromide

Note that the bromination of toluene here yields a different product than the bromination of toluene as shown in Section 13-3.1. This illustrates the fact that different reaction conditions (ultraviolet light versus $FeBr_3$) can produce different products.

Evidence indicates that chlorine atoms and other carbon radicals derived from the Freons are involved in the breakdown of the ozone (O_3) layer by Freons. The Freons are organic molecules with very few carbons and hydrogens that consist primarily of fluorine and chlorine. They are odorless and noncombustible gases, and have been used as propellants for aerosol cans and as refrigerants. Some of the more common Freons are shown in Table 13.1.

The use of Freons as aerosols has caused great controversy. The Freons are gases that, when released as aerosols into the atmosphere, drift upward into the troposphere (16 km above ground level). The bulk of the Freons remain there, but some drift further upward into the stratosphere (16–50 km above ground level). Scientists have expressed concern over the environmental effects of the Freons on the earth's protective ozone layer. The ozone layer shields the earth's surface from strong ultraviolet radiation. This radiation can kill crops, cause skin cancer, and

Table 13.1   Typical Freons

Structure	IUPAC Name	Commercial Name
CCl_3F	Trichlorofluoromethane	Freon 11
CCl_2F_2	Dichlorodifluoromethane	Freon 12
$CHClF_2$	Chlorodifluoromethane	Freon 22
$CClF_2CClF_2$	Dichlorotetrafluoroethane	Freon 114

change weather. According to a recent study, the Freons absorb ultraviolet radiation and decompose to chlorine atoms and other products.

$$\text{Freons} \xrightarrow{\text{ultraviolet radiation}} Cl\cdot + \text{other products}$$

The chlorine atoms then catalytically attack ozone to form oxygen and another radical.

$$Cl\cdot + O_3 \longrightarrow O_2 + ClO\cdot$$

Although this theory is still controversial because the laboratory results as described here have not been documented in the stratosphere, many manufacturers have restricted, and even ended, the use of Freons as aerosols.

PROBLEM

3. Write the structure and the name of the product of each reaction that represents the formation of an organic halide.

(a) $CH_3CH{=}CHCH_3 + HI \longrightarrow$

(b)  $+ Cl_2 \longrightarrow$

(c) 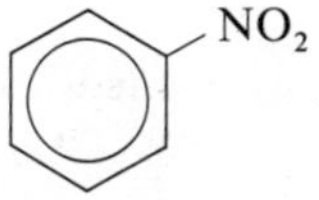 $\xrightarrow[\text{FeBr}_3]{\text{Br}_2}$

(d) $CH_3{-}\overset{\displaystyle CH_3}{\underset{\displaystyle CH_3}{\overset{|}{\underset{|}{C}}}}{-}CH_3 + Cl_2 \xrightarrow{\text{ultraviolet light}}$

(e) $CH_3C{\equiv}CH \xrightarrow[\text{1 mole}]{Cl_2} ? \xrightarrow{HBr}$

ANSWERS TO PROBLEM **3. (a)** $CH_3CH{-}CH_2CH_3$, *sec*-butyl iodide or 2-iodobutane
$\quad\quad\quad\quad\quad\quad\quad\quad\quad\quad\quad\quad\quad\quad\quad\quad\quad\quad\quad\quad |$
$\quad\quad\quad\quad\quad\quad\quad\quad\quad\quad\quad\quad\quad\quad\quad\quad\quad\quad\quad\quad I$

(b) 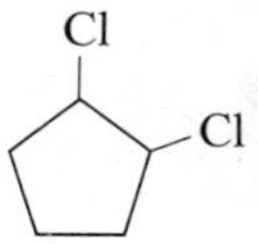 1,2-dichlorocyclopentane **(c)** 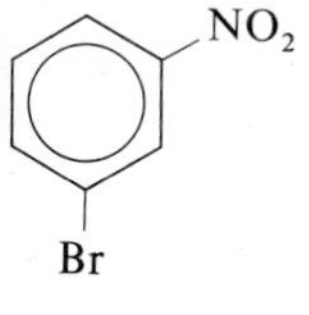 *m*-nitrobromo-

benzene **(d)** $CH_3\overset{\displaystyle CH_3}{\underset{\displaystyle CH_3}{\overset{|}{\underset{|}{C}}}}CH_2Cl$; 2,2-dimethyl-1-chloropropane

(e) $CH_3\underset{\underset{Cl}{|}}{C}{=}\underset{\underset{Cl}{|}}{C}H \xrightarrow{HBr} CH_3\underset{\underset{Cl}{|}}{\overset{\overset{Br}{|}}{C}}{-}\underset{\underset{Cl}{|}}{C}H_2$

1,2-dichloro-2-bromopropane

13-4 REACTIONS OF ORGANIC HALIDES

LEARNING OBJECTIVES

1. Given a substitution, or elimination, reaction for an organic halide, write the structure of the product.

2. Define and recognize examples of nucleophiles.

3. Write the general name of the product from substitution and elimination reactions of organic halides.

4. Given a dehydrohalogenation reaction of an organic halide, write the structure of the alkene formed.

5. Define detoxification and explain this process as an example of substitution and elimination reactions.

Alkyl halides are comparatively more reactive than aryl halides. Alkyl halides typically undergo two kinds of reaction. These include substitution and elimination. Furthermore, these two reactions often compete with one another. Aryl halides undergo substitution also, but we will be interested in aryl halides because of their use in biochemistry.

13-4.1 SUBSTITUTION AND ELIMINATION: WHAT ARE THEY?

Alkyl halides undergo **substitution** and **elimination reactions** (see Section 10-4.2) in the presence of a nucleophile. In a substitution reaction, the nucleophile (symbolized in the equations as Nu$^-$) attacks the carbon atom that has the halogen atom and replaces (or substitutes for) the halogen atom. The halogen atom is said to be the leaving group.

A substitution reaction is a reaction in which one group is replaced by another. An elimination reaction is one in which an organic molecule gives up something.

Substitution

$$CH_3CH_2Br \quad + \quad Nu^- \longrightarrow CH_3CH_2Nu + Br^-$$

leaves

nucleophile attacks carbon

In an elimination reaction, the nucleophile attacks the hydrogen atom on the carbon atom adjacent to the carbon atom that has the halogen atom. A double bond forms between the carbon atom bearing the halogen atom and the adjacent carbon atom. In elimination both the halogen atom and the hydrogen on the adjacent carbon atom are leaving groups.

Elimination

$$CH_2-CH_2Br \quad + \quad Nu^- \longrightarrow CH_2=CH_2 + HNu + Br^-$$
$$|$$
$$H$$

leaves

nucleophile attacks hydrogen

A **nucleophile** is a basic, electron-rich reagent. Generally, a nucleophile is a negatively charged species and has at least one unshared pair of electrons that it is willing to share with another atom. In substitution it will share its electrons with carbon; in elimination it will share its electrons with hydrogen. There are a

*A **nucleophile** is an electron-rich reagent that is willing to share a pair of electrons with another atom.*

Table 13.2 Some Common Nucleophiles

in vitro		*in vivo*	
:OH⁻	hydroxide	HÖCH$_2$CHCOOH	hydroxyl group in the
:OCH$_3$⁻	methoxide	$\quad\quad$ |	amino acid serine
:SH⁻	bisulfide	$\quad\quad$ NH$_2$	
:CN⁻	cyanide		
:I⁻	iodide	HSCH$_2$CHCOOH	thiol group in the
		$\quad\quad$ |	amino acid cysteine
		$\quad\quad$ NH$_2$	
		H$_2$N̈(CH$_2$)$_4$CHCOOH	amino group in the
		$\quad\quad\quad\quad$ |	amino acid lysine
		$\quad\quad\quad\quad$ NH$_2$	

variety of nucleophiles (see Table 13.2) that can cause substitution and elimination reactions. They are generally used as their sodium or potassium salts. Some typical substitution reactions are:

$$CH_3I + OH^- \longrightarrow CH_3OH + I^-$$

methyl iodide   methyl alcohol

$$CH_3-\overset{\overset{\displaystyle CH_3}{|}}{C}HCl + CH_3O^- \longrightarrow CH_3-\overset{\overset{\displaystyle CH_3}{|}}{C}HOCH_3 + Cl^-$$

isopropyl chloride   isopropyl methyl ether

$$CH_3-\overset{\overset{\displaystyle CH_3}{|}}{\underset{\underset{\displaystyle CH_3}{|}}{C}}-Br + I^- \longrightarrow CH_3-\overset{\overset{\displaystyle CH_3}{|}}{\underset{\underset{\displaystyle CH_3}{|}}{C}}-I + Br^-$$

tert-butyl bromide   *tert*-butyl iodide

Some typical elimination reactions are:

$$\underset{\underset{\displaystyle H}{|}}{CH_2}-\underset{\underset{\displaystyle Br}{|}}{CH_2} + OH^- \longrightarrow CH_2{=}CH_2 + HOH + Br^-$$

ethyl bromide   ethylene

$$CH_3-\overset{\overset{\displaystyle H}{|}}{\underset{\underset{\displaystyle I}{|}}{C}}-\underset{\underset{\displaystyle H}{|}}{CH_2} + OH^- \longrightarrow CH_3-\overset{\overset{\displaystyle H}{|}}{C}{=}CH_2 + HOH + I^-$$

2-iodopropane   propene

In each of these reactions the reactivity of the alkyl halide depends upon the nature of the alkyl group. In actuality there are two kinds of substitution reactions and two kinds of elimination reactions. These four reactions compete with one another and often occur at the same time. After we take a brief look at each type we will look at a competitive substitution-elimination reaction in living systems: detoxification.

13-4.2 SUBSTITUTION AND ELIMINATION: A CLOSER LOOK

Substitution

In a substitution reaction of an alkyl halide the nature of the organic product formed depends on the nucleophile used in the reaction. Table 13.2 lists some of the more common nucleophiles. For each nucleophile shown in Table 13.2 the organic product is as follows.

1. The nucleophile hydroxide (OH^-) converts an alkyl halide to the corresponding alcohol (see Table 10.5).

$$CH_3Cl \quad + \quad OH^- \quad \longrightarrow \quad CH_3OH \quad + \quad Cl^-$$

methyl chloride   hydroxide   methanol
an alcohol

2. The nucleophile methoxide (CH_3O^-) converts an alkyl halide to the corresponding ether (see Table 10.5).

$$CH_3-CHBr \quad + \quad CH_3O^- \quad \longrightarrow \quad CH_3CHOCH_3 \quad + \quad Br^-$$
$$\qquad\;\; | \qquad\qquad\qquad\qquad\qquad\qquad\qquad\;\; |$$
$$\qquad CH_3 \qquad\qquad\qquad\qquad\qquad\qquad\quad CH_3$$

isopropyl   methoxide   methyl isopropyl ether
bromide   *an ether*

3. The nucleophile bisulfide (SH^-) converts an alkyl halide to the corresponding thiol (see Table 10.5).

$$CH_3CH_2I \quad + \quad SH^- \quad \longrightarrow \quad CH_3CH_2SH \quad + \quad I^-$$

ethyl iodide   bisulfide   ethyl mercaptan
a thiol

4. The nucleophile cyanide (CN^-) converts an alkyl halide to the corresponding nitrile (see Table 10.5).

$$CH_3CH_2CH_2Cl \quad + \quad CN^- \quad \longrightarrow \quad CH_3CH_2CH_2CN \quad + \quad Cl^-$$

propyl chloride   cyanide   propyl cyanide
a nitrile

5. The nucleophile iodide (I^-) converts an alkyl halide to the corresponding alkyl iodide.

$$\qquad\qquad CH_3 \qquad\qquad\qquad\qquad\qquad CH_3$$
$$\qquad\qquad\;\; | \qquad\qquad\qquad\qquad\qquad\quad |$$
$$CH_3-C-Br \quad + \quad I^- \quad \longrightarrow \quad CH_3-C-I \quad + \quad Br^-$$
$$\qquad\qquad\;\; | \qquad\qquad\qquad\qquad\qquad\quad |$$
$$\qquad\qquad CH_3 \qquad\qquad\qquad\qquad\qquad CH_3$$

tert-butyl bromide   iodide   *tert*-butyl iodide
an alkyl iodide

Substitution of the halogen in an alkyl halide occurs in one of two ways. For some alkyl halides the nucleophile attacks the carbon atom from the rear, or the back side, relative to the halogen atom as in the case shown.

substitution by back-side attack

$$\text{HO}^- + \overset{\overset{\displaystyle CH_3}{|}}{CH_2} - Br \longrightarrow \text{HO} - \overset{\overset{\displaystyle CH_3}{|}}{CH_2} + Br^-$$

For other alkyl halides the alkyl halide first ionizes to form a carbonium ion (step 1, below), and the carbonium ion then undergoes attack by the nucleophile (step 2, below).

substitution by

first (1) ionization

$$CH_3 - \overset{\overset{\displaystyle CH_3}{|}}{\underset{\underset{\displaystyle Br}{|}}{C}} - CH_3 \longrightarrow CH_3 - \overset{\overset{\displaystyle CH_3}{|}}{\underset{+}{C}} - CH_3 + Br^-.$$

then, (2) nucleophilic attack

$$CH_3 - \overset{\overset{\displaystyle CH_3}{|}}{\underset{+}{C}} - CH_3 + OH^- \longrightarrow CH_3 - \overset{\overset{\displaystyle CH_3}{|}}{\underset{\underset{\displaystyle OH}{|}}{C}} - CH_3$$

Tertiary halides react with nucleophiles after ionizing to form a carbonium ion. For methyl and primary halides ionization does not occur and the nucleophile attacks from the rear.

Organic halides are rare in living systems but nucleophilic substitutions are common. Most nucleophilic substitutions in the living system involve the reactions of ethers (see Chapter 14) and esters (see Chapter 16). Furthermore, the nucleophilic reagents in the living system (*in vivo*) are more complex derivatives of the *in vitro* species shown in Table 13.2. Since nucleophilic substitutions in living systems are more common among esters and ethers, we will not discuss them any further here.

Elimination

Elimination reactions of alkyl halides leads to alkenes. In alkyl halides, elimination is the overall loss of a hydrogen and a halogen from adjacent carbons. For example,

$$\overset{\overset{\displaystyle }{|}}{\underset{\underset{\displaystyle H}{|}}{CH_2}} - \overset{\overset{\displaystyle }{|}}{\underset{\underset{\displaystyle Br}{|}}{CH_2}} + OH^- \longrightarrow CH_2 {=} CH_2 + HOH + Br^-$$

ethyl bromide   ethylene

$$\underset{\substack{\text{2-iodopropane}}}{CH_3-\overset{\displaystyle H}{\underset{\displaystyle \underset{\displaystyle H}{I}}{C}}-CH_2} + OH^- \longrightarrow \underset{\substack{\text{propene}}}{CH_3-\overset{\displaystyle H}{C}=CH_2} + HOH + I^-$$

Elimination in alkyl halides or **dehydrohalogenation** as it is sometimes called, can produce more than one alkene. This situation usually occurs when the carbon with the halogen atom is attached to more than one carbon atom, each of which has a hydrogen atom. For example, 2-bromobutane will give rise to two alkenes because each of the carbon atoms adjacent to the one with the halogen atom has a hydrogen atom on it.

Dehydrohalogenation is an elimination reaction in which hydrogen and halogen atoms are lost from adjacent carbon atoms in an alkyl halide.

secondary hydrogen   H   *primary hydrogen*

$$\underset{\substack{\text{2-bromobutane}}}{CH_3-CH_2-\overset{\displaystyle H}{\underset{\displaystyle Br}{C}}-CH_3} + OH^- \longrightarrow \underset{\substack{\text{2-butene}}}{CH_3-CH=CH-CH_3} + \underset{\substack{\text{1-butene}}}{CH_3-CH_2-CH=CH_2} + Br^- + H_2O$$

If a primary hydrogen is lost along with the halogen, the product formed is 1-butene. If the secondary hydrogen is lost along with the halogen, the product formed is 2-butene.

There are several ways in which these elimination reactions can occur. In one case the nucleophile can form a bond to the hydrogen on the adjacent carbon. Simultaneously, the C—H electrons form a C—C pi bond and halogen (bromine) leaves with its pair of electrons.

$$\underset{\substack{\text{hydroxide}}}{HO^-} + \underset{\substack{\text{ethyl bromide}}}{CH_2-CH_2-Br} \longrightarrow HOH + \underset{\substack{\text{ethylene}}}{CH_2=CH_2} + Br^-$$

Generally, it is the primary halides that undergo this elimination pathway.

In other cases the alkyl halide can ionize and form a carbonium ion just as it did in the reaction that led to substitution. For substitution to occur, the carbonium ion must then react with the nucleophile. For elimination to occur, the carbonium ion must lose a proton to the nucleophile to become an alkene. Tertiary halides undergo elimination via the carbonium ion more easily than do other halides.

Substitution: nucleophile attacks carbon

$$CH_3-\overset{\displaystyle CH_3}{\underset{\displaystyle CH_3}{C}}-I \longrightarrow \underset{\substack{\textit{tert}\text{-butyl}\\ \text{carbonium ion}\\ +I^-}}{CH_3-\overset{\displaystyle CH_3}{\underset{\displaystyle CH_3}{C^+}}} \xleftarrow{OH^-} \underset{\substack{\textit{tert}\text{-butyl alcohol}}}{CH_3-\overset{\displaystyle CH_3}{\underset{\displaystyle CH_3}{C}}-OH}$$

Elimination: nucleophile attacks hydrogen

$$CH_3-\underset{\underset{CH_3}{|}}{\overset{\overset{I}{|}}{C}}-CH_3 \longrightarrow CH_3-\underset{\underset{CH_3}{|}}{\overset{+}{C}}-CH_2 \xrightarrow{-OH} CH_3\underset{\underset{CH_3}{|}}{C}=CH_2 + HOH$$

tert-butyl   2-methyl propene
carbonium ion
$+ I^-$

A number of elimination reactions occur in living systems. Elimination reactions in living systems lead to the introduction of the carbon–carbon double bond. One example is an elimination reaction known as **dehydration**. A dehydration reaction occurs in the synthesis of fatty acids in the living system as shown and is discussed further in Section 27-3.1.

Dehydration is an elimination reaction in which a molecule of water is lost.

$$CH_3-\underset{\overset{|}{OH}}{\overset{\overset{OH}{|}}{CH}}-\overset{\overset{H}{|}}{CH}-COOH \xrightarrow{enzyme} CH_3-CH=CH-COOH + HOH$$

unsaturated fatty acid

In actuality, both substitution and elimination are competitive reactions. When an alkyl halide reacts with a nucleophile, the alkyl halide could, and often does, undergo both reactions simultaneously. Consequently, a mixture of products would result. Generally chemists can control the type of reaction taking place to a certain extent. However, these factors are beyond the objectives of this text. Competitive substitution and elimination reactions also occur in living systems. **Detoxification** is an example.

Detoxification is generally understood to describe those biotransformations that occur when foreign organic compounds that are not normally found in the organism are metabolized, regardless of whether toxicity is decreased, increased, or remains the same. Detoxification occurs when animals ingest insecticides and then attempt to metabolize them. For example, insects and mammals degrade Lindane (γ-hexa-chlorocyclohexane) by enzymic loss of the elements of hydrogen and chlorine. This, of course, is an example of a naturally occurring dehydrochlorination or elimination reaction. Pentachlorocyclohexene has been detected as a minor metabolite of Lindane in flies.

Detoxification is the breakdown of a foreign substance by an animal.

γ-hexachlorocyclohexane   pentachlorocyclohexene

Another example of dehydrochlorination is the detoxification of DDT (2,2-bis-*p*-chlorophenyl-1,1,1-trichloroethane) to DDE (1,1-bis-*p*-chlorophenyl-2,2-dichloro-ethene) by houseflies. However, a substitution reaction appears to be the course of detoxification in other insects. Resistance to DDT in *Drosophila melanogaster* is associated with the formation of Kelthane (1,1-bis-*p*-chlorophenyl-2,2-dichloro-ethanol). Kelthane is six times less toxic to houseflies than DDT. This detoxification product is, however, an effective toxicant for mites.

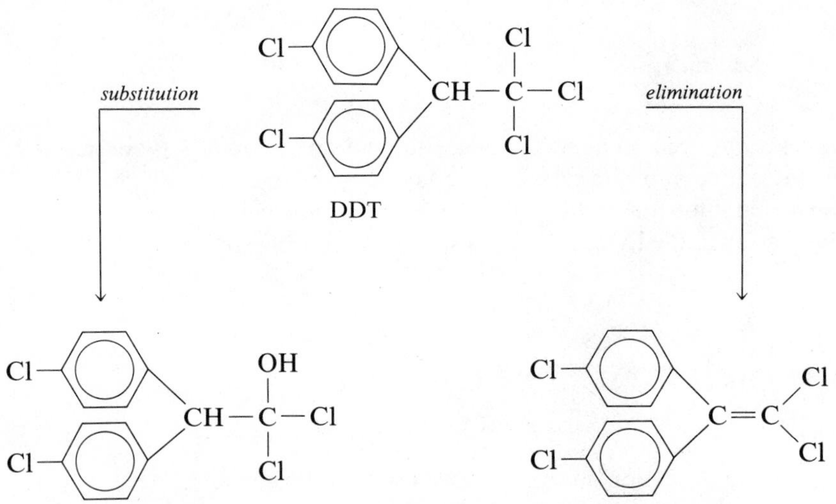

Kelthane
substitution product
in certain insects

DDE
elimination product
in houseflies

PROBLEM

4. Write the structure of the product of each reaction. Give the general name of the product of each reaction.

(a) $CH_3CH_2Br + NaCN \xrightarrow{\text{substitution}}$

(b) $CH_3\overset{\overset{CH_3}{|}}{C}HCH_2I + NaSH \xrightarrow{\text{substitution}}$

(c) $CH_3CH_2CH_2CH_2Cl + KOH \xrightarrow{\text{substitution}}$

(d) $CH_3CH_2CH_2CH_2Cl + KOH \xrightarrow{\text{elimination}}$

(e) $CH_3\underset{\underset{CH_3}{|}}{\overset{\overset{CH_3}{|}}{C}}{-}Br + NaOH \xrightarrow{\text{elimination}}$

ANSWERS TO PROBLEM **4. (a)** CH_3CH_2CN, nitrile **(b)** $(CH_3)_2CHCH_2SH$, thiol **(c)** $CH_3(CH)_2CH_2OH$, alcohol **(d)** $CH_3CH_2CH{=}CH_2$, alkene **(e)** $CH_3\overset{\overset{CH_3}{|}}{C}{=}CH_2$, alkene

13-5 ARYL HALIDES: A CHEMICAL HANDLE FOR AMINO ACID IDENTIFICATION

LEARNING OBJECTIVE

Given the structure of a nucleophile or an amino acid in a reaction with 2,4-dinitro-fluorobenzene, write the structure of the product.

Unsubstituted aryl halides do not react as easily with the nucleophilic reagents that are shown in Table 13.2 as do alkyl halides. For example, butyl chloride will be converted to butyl alcohol by refluxing the former with an aqueous solution of sodium hydroxide at temperatures less than 100°. Chlorobenzene can be converted to **phenol** by aqueous sodium hydroxide only under more strenuous conditions of temperatures and pressures in excess of 100° and 1 atmosphere pressure.

Phenol is a derivative of benzene in which the substituent is a hydroxyl group:

RCH_2—Cl *substitution here is easier* **but** *substitution here is difficult*

an alkyl halide   an aryl halide

However, if strongly electron-withdrawing groups (such as nitro groups) are positioned ortho and/or para to the halogen atom, nucleophilic substitution of the halogen does occur.

substitution here is difficult **but** *substitution here is easier*

For example,

2,4-dinitrochlorobenzene   2,4-dinitrophenol

2,4-dinitrofluorobenzene   2,4-dinitroaniline

2,4-Dinitrofluorobenzene (often abbreviated DNFB) will also react with the amino group of an **amino acid** as shown. This reaction has been utilized in determining the amino acid(s), in a protein, with a free amino group.

An **amino acid** is an organic compound that contains both an amino group and a carboxyl group (see Table 10.5).

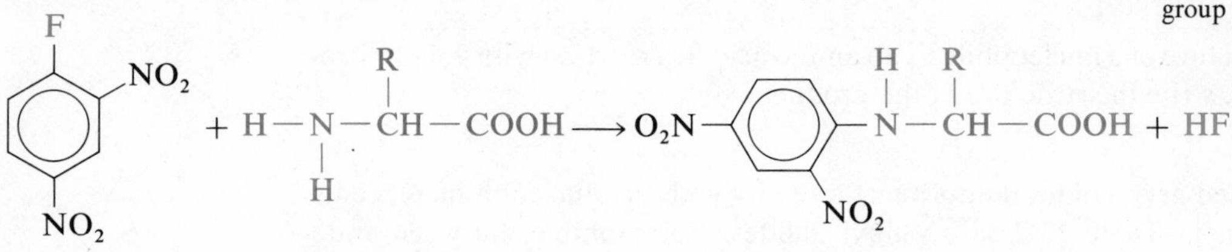

DNFB   amino acid

PROBLEM

5. Write the structure of the nucleophilic substitution product of each aryl halide.

(a) 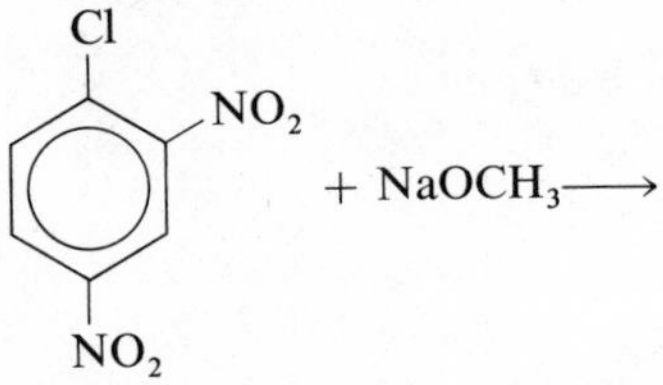 + NaOCH$_3$ ⟶

(b) 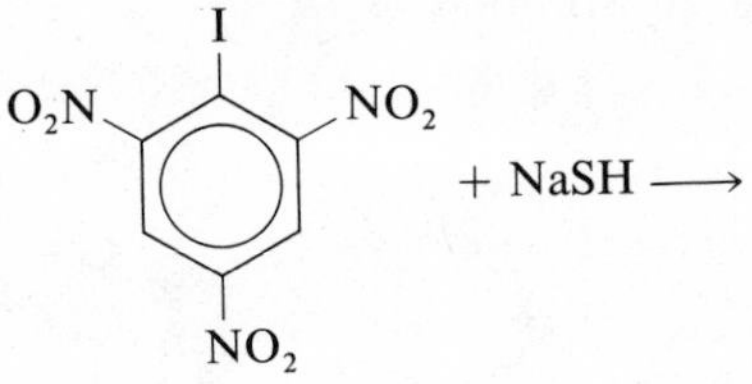 + NaSH ⟶

(c) 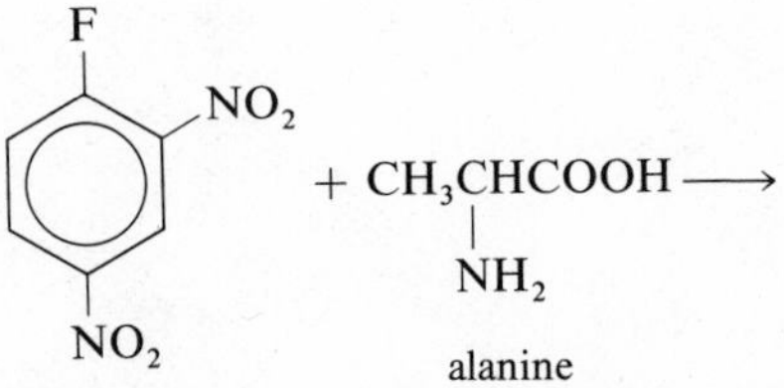 + CH$_3$CHCOOH ⟶
                        |
                       NH$_2$

alanine
*an amino acid
commonly found in proteins*

ANSWERS TO PROBLEM

5. (a) 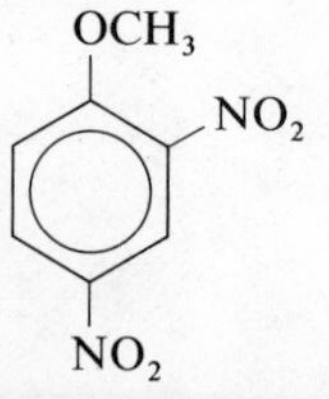   (b) 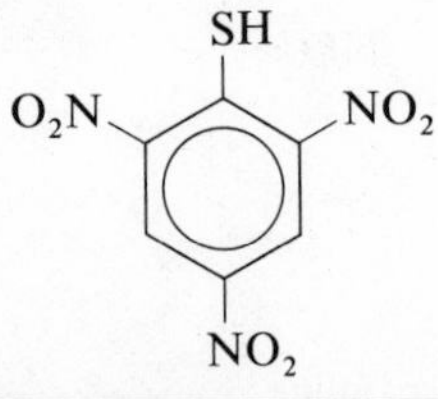   (c)

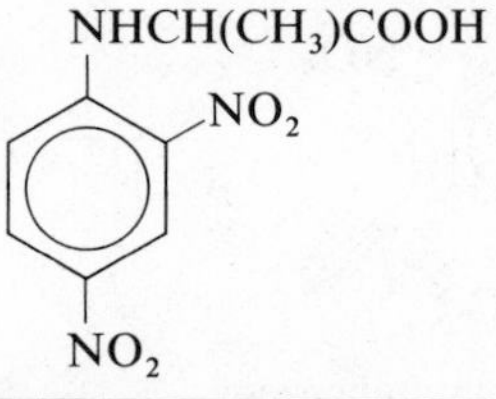

SUMMARY

Organic halides have been shown to have a variety of uses, particularly in medicine. In this chapter we learned the distinction between the two common organic halides, alkyl halides and aryl halides. Furthermore, we learned to classify alkyl halides as primary, secondary, and tertiary. We also learned how to name alkyl and aryl halides and how to draw a structure given the name of the compound.

$$R-X$$

an alkyl halide

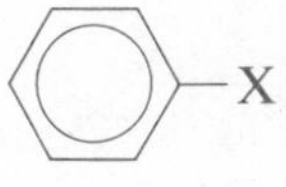

an aryl halide

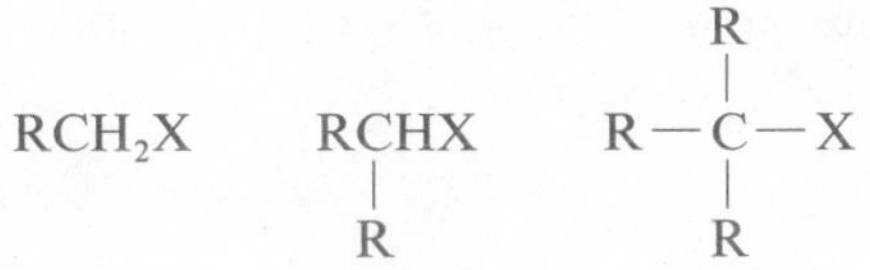

$$RCH_2X \qquad \underset{\underset{R}{|}}{RCHX} \qquad \underset{\underset{R}{|}}{\overset{\overset{R}{|}}{R-C-X}}$$

primary   *secondary*   *tertiary*

Alkyl halides can be formed from alkenes, from benzene, and from alkanes, as shown. The last equation is an example of free radical substitution.

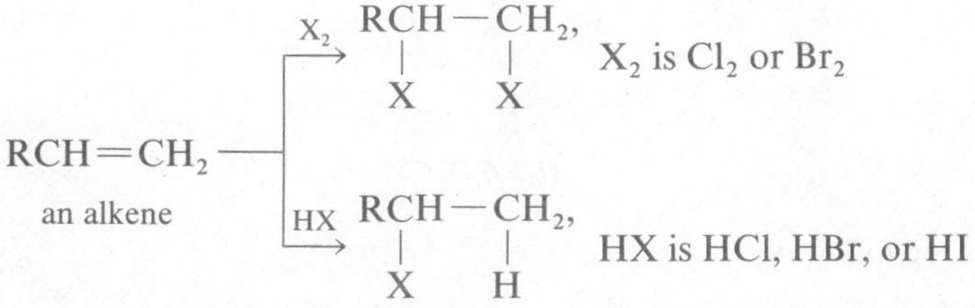

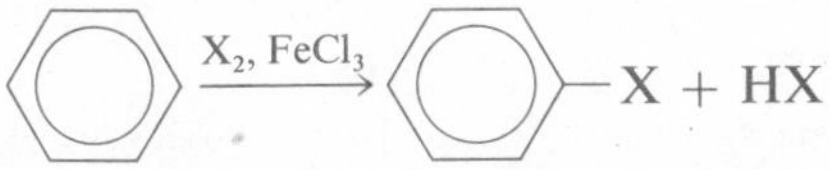

$$RCH=CH_2 \quad \begin{cases} \xrightarrow{X_2} \underset{\underset{X \quad X}{|\quad\;|}}{RCH-CH_2,} & X_2 \text{ is } Cl_2 \text{ or } Br_2 \\\\ \xrightarrow{HX} \underset{\underset{X \quad H}{|\quad\;|}}{RCH-CH_2,} & HX \text{ is } HCl, HBr, \text{ or } HI \end{cases}$$

an alkene

$\xrightarrow{X_2, FeCl_3}$ $-X + HX$

benzene

$$RCH_3 \xrightarrow{X_2, \text{ UV}} RCH_2X + HX$$

an alkane

The characteristic reactions of alkyl halides are substitution and elimination. The type of substitution product depends on the nature of the nucleophile. The elimination product of an alkyl halide, regardless of the nucleophile, is an alkene. A nucleophile is an electron-rich species such as OH^-, CH_3O^-, and SH^-. Substitution and elimination reactions can compete with one another and some insects detoxify DDT in reactions that are both substitution and elimination.

substitution

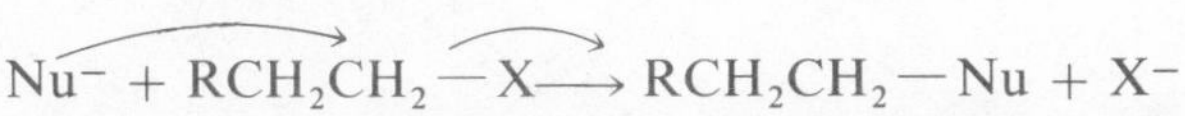

$$Nu^- + RCH_2CH_2-X \longrightarrow RCH_2CH_2-Nu + X^-$$

elimination

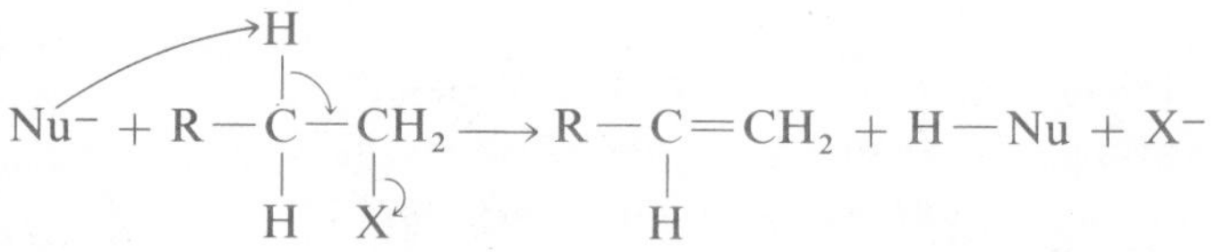

$$\text{Nu}^- + \text{R}-\underset{\underset{\text{H}}{|}}{\overset{\overset{\text{H}}{|}}{\text{C}}}-\underset{\underset{\text{X}}{|}}{\text{CH}_2} \longrightarrow \text{R}-\underset{\underset{\text{H}}{|}}{\text{C}}=\text{CH}_2 + \text{H}-\text{Nu} + \text{X}^-$$

Aryl halides are very reactive in substitution reactions when they contain nitro groups in the benzene ring. Such aryl halides, particularly 2,4-dinitrofluoroben-zene, have been used to identify amino acids in proteins that have free amino groups.

ADDITIONAL PROBLEMS

6. Write structures for each of the following.
   (a) isobutyl bromide
   (b) *o*-chlorotoluene
   (c) 4-iodo-1-hexene
   (d) *p*-chloronitrobenzene
   (e) benzyl bromide
   (f) 1,1-dichloro-2-pentyne
   (g) cyclohexyl iodide
   (h) 1,2-dibromocyclohexene
   (i) 2-chloro-1-phenylpropane
   (j) 1,4-dichlorobutane

7. Give the correct name for each of the following.

   (a) CH_3Cl

   (b)

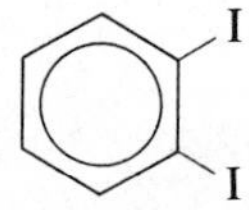

   (c)

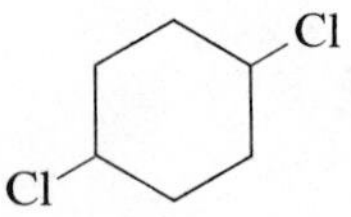

   (d)

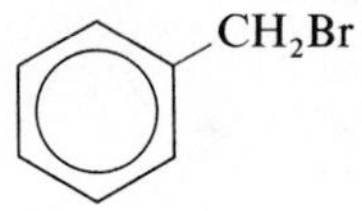

   (e) $CH_3CHCH=CHCH_2Br$
       |
       Br

(f) $HC{\equiv}CCH_2Cl$

(g)

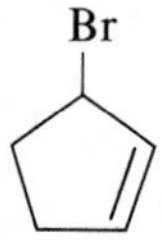

(h) CH_3CHCH_2I
        |
        CH_2
        |
        CH_3

(i)

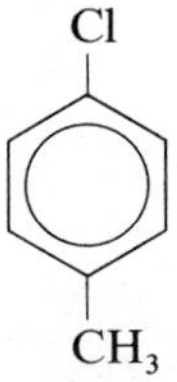

(j) $(CH_3)_3CBr$

8. Write the structure of the product of each reaction.
   (a) $CH_3CH_2CH{=}CH_2 + Br_2 \longrightarrow$
   (b) $CH_3CH_3 + Br_2 \xrightarrow[\text{light}]{\text{ultraviolet}}$
   (c) 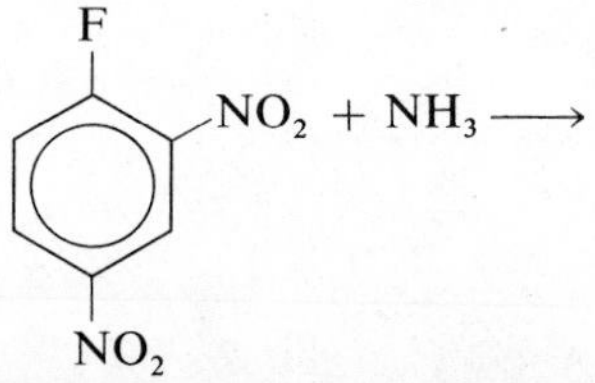 $+ NH_3 \longrightarrow$

(d)

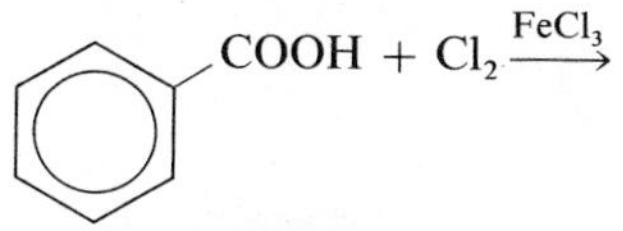

$$\text{C}_6\text{H}_5\text{COOH} + \text{Cl}_2 \xrightarrow{\text{FeCl}_3}$$

(e) $\text{CH}_3\text{CH}_2\text{CH}_2\text{Br} + \text{NaCN} \longrightarrow$

(f) 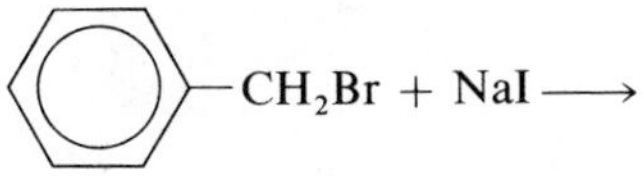 $\text{—CH}_2\text{Br} + \text{NaI} \longrightarrow$

(g) $\text{CH}_3\text{C}=\text{CH}_2 + \text{HBr} \longrightarrow$
    $\quad\quad\; |$
    $\quad\quad\; \text{CH}_3$

(h) $\text{CH}_3\text{CH}_2\text{Br} + \text{NaOH} \xrightarrow{\text{substitution}}$

(i)

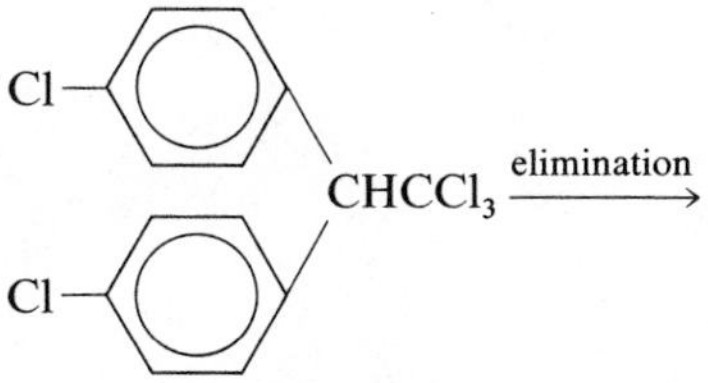

9. Draw and name all the isomers of $\text{C}_4\text{H}_9\text{Br}$.

10. Complete the following conversions by supplying the reagent and conditions necessary to cause each conversion.

    (a) $\text{CH}_2=\text{CH}_2 \longrightarrow \text{CH}_3\text{CH}_2\text{Br} \longrightarrow \text{CH}_3\text{CH}_2\text{SH}$

    (b)

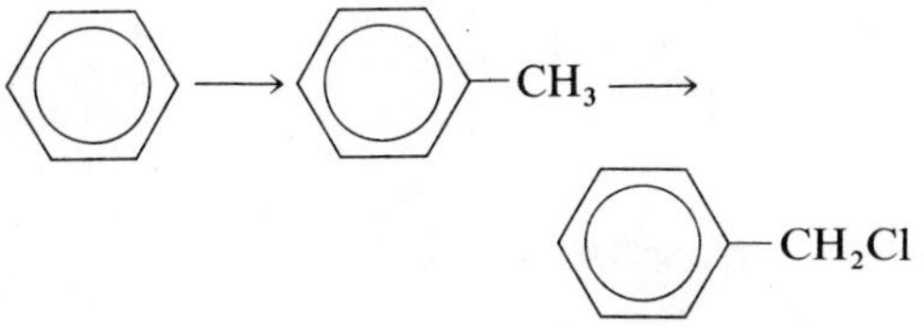

    (c) $\text{CH}_3\text{CH}_2\text{CH}_2\text{Br} \longrightarrow \text{CH}_3\text{CH}=\text{CH}_2 \longrightarrow$
    $$\quad\quad\quad\quad\quad\quad\quad\quad\quad\quad\quad\quad \text{CH}_3\text{CHCH}_3$$
    $$\quad\quad\quad\quad\quad\quad\quad\quad\quad\quad\quad\quad\quad\quad\; |$$
    $$\quad\quad\quad\quad\quad\quad\quad\quad\quad\quad\quad\quad\quad\; \text{Br}$$

    (d) $\text{HC}\equiv\text{CH} \longrightarrow \text{H}_2\text{C}=\text{CHBr} \longrightarrow$
    $$\quad\quad\quad\quad\quad\quad\quad\quad\quad\quad\quad \text{CH}_2\text{ClCHClBr}$$

11. Briefly discuss how detoxification of DDT exemplifies the competition between substitution and elimination.

12. Name four health-related uses of organic halides.

13. Choose the compound from A or B that correctly fits the description.

	A	B
(a) more reactive halide in nucleophilic substitution	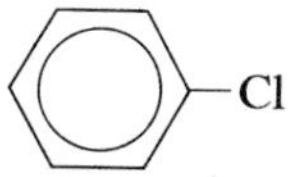 —Cl	$\text{CH}_3\text{CH}_2\text{CH}_2\text{Cl}$
(b) elimination product in reaction of propyl bromide with base (OH^-)	$\text{CH}_3\text{CH}=\text{CH}_2$	$\text{CH}_3\text{CH}_2\text{CH}_2\text{OH}$
(c) primary halide	$\text{CH}_3\text{CH}_2\text{CH}_2\text{Br}$	CH_3CHCH_3 with Br
(d) free radical	$\text{CH}_3\text{CH}_2\cdot$	CH_3CH_2^+
(e) Freon	CHClF_2	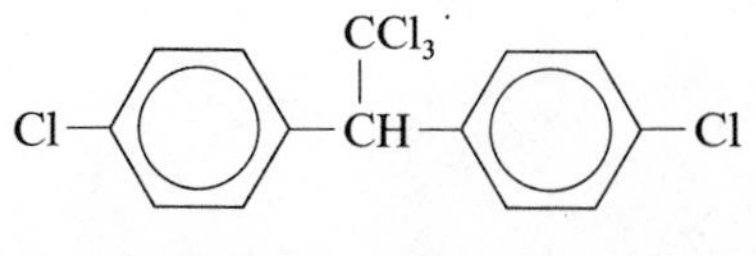

A   B

(f) intermediate in the substitution reaction

$$CH_3-\underset{\underset{CH_3}{|}}{\overset{\overset{CH_3}{|}}{C}}-I \xrightarrow{OH^-} CH_3-\underset{\underset{CH_3}{|}}{\overset{\overset{CH_3}{|}}{C}}-OH$$

A:
$$CH_3-\underset{\underset{CH_3}{|}}{\overset{\overset{CH_3}{|}}{C^+}}$$

B:
$$CH_3-\underset{\underset{CH_3}{|}}{\overset{\overset{CH_3}{|}}{C}}\cdot$$

(g) product of reaction

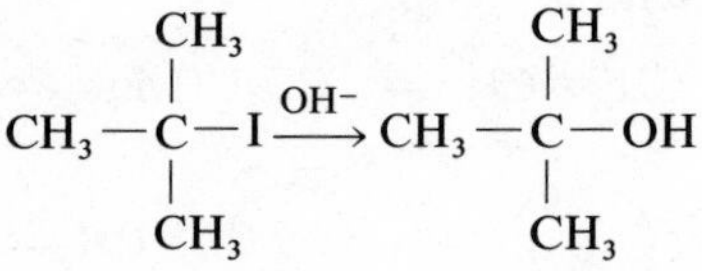

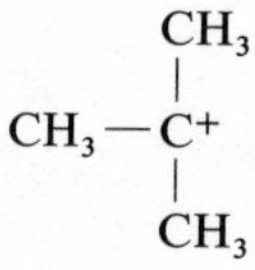

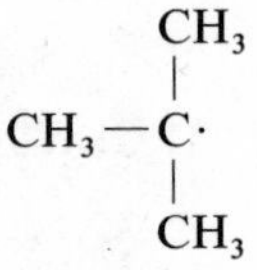

$$\text{(benzene ring)}-CH_3 + Cl_2 \xrightarrow[\text{light}]{\text{ultraviolet}}$$

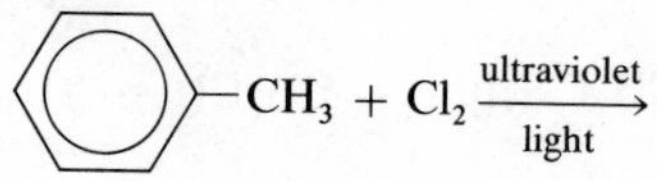

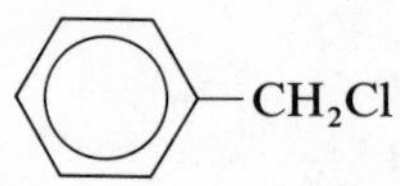

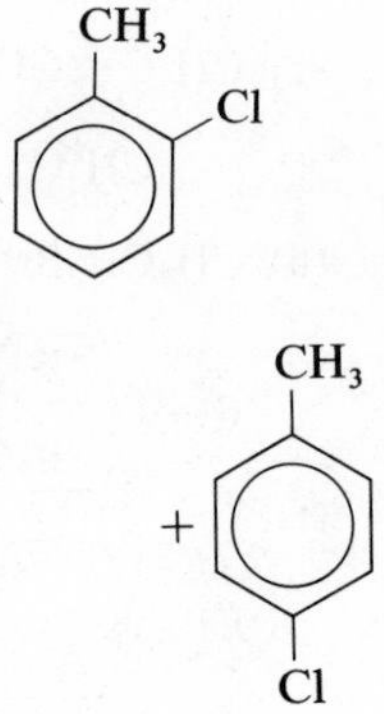

A: (benzene ring)$-CH_2Cl$

B: (toluene ring with Cl ortho) $+$ (toluene ring with Cl para)

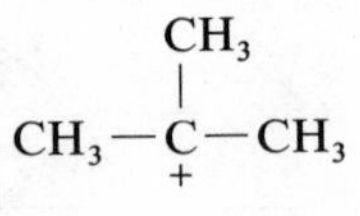

(h) nucleophile

A: $H-\overset{\cdot\cdot}{\underset{\cdot\cdot}{O}}:^-$

B:
$$CH_3-\underset{\underset{+}{\overset{\overset{CH_3}{|}}{C}}}{}-CH_3$$

(i) aryl halide

A:
$$CH_3\underset{\underset{CH_3}{|}}{CHBr}$$

B: (benzene ring)$-Cl$

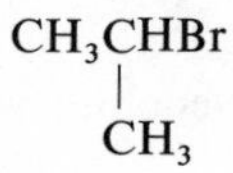

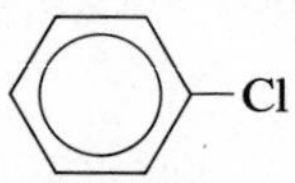

(j) substitution reaction that forms an alcohol

A: $CH_3CH_2Br + OH^- \longrightarrow$

B: $CH_3CH_2Br + CH_3O^- \longrightarrow$

CHAPTER 14

ALCOHOLS, PHENOLS, AND ETHERS

Alcohols and ethers are common substances of our daily lives. Ethyl alcohol is the alcohol of alcoholic beverages, formed by the fermentation of sugar by yeast or bacteria. Isopropyl alcohol is known to us as rubbing alcohol. We usually associate the word *ether* with the anesthetic, diethyl ether. Certain phenols and their ethers are found in oil of cloves and oil of nutmeg. Alcohols and phenols, and to a lesser extent ethers, are abundant in nature. A wide variety of these compounds have been identified in animal and plant sources.

Alcohols, phenols, and ethers may be considered to be derived from water by replacing one or both hydrogen atoms. Alcohols are derived from water by replacing one of the hydrogen atoms by an alkyl group. Phenols are derived from water by replacing one of the hydrogen atoms by an aromatic ring. Ethers are derived from water by replacing both hydrogen atoms by alkyl groups.

$$H-O-H \qquad R-O-H \qquad \langle\bigcirc\rangle-O-H \qquad R-O-R$$

水   *water*   *an alcohol*   *a phenol*   *an ether*

The characteristic functional group in an alcohol or phenol is the hydroxyl group, $-O-H$. In phenols the hydroxyl group is attached to an aromatic ring, while in alcohols it is attached to an alkyl group. Ethers do not contain the hydroxyl group but they are recognized by the presence of carbon–oxygen–carbon bonds. Ethers are isomers of alcohols.

Alcohols are more reactive than ethers. In fact, ethers, particularly ethyl ether, are routinely used as solvents in chemical reactions because of their inertness.

Because phenols contain the hydroxyl group attached to an aromatic ring, their chemistry is often different from that of the alcohols. Alcohols, then, are far more important to the organic chemist than ethers. In this chapter we will explore the chemistry of introducing the hydroxyl group into molecules and see how the hydroxyl group can be transformed into other functional groups, both in the laboratory and in the living system. Furthermore, we will consider the effect of some of these transformations when they occur in the living system.

14-1 NOMENCLATURE OF ALCOHOLS, PHENOLS, AND ETHERS

LEARNING OBJECTIVES

1. Given organic compounds, identify alcohols, phenols, ethers, thiols, thioethers, and disulfides.

2. Given the common or IUPAC name of an alcohol or phenol, write its structural formula.

3. Given the common name of an ether, thiol, thioether, or disulfide, write its structural formula.

4. Given the structural formula of an alcohol or phenol, write its common or IUPAC name.

5. Given the structural formula of an ether, thiol, thioether, or disulfide, write its common name.

The alcohols containing simple alkyl groups (see Table 11.4) are named by specifying the name of the alkyl group followed by the word *alcohol*. This combination is the common name of the alcohol. In discussing alcohols we often use the classification **primary, secondary,** and **tertiary alcohol,** just as we did with alkyl halides (see Section 13.1).

A **primary alcohol** is an alcohol in which the hydroxyl group is attached to a primary carbon (see Section 11-4.1). A **secondary alcohol** is an alcohol in which the hydroxyl group is attached to a secondary carbon. A **tertiary alcohol** is an alcohol in which the hydroxyl group is attached to a tertiary carbon.

CH_3CH_2OH — *a primary carbon*

CH_3CHCH_3 — *a secondary carbon*
|
OH

$CH_3 - \overset{\overset{\displaystyle CH_3}{|}}{\underset{\underset{\displaystyle OH}{|}}{C}} - CH_3$ — *a tertiary carbon*

ethyl alcohol
a primary alcohol

isopropyl alcohol
a secondary alcohol

tert-butyl alcohol
a tertiary alcohol

Whereas the common names are generally reserved for alcohols of simple structure, the IUPAC nomenclature system is applicable to a wide variety of structures. In the IUPAC system the longest continuous chain of carbon atoms that contains the OH group is selected to be given the parent name. The parent name is derived from the name of the corresponding alkane by replacing the *-e* of the corresponding alkane name by *-ol*. The OH group is given the lowest number and the number indicating the location of the OH is placed in front of the parent name. If the compound also contains other groups, their names and positions are also indicated in the name of the alcohol. For example,

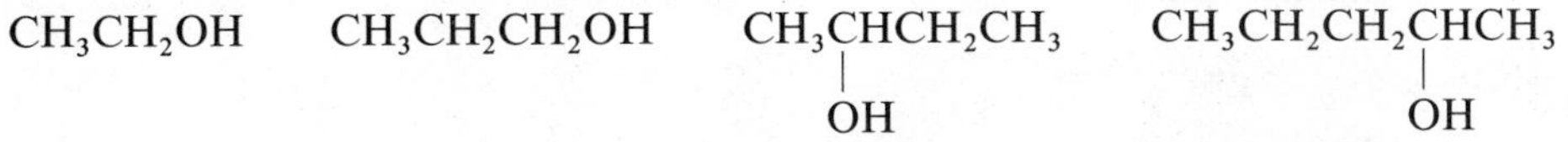

CH_3CH_2OH $\qquad$ $CH_3CH_2CH_2OH$ $\qquad$ $CH_3\underset{\underset{\displaystyle OH}{|}}{C}HCH_2CH_3$ $\qquad$ $CH_3CH_2CH_2\underset{\underset{\displaystyle OH}{|}}{C}HCH_3$

ethanol $\qquad$ 1-propanol $\qquad$ 2-butanol $\qquad$ 2-pentanol
$\qquad\qquad\qquad\qquad\qquad\qquad\qquad\qquad\qquad\qquad$ (*not* 4-pentanol)

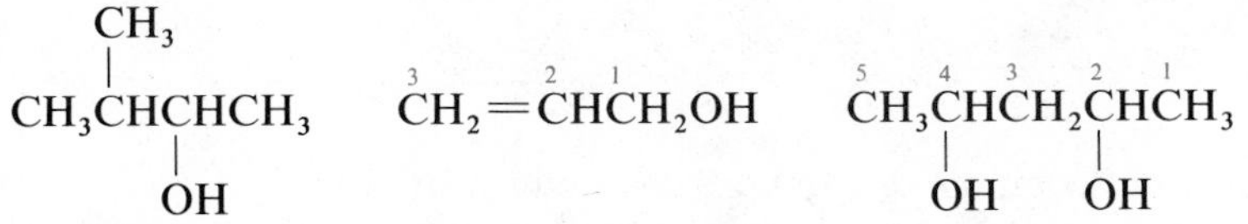

$CH_3\underset{\underset{\displaystyle OH}{|}}{\overset{\overset{\displaystyle CH_3}{|}}{C}}HCHCH_3$ $\qquad$ $\overset{3}{C}H_2{=}\overset{2}{C}H\overset{1}{C}H_2OH$ $\qquad$ $\overset{5}{C}H_3\overset{4}{\underset{\underset{\displaystyle OH}{|}}{C}}H\overset{3}{C}H_2\overset{2}{\underset{\underset{\displaystyle OH}{|}}{C}}H\overset{1}{C}H_3$

3-methyl-2-butanol $\qquad$ 2-propen-1-ol $\qquad$ 2,4-pentanediol
(*not* 2-methyl-3-butanol)

$CH_2{-}CH{-}CH_2$
$\;\;|\qquad|\qquad\;|$
$\;OH\;\;OH\;\;OH$

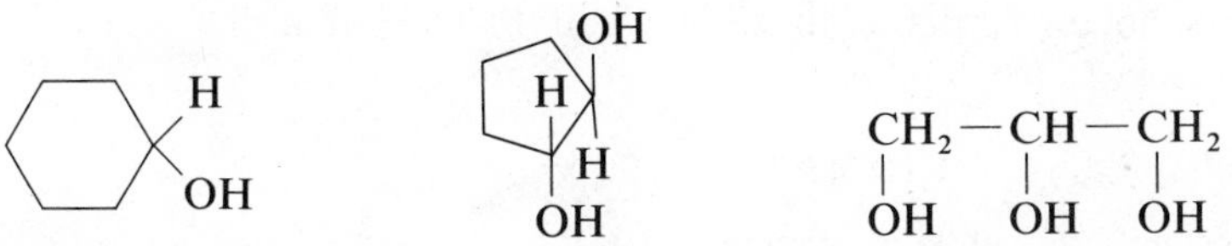

cyclohexanol $\qquad$ *trans*-1,2-cyclopentanediol $\qquad$ 1,2,3-propanetriol
$\qquad\qquad\qquad\qquad\qquad\qquad\qquad\qquad\qquad\qquad$ (commonly called glycerol)

Most phenols are named as derivatives of the parent compound, phenol.

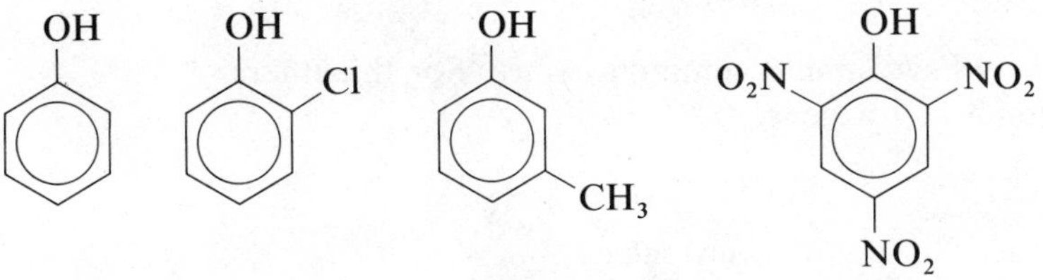

phenol $\qquad$ *o*-chlorophenol $\qquad$ *m*-methylphenol $\qquad$ 2,4,6-trinitrophenol

Ethers are usually named by common names. This is accomplished by designating the name of each alkyl group followed by the word *ether*. If the alkyl groups are identical, the alkyl group is named once and the prefix *di-* is used to indicate both alkyl groups.

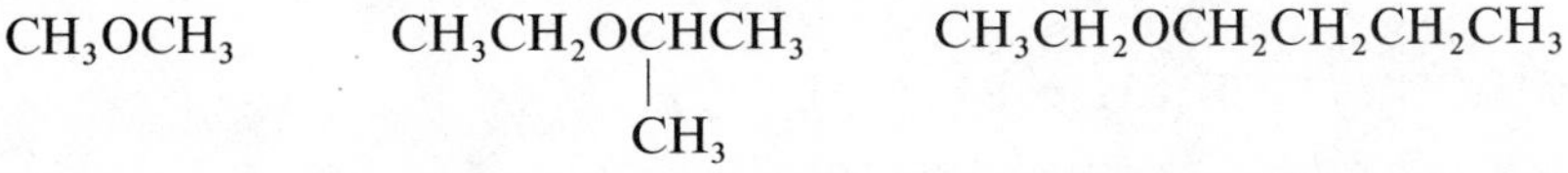

CH_3OCH_3 $\qquad$ $CH_3CH_2O\underset{\underset{\displaystyle CH_3}{|}}{C}HCH_3$ $\qquad$ $CH_3CH_2OCH_2CH_2CH_2CH_3$

dimethyl ether $\qquad$ ethyl isopropyl ether $\qquad$ ethyl butyl ether

There are many compounds containing carbon–sulfur single bonds that are analogs of the oxygen-containing compounds described above. Organic compounds containing sulfur are sometimes referred to as organosulfur compounds.

Table 14.1   Oxygen and Organosulfur Compounds

Oxygen Compound	Organosulfur Compound
R—OH alcohol	R—SH thiol (or mercaptan)
R—O—R ether	R—S—R thioether (or sulfide)
R—O—O—R peroxide	R—S—S—R disulfide

The similarity of these two classes of compounds can be accounted for by sulfur's position in the periodic table. Sulfur is just below oxygen in Group VIA. The organosulfur compounds in which we are interested are thiols (or mercaptans), thioethers (or sulfides), and disulfides; and these are shown in Table 14.1.

Thiols are given common names that are similar to those for alcohols. The name of the alkyl group is followed by the word *mercaptan*.

$$CH_3OH \qquad CH_3SH$$

methyl alcohol   methyl mercaptan
an alcohol   *a thiol*

$$CH_3CH_2CH_2OH \quad CH_3CH_2CH_2SH$$

propyl alcohol   propyl mercaptan
an alcohol   *a thiol*

Sulfides (thioethers) and ethers have similar common names. For thioethers, the name of the alkyl group is followed by the word *sulfide*.

$$CH_3OCH_3 \qquad CH_3SCH_3$$

dimethyl ether   dimethyl sulfide
an ether   *a sulfide*

$$CH_3OCH_2CH_3 \qquad CH_3SCH_2CH_3$$

methyl ethyl ether   methyl ethyl sulfide
an ether   *a sulfide*

Disulfides are named by giving the name of each alkyl group followed by the word *disulfide*.

$$CH_3-S-S-CH_2CH_3 \qquad CH_3CH_2-S-S-CH_2CH_3$$

methyl ethyl disulfide   diethyl disulfide

Many molecules containing the carbon–oxygen and carbon–sulfur single bond occur naturally and have a variety of uses. Table 14.2 lists the use or occurrence of some typical compounds. Phenols and phenol-related compounds are known for their antiseptic uses. Examples of phenols used as antiseptics are shown in Table 14.3.

Table 14.2 Some Naturally Occurring Alcohols, Phenols, Ethers, Thiols, and Thioethers.

Compounds	Structure	Compound Type	Occurrence or Use
Glycerol	$HOCH_2CHCH_2OH$ $\quad\quad\;\; OH$	Alcohol	Moisturizer
Menthol	$(CH_3)_2CH$ on cyclohexane ring with OH and CH_3	Alcohol	Cough drops, shaving lotion, mentholated cigarettes
Linalool	$(CH_3)_2C{=}CH(CH_2)_2CCH{=}CH_2$ with CH_3 above and OH below central C	Alcohol	Oil of lavender (perfume)
Eugenol	benzene ring with OH, OCH_3, and $CH_2CH{=}CH_2$	Phenol and ether	Oil of cloves
Propyl mercaptan	$CH_3CH_2CH_2SH$	Thiol	Freshly chopped onions
Isoamyl mercaptan	$CH_3CHCH_2CH_2SH$ $\quad\;\; CH_3$	Thiol	Component of skunk odor
Anethole	benzene ring with OCH_3 and $CH{=}CHCH_3$	Ether	Oil of aniseed
Tetrahydrocannabinol	fused ring structure with CH_3, OH, CH_3, O, C_5H_{11}	Ether and phenol	Constituent of marijuana
Methionine	$CH_3SCH_2CH_2CHCOOH$ $\quad\quad\quad\quad\quad\;\; NH_2$	Thioether	Component of proteins

Table 14.3 Phenols Used As Antiseptics and Disinfectants

Name	Structure	Germicidal Characteristic
Phenol		First widely used antiseptic. Causes severe burns on skin and is highly toxic. Serves today as a standard in establishing effectiveness of a substance as a germicide.
4-Chloro-3,5-dimethylphenol		Nonirritating topical antiseptic.
4-Hexylresorcinol		Active ingredient in several mouth washes. Has fewer side effects than phenol.
o-Phenylphenol		Ingredients in Lysol disinfectant. Used for disinfecting floors, walls, and furniture in hospitals, and surgical instruments.
2-Benzyl-4-chlorophenol (Chlorophene)		
Hexachlorophene		Ingredient of PhisoHex solution. Usefulness restricted to intact skin and not used on open wounds and skin burns because it is easily absorbed and is toxic. Currently banned for over-the-counter sale, but available by prescription and in hospitals.

PROBLEM

1. Identify each compound as an alcohol, phenol, ether, thiol, thioether (sulfide), or disulfide. Give a correct name for each.

(a) $CH_3CH_2CH_2CH_2OH$ (b)   (c)

(d) $CH_3CH_2OCH_2CH_3$ (e) CH_3CH_2SH (f) $CH_3CH_2SCH_2CH_3$

ANSWERS TO PROBLEM **1. (a)** alcohol; common name: butyl alcohol, IUPAC name: 1-butanol **(b)** phenol; *o*-nitrophenol **(c)** alcohol; 2-chlorocyclopentanol **(d)** ether; diethyl ether **(e)** thiol; ethyl mercaptan **(f)** thioether; diethyl sulfide

---

14-2 FORMATION OF ALCOHOLS AND ETHERS

LEARNING OBJECTIVES

1. Define fermentation.
2. Using equations, summarize the fermentation process for the production of ethyl alcohol.
3. Given the structure of an alkyl halide, write the structure of the alcohol that is formed by hydrolysis.
4. Given the structure of an alkyl halide and a sodium alkoxide, write the structure of the ether that is formed as product.
5. Explain the importance of biological hydroxylations using microorganisms.

In this section we will see how ethyl alcohol is obtained by the fermentation procedure and then look at some methods of preparing alcohols and ethers in the laboratory.

14-2.1 ETHYL ALCOHOL BY FERMENTATION

Ethyl alcohol is used as a solvent in toiletries, cosmetics, and pharmaceuticals and as a raw material that can be converted into other products, such as acetic acid in vinegar production. Ethyl alcohol is also the constituent of alcoholic beverages. For industrial purposes, ethyl alcohol is obtained by the ethanolic fermentation of materials containing sugar, such as molasses, or a substance convertible into sugar, such as the starch in grains. Ethyl alcohol is also obtained from petrochemical processes, such as the hydration of ethene in the vapor phase according to the following equation (see Section 11-5.2):

Ethanolic fermentation is the biological breakdown of sugar or starch materials to ethyl alcohol.

$$CH_2 {=\!=} CH_2 + H_2O \xrightarrow{\text{catalyst}} CH_3CH_2OH$$

In the production of ethyl alcohol for beverage purposes, cereal grains (corn, rye, barley, and malt) are the principal types of raw materials used.

Starch in grains is converted to sugar (maltose) by amylase, the active ingredient in malt. The maltose is subsequently hydrolyzed by another enzyme (maltase)
to glucose. Glucose is then fermented by microorganisms in yeast to ethyl alcohol
and CO_2. Ethanolic fermentation represents a series of complicated reactions that
can be summarized as follows:

*Conversion of starch
  to grain sugar:* $\text{starch} \xrightarrow{\text{amylase}} \text{maltose} \xrightarrow{\text{maltase}} \text{glucose } (C_6H_{12}O_6)$

Fermentation: $C_6H_{12}O_6 \xrightarrow{\text{yeast}} C_2H_5OH + CO_2$

The nature and flavor of the alcohol obtained depends on the source of starch
(for example, bourbon is obtained from corn and Scotch whiskey from barley),
the secondary products formed and retained during the subsequent distillation,
and the length of time the product is stored in oak barrels to attain the desired
ripeness or maturity.

As described here, the process by which enzymes in yeast convert glucose to
ethyl alcohol is called *fermentation.* A similar process during which glucose is converted to lactic acid occurs in muscle cells. This process is called *glycolysis* and is
described in Chapter 24.

The classical definition of fermentation is the *breakdown of organic compounds
by the action of microorganisms to products,* as described for the formation of ethyl
alcohol. A broader definition of the word is the *microbial action, controlled by
man, that makes products useful to man.* The formation of 11-hydroxy progesterone
from the action of *Rhizopus nigricans* on progesterone and its subsequent conversion to cortisone, a useful anti-inflammatory agent, is described in Section 14-2.3.

Alcohols containing a higher carbon content than ethyl alcohol are not obtained
by fermentation, but (on an industrial scale) by the reduction of lipids. Lipids are
animal fat or vegetable oils. For example, the reduction of coconut oil gives the
mixture of alcohols shown.

$$
\begin{array}{c}
\underset{\text{coconut oil}}{
\left.
\begin{array}{l}
CH_2-O-\overset{\displaystyle O}{\overset{\|}{C}}(CH_2)_{10}CH_3 \\[4pt]
CH-O-\overset{\displaystyle O}{\overset{\|}{C}}(CH_2)_{12}CH_3 \\[4pt]
CH_2-O-\overset{\displaystyle O}{\overset{\|}{C}}(CH_2)_7CH=CH(CH_2)_7CH_3
\end{array}
\right\}
}
\; + \; 7H_2 \longrightarrow
\underset{\text{glycerol}}{
\begin{array}{l}
CH_2OH \\
CHOH \\
CH_2OH
\end{array}
}
\; + \;
\begin{array}{l}
CH_3(CH_2)_{10}CH_2OH \\
\text{lauryl alcohol} \\
+ \\
CH_3(CH_2)_{12}CH_2OH \\
\text{myristyl alcohol} \\
+ \\
CH_3(CH_2)_{16}CH_2OH \\
\text{stearyl alcohol}
\end{array}
$$

Ethyl alcohol and other alcohols are important raw materials for the synthesis of
many other classes of compounds. We will see why this is so as we study the section
on the reactions of alcohols.

14-2.2 HYDROLYSIS OF ALKYL HALIDES: ALCOHOLS

Hydrolysis of alkyl halides to form alcohols is a substitution reaction, as shown in the general reaction below. Substitution reactions were discussed in Chapter 13. In this substitution reaction, the hydroxyl group is substituted for the halogen (represented by X).

A **hydrolysis reaction** is one in which water is one of the reactants in a bond-cleaving process.

$$RX + HOH \longrightarrow ROH + HX$$

an alkyl   *an alcohol*
halide

Some examples are the hydrolysis of *tert*-butyl bromide and of benzyl chloride.

$$\underset{\substack{| \\ Br}}{\overset{\substack{CH_3 \\ |}}{CH_3CCH_3}} + HOH \longrightarrow \underset{\substack{| \\ OH}}{\overset{\substack{CH_3 \\ |}}{CH_3CCH_3}} + HBr$$

tert-butyl bromide   *tert*-butyl alcohol

$$\langle\!\!\bigcirc\!\!\rangle\!-CH_2Cl + HOH \longrightarrow \langle\!\!\bigcirc\!\!\rangle\!-CH_2OH + HCl$$

benzyl chloride   benzyl alcohol

14-2.3 BIOLOGICAL HYDROXYLATIONS: ALCOHOLS

There is currently a large demand for cortisone (see structure) and cortisone-type drugs. These drugs have been found to be useful as anti-inflammatory agents in the treatment of arthritis. Efficient laboratory methods for large scale production of these materials have not been developed. The equation below outlines the basic steps for the conversion of the plant steroid diosgenin to cortisone. Note that a functional group is added at C-11. The use of chemical reagents to introduce the necessary functional group at C-11 is difficult in light of the fact that other points of substitution are possible. An efficient method involves introducing a hydroxyl group at C-11 by microbial hydroxylation. The most useful starting material is the plant steroid diosgenin, which is readily available in the Mexican yam. In the series of equations below, diosgenin is converted to progesterone. The hydroxylation of C-11 of progesterone is carried out enzymatically with the mold *Rhizopus nigricans* with more than 90% yield. Finally the 11-hydroxy progesterone is converted to cortisone. The importance to the pharmaceutical industry of the microbial hydroxylation step at C-11 is great. This hydroxylation is a crucial step in a sequence that allows the conversion of a readily available plant steroid to cortisone. Furthermore, this process illustrates that biochemical reactions that are enzyme-catalyzed can show remarkable selectivity (as here at C-11). We have seen already the same selectivity in the formation of L-dopa from tyrosine in Section 12-4.

A **steroid** is an organic molecule which contains four rings fused: three 6-membered rings and one 5-membered ring.

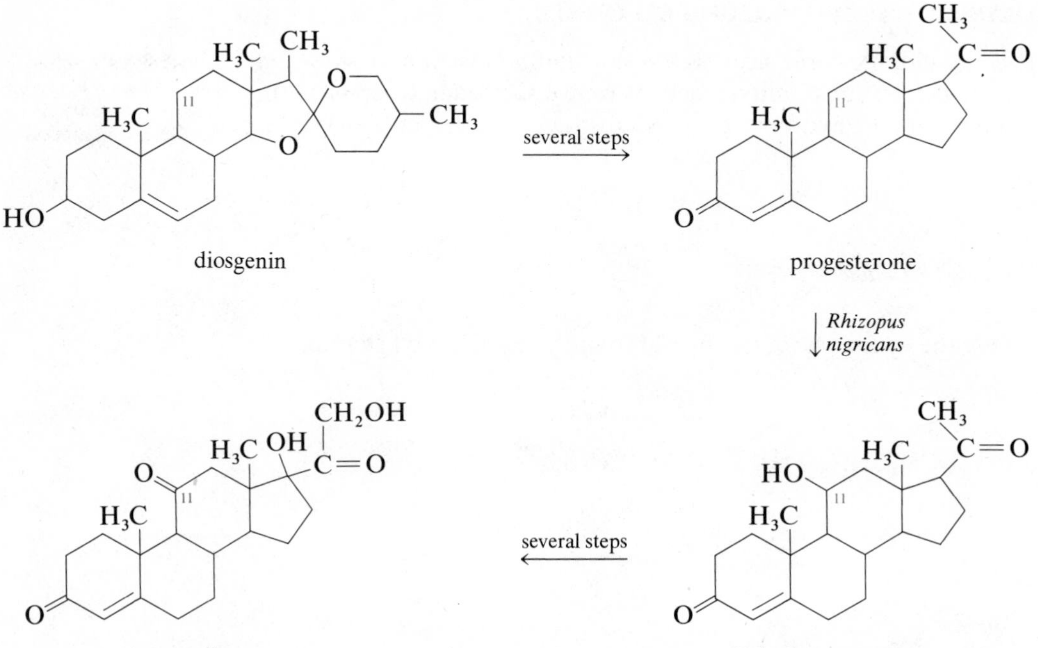

diosgenin

several steps →

progesterone

Rhizopus
nigricans

cortisone

← several steps

11-hydroxy progesterone

The preparation of phenols by fermentation or hydrolysis of the corresponding aryl halides does not occur. Furthermore, phenols are produced in the laboratory from other aromatic compounds only with difficulty. We will not consider these procedures here.

14-2.4 WILLIAMSON REACTION: ETHERS

Ethers can be formed in the lab by the **Williamson reaction**. This method involves the reaction of a sodium alkoxide with a methyl halide or a primary alkyl halide. A sodium alkoxide is a sodium salt of an alcohol. The Williamson reaction is a substitution reaction. It belongs to a class of reactions called *nucleophilic substitution* (see Section 13-4.2). The attacking nucleophile in this reaction is the alkoxide anion. The alkoxide anion replaces the halide anion. A carbon–oxygen single bond is formed.

> The **Williamson reaction** is a substitution reaction in which an alkoxide ion replaces a halogen in a methyl or primary alkyl halide to form an ether.

$$\text{Na}^+\text{OR} \quad + \quad \text{RX} \quad \longrightarrow \quad \text{ROR} \quad + \quad \text{Na}^+\text{X}^-$$

| a sodium alkoxide | a primary alkyl or methyl halide | ether | a sodium halide |

The following reactions illustrate the Williamson reaction. The student should notice that although several combinations of reactants (alkoxide and halide) are possible, only those reactions with a methyl halide or a primary alkyl halide will proceed to ether formation. If the alkyl halide is secondary or tertiary, elimination rather than substitution occurs (see Section 13-4.2).

$$CH_3O^-Na^+ + CH_3CH_2CH_2Br \longrightarrow CH_3CH_2CH_2OCH_3 + Na^+Br^-$$

sodium   propyl   propyl methyl
methoxide   bromide   ether

$$CH_3CH_2O^-Na^+ + CH_3CH_2Cl \longrightarrow CH_3CH_2OCH_2CH_3 + Na^+Cl^-$$

sodium ethoxide   ethyl chloride   diethyl ether

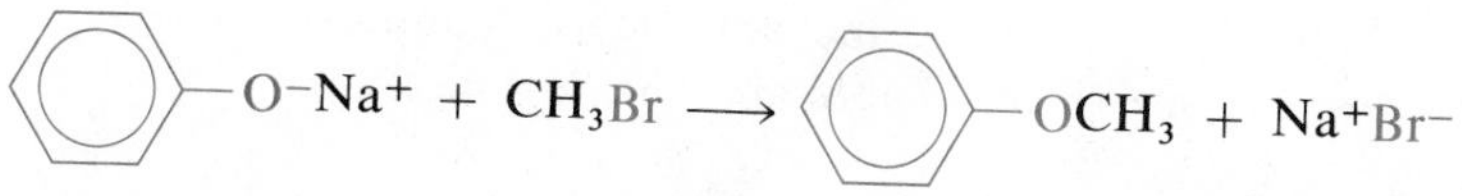

sodium phenoxide   methyl bromide   methyl phenyl ether

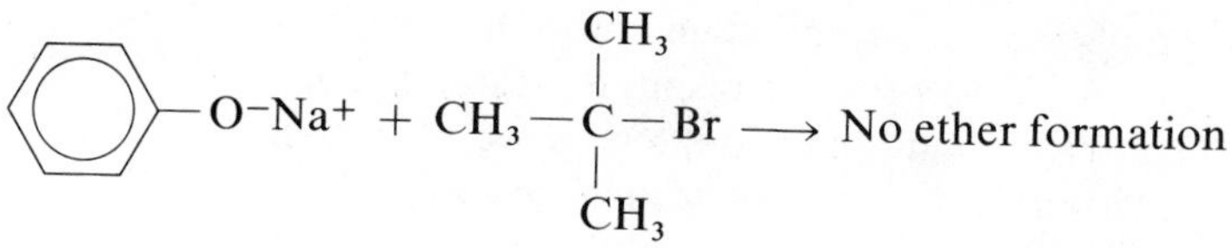

a tertiary halide

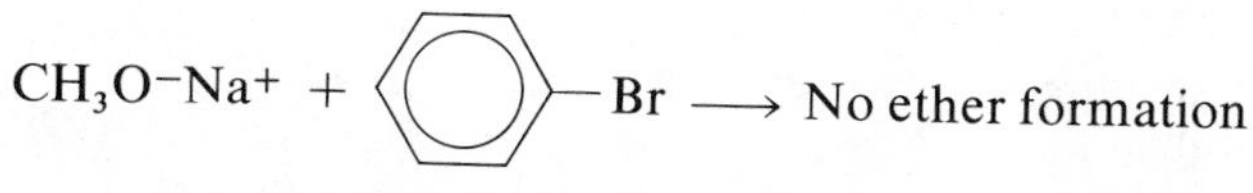

an aryl halide

Ethyl ether is the most commonly known ether because of its versatility. It can be prepared in the laboratory by the Williamson reaction as shown above. Ethyl ether is routinely used by organic chemists as a solvent for organic reactions. It is frequently used when working up a reaction mixture. In this procedure the reaction mixture is diluted with water and extracted with ethyl ether. Ethyl ether is immiscible with water and two separate liquid layers result. The organic reaction product is distributed largely into the ethyl ether layer. Inorganic by-products such as acids, bases, and salts remain in the water layer. The water layer is then separated and usually discarded. The organic reaction product is then recovered from the ethyl ether layer.

Until recently, ethyl ether was routinely used as an anesthetic. An *anesthetic* is a drug that depresses the central nervous system to the extent that sensitivity to pain is abolished and the individual suffers an interruption of consciousness. When used for surgical purposes, an anesthetic produces analgesia (absence of sense of pain), loss of consciousness, diminished reflex activity, and muscular relaxation. There is a minimal depression of the vital functions. Many of the common anesthetics are gaseous compounds that enter the blood stream by way of the lungs and quickly reach the central nervous system. Nitrous oxide (N_2O), ethyl ether, and halothane are three typical anesthetics.

Nitrous oxide (N_2O), the only inorganic gas that is practical for clinical anesthesia, is a very weak anesthetic agent. It is usually mixed with oxygen at either 60% N_2O to 40% O_2 or 50% N_2O and 50% O_2. The ratio is important since a high concentration of N_2O is needed to produce the desired effect as an additional anesthetic agent. Oxygen in a concentration of at least 40% is used to prevent hypoxia (inadequate oxygen for the needs of the body). The problem associated with too much N_2O is that not enough oxygen can be given. Induction of N_2O is rapid and relatively pleasant. Recovery is quick and without unpleasant after effects because N_2O lacks irritating properties. The use of N_2O has remained high in recent years in spite of the discovery of the newer anesthetic agents, particularly halothane. N_2O is used with halothane to decrease the amount of halothane required to produce the desired level of anesthesia. There are very few instances where N_2O is not used with halothane.

Ethyl ether is a complete anesthetic and has an excellent margin of safety. (A margin of safety is an amount allowed beyond what is actually assumed to be necessary.) An inhaled concentration of 6–8% in air maintains surgical anesthesia and a concentration of 12–15% is required to produce respiratory arrest. Ethyl ether causes excellent muscular relaxation without undue depression of the central nervous system. Induction with ether is unpleasant and slow, and recovery is prolonged. Ethyl ether is rarely used today because of its flammability, disagreeable odor, and tendency to produce extended nausea and vomiting. A small amount of ether undergoes biodegradation in the body.

A number of new anesthetic agents are currently in use. These include halothane (Fluothane) and enflurane (Ethrane). Chemically halothane is 1-bromo-1-chloro-2,2,2-trifluoroethane and enflurane is 2-chloro-1,1,2-trifluorethyl difluoromethyl ether. Both are nonflammable. Enflurane is less widely used than halothane and its use will not be discussed here.

halothane
a halogenated hydrocarbon

enflurane
a halogenated ether

Halothane is nonflammable and nonexplosive. It is generally used in concentration of 0.5–1.5% with nitrous oxide and oxygen. Muscle relaxants are used with halothane for abdominal surgery since halothane does not provide enough muscular relaxation. Induction and recovery are moderately slow because of the high blood solubility of halothane. The major disadvantage of halothane is its potent respiratory and cardiovascular depression potential. Unlike ethyl ether, halothane decreases respiration, pulse rate, and blood pressure in a dose-dependent manner. That is, the concentration required to maintain surgical anesthesia is only slightly different than that needed to cause respiratory arrest. Hence, halothane has a narrow margin of safety. Halothane is primarily excreted by the lungs; however, 12–20% is metabolized and excreted in the urine.

Today anesthesia begins with a pre-anesthetic evaluation by the anesthesiologist that includes history, physical examination, and review of pertinent laboratory tests and chest X-ray. At this time the anesthesiologist determines the best approach to anesthesia for the patient's safety and to meet the needs of the surgeon. The morning of surgery, the patient is given pre-anesthesia medication usually consisting of meperidine (Demerol analgesic) or morphine and a tranquilizer, such as diazepam (Valium muscle relaxant). This medication is intended to relieve the patient's anxiety and to make the patient drowsy. In the operating room, before anesthesia is induced, the patient is attached to various monitoring devices including a blood pressure cuff, electrocardiogram, and chest stethoscope. Anesthesia is started with an intravenous injection of sodium thiopental (Pentothal anesthetic) for the rapid induction of sleep. Then, a plastic tube is inserted into the trachea. Through this tube a mixture of halothane, nitrous oxide, and oxygen is fed to the patient. When an adequate depth of anesthesia is reached, the anesthesiologist allows the surgeon to begin the operation. During the operation the anesthesiologist watches for and treats any complications that may arise. At the end of the operation, the anesthetic gases are discontinued and the patient is allowed to breathe oxygen to wash the anesthetic gases out of the lungs. The patient awakens as the anesthetic concentration decreases to a low enough level.

PROBLEMS

2. Define fermentation.
3. Describe the reaction and write the structure of the alcohol or ether formed as the product of each reaction.

(a) $C_6H_{12}O_6 \xrightarrow{\text{fermentation}}$

(b) $CH_3\overset{\overset{\displaystyle CH_3}{|}}{\underset{\underset{\displaystyle I}{|}}{C}}CH_2CH_3 + H_2O \longrightarrow$

(c) $CH_3CH_2CH_2O^-Na^+ + CH_3I \longrightarrow$

(d) $CH_3CH_2O^-Na^+ + CH_3CH_2I \longrightarrow$

ANSWERS TO PROBLEMS   **2.** Fermentation is the biologic breakdown of sugar or starch materials to ethyl alcohol.
3. (a) Fermentation process, C_2H_5OH **(b)** hydrolysis of an alkyl halide,

$CH_3\overset{\overset{\displaystyle CH_3}{|}}{\underset{\underset{\displaystyle OH}{|}}{C}}CH_2CH_3$ **(c)** Williamson reaction, $CH_3CH_2CH_2OCH_3$

(d) Williamson reaction, $CH_3CH_2OCH_2CH_3$

14-3 REACTIONS OF ALCOHOLS AND PHENOLS

LEARNING OBJECTIVES

1. Given the reaction between an alcohol and a halogenating reagent, write the structure of the alkyl halide formed.
2. Given an alcohol and the conditions for dehydration, write the structure of the alkene formed.
3. Given the reaction between an alcohol or a phenol and a carboxylic acid or an inorganic acid, write the structure of the ester formed.
4. Write the structure of the phenoxide formed in the reaction of a phenol with a base.
5. Given a primary or secondary alcohol, write the product resulting from an oxidation reaction of it.
6. Explain the differences between the metabolisms of methyl alcohol and ethyl alcohol in the body.
7. Given the oxidation reaction of a thiol, write the structure of the disulfide formed.
8. Given the reduction reaction of a disulfide, write the structure of the thiol formed.
9. Describe the chemistry of permanent waving in terms of the thiol-disulfide interconversion.

The hydroxyl group in alcohols is much more reactive than the hydroxyl group in phenols. In alcohols, the reaction can take place at either the $C—O$ bond or the $O—H$ bond. In phenols, the reaction takes place only at the $O—H$ bond.

an alcohol　　　　　*a phenol*

Unlike alcohols and phenols, ethers are relatively unreactive compounds. With the exception of oxidation, their reactions will not be discussed in detail here. The two more common reactions of an alcohol in which reaction occurs at the $C—O$ bond are the formation of alkyl halides and the dehydration to alkenes. Alcohols and phenols react with both organic and inorganic acids to form esters. Additionally, the aromatic ring confers a degree of acidity on the hydrogen of phenol. In both ester formation and acidity, the oxygen–hydrogen bond is broken. We will discuss both carbon–oxygen and oxygen–hydrogen bond-breaking reactions in this section.

14-3.1 REACTION AT CARBON–OXYGEN BOND: FORMATION OF ALKYL HALIDES

The hydroxyl group in alcohols undergoes substitution by several halogenating reagents to form alkyl halides. These reagents include the hydrogen halides (HX), phosphorus tribromide (PBr_3), and thionyl chloride ($SOCl_2$). X is Cl, Br, or I.

$$R—OH \xrightarrow{\text{HX, PBr}_3, \text{ or SOCl}_2} R—X$$

an alcohol　　　　　　　*an alkyl halide*

For example,

$$CH_3CH_2CH_2OH + HI \longrightarrow CH_3CH_2CH_2I + HOH$$

propyl alcohol   propyl iodide

$$CH_3\overset{\displaystyle CH_3}{\underset{\displaystyle OH}{C}}CH_3 + SOCl_2 \longrightarrow CH_3\overset{\displaystyle CH_3}{\underset{\displaystyle Cl}{C}}CH_3 + SO_2 + HCl$$

tert-butyl alcohol   *tert*-butyl chloride

testosterone   bromide of testosterone
(male sex hormone)

Phenols do not react with these reagents to form organic halides.

14.3-2 REACTION AT CARBON–OXYGEN BOND: DEHYDRATION

Dehydration of an alcohol results in the formation of an alkene. In dehydration, an alcohol loses a molecule of water and a carbon–carbon double bond is formed. Dehydrations are usually catalyzed by acids such as sulfuric acid (H_2SO_4) or phosphoric acid (H_3PO_4) and the reaction is heated. The elements of water (H and OH) are lost from adjacent carbon atoms. As a result of this more than a single alkene can be formed in some cases. This is illustrated by the dehydration of 2-butanol. The loss of the hydroxyl group and the adjacent primary hydrogen gives 1-butene. The loss of the hydroxyl group and the adjacent secondary hydrogen gives 2-butene.

$$\overset{\displaystyle}{\underset{\displaystyle H}{C}H_2CH_2OH} + H_2SO_4 \xrightarrow{\text{heat}} CH_2{=}CH_2 + H_2O$$

ethyl alcohol   ethene

cyclopentanol   cyclopentene

a secondary   *a primary*
hydrogen   *hydrogen*

$$CH_3CH_2CHCH_3 + H_2SO_4 \xrightarrow{heat} CH_3CH_2CH{=}CH_2 + CH_3CH{=}CHCH_3 + H_2O$$
          |
          OH

    2-butanol   1-butene   2-butene

Dehydration also occurs in living systems. An example of dehydration was presented in Section 13-4.2 to illustrate an elimination reaction in the living system. Phenols and ethers do not undergo dehydration reaction.

$$ROR \;\; or \;\; \underset{}{\bigcirc}\!\!-\!OH + H_2SO_4 \xrightarrow{heat} \text{no dehydration}$$

14-3.3 REACTION AT OXYGEN–HYDROGEN BOND: FORMATION OF ESTERS

If an alcohol or a phenol reacts with an organic acid such as a carboxylic acid (an acid that contains a —COOH group), a compound called an organic ester is formed. On the other hand, inorganic esters are formed when alcohols react with inorganic acids such as sulfuric acid, phosphoric acid, or nitric acid. Some general examples of the formation of organic and inorganic esters are as follows.

$$\underset{\text{carboxylic acid}}{R'-\overset{\overset{\displaystyle O}{\|}}{C}-OH} + HOR \longrightarrow \underset{\textit{a carboxylate ester}}{R'-\overset{\overset{\displaystyle O}{\|}}{C}-OR} + HOH$$

$$\underset{\text{sulfuric acid}}{HO-\overset{\overset{\displaystyle O}{\|}}{\underset{\underset{\displaystyle O}{\|}}{S}}-OH} + HOR \longrightarrow \underset{\textit{a sulfate ester}}{HO-\overset{\overset{\displaystyle O}{\|}}{\underset{\underset{\displaystyle O}{\|}}{S}}-OR} + HOH$$

$$\underset{\text{phosphoric acid}}{HO-\overset{\overset{\displaystyle O}{\|}}{\underset{\underset{\displaystyle OH}{|}}{P}}-OH} + HOR \longrightarrow \underset{\textit{a phosphomonoester}}{HO-\overset{\overset{\displaystyle O}{\|}}{\underset{\underset{\displaystyle OH}{|}}{P}}-OR} + HOH$$

$$\underset{\text{nitric acid}}{O{=}\overset{\overset{\displaystyle O}{\|}}{N}-OH} + HOR \longrightarrow \underset{\textit{a nitrate ester}}{O{=}\overset{\overset{\displaystyle O}{\|}}{N}-OR} + HOH$$

Actual reactions that occur, resulting in organic and inorganic esters, are shown in the following equations.

$$\text{benzoic acid} + \text{HOCH}_2\text{CH}_3 \longrightarrow \text{ethyl benzoate} + \text{HOH}$$

benzoic acid   ethyl alcohol   ethyl benzoate
a carboxylate ester

$$\text{HO}-\overset{\displaystyle O}{\underset{\displaystyle OH}{\overset{\|}{P}}}-\text{OH} \;+\; \text{CH}_3\text{OH} \longrightarrow \text{HO}-\overset{\displaystyle O}{\underset{\displaystyle OH}{\overset{\|}{P}}}-\text{OCH}_3 \;+\; \text{HOH}$$

phosphoric acid   methyl alcohol   monomethyl phosphate
a phosphomonoester

$$O{=}N-\text{OH} + \text{HOCH}_2\text{CH}_2\text{CH}_2\text{CH}_2\text{CH}_3 \longrightarrow O{=}N-\text{OCH}_2\text{CH}_2\text{CH}_2\text{CH}_2\text{CH}_3 + \text{HOH}$$

nitrous   amyl alcohol   amyl nitrite
acid   (antispasmodic in
                                                   angina pectoria)
                                                   a nitrite ester

14-3.4 REACTION AT OXYGEN–HYDROGEN BOND: REACTION WITH BASES

The acidity of a substance is defined as its ability to lose a proton (see Chapter 8). Alcohols ionize to form a proton and an alkoxide anion; phenols ionize to form a proton and a phenoxide anion.

$$\text{RO}-\text{H} \xrightarrow{\text{ionization}} \text{H}^+ + \text{RO}^-$$

an alcohol   *an alkoxide anion*

phenol ionization:

phenol   *phenoxide anion*

Phenols are stronger acids than alcohols. Compared to acetic acid (a component of vinegar) however, phenols are weaker acids.

$$\textit{acidity:}\ \text{CH}_3\text{COOH} > \text{C}_6\text{H}_5\text{OH} > \text{ROH} > \text{HOH}$$

$$\underset{\xrightarrow{\hspace{3cm}}}{\text{decreasing acidity}}$$

Although phenols show acidic properties, they are relatively weak acids. Alcohols are even weaker acids. The aromatic ring is primarily responsible for increasing the acidity of a phenol over an alcohol. The aromatic ring is capable of delocalizing the negative charge in phenoxide (through resonance, Chapter 12) thus increasing the stability of phenoxide relative to phenol. A similar stabilization through de-localization is not possible for either ROH or RO$^-$. Ionization of phenol forms H$^+$

in solution. Since more protons are formed from the phenol than from the alcohol, phenol is the stronger acid. Phenols, but not alcohols, react with bases such as NaOH to form salts. A salt of phenol is called a *phenoxide.*

$$
\underset{\text{phenol}}{\text{C}_6\text{H}_5\text{OH}} + \text{NaOH} \longrightarrow \underset{\text{sodium phenoxide}}{\text{C}_6\text{H}_5\text{O}^-\text{Na}^+} + \text{H}_2\text{O}
$$

Phenol was the first widely used antiseptic. However, it causes severe burns when applied to the skin. Consequently, safer antiseptics have been found and some of these are described in Table 14.3.

14-3.5 OXIDATION OF ALCOHOLS, PHENOLS, AND ETHERS

The final reaction we will discuss is oxidation. In this section we will look at the oxidation of alcohols, phenols, and ethers and examine some applications of the oxidation of ethyl alcohol, methyl alcohol, and a phenol-related compound found in the respiratory chain.

Alcohols

In oxidation, the alcohol loses two hydrogen atoms. One of the hydrogens is lost from the carbon with the hydroxyl group and the other is lost from the oxygen atom. These two hydrogens will reappear in the water that is formed as a product. The carbon–oxygen single bond is replaced by a carbon–oxygen double bond.

$$
\text{H}-\overset{|}{\underset{|}{\text{C}}}-\text{O}-\text{H} \xrightarrow{\text{oxidizing agent}} \text{C}=\text{O} + \text{H}_2\text{O}
$$

a single bond   a double bond

Let us look more closely at this oxidation reaction. In actuality the alcohol loses the two hydrogens in a special way: as $\text{H}:^-$ and H^+. The oxygen that appears in the water molecule arises from the oxidizing agent. The symbol (O) in the equation below represents the oxidizing agent.

$$
-\overset{|}{\underset{|}{\text{C}}}-\text{O}-\text{H} \xrightarrow{\text{(O)}} \text{C}=\text{O} + \text{H}-\text{O}-\text{H}
$$

where $\text{C}-\text{O}-\text{H}$ gives H^+ (a proton)

and $\text{C}-\text{H}$ gives $\text{H}:^-$ (a hydride ion)

and $\text{H}^+ + \text{H}:^- + \text{(O)} \longrightarrow \text{H}-\text{O}-\text{H}$

In living systems a similar reaction occurs but with oxidizing agents that are not the same as those we use in the laboratory.

Table 14.4 Compounds Used to Oxidize Alcohols

Oxidizing Agent	In Vitro Name	In Vivo Oxidizing Agent*	In Vivo Name
CrO_3	Chromium trioxide	(structure with $CONH_2$, pyridine ring, N_+, R)	Nicotine adenine dinucleotide
$K_2Cr_2O_7$	Potassium dichromate	(flavin structure with H_3C, R, N, O, H)	Flavin adenine dinucleotide

*R as used here is not an alkyl group. See Sections 24-2.1 and 25-1 for structures of R.

In the laboratory we oxidize alcohols with such oxidizing agents as chromium trioxide and potassium dichromate. The oxidizing agent of alcohols in the living system is a derivative of the vitamin niacin. It is the aromatic compound nicotine adenine dinucleotide (NAD^+), which we discussed in Chapter 12. Another oxidizing agent in the living system is a derivative of another vitamin riboflavin and is called flavin adenine dinucleotide (FAD). The structures of the oxidizing agents used in the laboratory and in the living system are shown in Table 14.4.

The oxidation product of a primary alcohol depends on the strength of the oxidizing agent used in the reaction. If a mild oxidizing agent is used, an **aldehyde** is the product. A strong oxidizing agent, on the other hand, results in a **carboxylic acid** being formed.

An **aldehyde** is an organic molecule with the $-\overset{\overset{\text{O}}{\|}}{\text{C}}-\text{H}$ functional group. A **carboxylic acid** is an organic molecule with the $-\overset{\overset{\text{O}}{\|}}{\text{C}}-\text{OH}$ functional group.

$$RCH_2OH \text{ (a primary alcohol)} \xrightarrow{\text{mild oxidizing agent}} R-\overset{\overset{\text{O}}{\|}}{\text{C}}-H + H_2O \text{ (an aldehyde)}$$

$$RCH_2OH \text{ (a primary alcohol)} \xrightarrow{\text{strong oxidizing agent}} R-\overset{\overset{\text{O}}{\|}}{\text{C}}-OH + H_2O \text{ (a carboxylic acid)}$$

The oxidation of a primary alcohol by a strong oxidizing agent actually forms the aldehyde. However, if the aldehyde is not removed as rapidly as it is formed, it is oxidized further by the unchanged oxidizing agent to a carboxylic acid. Aldehydes, then, are more easily oxidized than are alcohols.

$$\underset{\substack{\displaystyle | \\ H}}{\overset{\substack{H \\ \displaystyle |}}{R-C-OH}} \xrightarrow{\text{strong oxidizing agent}} \underset{\substack{ \\ }}{\overset{\substack{O \\ \displaystyle \|}}{R-C-H}} \xrightarrow{\text{strong oxidizing agent}} \overset{\substack{O \\ \displaystyle \|}}{R-C-OH}$$

a primary alcohol   *an aldehyde*   *a carboxylic acid*

Consequently, when an organic chemist oxidizes a primary alcohol, a mild oxidizing agent is used if the desired product is an aldehyde and a strong oxidizing agent is used if a carboxylic acid is the desired product.

One of the two *in vitro* oxidizing agents shown in Table 14.4, chromium trioxide, is a mild oxidizing agent when dissolved in the solvent pyridine. Pyridine is a heterocyclic amine and a weak base (see Chapter 17). Chromium trioxide in pyridine is used to convert a primary alcohol to an aldehyde. Potassium dichromate in sulfuric acid is a strong oxidizing agent and is used to convert a primary alcohol to a carboxylic acid. A specific example is the oxidation of propyl alcohol to propionaldehyde or propionic acid, depending on the choice of oxidizing agent.

$$CH_3CH_2CH_2OH \begin{cases} \xrightarrow{CrO_3,\ \text{pyridine}} \overset{\substack{O \\ \|}}{CH_3CH_2C}-H + H_2O \\ \\ \xrightarrow{K_2Cr_2O_7,\ H_2SO_4} \overset{\substack{O \\ \|}}{CH_3CH_2C}-OH + H_2O \end{cases}$$

propyl alcohol   propionaldehyde

propionic acid

The oxidation of a secondary alcohol forms a ketone. Ketones, unlike aldehydes, are not oxidized further and are produced in good yield and without the formation of side products.

$$\underset{\substack{| \\ OH}}{\overset{\substack{H \\ |}}{R-C-R}} \xrightarrow{\text{oxidizing agent}} \underset{\substack{\| \\ O}}{R-C-R} + H_2O$$

a secondary alcohol   *a ketone*

The oxidizing agent used to convert a secondary alcohol to a ketone is potassium dichromate in sulfuric acid. For example,

$$\underset{\substack{| \\ OH}}{\overset{\substack{H \\ |}}{CH_3-C-CH_3}} \xrightarrow{K_2Cr_2O_7,\ H_2SO_4} \underset{\substack{\| \\ O}}{CH_3-C-CH_3} + H_2O$$

isopropyl alcohol   acetone

cyclohexyl alcohol   cyclohexanone

Tertiary alcohols, under neutral conditions, are not oxidized because the carbon with the hydroxyl group does not also have a hydrogen it can lose.

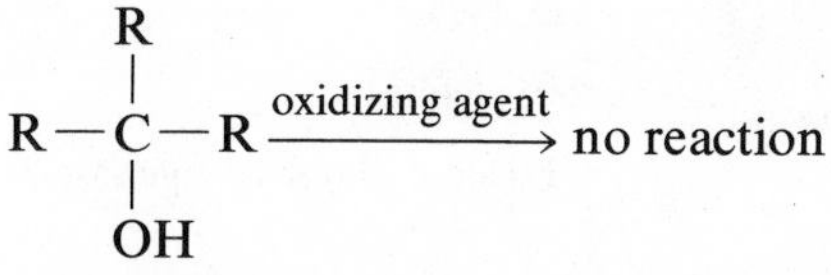

$$R - \underset{\underset{OH}{|}}{\overset{\overset{R}{|}}{C}} - R \xrightarrow{\text{oxidizing agent}} \text{no reaction}$$

a tertiary alcohol

As we said above, oxidation also occurs in the living system. For example, when ethyl alcohol has been ingested, one of the first reactions in its metabolism is the oxidation to acetaldehyde.

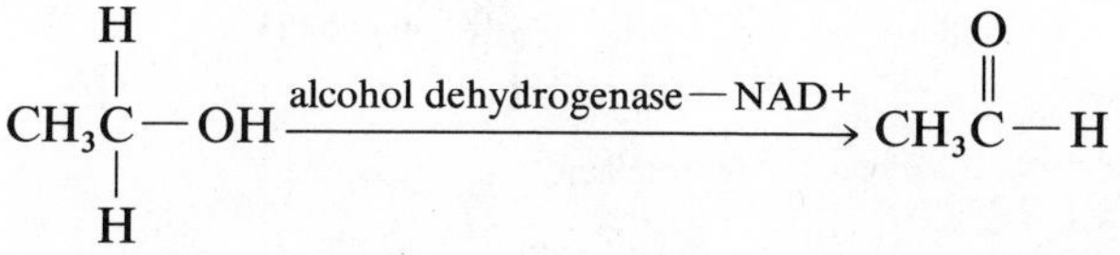

$$CH_3\underset{\underset{H}{|}}{\overset{\overset{H}{|}}{C}} - OH \xrightarrow{\text{alcohol dehydrogenase} - NAD^+} CH_3\overset{\overset{O}{\|}}{C} - H$$

ethyl alcohol   acetaldehyde

A *dehydrogenase* is an enzyme that converts a hydroxyl group to a carbonyl group. The hydrogen acceptor for this reaction is NAD^+ as shown below.

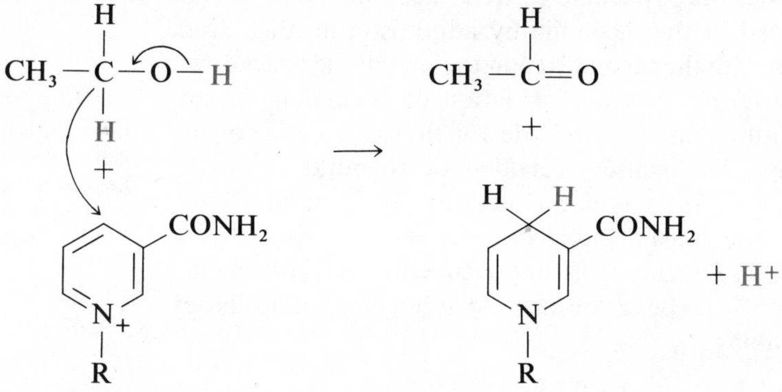

NAD⁺   NADH

The remainder of the reactions involved in metabolism of ethyl alcohol by the living system are illustrated in Figure 14.1.

The series of reactions in Figure 14.1 shows that after the ethyl alcohol is oxidized to acetaldehyde (step 1), the latter is converted to acetyl coenzyme A (step 2). The reaction is catalyzed by acetaldelyde dehydrogenase and NAD^+. The acetyl coenzyme A is then converted to carbon dioxide and water, step 3. Thus, the complete oxidation of ethyl alcohol in the living system forms carbon dioxide and water.

In some cases the concentration of acetaldehyde in a person's blood is abnormally increased after consuming alcohol and severe discomfort is experienced. The face becomes red, the heartbeat increases, and the blood pressure is markedly reduced. Nausea, vomiting, sweating, considerable uneasiness, and confusion is characteristic of this state, known medically as the *acetaldehyde syndrome*. To avoid such discomfort, the person is forced to avoid the use of alcohol.

Figure 14.1 A comparison of the metabolism of ethyl alcohol and of methyl alcohol.

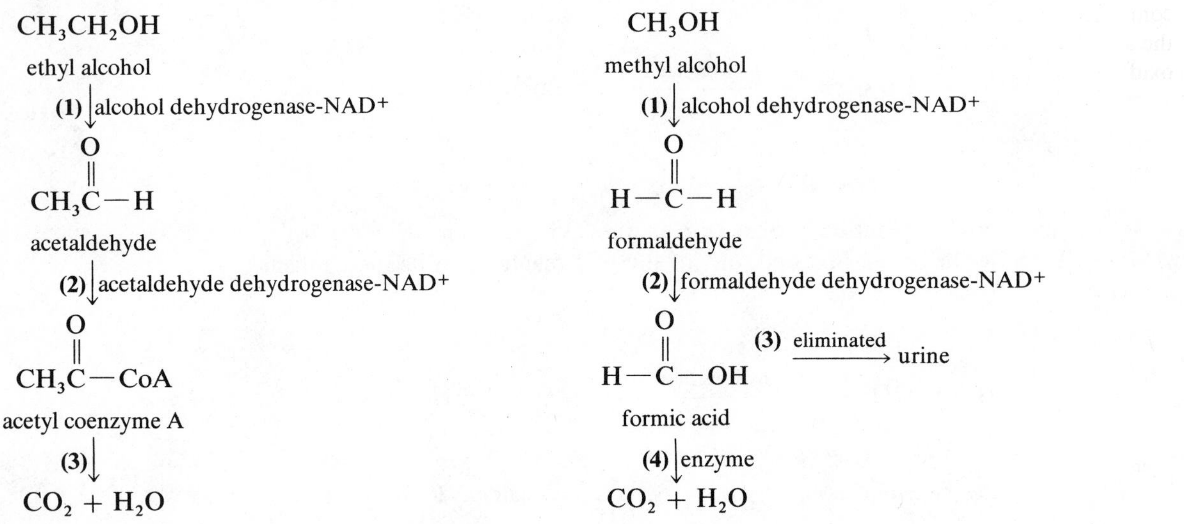

CH_3CH_2OH

ethyl alcohol

(1) $\Big\downarrow$ alcohol dehydrogenase-NAD$^+$

$$CH_3\overset{\overset{\displaystyle O}{\|}}{C}-H$$

acetaldehyde

(2) $\Big\downarrow$ acetaldehyde dehydrogenase-NAD$^+$

$$CH_3\overset{\overset{\displaystyle O}{\|}}{C}-CoA$$

acetyl coenzyme A

(3) $\Big\downarrow$

$CO_2 + H_2O$

CH_3OH

methyl alcohol

(1) $\Big\downarrow$ alcohol dehydrogenase-NAD$^+$

$$H-\overset{\overset{\displaystyle O}{\|}}{C}-H$$

formaldehyde

(2) $\Big\downarrow$ formaldehyde dehydrogenase-NAD$^+$

$$H-\overset{\overset{\displaystyle O}{\|}}{C}-OH \quad \xrightarrow[\text{eliminated}]{\text{(3)}} \text{urine}$$

formic acid

(4) $\Big\downarrow$ enzyme

$CO_2 + H_2O$

Doctors have used the acetaldehyde syndrome to treat alcoholics. The acetaldehyde syndrome can be induced in the alcoholic by administering disulfiram (Antabuse). Disulfiram interferes with the second step in the metabolic breakdown of alcohol. It functions by interfering with the oxidation of acetaldehyde, apparently by competing successfully with acetaldehyde for the active sites of the enzyme aldehyde dehydrogenase. This causes acetaldehyde to build up in the blood (see Figure 14.2). Since disulfiram is eliminated slowly from the body, therapy is long-lasting. The alcoholic must avoid alcohol for several days or suffer the accompanying severe reaction. Disulfiram is not a cure for alcoholism but, under medical supervision, patients can be encouraged to avoid alcohol until they can voluntarily abstain from drinking.

Active site is that portion of the enzyme to which the acetaldehyde binds when it is oxidized to acetyl coenzyme A.

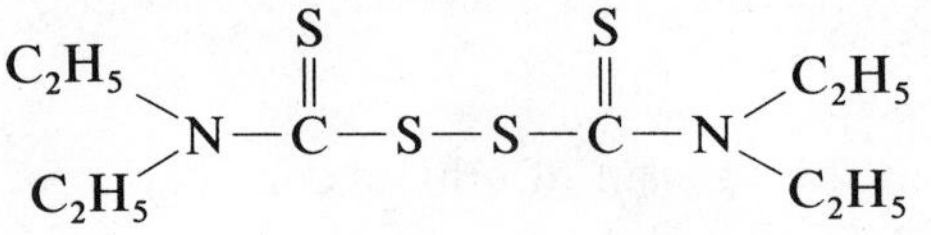

$$\begin{array}{ccc} C_2H_5 & \overset{\displaystyle S}{\underset{\displaystyle \|}{}} & \overset{\displaystyle S}{\underset{\displaystyle \|}{}} & C_2H_5 \\ {\diagdown} & {\|} & {\|} & {\diagup} \\ N-C-S-S-C-N \\ {\diagup} & & & {\diagdown} \\ C_2H_5 & & & C_2H_5 \end{array}$$

disulfiram

Let us now compare the oxidation of methyl alcohol with ethyl alcohol by the living system. As illustrated in Figure 14.1, the oxidation of methyl alcohol, like that of ethyl alcohol, ultimately forms carbon dioxide and water. Yet methyl alcohol is extremely toxic to living systems, especially humans'. During Prohibition many people died from drinking bootleg liquor contaminated with methyl alcohol. Furthermore, there is a report of a factory worker blinded as a result of spilling large amounts of methyl alcohol on his leg and failing to change his wet clothing. What is there about methyl alcohol that makes it harmful to humans?

As shown in Figure 14.1, methyl alcohol is first oxidized to formaldehyde, step 1. Formaldehyde is further oxidized to formic acid, step 2. Both of these steps require

Figure 14.2 Schematic drawing showing that the active site on the enzyme acetaldehyde dehydrogenase can accommodate disulfiram easier than acetaldehyde because of the complementary shape of disulfiram and the active site. Consequently, disulfiram occupies the active site of the enzyme. Since acetaldehyde cannot occupy the active site, it cannot be oxidized and its concentration will increase as long as it is produced.

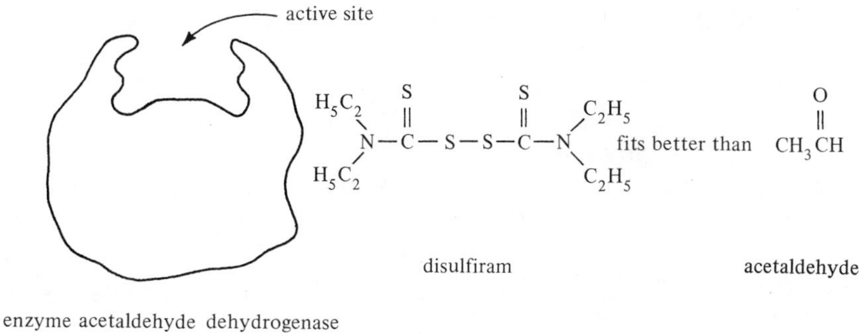

— active site

$$H_5C_2 \diagdown \underset{H_5C_2 \diagup}{N} - \overset{\overset{S}{\|}}{C} - S - S - \overset{\overset{S}{\|}}{C} - \underset{\diagdown C_2H_5}{\overset{\diagup C_2H_5}{N}}$$

fits better than

$$CH_3\overset{\overset{O}{\|}}{C}H$$

disulfiram   acetaldehyde

enzyme acetaldehyde dehydrogenase

the combination dehydrogenase-NAD$^+$—oxidizing system, as does ethyl alcohol. Formic acid either is eliminated in the urine, step 3, or undergoes further oxidation to carbon dioxide, step 4.

The occurrence of blindness in people who have ingested methanol has been attributed to the formation of formaldehyde in the retina. Studies indicate that the damaging effect of the aldehyde is a consequence of its forming directly in the retina by an alcohol-dehydrogenase-catalyzed oxidation of methyl alcohol. Alcohol dehydrogenase is produced in the retina and plays a role in the visual cycle by oxidizing *cis*- and *trans*- vitamin A (retinol) to the corresponding aldehydes (see Section 15-3.1).

It appears that there are two explanations for formaldehyde-caused blindness. One study postulates that blindness occurs in victims of methyl alcohol poisoning because formaldehyde inhibits formation of ATP. ATP is a compound which, through its enzymatic breakdown, provides energy for muscular contraction. The lack of formation of ATP causes degeneration of retinal cells by depriving these cells of the energy they need to function. The other study postulates that formic acid damages retinal cells. This explanation is based on the fact that partial and complete losses of vision occur after the onset of severe acidosis, a metabolic condition in which the pH of the blood and of body tissues falls. In victims of methyl alcohol poisoning, acidosis is caused by formic acid that apparently inactivates iron-containing enzymes responsible for transporting oxygen to the body's organs by complexing the iron in these enzymes. Because retinal tissue needs large amounts of oxygen, it is one of the first areas of the body to suffer damage (also see Chapter 8).

Acidosis also may lead to coma. This is followed by shock, convulsions, and, ultimately, death, resulting from respiratory failure. Methyl alcohol is metabolized slowly in the body. This can benefit the poison victim because the condition can

Acidosis is a condition in which the acidity of the blood is increased and its pH is decreased.

be treated successfully if it is recognized before toxic amounts of methyl alcohol metabolites accumulate. In most instances, ethyl alcohol is the major therapeutic agent for methyl-alcohol poisoning. Ethyl alcohol is metabolized by alcohol dehydrogenase much more rapidly than is methyl alcohol. Because alcohol dehydrogenase has a greater affinity for ethyl alcohol, metabolism of methyl alcohol is suppressed to the point where it is excreted from the body essentially unchanged. This example of an enzyme having greater affinity for one alcohol over another is an example of competitive inhibition, which will be dealt with again in Chapter 23.

Phenols

Phenols can also be oxidized. The hydroxyl group activates the aromatic ring and consequently makes the phenol easily susceptible to oxidation. *p*-Dihydroxy benzene (a hydroquinone), for example, can be oxidized to *p*-benzoquinone (a quinone).

p-dihydroxybenzene   *p*-benzoquinone
a hydroquinone   *a quinone*

In this oxidation the —OH group is converted to a carbonyl group ($C=O$) and the aromatic character of the original phenol is lost. The quinone is obtained in low yield, however, because it is susceptible to further oxidation and products of higher molecular weight are formed. An analogous oxidation of a phenol occurs in living systems. Unlike the *in vitro* oxidation, however, the *in vivo* oxidation is a controlled oxidation. In the respiratory chain (see Section 25-2), a series of reactions essential to life, a hydroquinone is oxidized to a quinone.

a hydroquinone   *a quinone*

Ethers

Ethers are also capable of being oxidized, particularly the lower-molecular-weight ethers. Oxidation of ethers usually occurs on standing in contact with air. Most ethers are converted to unstable peroxides. Peroxides are usually present only in low concentrations, but they are very dangerous. Their instability usually results in an explosion. In addition, the lower-molecular-weight ethers are very volatile and flammable. Because of these chemical characteristics, hospitals take numerous precautions when using ether or have changed to other, less chemically dangerous, anesthetics (see Section 14-2.4).

$$ROR \xrightarrow{\text{oxidation}} ROOR$$

an ether   *a peroxide*

14-3.6 OXIDATION OF THIOLS

The most important reaction of thiols is their oxidation to disulfides. The disulfides can be easily reduced to regenerate the thiol.

$$R-SH \underset{\text{reducing agent}}{\overset{\text{oxidizing agent}}{\rightleftharpoons}} R-S-S-R$$

a thiol   *a disulfide*

For example,

$$2\,CH_3SH \xrightarrow{\text{oxidation}} CH_3-S-S-CH_3 + H_2O$$

methyl mercaptan   dimethyl disulfide
a thiol   *a disulfide*

and

$$CH_3CH_2CH_2-S-S-CH_2CH_2CH_3 \xrightarrow{\text{reduction}}$$

dipropyl disulfide
a disulfide

$$CH_3CH_2CH_2SH + HSCH_2CH_2CH_3$$

propyl mercaptan
a thiol

As another example, the two amino acids cysteine and cystine undergo such a transformation.

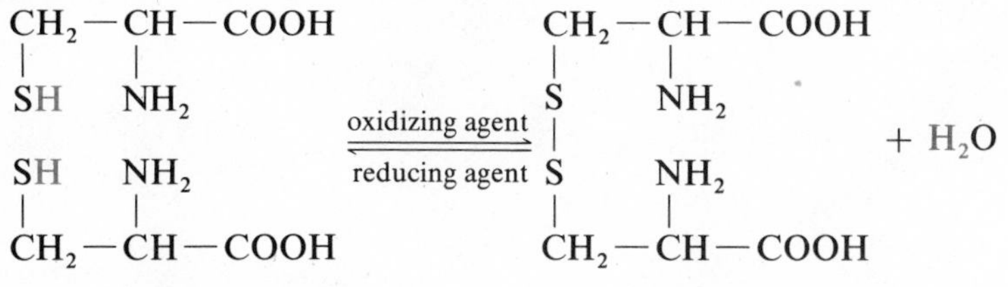

$$
\begin{array}{l}
CH_2-CH-COOH \\
\;|\qquad\;| \\
SH\quad NH_2 \\
\\
SH\quad NH_2 \\
\;|\qquad\;| \\
CH_2-CH-COOH
\end{array}
\underset{\text{reducing agent}}{\overset{\text{oxidizing agent}}{\rightleftharpoons}}
\begin{array}{l}
CH_2-CH-COOH \\
\;|\qquad\;| \\
S\quad\; NH_2 \\
\;| \\
S\quad\; NH_2 \\
\;|\qquad\;| \\
CH_2-CH-COOH
\end{array}
+ H_2O
$$

cysteine   cystine

The structural proteins of hair and skin have a high content of cystine as cross-links between protein chains. Permanent waves or curls in hair are achieved by the reduction and reforming of disulfide cross-links in the protein of hair. The permanent waving technique is illustrated in Figure 14.3. A reducing agent is used to cleave the disulfide linkages. The reduced, disordered hair is then placed on curlers and set in the desired pattern. Reduction allows the molecules to slip over one another. An oxidizing agent is then added to reform the disulfide linkages, but this time between different thiol groups. In effect, the hair has been molecularly set into a new pattern.

Figure 14.3   Permanent waving.

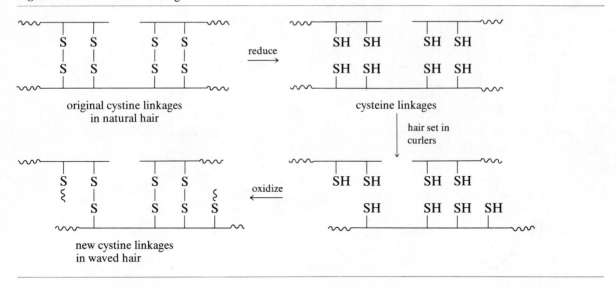

original cystine linkages
in natural hair

cysteine linkages

reduce

hair set in
curlers

new cystine linkages
in waved hair

oxidize

PROBLEM

4. Write the structure of the product of each reaction and give its general name:

(a) $CH_3CH_2CH_2OH + PBr_3 \longrightarrow$

(b)
$$CH_3\overset{\overset{\displaystyle CH_3}{\displaystyle |}}{\underset{\underset{\displaystyle OH}{\displaystyle |}}{C}}CH_3 + HCl \longrightarrow$$

(c) $HO-\bigcirc-CH_3 + H_2SO_4 \xrightarrow{\text{heat}}$

(d) $CH_3CH_2OH + CH_3\overset{\overset{\displaystyle O}{\displaystyle \|}}{C}OH \longrightarrow$

(e) $HO-\overset{\overset{\displaystyle O}{\displaystyle \|}}{\underset{\underset{\displaystyle O}{\displaystyle \|}}{S}}-OH + CH_3CH_2CH_2OH \longrightarrow$

(f)
$$CH_3\underset{\underset{\displaystyle CH_3}{\displaystyle |}}{CH}CH_2OH \xrightarrow{CrO_3,\ \text{pyridine}}$$

(g) $\overset{\overset{\displaystyle H}{}}{\underset{\underset{\displaystyle OH}{}}{\bigcirc}} \xrightarrow{K_2Cr_2O_7,\ H_2SO_4}$

5. Write the formula of the product of each reaction and identify the product as a thiol or a disulfide.

(a)

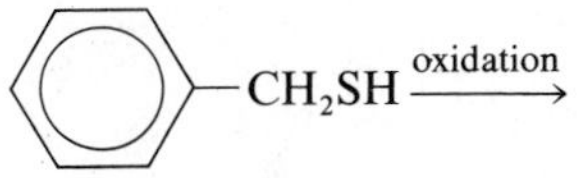

—CH_2SH $\xrightarrow{\text{oxidation}}$

(b) CH_3CHCH_2—S—S—CH_2CHCH_3 $\xrightarrow{\text{reduction}}$
    | |
    CH_3 CH_3

(c) CH_3—S—S—$CH_2CH_2CH_3$ $\xrightarrow{\text{reduction}}$

ANSWERS TO PROBLEM **4. (a)** An alkyl halide, $CH_3CH_2CH_2Br$

                    CH_3
                    |
(b) An alkyl halide, CH_3—C—CH_3 **(c)** An alkene, —CH_3
                    |
                    Cl

                    O
                    ||
(d) A carboxylate ester, CH_3C—O—CH_2CH_3

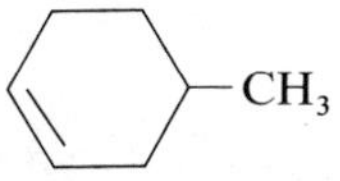

                                                      O
                                                      ||
(e) A sulfate ester, HO—S—$OCH_2CH_2CH_3$ **(f)** An aldehyde, CH_3CHCH
                          || |
                          O CH_3

(g) A ketone, 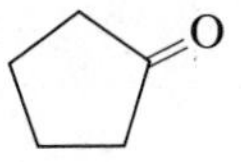   **5. (a)** 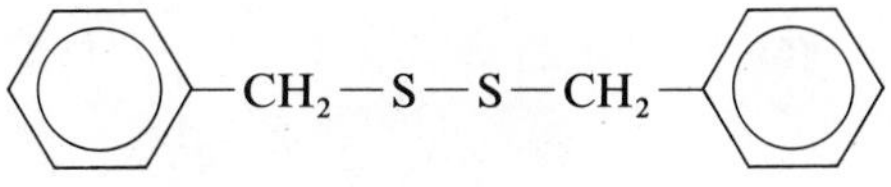 —CH_2—S—S—CH_2—

                                              a disulfide

(b) CH_3CHCH_2SH, a thiol   **(c)** CH_3SH and $CH_3CH_2CH_2SH$, thiols
        |
        CH_3

SUMMARY

The important carbon–oxygen single bond compounds are: alcohol, ROH; phenol, C_6H_5OH; and ether, ROR. The important organosulfur compounds are: thiol (mercaptan), R—SH; thioether (sulfide), R—S—R; and disulfide, R—S—S—R. These compounds have common names and are also systematically named by the IUPAC system. In the IUPAC system alcohols are named by indicating the number of carbons in the longest continuous chain containing the —OH group. The characteristic ending indicating the —OH group is -ol. These compounds occur naturally and have various uses (see Table 14.2).

Ethanol is produced by fermentation of sugar or starch-containing materials. Alcohols of higher carbon content are produced industrially by the reduction of animal fat or vegetable oils.

Alcohols, in general, can be formed by the hydrolysis of alkyl halides. Ethers

$$RX + H_2O \longrightarrow ROH + HX$$

are produced in the lab through the reaction of a sodium alkoxide and a methyl or primary alkyl halide. Ethyl ether has declined in use as an anesthetic because unstable compounds can form when it is allowed to stand in air and because it is flammable.

The reactions of alcohols involve breaking either the oxygen–hydrogen or the carbon–oxygen bond and include the following.

$$ROH \xrightarrow{\text{HX or PBr}_3 \text{ or SOCl}_2} RX \qquad (14\text{-}1)$$

an alkyl halide

$$RCH_2CH_2OH + H_2SO_4 \xrightarrow{\text{heat}} RCH{=}CH_2 + H_2O \qquad (14\text{-}2)$$

an alkene

$$ROH + HO-\overset{\displaystyle O}{\overset{\displaystyle \|}{C}}-R' \longrightarrow RO-\overset{\displaystyle O}{\overset{\displaystyle \|}{C}}-R' + H_2O \qquad (14\text{-}3)$$

a carboxylate ester

$$ROH + HO-\overset{\displaystyle O}{\underset{\displaystyle \underset{\displaystyle OH}{|}}{\overset{\displaystyle \|}{P}}}-OH \longrightarrow RO-\overset{\displaystyle O}{\underset{\displaystyle \underset{\displaystyle OH}{|}}{\overset{\displaystyle \|}{P}}}-OH + H_2O \qquad (14\text{-}4)$$

a phosphomonoester

$$RCH_2OH \xrightarrow{\text{oxidizing agent}} R-\overset{\displaystyle O}{\overset{\displaystyle \|}{C}}-H \quad \text{or} \quad R-\overset{\displaystyle O}{\overset{\displaystyle \|}{C}}-OH \qquad (14\text{-}5)$$

a primary alcohol   *an aldehyde*   *a carboxylic acid*

$$R-\overset{\displaystyle OH}{\overset{\displaystyle |}{C}}H-R \xrightarrow{\text{oxidizing agent}} R-\overset{\displaystyle O}{\overset{\displaystyle \|}{C}}-R \qquad (14\text{-}6)$$

a secondary alcohol   *a ketone*

Phenols undergo only reactions 3 and 6. In addition, phenols are more acidic than alcohols and form phenoxides with bases such as NaOH.

The normal metabolism (oxidation) of ethanol forms CO_2 and H_2O. In some cases a high concentration of acetaldehyde builds up in the person and this causes discomfort. This abnormal build-up of acetaldehyde is the basis of the medical treatment of alcoholism. Disulfiram causes the abnormal build-up of acetaldehyde, and the discomfort experienced by the alcoholic is enough to encourage the alcoholic to give up alcohol.

The metabolism of methanol forms formaldehyde. Formaldehyde is formed in the retina and its formation inhibits formation of ATP. The lack of formation of ATP causes degeneration of retinal cells and leads to blindness.

The oxidation–reduction chemistry associated with thiols and disulfides is the basis of permanent waving procedure.

ADDITIONAL PROBLEMS

6. Give a correct name for each of the following and identify each as an alcohol, phenol, ether, thiol, thio-ether (sulfide), or disulfide.

(a) [structure: benzene ring with OH and Br substituents]

(b) [structure: cyclopentane ring with H and OH]

(c) $CH_3CH_2SCH_2CH_3$

(d) $CH_3CHCH_2CH_2CH_3$ with OH

(e) $C_6H_5OCH_3$

(f) CH_3CH_2CHOH with CH_3

(g) $CH_3CH_2CH_2OCHCH_3$ with CH_3

(h) $CH_3CH_2CH_2SH$

(i) $HOCH_2CH_2OH$

(j) $HC{\equiv}CCH_2CH_2CH_2CH_2OH$

7. Write a complete structure for each of the following.
   (a) glycerol
   (b) diethyl ether
   (c) cyclopentyl alcohol
   (d) benzyl alcohol
   (e) 2,4-dinitrophenol
   (f) *tert*-butyl alcohol
   (g) 2-isopropyl-5-methylcyclohexanol
   (h) methyl isopropyl sulfide
   (i) 2,3-pentanediol
   (j) *cis*-1,2-cyclopentanediol

8. Write structures and give common names and IUPAC names for all the isomeric alcohols of molecular formula C_4H_9OH. Identify each alcohol as primary, secondary, or tertiary.

9. Write the product of the reaction of *sec*-butyl alcohol with the following reagents.
   (a) HI
   (b) PBr_3
   (c) $SOCl_2$
   (d) H_2SO_4, heat
   (e) H_3PO_4 (ester formation)
   (f) $CH_3CH_2C{-}OH$ with O above
   (g) $K_2Cr_2O_7$, H_2SO_4

10. Write the structure of the organic product of each reaction.

(a) $C_6H_{12}O_6 \xrightarrow{\text{fermentation}}$

(b) $CH_3CCH_3 + HOH \longrightarrow$ with CH_3 above and Br below

(c) $CH_3CH_2CHCH_2OH \xrightarrow{\text{CrO}_3,\text{ pyridine}}$ with CH_3 below

(d) $CH_3O^-Na^+ + CH_3CH_2CH_2CH_2Cl \longrightarrow$

(e) [benzene ring]$-CH_2OH + HI \longrightarrow$

(f) [steroid structure with CH_3, C_8H_{17} labels, HO group] $+ PBr_3 \longrightarrow$

(g) $CH_3CHCH_2CH_2CH_2OH + H_2SO_4 \xrightarrow{\text{heat}}$ with CH_3 below

(h) $CH_3CH_2CH_2OH + HO{-}CCH_3 \longrightarrow$ with O above

(i) [structure: benzene ring with CH_3 and $CH(CH_3)_2$ and OH] $\xrightarrow[\text{acid}]{\text{K}_2\text{Cr}_2\text{O}_7}$ menthone ($C_{10}H_{18}O$)

(j) $CH_3CH_2OH \xrightarrow{\text{dehydrogenase}}$

11. What is the structural difference between the following?
    (a) an alcohol and a thiol

(b) ether and thioether

(c) thioether and sulfide

12. From what you have learned in this chapter, do enzymes known as dehydrogenases cause dehydration or oxidation–reduction reactions? Give an example from this chapter.

13. What is the Williamson reaction? What is the nature of the reactants in the reaction?

14. To what class of organic compounds do sulfates, phosphates, nitrates, and carboxylates belong?
    Draw the structure of a typical example of each using methanol as the compound.

15. Glycerol reacts with 3 mol nitric acid to form nitroglycerin. Nitroglycerin is used as an explosive and to relieve the pain due to attacks of angina pectoris. Write the equation for the reaction between glycerol and nitric acid.

16. Is phenol or ethanol more acidic? Why?

17. Briefly outline the normal and abnormal metabolism of ethyl alcohol. How can disulfiram help the alcoholic?

18. Explain how the normal metabolism of methanol causes blindness.

19. What is the structural relationship between ethers and peroxides?

20. Lipoic acid is a growth factor that is required by many microorganisms. It functions in the living system by catalyzing certain oxidation reactions and is reduced to a thiol in the process. Write the structure of the reduced form of lipoic acid.

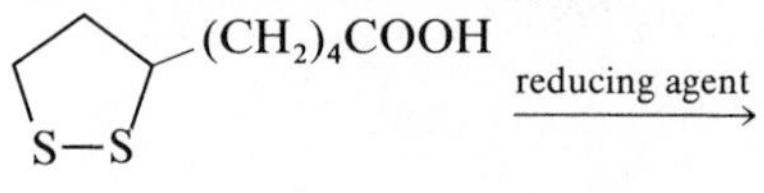

$$\text{(structure)} \quad -(CH_2)_4COOH \qquad \xrightarrow{\text{reducing agent}}$$

S—S

lipoic acid

a disulfide

CHAPTER 15

THE CARBONYL COMPOUNDS: ALDEHYDES AND KETONES

The carbonyl compounds are aldehydes and ketones. Aldehydes and ketones are recognized by the functional group of a carbon–oxygen double bond (C=O), known as a *carbonyl group*. Many carbonyl compounds occur in plants and animals and many have been used as perfume and flavoring ingredients. Carbonyl groups also exist in the male and female sex hormones. Sometimes the carbonyl group occurs in a combined form, as in carbohydrates. Carbohydrates are polyhydroxy aldehydes and ketones (see Chapter 19). Furthermore, carbohydrates are the major structural support material in many plants, as well as a major source of energy in living systems. An understanding of the fundamental reactions of the carbonyl group is required for an understanding of the mechanism by which energy is derived from carbohydrates. In this chapter we will consider some of the typical chemistry of aldehydes and ketones.

15-1 STRUCTURE, NOMENCLATURE, AND USES

LEARNING OBJECTIVES

1. Compare and contrast the features of the carbonyl group and carbon–carbon double bond: nature of the double bonds, bond angles, and degree of sharing of electrons between atoms of the double bonds.

2. Given the common or IUPAC name of an aldehyde or a ketone, write its structural formula.

3. Given the structural formula of an aldehyde or a ketone, write its common or IUPAC name.

The carbonyl group is the functional group characteristic of aldehydes and ketones. The carbonyl group consists of a carbon atom doubly bonded to an oxygen atom as shown. In many respects it is similar to the carbon–carbon double bond (C=C) found in an alkene (see Chapter 11). The atoms attached to the carbonyl carbon and the oxygen atom of the carbonyl group all lie in the same plane, 120° apart. Similarly, the atoms attached to the C=C double bond of an alkene lie in the same plane, 120° apart.

carbonyl group   *carbon–carbon double bond*

The carbonyl C=O double bond consists of four electrons. Two of these electrons occupy the space along the bond axis, the sigma bond. The second pair of electrons occupies the regions of space above and below the plane of the molecule. This pair of electrons then forms a pi bond. The carbonyl C=O double bond, like the C=C double bond, consists of a sigma and a pi bond. However, unlike the electrons of the C=C double bond, the electrons of the C=O double bond are not equally shared and the bond is polarized. The electrons are attracted toward the more electronegative oxygen. The symbol δ (delta) represents the partial charge resulting from a distortion of the electrons in the C=O bond toward the more electronegative oxygen atom. Consequently, the chemistry of the C=O double bond will be different from that of the C=C double bond.

If one or both of the atoms attached to the carbonyl carbon are hydrogen, the carbonyl compound is an **aldehyde**. On the other hand, if none of the atoms is hydrogen, the carbonyl compound is a **ketone**.

polarized carbonyl group

An **aldehyde** has a carbonyl group bonded to at least one hydrogen atom. A **ketone** has a carbonyl group bonded to two carbon atoms.

$$\overset{O}{\overset{\|}{R-C-H}} \quad \text{or} \quad R\text{CHO} \qquad \overset{O}{\overset{\|}{R-C-R}} \quad \text{or} \quad R\text{COR}, \qquad R = \text{alkyl or cyclic}$$

an aldehyde   *a ketone*

Carbonyl compounds have both common and IUPAC names. Table 15.1 lists some typical carbonyl compounds and their common names. The common names of ketones consist of the name for each alkyl group (see Table 11.4) attached to the carbonyl group and followed by the word *ketone*.

In the IUPAC system of nomenclature aldehydes are named using the following method. The longest chain carrying the carbonyl group is given the parent name derived from the corresponding alkane. The *-e* of the alkane is replaced by *-al*. The carbonyl carbon in aldehydes is always carbon number one and the position of other groups is also indicated. Examples are:

$$\overset{O}{\overset{\|}{CH_3CH_2C-H}} \qquad \overset{O}{\overset{\|}{CH_3CH_2CH_2CH_2C-H}}$$

propanal   pentanal

$$\overset{O}{\underset{\overset{|}{CH_3}}{\overset{\|}{CH_3CHCH_2CH_2C-H}}} \qquad \underset{\overset{|}{}}{\overset{Cl \qquad CH_3 \quad O}{CH_3CHCH_2CH_2CHCH_2C-H}}$$

4-methylpentanal   3-methyl-6-chloroheptanal

Table 15.1 Typical Carbonyl Compounds

Structure of the Carbonyl Compound	Common Name	IUPAC Name
Aldehydes H—C—H (O above C)	formaldehyde	methanal
$\text{CH}_3\text{C—H}$ (O above C)	acetaldehyde	ethanal
$\text{CH}_3\text{CH}_2\text{C—H}$ (O above C)	propionaldehyde	propanal
$\text{CH}_3\text{CH}_2\text{CH}_2\text{C—H}$ (O above C)	butyraldehyde	butanal
$\text{CH}_3\text{CHC—H}$ (O above C, CH_3 below)	isobutyraldehyde	2-methylpropanal
Ketones CH_3CCH_3 (O above C)	dimethyl ketone (acetone)	propanone
$\text{CH}_3\text{CCH}_2\text{CH}_3$ (O above C)	methyl ethyl ketone	2-butanone
$\text{CH}_3\text{CH}_2\text{CCH}_2\text{CH}_3$ (O above C)	diethyl ketone	3-pentanone
$\text{CH}_3\text{CH}_2\text{CH}_2\text{CCH}_3$ (O above C)	methyl propyl ketone	2-pentanone
cyclohexanone ring with =O	cyclohexanone	cyclohexanone

The simplest aromatic aldehyde is benzaldehyde. Substituted benzaldehydes are named as derivatives of benzaldehyde. For example, *m*-nitrobenzaldehyde is a substituted benzaldehyde.

benzaldehyde   *m*-nitrobenzaldehyde

In naming ketones by the IUPAC system, the longest chain carrying the carbonyl group is given the parent name derived from the corresponding alkane by replacing *-e* with *-one*. The carbonyl carbon is given the smallest number and the position of other groups is indicated. For example, 2-butanone, 5-methyl-3-hexanone, and 4,4-dimethyl-2-pentanone are ketones.

$$\overset{\displaystyle O}{\overset{\displaystyle \|}{CH_3CH_2CCH_3}}$$

2-butanone
(*not* 3-butanone)

$$\overset{\displaystyle CH_3 \quad O}{\overset{\displaystyle | \quad\quad \|}{CH_3CHCH_2CCH_2CH_3}}$$

$$\overset{\displaystyle CH_3 \quad O}{\overset{\displaystyle | \quad\quad \|}{\underset{\displaystyle \underset{\displaystyle CH_3}{|}}{CH_3CCH_2CCH_3}}}$$

5-methyl-3-hexanone
(*not* 2-methyl-4-hexanone)

4,4-dimethyl-2-pentanone
(*not* 2,2-dimethyl-
4-pentanone)

The simplest aromatic ketone is acetophenone and substituted acetophenones are named as derivatives of acetophenone. For example, 2,3-dichloroacetophenone and *o*-iodoacetophenone are substituted acetophenones.

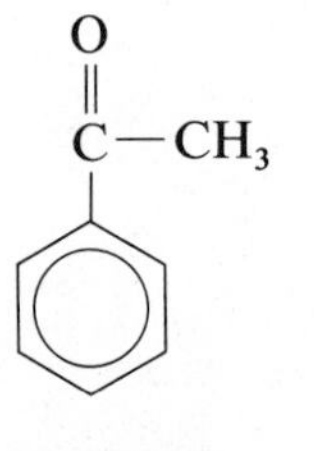

acetophenone

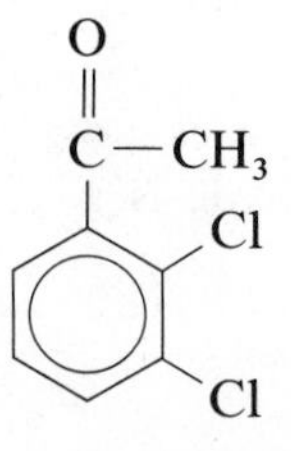

2,3-dichloroacetophenone

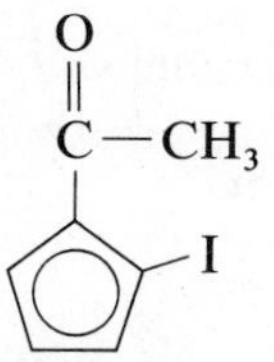

o-iodoacetophenone

Carbonyl compounds are abundant in nature and they demonstrate a wide variety of uses and sources. Some of these are given in Table 15.2 and in Table 15.3.

Table 15.2   Uses and Sources of Aldehydes

Structure of Aldehyde	Name	Occurrence or Use
HCHO	Formaldehyde	Preservative of specimens; possesses antibacterial properties
$CH_3(CH_2)_9CHO$	Undecanal	Pheromone of the wax moth
(benzaldehyde structure) CHO	Benzaldehyde	Chief constituent of oil of bitter almond
(vanillin structure) CHO, HO, OCH_3	Vanillin	Chief constituent of vanilla bean
(cinnamaldehyde structure) CH=CHCHO	Cinnamaldehyde	Chief constituent of oil of cinnamon
$(CH_3)_2C=CHCH_2CH_2\overset{\underset{\displaystyle CH_3}{\mid}}{C}=CHCHO$	Citral	Chief constituent of rinds of lemons, limes, and oranges
(pyridoxal phosphate structure) CHO, HO, $CH_2OPO_3H_2$, H_3C, N	Pyridoxal phosphate	Active form of vitamin B_6 which functions in metabolic reactions of amino acids
CHO H—C—OH HO—C—H H—C—OH H—C—OH CH_2OH	Glucose	Most abundant carbohydrate in nature; provides more than half the energy required by the human body.

Table 15.3   Uses and Sources of Ketones

Structure of Ketone	Name	Occurrence or Use
$CH_3CCH_2CH_3$ (with O double bond)	Methyl ethyl ketone (2-butanone)	Solvent, such as fingernail polish remover
CH_3C-CCH_3 (with two O double bonds)	Diacetyl (2,3-butanedione)	Diketone; gives butter its characteristic odor
(cyclohexanone ring with CH_3 and $CH(CH_3)_2$ substituents)	Menthone	Used as a peppermint flavoring
(benzene ring with CCH_2Cl and O)	Chloroacetophenone	Component of Mace and used as a tear gas
(large ring: $(CH_2)_{12}$, CH_2, CH, CH_3, C=O)	Muscone	Active ingredient of musk oil from a small male deer of Asia
(benzene ring with $COCH_3$, H_3C, CH_3, O_2N, NO_2, $C(CH_3)_3$)	4-t-Butyl-2,6-dimethyl-3,5-dinitroacetophenone (musk ketone)	Synthetic musk and a common constituent of musk colognes
(cyclohexene ring with CH_3 CH_3, $CH=CHCOCH_3$, CH_3)	β-Ionone	Fragrance responsible for odor of violets
CH_2OH / $C=O$ / $HO-C-H$ / $H-C-OH$ / $H-C-OH$ / CH_2OH	Fructose	Released (with glucose) upon hydrolysis of table sugar

PROBLEMS

1. Identify each of the following as an aldehyde or ketone and give the correct name for each.

(a) $CH_3CH_2CH_2\overset{\displaystyle O}{\overset{\displaystyle \|}{C}}CH_2CH_3$ (b) $CH_3CH{=}CH\overset{\displaystyle O}{\overset{\displaystyle \|}{C}}{-}H$ (c)

(d) $H{-}\overset{\displaystyle O}{\overset{\displaystyle \|}{C}}CH_2\overset{\displaystyle Br}{\overset{\displaystyle |}{C}}HCH_2\overset{\displaystyle CH_3}{\overset{\displaystyle |}{C}}HCH_3$ (e)

2. Draw a complete structure for the following.
   (a) ethanal   (b) dipropylketone   (c) butyraldehyde
   (d) 2-pentanone   (e) methyl isopropyl ketone   (f) cyclohexanone

---

ANSWERS TO PROBLEMS   **1. (a)** ketone; common, ethyl propyl ketone; IUPAC, 3-hexanone
(b) aldehyde; 2-butenal   **(c)** ketone; 2-methylcyclopentanone   **(d)** aldehyde; 3-bromo-5-methylhexanal   **(e)** ketone; *p*-methylacetophenone

2. (a) $CH_3\overset{\displaystyle O}{\overset{\displaystyle \|}{C}}{-}H$ **(b)** $CH_3CH_2CH_2\overset{\displaystyle O}{\overset{\displaystyle \|}{C}}CH_2CH_2CH_3$ **(c)** $CH_3CH_2CH_2\overset{\displaystyle O}{\overset{\displaystyle \|}{C}}{-}H$

(d) $CH_3\overset{\displaystyle O}{\overset{\displaystyle \|}{C}}CH_2CH_2CH_3$ **(e)** $CH_3\overset{\displaystyle O}{\overset{\displaystyle \|}{C}}\overset{\displaystyle |}{C}HCH_3$ **(f)**
$\qquad\qquad\qquad\qquad\qquad\quad CH_3$

15-2 METHODS OF INTRODUCING THE CARBONYL GROUP IN MOLECULES

LEARNING OBJECTIVES

1. Given the oxidation reaction of a primary alcohol, write the structure of the aldehyde formed.
2. Given the oxidation reaction of a secondary alcohol, write the structure of the ketone formed.

3. Given the reaction of an aromatic compound with an acid chloride in the presence of $AlCl_3$, write the structure of the ketone formed.

15-2.1 OXIDATION OF PRIMARY AND SECONDARY ALCOHOLS

In the laboratory aldehydes are prepared by the oxidation of primary alcohols (see Section 14-3.5). The oxidizing medium is generally chromium trioxide (CrO_3) in pyridine (C_5H_5N). Pyridine is a weak heterocyclic base. Examples of these reactions are the oxidation of ethyl alcohol and benzyl alcohol as shown.

$$R{-}CH_2OH \xrightarrow{\ CrO_3,\ pyridine\ } R{-}\overset{\displaystyle O}{\overset{\displaystyle \|}{C}}{-}H + H_2O$$

a primary alcohol   *an aldehyde*

$$CH_3CH_2OH \xrightarrow{CrO_3, \text{ pyridine}} CH_3\overset{\displaystyle O}{\overset{\|}{C}}-H + H_2O$$

ethyl alcohol   acetaldehyde

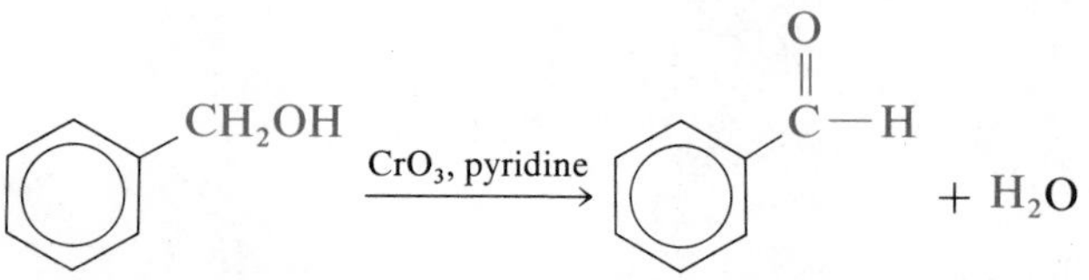

benzyl alcohol   benzaldehyde

Ketones, on the other hand, are prepared from secondary alcohols (see Section 14-3.5) in an oxidizing medium. The oxidizing medium is usually potassium dichromate ($K_2Cr_2O_7$) in sulfuric acid. Examples of these reactions are the oxidation of isopropyl alcohol and cyclohexanol.

$$R-\underset{\underset{\displaystyle OH}{|}}{C}H-R \xrightarrow{K_2Cr_2O_7, H_2SO_4} R-\underset{\underset{\displaystyle O}{\|}}{C}-R + H_2O$$

a secondary alcohol   *a ketone*

$$CH_3\underset{\underset{\displaystyle OH}{|}}{C}HCH_3 \xrightarrow{K_2Cr_2O_7, H_2SO_4} CH_3\underset{\underset{\displaystyle O}{\|}}{C}CH_3 + H_2O$$

isopropyl alcohol   acetone

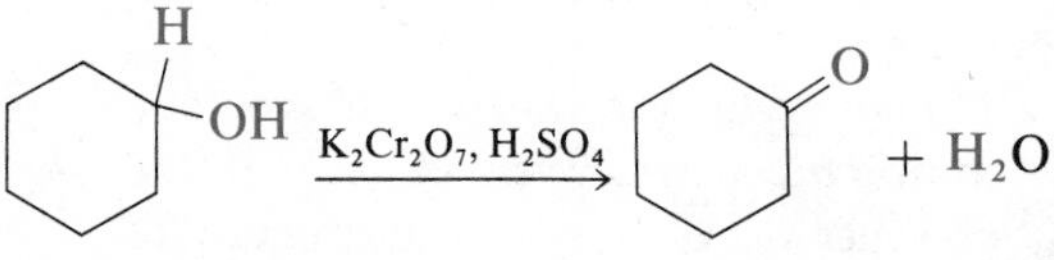

cyclohexanol   cyclohexanone

There are several examples of reactions in the living system whose overall reaction represents the oxidation of a secondary alcohol to a ketone. The chemistry involved in the condition known as ketosis and the formation of the male sex hormone testosterone from cholesterol are examples.

In addition to the formation of acetone from the *in vitro* oxidation of isopropyl alcohol as shown above, acetone can also be formed *in vivo* by the oxidation of the secondary hydroxy group in 3-hydroxybutanoic acid (also called β-hydroxybutyric acid) to the ketone group in acetoacetic acid. This reaction is followed by the subsequent **decarboxylation** of acetoacetic acid to acetone. The reactions shown are first an oxidation of the secondary alcohol and then decarboxylation of the acetoacetic acid.

Decarboxylation is the chemical loss of carbon dioxide.

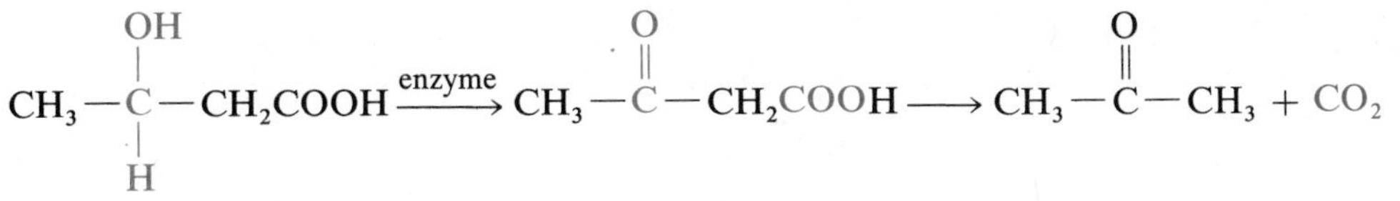

$$CH_3-\underset{\underset{H}{|}}{\overset{\overset{OH}{|}}{C}}-CH_2COOH \xrightarrow{\text{enzyme}} CH_3-\overset{\overset{O}{||}}{C}-CH_2COOH \longrightarrow CH_3-\overset{\overset{O}{||}}{C}-CH_3 + CO_2$$

3-hydroxybutanoic acid   acetoacetic acid   acetone
(β-hydroxybutyric acid)

Acetoacetic acid, 3-hydroxybutanoic acid, and acetone are collectively referred to as *ketone bodies.* If they accumulate in the blood (in excess of 1 mg/100 mL) and urine a condition known as **ketosis** occurs. Acetone is a fairly volatile compound and is often expelled in the breath. Acetone breath is a good indication of ketosis. Ketosis, a consequence of restricted carbohydrate metabolism and increased lipid metabolism of storage fat, is characteristic of diabetes. Because both 3-hydroxybutanoic acid and acetoacetic acid are acids, their presence in excess amounts decreases the pH of the blood. Such an effect is more serious than ketosis and can result in a condition known as **acidosis** (see Chapters 8 and 14). Acidosis interferes with oxygen transport by hemoglobin and unconsciousness can result. Insulin administration alleviates both ketosis and acidosis in diabetics.

> **Ketosis** is a condition in which certain ketones accumulate in the blood.

> **Acidosis** is a condition in which certain acids accumulate in the blood.

Another good example of the biosynthesis of a ketone is the formation of the ketone testosterone from cholesterol, shown below. The series of steps in the reaction is complicated and, in some steps, the mechanisms are not known. Testosterone functions in developing the male reproductive organs and maintaining the secondary male characteristics. These secondary characteristics include development of deep voice, growth and pattern of facial and body hair, and the male type of skeletal muscular development.

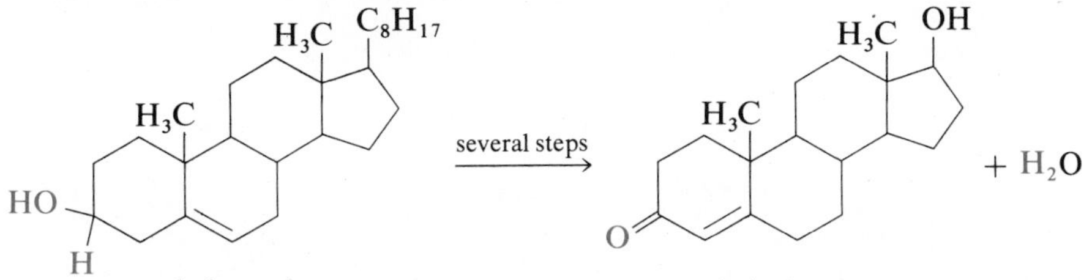

cholesterol   several steps   testosterone $+ H_2O$

15-2.2 ACYLATION OF AROMATIC COMPOUNDS

In addition to oxidation of secondary alcohols, ketones can be prepared by an **acylation reaction.** In this reaction an aromatic compound such as benzene is reacted with an **acid chloride** in the presence of aluminum chloride. The reaction is a substitution reaction and forms an aromatic ketone.

> An **acylation reaction** is one in which a ketone is formed from an aromatic compound and an acid chloride in the presence of $AlCl_3$. An **acid chloride** has the general formula
> $$R-\overset{\overset{O}{||}}{C}-Cl.$$

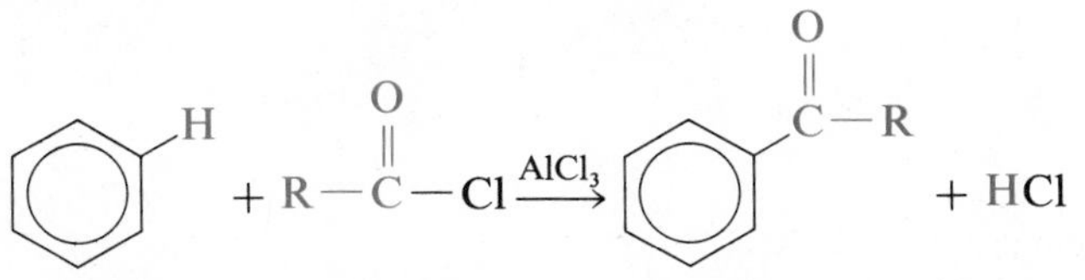

benzene   *an acid chloride*   *a ketone*

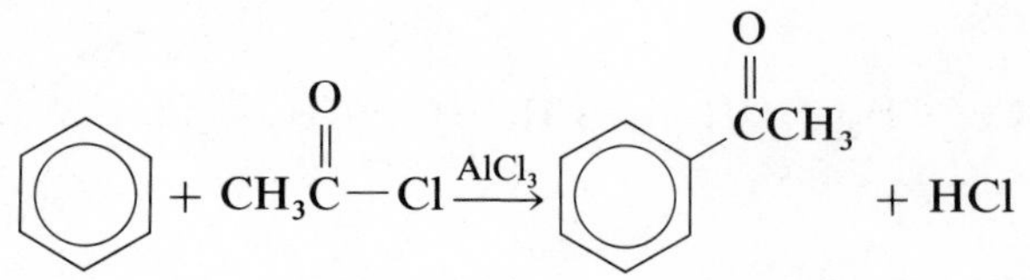

acetyl chloride   acetophenone

In this reaction the acid chloride may be thought of as transferring an acetyl group to the benzene molecule. Substances in the living system also transfer acetyl groups to other compounds. One such substance is acetyl coenzyme A. Acetyl coenzyme A is a substance that functions in reactions *in vivo* in which an acetyl group is transferred.

$$CH_3-\overset{\overset{\displaystyle O}{\|}}{C}-Cl \qquad\qquad CH_3-\overset{\overset{\displaystyle O}{\|}}{C}-S-CoA$$

acetyl chloride   acetyl coenzyme A
transfers acetyl groups in vitro   *transfers acetyl groups in vivo*

PROBLEM

3. Describe the reaction and write the structure of the product for each of the following.

(a) $CH_3CH_2CH_2OH \xrightarrow{\text{CrO}_3,\ \text{pyridine}}$

(b) 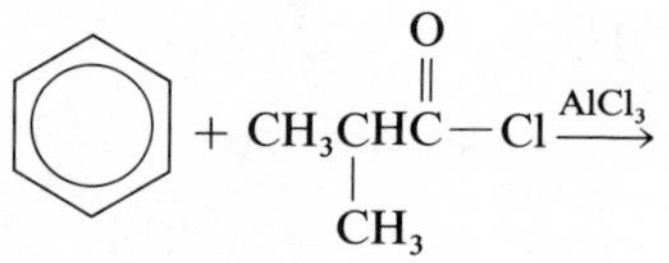 $+ CH_3\underset{\underset{\displaystyle CH_3}{|}}{CH}\overset{\overset{\displaystyle O}{\|}}{C}-Cl \xrightarrow{\text{AlCl}_3}$

(c) $CH_3CH_2\underset{\underset{\displaystyle CH_3}{|}}{C}HOH \xrightarrow{\text{K}_2\text{Cr}_2\text{O}_7,\ \text{H}_2\text{SO}_4}$

ANSWERS TO PROBLEM   **3. (a)** oxidation of a primary alcohol, product is $CH_3CH_2\overset{\overset{\displaystyle O}{\|}}{C}-H$

(b) an acylation of benzene, product is a ketone 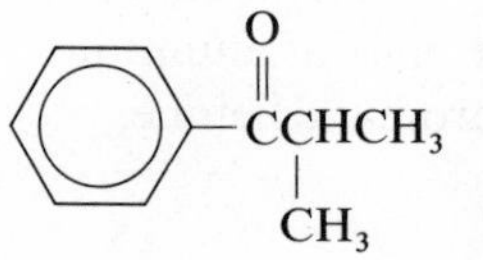 **(c)** oxidation of a

secondary alcohol, product is a ketone, $CH_3CH_2\overset{\overset{\displaystyle O}{\|}}{C}CH_3$

15-3 REACTIONS OF THE CARBONYL GROUP

LEARNING OBJECTIVES

1. Given the reaction between a carbonyl compound and sodium cyanide, write the structure of the cyanohydrin formed.
2. Given the reaction between a carbonyl compound and an amine, write the structure of the imine formed.
3. Given the reaction between a carbonyl compound and an ammonia derivative, write the structure of the oxime, hydrazone, or phenylhydrazone formed.
4. Given the reaction between a carbonyl compound and an alcohol, write first the structure of the hemiacetal

formed and then the structure of the acetal formed.
5. Identify an alpha hydrogen and a beta hydroxyl group in a carbonyl compound.
6. Given a carbonyl compound with an alpha hydrogen, write the structure of the beta-hydroxycarbonyl compound formed in an aldol condensation.
7. Given the reduction reaction of a carbonyl compound, write the structure of the alcohol formed.
8. Given the oxidation reaction of an aldehyde, write the structure of the carboxylic acid formed.

15-3.1 ADDITION REACTIONS OF CARBONYL COMPOUNDS

The pi electrons in the bond between carbon and oxygen of the carbonyl group are not equally shared. Because oxygen is more electronegative than carbon, the pi bond is polarized, with the electrons attracted more by oxygen than by carbon (see Section 15-1). Consequently, the carbonyl carbon is relatively positive and susceptible to attack by nucleophiles. As shown, a nucleophile (Nu:$^-$) attacks the carbonyl carbon. The intermediate is unstable and so it picks up a proton from the solvent to form a neutral product.

$$Nu:^- + \ \underset{/}{\overset{\backslash}{C}}{=}O \longrightarrow Nu-\overset{|}{\underset{|}{C}}-O^- \xrightarrow{\text{solvent}} Nu-\overset{|}{\underset{|}{C}}-OH$$

To increase the electrophilic nature of the carbonyl carbon, reactions of aldehydes and ketones are often carried out in acidic solutions. The proton, supplied by the acid, attacks the oxygen of the carbonyl group, resulting in a species in which carbon bears the full positive charge.

$$\underset{/}{\overset{\backslash}{C}}{=}O \ + H^+ \longrightarrow \ ^+\overset{|}{\underset{|}{C}}-O-H$$

Nucleophilic reagents are then easily attracted to the electrophilic carbon. The overall reaction of carbonyl compounds is referred to as *nucleophilic addition*.

$$Nu:^- + \ ^+\overset{|}{\underset{|}{C}}-O-H \longrightarrow Nu-\overset{|}{\underset{|}{C}}-OH$$

$$\underset{/}{\overset{\backslash}{C}}{\overset{\pi}{\underset{\sigma}{=}}}O$$

carbonyl group

$$\underset{/}{\overset{\backslash\delta+}{C}}{=}\overset{\delta-}{O}$$

polarized carbonyl group

In this section we will consider four important nucleophilic addition reactions of carbonyl compounds and, in particular, their relation to the chemistry of biomolecules. These reagents include (1) aqueous sodium cyanide in acid solution, (2) primary amines, (3) reagents that are classified as derivatives of ammonia, and (4) alcohols.

Nucleophilic Addition of Aqueous Sodium Cyanide

The addition of aqueous sodium cyanide (NaCN) in acid solution to a carbonyl compound gives a cyanohydrin. The cyanohydrin is formed by the attack of the nucleophilic cyanide ($^-$:C≡N) on the protonated carbonyl group ($^+$C—OH).

A **cyanohydrin** is an organic compound derived from a carbonyl compound and has the structure

$$\begin{array}{c} OH \\ | \\ -C-CN. \\ | \end{array}$$

$$\begin{array}{ccc} \diagdown & & OH \\ C=O + NaCN & \xrightarrow{H^+} & | \\ \diagup & & -C-CN \\ & & | \end{array}$$

a carbonyl compound   *a cyanohydrin*

benzaldehyde   benzaldehyde cyanohydrin

$$\begin{array}{ccc} O & & OH \\ \| & & | \\ CH_3C-H + NaCN & \xrightarrow{H^+} & CH_3CHCN \end{array}$$

acetaldehyde   acetaldehyde cyanohydrin

Cyanohydrins occur in nature. For example, the cyanohydrin of benzaldehyde is the principal component of the defense secretion of certain millipedes. An approaching enemy is sprayed with this compound and in the process the spray is mixed with a catalyst that rapidly decomposes the cyanohydrin into hydrogen cyanide and benzaldehyde. Because hydrogen cyanide is toxic, most predators will retreat.

benzaldehyde   benzaldehyde   hydrogen cyanide
cyanohydrin   *toxic*

The cyanohydrin of benzaldehyde also occurs in nature in some fruit seeds such as cherries, plums, peaches, and almonds. These seeds contain the benzaldehyde cyanohydrin as a derivative of glucose in a compound known as amygdalin. The structure of amygdalin is shown.

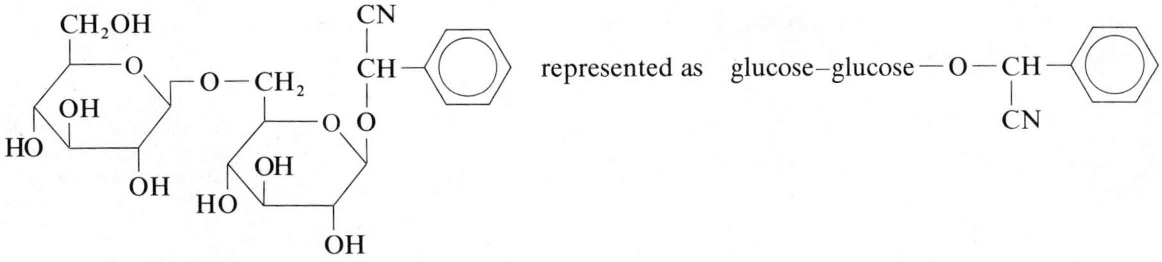

There have been incidences of persons chewing these fruit seeds and suddenly becoming cyanotic. Apparently what happens is that the amygdalin is enzymatically degraded to glucose and benzaldehyde cyanohydrin. The latter is further decomposed enzymatically to benzaldhyde and toxic hydrogen cyanide.

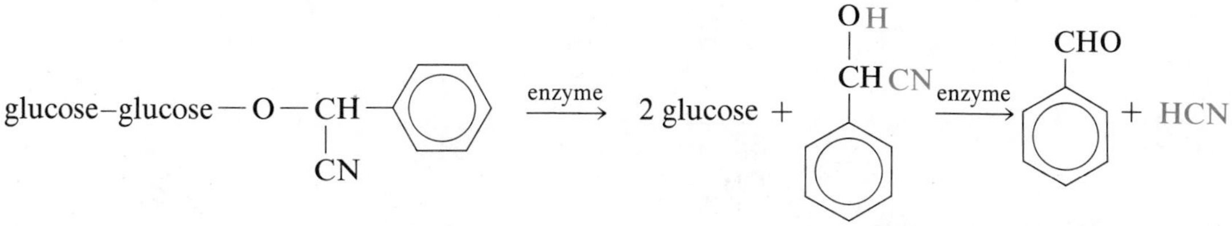

amygdalin
short-hand notation

benzaldehyde
cyanohydrin

benzaldehyde

A final example of a cyanohydrin is the controversial compound Laetrile. Laetrile has the structure shown and consists of a glucose derivative, benzaldehyde, and cyanide. Some Laetrile users claim that the drug is effective in the treatment of cancer. However, the distribution of Laetrile for cancer treatment never has been legal in this country in spite of the large number of Americans now using it. Proponents of Laetrile argue that it not only relieves the physical pain associated with cancer but also offers a cure. The fact is that there is no scientific evidence that Laetrile is chemically effective or that it has antitumor effects.

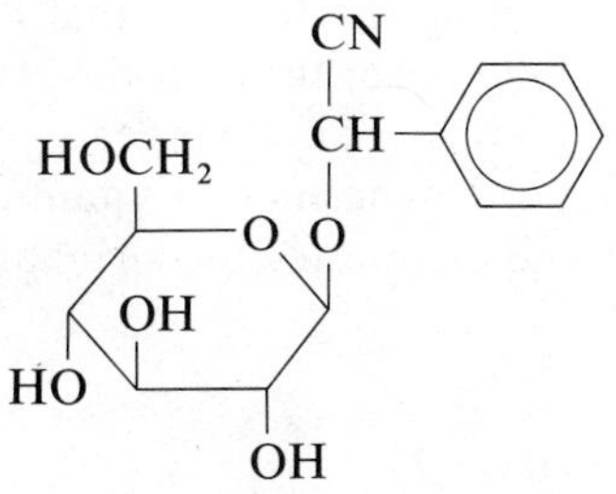

Laetrile

Nucleophilic Addition of Primary Amines

Primary amines (R$\ddot{\text{N}}$H$_2$) like cyanide anions, are reactive nucleophiles and readily attack the carbonyl carbon atom of an aldehyde and a ketone. The product of this reaction is an imine. These compounds are identified by their carbon–nitrogen double bond and result from the loss of water by the initial adduct.

A **primary amine** is an organic compound derived from ammonia by the replacement of one hydrogen atom by a carbon-containing group.

An **imine** is an organic compound derived from a carbonyl compound and characterized by a carbon–nitrogen double bond. It has the structure
$$-\overset{|}{\text{C}}=\text{N}-\text{R}.$$

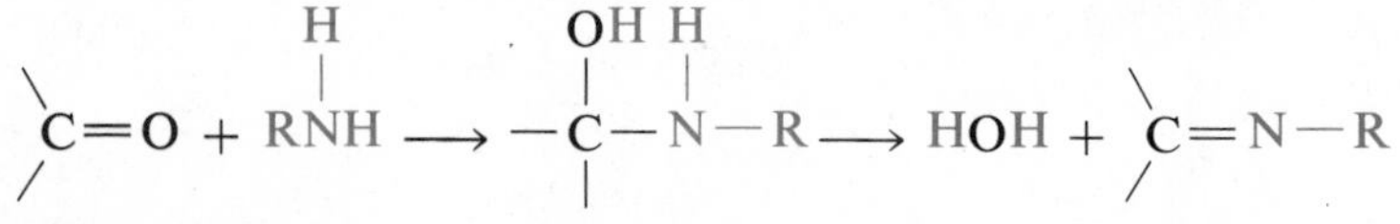

$$\underset{\substack{a\ carbonyl \\ compound}}{\overset{}{\text{C}}=\text{O}} + \underset{\substack{a\ primary \\ amine}}{\text{RNH}} \longrightarrow -\overset{\text{OH}}{\underset{}{\text{C}}}-\overset{\text{H}}{\underset{}{\text{N}}}-\text{R} \longrightarrow \text{HOH} + \underset{an\ imine}{\overset{}{\text{C}}=\text{N}-\text{R}}$$

$$\underset{acetaldehyde}{\text{CH}_3\overset{\text{O}}{\overset{||}{\text{C}}}-\text{H}} + \underset{ethylamine}{\text{CH}_3\text{CH}_2\text{NH}_2} \longrightarrow \text{CH}_3\overset{\text{OH}}{\underset{\text{H}}{\text{C}}}-\text{NCH}_2\text{CH}_3 \longrightarrow \text{HOH} + \underset{an\ imine}{\text{CH}_3\overset{}{\text{C}}=\text{NCH}_2\text{CH}_3}$$

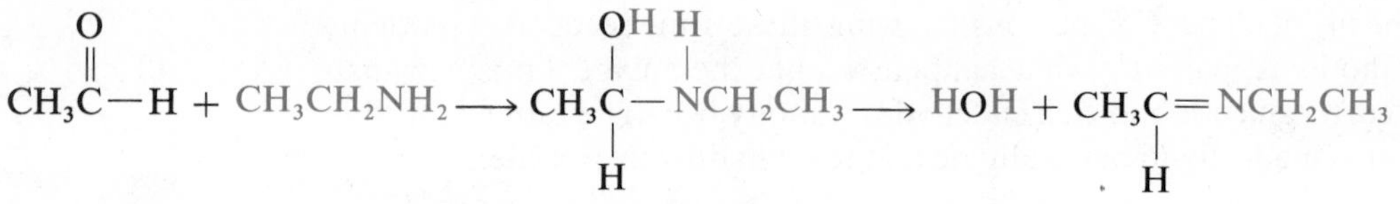

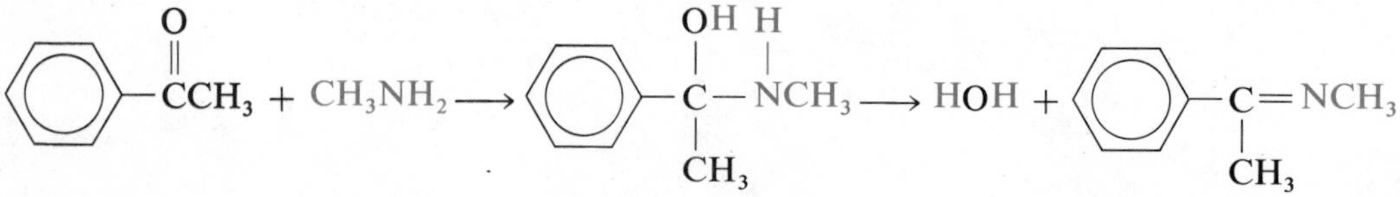

$$\underset{acetophenone}{\bigcirc-\overset{\text{O}}{\overset{||}{\text{C}}}\text{CH}_3} + \underset{methylamine}{\text{CH}_3\text{NH}_2} \longrightarrow \bigcirc-\overset{\text{OH}}{\underset{\text{CH}_3}{\text{C}}}-\text{NCH}_3 \longrightarrow \text{HOH} + \underset{an\ imine}{\bigcirc-\overset{}{\text{C}}=\text{NCH}_3}$$

An imine is known to be involved in the chemistry of vision. The reactions occurring in the vision sequence are summarized in Figure 15.1. The light sensitive pigment in the rods, rhodopsin, is an imine of the enzyme opsin and cis retinal. This imine is thought to be formed between an amino group in the opsin and the aldehyde group in cis retinal in the manner shown in the examples above and as illustrated in step 1 in Figure 15.1. The geometry of the resulting cis isomer is such that it fits tightly into a cavity on the rhodopsin.

When light strikes rhodopsin, the cis isomer becomes a trans isomer as shown in step 2. The trans isomer is not stable with respect to forming a good fit in the enzyme cavity and dissociates, releasing trans retinal and opsin (step 3). As this dissociation occurs a message is sent to the brain indicating that light has been detected. A series of chemical reactions occur, converting the trans retinal back to the cis retinal (step 4). The cis retinal rebinds to the opsin through the imine and forms again the light sensitive rhodopsin (step 1). Thus, the cycle repeats itself.

cis Retinal is formed in the body from vitamin A. A deficiency of vitamin A leads to a condition of nightblindness. Nightblindness is manifested by the inability of a person to see in darkness.

Nucleophilic Addition of Some Derivatives of Ammonia

Another important nucleophilic addition reaction of carbonyl compounds is that involving reagents considered to be derivatives of ammonia (NH$_3$). The derivatives

Figure 15.1 The chemistry of vision. (Adapted from *Contemporary Organic Chemistry* by Marion H. O'Leary, p. 50. Copyright © 1976 by McGraw-Hill, Inc. Used with permission of McGraw-Hill Book Company.

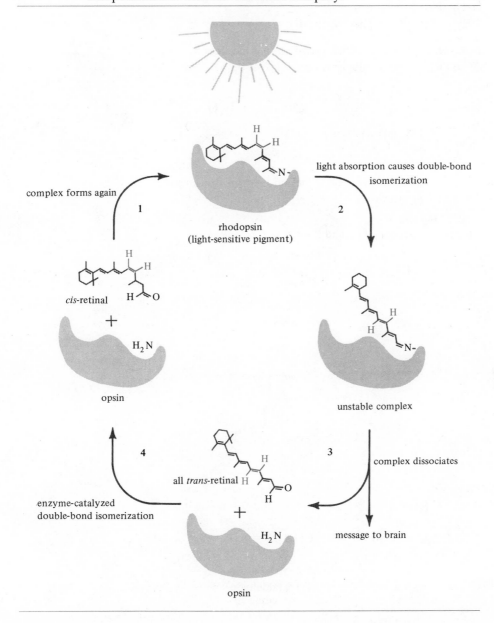

of ammonia considered here are those in which one of the hydrogen atoms is replaced by another group or grouping, G. Of interest to us are hydroxylamine, hydrazine, and phenylhydrazine. The nucleophilic addition these reagents undergo with carbonyl compounds is catalyzed by acid. It is the nitrogen atom with the

$\overset{\cdot\cdot}{N}H_2-G$	Name
$\overset{\cdot\cdot}{N}H_2-OH$	hydroxylamine
$\overset{\cdot\cdot}{N}H_2-NH_2$	hydrazine
$\overset{\cdot\cdot}{N}H_2-NHC_6H_5$	phenylhydrazine

unshared electron pair shown in the structures above that is nucleophilic. The initial adduct is unstable and loses a molecule of water as shown below.

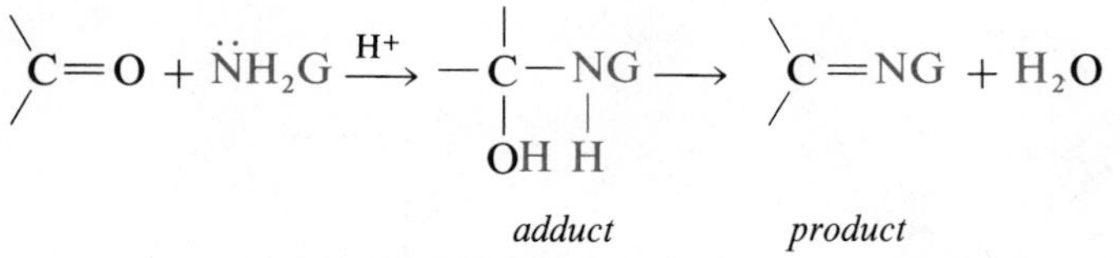

$$\overset{\diagdown}{\underset{\diagup}{}}C=O + \overset{\cdot\cdot}{N}H_2G \xrightarrow{H^+} -\overset{|}{\underset{|}{C}}-NG \longrightarrow \overset{\diagdown}{\underset{\diagup}{}}C=NG + H_2O$$
$$\qquad\qquad\qquad\qquad\quad OH\ H$$

$$\qquad\qquad\quad adduct \qquad\qquad product$$

The product of the reaction has a unique name depending on the ammonia derivative used in the reaction. For example,

if the ammonia derivative is	then	the structure of the product is	and	the product is called
NH_2OH		$\overset{\diagdown}{\underset{\diagup}{}}C=NOH$		an oxime
NH_2NH_2		$\cdot\ \overset{\diagdown}{\underset{\diagup}{}}C=NNH_2$		a hydrazone
$NH_2NHC_6H_5$		$\overset{\diagdown}{\underset{\diagup}{}}C=NNHC_6H_5$		a phenylhydrazone

Some examples are:

$$\overset{O}{\overset{\|}{CH_3C}}-H \quad + \quad H_2NOH \xrightarrow{H^+}$$

acetaldehyde   hydroxylamine

$$\overset{OH\ H}{\overset{|\quad|}{CH_3C}}-NOH \longrightarrow CH_3CH=NOH \quad + \quad H_2O$$
$$\overset{|}{H}$$

$$\qquad\quad adduct \qquad\qquad\qquad\qquad acetaldehyde$$
$$\qquad\qquad\qquad\qquad\qquad\qquad\quad oxime$$

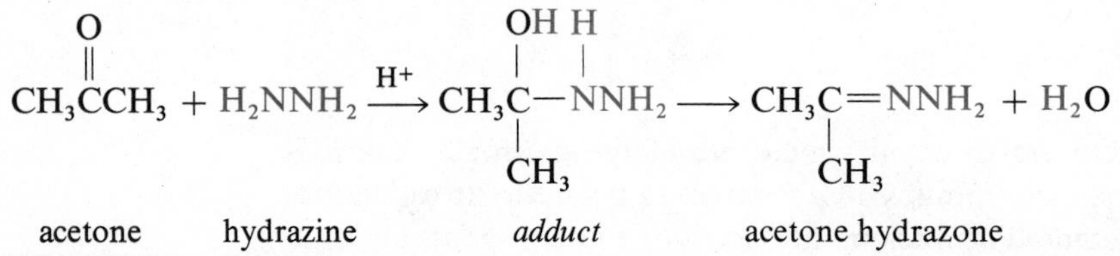

$$\overset{O}{\overset{\|}{CH_3CCH_3}} + H_2NNH_2 \xrightarrow{H^+} \overset{OH\ H}{\overset{|\quad|}{CH_3C}}-NNH_2 \longrightarrow CH_3C=NNH_2 + H_2O$$
$$\qquad\qquad\qquad\qquad\qquad\quad \overset{|}{CH_3} \qquad\qquad \overset{|}{CH_3}$$

acetone   hydrazine   adduct   acetone hydrazone

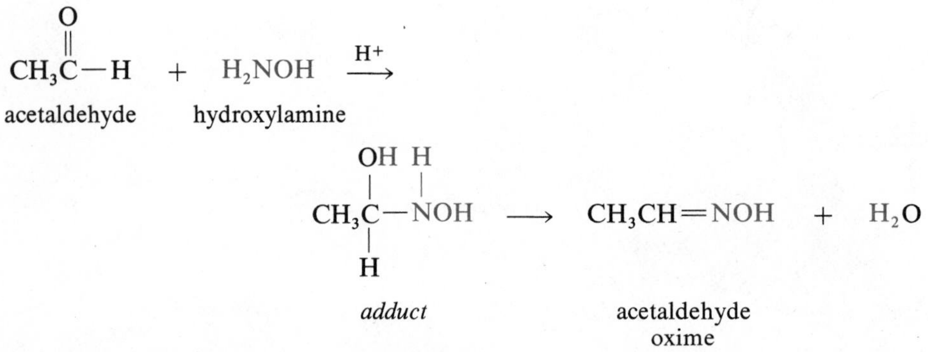

$$
\underset{\text{2-butanone}}{CH_3CH_2\overset{\overset{\textstyle O}{\|}}{C}CH_3} + \underset{\text{phenylhydrazine}}{H_2NNHC_6H_5} \xrightarrow{H^+} \underset{\text{adduct}}{CH_3CH_2\overset{\overset{\textstyle OH}{|}}{\underset{\underset{\textstyle CH_3}{|}}{C}}-NHNHC_6H_5} \longrightarrow
$$

$$
\underset{\text{2-butanone phenylhydrazone}}{CH_3CH_2\overset{}{\underset{\underset{\textstyle CH_3}{|}}{C}}=NNHC_6H_5} + H_2O
$$

The phenylhydrazones formed from the reactions of carbonyl compounds with phenylhydrazine are bright yellow, crystalline compounds. These crystalline compounds often serve as a means of identifying a particular carbonyl compound because of its characteristic melting point.

Nucleophilic Addition of Alcohols

A fourth addition reaction of carbonyl compounds is the addition reaction that occurs with alcohols. One mole of alcohol in the presence of anhydrous HCl reacts with an aldehyde to form a **hemiacetal** and two moles of alcohol react with an aldehyde to form an **acetal**. The reaction of an aldehyde with an alcohol first forms a hemiacetal. The reaction is the typical reaction of an aldehyde: *nucleophilic addition*. The hemiacetal then reacts with another alcohol to form an acetal. The reaction in this step is a *substitution reaction*.

A **hemiacetal** is an organic compound derived from a carbonyl compound and has the structure $-\overset{\overset{\textstyle OH}{|}}{\underset{|}{C}}-OR$.

An **acetal** is an organic compound derived from a carbonyl compound and has the structure $-\overset{\overset{\textstyle OR}{|}}{\underset{|}{C}}-OR$.

$$
\underset{\text{an aldehyde}}{R-\overset{\overset{\textstyle O}{\|}}{C}-H} + R'OH \xrightarrow[\substack{\text{an addition} \\ \text{reaction}}]{\text{dry HCl}} \underset{\text{a hemiacetal}}{R-\overset{\overset{\textstyle OH}{|}}{\underset{\underset{\textstyle H}{|}}{C}}-OR'} \xrightarrow[\substack{\text{a substitution} \\ \text{reaction}}]{R'OH} \underset{\text{an acetal}}{R-\overset{\overset{\textstyle OR'}{|}}{\underset{\underset{\textstyle H}{|}}{C}}-OR'} + HOH
$$

Ketones react in a similar fashion to form hemiacetals and acetals as shown.

$$
\underset{\text{a ketone}}{R-\overset{\overset{\textstyle O}{\|}}{C}-R} + R'OH \xrightarrow[\substack{\text{an addition} \\ \text{reaction}}]{\text{dry HCl}} \underset{\text{a hemiacetal}}{R-\overset{\overset{\textstyle OH}{|}}{\underset{\underset{\textstyle R}{|}}{C}}-OR'} \xrightarrow[\substack{\text{a substitution} \\ \text{reaction}}]{R'OH} \underset{\text{an acetal}}{R-\overset{\overset{\textstyle OR'}{|}}{\underset{\underset{\textstyle R}{|}}{C}}-OR'} + HOH
$$

In a hemiacetal the original carbonyl carbon is bonded to a hydroxyl and an alkoxyl group ($-OR$). In an acetal the original carbonyl carbon is bonded to two alkoxyl groups. The alkyl groups of the alkoxyl group arise from the alcohol. Examples shown are the hemiacetal and acetal from acetaldehyde and methanol and the acetal from acetone and methyl alcohol.

$$CH_3\overset{\overset{O}{\|}}{C}-H + CH_3OH \xrightarrow{\text{dry HCl}} CH_3\overset{\overset{OH}{|}}{\underset{|}{C}}-OCH_3 \xrightarrow{CH_3OH} CH_3\overset{\overset{OCH_3}{|}}{\underset{|}{C}}-OCH_3 + HOH$$
$$\qquad\qquad\qquad\qquad\quad H\qquad\qquad\qquad\qquad H$$

acetaldehyde   methanol   methyl hemiacetal   dimethylacetal
                       of acetaldehyde   of acetaldehyde

$$CH_3\overset{\overset{O}{\|}}{C}CH_3 + 2\,CH_3OH \xrightarrow{\text{dry HCl}} CH_3\overset{\overset{OCH_3}{|}}{\underset{\underset{CH_3}{|}}{C}}OCH_3 + H_2O$$

acetone   methanol   dimethylacetal of acetone

The reaction of an aldehyde with an alcohol such as methanol actually provides us with chemical insight into the nature of the structure of simple carbohydrates such as glucose and other **monosaccharides**. A simple aldehyde will react with two moles of methanol to form an acetal. Glucose reacts with a single mole of methanol to form an acetal. This is interpreted to mean that glucose already exists as a hemiacetal.

A **monosaccharide** is a polyhydroxy aldehyde or ketone.

Glucose forms a hemiacetal through the reaction of the aldehyde group and the C-5 hydroxy group. A reaction such as this, which takes place between groups within the same molecule, is called an *intramolecular reaction.*

glucose   hemiacetal of   hemiacetal
a polyhydroxy aldehyde   glucose   grouping

Glucose, in the hemiacetal form, can now react with an alcohol to form an acetal as shown.

glucose   acetal of   acetal
                           glucose   grouping

Sucrose, a carbohydrate, has an acetal grouping rather than a hemiacetal grouping. This distinction between glucose and sucrose will be discussed further in Chapter 19 because it serves as a basis for distinguishing the two carbohydrates.

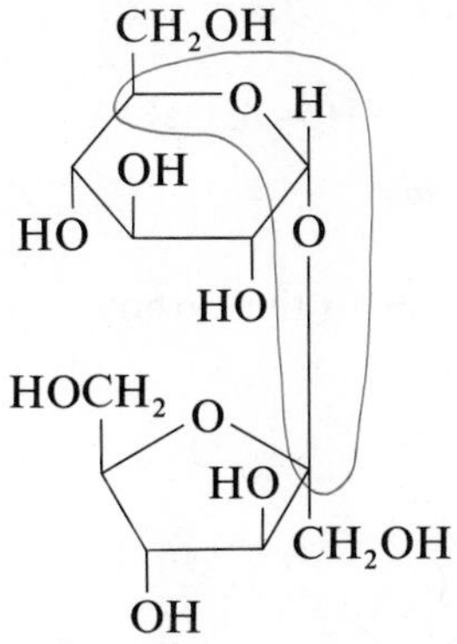

sucrose

15-3.2 THE ALDOL CONDENSATION

The aldol condensation is a base-catalyzed reaction of a carbonyl compound that possesses an **alpha hydrogen**. Under the influence of a base, usually dilute sodium

The **alpha (α) hydrogen** is the hydrogen atom on the carbon directly bonded to the carbonyl carbon.

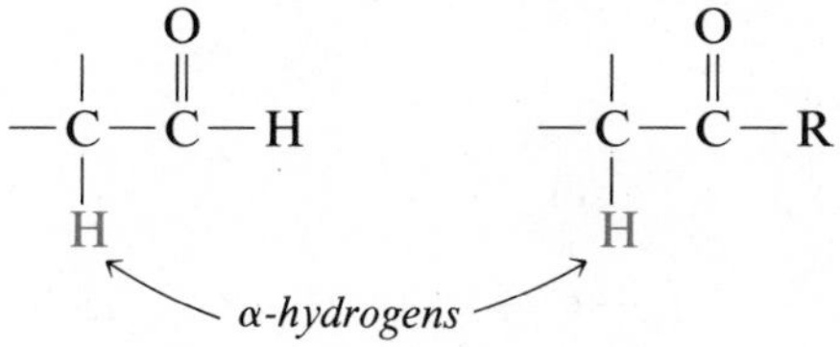

α-hydrogens

hydroxide, a carbonyl compound with at least one alpha hydrogen will condense with itself or another carbonyl compound. The product of such a reaction is a **β-hydroxyaldehyde** or a **β-hydroxyketone**.

A **β-hydroxyaldehyde** or a **β-hydroxyketone** is an aldehyde or ketone in which the hydroxyl group is attached to the number three carbon with the carbonyl carbon designated as C-1.

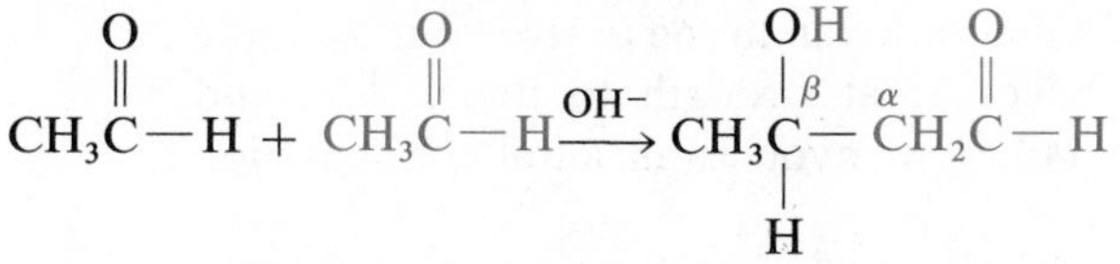

acetaldehyde β-hydroxybutyraldehyde
a β-hydroxyaldehyde

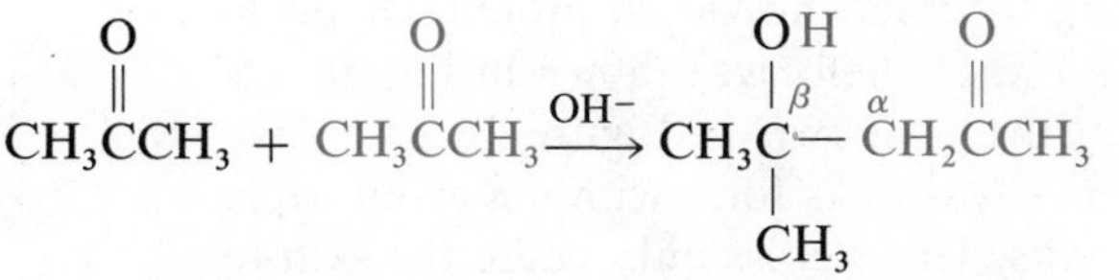

acetone   4-hydroxy-4-methyl-2-pentanone
a β-hydroxyketone

The aldol condensation of acetaldehyde with itself occurs in several steps. The first step involves the abstraction of the α-hydrogen by the base (OH⁻), resulting in the formation of an intermediate called a **carbanion.**

A **carbanion** is a carbon-containing species in which carbon has an unshared pair of electrons and bears a negative charge.

$$CH_2-\overset{\overset{\displaystyle O}{\|}}{C}-H + OH^- \longrightarrow \; ^-:CH_2-\overset{\overset{\displaystyle O}{\|}}{C}-H + HOH \qquad (15\text{-}1)$$
$$\underset{\displaystyle H}{|}$$

a carbanion

In the second step, the nucleophilic carbanion adds to the carbonyl group of another acetaldehyde molecule.

$$^-:CH_2-\overset{\overset{\displaystyle O}{\|}}{C}-H + CH_3-\overset{\overset{\displaystyle O}{\|}}{C}-H \longrightarrow CH_3-\overset{\overset{\displaystyle O^-}{|}}{\underset{\underset{\displaystyle H}{|}}{C}}-CH_2-\overset{\overset{\displaystyle O}{\|}}{C}-H \qquad (15\text{-}2)$$

Finally, the intermediate species from the previous step removes a proton from the solvent (water), forming the β-hydroxy carbonyl compound. Thus, the aldol condensation is really another example of the characteristic reaction of a carbonyl compound: *nucleophilic addition.*

$$CH_3-\overset{\overset{\displaystyle O^-}{|}}{\underset{\underset{\displaystyle H}{|}}{C}}-CH_2-\overset{\overset{\displaystyle O}{\|}}{C}-H + HOH \longrightarrow$$

$$CH_3-\overset{\overset{\displaystyle OH}{|}}{\underset{\underset{\displaystyle H}{|}}{C}}-CH_2-\overset{\overset{\displaystyle O}{\|}}{C}-H + OH^- \qquad (15\text{-}3)$$

β-hydroxybutyraldehyde

Let us now look at an aldol condensation that occurs in the living system. This will also be a good place to review the formation of an imine *in vivo.* Our example concerns the protein collagen, which gives great strength to bone, skin, and cartilage. The strength associated with collagen involves an aldol condensation and the formation of an imine.

More than 30% of the protein in the human body is collagen. Collagen gives structure to many tissues. The collagen content is especially high in structural components such as bone (88%), tendons (86%), and skin (72%). Collagen consists of a triple helix that is formed by twisting together individual protein chains as shown in Figure 15.2(a). Remember that a single helix was shown in Figure 5.5. Cross links form between the protein chains as shown in Figure 15.2(b). Two kinds of cross links are formed in collagen. Cross links form within a given triple helix and between different triple helix units. These cross links cause the skin to stiffen and wrinkle. They also cause other tissues like blood vessels to stiffen. The

Figure 15.2   (a) Triple helix such as is found in collagen. (b) Triple helix with cross links within each triple helix unit and between different triple helix units.

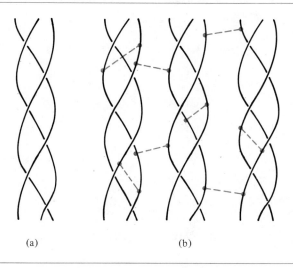

(a)   (b)

number of cross links increases with age and causes some of the medical problems of older persons.

It is the cross links (actually, covalent bonds) that form between particular residues between protein chains that we want to consider here. The cross links form because an enzyme forms aldehyde residues on collagen molecules as shown.

$$
\begin{array}{c}
\qquad\qquad\;\;\overset{\displaystyle O}{\underset{\displaystyle \|}{}} \\
-NH-CH-C\!\sim\!\text{protein}\!\sim \\
\quad\;\;\; | \\
\quad\;\;\; CH_2 \\
\quad\;\;\; | \\
\quad\;\;\; CH_2 \\
\quad\;\;\; | \\
\quad\;\;\; CH_2 \\
\quad\;\;\; | \\
\quad\;\;\; CH_2 \\
\quad\;\;\; | \\
\quad\;\;\; NH_2
\end{array}
\qquad\longrightarrow\qquad
\begin{array}{c}
\qquad\qquad\;\;\overset{\displaystyle O}{\underset{\displaystyle \|}{}} \\
-NH-CH-C\!\sim\!\text{protein}\!\sim \\
\quad\;\;\; | \\
\quad\;\;\; CH_2 \\
\quad\;\;\; | \\
\quad\;\;\; CH_2 \\
\quad\;\;\; | \\
\quad\;\;\; CH_2 \\
\quad\;\;\; | \\
\quad\;\;\; HC\!=\!O
\end{array}
$$

(collagen)   (collagen)

These aldehyde residues undergo the two reactions we have discussed in this chapter: aldol condensation and imine formation. As shown in Figure 15.3(a), the aldehyde residues react with each other to form the product of an aldol condensation, namely, a β-hydroxyaldehyde. They may also react with an amine to form an imine, as shown in Figure 15.3(b). The formation of these cross links and the structure imparted by the triple helix are responsible for the properties of collagen: strength, stretch resistance, and rigidity.

Figure 15.3   (a) Aldol condensation between two aldehyde residues. (b) Imine formation between amine and aldehyde residues.

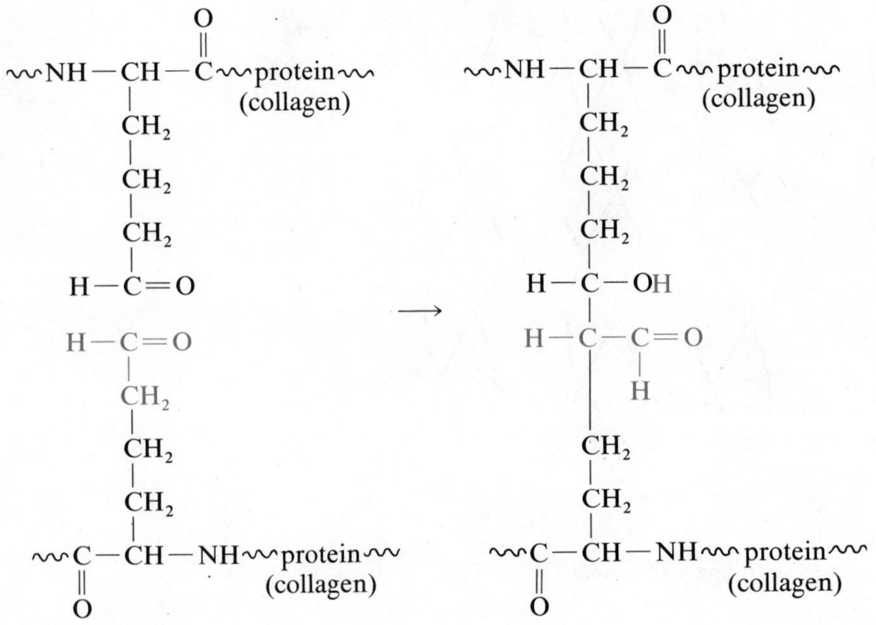

(a)

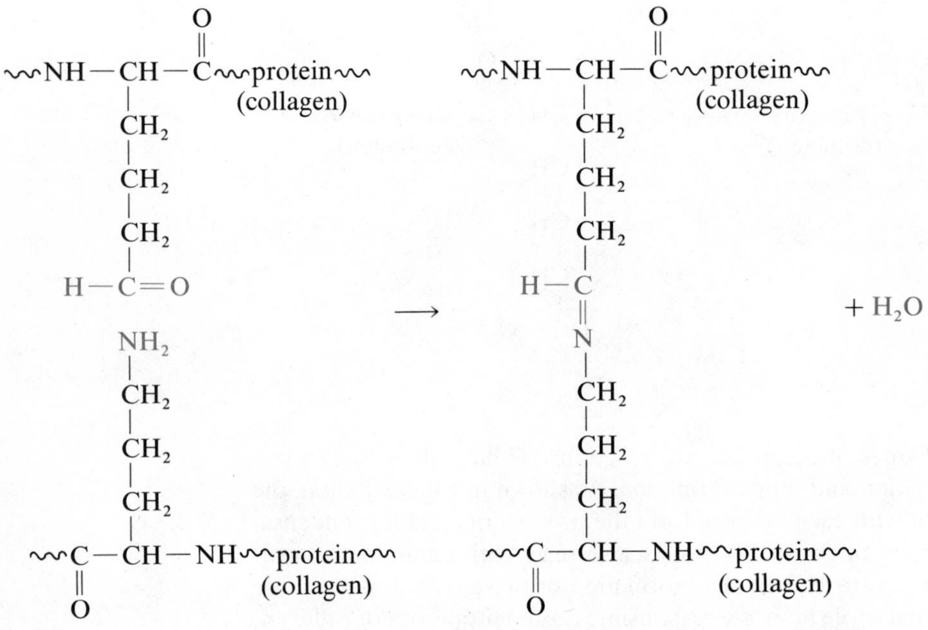

(b)

15-3.3 REDUCTION OF CARBONYL COMPOUNDS

The reduction product of a carbonyl compound is an alcohol. The class of alcohol formed depends upon the carbonyl compound reduced. If an aldehyde is reduced, a primary alcohol is the product. If a ketone is reduced, a secondary alcohol is the product. This reduction reaction is just the reverse of the oxidation reaction of alcohols that we discussed in Section 14-3.5.

$$R-\overset{\overset{\displaystyle O}{\|}}{C}-H \xrightarrow{\text{reduction}} R-CH_2OH$$

an aldehyde   *a primary alcohol*

$$R-\overset{\overset{\displaystyle O}{\|}}{C}-R \xrightarrow{\text{reduction}} R-\overset{\overset{\displaystyle OH}{|}}{\underset{\underset{\displaystyle H}{|}}{C}}-R$$

a ketone   *a secondary alcohol*

The overall reduction may be considered to be the addition of hydrogen to a pi bond. Reduction of carbonyl compounds may be accomplished by a number of reagents; for example, hydrogen in the presence of a nickel catalyst, lithium aluminum hydride ($LiAlH_4$), or sodium borohydride ($NaBH_4$).

$$CH_3\underset{\underset{\displaystyle CH_3}{|}}{CH}\overset{\overset{\displaystyle O}{\|}}{C}-H + H_2 \xrightarrow{Ni} CH_3\underset{\underset{\displaystyle CH_3}{|}}{CH}CH_2OH$$

isobutyraldehyde   isobutyl alcohol
an aldehyde   *a primary alcohol*

$$\text{acetophenone} \xrightarrow{LiAlH_4} \text{1-phenylethanol}$$

acetophenone   1-phenylethanol
a ketone   *a secondary alcohol*

$$\text{glucose} + NaBH_4 \longrightarrow \text{glucitol}$$

glucose   glucitol
an aldehyde   *a primary alcohol*
used as a moisture retainer
in tobacco

An example of a biochemical process involving the reduction of a carbonyl compound occurs in fermentation (see Section 14-2.1). In this reaction, acetaldehyde is reduced to ethyl alcohol with NADH (the reduced form of NAD^+ as shown in Section 12-5.2).

| acetaldehyde | NADH | NAD+ | ethyl alcohol |
| | *reducing agent* | | |

The reduction of a carbonyl group by lithium aluminum hydride, sodium borohydride, and NADH are similar mechanistically. Each reaction is a nucleophilic addition. The addition involves the transfer of a hydrogen with a pair of electrons, such as the hydride ion ($H:^-$), from the reducing agent to the carbonyl carbon.

Lithium aluminum hydride and sodium borohydride are metal hydrides. We can represent these substances by the general formula M-H. The metal hydride acts as a hydride donor and the carbonyl compound is the acceptor of the hydride ion. This reaction results in the formation of the salt of an alcohol.

a salt of an alcohol

In the presence of water the alcohol salt picks up a proton to form an alcohol. The overall reaction may be thought of as the addition of the elements of hydrogen to the carbonyl group.

| *alcohol salt* | *alcohol* |

The reduction of a carbonyl compound by NADH is similar to the reduction of a carbonyl compound by M-H. After the initial protonation of the carbonyl oxygen, a hydride ion from NADH is transferred to the electrophilic carbonyl carbon.

acetaldehyde   ethyl alcohol

15-3.4 OXIDATION OF ALDEHYDES

Aldehydes, but not ketones, are oxidized to carboxylic acids by several oxidizing agents, generally: (1) Tollens' reagent [silver nitrate ($AgNO_3$) dissolved in aqueous ammonia], (2) Benedict's solution [an alkaline solution of cupric sulfate ($CuSO_4$), sodium citrate, and sodium carbonate], or (3) potassium dichromate (see Section 14-3.5).

$$R - \overset{\overset{\text{O}}{\|}}{C} - H \xrightarrow{\text{oxidizing agent}} R - \overset{\overset{\text{O}}{\|}}{C} - OH$$

an aldehyde   *a carboxylic acid*

These oxidizing media have the advantage of producing an observable result. In the case of *Tollens' reagent,* a silver mirror is formed on the inner walls of the vessel in which the reaction is performed. A red precipitate of cuprous oxide (Cu_2O) is formed with *Benedict's reagent.*

$$R - \overset{\overset{\text{O}}{\|}}{C} - H \xrightarrow{\textit{Tollens' reagent: } AgNO_3 + NH_3 \text{ (aq)}} R - \overset{\overset{\text{O}}{\|}}{C} - OH + Ag$$

$$R - \overset{\overset{\text{O}}{\|}}{C} - H \xrightarrow{\textit{Benedict's reagent: } Cu^{2+}, \text{ citrate, } OH^-} R - \overset{\overset{\text{O}}{\|}}{C} - OH + Cu_2O$$

For example,

$$CH_3CH_2CH_2\overset{\overset{\text{O}}{\|}}{C} - H \xrightarrow{\text{Tollens' reagent}} CH_3CH_2CH_2\overset{\overset{\text{O}}{\|}}{C} - OH + Ag$$

butyraldehyde   butyric acid

$$CH_3\underset{\underset{\text{CH}_3}{|}}{CH}\overset{\overset{\text{O}}{\|}}{C} - H \xrightarrow{\text{Benedict's reagent}} CH_3\underset{\underset{\text{CH}_3}{|}}{CH}\overset{\overset{\text{O}}{\|}}{C} - OH + Cu_2O$$

isobutyraldehyde   isobutyric acid

Tollens' or Benedict's reagent is often used as a means of identifying certain carbohydrates. Glucose, a polyhydroxy aldehyde, gives a positive Tollens' test and Benedict's test.

$$
\begin{array}{l}
\text{H} - \text{C} = \text{O} \\
\,\,\,|\\
\text{H} - \text{C} - \text{OH} \\
\,\,\,|\\
\text{HO} - \text{C} - \text{H} \\
\,\,\,|\\
\text{H} - \text{C} - \text{OH} \\
\,\,\,|\\
\text{H} - \text{C} - \text{OH} \\
\,\,\,|\\
\text{CH}_2\text{OH}
\end{array}
\xrightarrow{\text{Tollens' reagent or Benedict's reagent}}
\begin{array}{l}
\overset{\overset{\text{O}}{\|}}{\text{C}} - \text{OH} \\
\,\,\,|\\
\text{H} - \text{C} - \text{OH} \\
\,\,\,|\\
\text{HO} - \text{C} - \text{H} \\
\,\,\,|\\
\text{H} - \text{C} - \text{OH} \\
\,\,\,|\\
\text{H} - \text{C} - \text{OH} \\
\,\,\,|\\
\text{CH}_2\text{OH}
\end{array}
+ \text{Ag or } Cu_2O
$$

glucose   gluconic acid

There are many different kinds of reducing substances in urine. Glucose, a reducing sugar, is one that has pathological significance. A healthy, fasting individual has no glucose in the urine. The presence of glucose in the urine is an indication of abnormal glucose metabolism such as occurs with diabetes. Clinically the test for glucose commonly is performed with a BiliLabstix reagent strip although other reagent strips and tablets are available. BiliLabstix is a firm, plastic strip to which are affixed six separate reagent areas for testing for pH, protein, glucose, ketones, bilirubin (a red pigment present in bile, a fluid excreted by the liver), and blood in urine. The reagent test areas on the strip are ready for use from the bottle. The BiliLabstix is used with a dip-and-read technique. The color that develops on the strip after a given time is compared to a color chart provided by the manufacturer. This test is capable of detecting 0.1 g of glucose per 100 mL of urine and no substance excreted in the urine other than glucose is known to give a positive result. The chemistry involved in this test involves two separate enzyme reactions and is discussed further in Section 23-4.

PROBLEM

4. Describe the following reactions and write the structure of the product.

$$\text{(a)} \quad \underset{\displaystyle \overset{\displaystyle \text{O}}{\underset{\displaystyle \|}{}}}{\text{CH}_3\text{CH}_2\text{C}}\text{—H} + \text{NaBH}_4 \longrightarrow$$

$$\text{(b)} \quad \underset{\displaystyle \overset{\displaystyle \text{O}}{\underset{\displaystyle \|}{}}}{\text{CH}_3\text{CCH}_3} + \text{CH}_3\text{NH}_2 \longrightarrow$$

(c) 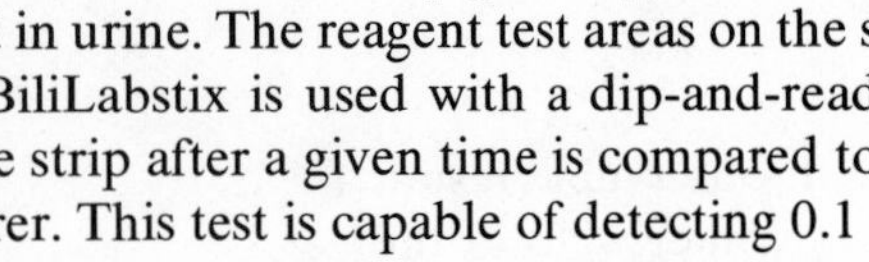 ⟨benzene ring⟩—$\overset{\displaystyle \text{O}}{\overset{\displaystyle \|}{\text{C}}}$—H + $\text{C}_6\text{H}_5\text{NHNH}_2 \xrightarrow{\text{H}^+}$

$$\text{(d)} \quad \underset{\displaystyle \overset{\displaystyle \text{O}}{\underset{\displaystyle \|}{}}}{\text{CH}_3\text{CCH}_3} + \text{dilute NaOH} \longrightarrow$$

(e) $\text{CH}_3\overset{\displaystyle \text{O}}{\underset{\displaystyle \underset{\displaystyle \text{CH}_3}{|}}{\overset{\displaystyle \|}{\text{CHC}}}}$—H + Tollens' reagent $\longrightarrow$

$$\text{(f)} \quad \underset{\displaystyle \overset{\displaystyle \text{O}}{\underset{\displaystyle \|}{}}}{\text{CH}_3\text{CH}_2\text{CCH}_3} + \text{excess CH}_3\text{OH} \longrightarrow$$

ANSWERS TO PROBLEM   4. (a) reduction of an aldehyde; product is a primary alcohol, $\text{CH}_3\text{CH}_2\text{CH}_2\text{OH}$ (b) nucleophilic addition of a primary amine to a ketone; product is an imine, $\text{CH}_3\text{C}{=}\text{NCH}_3$ (c) nucleophilic addition of an ammonia derivative, phenyl-
   |
   CH_3

hydrazine, to an aldehyde; product is a phenylhydrazone, $\text{C}_6\text{H}_5\text{CH}{=}\text{NNHC}_6\text{H}_5$ (d) aldol condensation of acetone; product is a β-hydroxy carbonyl compound, $CH_3\overset{\underset{\displaystyle |}{OH}}{C}CH_2\overset{\underset{\displaystyle |}{\overset{\displaystyle O}{\|}}}{C}CH_3$

$$\underset{CH_3}{|}$$

(e) oxidation reaction of an aldehyde; product is a carboxylic acid, $CH_3\overset{\underset{\displaystyle |}{\overset{\displaystyle O}{\|}}}{C}HC{-}OH$

$$\underset{CH_3}{|}$$

(f) nucleophilic addition of an alcohol to a ketone; product is an acetal, $CH_3CH_2\overset{\underset{\displaystyle |}{\overset{\displaystyle CH_3}{|}}}{C}{-}OCH_3$

$$\underset{OCH_3}{|}$$

SUMMARY

The carbonyl compounds are aldehydes and ketones. Carbonyl compounds contain the carbonyl group as the functional group. Aldehydes differ from ketones in that an aldehyde contains a hydrogen bonded to the carbonyl carbon.

$$\underset{\textit{a carbonyl group}}{\overset{\diagdown}{\underset{\diagup}{C}}{=}O} \qquad \underset{\textit{an aldehyde}}{R{-}\overset{\overset{\displaystyle O}{\|}}{C}{-}H} \qquad \underset{\textit{a ketone}}{R{-}\overset{\overset{\displaystyle O}{\|}}{C}{-}R}$$

Aldehydes and ketones have both common and IUPAC names. The common names of simple carbonyl compounds are listed in Table 15.1. The IUPAC names of these compounds consist of a parent name that is derived from the name of the corresponding alkane. The carbonyl group is included in the parent name. For aldehydes, the carbonyl carbon is always number 1; for ketones, the carbonyl carbon is given the lowest number. The suffix for an aldehyde is *-al* and that for a ketone is *-one.* Substituted benzaldehydes and acetophenones are named as derivatives of benzaldehyde and acetophenone, respectively. There are numerous occurrences of carbonyl compounds in nature and these compounds have a variety of uses.

The carbonyl group is introduced into molecules primarily through the oxidation of alcohols. Primary alcohols give aldehydes; secondary alcohols give ketones. Additionally, ketones are formed through an acylation reaction.

$$\underset{\textit{a primary alcohol}}{RCH_2OH} \xrightarrow{CrO_3,\ \text{pyridine}} \underset{\textit{an aldehyde}}{R{-}\overset{\overset{\displaystyle O}{\|}}{C}{-}H}$$

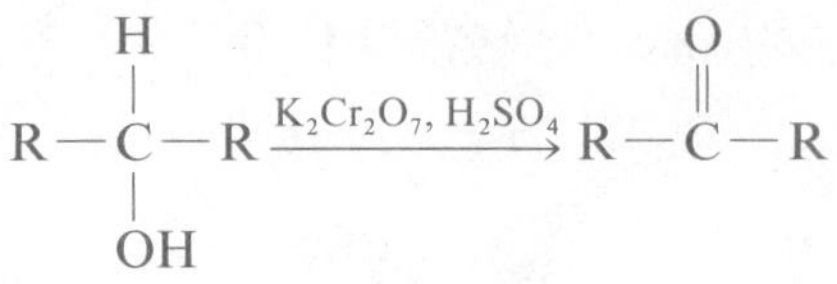

$$\underset{\substack{| \\ OH}}{\overset{\substack{H \\ |}}{R-C-R}} \xrightarrow{K_2Cr_2O_7,\ H_2SO_4} \overset{\substack{O \\ ||}}{R-C-R}$$

a secondary alcohol   *a ketone*

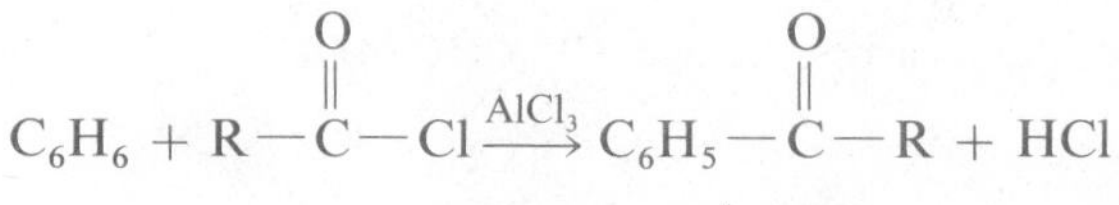

$$C_6H_6 + \overset{\substack{O \\ ||}}{R-C-Cl} \xrightarrow{AlCl_3} \overset{\substack{O \\ ||}}{C_6H_5-C-R} + HCl$$

a ketone

The reaction of carbonyl compounds is a nucleophilic addition to the pi bond of the carbonyl group. The nucleophiles of importance are sodium cyanide, primary amines, derivatives of ammonia, and alcohols. A variety of addition products is possible.

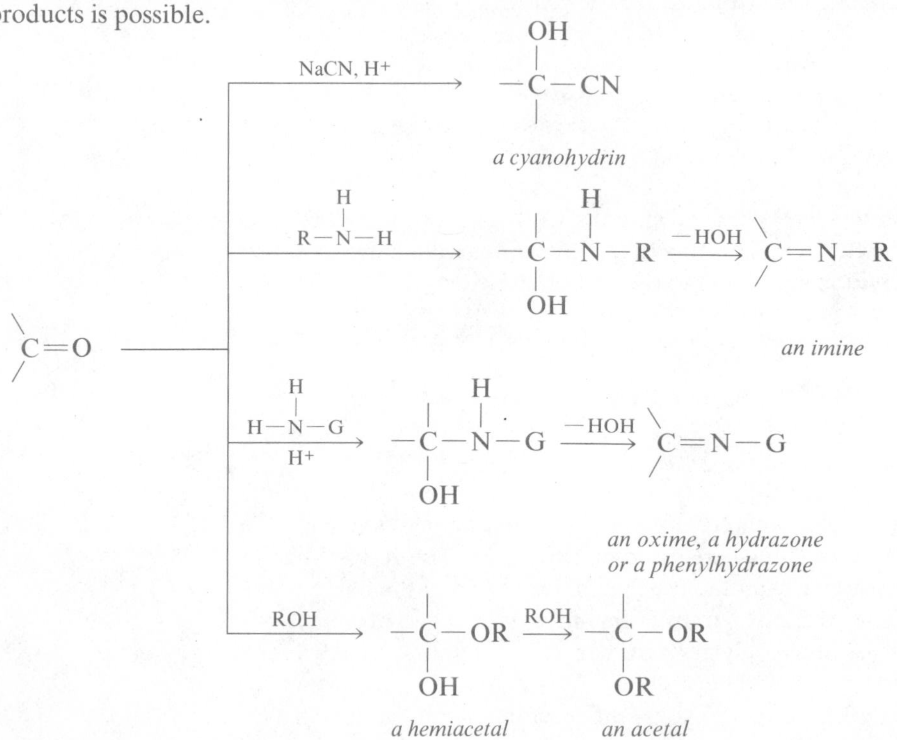

$$\underset{/}{\overset{\backslash}{C}}=O$$

$\xrightarrow{NaCN,\ H^+}$ $\underset{\substack{| \\ }}{\overset{\substack{OH \\ |}}{-C-CN}}$

a cyanohydrin

$\xrightarrow[R-N-H]{\overset{H}{|}}$ $\underset{\substack{| \\ OH}}{\overset{\substack{H \\ |}}{-C-N-R}}$ $\xrightarrow{-HOH}$ $\underset{/}{\overset{\backslash}{C}}=N-R$

an imine

$\xrightarrow[H^+]{\overset{H}{\underset{|}{H-N-G}}}$ $\underset{\substack{| \\ OH}}{\overset{\substack{H \\ |}}{-C-N-G}}$ $\xrightarrow{-HOH}$ $\underset{/}{\overset{\backslash}{C}}=N-G$

an oxime, a hydrazone or a phenylhydrazone

$\xrightarrow{ROH}$ $\underset{\substack{| \\ OH}}{\overset{\substack{| \\ }}{-C-OR}}$ $\xrightarrow{ROH}$ $\underset{\substack{| \\ OR}}{\overset{\substack{| \\ }}{-C-OR}}$

a hemiacetal   *an acetal*

Other important reactions of carbonyl compounds are the aldol condensation, reduction, and oxidation. The aldol condensation occurs only when the carbonyl compound contains an α-hydrogen. The product is a β-hydroxy carbonyl compound.

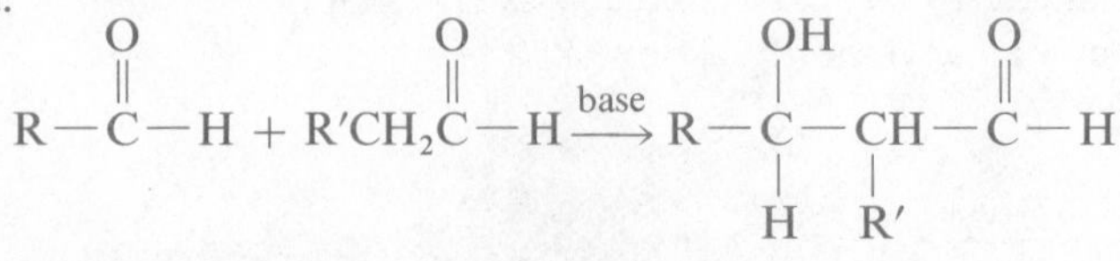

$$\overset{\substack{O \\ ||}}{R-C-H} + \overset{\substack{O \\ ||}}{R'CH_2C-H} \xrightarrow{base} \underset{\substack{| \quad | \\ H \quad R'}}{\overset{\substack{OH \quad O \\ | \qquad ||}}{R-C-CH-C-H}}$$

a β-hydroxy carbonyl compound

The reduction product of a carbonyl compound is an alcohol. Upon reduction aldehydes form primary alcohols, ketones form secondary alcohols.

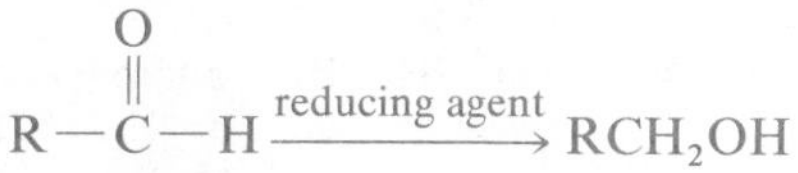

$$R-\overset{\overset{\displaystyle O}{\|}}{C}-H \xrightarrow{\text{reducing agent}} RCH_2OH$$

an aldehyde   *a primary alcohol*

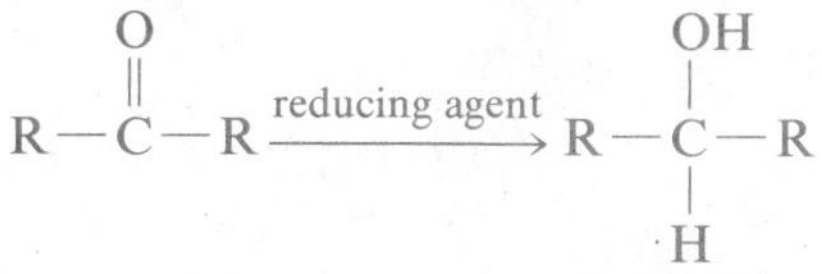

$$R-\overset{\overset{\displaystyle O}{\|}}{C}-R \xrightarrow{\text{reducing agent}} R-\overset{\overset{\displaystyle OH}{|}}{\underset{\underset{\displaystyle H}{|}}{C}}-R$$

a ketone   *a secondary alcohol*

Aldehydes can be oxidized by Tollens' reagent or Benedict's reagent to a carboxylic acid. Ketones are stable to these oxidizing agents.

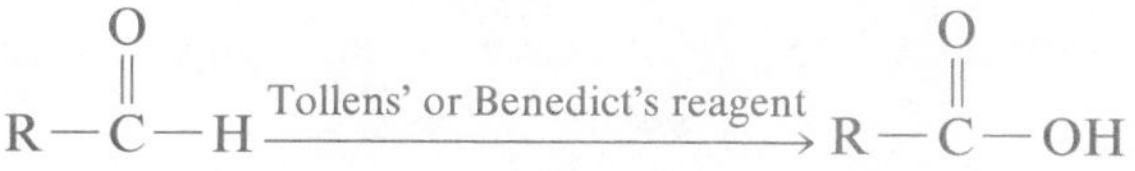

$$R-\overset{\overset{\displaystyle O}{\|}}{C}-H \xrightarrow{\text{Tollens' or Benedict's reagent}} R-\overset{\overset{\displaystyle O}{\|}}{C}-OH$$

an aldehyde   *a carboxylic acid*

ADDITIONAL PROBLEMS

5. Write a correct name for each of the following:

(a) 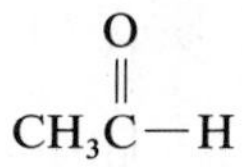 $CH_3\overset{\overset{\displaystyle O}{\|}}{C}-H$

(b) $CH_3CH_2\overset{\underset{\underset{\displaystyle CH_3}{|}}{}}{CH}\overset{\overset{\displaystyle O}{\|}}{C}CH_3$

(c) $CH_3\overset{\underset{\underset{\displaystyle CH_3}{|}}{}}{CH}\overset{\overset{\displaystyle O}{\|}}{C}-H$

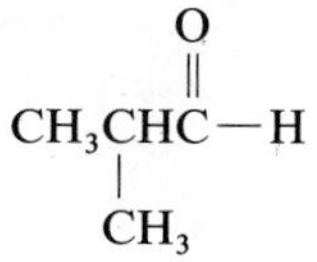

(d) 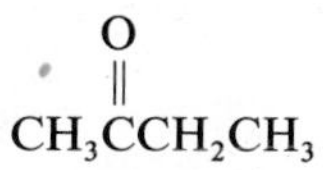 $CH_3\overset{\overset{\displaystyle O}{\|}}{C}CH_2CH_3$

(e) 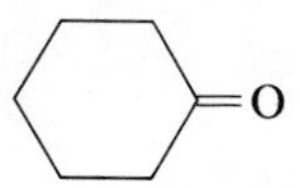 cyclohexanone structure with =O

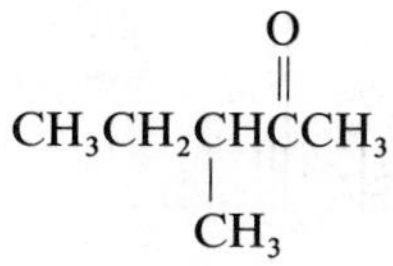

(f) 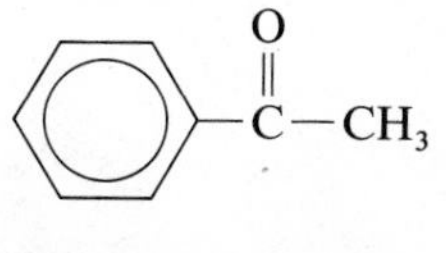 benzene ring $-\overset{\overset{\displaystyle O}{\|}}{C}-CH_3$

(g) $BrCH_2CH_2CH_2CH_2CHO$

(h) 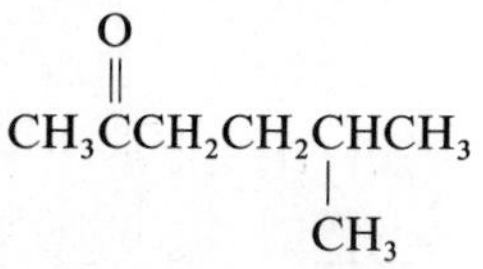 $CH_3\overset{\overset{\displaystyle O}{\|}}{C}CH_2CH_2\overset{\underset{\underset{\displaystyle CH_3}{|}}{}}{CH}CH_3$

(i) 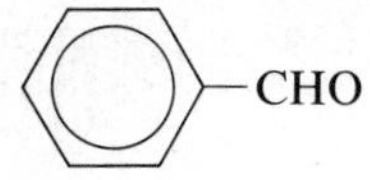 benzene ring $-CHO$

(j) 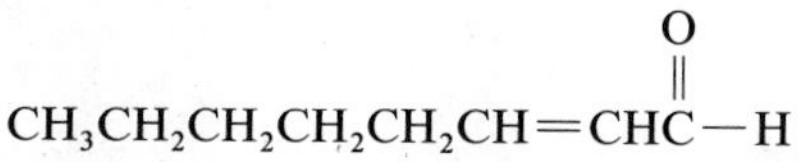 $CH_3CH_2CH_2CH_2CH_2CH=CH\overset{\overset{\displaystyle O}{\|}}{C}-H$

6. Write complete structures for each of the following:
   (a) acetone
   (b) propanal
   (c) 4-chlorobutanal
   (d) 2-hexanone
   (e) acetophenone
   (f) diphenyl ketone
   (g) *m*-chlorobenzaldehyde
   (h) cyclopentanone
   (i) 5-methyl-4-chloro-2-heptanone
   (j) propionaldehyde

7. Write the general structure of a compound that corresponds to the following term. Give an example of each.
(a) aldehyde
(b) ketone
(c) cyanohydrin
(d) hemiacetal
(e) acetal
(f) polyhydroxyaldehyde
(g) imine
(h) oxime
(i) phenylhydrazone
(j) hydrazone

8. Write the structure for the organic product of each reaction.

(a) 
$$\begin{array}{c} CHO \\ | \\ HOCH \\ | \\ HOCH \\ | \\ HCOH \\ | \\ HCOH \\ | \\ CH_2OH \end{array}$$ + Tollen's reagent $\longrightarrow$

mannose

(b) $CH_3CH_2CCH_2CH_3$ + $NH_2OH \xrightarrow{H^+}$

(c) 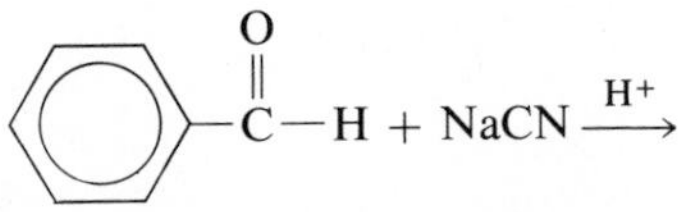 Ph—C(=O)—H + NaCN $\xrightarrow{H^+}$

(d) 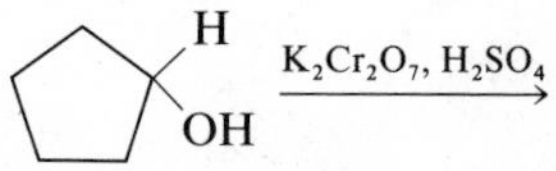 cyclopentanol with H, OH $\xrightarrow{K_2Cr_2O_7, H_2SO_4}$

(e) $CH_3CH_2CH_2CHO$ + 2 $CH_3OH \xrightarrow{dry\ HCl}$

(f) 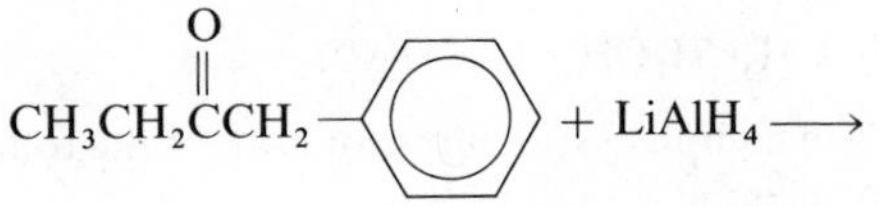 $CH_3CH_2CCH_2$—Ph + $LiAlH_4 \longrightarrow$

(g) 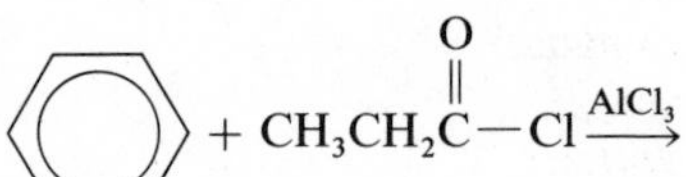 Ph + $CH_3CH_2C(=O)$—Cl $\xrightarrow{AlCl_3}$

(h) $CH_3OH \xrightarrow{CrO_3,\ pyridine}$

(i) CH_3CH(=O) + dilute NaOH $\longrightarrow$

(j) CH_3CH_2C(=O)—H + $NH_2NH_2 \xrightarrow{H^+}$

(k) CH_3CHC(=O)—H with CH_3 + $LiAlH_4 \longrightarrow$

(l) CH_3—C(=O)—CH_3 + $CH_3NH_2 \longrightarrow$

9. Using question 8 as a reference, identify those reactions that fit the following descriptions.
(a) oxidation
(b) reduction
(c) aldol condensation
(d) cyanohydrin formation
(e) acetal formation
(f) acylation
(g) formation of a silver mirror
(h) oxime formation

10. Glucose forms a hemiacetal through the reaction of the carbonyl group and the hydroxy group on C-5. Show the structure of the hemiacetal form of glucose.

11. What is the observable indication in a positive (a) Tollens' test and (b) Bendict's test for glucose?

12. Some insects use a cyanohydrin as a component of their defense mechanisms. Describe why the cyanohydrin is an effective defense against a threatening enemy.

13. In the living system, NAD+ is a coenzyme that functions in oxidation reactions. Predict the oxidation product of the following reaction.

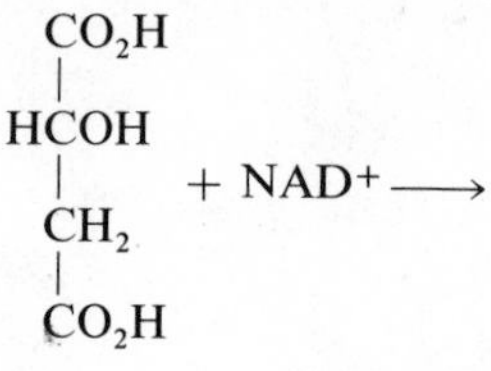

$$\begin{array}{c} CO_2H \\ | \\ HCOH \\ | \\ CH_2 \\ | \\ CO_2H \end{array}$$ + NAD+ $\longrightarrow$

CHAPTER 16

CARBOXYLIC ACIDS AND THEIR DERIVATIVES

Carboxylic acids are identified by the functional group called a carboxyl group. A carboxyl group is a combination of a carbonyl group ($C=O$) and a hydroxyl group (OH).* Although carboxylic acids are widely distributed in plants and animals, they occur primarily as the derivatives of carboxylic acids: acid anhydrides, esters, or amides. Many of these compounds are technologically important and are used in food additives, soaps and detergents, synthetic fibers (Nylon and Dacron), and medicines (aspirin, aspirin substitutes, and barbiturates). Proteins, the major structural component of animal tissue, are carboxylic acid derivatives. Carboxylic acids and their derivatives are important in the degradation of triglycerides. The triglycerides are a class of water-soluble esters that, when metabolized by the living system, provide even more energy than do the carbohydrates. The chemistry involved in the degradation (or synthesis) of triglycerides is another application of the chemistry of carboxylic acids and carboxylic acid derivatives. Another kind of acid derivative, the acid chloride, is also useful to the organic chemist. An acid chloride is very reactive and does not occur in nature; however, because of its high reactivity, it is used to make the other acid derivatives.

---

*However, the combination of these two groups provides some special properties and consequently the chemistry of the carboxyl group is treated in a separate chapter. For example, as we saw in Chapter 14, the hydroxyl group in alcohols is not acidic, but when present in the carboxyl group a greater acidity is imparted making a carboxylic acid a stronger acid than an alcohol.

16-1 STRUCTURE AND NOMENCLATURE

LEARNING OBJECTIVES

1. Given the structural formulas of organic compounds, identify those that are carboxylic acids, acid chlorides, anhydrides, esters, and amides.

2. Given the common or IUPAC name of a carboxylic acid or a carboxylic acid derivative, write its structural formula.

3. Given the structural formula of a carboxylic acid or carboxylic acid derivative, write its common or IUPAC name.

The **carboxyl group** in a carboxylic acid is represented by the general formulas as shown.

$$
\begin{array}{ccc}
\overset{\displaystyle O}{\overset{\displaystyle \|}{-C-OH}} & \overset{\displaystyle O}{\overset{\displaystyle \|}{R-C-OH}} \quad \text{or} \quad \text{RCOOH} \quad \text{or} \quad \text{RCO}_2\text{H}
\end{array}
$$

a carboxyl group   *a carboxylic acid*

A **carboxyl group** consists of a carbon atom doubly bonded to an oxygen atom and singly bonded to a hydroxyl group.

The carboxylic acid derivatives can be categorized as acid chlorides, anhydrides, esters, and amides. The general formula for each class of compounds is as shown.

$$
\overset{\displaystyle O}{\overset{\displaystyle \|}{R-C-Cl}} \qquad \text{or} \quad \text{RCOCl} \qquad \text{is an } \textit{acid chloride}
$$

$$
\overset{\displaystyle O}{\overset{\displaystyle \|}{R-C-O-}}\overset{\displaystyle O}{\overset{\displaystyle \|}{C-R}} \quad \text{or} \quad \text{RCOOCOR} \quad \text{is an } \textit{anhydride}
$$

$$
\overset{\displaystyle O}{\overset{\displaystyle \|}{R-C-OR'}} \qquad \text{or} \quad \text{RCOOR}' \qquad \text{is an } \textit{ester}
$$

$$
\overset{\displaystyle O}{\overset{\displaystyle \|}{R-C-NH_2}} \qquad \text{or} \quad \text{RCONH}_2 \qquad \text{is an } \textit{amide}
$$

With the exception of acid chlorides, carboxylic acids and their derivatives are commonly found in living systems. Acid chlorides are highly reactive and do not exist in living cells because they decompose in an aqueous medium to carboxylic acids with the release of hydrogen chloride. When hydrogen chloride dissolves in water, the primary component of the living system, hydrochloric acid is formed. Thus, acid chlorides cannot exist in the living system because this formation of hydrochloric acid would greatly alter the pH of the system with detrimental results.

Carboxylic acids are named by common and IUPAC names. These names, and the sources of some carboxylic acids, are shown in Table 16.1. The IUPAC names of carboxylic acids are derived by choosing the longest chain of carbon atoms including the carboxyl group. The *-e* in the name of the corresponding alkan*e* is dropped and *-oic acid* is added in its place.

Table 16.1 Common and IUPAC Names of Carboxylic Acids

Structure	Number of Carbons	Common Name	IUPAC Name	Occurrence
HCOOH	1	Formic acid	Methanoic acid	Red ants
CH₃COOH	2	Acetic acid	Ethanoic acid	Vinegar
CH₃CH₂COOH	3	Propionic acid	Propanoic acid	Some dairy products
CH₃CH₂CH₂COOH	4	Butyric acid	Butanoic acid	Rancid butter
CH₃(CH₂)₃COOH	5	Valeric acid	Pentanoic acid	Valerian root
CH₃(CH₂)₄COOH	6	Caproic acid	Hexanoic acid	Hair and secretions of goats
CH₃(CH₂)₁₂COOH	14	Myristic acid*	—	Nutmeg butter, sperm whale oil
CH₃(CH₂)₁₄COOH	16	Palmitic acid*	—	Palm oil
CH₃(CH₂)₁₆COOH	18	Stearic acid*	—	Beef and sheep tallow, cottonseed oil

*These acids are referred to as fatty acids because they are found in vegetable oils and animal fats. These fatty acids do have IUPAC names but common names are used more often.

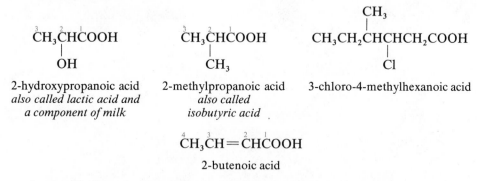

When other groups or substituents are present, the carboxyl carbon is the number one carbon as shown.

Aromatic acids are named as derivatives of benzoic acid.

There are also a number of important dicarboxylic acids. These contain two carboxyl groups. Some of the more important dicarboxylic acids are shown. In these examples the first name is the IUPAC name and the name in parentheses is the common name.

$$
\overset{O}{\underset{\|}{HOC}}-\overset{O}{\underset{\|}{COH}} \qquad \overset{O}{\underset{\|}{HOC}}CH_2\overset{O}{\underset{\|}{COH}} \qquad \overset{O}{\underset{\|}{HOC}}(CH_2)_2\overset{O}{\underset{\|}{COH}}
$$

ethanedioic acid   propanedioic acid   butanedioic acid
(oxalic acid)   (malonic acid)   (succinic acid)

$$
\overset{O}{\underset{\|}{HOC}}(CH_2)_3\overset{O}{\underset{\|}{COH}} \qquad \overset{O}{\underset{\|}{HOC}}(CH_2)_4\overset{O}{\underset{\|}{COH}} \qquad HOOC-\langle\bigcirc\rangle-COOH
$$

pentanedioic acid   hexanedioic acid
(glutaric acid)   (adipic acid)   (terephthalic acid)

Derivatives of carboxylic acids are named by common and IUPAC names. Common names rather than IUPAC names are usually the name of choice where possible and this method is illustrated in Table 16.2 for the four carboxylic acid derivatives derived from acetic acid. The common name of each carboxylic acid derivative can be obtained by first determining the common name of the acid from which the carboxylic acid derivative is derived and then applying the appropriate rule as in Table 16.2. Important acid derivatives are described throughout this chapter.

Table 16.2   Common Names of Acid Derivatives

Acid Derivative	Rule for Naming from Common Name of Acid	Example*
Acid chloride	Change *-ic acid* to *-yl chloride*	$CH_3\overset{O}{\underset{\|}{C}}-Cl$, acetyl chloride
Anhydride	Change *acid* to *anhydride*	$CH_3\overset{O}{\underset{\|}{C}}-O-\overset{O}{\underset{\|}{C}}CH_3$, acetic anhydride
Ester	Change *-ic acid* to *-ate*, preceded by name of group bonded to oxygen	$CH_3\overset{O}{\underset{\|}{C}}-OCH_2CH_3$, ethyl acetate
Amide	Change *-ic acid* to *amide*	$CH_3\overset{O}{\underset{\|}{C}}-NH_2$, acetamide

*The colored portion identifies the carboxylic acid from which the acid derivative is derived.

PROBLEMS

1. Name the following acids by common or IUPAC names.
   (a) CH_3CH_2COOH

(b) 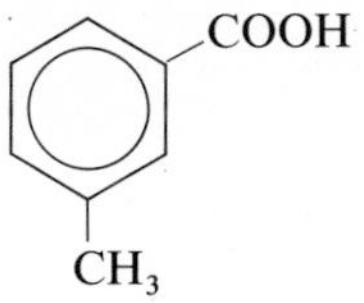

(c) 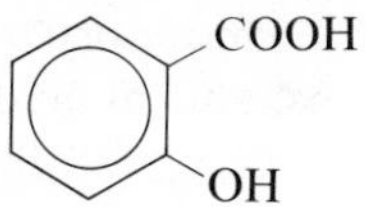

(d) $HC{\equiv}CCH_2COOH$

(e) $HOOCCH_2COOH$

2. Identify the following acid derivatives and name each one.

(a) $CH_3CH_2CH_2\overset{O}{\overset{\|}{C}}{-}NH_2$

(b) 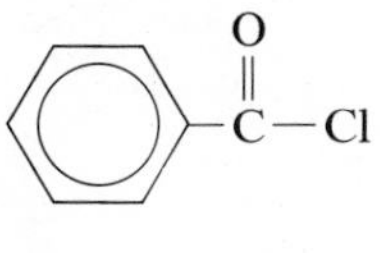

(c) $CH_3CH_2{-}\overset{O}{\overset{\|}{C}}{-}O{-}\overset{O}{\overset{\|}{C}}{-}CH_2CH_3$

(d) $CH_3CH_2CH_2\overset{O}{\overset{\|}{C}}{-}OCH_3$

3. Write the structural formula for each of the following.

(a) 2,4-dinitrobenzoic acid
(b) propionyl chloride
(c) butyl acetate
(d) 2-chloro-4-ethylheptanoic acid
(e) 4-pentynoic acid

ANSWERS TO PROBLEMS **1. (a)** propionic acid or propanoic acid **(b)** *m*-methylbenzoic acid **(c)** *o*-hydroxybenzoic acid or salicylic acid **(d)** 3-butynoic acid **(e)** malonic acid or propanedioic acid **2. (a)** amide of butyric acid, butyramide **(b)** acid chloride of benzoic acid, benzoyl chloride **(c)** anhydride of propionic acid, propionic anhydride **(d)** ester of butryic acid, methyl butyrate **3. (a)** 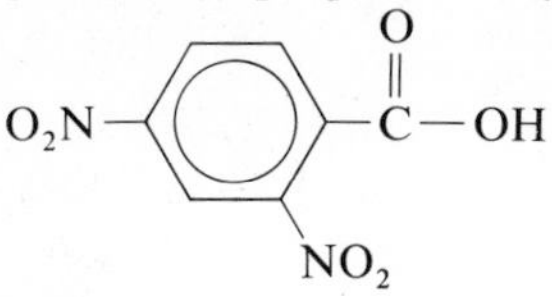

(b) $CH_3CH_2\overset{O}{\overset{\|}{C}}{-}Cl$ **(c)** $CH_3\overset{O}{\overset{\|}{C}}{-}OCH_2CH_2CH_2CH_3$

(d) $CH_3CH_2CH_2\underset{\underset{CH_3}{|}}{\underset{CH_2}{|}}{CH}CH_2\underset{Cl}{\underset{|}{CH}}\overset{O}{\overset{\|}{C}}{-}OH$ **(e)** $H{-}C{\equiv}CCH_2CH_2\underset{O}{\underset{\|}{C}}{-}OH$

16-2 PREPARATION OF CARBOXYLIC ACIDS

LEARNING OBJECTIVES

1. Given an oxidation reaction of a primary alcohol, an aldehyde, or an aromatic hydrocarbon, write the structure of the carboxylic acid formed.

2. Given a hydrolysis reaction of a nitrile, write the structure of the carboxylic acid formed.

Carboxylic acids usually are prepared in the laboratory by oxidation of primary alcohols, aldehydes, or aromatic hydrocarbons, and by hydrolysis of nitriles. We will discuss each of these processes.

16-2.1 OXIDATION OF PRIMARY ALCOHOLS, ALDEHYDES, AND AROMATIC HYDROCARBONS

Alcohols, aldehydes, and aromatic hydrocarbons are oxidized to carboxylic acids by potassium dichromate in acid solution (see Section 14-3.5). The alkyl group (R) in the aromatic hydrocarbon is converted to a carboxyl group regardless of the nature of the alkyl group. Therefore, the oxidation of an aromatic hydrocarbon gives benzoic acid or a derivative of benzoic acid.

$$RCH_2OH \xrightarrow{K_2Cr_2O_7,\ H_2SO_4} R-\overset{\overset{\displaystyle O}{\|}}{C}-OH$$

a primary alcohol   *a carboxylic acid*

$$R-\overset{\overset{\displaystyle O}{\|}}{C}-H \xrightarrow{K_2Cr_2O_7,\ H_2SO_4} R-\overset{\overset{\displaystyle O}{\|}}{C}-OH$$

an aldehyde   *a carboxylic acid*

$$\langle\bigcirc\rangle-R \xrightarrow{K_2Cr_2O_7,\ H_2SO_4} \langle\bigcirc\rangle-\overset{\overset{\displaystyle O}{\|}}{C}-OH$$

an aromatic hydrocarbon   *benzoic acid*

The oxidation of ethyl alcohol, propionaldehyde, and isopropylbenzene illustrates this process.

$$CH_3CH_2OH \xrightarrow{K_2Cr_2O_7,\ H_2SO_4} CH_3\overset{\overset{\displaystyle O}{\|}}{C}-OH$$

ethyl alcohol   acetic acid

$$CH_3CH_2\overset{\overset{\displaystyle O}{\|}}{C}-H \xrightarrow{K_2Cr_2O_7,\ H_2SO_4} CH_3CH_2\overset{\overset{\displaystyle O}{\|}}{C}-OH$$

propionaldehyde   propionic acid

$$\langle\bigcirc\rangle-\overset{\overset{\displaystyle CH_3}{|}}{C}HCH_3 \xrightarrow{K_2Cr_2O_7,\ H_2SO_4} \langle\bigcirc\rangle-\overset{\overset{\displaystyle O}{\|}}{C}-OH$$

isopropylbenzene   benzoic acid

16-2.2 HYDROLYSIS OF NITRILES

Hydrolysis of a nitrile (see Table 10.5), or *alkyl cyanide* as it is sometimes called, in the presence of hot aqueous acid yields a carboxylic acid.

$$R-C{\equiv}N + H_2O \xrightarrow{H^+,\ heat} RCOOH + NH_4^+$$

a nitrile   *a carboxylic acid*

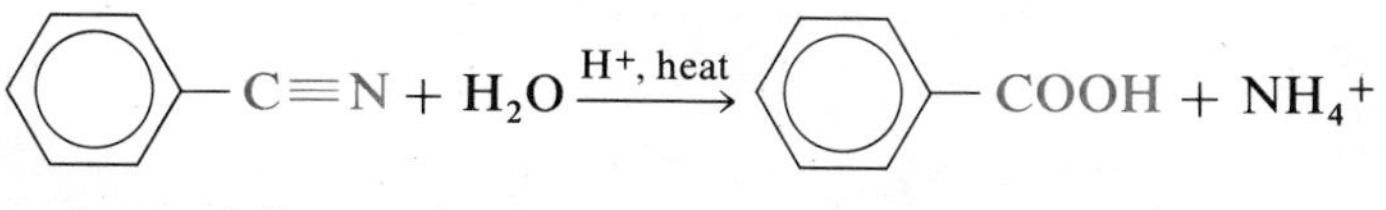

benzonitrile   benzoic acid

PROBLEM

4. Describe the following reactions and write the structure of the carboxylic acid which would be formed from each.

(a) $CH_3CH_2CH_2CH_2OH \xrightarrow{K_2Cr_2O_7,\ H_2SO_4}$

(b) 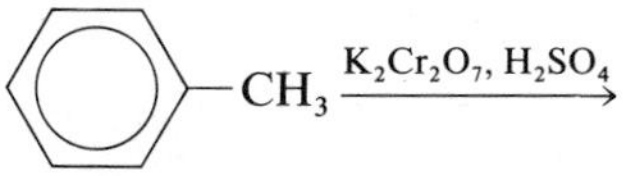 — $CH_3 \xrightarrow{K_2Cr_2O_7,\ H_2SO_4}$

(c) 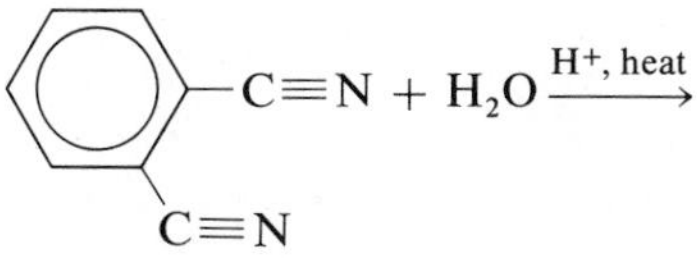 — $C{\equiv}N + H_2O \xrightarrow{H^+,\ heat}$
      $C{\equiv}N$

(d) 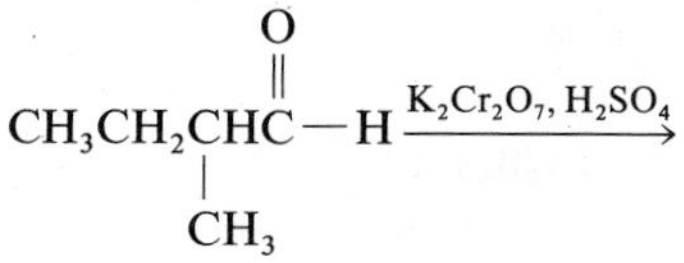 $CH_3CH_2\overset{\displaystyle O}{\overset{\|}{C}}HC-H \xrightarrow{K_2Cr_2O_7,\ H_2SO_4}$
         $\underset{CH_3}{|}$

ANSWERS TO PROBLEM   **4. (a)** oxidation of a primary alcohol, $CH_3CH_2CH_2COOH$

(b) oxidation of an aromatic hydrocarbon, $\overset{\displaystyle O}{\overset{\|}{C}}-OH$

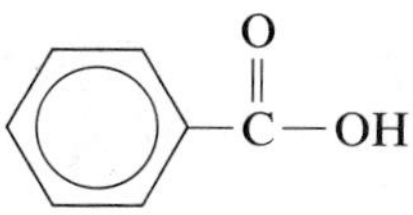

(c) hydrolysis of a nitrile, $\begin{array}{c} COOH \\ COOH \end{array}$

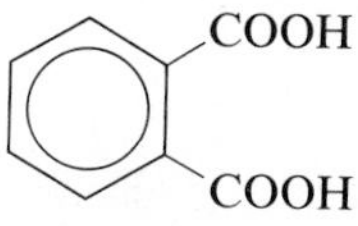

(d) oxidation of an aldehyde, $CH_3CH_2\overset{\displaystyle O}{\overset{\|}{C}}HC-OH$
                 $\underset{CH_3}{|}$

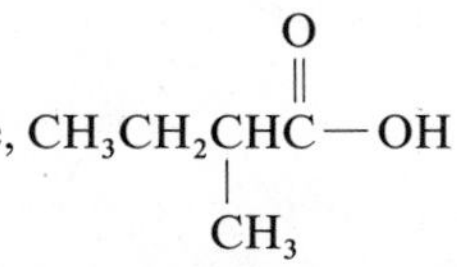

16-3 REACTIONS OF CARBOXYLIC ACIDS

LEARNING OBJECTIVES

1. Given the reaction of a carboxylic acid with sodium hydroxide or sodium bicarbonate, complete the equation.
2. Given the reaction between a carboxylic acid and thionyl chloride, another carboxylic acid, phosphoric acid, or an alcohol, write the structure of the carboxylic acid derivative formed.

3. Given a carboxylic acid, write the products of its decarboxylation.
4. List the commercial importance of zinc 10-undecylenate, the phosphotriesters, barbiturates, and Nylon 66.

16-3.1 ACIDITY: REACTION WITH BASE

We have already learned about the acidic nature of a carboxylic acid (see Chapter 8). Carboxylic acids are stronger acids than alcohols and phenols (see Chapter 14), but are weaker than acids such as sulfuric acid and hydrochloric acid.

$$acidity: \quad H_2SO_4 \text{ and } HCl > \text{carboxylic acids} > \text{phenols} > \text{alcohols}$$

$$\xrightarrow{\text{decreasing acidity}}$$

Carboxylic acids ionize to form a carboxylate anion and a proton.

$$\underset{\text{a carboxylic acid}}{R-\overset{\overset{\textstyle O}{\|}}{C}-OH} \longrightarrow \underset{\text{a carboxylate anion}}{R-\overset{\overset{\textstyle O}{\|}}{C}-O^-} + H^+$$

Carboxylic acids form carboxylate anions, or salts, upon reaction with bases such as sodium hydroxide (NaOH) and sodium bicarbonate (NaHCO$_3$). Sodium benzoate, the salt that results from the reaction of benzoic acid with sodium bicarbonate, occurs naturally in many foods, especially cranberries and prunes. It is used as a preservative in bakery products (see Section 16-5).

benzoic acid → sodium benzoate + H$_2$O + CO$_2$

$$\underset{\text{acetic acid}}{CH_3\overset{\overset{\textstyle O}{\|}}{C}-OH} \xrightarrow{NaOH} \underset{\text{sodium acetate}}{CH_3\overset{\overset{\textstyle O}{\|}}{C}-O^-Na^+} + H_2O$$

10-Undecylenic acid in the form of its zinc salt, zinc 10-undecylenate, is commonly used as an effective treatment of *athlete's foot*. Athlete's foot is a fungal infection of the feet resulting from the reaction of bacteria with excessive perspiration on the feet. This combination usually produces a foul odor, itching, and peeling of the white skin in the webs of the toes. One commercial product that contains zinc 10-undecylenate is Desenex. The structural formula of zinc 10-undecylenate is shown.

$$(CH_2=CHCH_2CH_2CH_2CH_2CH_2CH_2CH_2CH_2\overset{\displaystyle O}{\overset{\displaystyle \|}{C}}-O^-)_2Zn^{2+}$$

zinc 10-undecylenate

Phenols do not form phenoxide anions (salts) with $NaHCO_3$ and alcohols do not form alkoxide anions (salts) with either NaOH or $NaHCO_3$. Thus, these reactions can be useful in separating mixtures of alcohols, phenols, and carboxylic acids.

16-3.2 FORMATION OF FUNCTIONAL DERIVATIVES

The functional derivatives of carboxylic acids are the acid chlorides, anhydrides, esters, and amides. The general structures for these derivatives are shown in Section 16.1. These derivatives are formed by a common mechanism: *nucleophilic substitution*. Stated simply, one nucleophile, for example the —OH of the carboxyl group, is replaced by another nucleophile. These substituting nucleophiles may be considered to be one of the following nucleophiles. Note that each nucleophile has at least one unshared pair of electrons available to form a bond to carbon.

$$:\overset{..}{\underset{..}{Cl}}:^- \qquad R-\overset{\displaystyle O}{\overset{\displaystyle \|}{C}}-\overset{..}{\underset{..}{O}}:^- \qquad R\overset{..}{\underset{..}{O}}:^- \qquad H_2N:^-$$

The substitution for the formation of an acid chloride is shown.

$$R-\overset{\displaystyle O}{\overset{\displaystyle \|}{C}}-OH + :Cl^- \longrightarrow R-\overset{\displaystyle O}{\overset{\displaystyle \|}{C}}-Cl + :OH^-$$

OH *is replaced by* Cl

The substitution reactions leading to the formation of each of the other functional derivatives follow a similar pattern. As noted, the mechanism is slightly more complicated than described here. However, this explanation is complete enough to help you understand that the formation of each functional derivative from the carboxylic acid is another example of nucleophilic substitution. We will now discuss the formation of each carboxylic acid derivative in greater detail.

Acid chlorides

Carboxylic acids yield acid chlorides when treated with thionyl chloride ($SOCl_2$).

$$R\overset{\displaystyle O}{\overset{\displaystyle \|}{C}}-OH \overset{SOCl_2}{\longrightarrow} R\overset{\displaystyle O}{\overset{\displaystyle \|}{C}}-Cl + SO_2 + HCl$$

For example, acetic acid reacts with thionyl chloride to produce acetyl chloride.

$$\underset{\text{acetic acid}}{CH_3\overset{\displaystyle O}{\overset{\|}{C}}-OH} + SOCl_2 \longrightarrow \underset{\text{acetyl chloride}}{CH_3\overset{\displaystyle O}{\overset{\|}{C}}-Cl} + SO_2 + HCl$$

Anhydrides

Anhydrides can be formed from the condensation of a carboxylic acid with itself. The condensation results in the formation of water as a by-product.

$$R\overset{\displaystyle O}{\overset{\|}{C}}-OH + HO-\overset{\displaystyle O}{\overset{\|}{C}}R \overset{heat}{\longrightarrow} R\overset{\displaystyle O}{\overset{\|}{C}}-O-\overset{\displaystyle O}{\overset{\|}{C}}R + HOH$$

For example, butyric acid reacts with itself to produce butyric anhydride and water.

$$\underset{\text{butyric acid}}{CH_3CH_2CH_2\overset{\displaystyle O}{\overset{\|}{C}}-OH} + HO-\overset{\displaystyle O}{\overset{\|}{C}}CH_2CH_2CH_3 \overset{heat}{\longrightarrow}$$

$$\underset{\text{butyric anhydride}}{CH_3(CH_2)_2\overset{\displaystyle O}{\overset{\|}{C}}-O-\overset{\displaystyle O}{\overset{\|}{C}}(CH_2)_2CH_3} + HOH$$

Unlike acid chlorides, anhydrides are very common in living systems. The anhydrides of living systems, however, often contain some combination of an organic molecule with the inorganic acid phosphoric acid.

$$\underset{\text{a carboxylic acid}}{R\overset{\displaystyle O}{\overset{\|}{C}}-OH} + \underset{\text{phosphoric acid}}{HO-\overset{\displaystyle O}{\underset{\displaystyle OH}{\overset{\|}{P}}}-OH} \overset{heat}{\longrightarrow} \underset{\substack{\text{a mixed anhydride of a}\\\text{carboxylic and}\\\text{phosphoric acids}}}{R\overset{\displaystyle O}{\overset{\|}{C}}-O-\overset{\displaystyle O}{\underset{\displaystyle OH}{\overset{\|}{P}}}-OH} + HOH$$

The many metabolic reactions of biomolecules occur only when these organic substances are in the phosphorylated state. Glycolysis (see Chapter 24), the metabolic breakdown of glucose and an important source of energy, takes place in the presence of phosphate. An important anhydride in this metabolic scheme is the phosphoric acid anhydride of glyceric acid, 1,3-diphosphoglyceric acid.

$$HO-\overset{\displaystyle O}{\underset{\displaystyle OH}{\overset{\|}{P}}}-O-CH_2-\overset{\displaystyle H}{\underset{\displaystyle OH}{C}}-\overset{\displaystyle O}{\overset{\|}{C}}-O-\overset{\displaystyle O}{\underset{\displaystyle OH}{\overset{\|}{P}}}-OH$$

a mixed anhydride bond

1,3-diphosphoglyceric acid

Esters

Esters are produced by one of two methods. The first is the reaction between a carboxylic acid and an alcohol.

$$\underset{\text{a carboxylic acid}}{RC-OH} \;+\; \underset{\text{an alcohol}}{HOR'} \;\overset{H^+}{\rightleftharpoons}\; HOH \;+\; \underset{\text{an ester}}{RC-OR'}$$

This is a reversible reaction catalyzed by trace amounts of acid. It is reversible because the water formed can cause hydrolysis of the ester. Consequently, the preferred method for ester formation is to cause an acid chloride to react with an alcohol. The by-product is hydrogen chloride, a gas, which escapes and thereby forces the reaction in the direction of the ester.

$$\underset{\text{an acid chloride}}{R-C-Cl} \;+\; \underset{\text{an alcohol}}{H-OR'} \;\longrightarrow\; \underset{\text{an ester}}{R-C-OR'} \;+\; HCl$$

For example, when acetyl chloride reacts with ethyl alcohol, ethyl acetate and hydrogen chloride result.

$$\underset{\text{acetyl chloride}}{CH_3C-Cl} \;+\; \underset{\text{ethyl alcohol}}{H-OCH_2CH_3} \;\longrightarrow\; \underset{\text{ethyl acetate}}{CH_3C-OCH_2CH_3} \;+\; HCl$$

Anhydrides also yield esters upon reaction with alcohols.

$$\underset{\text{an anhydride}}{R-C-O-C-R} + \underset{\text{an alcohol}}{H-OR'} \longrightarrow \underset{\text{an ester}}{R-C-OR'} + \underset{\text{a carboxylic acid}}{R-C-OH}$$

The preparation of aspirin from salicylic acid and acetic anhydride is an example of this reaction. Although salicylic acid contains both a carboxyl group and a phenolic hydroxyl group, it is the latter that undergoes the reaction with the acetic anhydride.

$$\underset{\substack{\text{acetic}\\\text{anhydride}}}{CH_3C-O-CCH_3} + \underset{\substack{\text{salicylic}\\\text{acid}}}{HO-\text{(COOH)}} \overset{H^+}{\rightleftharpoons} HO-CCH_3 + \underset{\substack{\text{acetylsalicylic acid}\\\text{(aspirin)}}}{CH_3C-O-\text{(COO)}}$$

Aspirin is probably the best known analgesic and antipyretic. Many people can take it safely: only one out of every 500 people who take aspirin experiences any undesirable side effects. Possible side effects include skin rashes, asthma, and gastric or intestinal bleeding.

One common side effect of aspirin ingestion is the loss of small (0.5 to 2.0 mL) quantities of blood through the stomach and intestines. This effect takes place in all people, but for most individuals it is entirely inconsequential. In other individuals, aspirin can inhibit the normal blood clotting mechanism and thus cause prolonged bleeding. It appears that aspirin prevents blood from clotting normally by interfering with the aggregation (clustering) of **platelets**. Normally the platelets adhere to an injured area and, eventually, enough join together to form a plug on which the rest of the blood clot forms. In the absence of this plug, clotting takes place more slowly. During the clotting process the platelets release substances that promote continued aggregation. When two aspirin tablets (the recommended dosage) are swallowed, aggregation is prevented because release of the necessary chemicals is blocked. Hence, ruptured capillaries cannot clot quickly enough to prevent a small loss of blood. Generally, the amount of blood lost is insignificant, but some habitual aspirin users could develop an iron-deficiency anemia.

Platelets are small cells in the blood that are instrumental in the blood-clotting process.

There is evidence that people who take aspirin regularly have a significantly lower incidence of fatal heart attack and strokes. Since a heart attack is caused by a blood clot, medical scientists are studying the blood clotting—aspirin interference problem, in addition to concentrating on such factors as blood cholesterol and arteriosclerosis, in attempting to find a method for preventing heart attacks.

Aspirin is remarkably versatile. It is capable of lowering body temperature and relieves pains due to headache, sprain, toothache, and arthritis. Just how aspirin works is not known. Recent studies indicate that aspirin interferes with the synthesis of prostaglandins. The prostaglandins are a group of structurally related carboxylic acids that are believed to cause fever and are probably involved in the process of inflammation. The function of prostaglandins seems to be hormonal. Besides controlling body temperature, they also regulate the contraction of smooth muscles and reproduction. Two representative prostaglandins are shown.

prostaglandin E_1

prostaglandin $E_{1\alpha}$

Two aspirin substitutes are p-hydroxyacetanilide (acetaminophen, or Tylenol) and m-(2,4-difluorophenyl)salicyclic acid (Dolobid).

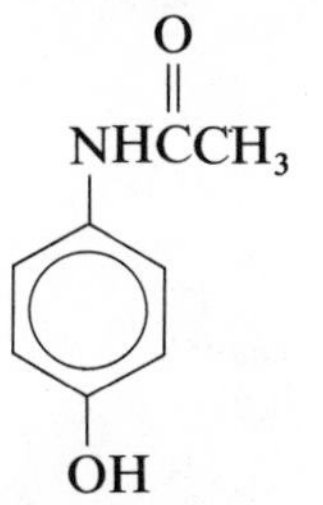

O
‖
NHCCH₃

OH

p-hydroxyacetanilide
an amide

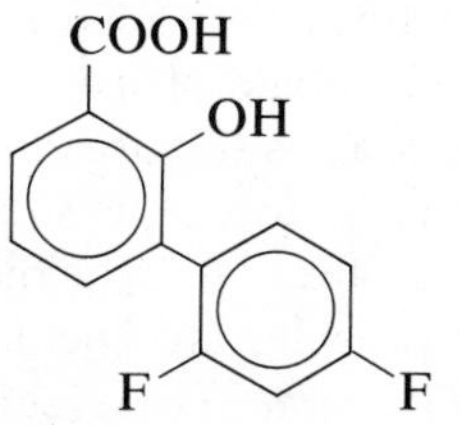

COOH

OH

F   F

m-(2,4-difluorophenyl) salicylic acid
a substituted benzoic acid

Esters also can be formed by the reaction between an alcohol and an inorganic acid (see Section 14-3.3). The esters of phosphoric acid are probably the most important compounds of this group. Depending on the number of moles of alcohol reacting with each phosphoric acid, it is possible to prepare a phosphomonoester, a phosphodiester, and a phosphotriester.

A **phosphomonoester** is an ester formed from the reaction of one mole each of an alcohol and phosphoric acid. A **phosphodiester** is an ester that requires two moles of alcohol and one mole of phosphoric acid. A **phosphotriester** is an ester that requires three moles of alcohol and one mole of phosphoric acid.

O
‖
HO—P—OH
|
OH

phosphoric acid

ROH →

O
‖
RO—P—OH + HOH
|
OH

a phosphomonoester

2 ROH →

O
‖
RO—P—OR + 2 HOH
|
OH

a phosphodiester

3 ROH →

O
‖
RO—P—OR + 3 HOH
|
OR

a phosphotriester

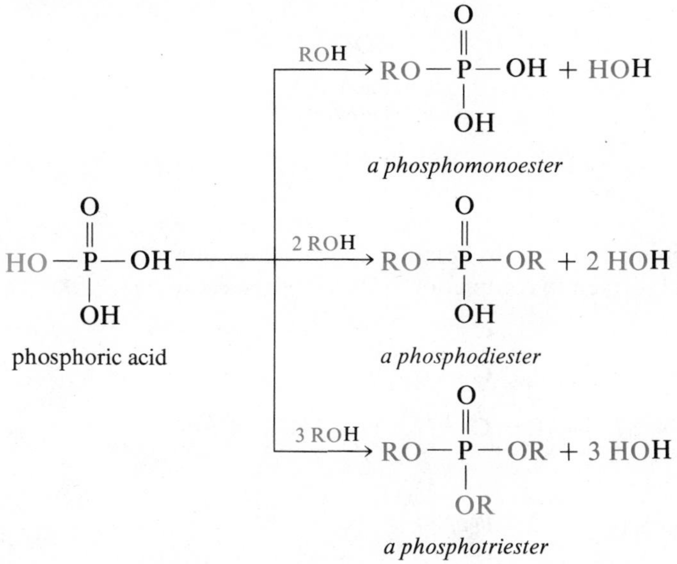

Glucose 6-phosphate, an intermediate in the metabolism of glucose, is an example of a phosphomonoester.

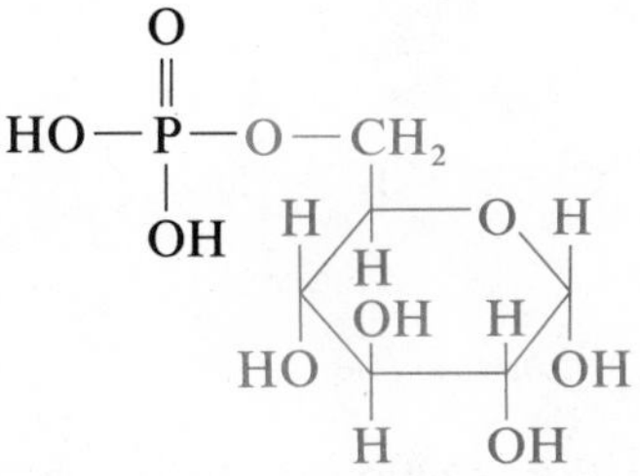

O
‖
HO—P—O—CH₂
|
OH   H   O   H
          H
          OH   H
HO   OH
      H   OH

glucose 6-phosphate

Some phosphotriesters are useful as insecticides. Although the phosphotriester insecticides are highly toxic compounds, they have largely replaced organochlorine insecticides such as DDT. The advantage of using phosphotriesters is that they do not last in the environment as long as the organochlorine insecticides. Usually phosphotriesters can be hydrolyzed to less toxic substances. The phosphotriester insecticides include esters in which the P=O of the ester has been replaced by P=S. Two examples of such insecticides are Dichlorovos, the active constituent in Shell's No-Pest Strip, and Parathion. These substances are extremely toxic and hazardous to living systems.

$$Cl_2C=CHO-\overset{\overset{\displaystyle O}{\|}}{\underset{\underset{\displaystyle OCH_3}{|}}{P}}-OCH_3$$

Dichlorovos
a phosphotriester

$$CH_3CH_2O-\overset{\overset{\displaystyle S}{\|}}{\underset{\underset{\displaystyle O}{|}}{P}}-OCH_2CH_3$$

Parathion
a phosphotriester

Amides

The most convenient way of obtaining amides is by treating an acid chloride with ammonia or amines. The reaction is driven to completion because of the escaping hydrogen chloride.

$$R-\overset{\overset{\displaystyle O}{\|}}{C}-Cl + H-\overset{\overset{\displaystyle H}{|}}{N}-H \longrightarrow R-\overset{\overset{\displaystyle O}{\|}}{C}-\overset{}{\underset{\underset{\displaystyle H}{|}}{N}}-H + HCl$$

an acid chloride   ammonia   *an amide*

Substituted amides can be made from the reaction of an acid chloride and a primary amine (see Section 15-3.1). A substituted amide is one in which —Cl has been replaced by the —NHR group.

$$R-\overset{\overset{\displaystyle O}{\|}}{C}-Cl + H-\overset{}{\underset{\underset{\displaystyle H}{|}}{N}}-R \longrightarrow R-\overset{\overset{\displaystyle O}{\|}}{C}-\overset{}{\underset{\underset{\displaystyle H}{|}}{N}}-R + HCl$$

an acid chloride   a primary amine   *a substituted amide*

For example, N-methylacetamide is made from the reaction of acetyl chloride and methyl amine.

$$CH_3\overset{\displaystyle O}{\overset{\|}{C}}-Cl \ + \ H-\underset{\underset{\displaystyle H}{|}}{N}-CH_3 \ \longrightarrow \ CH_3\overset{\displaystyle O}{\overset{\|}{C}}-\underset{\underset{\displaystyle H}{|}}{N}-CH_3 \ + \ HCl$$

acetyl chloride   methyl amine   N-methylacetamide

Amides are also important commercially. When adipic acid is treated with hexamethylenediamine, a polyamide known as *Nylon 66* is formed. The polyamide is called Nylon 66 because it is derived from two monomers each having six carbon atoms. It is possible to make other nylons by varying the number of carbons in either the acid or the amine. Nylon is a useful synthetic textile fiber. It is strong and can be drawn into fine strands. Permanent creases can be introduced and it is wrinkle resistant.

$$HO\overset{\displaystyle O}{\overset{\|}{C}}(CH_2)_4\overset{\displaystyle O}{\overset{\|}{C}}-OH \ + \ H-\underset{\underset{\displaystyle H}{|}}{N}(CH_2)_6NH_2 \ \overset{heat}{\longrightarrow} \ HO\overset{\displaystyle O}{\overset{\|}{C}}(CH_2)_4\overset{\displaystyle O}{\overset{\|}{C}}-\underset{\underset{\displaystyle H}{|}}{N}(CH_2)_6NH_2 \ + \ HOH$$

adipic acid   hexamethylenediamine   ↓ *further reaction*

$$\left[\ \sim\!\!\sim\!\!\sim O-\overset{\displaystyle O}{\overset{\|}{C}}(CH_2)_4\overset{\displaystyle O}{\overset{\|}{C}}-\underset{\underset{\displaystyle H}{|}}{N}(CH_2)_6\underset{\underset{\displaystyle H}{|}}{N}\!\!\sim\!\!\sim\!\!\sim \ \right]_n$$

Nylon 66
a polyamide

The *barbiturates* are amides of medical importance. Barbiturates are physiologically active hypnotics and sedatives. Barbiturates are formed by the reaction of the diethyl ester of malonic acid (a dicarboxylic acid) and urea. In living systems urea is the product of protein metabolism.

$$\begin{array}{c}\overset{\displaystyle O}{\overset{\|}{C}}-OC_2H_5\\ | \\ CH_2 \\ | \\ \underset{\underset{\displaystyle O}{\|}}{C}-OC_2H_5\end{array} \ + \ \begin{array}{c} H-N\\ \;\;\;\;\;\;\;\;\;\;\;\diagdown\\ \;\;\;\;\;\;\;\;\;\;\;\;C=O\\ \;\;\;\;\;\;\;\;\;\;\;\diagup\\ H-N\\ |\\ H \end{array} \ \longrightarrow \ 2\,C_2H_5OH \ + \ \begin{array}{c}\overset{\displaystyle O}{\overset{\|}{C}}-NH\\ \;\;\;\;\;\;\;\;\;\;\;\;\;\;\;\diagdown\\ CH_2\;\;\;\;\;\;\;\;\;\;C=O\\ \;\;\;\;\;\;\;\;\;\;\;\;\;\;\;\diagup\\ \underset{\underset{\displaystyle O}{\|}}{C}-NH\end{array}$$

diethyl malonate   urea   barbituric acid
*parent member of a
series of hypnotics
and sedatives*

Barbiturates depress the central nervous system. They are generally used for sedation and insomnia. A variety of barbiturates of varying activity are obtained using different substituent groups attached to the carbon atom bearing the two hydrogens. The potency of a barbiturate is determined by the length of time over which it depresses the central nervous system. Table 16.3 summarizes a variety of barbiturates according to their nature, action, and use.

Table 16.3   Some Common Barbiturates

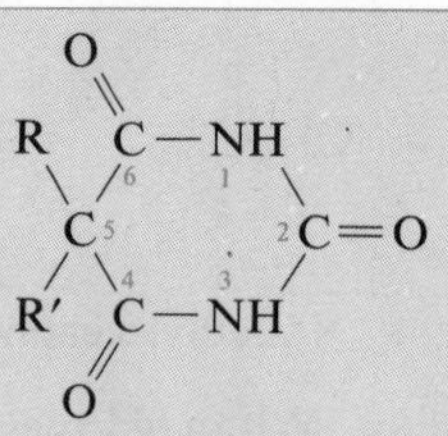

$$\begin{array}{c}
O \\
\parallel \\
R \quad C-NH \\
\diagdown \; {}_6 \qquad {}_1 \diagdown \\
C \; {}_5 \qquad \; {}_2 C=O \\
\diagup \; {}_4 \qquad {}_3 \diagup \\
R' \quad C-NH \\
\parallel \\
O
\end{array}$$

R	R'	Name (trade name)	Action	Onset	Duration	Use
CH_3CH_2-	$CH_3(CH_2)_2CH-$ $\mid$ CH_3	Thiopental sodium* (Pentothal)	Ultrashort	Seconds	Minutes	Intravenous anesthesia
CH_3CH_2-	$CH_3(CH_2)_2CH-$ $\mid$ CH_3	Pentobarbital (Nembutal)	Short	Minutes	4–8 hours	Brief hypnosis, preparative sedation, insomnia
CH_3CH_2-	$(CH_3)_2CH(CH_2)_2-$	Amobarbital (Amytal)	Intermediate	1 hour	6–8 hours	Insomnia
CH_3CH_2-	CH_3CH_2-	Barbital (Veronal)	Long	More than 1 hour	10–12 hours	Continuous sedation, hypertension, epilepsy
CH_3CH_2-	C_6H_5-	Phenobarbital (Luminal)				

*In the general structure shown above, carbon number 2 is doubly bonded to a sulfur atom.

16-3.3 DECARBOXYLATION

Decarboxylation is the loss of carbon dioxide. It is a reaction characteristic of some carboxylic acids both *in vitro* and *in vivo*. *In vitro* decarboxylation of β-keto acids, such as acetoacetic acid, when heated, gives rise to a ketone (acetone) and carbon dioxide. This decarboxylation occurs *in vivo* in the condition known as diabetes as was discussed in Section 15-2.1.

$$\underset{\substack{\text{acetoacetic acid} \\ \textit{a β-keto acid}}}{\overset{\displaystyle O \quad\;\; O}{\underset{\displaystyle \parallel \quad\;\; \parallel}{CH_3CCH_2C-OH}}} \xrightarrow{\text{heat}} \underset{\text{acetone}}{\overset{\displaystyle O}{\overset{\displaystyle \parallel}{CH_3CCH_3}}} + CO_2$$

In Chapter 17 we will look at the decarboxylation of amino acids *in vivo*. This process provides a variety of physiologically active amines.

PROBLEM

5. Each of the following reactions forms a functional derivative of a carboxylic acid. Give the general name and write the structure of the product of each reaction.

(a)

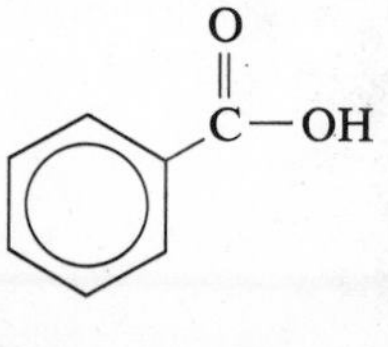

$$\begin{array}{c} O \\ \parallel \\ C-OH \end{array}$$

$+ \; SOCl_2 \longrightarrow$

(b) $CH_3\overset{\overset{\displaystyle O}{\|}}{C}OH + HO\overset{\overset{\displaystyle O}{\|}}{\underset{\underset{\displaystyle OH}{|}}{P}}OH \xrightarrow{\text{heat}}$

(c) $CH_3CH_2CH_2\overset{\overset{\displaystyle O}{\|}}{C}-OH + CH_3OH \longrightarrow$

(d)

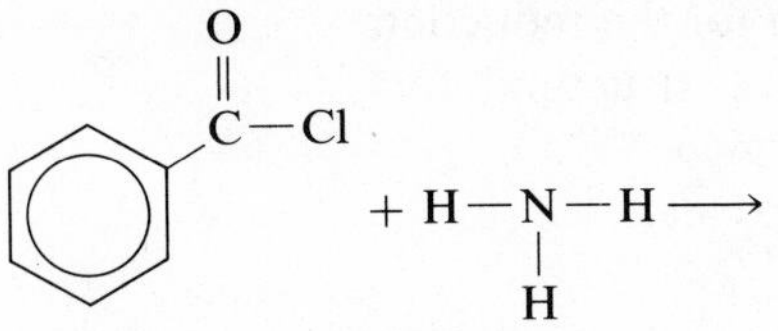

$\overset{\overset{\displaystyle O}{\|}}{C}-Cl$ (attached to benzene ring) $+ H-\underset{\underset{\displaystyle H}{|}}{N}-H \longrightarrow$

(e) $HO-\overset{\overset{\displaystyle O}{\|}}{\underset{\underset{\displaystyle OH}{|}}{P}}-OH + CH_3OH \ (2 \ mol) \longrightarrow$

(f) 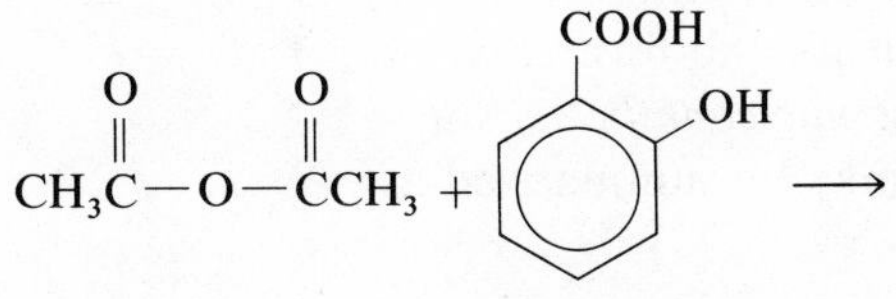 $CH_3\overset{\overset{\displaystyle O}{\|}}{C}-O-\overset{\overset{\displaystyle O}{\|}}{C}CH_3 + $ (benzene ring with COOH and OH) $\longrightarrow$

(g) $CH_3\overset{\overset{\displaystyle O}{\|}}{C}-\overset{\overset{\displaystyle O}{\|}}{C}-OH \xrightarrow{\text{pyruvic acid decarboxylase}}$

ANSWERS TO PROBLEM   5. (a) acid chloride, (benzene ring)$-\overset{\overset{\displaystyle }{}}{\underset{\underset{\displaystyle O}{\|}}{C}}-Cl$

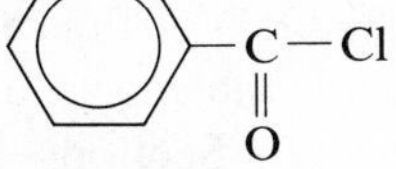

(b) anhydride, $CH_3\overset{\overset{\displaystyle O}{\|}}{C}-O-\overset{\overset{\displaystyle O}{\|}}{\underset{\underset{\displaystyle OH}{|}}{P}}-OH$

(c) ester, $CH_3CH_2CH_2\overset{\overset{\displaystyle O}{\|}}{C}-OCH_3$

(d) amide, (benzene ring)$-\overset{\overset{\displaystyle O}{\|}}{C}-NH_2$

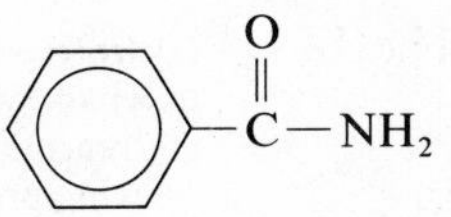

(e) phosphodiester, $CH_3O-\overset{\overset{\displaystyle O}{\|}}{\underset{\underset{\displaystyle OH}{|}}{P}}-OCH_3$

(f) ester (aspirin), (benzene ring with COOH and $O-\overset{\overset{\displaystyle O}{\|}}{C}-CH_3$)

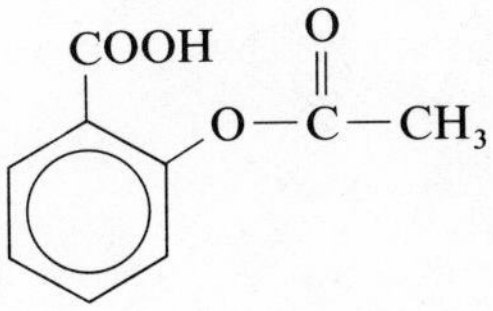

(g) $CH_3\overset{\overset{\displaystyle O}{\|}}{C}-H + CO_2$

16-4 REACTIONS OF CARBOXYLIC ACID DERIVATIVES

LEARNING OBJECTIVES

1. Define hydrolysis, saponification, soap, detergent, alcoholysis, and trans-esterification.

2. Given an equation for the hydrolysis of a carboxylic acid derivative, write the structure of the carboxylic acid formed.

3. Given an equation for the alcoholysis of a carboxylic acid derivative, write the structure of the ester formed.

4. Given the reaction for the reduction of an acid chloride, anhydride, or ester, write the structural formula of the alcohol formed.

5. Given the reaction for the reduction of an amide, write the structural formula of the amine formed.

6. Explain how a soap cleans.

The reactions of carboxylic acid derivatives are numerous. In this section we will learn about two of the more typical reactions of these functional derivatives and compare them to the reactions of the living system. One of the reactions of these functional derivatives follows the same mechanisms as that for the formation of functional derivatives from carboxylic acids. The reaction is *nucleophilic substitution*. The mechanism of reaction of functional derivatives may be summarized as shown.

$$\underset{\textit{functional derivative}}{R-\overset{\overset{\textstyle O}{\|}}{C}-Z} + :Y \longrightarrow R-\overset{\overset{\textstyle O}{\|}}{C}-Y + Z:$$

The group Z bonded to the carboxyl carbon is replaced by the group Y. Y can be a nucleophile such as H_2O or an alcohol (ROH). If the reactant Y is water, the reaction is called *hydrolysis;* if it is an alcohol, the reaction is called *alcoholysis.* Hydrolysis and alcoholysis will be discussed in Sections 16-4.1 and 16-4.2. A second reaction, reduction of carboxylic acid derivatives, will be discussed in Section 16-4.3.

16-4.1 HYDROLYSIS OF FUNCTIONAL DERIVATIVES

Each functional derivative undergoes hydrolysis to a carboxylic acid. The by-product is different in each case, however.

Hydrolysis is a reaction in which water is one of the reactants in a bond-cleaving process.

$$\underset{\substack{\textit{an acid} \\ \textit{chloride}}}{R-\overset{\overset{\textstyle O}{\|}}{C}-Cl} + \underset{\text{water}}{HOH} \longrightarrow \underset{\substack{\textit{a carboxylic} \\ \textit{acid}}}{R-\overset{\overset{\textstyle O}{\|}}{C}-OH} + \underset{\substack{\text{hydrogen} \\ \text{chloride}}}{HCl}$$

$$\underset{\textit{an anhydride}}{R-\overset{\overset{\textstyle O}{\|}}{C}-O-\overset{\overset{\textstyle O}{\|}}{C}-R} + \underset{\text{water}}{HOH} \longrightarrow \underset{\textit{a carboxylic acid}}{R-\overset{\overset{\textstyle O}{\|}}{C}-OH} + \underset{\text{carboxylic acid}}{R-\overset{\overset{\textstyle O}{\|}}{C}-OH}$$

$$\underset{\text{an ester}}{R-\overset{\displaystyle O}{\overset{\|}{C}}-OR'} + \underset{\text{water}}{HOH} \xrightarrow{H^+} \underset{\text{a carboxylic acid}}{R-\overset{\displaystyle O}{\overset{\|}{C}}-OH} + \underset{\text{alcohol}}{R'OH}$$

$$\underset{\text{an amide}}{R-\overset{\displaystyle O}{\overset{\|}{C}}-NH_2} + \underset{\text{water}}{HOH} \xrightarrow{H^+} \underset{\text{a carboxylic acid}}{R-\overset{\displaystyle O}{\overset{\|}{C}}-OH} + \underset{\text{ammonium ion}}{NH_4^+}$$

These functional derivatives do not have the same reactivity with water. Acid chlorides react rapidly and explosively. Anhydrides are moderately reactive. Esters and amides are less reactive.

hydrolysis reactivity: acid chlorides > anhydrides > esters > amides

$$\xrightarrow{\text{decreasing reactivity}}$$

For example, an acid chloride undergoes hydrolysis to the corresponding carboxylic acid. An anhydride is less reactive than an acid chloride in hydrolysis, but it also can be hydrolyzed to a carboxylic acid.

$$\underset{\text{acetyl chloride}}{CH_3\overset{\displaystyle O}{\overset{\|}{C}}-Cl} + HOH \longrightarrow \underset{\text{acetic acid}}{CH_3\overset{\displaystyle O}{\overset{\|}{C}}-OH} + HCl$$

$$\underset{\text{benzoic anhydride}}{\bigcirc\!\!\!-\overset{\displaystyle O}{\overset{\|}{C}}-O-\overset{\displaystyle O}{\overset{\|}{C}}-\!\!\!\bigcirc} + HOH \longrightarrow \underset{\text{benzoic acid}}{\bigcirc\!\!\!-\overset{\displaystyle O}{\overset{\|}{C}}-OH} + \underset{\text{benzoic acid}}{HO-\overset{\displaystyle O}{\overset{\|}{C}}-\!\!\!\bigcirc}$$

Because esters and amides are not very reactive with water, their hydrolysis requires an acidic or basic solution. A typical example of an ester hydrolysis in basic solution is **saponification**. *Saponification* has historical meaning. This reaction was once the method for making soap from animal fat. In this reaction the ester is converted to an alcohol and a carboxylic acid salt (a carboxylate).

Saponification is the hydrolysis of an ester in basic solution.

$$\underset{\text{an ester}}{R\overset{\displaystyle O}{\overset{\|}{C}}-OR'} + HOH \xrightarrow{NaOH} \underset{\substack{\text{a sodium} \\ \text{carboxylate}}}{R\overset{\displaystyle O}{\overset{\|}{C}}-O^-Na^+} + \underset{\text{an alcohol}}{R'OH}$$

When the ester is derived from animal fat, the resulting carboxylic acid salt is called **soap**. Tristearin is a typical ester of animal fat. Saponification of tristearin forms glycerol, an alcohol used in hand lotions as a moisturizer, and sodium stearate. The sodium salt obtained in this saponification is a soap.

A **soap** is any salt of a fatty acid.

$$
\begin{array}{c}
\overset{\displaystyle O}{\overset{\displaystyle \|}{CH_2OC(CH_2)_{16}CH_3}} \\
| \quad \overset{\displaystyle O}{\overset{\displaystyle \|}{}} \\
CHOC(CH_2)_{16}CH_3 \\
| \quad \overset{\displaystyle O}{\overset{\displaystyle \|}{}} \\
CH_2OC(CH_2)_{16}CH_3
\end{array}
\;+\; 3\,HOH \xrightarrow{\text{NaOH}}\;
\begin{array}{c}
CH_2OH \\
| \\
CHOH \\
| \\
CH_2OH
\end{array}
\;+\; 3\,CH_3(CH_2)_{16}\overset{\displaystyle O}{\overset{\displaystyle \|}{C}}-O^-Na^+
$$

tristearin		glycerol	sodium stearate
an ester			*a soap*

In the past, the making of soap was a simple process. Fat, usually beef tallow, was boiled in the presence of water and lye (sodium hydroxide). The process is the same today but modern technology is more efficient than home production. In addition, certain ingredients are added: perfumes, to make fragrant soaps; antiseptics, to make deodorants; and dyes, to give a soap color. Although bar and laundry soaps are sodium salts of various fatty acids, potassium soaps are sometimes desirable. Potassium salts are softer soaps and are more soluble, but they are more expensive to produce. These soaps are often used in liquid soaps. Lithium soaps are lubricants.

How does soap clean? A soap molecule consists of a *hydrophobic* (water-insoluble) hydrocarbon end and a *hydrophilic* (water-soluble) carboxylate end.

$$
CH_3CH_2CH_2CH_2CH_2CH_2CH_2CH_2CH_2CH_2CH_2CH_2CH_2CH_2CH_2CH_2CH_2C \overset{\displaystyle O}{\underset{\displaystyle O^-Na^+}{\diagup\diagdown}}
$$

hydrophobic	*hydrophilic*
sodium stearate	

Oils and greases (dirt) are non-polar, water-insoluble substances. The hydrophobic end of the soap molecule attaches itself to the surface of the grease particle. The hydrophilic end of the soap molecule is soluble in the water. As Figure 16.1 shows, many soap molecules arrange themselves around the surface of the grease particle. A grease particle surrounded by soap molecules is called a *micelle*. Because the negatively charged surfaces of these micelles repel each other, the micelles remain suspended in the water solution and are eventually washed away.

There is one significant problem associated with soaps. Many ions frequently found in water, particularly Mg^{2+} and Ca^{2+}, form insoluble soaps. Water containing such ions is called hard water. The insoluble soaps in the water appear as an unsightly scum. The formation of these insoluble soaps diminishes the cleansing ability of soap. To some extent soaps have been replaced by synthetic detergents (syndets) because of the improved cleansing ability of the latter. Like soaps, syndets contain a hydrophobic hydrocarbon end and a hydrophilic end. It is the hydro-

Hard water is water that contains Mg^{2+} and Ca^{2+} and causes the precipitation of soap.

Syndet (synthethic detergent) is a substance in which the carboxylated group of a soap molecule has been replaced by a sulfate group or an ammonium group.

Figure 16.1   Cleansing action of soap. The soap molecules orient their hydrophobic ends toward grease and oil particles that rest on the surface (a). The soap molecules attach themselves to the grease molecules (b) and disrupt the grease molecules from the surface (c). The hydrophilic ends of the soap molecules protrude into the aqueous medium and stabilize the grease particles in aqueous solution so that they can be washed away (c).

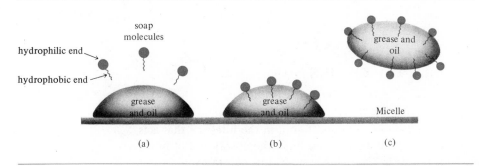

philic end that differs in soaps and syndets. Syndets that have a negatively charged sulfate group are called *anionic syndets* and those with a positively charged ammonium group are called *cationic syndets.* As shown in Table 16.4 the cationic syndets are useful as germicides.

Table 16.4   Some Ammonium Compounds of Germicidal Use

Name	Structure	Germicidal Use
Benzethonium chloride	$\left[C_8H_{17}\text{—}\bigcirc\text{—}OCH_2CH_2OCH_2CH_2\text{—}\overset{\displaystyle CH_3}{\underset{\displaystyle CH_3}{N}}\text{—}CH_2\text{—}\bigcirc \right]^+ \; Cl^-$ (with CH_3 on ring)	Skin disinfectant
Cetylpyridinium chloride	$\left[\bigcirc_{N}\text{—}CH_2(CH_2)_{14}CH_3 \right]^+ \; Cl^-$	Mucous membrane disinfectant, milder than benzethonium chloride
Methylbenzethonium chloride (Diaparene)	$\left[C_8H_{17}\text{—}\bigcirc\text{—}OCH_2CH_2OCH_2CH_2\text{—}\overset{\displaystyle CH_3}{\underset{\displaystyle CH_3}{N}}\text{—}CH_2\text{—}\bigcirc \right]^+ \; Cl^-$	Disinfectant for clothes, especially diapers

$$CH_3CH_2CH_2CH_2CH_2CH_2CH_2CH_2CH_2\overset{\overset{\displaystyle O}{\|}}{C}-O^- \quad \textit{a soap}$$

$$CH_3CH_2CH_2CH_2CH_2CH_2CH_2CH_2CH_2CH_2O-\overset{\overset{\displaystyle O}{\|}}{\underset{\underset{\displaystyle O}{\|}}{S}}-O^- \quad \textit{an anionic syndet}$$

$$CH_3CH_2CH_2CH_2CH_2CH_2CH_2CH_2CH_2CH_2\overset{\overset{\displaystyle CH_3}{|}}{\underset{\underset{\displaystyle CH_3}{|}}{\overset{+}{N}}}-CH_3 \quad \textit{a cationic syndet}$$

Cholic acid (see Section 27-1.2) is a compound found in the intestine that acts like a soap and a syndet. Its main function is to facilitate the absorption of lipids (water insoluble organic substances in the living system) through the wall of the intestine. The sodium salt of cholic acid has a large hydrophobic hydrocarbon end and a hydrophilic carboxylate end and is, therefore, soaplike. Cholic acid disrupts some of the hydrophobic bonds of the ingested lipid molecules that hold the lipid molecules together. It converts the lipids into smaller globules that are more readily digested and more easily transported.

hydrophobic

cholic acid
a naturally occurring syndet

Amides undergo hydrolysis in acid solution to yield carboxylic acids and ammonium salts.

$$R-\overset{\overset{\displaystyle O}{\|}}{C}-NH_2 + HOH + HCl \longrightarrow R-\overset{\overset{\displaystyle O}{\|}}{C}-OH + NH_4{}^+Cl^-$$

an amide   *a carboxylic acid*

$$CH_3CH_2\overset{\overset{\displaystyle O}{\|}}{C}-NH_2 + HOH + HCl \longrightarrow CH_3CH_2\overset{\overset{\displaystyle O}{\|}}{C}-OH + NH_4{}^+Cl^-$$

propionamide   propionic acid

The amide linkage (shown below in color in the formula of the dipeptide) is the characteristic functional group found in **peptides.** Hydrolysis of a peptide yields amino acids. For example, hydrolysis of a dipeptide yields two amino acids as shown. This hydrolysis reaction of a dipeptide and peptides in general is important to the living system. Peptides are a major part of our diets and digestion (hydrolysis) of peptides provides some amino acids which we are incapable of synthesizing *in vivo.* These amino acids are called essential amino acids and will be discussed further in Chapter 28. Peptides are examples of substituted amides. We will discuss substituted amides again in Section 17-4.1.

A **peptide** is two or more amino acids that are joined covalently by an amide bond.

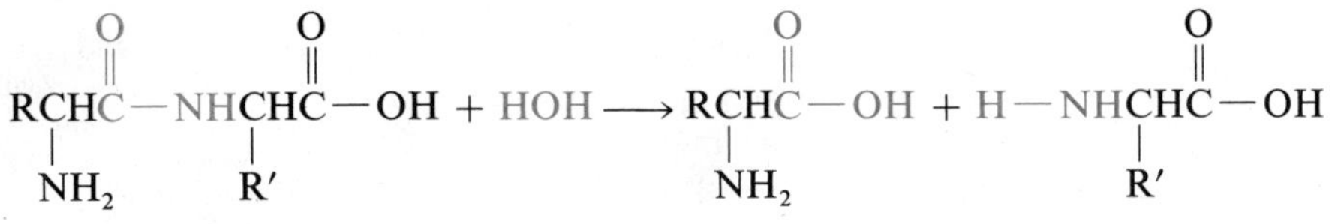

$$
\underset{\text{a dipeptide}}{\overset{\text{O}\qquad\qquad\text{O}}{\text{RCHC}-\text{NHCHC}-\text{OH}}} + \text{HOH} \longrightarrow \underset{\text{an amino acid}}{\overset{\text{O}}{\text{RCHC}-\text{OH}}} + \underset{\text{an amino acid}}{\overset{\text{O}}{\text{H}-\text{NHCHC}-\text{OH}}}
$$

a dipeptide (with NH₂ and R′ substituents) + HOH → an amino acid (NH₂) + an amino acid (R′)

16-4.2 ALCOHOLYSIS

In **alcoholysis,** the functional derivative reacts with an alcohol. The major organic product is an ester, but the by-product is different in each case.

Alcoholysis is a reaction in which an alcohol is one of the reactants in a bond-cleaving process.

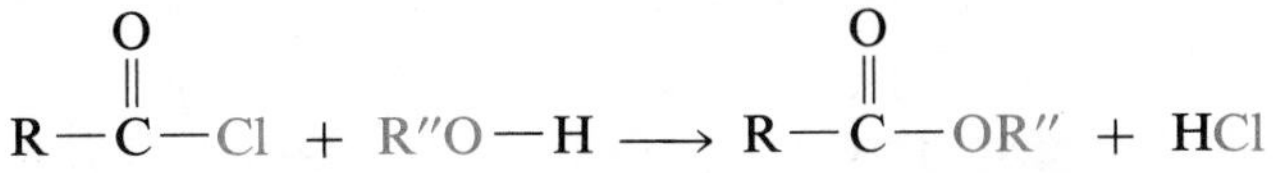

$$
\underset{\text{an acid chloride}}{\overset{\text{O}}{\text{R}-\text{C}-\text{Cl}}} + \underset{\text{an alcohol}}{\text{R}''\text{O}-\text{H}} \longrightarrow \underset{\text{an ester}}{\overset{\text{O}}{\text{R}-\text{C}-\text{OR}''}} + \underset{\substack{\text{hydrogen}\\\text{chloride}}}{\text{HCl}}
$$

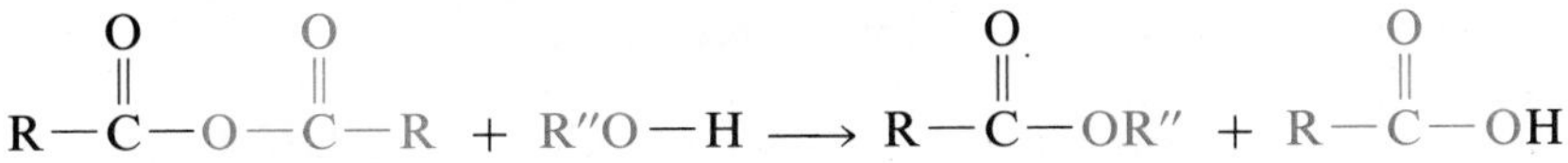

$$
\underset{\text{an anhydride}}{\overset{\text{O}\qquad\text{O}}{\text{R}-\text{C}-\text{O}-\text{C}-\text{R}}} + \underset{\text{an alcohol}}{\text{R}''\text{O}-\text{H}} \longrightarrow \underset{\text{an ester}}{\overset{\text{O}}{\text{R}-\text{C}-\text{OR}''}} + \underset{\text{carboxylic acid}}{\overset{\text{O}}{\text{R}-\text{C}-\text{OH}}}
$$

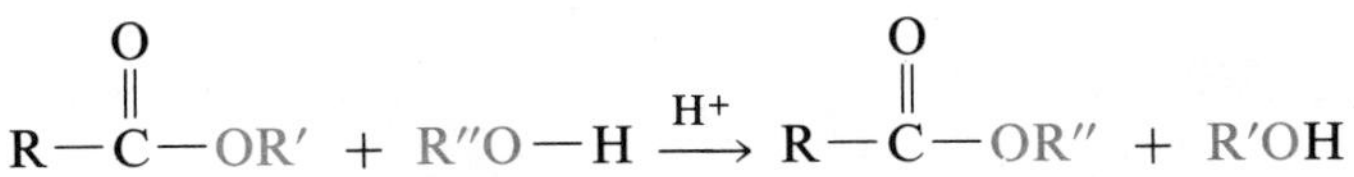

$$
\underset{\text{an ester}}{\overset{\text{O}}{\text{R}-\text{C}-\text{OR}'}} + \underset{\text{an alcohol}}{\text{R}''\text{O}-\text{H}} \overset{\text{H}^+}{\longrightarrow} \underset{\text{an ester}}{\overset{\text{O}}{\text{R}-\text{C}-\text{OR}''}} + \underset{\text{alcohol}}{\text{R}'\text{OH}}
$$

The order of reactivity of the functional derivatives in alcoholysis is the same as in hydrolysis.

alcoholysis reactivity: acid chlorides > anhydrides > esters

decreasing reactivity
→

For example, acetyl chloride reacts with methyl alcohol to form the ester methyl acetate and benzoic anhydride reacts with ethyl alcohol to form ethyl benzoate.

$$CH_3\overset{\displaystyle O}{\overset{\|}{C}}-Cl + CH_3O-H \longrightarrow CH_3\overset{\displaystyle O}{\overset{\|}{C}}-OCH_3 + HCl$$

acetyl chloride   methyl alcohol   methyl acetate

$$\bigcirc -\overset{\displaystyle O}{\overset{\|}{C}}-O-\overset{\displaystyle O}{\overset{\|}{C}}-\bigcirc + CH_3CH_2O-H \longrightarrow \bigcirc -\overset{\displaystyle O}{\overset{\|}{C}}-OCH_2CH_3 + \bigcirc -\overset{\displaystyle O}{\overset{\|}{C}}-OH$$

benzoic anhydride   ethyl alcohol   ethyl benzoate

The product of the reaction of an ester with an alcohol is another ester. This process is called **transesterification**. For example, methyl acetate reacts with isopropyl alcohol to form isopropyl acetate. In effect, the methyl group of the original ester has been replaced by the isopropyl group of the reacting alcohol.

Transesterification is the reaction of an ester with an alcohol. The product is a new ester in which the reacting alcohol replaces the original alcohol group.

$$CH_3\overset{\displaystyle O}{\overset{\|}{C}}-OCH_3 + CH_3\underset{\underset{\displaystyle CH_3}{|}}{CH}O-H \longrightarrow CH_3\overset{\displaystyle O}{\overset{\|}{C}}-O\underset{\underset{\displaystyle CH_3}{|}}{CH}CH_3 + CH_3O-H$$

methyl acetate   isopropyl   isopropyl   methyl alcohol
                alcohol   acetate

Transesterification is a commercially important reaction. When the dimethyl ester of terephthalic acid (dimethyl terephthalate) is treated with ethylene glycol, *Dacron* polyester fiber is formed. Dacron is a synthetic fiber known for its remarkable resistance to creasing and wrinkling. Dacron fibers are unusually stiff and harsh, so they are blended with wool and cotton in fabrics. The fibers also can be set by heat. This means that permanent creases can be introduced in garments manufactured from these fibers. Thus, the garments may be used repeatedly with no ironing.

$$CH_3O\overset{\displaystyle O}{\overset{\|}{C}}-\bigcirc-\overset{\displaystyle O}{\overset{\|}{C}}OCH_3 + HOCH_2CH_2OH \longrightarrow$$

dimethyl terephthalate   ethylene glycol

$$\left(\!\!\!\sim OCH_2CH_2O-\overset{\displaystyle O}{\overset{\|}{C}}-\bigcirc-\overset{\displaystyle O}{\overset{\|}{C}}OCH_2CH_2O\overset{\displaystyle O}{\overset{\|}{C}}-\bigcirc-\overset{\displaystyle O}{\overset{\|}{C}}\sim\!\!\!\right)_n$$

Dacron

In addition to its use in wash-and-wear fabrics, Dacron is used in a variety of ways in medicine and surgery. When a wound is repaired, the edges must be joined and firmly secured until the tissue heals and regains its mechanical strength. Dacron sutures are often used to accomplish this and then are removed after the wound heals. Occasionally, as a result of blood clots or arteriosclerosis, it is neces-

Table 16.5 Polymers Useful in Medicine and Surgery

Polymer	Medical or Surgical Use
Poly(lactide-co-glycolide)	Absorbable sutures
Poly(methyl methacrylate)	Intraocular (within the eye) replacement for lenses during cataract surgery
Poly(alkyl cyanoacrylate)	Tissue adhesives, protective coatings in treating burns
Teflon	Applied as a paste to vocal cords to correct unilateral paralysis and restore voice
Silicones	Protective hand creams, for tear ducts to control glaucoma, corrective plastic surgery, oxygenating membrane for heart lung machine

sary to replace sections of blood vessels that are blocked or malfunctioning. After the damaged or diseased sections have been removed, it may not be possible to stretch the remaining portion enough to allow reconnection. Flexible tubing made of Dacron has been used to connect large veins. Dacron fabric coated with silicone rubber has been considered as a replacement for the trachea, the air duct to the lungs, and the esophagus, through which food travels to the stomach. The versatility of other polymers and their potentials for medical use are summarized in Table 16.5.

16-4.3 REDUCTION

Carboxylic acids and their functional derivatives can be reduced. The reagent lithium aluminum hydride ($LiAlH_4$) gives the best results and the product in each case, except amides, is an alcohol. Amides are reduced to amines.

$$R-\overset{\overset{\textstyle O}{\|}}{C}-Cl$$
an acid chloride

$$R-\overset{\overset{\textstyle O}{\|}}{C}-O-\overset{\overset{\textstyle O}{\|}}{C}-R$$
an anhydride

$$R-\overset{\overset{\textstyle O}{\|}}{C}-OR$$
an ester

$$R-\overset{\overset{\textstyle O}{\|}}{C}-OH$$
a carboxylic acid

$$\xrightarrow{\;LiAlH_4\;} RCH_2OH$$
an alcohol

$$R-\overset{\overset{\textstyle O}{\|}}{C}-NH_2 \xrightarrow{\;LiAlH_4\;} RCH_2NH_2$$
an amide   *an amine*

For example, benzoyl chloride is reduced to benzyl alcohol and acetic anhydride becomes ethyl alcohol.

benzoyl chloride   benzyl alcohol

acetic anhydride   ethyl alcohol

Functional derivatives are more easily reduced than carboxylic acids. The living system takes advantage of the ease of reduction of functional derivatives over the carboxylic acids. In living systems, the carboxyl group or its derivatives can be reduced to the carbonyl group rather than an alcohol.

a carboxyl group   a carbonyl group

For example, 3-phosphoglyceric acid (a carboxylic acid) is converted to the anhydride, 1,3-diphosphoglyceric acid, which is subsequently reduced to 3-phosphoglyceraldehyde (a carbonyl compound) rather than to the carbonyl group in a one-step reduction.

3-phosphoglyceric acid
a carboxylic acid

1,3-diphosphoglyceric acid
an anhydride

3-phosphoglyceraldehyde
an aldehyde

PROBLEM

6. Describe the following and write the product of the reaction of each functional derivative.

(a)

(b) $CH_3C-O-CCH_3 + H_2O \longrightarrow$

(c)

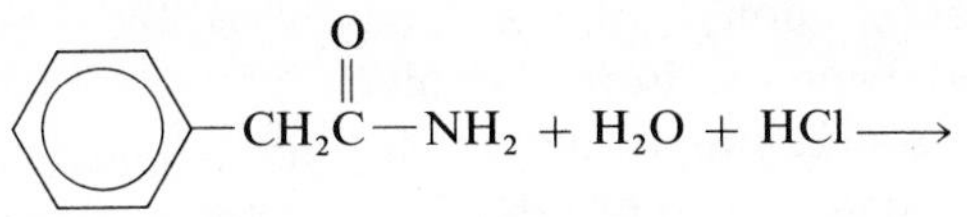

$$\text{(benzyl ring)} - CH_2\overset{\displaystyle O}{\overset{\|}{C}} - NH_2 + H_2O + HCl \longrightarrow$$

(d) $CH_3CH_2\overset{\displaystyle O}{\overset{\|}{C}} - Cl + CH_3OH \longrightarrow$

(e) 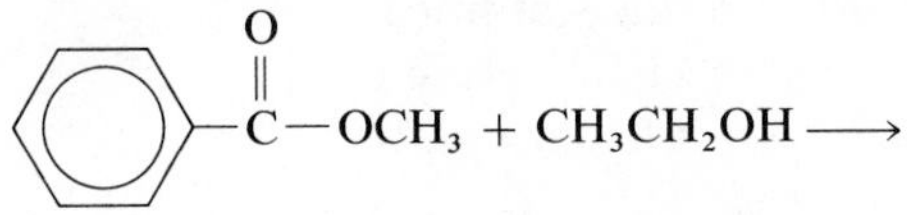 $\text{(benzyl ring)} - \overset{\displaystyle O}{\overset{\|}{C}} - OCH_3 + CH_3CH_2OH \longrightarrow$

(f) $CH_3CH_2\overset{\displaystyle O}{\overset{\|}{C}} - NH_2 \xrightarrow{\text{LiAlH}_4}$

(g) $CH_3\overset{\displaystyle O}{\overset{\|}{C}} - OCH_2CH_3 \xrightarrow{\text{LiAlH}_4}$

ANSWERS TO PROBLEM   **6. (a)** hydrolysis of an acid chloride, $\text{(benzyl ring)} - \overset{\displaystyle O}{\overset{\|}{C}} - OH$

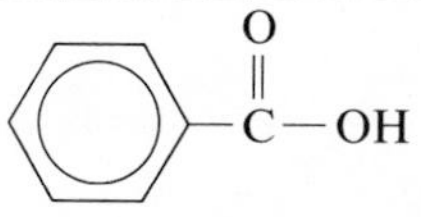

(b) hydrolysis of an anhydride, $2\ CH_3\overset{\displaystyle O}{\overset{\|}{C}} - OH$ **(c)** hydrolysis of an amide,

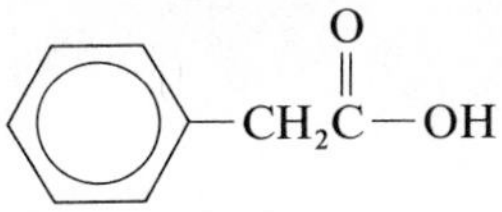

$\text{(benzyl ring)} - CH_2\overset{\displaystyle O}{\overset{\|}{C}} - OH$ **(d)** alcoholysis of an acid chloride,

$CH_3CH_2\overset{\displaystyle O}{\overset{\|}{C}} - OCH_3$ **(e)** transesterification, $\text{(benzyl ring)} - \overset{\displaystyle O}{\overset{\|}{C}}OCH_2CH_3 + CH_3OH$

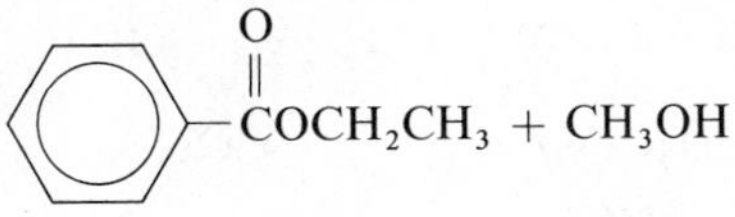

(f) reduction of an amide, $CH_3CH_2CH_2NH_2$ **(g)** reduction of an ester, CH_3CH_2OH

16-5 FOOD ADDITIVES

LEARNING OBJECTIVES

1. Define food additive.
2. List the five categories of food additives.
3. Summarize the various risk categories of food additives.

Carboxylic acids and carboxylic acid derivatives are commonly used as food additives. Additives have economic and health-related benefits. For example, vitamin D is routinely added to milk to prevent rickets, a softening of the bones, and calcium propionate is added to bakery products to extend their shelf-life.

Food additives are usually categorized as follows:

1. *Flavor enhancers* are usually artificial and include coloring and flavoring compounds that provide a desirable taste and appearance. In many cases, the use of imitation fruit flavorings permits the marketing of useful, tasty products at prices that the majority of people can afford. Esters blended with other chemicals and essential oils produce a fruity flavor and are commonly used in fruit flavorings. In Table 16.6 some esters used in fruit flavorings are listed.

Food additives are chemicals incorporated in foods and food preparations for the purpose of enhancing their flavor or attractiveness and prolonging their lifetime on the shelf.

Table 16.6 Esters Used As Food Additives

Structure	Name of Ester	Flavor
$CH_3COO(CH_2)_4CH_3$	Amyl acetate*	Banana
$o\text{-}H_2NC_6H_4COOCH_3$	Methyl o-aminobenzoate	Grape
$CH_3(CH_2)_2COOCH_3$	Methyl butyrate	Apple
$CH_3COO(CH_2)_7CH_3$	Octyl acetate	Orange
$CH_3COOCH_2CH{=}C(CH_3)_2$	3-Methyl-2-butenyl ethanoate	Juicy Fruit gum
$CH_3(CH_2)_2COO(CH_2)_4CH_3$	Amyl butyrate*	Apricot

*The pentyl group, $CH_3CH_2CH_2CH_2CH_2-$, is often called *amyl*.

2. *Acidity-controlling agents* are used to augment fruit flavors and impart a tartness to foods. Examples are citric acid and tartaric acid.

citric acid   tartaric acid

3. *Antioxidants* keep fatty products (e.g., butter and cheese) from smelling or tasting bad. Bad odor and taste are a common type of spoilage in foods that is known as *rancidity*. Vitamin C (ascorbic acid) and BHA (butylated hydroxyanisole) are examples of antioxidants.

ascorbic acid   BHA

4. *Color and flavor stabilizers* are preservatives that retard spoilage and mold formation in jams, jellies, breads, and cheeses and they prevent damage to flavor and color in foods. Calcium propionate and sodium benzoate, salts of carboxylic acids, are typical preservatives.

$$(CH_3CH_2\overset{\overset{\displaystyle O}{\|}}{C}-O^-)_2Ca^{2+}$$

calcium propionate

$$\langle\hspace{-0.3em}\bigcirc\hspace{-0.3em}\rangle-\overset{\overset{\displaystyle O}{\|}}{C}-O^-Na^+$$

sodium benzoate

5. *Emulsifiers and thickeners* keep foods from separating and add thickness to the consistency of foods. Propylene glycol monostearate, an ester, is an emulsifier found in Dream Whip food topping.

$$CH_3(CH_2)_{16}\overset{\overset{\displaystyle O}{\|}}{C}OCH_2\overset{\overset{\displaystyle OH}{|}}{C}HCH_3$$

propylene glycol monostearate

Whether or not the use of a chemical additive is dangerous is indicated by the risk category in which the chemical additive is placed. Food additives fall into one of the following risk categories.

1. *Chemicals that occur in nature and are easily metabolized by the body.* These are the safest additives. They include proteins, which tenderize meat and speed the brewing of beer; vitamins and minerals, which increase the nutritional value of foods; starches such as cornstarch, a wholesome thickening agent; and sugars. If taken in excess, however, any of these chemicals can be dangerous to life. A high sugar intake, for example, can add fattening calories, cause dental caries (tooth decay), and lead to high blood sugar levels.

2. *Chemicals that are not absorbed by the body.* These are complex chemicals that cannot be digested by enzymes and therefore do not enter the bloodstream, where they can damage the vital organs. Cellulose, a major component of broccoli, lettuce, and other vegetables, is typical of these chemicals. These are generally safe and act as natural laxatives.

3. *Chemicals that occur naturally and are known to affect the body's operation.* This category includes caffeine, a powerful stimulant that occurs naturally in coffee, tea, and chocolate and is added to cola soft drinks. Quinine, a soft drink flavoring, can, in large doses, control malaria and also cause blindness and severe heart damage. Scientists are unsure of the long-term effect of these chemicals on the human body.

caffeine

4. *The synthetics, that is, chemicals that do not occur in nature and are absorbed into the body.* The use of these chemicals may involve high risk. They include many artificial colors and some preservatives. These chemicals are transported to the liver, where they are thought to be detoxified. Examples of compounds in this category are sodium cyclamate, a synthetic sweetener, and methyl tetra-O-methyl carminate, the yellow dye found in margarine.

sodium cyclamate   methyl tetra-O-methyl carminate

Ideally, every additive should be tested to determine its exact chemical makeup, how the body metabolizes it, and whether or not long-term ingestion of it can have harmful effects. However, this is a time-consuming and expensive task. It does not follow that if a chemical harms an animal, it will harm a human being. The animal could differ from human beings in some crucial metabolic reaction, so that the animal could be more (or less) sensitive to an additive; thus a false result for human beings could be obtained. This hazard is minimized if the chemical is fed to more than one species, but still the question of similarity to human beings remains.

SUMMARY

Carboxylic acids contain the carboxyl group. They have common names (see Table 16.1) even if R is not a simple alkyl group. In the IUPAC system the carboxyl carbon is number one and the longest continuous chain containing the carboxyl carbon is the parent structure. The appropriate suffix is *-oic acid*.

$$R-\overset{\displaystyle O}{\overset{\|}{C}}-OH \qquad -\overset{\displaystyle O}{\overset{\|}{C}}-OH$$

a carboxylic acid   *a carboxyl group*

The carboxylic acid derivatives are acid chlorides, anhydrides, ester and amides.

$$R-\overset{\displaystyle O}{\overset{\|}{C}}-Cl \quad R-\overset{\displaystyle O}{\overset{\|}{C}}-O-\overset{\displaystyle O}{\overset{\|}{C}}-R \quad R-\overset{\displaystyle O}{\overset{\|}{C}}-OR \quad R-\overset{\displaystyle O}{\overset{\|}{C}}-NH_2$$

an acid chloride   *an anhydride*   *an ester*   *an amide*

These derivatives are named as summarized in Table 16.2.

Carboxylic acids are prepared by oxidation of primary alcohols, aldehydes, or aromatic hydrocarbons and hydrolysis of nitriles.

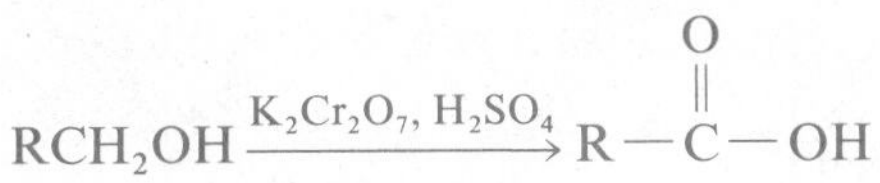

$$RCH_2OH \xrightarrow{K_2Cr_2O_7,\ H_2SO_4} R-\overset{\overset{\displaystyle O}{\|}}{C}-OH$$

a primary alcohol

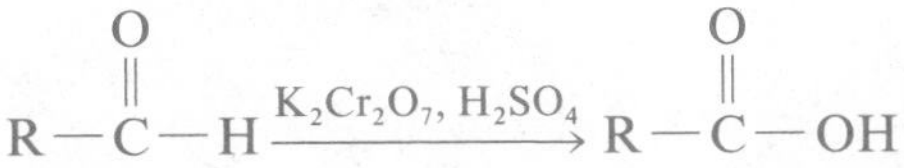

$$R-\overset{\overset{\displaystyle O}{\|}}{C}-H \xrightarrow{K_2Cr_2O_7,\ H_2SO_4} R-\overset{\overset{\displaystyle O}{\|}}{C}-OH$$

an aldehyde

$$C_6H_5R \xrightarrow{K_2Cr_2O_7,\ H_2SO_4} C_6H_5\overset{\overset{\displaystyle O}{\|}}{C}-OH$$

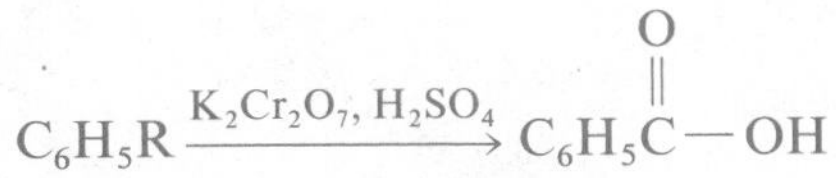

an aromatic hydrocarbon

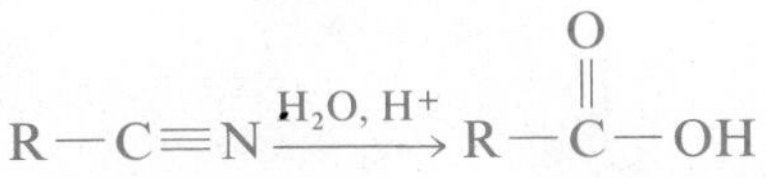

$$R-C\equiv N \xrightarrow{H_2O,\ H^+} R-\overset{\overset{\displaystyle O}{\|}}{C}-OH$$

a nitrile

Carboxylic acids are stronger acids than phenols or alcohols. They react with bases (NaOH or NaHCO$_3$) to form carboxylates.

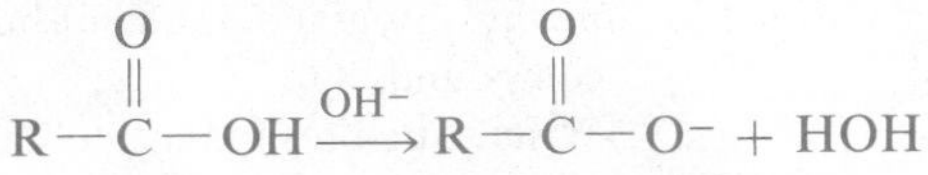

$$R-\overset{\overset{\displaystyle O}{\|}}{C}-OH \xrightarrow{OH^-} R-\overset{\overset{\displaystyle O}{\|}}{C}-O^- + HOH$$

a carboxylic acid   *a carboxylate anion*

Carboxylic acids form a variety of carboxylic acid derivatives. Each of these reactions is an example of nucleophilic substitution.

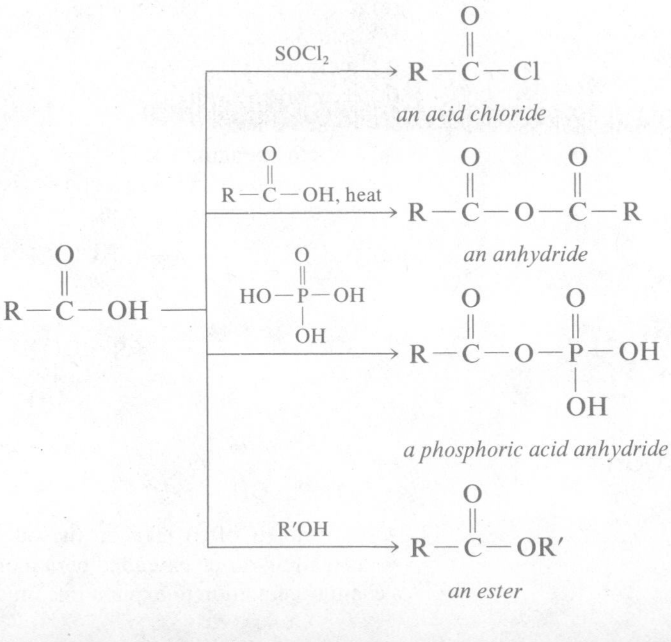

$$R-\overset{\overset{\displaystyle O}{\|}}{C}-OH$$

$\xrightarrow{SOCl_2}$ R$-\overset{\overset{\displaystyle O}{\|}}{C}-$Cl

an acid chloride

$\xrightarrow{R-\overset{\overset{\displaystyle O}{\|}}{C}-OH,\ heat}$ R$-\overset{\overset{\displaystyle O}{\|}}{C}-O-\overset{\overset{\displaystyle O}{\|}}{C}-$R

an anhydride

$\xrightarrow{HO-\overset{\overset{\displaystyle O}{\|}}{\underset{\underset{\displaystyle OH}{|}}{P}}-OH}$ R$-\overset{\overset{\displaystyle O}{\|}}{C}-O-\overset{\underset{\underset{\displaystyle OH}{|}}{P}}{}-$OH

a phosphoric acid anhydride

$\xrightarrow{R'OH}$ R$-\overset{\overset{\displaystyle O}{\|}}{C}-OR'$

an ester

Carboxylic acid derivatives undergo hydrolysis, alcoholysis, and reduction. Hydrolysis gives carboxylic acids, alcoholysis gives esters, and reduction gives alcohols (with the exception of an amide, which gives an amine).

$$
\underset{\text{an ester}}{R-\overset{\overset{\displaystyle O}{\|}}{C}-OR''} \xleftarrow[\text{alcoholysis}]{R''OH}
\begin{bmatrix}
R-\overset{\overset{\displaystyle O}{\|}}{C}-Cl \\
\text{or} \\
R-\overset{\overset{\displaystyle O}{\|}}{C}-O-\overset{\overset{\displaystyle O}{\|}}{C}-R \\
\text{or} \\
R-\overset{\overset{\displaystyle O}{\|}}{C}-OR' \\
\text{or} \\
R-\overset{\overset{\displaystyle O}{\|}}{C}-NH_2
\end{bmatrix}
\xrightarrow[\text{hydrolysis}]{HOH}
\underset{\text{a carboxylic acid}}{R-\overset{\overset{\displaystyle O}{\|}}{C}-OH}
$$

ADDITIONAL PROBLEMS

7. Name the following carboxylic acids.

(a) $C_6H_5-\overset{\overset{\displaystyle O}{\|}}{C}-OH$

(b) $CH_3CH_2CH_2CH_2\overset{\overset{\displaystyle O}{\|}}{C}-OH$

(c) $HO-\overset{\overset{\displaystyle O}{\|}}{C}CH_2CH_2\overset{\overset{\displaystyle O}{\|}}{C}-OH$

(d) $CH_3\underset{\underset{\displaystyle Cl}{|}}{C}HCH_2\overset{\overset{\displaystyle O}{\|}}{C}-OH$

(e) $CH_3\underset{\underset{\displaystyle OH}{|}}{C}H\overset{\overset{\displaystyle O}{\|}}{C}-OH$

(f) $H_3C-C_6H_4-COOH$

(g) $CH_3CH_2CH=CHCH_2CH_2COOH$

(h) $HCOOH$

(i) Br_2CHCH_2COOH

(j) $ClCH_2C\equiv CCH_2COOH$

8. Write the structure of the acid chloride, anhydride, methyl ester, and amide of butyric acid. Name each compound.

9. Write structures for each of the following carboxylic acids.
   (a) butyric acid
   (b) 3,5-dinitrobenzoic acid
   (c) 2-pentenoic acid
   (d) propanoic acid
   (e) 2-iodo-3-methylhexanoic acid
   (f) salicylic acid
   (g) malonic acid
   (h) terephthalic acid
   (i) octanoic acid
   (j) 3-methylpentanoic acid

10. Write the product of the reaction of acetic acid with each of the following reagents.

    (a) NaOH   (b) CH_3CH_2OH

    (c) $SOCl_2$ (d) $HO\underset{\underset{\displaystyle OH}{|}}{\overset{\overset{\displaystyle O}{\|}}{P}}OH$

    (e) $CH_3\overset{\overset{\displaystyle O}{\|}}{C}-OH$ + heat

11. Aspirin tablets often take on the odor of vinegar after standing for an extended period of time. Write a chemical equation to explain this situation.

12. What are food additives? Describe each of the categories for classifying food additives. Briefly describe each risk category.
13. We mentioned or discussed each of the following substances in this chapter. Draw the structure, or the major component, of each and identify the functional group of each discussed in this chapter.
    (a) vinegar (b) nutmeg
    (c) aspirin (d) Juicy Fruit gum
    (e) soap (f) Dacron
    (g) prostaglandin
    (h) 1,3-disphosphoglyceric acid
    (i) dipeptide (j) Nylon 66
    (k) amytal (l) tristearin
14. Pineapple flavoring contains the following esters: amyl propionate (see Table 16.6 for the structure of the amyl group), ethyl butyrate, and ethyl caproate. Draw the structure for each ester.
15. Write the structure for the product of each of the following reactions.

(a) 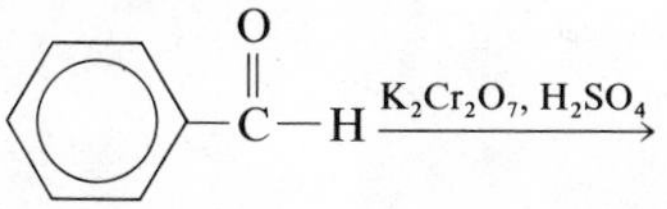

(b) $CH_3CH_2CH_2C{\equiv}N + H_2O \xrightarrow{H^+,\ heat}$

(c) $CH_3C(=O)-OH + NaOH \longrightarrow$

(d) $CH_3CH_2CH_2CH_2C(=O)-OH \xrightarrow{SOCl_2}$

(e) $CH_3C(=O)-OH + HO-P(=O)(OH)-OH \longrightarrow$

(f) 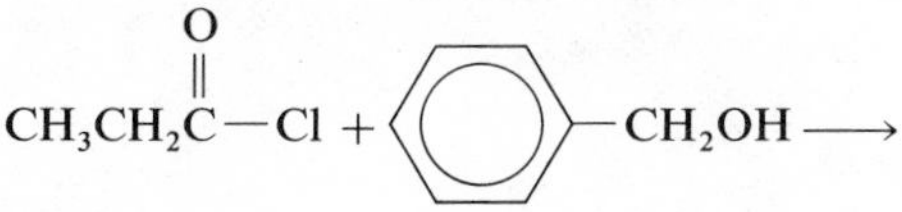

(g) $CH_3C(=O)CH_2C(=O)-OH \xrightarrow{\text{acetoacetic acid decarboxylase}}$

(h) $CH_2O-C(=O)(CH_2)_{12}CH_3$ / $CHO-C(=O)(CH_2)_{12}CH_3$ / $CH_2O-C(=O)(CH_2)_{12}CH_3$ $+ H_2O \xrightarrow{NaOH}$

(i) $CH_3CH(CH_3)CH_2C(=O)-NH_2 + H_2O + HCl \longrightarrow$

(j) $CH_3C(=O)OCH_2CH_3 + CH_3CH(CH_3)OH \longrightarrow$

(k) 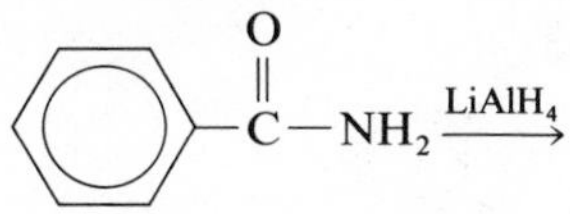

(l) $CH_3OH + HO-P(=O)(OH)-OH \longrightarrow$

16. Lysergic acid comes from the parasitic fungus ergot, which grows on rye and wheat. LSD, the potent hallucinogen, is the amide of lysergic acid. It does not occur in nature, but is synthesized in the lab by the following procedure. Identify the compounds A and B needed to perform the two transformations.

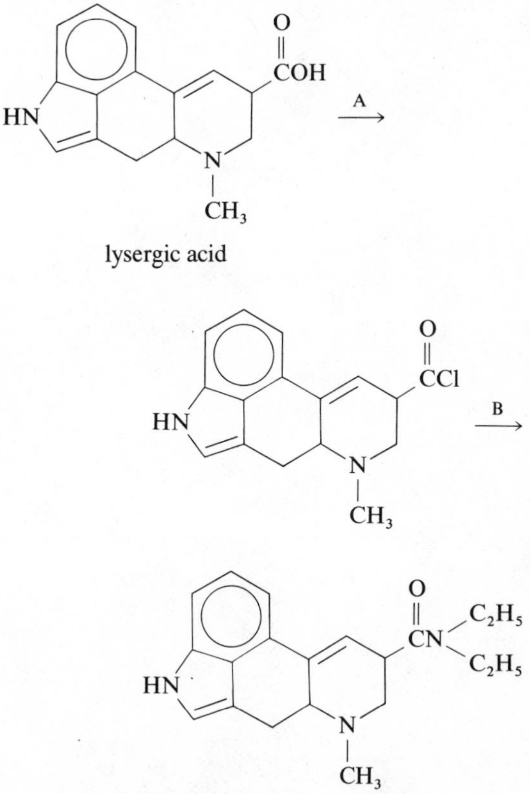

lysergic acid

lysergic acid diethylamide

17. Luminal, or phenobarbital, is a long-acting barbiturate. The R groups (see structure in Table 16.3) are ethyl and phenyl. Draw the structure of luminal.

18. Draw the structures of the monoethyl, diethyl, and triethyl esters of phosphoric acid.

19. 3,5-Cyclic AMP is an internal ester formed from the reaction of the acid grouping on C-5 and the alcohol group on C-3 in the following molecule. Show the structure of 3,5-cyclic AMP.

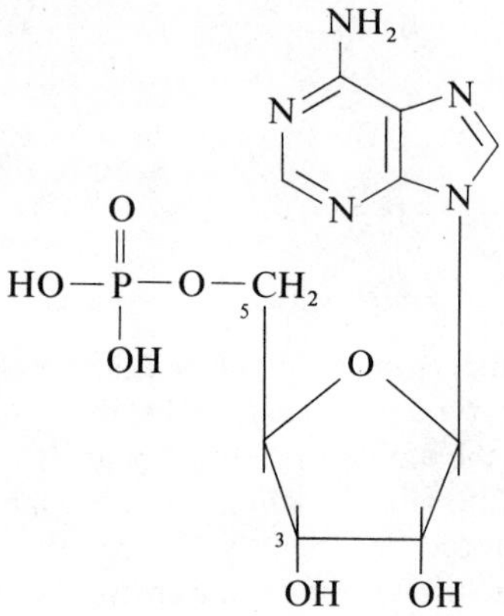

20. Kodel is a polyester used as a fiber in many garments. It is made from the alcohol and acid shown. Draw a portion of the Kodel polyester molecule.

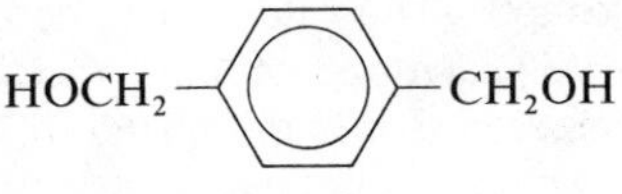

alcohol

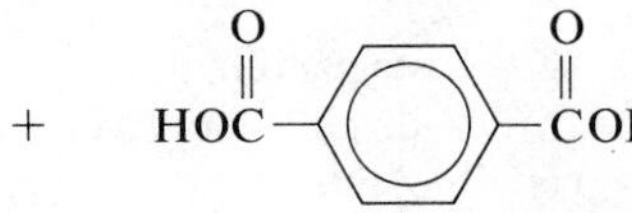

acid

21. Distinguish among the terms *hydrolysis, alcoholysis, transesterification,* and *saponification.* Using $CH_3CH_2COOCH_3$ as your example, write an equation to illustrate each term.

CHAPTER 17

ORGANIC NITROGEN COMPOUNDS: AMINES

There are many organic compounds containing nitrogen. We have considered several of these in previous chapters: heterocyclic aromatic compounds, nitriles or cyanides, and imines. In this chapter we will turn our attention to a class of organic nitrogen compounds called amines. In general, amines are recognized by their obnoxious odors. The lower molecular weight amines have an ammonia-like odor, while other amines have a characteristic fishy odor.

Amines are abundant in nature, having a variety of sources. The amino acids that make up proteins and the purine and pyrimidine bases of DNA and RNA are amine derivatives. Many amines have physiological properties. Among these amines are the narcotics, the phenethylamines, and the hallucinogens. As we will see in this chapter, the physiological property of these amines arises from their structural similarity to naturally occurring nitrogen compounds called enkephalins, adrenaline, and serotonin, respectively.

Since amines are derivatives of ammonia, they behave as bases and are capable of forming ammonium salts. Amines are the starting point for a variety of other important compounds such as those used as color dyes in foods and clothing and antibacterials such as the sulfa drugs and penicillins.

17-1 STRUCTURE AND NOMENCLATURE OF AMINES

LEARNING OBJECTIVES

1. Given organic compounds, identify them as primary amines, secondary amines, tertiary amines, amine salts, and heterocyclic amines.
2. Given the structural formula of an amine or an amine salt, write its common name or IUPAC name.
3. Given the common or IUPAC name of an amine or amine salt, write its structural formula.

Amines are derived from ammonia by the replacement of one, two, or all three of the hydrogen atoms by an alkyl or cyclic group. If one of the hydrogens has been replaced, the amine is classified as a *primary amine*. If two hydrogens have been replaced, the amine is classified as a *secondary amine*. If all three hydrogens are replaced, the amine is classified as a *tertiary amine*. Note that this is different from the designation of alcohols as primary, secondary, or tertiary (see Section 14-1) because the classification of the alcohols is based on whether the carbon having the hydroxyl group is primary, secondary, or tertiary.

A **primary amine** is an amine in which the nitrogen atom is bonded to one carbon atom. A **secondary amine** is an amine in which the nitrogen atom is bonded to two carbon atoms. A **tertiary amine** is an amine in which the nitrogen is bonded to three carbon atoms.

$$H-\underset{\underset{H}{|}}{N}-H \qquad R-\underset{\underset{H}{|}}{N}-H \qquad R-\underset{\underset{H}{|}}{N}-R \qquad R-\underset{\underset{R}{|}}{N}-R$$

ammonia   *a primary amine*   *a secondary amine*   *a tertiary amine*

Simple amines are named by naming the alkyl groups (see Table 11.4) attached to the nitrogen atom followed by the word *amine*. Thus, methyl amine, methyl ethyl amine, and trimethylamine are simple amines.

$$CH_3-\underset{\underset{H}{|}}{N}-H \qquad CH_3-\underset{\underset{H}{|}}{N}-CH_2CH_3 \qquad CH_3-\underset{\underset{CH_3}{|}}{N}-CH_3$$

methyl amine   methyl ethyl amine   trimethylamine
a primary amine   *a secondary amine*   *a tertiary amine*

In the IUPAC system, the $-NH_2$ group is named *amino*. The following are examples of the IUPAC names of amines.

$$\overset{5}{C}H_3\overset{4}{C}H\overset{3}{C}H_2\overset{2}{C}H_2\overset{1}{C}H_2NH_2 \qquad \overset{2}{H_2N}CH_2\overset{1}{C}H_2OH$$
$$\underset{CH_3}{|}$$

4-methyl-1-aminopentane   2-aminoethanol

—$CH_2CH_2CH_2NH_2$ $H_2NCH_2CH_2CH_2CH_2NH_2$

3-phenyl-1-aminopropane   1,4-diaminobutane
                                   (putrescine)

Amines that contain a phenyl group attached to nitrogen are *aromatic amines*. The simplest aromatic amine is aniline. Many aromatic amines are named as derivatives of aniline, such as *m*-bromoaniline, 2,4-dinitroaniline, N-methylaniline, and *o*-methylaniline.

aniline   *m*-bromoaniline   2,4-dinitroaniline

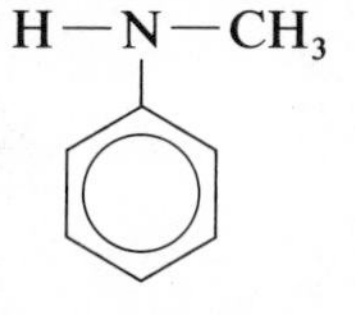

H—N—CH$_3$ NH$_2$

                         CH$_3$

N-methylaniline 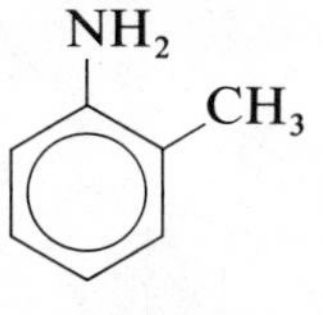   o-methylaniline

Nitrogen in amines has an unshared pair of electrons, so it can form an additional covalent bond with a hydrogen atom, an alkyl group, or a cyclic group. These compounds are similar to ammonium salts and are known collectively as *salts of amines*. There are four different salts of amines, depending on the number of hydrogen atoms that have been replaced. If all four hydrogen atoms have been replaced, the ion is called a quaternary ammonium ion. Compounds derived from this ion are called quaternary ammonium salts.

A **quaternary ammonium salt** is an amine salt in which there are four R groups attached to the nitrogen.

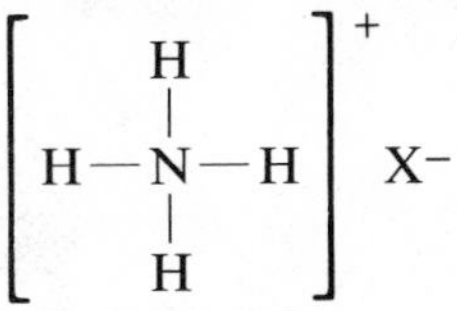

$$\left[\begin{array}{c} H \\ | \\ H-N-H \\ | \\ H \end{array}\right]^{+} X^{-} \quad \left[\begin{array}{c} H \\ | \\ R-N-H \\ | \\ H \end{array}\right]^{+} X^{-} \quad \left[\begin{array}{c} R \\ | \\ R-N-H \\ | \\ H \end{array}\right]^{+} X^{-} \quad \left[\begin{array}{c} R \\ | \\ R-N-R \\ | \\ H \end{array}\right]^{+} X^{-}$$

ammonium salt 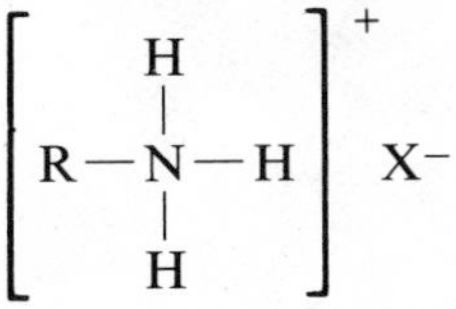   *an amine salt* 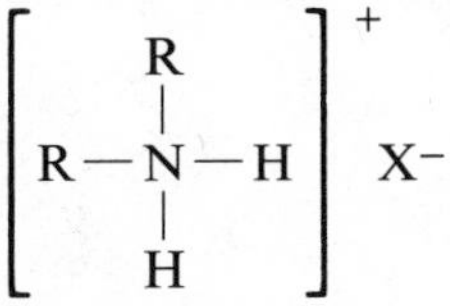   *an amine salt* 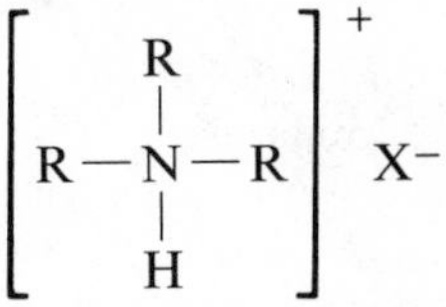   *an amine salt*

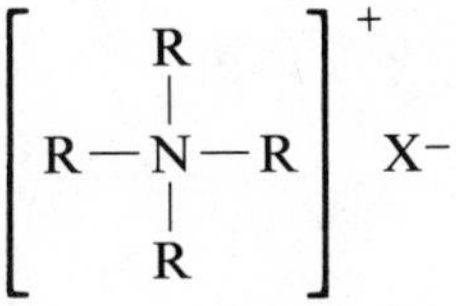

$$\left[\begin{array}{c} R \\ | \\ R-N-R \\ | \\ R \end{array}\right]^{+} X^{-}$$

a quaternary ammonium salt

Salts of amines are named by replacing the word *amine* by *ammonium* and adding the name of the anion (for example, chloride, sulfate, nitrate, and so forth).

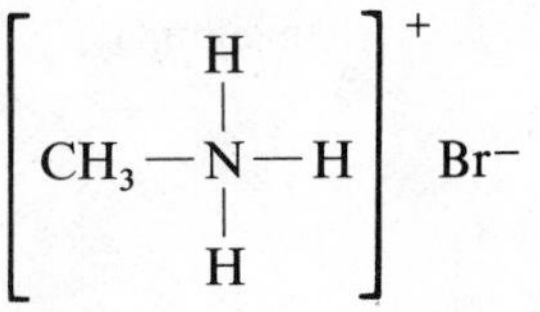

$$\left[\begin{array}{c} H \\ | \\ CH_3-N-H \\ | \\ H \end{array}\right]^{+} Br^{-} \quad \left[\begin{array}{c} CH_3 \\ | \\ CH_3-N-CH_3 \\ | \\ CH_3 \end{array}\right]^{+} NO_3^{-}$$

methyl ammonium bromide 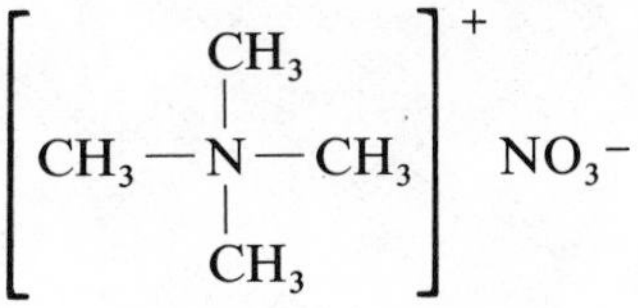   tetramethyl ammonium nitrate
                                 a quaternary ammonium salt

Many amines that occur in the living system and in nature contain the nitrogen atom as part of a ring structure. Such amines are called heterocyclic amines. Many heterocyclic amines occur naturally and many have uses in medicine. Some examples are shown in Table 17.1.

A **heterocyclic amine** is an amine which contains one or more nitrogen atoms as part of the ring structure.

Table 17.1   Heterocyclic Amines

| *Heterocyclic Ring System* | | *Example* |
Name	Structure	Name
Pyrrole		Heme
Imidazole		Histamine
Pyrrolidine		Nicotine
Thiazole		Penicillin G
Pyridine		Niacin (nicotinic acid)

Structure *Example*	Occurrence or Medical Use
HOOC—CH$_2$—CH$_2$ CH$_2$—CH$_2$—COOH H$_3$C CH$_3$ Fe H$_2$C=CH CH$_3$ CH$_3$ CH=CH$_2$	Part of the iron porphyrin complex, the red pigment in blood
CH$_2$CH$_2$NH$_2$ N N | H	Derived from the amino acid histidine. Causes dilation of capillaries, decrease in blood pressure, flushing, and headache. Antihistamines block the effect of histamines.
N | CH$_3$	Constituent of tobacco leaves. Oxidation produces niacin (see below). Also contains a pyridine ring.
O ‖ CH$_2$—CNH S CH$_3$ CH$_3$ O N COOH	An antibiotic, a compound produced by microorganisms that inhibits the growth of other microorganisms.
O ‖ C—OH N	Prevents a skin disease known as pellagra or blacktongue. Component of the coenzyme NAD$^+$.

Table 17.1   Heterocyclic Amines (*cont'd*)

| *Heterocyclic Ring System* | | *Example* |
Name	Structure	Name
Piperidine		Piperine
Pyrimidine		Sulfadiazine
Indole		Reserpine
Purine		Caffeine
Quinoline		Quinine

Example	
Structure	**Occurrence or Medical Use**

Main constituent of black pepper.

One of several sulfa drugs (see Section 17-4.2).

Used in treatment of high blood pressure (hypertension) and as a tranquilizer.

Occurs in tea plant and coffee bean and is responsible for stimulation derived from tea, coffee, and many soft drinks.

An antimalarial drug

PROBLEMS

1. Identify each of the following as a primary, secondary, or tertiary amine or amine salt. Give the correct name for each.

   (a) $(CH_3CH_2)_2NH$

   (b) 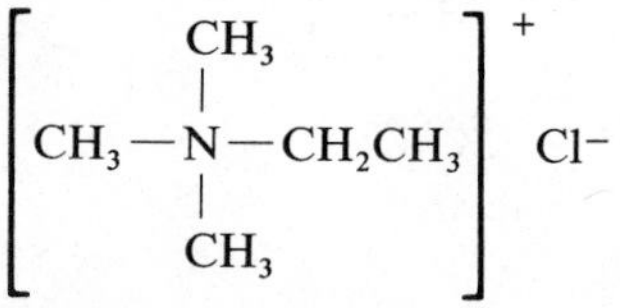 $\left[\begin{array}{c} CH_3 \\ | \\ CH_3-N-CH_2CH_3 \\ | \\ CH_3 \end{array}\right]^+ Cl^-$

   (c) $H_2NCH_2CH_2NH_2$

   (d) 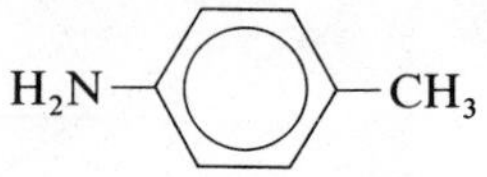 $H_2N-\bigcirc-CH_3$

   (e) 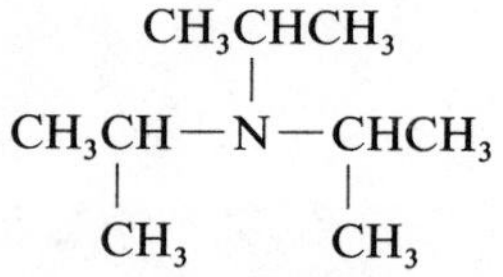 $\begin{array}{c} CH_3CHCH_3 \\ | \\ CH_3CH-N-CHCH_3 \\ | \qquad\quad | \\ CH_3 \qquad CH_3 \end{array}$

2. Tryptophan is an amino acid whose structure is shown below.

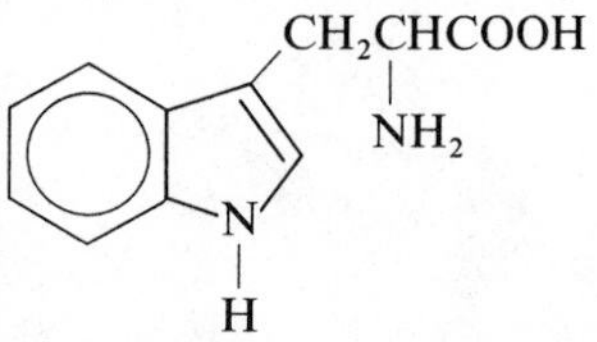

   $$CH_2CHCOOH$$
   $$| $$
   $$NH_2$$

   (a) Classify each kind of amino nitrogen.   (b) Identify the heterocyclic system.

ANSWERS TO PROBLEMS   **1. (a)** secondary amine, diethyl amine   **(b)** amine salt; trimethylethylammonium chloride   **(c)** primary amine, 1,2-diaminoethane   **(d)** primary aromatic amine, *p*-methylaniline   **(e)** tertiary amine, triisopropylamine   **2. (a)** a heterocyclic secondary amine and a primary amine   **(b)** similar to that in indole

---

17-2 PREPARATION OF AMINES

LEARNING OBJECTIVES

1. Given the reduction reaction of a nitro compound, a nitrile, an amide, or a carbonyl compound (after its reaction with ammonia), write the structure of the amine formed.

2. Given a decarboxylation reaction of an amino acid, write the structure of the amine formed.

In this section we will learn how a variety of organic molecules can be converted to amines by a single chemical technique—reduction. We also will see how the living system converts amino acids to amines, many of which have physiological importance.

17-2.1 REDUCTION OF NITRO COMPOUNDS, NITRILES, AMIDES, AND CARBONYL COMPOUNDS

A variety of organic functional groups can be reduced to amines. These include nitro compounds, nitriles, amides, and carbonyl compounds. Aromatic nitro compounds may be reduced with metals in acid solutions. For example, nitrobenzene reduces to aniline and *p*-nitrotoluene reduces to *p*-aminotoluene.

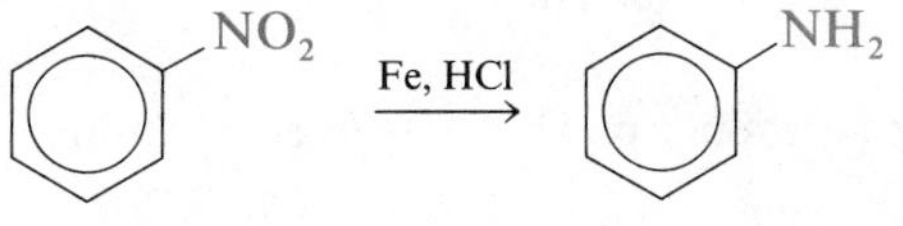

nitrobenzene → aniline

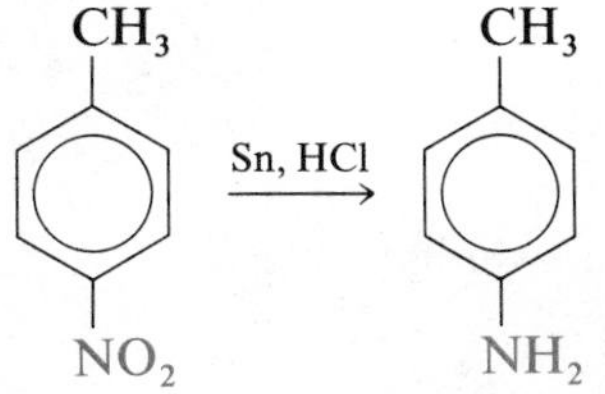

Fe, HCl

p-nitrotoluene   *p*-aminotoluene

Sn, HCl

Nitriles are conveniently reduced to amines with lithium aluminum hydride. For example, propionitrile reduces to propylamine and benzonitrile reduces to benzyl amine.

$$CH_3CH_2C{\equiv}N \xrightarrow{\text{LiAlH}_4} CH_3CH_2CH_2NH_2$$

propionitrile   propylamine

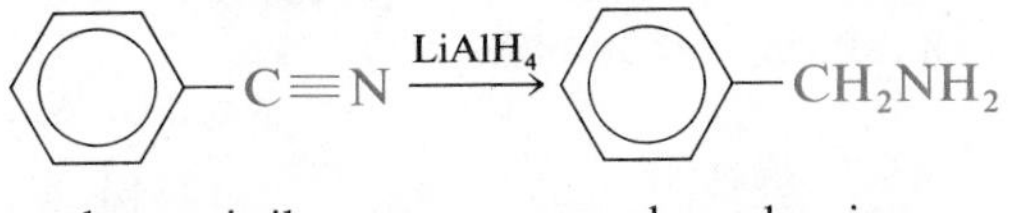

benzonitrile   benzyl amine

Amides, like nitriles, are reduced to amines by lithium aluminum hydride.

acetamide   ethylamine

N,N-dimethylbenzamide   dimethylbenzylamine

The reduction of the carbonyl compounds, aldehydes and ketones, occurs in the presence of ammonia, hydrogen, and a metal catalyst. This reaction is illustrated for the conversion of acetaldehyde to ethyl amine. In this reaction an imine (see Section 15-3.1) is first formed between the carbonyl compound, acetaldehyde, and ammonia.

$$\underset{\text{acetaldehyde}}{CH_3\overset{\displaystyle O}{\overset{\|}{C}}-H} + NH_3 \longrightarrow HOH + \underset{\textit{an imine}}{CH_3CH{=}NH}$$

The imine is then reduced by hydrogen in the presence of a catalyst to ethyl amine.

$$\underset{\textit{an imine}}{CH_3CH{=}NH} + H_2 \overset{Ni}{\longrightarrow} \underset{\text{ethyl amine}}{CH_3\underset{\displaystyle H}{\overset{\displaystyle |}{C}}H-\underset{\displaystyle H}{\overset{\displaystyle |}{N}}H}$$

Cyclopentanone is converted to cyclopentyl amine in a similar way.

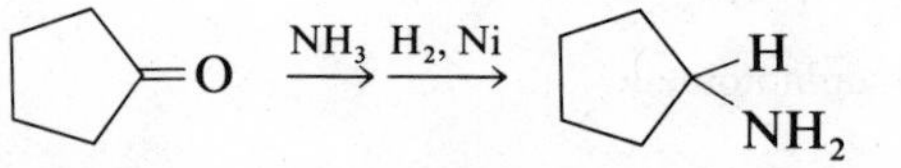

cyclopentanone   cyclopentylamine

This reaction, as illustrated above in the examples, represents the overall replacement of the doubly bonded oxygen of a carbonyl group by an amino group.

When such a transfer of an amino group occurs in the living system, the reaction is referred to as a **transamination**. For example, as shown below and discussed further in Section 28-2.1, the carbonyl oxygen of pyruvic acid is replaced by the amino group of glutamic acid to form α-ketoglutaric acid and alanine.

> **Transamination** is the biological substitution of an amino group for a carbonyl oxygen.

$$\underset{\text{pyruvic acid}}{CH_3\overset{\displaystyle O}{\overset{\|}{C}}CO_2H} + \underset{\text{glutamic acid}}{HO_2CCH_2CH_2\overset{\displaystyle NH_2}{\overset{\displaystyle |}{C}}HCO_2H} \rightleftharpoons \underset{\text{alanine}}{CH_3\overset{\displaystyle NH_2}{\overset{\displaystyle |}{C}}HCO_2H} + \underset{\text{α-ketoglutaric acid}}{HO_2CCH_2CH_2\overset{\displaystyle O}{\overset{\|}{C}}COH_2H}$$

17-2.2 *IN VIVO* DECARBOXYLATION

A variety of physiologically active amines can be synthesized by the living system. Many of these amines are formed by *decarboxylation* of amino acids. During decarboxylation, the —COOH group of the amino acid is replaced by a single hydrogen atom (see Section 16-3.3). Carbon dioxide is the by-product of this reaction. Decarboxylation reactions are catalyzed by enzymes that are known collectively as decarboxylases.

The decarboxylation of dopa to dopamine is a typical example of this type of reaction.

$$\underset{\substack{\text{dopa}}}{\text{HO} \diagdown \bigcirc \diagup \overset{\displaystyle\underset{\underset{\displaystyle NH_2}{|}}{CH_2CHCOH}}{\overset{\overset{\displaystyle O}{\|}}{}} } \xrightarrow{\text{dopa decarboxylase}} \underset{\substack{\text{dopamine}}}{\text{HO} \diagdown \bigcirc \diagup CH_2CH_2NH_2} + CO_2$$

Dopamine is an important transmitter in the central nervous system. The lack of dopamine in the central nervous system causes a disease known as *Parkinson's disease*. Unfortunately, the disease cannot be remedied by administering dopamine to the victim because dopamine cannot pass the **blood-brain barrier.** However, individuals with Parkinson's disease are usually given large amounts of dopa. This compound, unlike dopamine, can pass from the digestive system into the blood stream and through the blood-brain barrier. After passing through the blood-brain barrier, dopa is decarboxylated into dopamine. This discovery has provided many stricken individuals with a "normal" life.

A characteristic feature of amines is their pronounced odors. The odors of amines are less pungent than those of ammonia and more fishy. Dimethylamine and trimethylamine are amines found in some fish. The name of the decarboxylation product in decaying flesh is indicative of its unpleasant odor. For example, the decomposition of the amino acid lysine by a decarboxylase enzyme results in cadaverine (1,5-diaminopentane). The structure of putrescine was shown in Section 17-1.

The **blood-brain barrier** is the obstruction, presumably the walls of the blood vessels of the central nervous system and the surrounding membranes, that prevents substances from passing from the blood to the brain and cerebrospinal fluid.

$$\underset{\substack{\text{lysine}}}{\underset{\substack{| \\ NH_2}}{H_2NCH_2CH_2CH_2CH_2CHC}} \overset{\overset{\displaystyle O}{\|}}{} \text{—OH} \xrightarrow{\text{lysine decarboxylase}} \underset{\substack{\text{1,5-diaminopentane} \\ \text{(cadaverine)}}}{H_2NCH_2CH_2CH_2CH_2CH_2NH_2} + CO_2$$

PROBLEMS

3. Give the structure of each compound that will form the amine product on reduction.

(a) amide $\xrightarrow{\text{LiAlH}_4}$ ⬡—CH₂NH₂

(b) nitro compound $\xrightarrow{\text{Sn, HCl}}$ ⬡ with OCH₃ and NH₂

(c) nitrile $\xrightarrow{\text{LiAlH}_4}$ $\underset{\substack{| \\ CH_3}}{CH_3CHCH_2NH_2}$

(d) carbonyl compound $\xrightarrow{\text{NH}_3}$ $\xrightarrow{\text{H}_2, \text{Ni}}$

(benzene ring with CH_2NH_2 substituent)

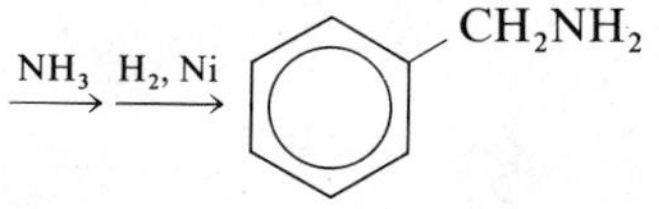

4. Draw the structure of histamine, the amine formed by the decarboxylation of histidine. Histidine is an amino acid and histamine is known to relax and dilate blood vessels. Antihistamines are sometimes used to counteract these effects.

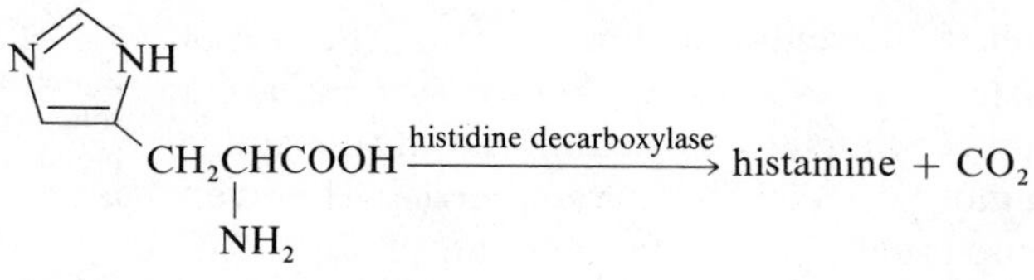

(imidazole ring with N, NH)

$CH_2CHCOOH \xrightarrow{\text{histidine decarboxylase}} \text{histamine} + CO_2$

$\underset{NH_2}{|}$

histidine

ANSWERS TO PROBLEMS

3. (a) 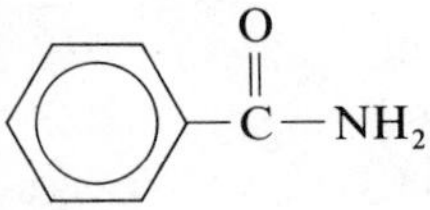 (benzene ring) $-\overset{O}{\overset{||}{C}}-NH_2$ (b) 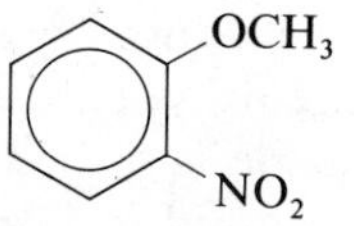 (benzene ring with OCH_3 and NO_2 substituents)   (c) CH_3CHCN
$\underset{CH_3}{|}$

(d) 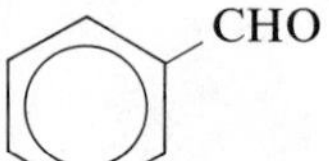 (benzene ring with CHO substituent)   4. see Table 17.1.

17-3 BASICITY OF AMINES

LEARNING OBJECTIVES

1. Given primary, secondary, or tertiary amines; saturated heterocyclic amines; aniline; and heterocyclic aromatic amines; rank them as stronger or weaker bases than ammonia.

2. Given the reaction of an amine with an acid, write the formula of the amine salt formed.

Amines, like ammonia, have a nitrogen atom with an unshared pair of electrons to share with another atom or group. For example, amines, like ammonia, will react with a variety of acids to form amine salts as shown.

$$\ddot{N}H_3 + HCl \longrightarrow \left[H-\overset{\overset{H}{|}}{\underset{\underset{H}{|}}{N}}-H \right]^+ Cl^-$$

ammonia   ammonium chloride

$$CH_3\ddot{N}H_2 + HI \longrightarrow \begin{bmatrix} & H \\ & | \\ CH_3-&N-H \\ & | \\ & H \end{bmatrix}^+ I^-$$

methylamine   methyl ammonium iodide
a primary amine

$$\text{pyrrolidine} + HNO_3 \longrightarrow \begin{bmatrix} & N & \\ & / \backslash & \\ H & & H \end{bmatrix}^+ NO_3^-$$

pyrrolidine   pyrrolidinium nitrate
a secondary amine

$$2\,(CH_3)_3\ddot{N} + H_2SO_4 \longrightarrow \begin{bmatrix} & CH_3 & \\ & | & \\ CH_3-&N-CH_3 \\ & | & \\ & H & \end{bmatrix}_2^+ SO_4^{2+}$$

trimethylamine   trimethylammonium sulfate
a tertiary amine

The basicity of an amine is a measure of the amine's ability to accept a proton.

$$R-\overset{\displaystyle ..}{\underset{\displaystyle H}{N}}-H + H^+ \longrightarrow \begin{bmatrix} & H \\ & | \\ R-&N-H \\ & | \\ & H \end{bmatrix}^+$$

The basicity constant, K_b, can be formulated as:

$$K_b = \frac{[RNH_3{}^+]}{[RNH_2][H^+]}$$

Thus, the numerical value of K_b is a measure of the amine's ability to accept the proton. As the numerical value of K_b increases, the strength of the amine as a base increases. The measure of the basicity of amines usually is related to ammonia. Ammonia has a K_b of 1.76×10^{-5} (see Table 8.3). If an amine has a K_b larger than this value, the amine has a greater tendency to accept a proton than does ammonia. Conversely, if the amine has a K_b value smaller than this value, the amine has less tendency to accept a proton than does ammonia.

A **saturated heterocyclic amine** is a cyclic amine in which carbon is bonded to the maximum of four atoms and nitrogen is bonded to the maximum of three atoms.

Primary, secondary, and tertiary amines and **saturated heterocyclic amines** are stronger bases than ammonia.

$CH_3\ddot{N}H_2$ $(CH_3)_2\ddot{N}H$ $(CH_3CH_2)_3\ddot{N}$ are stronger bases than NH_3
$K_b = 1.76 \times 10^{-5}$

$K_b = \sim 10^{-4}$ $K_b = \sim 10^{-3}$

primary, secondary, and tertiary amines and saturated heterocyclic amines

Aromatic amines, such as aniline, and aromatic heterocyclic amines, such as pyridine, are weaker bases than ammonia.

$\overset{..}{N}H_3$ is a stronger base than

$K_b = 1.76 \times 10^{-5}$

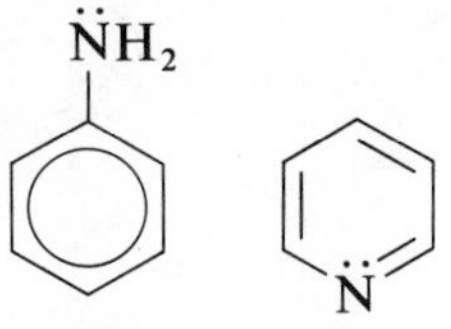

aniline   pyridine

$K_b = \sim 10^{-9}$

aromatic amines

In summary, the basicity of amines follows the general order:

primary, secondary and tertiary amines,   aniline and aromatic
and saturated heterocyclic amines $>$ NH_3 $>$ heterocyclic amines

decreasing basicity

Many amines often are used as drugs. However, the amine usually is not administered in the free-amine form. Because of the buffering action of the digestive system, the free amine would be rapidly converted into the amine salt form. Apparently, the absorption of protons by the free amine can cause local irritation and discomfort due to slight pH changes in the gastrointestinal tract. Hence, amine salts are usually administered. For example, cocaine, a local anesthetic, is administered to the patient as the amine salt, cocaine hydrochloride.

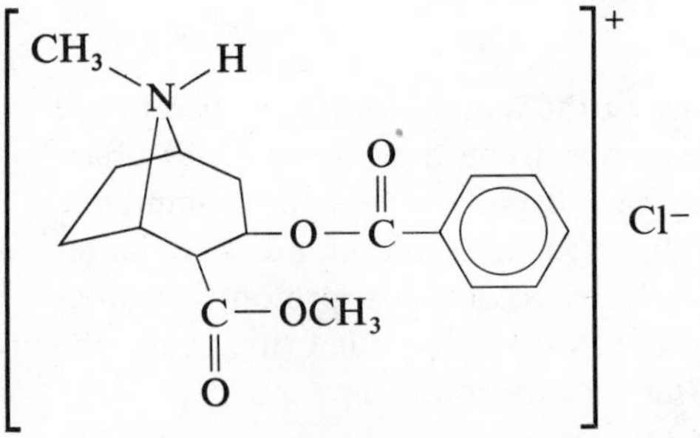

cocaine hydrochloride

Other amine salts are common in living systems. Acetylcholine is possibly one of the most important. It functions in the transmission of nerve impulses in the body. Stimulation of a nerve cell in a muscle or gland releases acetylcholine. The acetylcholine migrates to an adjacent nerve cell and transmits a nerve impulse to the nerve cell. After the transmission of the nerve impulse is complete (about 30–50 microseconds), the acetylcholine must be deactivated. If it is not, continual stimulation of a muscle or gland can occur and paralysis or death eventually will result. In Chapter 23 you will learn how this nerve transmitter is deactivated and how poisons or nerve gases can interfere with the deactivation process.

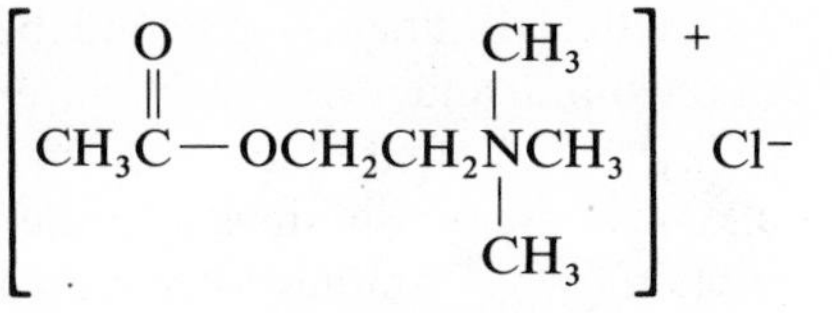

$$\left[\begin{array}{c} \overset{O}{\underset{\|}{CH_3C}} - OCH_2CH_2\overset{CH_3}{\underset{\underset{CH_3}{|}}{\overset{|}{N}}}CH_3 \end{array} \right]^+ \quad Cl^-$$

acetylcholine

PROBLEMS

5. Anabasine is an amine, found in tobacco, that contains pyridine and piperidine rings (see Table 17.1). Which of the nitrogen atoms is more basic?

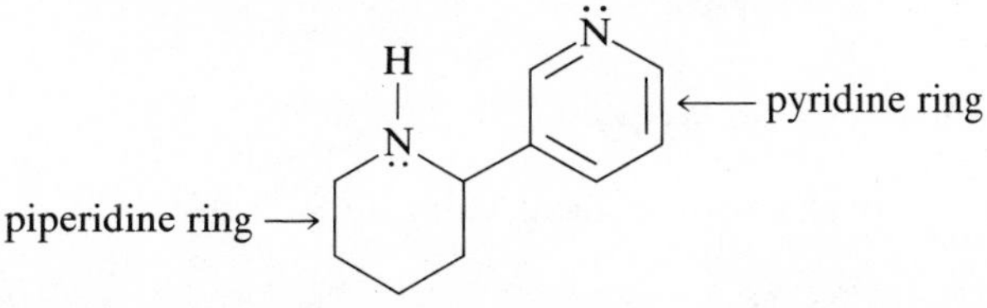

piperidine ring ⟶   ⟵ pyridine ring

anabasine

6. A 0.5% solution of clopane is commonly used as a nasal decongestant. Clopane is used in this preparation as the hydrochloride. Draw the structure of the hydrochloride.

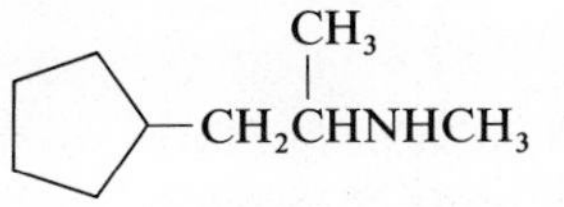

$$\overset{CH_3}{\underset{|}{-CH_2CHNHCH_3}}$$

clopane

ANSWERS TO PROBLEMS   **5.** Saturated heterocyclic amines (such as piperidine) are stronger bases than aromatic heterocyclic amines (such as pyridine).

6. 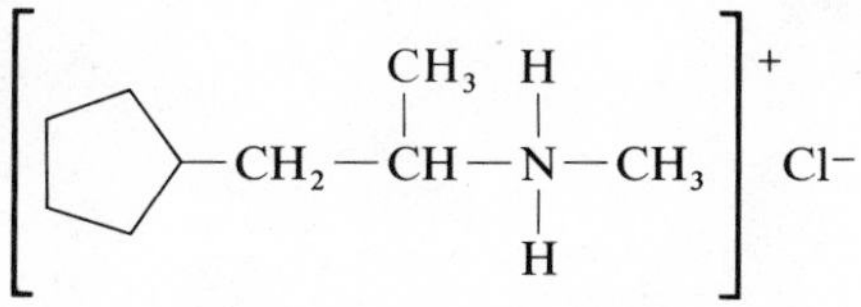 $\left[\begin{array}{c} -CH_2 - \overset{CH_3}{\underset{}{CH}} - \overset{H}{\underset{H}{N}} - CH_3 \end{array} \right]^+ \quad Cl^-$

17-4 REACTION OF AMINES

LEARNING OBJECTIVES

1. Given the reaction between an acid chloride and an amine, write the structure of the amide formed.
2. Given the reaction beween an amine and benzenesulfonyl chloride,

write the structure of the sulfonamide formed.
3. Explain how sulfa drugs act as antibacterial agents.
4. Given the reaction between a

primary aromatic amine and nitrous acid, write the structure of the diazonium ion formed.
5. Given the reaction between a secondary amine and nitrous acid, write the structure of the nitrosoamine formed.
6. Given the substitution reaction between an amine and a methyl or primary alkyl halide, write the structure of the quaternary ammonium salt formed.

7. Given the elimination reaction between an amine and a secondary or tertiary alkyl halide, write the structure of the alkene and ammonium salt formed.
8. Explain why amines behave as nucleophiles.
9. Given an organic compound, identify it as a diazonium ion, an azo compound, a nitrosoamine, an amide, sulfonamide, or a sulfa drug.

17-4.1 CONVERSION TO AMIDES: THE PEPTIDE BOND

An amide is formed whenever ammonia reacts with an acid chloride (see Section 16-3.2).

$$R-\overset{\overset{\displaystyle O}{\|}}{C}-Cl + H-\underset{\underset{\displaystyle H}{|}}{N}-H \longrightarrow R-\overset{\overset{\displaystyle O}{\|}}{C}-\underset{\underset{\displaystyle H}{|}}{N}-H + HCl$$

an acid chloride   ammonia   an amide

Primary and secondary amines also react with acid chlorides to form substituted amides. For example, aniline reacts with acetyl chloride to produce **acetanilide** and dimethylamine reacts with benzoyl chloride to form N,N-dimethylbenzamide.

Acetanilide is the acetic acid amide of aniline.

$$CH_3\overset{\overset{\displaystyle O}{\|}}{C}-Cl + H-\underset{\underset{\displaystyle H}{|}}{N}-\bigcirc \longrightarrow CH_3\overset{\overset{\displaystyle O}{\|}}{C}-\underset{\underset{\displaystyle H}{|}}{N}-\bigcirc + HCl$$

acetyl chloride   aniline   acetanilide

$$\bigcirc-\overset{\overset{\displaystyle O}{\|}}{C}-Cl + HN(CH_3)_2 \longrightarrow \bigcirc-\overset{\overset{\displaystyle O}{\|}}{C}-N(CH_3)_2 + HCl$$

benzoyl chloride   dimethylamine   N,N-dimethylbenzamide

Substituted amides are composed of an acid part and an amine part.

$$R-\overset{\overset{\displaystyle O}{\|}}{C}-\underset{\underset{\displaystyle H}{|}}{N}-R'$$

acid part   amine part

substituted amide

Table 17.2 Natural and Synthetic Penicillins

Penicillins (see Table 17.2) are substituted amides that differ in the nature of the R group in the acid portion. They are well known as antibiotics. Benzylpenicillin (Penicillin G) and phenoxymethylpenicillin (Penicillin V) are of microbial origin. Others, for example ampicillin, are now made synthetically.

Peptides and proteins are biomolecules composed of amino acids joined by an amide bond. Many peptides and proteins are substituted amides. When the amide bond is present in a peptide or protein it is called a *peptide bond*. For example, a peptide known to influence transmission in some parts of the brain and to mimic some of the properties of morphine (see Section 17-5) is the substituted amide shown.

An **amide bond** is the
$$\begin{array}{c} O \\ \parallel \\ \text{grouping} \quad -C-N- . \\ \quad\quad\quad\quad | \end{array}$$

a substituted amide and a peptide

17-4.2 CONVERSION TO SULFONAMIDES: SULFA DRUGS

Sulfonamides are amides of benzenesulfonic acid.

benzenesulfonic acid   *a sulfonamide*

These compounds are well-known for their antibacterial properties and are often called sulfa drugs. Sulfanilamide, the parent compound, is prepared as shown from acetanilide. The reagent $ClSO_3H$ is chlorosulfonic acid.

Sulfa drugs are amides of benzenesulfonic acid that possess antibacterial properties.

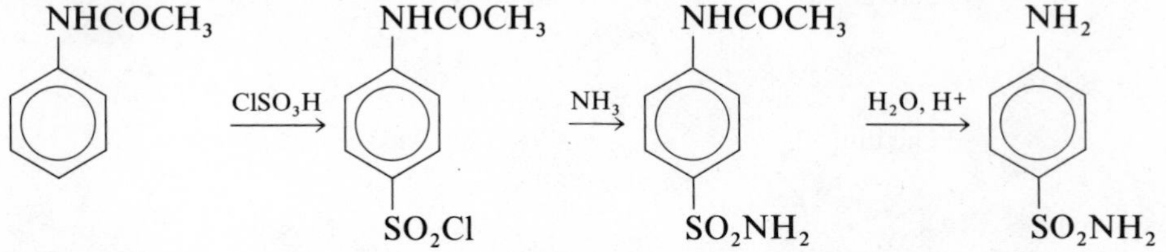

acetanilide   sulfanilamide

The activity and effectiveness of sulfanilamide can be increased by replacing one of the two hydrogens on the amide nitrogen by other groups (R). Some examples of the resultant compound are sulfathiazole and sulfapyrazine.

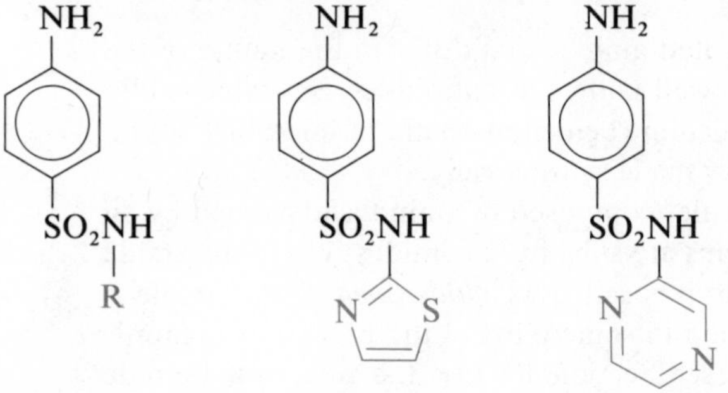

sulfathiazole   sulfapyrazine

Many microorganisms require *p*-aminobenzoic acid (PABA). PABA is incorporated into folic acid which is required for growth and function. Sulfanilamide resembles PABA structurally and interferes with the enzyme-controlled step involving the incorporation of the PABA molecule into folic acid. As a result, no folic acid is produced and the microorganism dies. Humans also require folic acid. However, we do not synthesize it, but get it from such foods as liver, mushrooms, and green leaf vegetables.

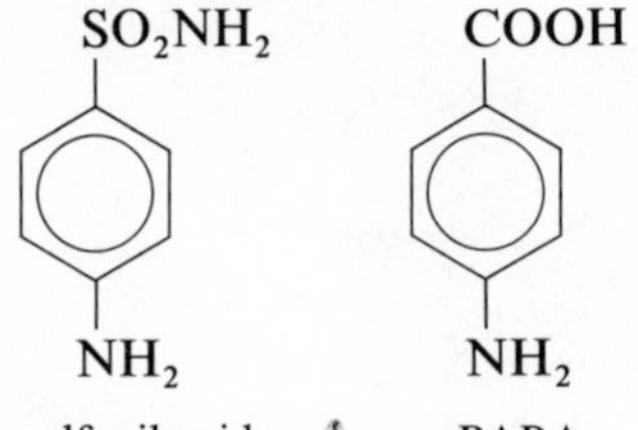

sulfanilamide   PABA

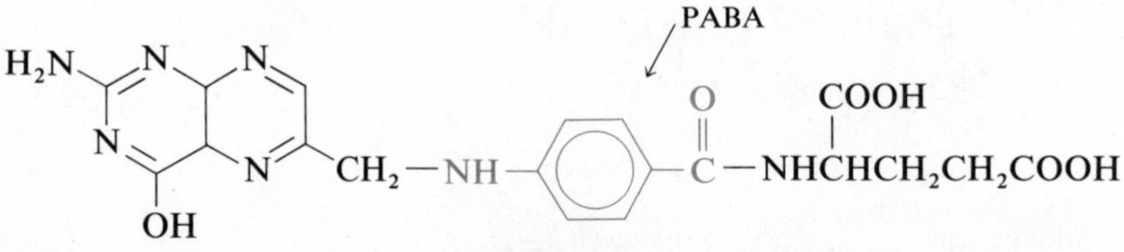

folic acid

Table 17.3 Sulfonamides Currently Used As Urinary Antiseptics

$$H_2N-\bigcirc-SO_2N\overset{H}{\underset{|}{N}}-R$$

—R	Name	Trade Name	Potency
(pyrimidinone, N—C₂H₅)	Sulfacytine	Renoquid	very short acting
(isoxazole, O—N, CH₃, CH₃)	Sulfisoxazole	Gantrisin	short acting
(isoxazole, N—O, CH₃)	Sulfamethoxazole	Gantanol	long acting

The sulfa drugs are biologically active and effective. They have been used to treat pneumonia, tuberculosis, gangrene, and gonorrhea. However, the sulfa drugs cause side reactions in some people, particularly infants, and some bacterial strains are resistant to them. Although the use of sulfa drugs has been largely replaced by the antibiotics such as the penicillins and tetracyclines, some are still routinely used in treating certain bacterial infections, such as those occurring in the urinary tract. Table 17.3 lists several common sulfa drugs in use today as urinary antiseptics (see Section 13-2) that have been shown to be effective in killing many microorganisms. Apparently the sulfa drugs are more widely used in urinary tract infections than the penicillins and tetracyclines because of their stability in the digestive tract and their greater solubility in urine which allows for a higher concentration in the urinary tract.

17-4.3 REACTIONS OF AMINES WITH NITROUS ACID

Unlike hydrochloric acid (HCl), hydroiodic acid (HI), nitric acid (HNO_3) and sulfuric acid (H_2SO_4), nitrous acid (HNO_2) does not react with amines to form ammonium salts. Instead, nitrous acid is commonly used to distinguish primary, secondary, and tertiary amines because each class of amine reacts differently. In this section we will learn the reactions of primary and secondary amines with nitrous acid. Nitrous acid is not stable and usually is generated *in situ* from sodium nitrite ($NaNO_2$) and hydrochloric acid. However, in the reactions using nitrous acid, we will use the formula HNO_2.

Formation of Azo Compounds: Dyes

The reaction of a primary aromatic amine with nitrous acid forms a *diazonium ion*. A diazonium ion contains a diazo group.

A **diazo group** is a functional group that is characterized by a nitrogen–nitrogen triple bond, $-\overset{+}{N}\equiv N$.

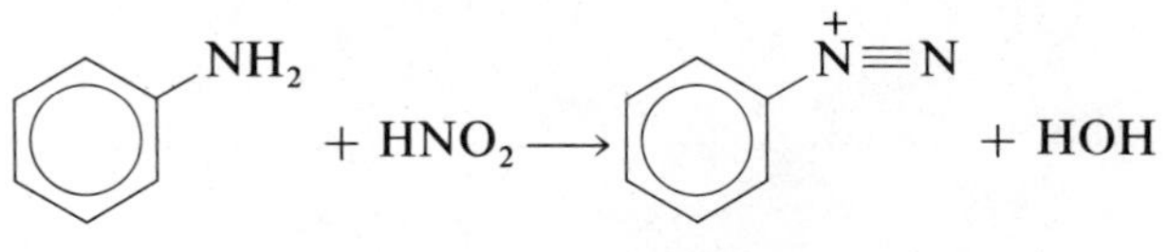

aniline
a primary aromatic amine

benzenediazonium ion
a diazonium ion

Diazonium ions are unstable and are seldom isolated once they are formed. They can react with a variety of aromatic compounds. The products of these reactions are called azo compounds. Depending on the aromatic compound used in the reaction, a variety of colors and shades can be produced. These colored azo compounds constitute a large class of synthetic commercial dyes. For example, the diazonium ion formed from *p*-nitroaniline and nitrous acid will be converted to para red when reacted with β-naphthol. Para red is a water insoluble azo compound and dye.

An **azo compound** is a molecule that contains the azo group in which two nitrogen atoms are doubly bonded, $(-N=N-)$.

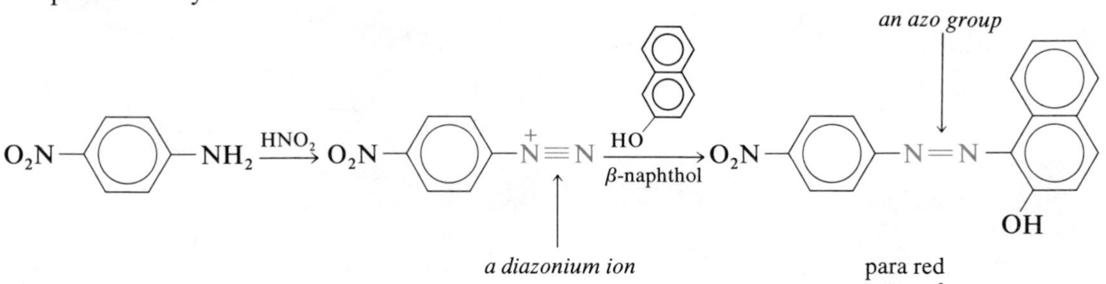

a diazonium ion

para red
a cotton dye

Azo compounds are also used as indicators in acid-base titrations. Methyl orange is a commonly used indicator. It is red at pH 3 and yellow-orange at pH 4.

An **indicator** is an organic molecule with the ability to change color in dilute solution when the hydrogen ion concentration (pH) of the solution attains a definite value.

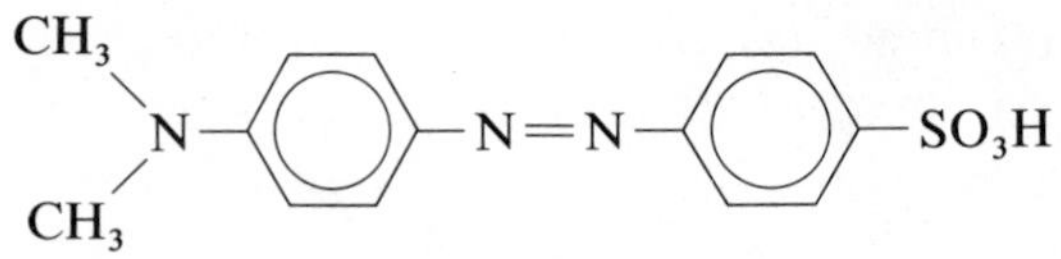

methyl orange
an indicator

Another dye is Red No. 2. At one time, this dye was used in almost every processed food with a reddish or brownish color, including canned fruits, gelatins, candy bars, salad dressings, cereals, frankfurters, ice creams, and cake mixes, as well as a variety of drugs and cosmetics. However, Red No. 2 has been banned from use as a coloring agent by the Food and Drug Administration (FDA) because after years of laboratory testing, it had not been proven safe for human consumption.

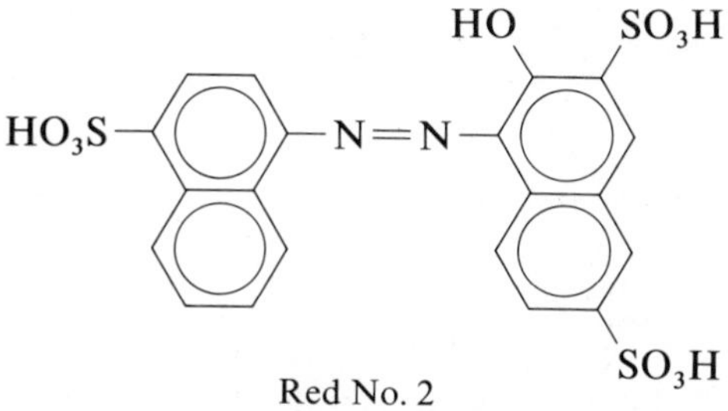

$$HO_3S-\left\langle\bigcirc\right\rangle-N=N-\quad\text{Red No. 2}$$

Red No. 2

In high doses, Red No. 2 has caused tumors in rats. The FDA is continually reviewing a multitude of compounds including other coloring agents to prove that food and drug additives are safe for human use. The studies so far have shown some dyes to be dangerous to test animals. The FDA can suspend use on this basis without proof of actual dangers to humans.

Due to the potential danger in using azo compounds as dyes, a novel approach by the food-dye industry is currently being considered. Scientists propose to develop large dye molecules that are, in effect, polymers (see Section 11-5.2). These polymers are so large that they will not break into smaller pieces during cooking, food storage, or passage through the body. The polymers must be able to withstand the acidity of soft drinks, the high temperature of baking and candy making, and the attacks of intestinal enzymes and microbes. In effect, these dye molecules must be able to color foods and, upon ingestion, pass directly through the digestive system without being absorbed through the intestinal wall.

The dye molecules being considered have a structural backbone of carbon molecules. They contain a chromophore group and solubilizing groups. The chromophore is the part of the molecule that absorbs light and bestows color. Solubilizing groups are incorporated because the product has to be soluble in water. As shown below, the two major portions of the molecules (in parentheses) are repeated many times so that the final dye molecules have backbones about 600 carbons long. In this structure, Ch indicates chromophore and S indicates solubilizing group. Note that the polymer itself is the dye and not a dye attached to a polymer.

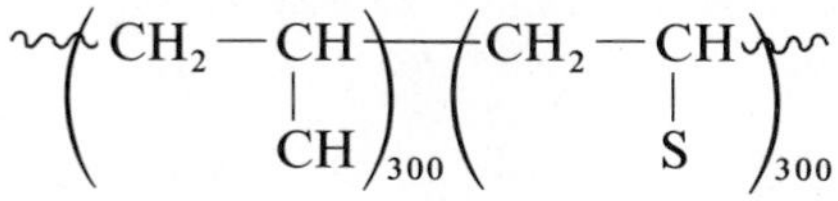

$$\sim\!\!\left(\begin{array}{c}CH_2-CH \\ | \\ CH\end{array}\right)_{\!300}\!\!\left(\begin{array}{c}CH_2-CH \\ | \\ S\end{array}\right)_{\!300}\!\!\sim$$

Nitrosamines

Nitrosamines are commonly prepared in the laboratory from the reaction of nitrous acid with a secondary amine as shown.

$$\begin{array}{ccc} R-\underset{\underset{R}{|}}{N}-H + HNO_2 & \longrightarrow & R-\underset{\underset{R}{|}}{N}-NO + HOH \\ \textit{a secondary amine} & & \textit{a nitrosamine} \end{array}$$

The equation for the formation of N-nitrosodimethylamine, a nitrosamine found in canned fish, cooked bacon, and frankfurters, is shown.

$$CH_3-\underset{\underset{CH_3}{|}}{N}-H \xrightarrow{HNO_2} CH_3-\underset{\underset{CH_3}{|}}{N}-NO + H_2O$$

    dimethylamine   N-nitrosodimethylamine

Nitrosamines have been found to be potent carcinogens in a wide variety of animal species. Although there is no direct evidence to associate human cancer with nitrosamines, every type of animal that has been tested has been found to be susceptible to nitrosamine carcinogenesis. Furthermore, it is now widely accepted that nitrosamines are probably carcinogenic to humans. How are these carcinogens formed?

As shown, the amine–nitrous acid reaction proceeds under acidic conditions. These acidic conditions are very near the pH of human gastric juice. In addition, a number of amines occur naturally in foods. These two facts suggest the possibility of nitrosamine formation from amines and nitrite ion when these reagents are both present in the human stomach. The level of nitrite ion in the body is dependent on several factors. These factors include local water supplies, consumption of high levels of nitrate ion that can be reduced by salivary microorganisms to nitrite ion, and the nature of certain foods after prolonged storage. In the latter case, for example, the nitrite level in fresh spinach can increase to sixty times its original amount after two weeks of refrigeration.

Nitrite ion is widely used to preserve and enhance the color and flavor of meats such as frankfurters and bacon. A high level of nitrosamines is generally found in these meat products, particularly bacon, after cooking. In addition to N-nitrosodimethylamine (shown above), N-nitrosodiethylamine and N-nitrosopyrrolidine have been detected in cooked bacon. The latter nitrosamine is typically the most abundant nitrosamine in bacon. The structures of N-nitrosodiethylamine and N-nitrosopyrrolidine are shown.

The formation of N-nitrosopyrrolidine is not well understood because N-nitrosopyrrolidine is found only in cooked bacon; the precursor amine, pyrrolidine, is not present in raw bacon. However, a possible pathway of N-nitrosopyrrolidine in cooked bacon is shown in Figure 17.1. The natural occurrence of the secondary amine and amino acid proline in foods constitutes a potential source of N-nitro-

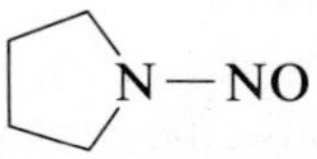

$$CH_3CH_2-\underset{\underset{CH_3CH_2}{|}}{N}-NO$$

N-nitrosodiethylamine

N—NO

N-nitrosopyrrolidine

Figure 17.1   Proposed reaction pathway for the formation of N-nitrosopyrrolidine in cooked bacon. Proline, an amino acid, is decarboxylated to pyrrolidine which undergoes reaction to form the nitrosamine, N-nitrosopyrrolidine.

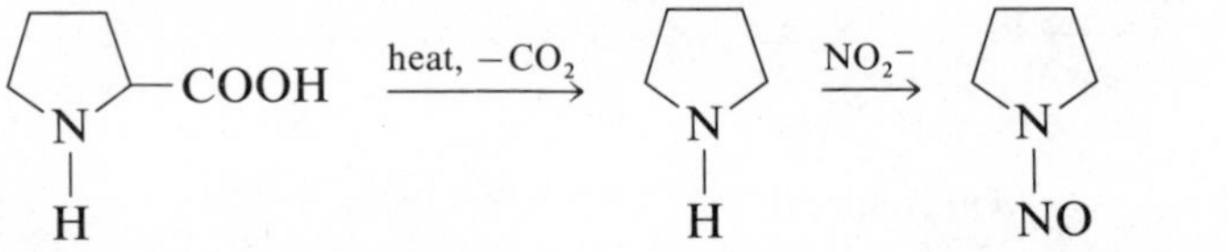

proline   pyrrolidine   N-nitrosopyrrolidine

sopyrrolidine. The biological interactions through which nitrosamines initiate tumors are not well understood. Although considerable controversy has been generated over whether or not the nitrosamine levels found in food are sufficiently high to pose a true hazard for humans, it is widely believed that nitrosamines are potentially hazardous and are probably carcinogenic to humans.

17-4.4 AMINES AS NUCLEOPHILES

Amines can act as nucleophiles, with nitrogen furnishing a pair of electrons. If the nitrogen atom furnishes the pair of electrons for a bond between the nitrogen atom and a carbon atom, substitution occurs. On the other hand, if the nitrogen atom furnishes the pair of electrons for a bond between the nitrogen atom and a hydrogen atom, elimination occurs. Substitution and elimination reactions in which amines act as nucleophiles occur with alkyl halides. Substitution occurs between amines and methyl or primary alkyl halides. The product of the reaction is a quaternary ammonium salt.

$$CH_3CH_2\overset{..}{N}H_2 + 3\,CH_3-Br \longrightarrow \left[CH_3CH_2-\overset{\overset{\displaystyle CH_3}{|}}{\underset{\underset{\displaystyle CH_3}{|}}{N}}-CH_3 \right]^+ Br^-$$

ethyl amine	methyl bromide	trimethyl ethyl ammonium bromide

$$HOCH_2CH_2\overset{..}{N}H_2 + 3\,CH_3I \longrightarrow \left[HOCH_2CH_2-\overset{\overset{\displaystyle CH_3}{|}}{\underset{\underset{\displaystyle CH_3}{|}}{N}}-CH_3 \right]^+ I^-$$

ethanolamine	methyl iodide	choline

Elimination occurs between amines and secondary or tertiary alkyl halides. The product of the reaction is an alkene and an amine salt.

$$CH_3CH_2\overset{..}{N}H_2 + \quad \longrightarrow \quad CH_3CH_2NH_3{}^+Br^- \; + \quad$$

ethylamine	cyclohexyl bromide	ethyl ammonium bromide	cyclohexene

Nucleotides are the monomeric units found in the nucleic acids RNA and DNA. Living systems can synthesize nucleotides using derivatives of ribose (in the case of RNA) and pyrimidine bases. The reaction involves nucleophilic substitution of amines on the carbon of the ribose derivative.

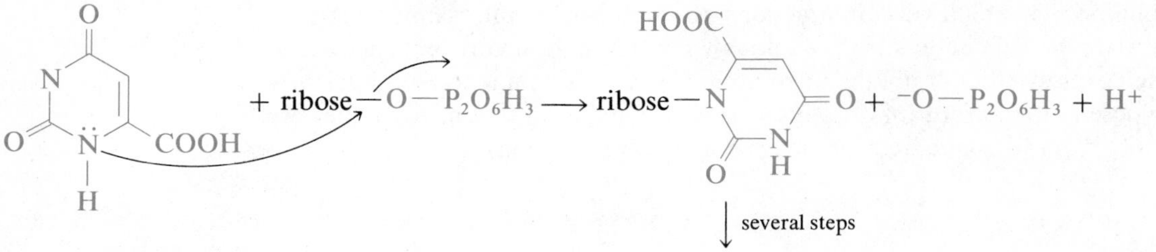

a pyrimidine base   *a nucleotide*

PROBLEMS

7. Write the structures of the amines 2-aminothiazole and 2-aminopyrazine which yield sulfathiazole and sulfapyrazine, respectively.

8. Write the structure of the product of each reaction illustrating the effect of nitrous acid on the type of amine shown.

(a) 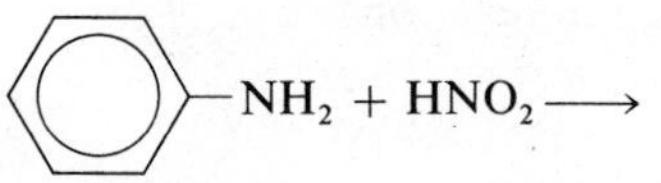 $-NH_2 + HNO_2 \longrightarrow$

(b) $CH_3CH_2N-H + HNO_2 \longrightarrow$
     $\quad\quad\quad\; |$
     $\quad\quad\quad CH_3$

9. Write the structure of the product of each reaction.

(a) $CH_3CHC-Cl + NH_3 \longrightarrow$ (with O double-bonded to C, and CH_3 below)

(b) 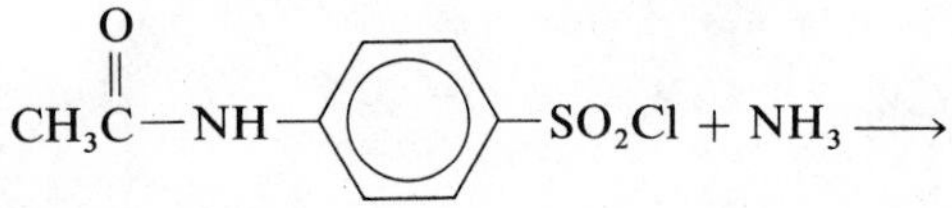 CH_3C-NH- $-SO_2Cl + NH_3 \longrightarrow$ (with O double-bonded to C)

(c) $CH_3CH_2NH_2 + 3\ CH_3CH_2Br \longrightarrow$

ANSWERS TO PROBLEMS

7.

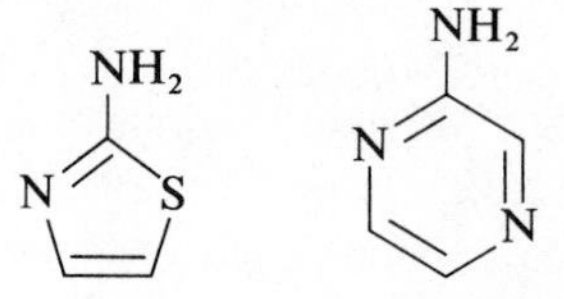

2-aminothiazole   2-aminopyrazine

8. (a) primary aromatic

amines form diazonium ion (salts), $-\overset{+}{N}\equiv N + H_2O$ (b) secondary amines

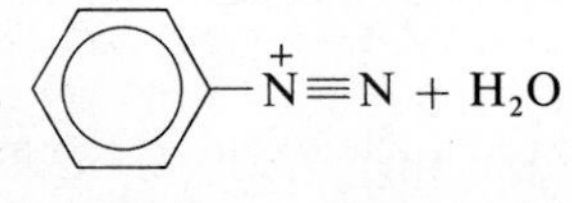

form nitrosamines, $CH_3CH_2N-NO + H_2O$ 9. (a) Formation of an amide,
                    $\quad\quad\quad\quad |$
                    $\quad\quad\quad\; CH_3$

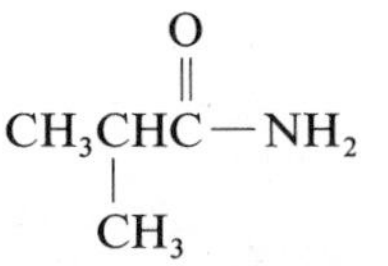

$$\text{CH}_3\text{CHC}-\text{NH}_2 \quad \text{(b) formation of a sulfonamide,} \quad \text{CH}_3\text{C}-\text{NH}-\bigcirc-\text{SO}_2\text{NH}_2$$

with the O double bonds shown above each carbonyl carbon and a CH_3 group below the first carbon.

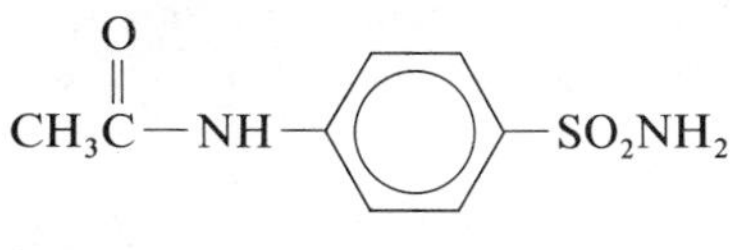

(c) formation of a quaternary ammonium salt, $(\text{CH}_3\text{CH}_2)_4\text{N}^+\text{Br}^-$

17-5 THE PHYSIOLOGICALLY ACTIVE AMINES

LEARNING OBJECTIVES

1. Identify from a list the three broad classes of physiologically active amines: narcotics, β-phenethylamines, and hallucinogens.

2. Given the name of any class of physiologically active amine, state whether it is a stimulant or depressant.

Many amines are characterized by their marked biological activity. Among the physiologically active amines, the most common are the narcotics, the phenethylamines, and the hallucinogens. These amines can cause a psychological, as well as a physical, dependence. *Dependence* is a condition in which the body needs a particular compound to the extent that deprivation produces withdrawal symptoms. Furthermore, many of these amines can also cause tolerance. *Tolerance* means that the user must take increasingly larger doses of the drug to achieve the effect obtained with the initial dose.

17-5.1 NARCOTICS

The narcotics are depressants. The short-term depressant actions include analgesia, sedation (freedom from anxiety, muscular relaxation, and decreased motor activity), hypnosis (drowsiness and lethargy), and euphoria (a sense of well-being and contentment). The long-term effects include constipation, loss of appetite and weight, temporary impotency, habituation, and addiction with unpleasant and painful withdrawal illness. Examples of narcotic drugs are morphine, codeine, heroin, demerol, nalorphine, and methadone.

Chemically, all narcotics are saturated, heterocyclic, tertiary amines that contain one aromatic ring. Only morphine is found in nature; codeine and heroin are prepared from morphine. Morphine is the principal active component (about 10%) of the opium poppy. The milky juice that oozes from the cut pods contains the morphine. Heating morphine with acetic acid produces the more potent drug, heroin. Converting the 3-hydroxyl group of morphine to a methyl ether produces codeine, a common component of cough medicines. Other synthetic opium derivatives are demerol, nalorphine, and methadone. Because the withdrawal symptoms of methadone are relatively mild, it is used in the withdrawal treatment of chronic opium users. Nalorphine has the ability to counteract nearly all of the effects of morphine and other narcotics. It is an important antidote in cases of narcotic overdoses. In the addict, an injection of nalorphine can initiate an im-

Figure 17.2 The chemical structures of some narcotics. All compounds are amines. The structures of demerol and methadone are drawn to suggest similarity to morphine.

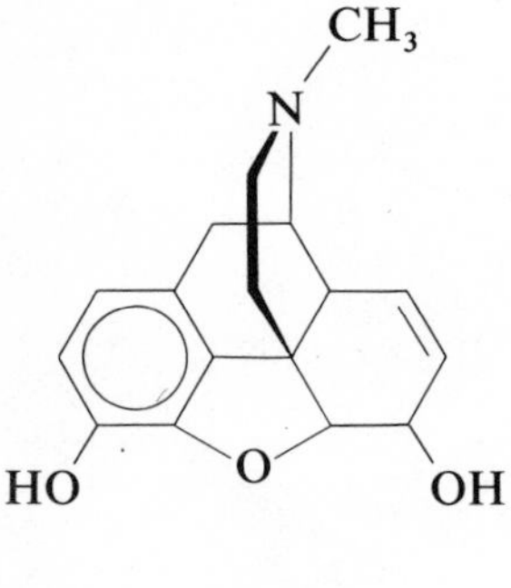

morphine

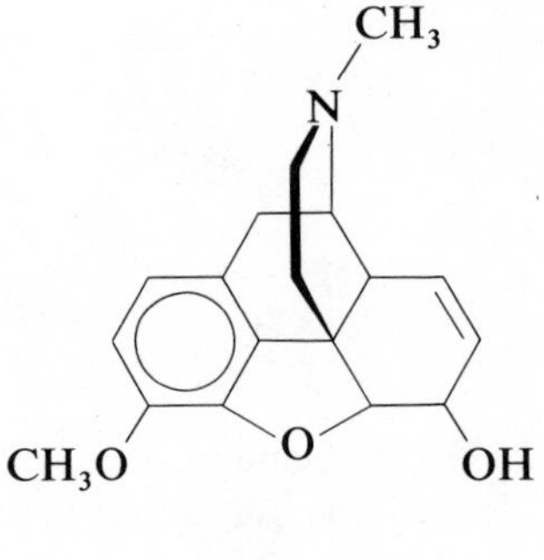

codeine

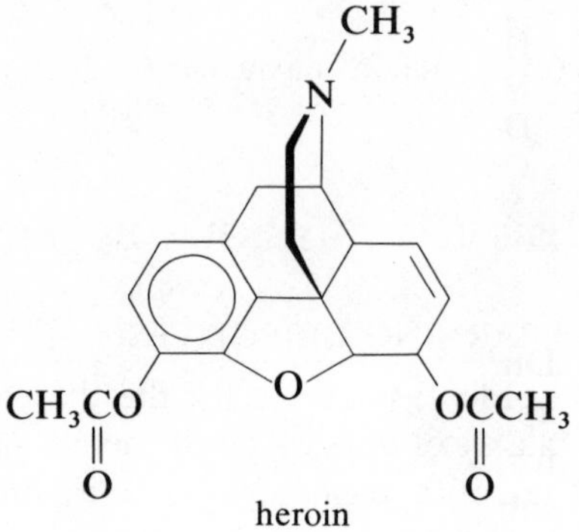

heroin

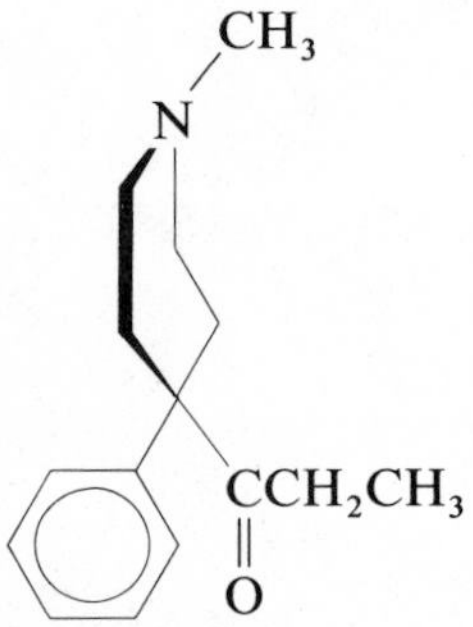

demerol

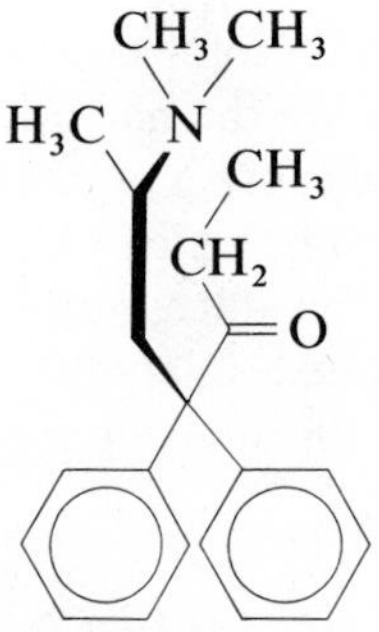

methadone

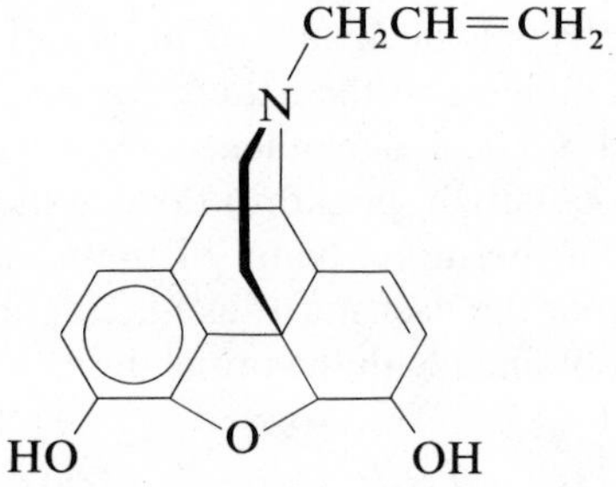

nalorphine

mediate onset of withdrawal symptoms. The structures of these compounds are shown in Figure 17.2.

For years, morphine was the only effective pain killer available, and it became indispensable in medicine and surgery. However, addiction to morphine and other narcotic amines is such a serious social problem that the sale or possession of them is now controlled by government agencies. Morphine and morphine-like drugs literally trap the abuser. First, the abuser develops a physical dependence. This can be initiated by repeated small doses of the drug. Withdrawal begins after about twelve hours and the abuser generally experiences an unpleasant feeling. Repeated use causes the body to adapt to the drug (tolerance). Now the abuser requires increasing quantities of the drug to produce the effects felt initially and to avoid the intense discomforts of withdrawal. Eventually the abuser takes doses that ordinarily would be lethal. With this increased dosage the degree of dependence on, and tolerance of, the drug also increases. The abuser is trapped. Continued use can result in death from respiratory failure.

Withdrawal can be painful and psychologically damaging. Abrupt withdrawal (*cold turkey*) from the drug is a cruel experience and not medically approved.

Table 17.4 Pain Killing Rating of Analgesics

Analgesic	Relative Rating Compared with Aspirin
Heroin	300
Morphine	100
Demerol	20–30
Codeine	15
Aspirin	1

During withdrawal the person has a consuming craving for the drug. A better approach is gradual withdrawal: the person is kept on minimal doses of morphine for a few days under medical supervision. Then, for about a week, methadone is substituted. Methadone, unlike morphine and morphine-type drugs, does not normally induce euphoria, but it eliminates the hunger for the drug and the withdrawal pains that accompany the giving up of the use of morphine. The dose of methadone is gradually reduced, then discontinued. However, the psychological dependence on the drug may linger for the rest of the person's life.

The chief medical use of morphine and morphine-like drugs is as an analgesic for the relief of pain. The pain-killing power of some analgesic drugs is shown in Table 17.4. The central nervous system contains receptor sites for naturally occurring modulators of pain responses. It is now believed that morphine produces its effect by binding to these receptor sites. Morphine binds to the receptor sites because it has a structural feature in common with the naturally occurring modulators. Recent studies indicate that these naturally occurring modulators are two peptides. These naturally occurring peptides to which morphine bears a structural similarity are called *enkephalins*. Figure 17.3 shows the structural relationship between morphine and one of these peptides.

Figure 17.3 The structural similarity between morphine and a naturally occurring peptide. That portion of the peptide is colored to suggest its similarity to morphine.

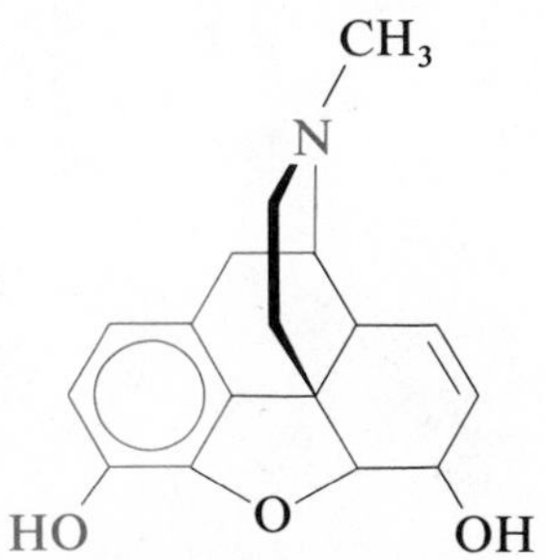

morphine 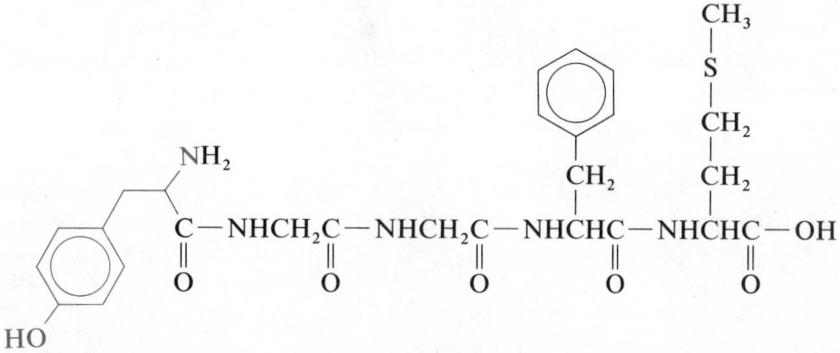   an enkephalin

17-5.2 β-PHENETHYLAMINES

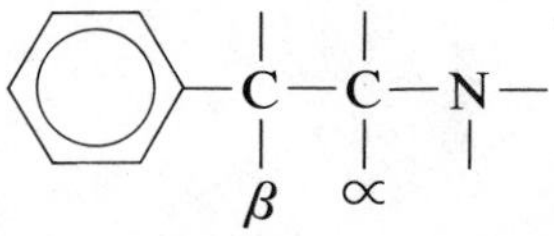

The β-phenethylamines are compounds that have a phenyl group on a carbon atom that is beta to an amino group or a substituted amino group. The most common β-phenethylamines are amphetamine, isoproterenol, phenylephrine, isoxsuprine, and mescaline. We will discuss amphetamine below and the hallucinogen, mescaline, is mentioned in the next section. The others are summarized in Table 17.5. All of these compounds stimulate the central nervous system.

The amphetamines are synthetic amines. Amphetamine (or benzedrine) and methamphetamine are the most common. Their chemical structure is similar to adrenaline, the hormone released by the adrenal gland at times of stress. Adrenaline causes stimulation of the central nervous system, and since amphetamines mimic the action of adrenaline on the central nervous system, they are commonly referred to as *speed, uppers,* or *pep pills.* The use of the other phenethylamines is described in Table 17.5.

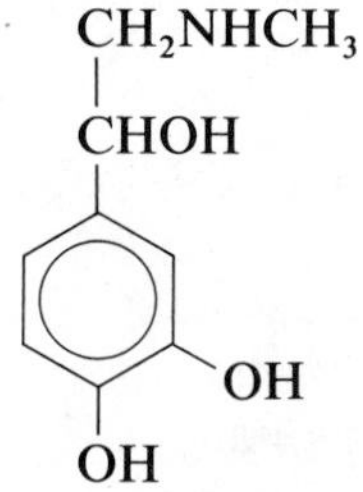

adrenaline

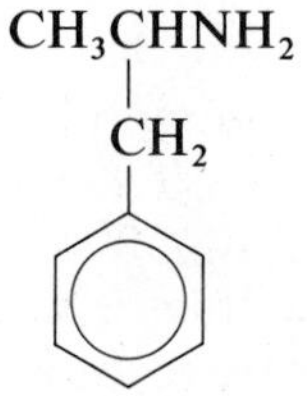

amphetamine (benzedrine)

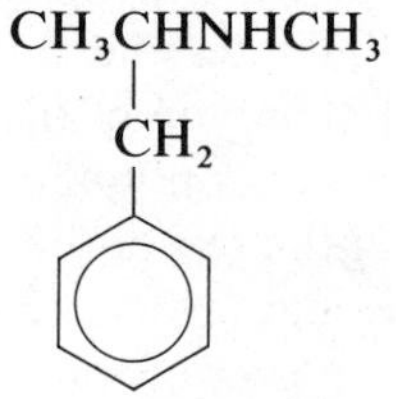

methamphetamine

Table 17.5   Some Physiologically Active β-Phenethylamines

Name	Structure	Use
Isoproterenol	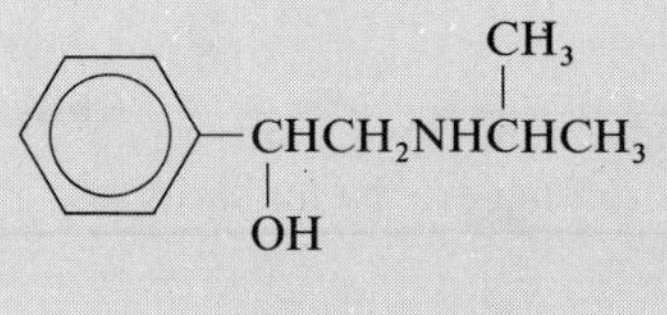	Most widely used bronchodilator
Phenylephrine	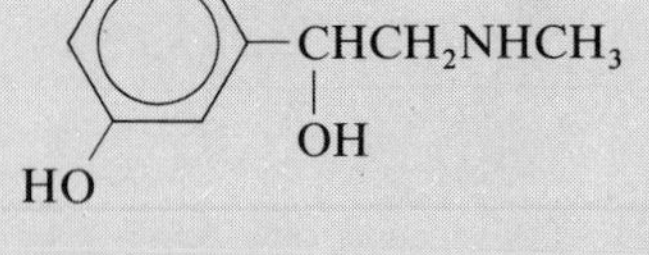	Widely used as a nasal spray
Isoxsuprine	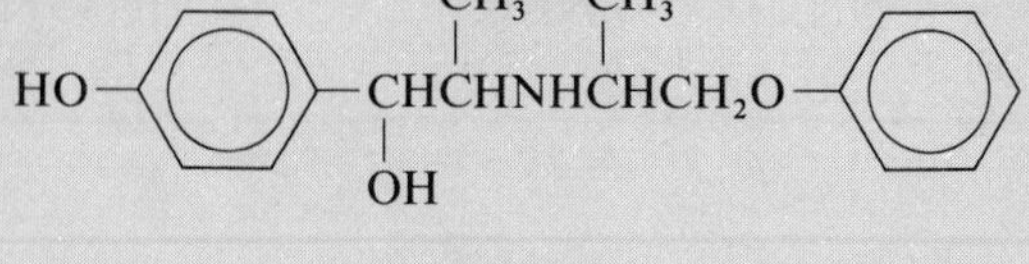	Very good relaxant for uterine muscle. Used when there is pain during menstruation and to prevent spasms and premature birth.

When taken under medical supervision and in low dosages, amphetamines have been found useful for treating a variety of conditions. These include control of hyperactive children, treatment of narcolepsy (compulsive tendency to fall asleep), obesity, fatigue, and minor depressions. But in all of these examples, physicians are doubtful as to the long-term effectiveness of amphetamines. The American Medical Association has been outspoken over the use of amphetamines by people who are trying to control their weight.

Amphetamines produce a general increase in alertness, a sense of well-being, wakefulness, and decreased feelings of fatigue. They may be used in low oral doses for limited periods by long distance truck drivers and athletes requiring extreme physical endurance. However, with prolonged use tolerance develops and the same amount of amphetamine no longer induces the desired effect. Thus, the dosage may be gradually increased. This can lead to periods of sleeplessness, loss of weight, and severe paranoia. Some uses, particularly intravenous, of amphetamines are associated with inclinations toward agression and violence and with greater tendency to take action quickly.

17-5.3 HALLUCINOGENS

Hallucinogens are compounds that stimulate sensory perceptions that have no physical reality. This may involve our senses of taste, smell, sight, hearing, or feeling. Frequently hallucinogens are taken for their visual effects. Scientific interest in the hallucinogens lies in their use to produce experimental psychoses .

Psychosis is a severe form of mental disorder.

The compounds in this class of hallucinogens are obtained from plants. Mescaline comes from the peyote cactus, psilocybin is derived from certain species of mushroom, and lysergic acid comes from the parasitic fungus ergot which grows in rye and wheat. LSD (the amide of lysergic acid) is chemically synthesized in the lab. All of these compounds are amines. Psilocybin and mescaline are not used medically and LSD is a compound used in psychotherapeutic studies.

psilocybin

lysergic acid diethylamide
(LSD)

mescaline

serotonin

It is thought that hallucinogens affect the serotonin level in the brain. Serotonin is an amine that is structurally similar to psilocybin, LSD, and mescaline. It is widely distributed in plants and animals and affects a number of bodily functions in humans including blood pressure. Furthermore, variations in the serotonin level in the blood have been correlated with the incidence of schizophrenia.

LSD is the most common and representative example of the hallucinogens. About one-millionth of an ounce of LSD will produce a series of sequential events in the user. The first effects include dizziness, weakness, nausea, and dilation of the pupils. The second series of changes are perceptual alterations. Among these are a distortion of sense of time, visual aberrations, and an uncontrolled psychedelic state in which all sensory input loses customary meaning (for example, "hearing colors" and "seeing sounds"). These effects are followed by, and in some cases, overlap, the psychic symptoms: rapid changes in mood, depersonalization, distortions in body image, and dissociation of the self from external reality.

Unlike the users of marijuana, users of LSD can experience severe panic reactions (*bad trips*) and flashbacks. Bad trips are more common with high doses, which can occur accidentally because the purity of LSD varies widely. Flashbacks are sudden, unexpected perceptual distortions and bizarre memories of an LSD trip that occur after the pharmacological effects of the drug have worn off. Flashbacks can occur as long as a month or a year after the last ingestion of LSD.

Available evidence indicates that a moderate dose of *pure* LSD does not damage chromosomes *in vivo*, does not cause detectable genetic damage, and is not a carcinogen in humans. Further, despite all claims to the contrary, LSD does not enhance creativity. LSD, in fact, impairs the critical synthetic operations of the creative process; that is, the sequential coordination of original responses into a pattern that is relevant to the problem or endeavor at hand.

The effective oral dose of LSD is about 100 micrograms. LSD does not act for about 45 minutes; then, about four hours after consumption, the effects begin to decrease and are usually gone in 6–12 hours. Tolerance to LSD is rapidly developed. Unlike opium tolerance, LSD tolerance may be developed in a few days and is usually lost in 2 or 3 days. Over a period of days, therefore, some users build up an LSD tolerance of 1000–2000 micrograms. For the original effect, therefore, massive doses may be taken once a tolerance has been built up. There is no evidence of a physical LSD dependence.

SUMMARY

Amines are a class of organic compounds that contain a nitrogen atom. Amines can be classified as primary, secondary, or tertiary amines. Amine salts contain a fourth atom covalently bonded to nitrogen through the unshared electron pair on nitrogen. Amines with simple alkyl groups are named by common names. Otherwise, IUPAC names are used and the $-NH_2$ group is called the amino group.

$$
\begin{array}{cccc}
\text{R}-\overset{\displaystyle |}{\underset{\displaystyle |}{\text{N}}}-\text{H} &
\text{R}-\overset{\displaystyle |}{\underset{\displaystyle |}{\text{N}}}-\text{R} &
\text{R}-\overset{\displaystyle |}{\underset{\displaystyle |}{\text{N}}}-\text{R} &
\left[\text{R}-\overset{\displaystyle \text{H} |}{\underset{\displaystyle |\text{H}}{\text{N}}}-\text{H}\right]^{+} \text{X}^{-} \\
\text{H} & \text{H} & \text{R} & \\
\textit{primary amine} & \textit{secondary amine} & \textit{tertiary amine} & \textit{amine salt}
\end{array}
$$

Amines are prepared *in vitro* by the reduction of nitro compounds, nitriles, amides, and carbonyl compounds.

$$R-NO_2 \xrightarrow{\text{reducing agent}} RNH_2$$

a nitro compound   *an amine*

$$R-C\equiv N \xrightarrow{\text{reducing agent}} RCH_2NH_2$$

a nitrile   *an amine*

$$R-\overset{\overset{\displaystyle O}{\|}}{C}-NH_2 \xrightarrow{\text{reducing agent}} RCH_2NH_2$$

an amide   *an amine*

$$R-\overset{\overset{\displaystyle O}{\|}}{C}-H \xrightarrow{NH_3} \xrightarrow{H_2,Ni} RCH_2NH_2$$

an aldehyde   *an amine*

In vivo, amines are often the products of decarboxylation of amino acids.

$$\underset{\underset{\displaystyle NH_2}{|}}{R\overset{\overset{\displaystyle O}{\|}}{C}HCOH} \longrightarrow CO_2 + RCH_2NH_2$$

an amino acid   *an amine*

The aliphatic primary, secondary, and tertiary amines and the saturated heterocyclic amines are stronger bases than ammonia.

basicity: primary, secondary, and tertiary   aniline and aromatic
amines, saturated heterocyclic amines $>$ NH_3 $>$ heterocyclic amines

Amines react with acids and form amine salts. Amines, used as therapeutic medicines, are administered as amine salts.

$$RNH_2 + HX \longrightarrow [RNH_3]^+X^-$$

an amine   *an amine salt*

The reaction of primary aromatic amines with nitrous acid gives a diazonium ion. These ions, upon reaction with aromatic compounds, give a variety of colored dyes.

a primary aromatic amine   *a diazonium ion*

Secondary amines react with nitrous acid to form nitrosamines. Nitrosamines are suspected carcinogens.

$$R-\underset{\underset{R}{|}}{N}-H + HNO_2 \longrightarrow R-\underset{\underset{R}{|}}{N}-NO + HOH$$

a secondary amine   *a nitrosamine*

Primary and secondary amines react with acid chlorides to form substituted amides. The amide grouping is the characteristic functional group in peptides and proteins.

$$R-\overset{\overset{O}{||}}{C}-Cl \begin{cases} \xrightarrow{RNH_2} R-\overset{\overset{O}{||}}{C}-\underset{\underset{H}{|}}{N}-R + HCl \\ \\ \xrightarrow{R_2NH} R-\overset{\overset{O}{||}}{C}-\underset{\underset{R}{|}}{N}-R + HCl \end{cases}$$

an acid chloride   *substituted amides*

Amines react with benzenesulfonyl chloride to form benzenesulfonamides. The latter compounds are useful in forming a class of antibacterials known as sulfa drugs.

$$RNH_2 + C_6H_5SO_2Cl \longrightarrow C_6H_5SO_2\underset{\underset{}{}}{\overset{\overset{H}{|}}{N}}-R \xrightarrow{\text{several steps}} \text{sulfa drugs}$$

an amine   *a benzenesulfonamide*

Amines are nucleophilic and are capable of causing substitution reactions in methyl and primary halides, but elimination in secondary and tertiary halides. The physiologically active amines are the narcotics, the β-phenethylamines, and the hallucinogens. Narcotics are depressants and β-phenethylamines and hallucinogens are stimulants.

ADDITIONAL PROBLEMS

10. Give a correct name for each of the following amines.

(a) $CH_3CH_2CH_2NH_2$

(b) $CH_3\underset{\underset{CH_3}{|}}{N}CH_3$

(c) $CH_3CH{=}CHCH_2NH_2$

(d)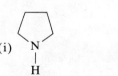

(e) $CH_3CH_2NHCH_2CH_3$

(f) $(CH_3)_3CNH_2$

(g)

(h)

(i)

(j)

11. Using the amines in preceding questions, which amine is an example of each of the following descriptions?
    (a) primary amine
    (b) aromatic heterocyclic amine
    (c) saturated heterocyclic amine
    (d) tertiary amine
    (e) substituted aniline
    (f) secondary amine
12. Write structures for each of the following.
    (a) diethylamine
    (b) isopropylamine
    (c) 2-aminopentane
    (d) 3-methyl-1,6-diaminohexane
    (e) o-bromoaniline
    (f) pyrrole
    (g) alanine (an amino acid); 2-aminopropanoic acid
    (h) tetraethyl ammonium bromide
    (i) cyclohexyl amine
    (j) p-aminobenzoic acid
13. Write the product of the reaction of methyl amine with each of the following.

    (a) HCl

    (b) $CH_3\overset{\displaystyle O}{\overset{\|}{C}}-Cl$

    (c) CH_3I (3 moles)

    (d) $ClO_2S-\langle\bigcirc\rangle-NH\overset{\displaystyle O}{\overset{\|}{C}}-CH_3$

14. Write the structure of the product of each reaction.

    (a) $\langle\bigcirc\rangle-NO_2 \xrightarrow{Fe, HCl}$

    (b) $CH_3C\equiv N \xrightarrow{LiAlH_4}$

    (c) $H_2NCH_2CH_2CH_2\overset{\displaystyle}{\underset{\displaystyle NH_2}{CH}}\overset{\displaystyle O}{\overset{\|}{C}}-OH \xrightarrow{decarboxylase} CO_2 +$

    (d) $CH_3-\overset{\displaystyle CH_3}{\underset{\displaystyle CH_3}{\overset{\displaystyle |}{\underset{\displaystyle |}{C}}}}-NH_2 + HCl \longrightarrow$

(e) $CH_3-\langle\bigcirc\rangle-NH_2 \xrightarrow{HNO_2}$

(f) $\langle\bigcirc\rangle-\overset{\displaystyle O}{\overset{\|}{C}}-NH_2 \xrightarrow{LiAlH_4}$

(g) $CH_3\overset{\displaystyle O}{\overset{\|}{C}}NH-\langle\bigcirc\rangle-SO_2Cl + H_2N-\langle\overset{S}{\underset{N}{\rangle}} \longrightarrow$

(h) $HOCH_2CH_2NH_2 + 3\ CH_3I \longrightarrow$

(i) $CH_3CH_2CH_2\overset{\displaystyle}{\underset{\displaystyle CH_3}{N}}-H \xrightarrow{HNO_2}$

15. Benzedrine, or amphetamine sulfate, is a central nervous stimulant used to produce wakefulness, to reverse fatigue, and to treat depressed psychic states. Chemically, it is an amine sulfate derived from 1-phenyl-2-amino propane. Draw the structure of the amine sulfate.
16. How are dyes produced chemically? What kinds of functional groups occur in dyes? Why does the FDA ban some dyes and review others? What approach to making dyes safe for human consumption is currently being considered?
17. Write the general structure of a sulfa drug. Why were the sulfa drugs found to be effective against microorganisms? Why have the sulfa drugs been largely replaced by antibiotics such as penicillins?
18. Lidocaine (Xylocaine) is frequently used as a local anesthetic for minor surgery. It may be injected into the site of the surgery or applied topically to the surface of the membranes of the eye, nose, and throat. Identify the functional group (colored) in lidocaine. What two organic compounds would you use to make lidocaine?

lidocaine

CHAPTER 18

STEREOCHEMISTRY: INTRODUCTION TO BIOCHEMISTRY

Stereochemistry deals with structure in three dimensions. An important part of stereochemistry is stereoisomers. *Stereoisomers* are isomeric compounds that are different from one another only in the way the atoms are oriented in space. From Chapter 11 we recall that our sense of smell was related to receptor sites of varying size and shape. In Chapter 17 we saw that the analgesic property of morphine was due to its structural similarity to a naturally occurring pain modulator. Similarly, foods and medicines must have the proper molecular orientation if they are to be beneficial to living systems. The orientation of biomolecules (see Section 10-3) is very important to living systems. Biological reactions are catalyzed by enzymes that can assume a particular molecular orientation. The differences in structural orientation between stereoisomers is subtle, but this difference is very important to life. To gain a better understanding of how living systems recognize subtle differences of shape and structure we will study the terminology of stereochemistry and its relationship to biomolecules.

18-1 ISOMERISM: A REVIEW

LEARNING OBJECTIVE

Given a molecular formula, write structural formulas for its isomers.

In Section 11-4 we considered some of the aspects of isomerism. An *isomer* is any one of a group of compounds that have the same molecular formula but different structural formulas and that display different physical, chemical, or physiological properties. Table 18.1 lists some of the isomers for the molecular formulas

Table 18.1 Comparison of Physical, Chemical, and Physiological Properties of Isomers

Molecular Formula	Structural Formula of Isomer and Properties			
	Isomer	Physical Property	Chemical Property	Physiological Chemistry
C_4H_{10}	CH_3 | CH_3CHCH_3 2-methylpropane	Boiling point: $-12°$	Burns smoothly in oxygen	Toxic
	$CH_3CH_2CH_2CH_3$ butane	Boiling point: $0°$	Burns less smoothly in oxygen	Toxic
C_2H_6O	CH_3CH_2OH ethyl alcohol	Boiling point: $78°$	Reacts with PBr_3 to form ethyl bromide	Can be metabolized to CO_2 and H_2O in suitable dilution
	CH_3OCH_3 methyl ether	Boiling point: $-24°$	Does not react with PBr_3 to form an alkyl halide	May be a narcotic in high concentrations

C_4H_{10} and C_2H_6O, and compares their physical, chemical, and physiological properties. In the case of C_4H_{10}, the isomers belong to the same class of organic compounds (for example, 2-methylpropane and butane are alkanes); in C_2H_6O, the isomers have different functional groups (ethyl alcohol and dimethyl ether).

The concept of geometric isomerism was also presented in Chapter 11. Geometric isomers are commonly found among alkenes and cyclic compounds because of lack of free rotation about certain bonds. For alkenes, the pi bond of the carbon–carbon double bond restricts rotation about the carbon–carbon double bond. The isomer that has the like groups on the same side of the molecule is called the *cis isomer,* and the isomer with the like groups on opposite sides is the *trans isomer.* If one of the carbon atoms of the carbon–carbon double bond has like groups on the same carbon atom, no geometric isomers are possible (see Section 11-4.2). Table 18.2 illustrates geometric isomerism in 2-butene and compares the properties of cis-trans isomers of 2-butene. For cyclic compounds, the ring structure restricts rotation about the carbon–carbon single bond and geometric isomers are possible. For example, as shown in Table 18.2 for 1,2-dichlorocyclohexane, two groups can lie on the same side of the planar cyclohexane ring (cis isomer) or on opposite sides of the planar cyclohexane ring (trans isomer).

Geometric isomerism manifests itself in living systems. For example, a typical honey bee colony consists of a queen (the only fully developed female), several hundred drones (the males), and thousands of workers (underdeveloped females). The queen bee secretes a compound known as a pheromone (see Section 11-4.2) that prevents construction of cells for rearing new queens. This pheromone also prevents ovarian development in workers. Chemically, the pheromone is an

Table 18.2

Structural Formula	cis Isomer		trans Isomer	
$CH_3CH=CHCH_3$	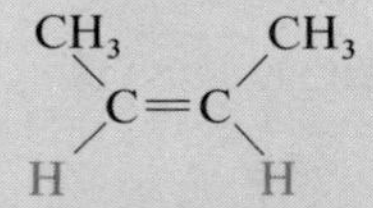	boiling point: 4° melting point: −130° density: 0.667 g/mL	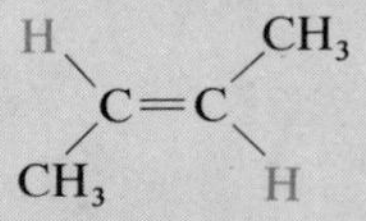	boiling point: 1° melting point: −106° density: 0.649 g/mL
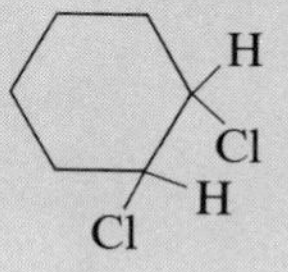	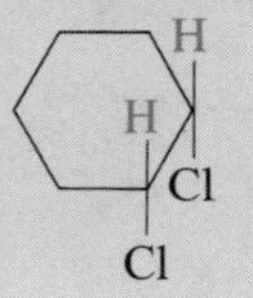	boiling point: 206° melting point: −1.5° density: 1.20 g/mL	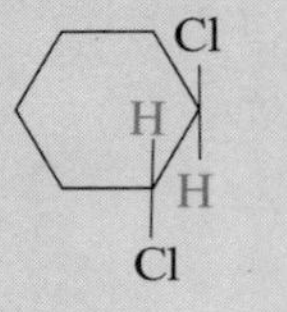	boiling point: 189° melting point: −6.3° density: 1.18 g/mL

unsaturated carboxylic acid that contains a carbonyl group. It is interesting that the trans isomer of the carboxylic acid, not the cis isomer, is the physiologically active isomer.

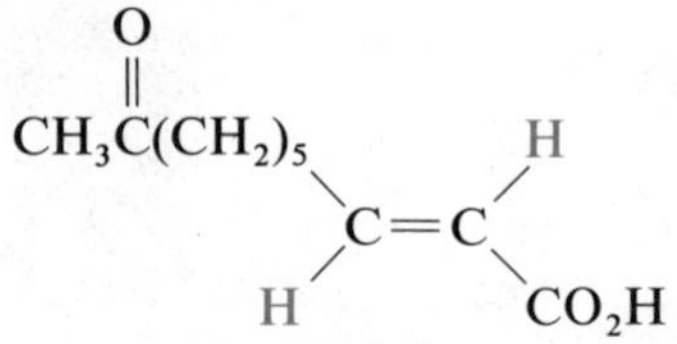

trans isomer
physiologically active

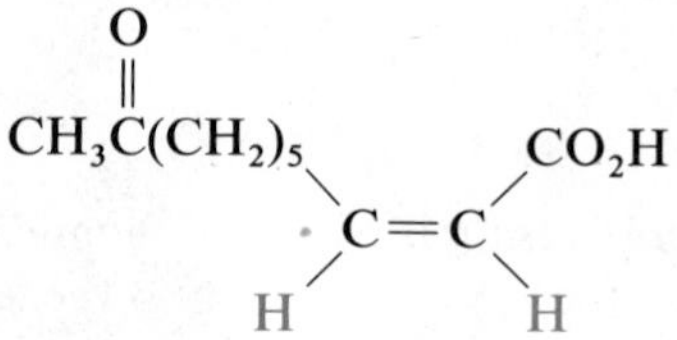

cis isomer
physiologically inactive

PROBLEMS

1. Draw structural isomers for C_5H_{12} and C_4H_9Cl.
2. Grandisol, found in the male boll weevil, acts as a sex attractant. Can grandisol exist as cis-trans isomers about the carbon–carbon double bond? Can it exist as cis-trans isomers about the cyclobutane ring? Draw the structures of any cis-trans isomers. The cis isomer is physiologically active. Which structure corresponds to the cis isomer?

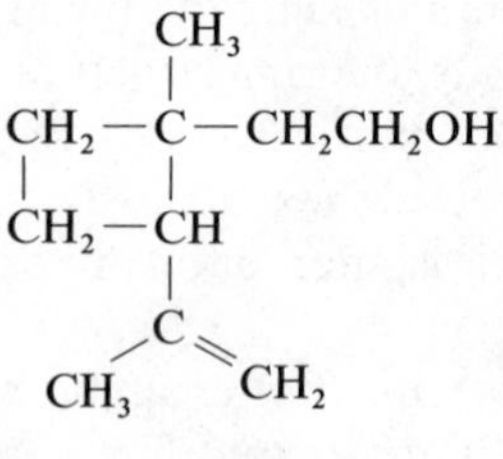

grandisol

ANSWERS TO PROBLEMS   **1.** C_5H_{12}: $CH_3(CH_2)_3CH_3$, $CH_3CHCH_2CH_3$, CH_3CCH_3;

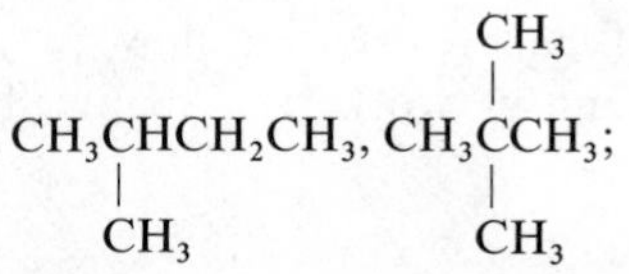

C_4H_9Cl: $CH_3(CH_2)_2CH_2Cl$, $CH_3CHCH_2CH_3$, CH_3CHCH_2Cl, CH_3CCH_3

with substituents Cl, CH₃, Cl, and CH₃ respectively below/above as shown.

2. Cis-trans isomers are not possible about the carbon–carbon double bond, but cis-trans isomers are possible about the cyclobutane ring structure. The cis and trans isomers are shown here.

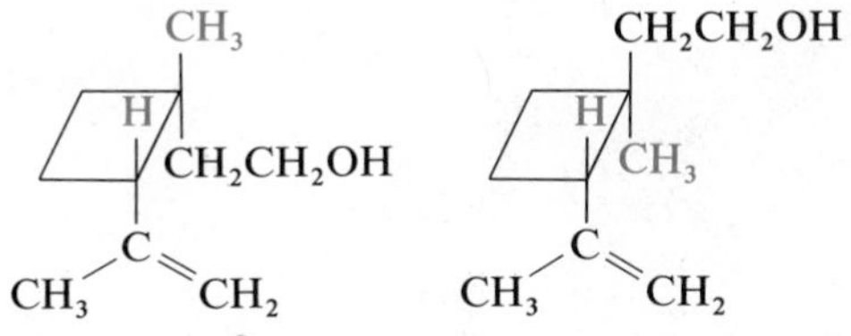

cis isomer   trans isomer

18-2 THE FISCHER PROJECTION

LEARNING OBJECTIVES

1. Given an organic molecule, write the Fischer projection.

2. Given the tetrahedral model of a molecule, write the Fischer projection.

A carbon atom that is bonded to four other atoms is tetrahedrally oriented. A common notation for tetrahedral carbon is the **perspective formula**.

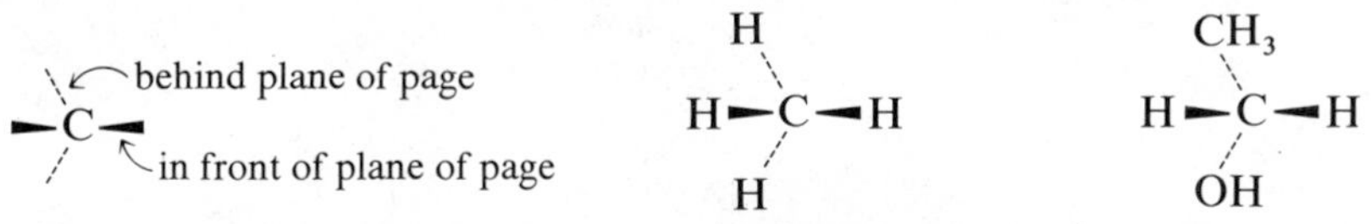

behind plane of page

in front of plane of page

perspective formula
for tetrahedral carbon

$H \blacktriangleright C \blacktriangleleft H$ with H above and H below

perspective formula
for methane

CH_3 above, $H \blacktriangleright C \blacktriangleleft H$, OH below

perspective formula
for ethanol

A **perspective formula** is a formula in which the four groups attached to a carbon atom that exists in the plane of the page are indicated as going behind the plane of the page by a dashed bond and coming out of the page to the reader by a wedged-shaped bond.

Chemists often use another notation for the tetrahedral model called the **Fischer projection**. The relationship between the perspective formulas and the corresponding Fischer projection is shown in the examples below. The spatial arrangement, or orientation, of atoms or groups about a carbon atom is an important aspect of stereochemistry. We will discuss this aspect in the following sections.

A **Fischer projection** is a planar representation of the perspective formula. In the Fischer projection, vertical lines are used to represent bonds going away from the reader and horizontal lines represent bonds coming toward the reader out of the plane of the page.

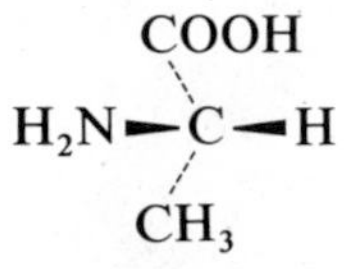

COOH
$H_2N \blacktriangleright C \blacktriangleleft H$
CH_3
alanine
an amino acid

COOH — vertical lines indicate groups behind the page
$H_2N — C — H$
CH_3 — horizontal lines indicate groups in front of the page

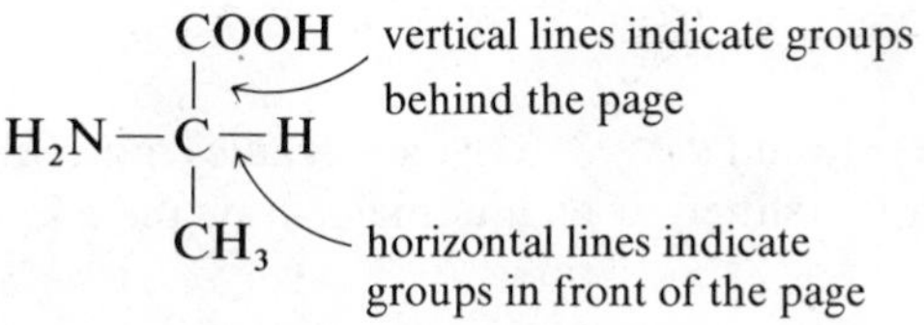

Fischer projection of alanine

$$\begin{array}{c} \text{COOH} \\ \text{HO} - \text{C} - \text{H} \\ \text{CH}_3 \end{array}$$

lactic acid
a component of muscle tissue

$$\begin{array}{c} \text{COOH} \\ \text{HO} - \text{C} - \text{H} \\ \text{CH}_3 \end{array}$$

Fischer projection of lactic acid

$$\begin{array}{c} \text{CHO} \\ \text{H} - \text{C} - \text{OH} \\ \text{H} - \text{C} - \text{OH} \\ \text{CH}_2\text{OH} \end{array}$$

erythrose
a carbohydrate

$$\begin{array}{c} \text{CHO} \\ \text{H} - \text{C} - \text{OH} \\ \text{H} - \text{C} - \text{OH} \\ \text{CH}_2\text{OH} \end{array}$$

Fischer projection of erythrose

PROBLEM

3. Write Fischer projection formulas for each of the following.

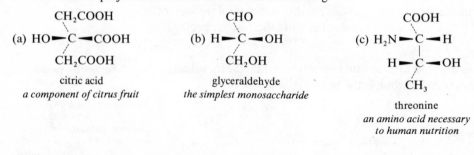

(a)
$$\begin{array}{c} \text{CH}_2\text{COOH} \\ \text{HO} - \text{C} - \text{COOH} \\ \text{CH}_2\text{COOH} \end{array}$$

citric acid
a component of citrus fruit

(b)
$$\begin{array}{c} \text{CHO} \\ \text{H} - \text{C} - \text{OH} \\ \text{CH}_2\text{OH} \end{array}$$

glyceraldehyde
the simplest monosaccharide

(c)
$$\begin{array}{c} \text{COOH} \\ \text{H}_2\text{N} - \text{C} - \text{H} \\ \text{H} - \text{C} - \text{OH} \\ \text{CH}_3 \end{array}$$

threonine
an amino acid necessary to human nutrition

ANSWERS TO PROBLEM

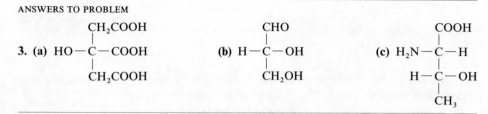

3. (a)
$$\begin{array}{c} \text{CH}_2\text{COOH} \\ \text{HO} - \text{C} - \text{COOH} \\ \text{CH}_2\text{COOH} \end{array}$$

(b)
$$\begin{array}{c} \text{CHO} \\ \text{H} - \text{C} - \text{OH} \\ \text{CH}_2\text{OH} \end{array}$$

(c)
$$\begin{array}{c} \text{COOH} \\ \text{H}_2\text{N} - \text{C} - \text{H} \\ \text{H} - \text{C} - \text{OH} \\ \text{CH}_3 \end{array}$$

18-3 ENANTIOMERS: THE AMINO ACID ALANINE

LEARNING OBJECTIVES

1. Given an object or a molecule and its mirror image, decide which are superimposable.
2. Define or recognize examples of enantiomers.

3. Given organic molecules that are enantiomers, draw their Fischer projections.

Stereoisomers, as defined at the beginning of this chapter, are isomeric compounds that are different from one another only in the way the atoms are oriented in space. The concept of stereoisomerism can be explained adequately in terms of mirror-image relationships. An object is superimposable on its mirror image if, when it is placed on its mirror image, the corresponding parts lie together. As shown in Figure 18.1, a scalpel, a culture dish, and a stethoscope are superimposable on their respective mirror images. On the other hand, a left-hand surgical glove and its mirror image (a right-hand surgical glove) are not superimposable. Likewise, a sterile container and a syringe have nonsuperimposable mirror images.

Figure 18.1 (a) Examples of objects that are not superimposable on their mirror image. (b) Examples of objects that have superimposable mirror images.

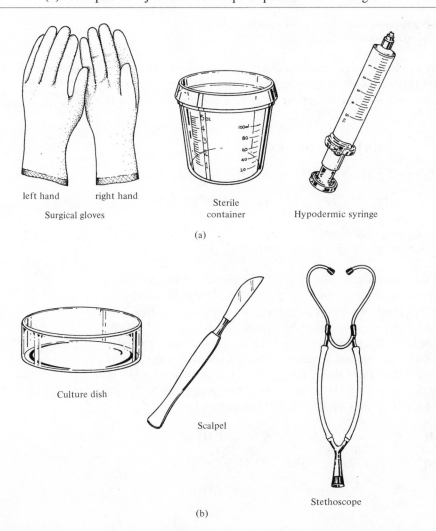

left hand   right hand

Surgical gloves

Sterile container

Hypodermic syringe

(a)

Culture dish

Scalpel

Stethoscope

(b)

These objects and the mirror image relationship of the surgical glove are also il-lustrated in Figure 18.1.

Similarly, two molecules may have a mirror image relationship. Furthermore, the two molecules may or may not be superimposable. If a molecule and its mirror image are superimposable, then they are two molecules of the same compound. As shown, the mirror image (II) of the Fischer projection of methane (I) is super-imposable on the Fischer projection (I). Thus, both projections of methane repre-sent two different *molecules* of the same compound.

mirror

H	H
|	|
H—C—H	H—C—H
|	|
H	H
I	II
Fischer projection	*mirror image*
of methane	*of* I

On the other hand, the Fischer projection of the amino acid alanine (III), and its mirror image (IV), are not superimposable. They represent two different *isomers* of alanine and are called **enantiomers**. It may appear to us that these mirror images

Enantiomers are stereo-isomers that are not superimposable mirror images of one another.

mirror

COOH	COOH
|	|
H_2N—C—H	H—C—NH_2
|	|
CH_3	CH_3
III	IV
Fischer projection	*mirror image*
of alanine	*of* III

are superimposable. If IV is taken out of the plane of the page, and flipped over, then the corresponding parts will superimpose on each other. However, this leads to the wrong conclusion. We must be careful in testing for superimposability. The rule to follow is this: *A Fischer projection must be rotated end-for-end and only in the plane of the page.* The resulting projection can then be moved over mentally in the plane of the page onto the original Fischer projection to test for the super-imposability of the groups. When IV is rotated end-for-end, V results. When V is superimposed on III (by sliding V onto III in the plane of the page), the NH_2 and H lie on their respective parts, but the COOH and CH_3 do not.

COOH		CH_3
|	*rotate 180°*	|
H—C—NH_2	*in plane* gives	H_2N—C—H
|	*of page*	|
CH_3		COOH
IV		V

Another example of enantiomers is illustrated by lactic acid, which is commonly produced in muscle tissue when work is done.

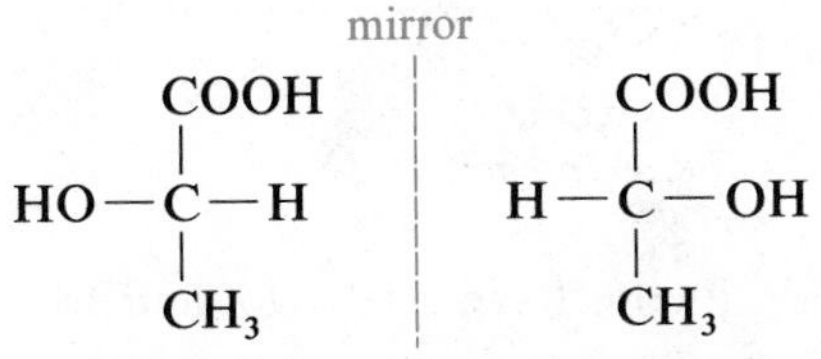

mirror

$$
\begin{array}{cc}
\text{COOH} & \text{COOH} \\
| & | \\
\text{HO}-\text{C}-\text{H} & \text{H}-\text{C}-\text{OH} \\
| & | \\
\text{CH}_3 & \text{CH}_3
\end{array}
$$

Fischer projection   *mirror image of*
of lactic acid   *Fischer projection*
                           of lactic acid

PROBLEM

4. Which of the following pairs represent mirror images that are not superimposable?

(a) 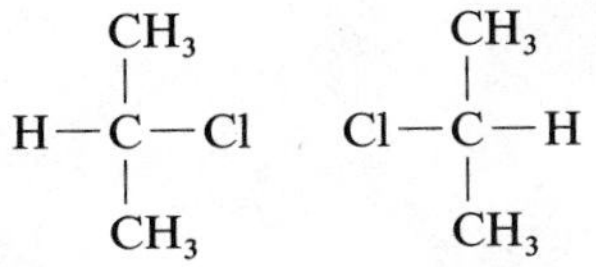
$$
\begin{array}{cc}
\text{CH}_3 & \text{CH}_3 \\
| & | \\
\text{H}-\text{C}-\text{Cl} & \text{Cl}-\text{C}-\text{H} \\
| & | \\
\text{CH}_3 & \text{CH}_3 \\
\text{A} & \text{B}
\end{array}
$$

(b) 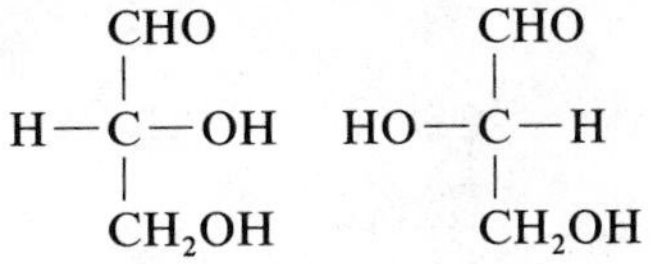
$$
\begin{array}{cc}
\text{CHO} & \text{CHO} \\
| & | \\
\text{H}-\text{C}-\text{OH} & \text{HO}-\text{C}-\text{H} \\
| & | \\
\text{CH}_2\text{OH} & \text{CH}_2\text{OH} \\
\text{A} & \text{B}
\end{array}
$$

(c) 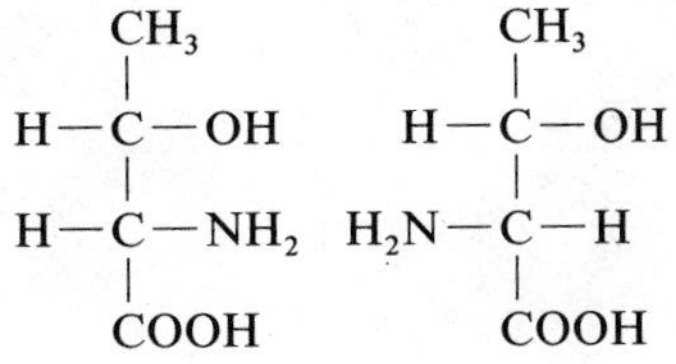
$$
\begin{array}{cc}
\text{CH}_3 & \text{CH}_3 \\
| & | \\
\text{H}-\text{C}-\text{OH} & \text{H}-\text{C}-\text{OH} \\
| & | \\
\text{H}-\text{C}-\text{NH}_2 & \text{H}_2\text{N}-\text{C}-\text{H} \\
| & | \\
\text{COOH} & \text{COOH} \\
\text{A} & \text{B}
\end{array}
$$

ANSWERS TO PROBLEM   **4. (a)** superimposable; rotate B end-for-end in plane of page and it becomes A   **(b)** not superimposable   **(c)** not superimposable and not mirror images

---

18-4 THE CHIRAL CARBON ATOM

LEARNING OBJECTIVE

Define or recognize a chiral carbon atom in a molecule.

Two molecules that are enantiomers, that is, that are not superimposable on their mirror images, are referred to as *chiral. In most cases,* a chiral molecule can be identified as that molecule that contains a carbon atom with four different groups bonded to it. A carbon atom with four different groups bonded to it is called a **chiral carbon**. Examples of molecules that have no chiral carbon are: methane, methyl chloride, and glycine.

A **chiral carbon** is a carbon atom bonded to four different groups.

$$CH_4 \qquad CH_3Cl \qquad H_2NCH_2COOH$$

methane   methyl chloride   glycine
                                an amino acid

On the other hand, lactic acid, threonine, and glucose have one, two, and four chiral carbons (colored), respectively. Lactic acid, with only one chiral carbon, exists as a pair of enantiomers. Threonine, which has two chiral centers, can exist as two pairs of enantiomers (four stereoisomers). Glucose has four chiral centers and exists as eight pairs of enantiomers (sixteen stereoisomers). Half of these stereoisomers are shown in Figure 19.1. In general, for n chiral centers a theoretical maximum of 2^n stereoisomers exist.

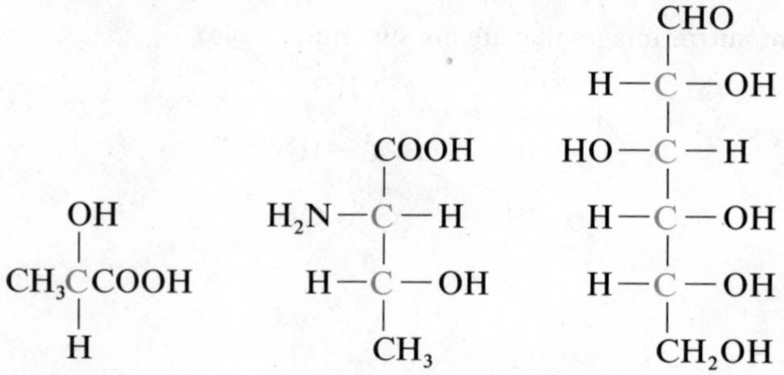

lactic acid
one chiral carbon

threonine
two chiral carbons

glucose
four chiral carbons

PROBLEM

5. Chloramphenicol is a naturally occurring antibiotic. It contains two chiral carbons. Indicate which of the carbons is chiral.

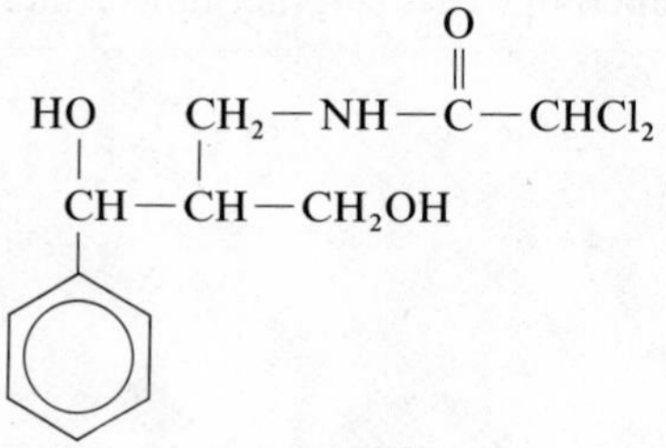

chloramphenicol

ANSWER TO PROBLEM   **5.** The secondary carbon bearing the OH group and the carbon bearing the CH_2OH group are chiral carbons.

18-5 OPTICAL ACTIVITY

LEARNING OBJECTIVES

1. Define optical activity, specific rotation, dextrorotatory, levorotatory, $(+)$, and $(-)$.

2. Explain the effect of molecules on plane-polarized light.

Molecules that are not superimposable on their mirror images have the ability to rotate plane-polarized light. *Plane-polarized light* is light consisting of waves that travel in a single plane, as opposed to ordinary light in which the waves travel in all random directions. If a molecule possesses the ability to rotate plane-polarized light it is said to be **optically active**. The ability of a substance to rotate plane-polarized light is measured by an instrument called a *polarimeter*. When plane-polarized light is passed through a substance (contained in the polarimeter) that is optically inactive, the emerging light will still travel in the same plane. However, when the substance is optically active, the emerging light will be oriented in a different plane. The polarimeter can measure the difference in rotation between the original and final planes of polarization. This difference is called the *observed rotation*.

An **optically active molecule** is a molecule which has the ability to rotate plane-polarized light.

The observed rotation of an optically active substance depends upon a variety of conditions: temperature, length of the tube containing the substance, wavelength of the light source, nature of the solvent, amount of the substance, and the structure of the substance. Because of the variety of these variables, chemists use a conventional rule to compare the optical activity of substances.

The optical activity of a substance is reported as the *specific rotation,* $[\alpha]$. The specific rotation of an optically active substance is a physical property like boiling point and density. The specific rotation is a numerical value and is determined as shown.

$$[\alpha] = \frac{\text{observed optical rotation}}{\text{length of sample tube} \times \text{concentration of sample}}$$

Enantiomers, mirror-image isomers that are not superimposable, are alike in all physical properties except the observed direction of rotation of plane-polarized light. One enantiomer rotates plane-polarized light in a clockwise direction and is said to be *dextrorotatory*. The symbol $(+)$ is used to indicate an optically active compound that is dextrorotatory. The other enantiomer rotates plane-polarized light in a counterclockwise direction and is said to be *levorotatory*. The symbol $(-)$ is used to indicate an optically active compound that is levorotatory.

Optical activity, then, indicates molecular structure, and slight changes in structure can cause dramatic biological effects. For example, the levorotatory isomer of monosodium glutamate (MSG) is used as a flavoring additive in food, whereas the dextrorotatory isomer cannot be recognized by our taste buds as having any flavoring property. The dextrorotatory form of amphetamine (see Section 17-5.2) is a stronger stimulant (four times more physiologically active) than the levorotatory isomer.

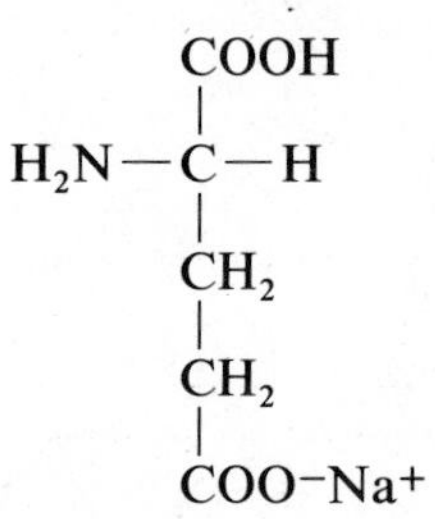

$$COOH$$
$$H_2N-C-H$$
$$CH_2$$
$$CH_2$$
$$COO^-Na^+$$

$$CH_3$$
$$H_2N-C-H$$
$$CH_2$$

monosodium glutamate 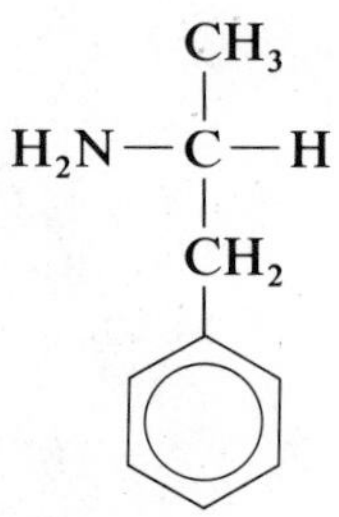   amphetamine

It is possible to determine whether a substance is optically active without the use of a polarimeter. If the substance and its mirror image are not superimposable, the substance will be optically active. Alanine, 2-butanol, threonine, and ribose are optically active because they are not superimposable on their mirror images.

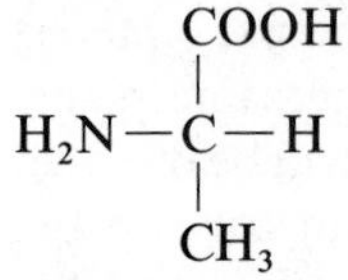

$$COOH$$
$$H_2N-C-H$$
$$CH_3$$

$$CH_3$$
$$H-C-OH$$
$$CH_2CH_3$$

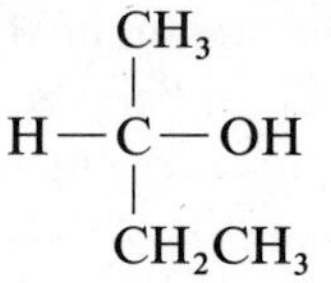

alanine   2-butanol
an amino acid   *a secondary alcohol*

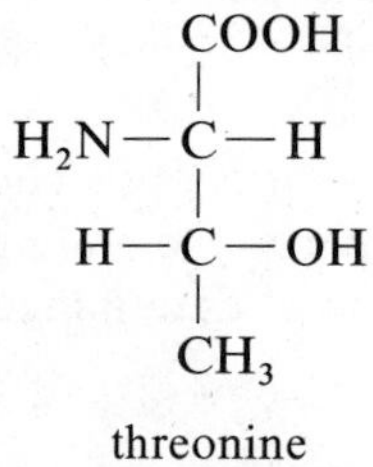

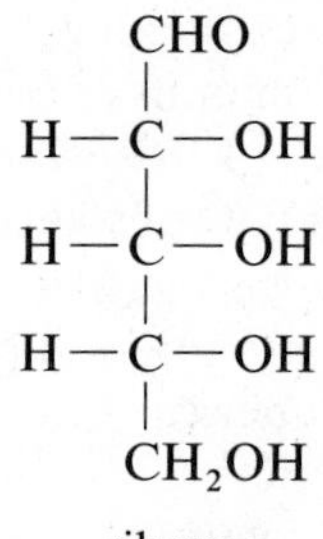

$$COOH$$
$$H_2N-C-H$$
$$H-C-OH$$
$$CH_3$$

$$CHO$$
$$H-C-OH$$
$$H-C-OH$$
$$H-C-OH$$
$$CH_2OH$$

threonine   ribose
an amino acid   *a carbohydrate*

18-6 CONFIGURATION: D- AND L-

LEARNING OBJECTIVE

Given a Fischer projection, designate configuration with the letters D- and L-, using D- and L-glyceraldehyde as standards.

Chemists often speak of the **configuration** of a molecule. The configurational designation is the use of the letters D- and L-. The reference compounds are D-glyceraldehyde and L-glyceraldehyde. By convention, the number one carbon in the IUPAC system is placed at the *top* in the Fischer projection. If the group on the *last* chiral carbon atoms lies to the right as in D-glyceraldehyde, the compound belongs to the D-series; if it lies to the left as in L-glyceraldehyde, it belongs to the L-series.

The **configuration** of a molecule is the arrangement in space of the atoms that can be changed only by the breaking and making of bonds.

$$\begin{array}{ccc}
& \text{COOH} & \text{COOH} \\
\text{CHO} & | & | \\
| & \text{HO}-\text{C}-\text{H} & \text{HO}-\text{C}-\text{H} \\
\text{H}-\text{C}-\text{OH} & | & | \\
| & \text{H}-\text{C}-\text{OH} & \text{H}-\text{C}-\text{OH} \\
\text{CH}_2\text{OH} & | & | \\
& \text{CH}_2\text{OH} & \text{COOH}
\end{array}$$

| D-glyceraldehyde | D-threose | D-tartaric acid |

$$\begin{array}{ccc}
& & \text{COOH} \\
& & | \\
\text{CHO} & \text{COOH} & \text{H}-\text{C}-\text{OH} \\
| & | & | \\
\text{HO}-\text{C}-\text{H} & \text{H}_2\text{N}-\text{C}-\text{H} & \text{HO}-\text{C}-\text{H} \\
| & | & | \\
\text{CH}_2\text{OH} & \text{CH}_3 & \text{COOH}
\end{array}$$

| L-glyceraldehyde | L-alanine | L-tartaric acid |

The letters D- and L- designate configuration and must not be confused with (+) and (−) which designate the rotation of plane-polarized light. The direction of rotation, a physical property, must be determined by the use of a polarimeter. D-Glyceraldehyde rotates plane-polarized light to the right (+). D-Glyceraldehyde is dextrorotatory. L-Alanine is dextrorotatory and D-lactic acid is levorotatory. The configuration of a molecule is determined by comparison with D- and L-glyceraldehyde.

The configuration of a molecule is important to living systems and can determine the physiological use of the molecule. D-Thyroxine is used for lowering the blood cholesterol level. L-Thyroxine, on the other hand, is used as therapy for thyroid deficiency. The amino acids used by the living system to make protein are exclusively L-amino acids. The glucose used as a source of energy by the living system is D-glucose.

$$\begin{array}{cc}
\text{COOH} & \text{COOH} \\
| & | \\
\text{H}-\text{C}-\text{NH}_2 & \text{H}_2\text{N}-\text{C}-\text{H} \\
| & | \\
\text{CH}_2 & \text{CH}_2 \\
\end{array}$$

| D-thyroxine | L-thyroxine |

PROBLEM

6. Designate the following molecules as D- or L-.

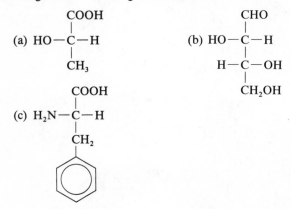

ANSWERS TO PROBLEM **6. (a)** L- **(b)** D- **(c)** L-

18-7 DIASTEREOMERS:
THE CARBOHYDRATES, ERYTHROSE,
AND THREOSE

LEARNING OBJECTIVES

1. Define or recognize examples of diastereomers.
2. Given organic molecules that are diastereomers, draw their Fischer projections.

Two stereoisomers that are not superimposable on their mirror image are enantiomers. However, it is possible that two stereoisomers may not have a mirror-image relationship. Such stereoisomers are called diastereoisomers or **diastereomers.** The monosaccharides (see Chapter 19), which are polyhydroxy carbonyl compounds, provide many examples of enantiomers and diastereomers. Two compounds that illustrate this definition are D-threose and D-erythrose.

Diastereomers are stereoisomers that are not mirror images of each other.

$$
\begin{array}{cc}
\text{CHO} & \text{CHO} \\
| & | \\
\text{H}-\text{C}-\text{OH} & \text{HO}-\text{C}-\text{H} \\
| & | \\
\text{H}-\text{C}-\text{OH} & \text{H}-\text{C}-\text{OH} \\
| & | \\
\text{CH}_2\text{OH} & \text{CH}_2\text{OH} \\
\text{D-erythrose} & \text{D-threose}
\end{array}
$$

These two monosaccharides are stereoisomers because they each have the same groups bonded to the same atoms. They differ in the orientation of the hydroxy groups on C-2 (colored). They do not possess a mirror image relationship. Hence, D-erythrose and D-threose are diastereomers. D-Erythrose and D-threose each have a mirror image. D-Erythrose and its mirror image, L-erythrose, are not superimposable. They are enantiomers.

$$
\begin{array}{cc}
\text{CHO} & \text{CHO} \\
| & | \\
\text{H}-\text{C}-\text{OH} & \text{HO}-\text{C}-\text{H} \\
| & | \\
\text{H}-\text{C}-\text{OH} & \text{HO}-\text{C}-\text{H} \\
| & | \\
\text{CH}_2\text{OH} & \text{CH}_2\text{OH} \\
\text{D-erythrose} & \text{L-erythrose}
\end{array}
$$

PROBLEM

7. Identify the following pairs of compounds as enantiomers, diastereomers, or identical.

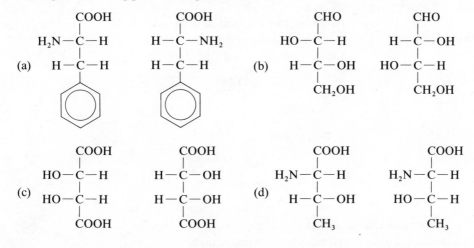

ANSWERS TO PROBLEM **7. (a)** enantiomers **(b)** enantiomers **(c)** identical **(d)** diastereomers

18-8 A MESO COMPOUND: TARTARIC ACID

LEARNING OBJECTIVES

1. Define or recognize examples of a meso compound.
2. Draw the Fischer projection for *meso*-tartaric acid.

3. Given organic molecules, draw Fischer projections of meso compounds.

A meso compound is recognized by the fact that one-half of the molecule is the mirror image of the other half. A meso compound is characterized by its plane of symmetry. The objects shown in Figure 18.1(b) have a plane of symmetry. Tartaric acid, often used as an acidity-controlling agent in foods (see Section 16-5), is a good example of a meso compound. Its purpose is to impart tartness to foods.

A **meso compound** is a molecule which has a plane of symmetry and cannot exhibit optical activity because it is superimposable on its mirror image.

$$
\begin{array}{c}
CO_2H \\
| \\
H\!-\!C\!-\!OH \\
| \\
H\!-\!C\!-\!OH \\
| \\
CO_2H
\end{array}
$$

plane of symmetry --------------------------   *this half is the mirror image*

of this half

*Fischer projection
of meso-tartaric acid*

A meso compound cannot exhibit optical activity because the effect of plane-polarized light on one-half of the molecule is exactly canceled by the opposite effect of plane-polarized light on the other half of the molecule.

meso-Tartaric acid has a mirror image; but *meso*-tartaric acid and its mirror image are superimposable. A molecule that is superimposable on its mirror image cannot be optically active.

$$
\begin{array}{c}
CO_2H \\
| \\
H\!-\!C\!-\!OH \\
| \\
H\!-\!C\!-\!OH \\
| \\
CO_2H
\end{array}
\qquad\qquad
\begin{array}{c}
CO_2H \\
| \\
HO\!-\!C\!-\!H \\
| \\
HO\!-\!C\!-\!H \\
| \\
CO_2H
\end{array}
$$

meso-tartaric acid   *superimposable
mirror images*   *meso*-tartaric acid

meso-Tartaric acid has two stereoisomers, and it is a diastereomer (see Section 18-7) of both of the stereoisomers. The two stereoisomers are enantiomers.

$$
\begin{array}{c}
CO_2H \\
| \\
H\!-\!C\!-\!OH \\
| \\
H\!-\!C\!-\!OH \\
| \\
CO_2H
\end{array}
\qquad
\begin{array}{c}
CO_2H \\
| \\
H\!-\!C\!-\!OH \\
| \\
HO\!-\!C\!-\!H \\
| \\
CO_2H
\end{array}
\qquad
\begin{array}{c}
CO_2H \\
| \\
HO\!-\!C\!-\!H \\
| \\
H\!-\!C\!-\!OH \\
| \\
CO_2H
\end{array}
$$

meso-tartaric acid   L-tartaric acid   D-tartaric acid

diastereomers   *enantiomers*

diastereomers

Table 18.3 lists some physical properties of the various forms of tartaric acid.

Table 18.3 Physical Properties of Various Forms of Tartaric Acid

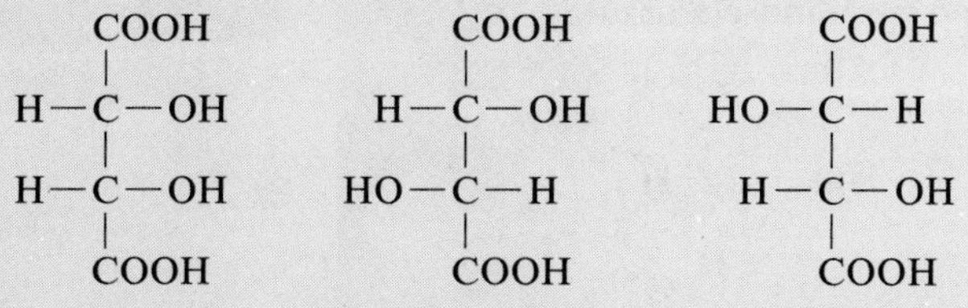

	COOH	COOH	COOH
	|	|	|
	H—C—OH	H—C—OH	HO—C—H
	|	|	|
	H—C—OH	HO—C—H	H—C—OH
	|	|	|
	COOH	COOH	COOH
	meso-tartaric acid	L(+)-tartaric acid	D(−)-tartaric acid

Physical Property	Meso Form	Dextrorotatory Form	Levorotatory Form
Melting point	140°	169°	169°
Solubility in 100 mL HOH	125 g	139 g	139 g
Density (g/mL)	1.67	1.76	1.76
Specific rotation	0°	+12°	−12°

As can be seen, diastereomers have different physical properties (compare *meso*-tartaric acid and L(+)-tartaric acid), but enantiomers have identical physical properties except for specific rotation (compare L(+)-tartaric acid and D(−)-tartaric acid).

PROBLEM

8. Each of the compounds shown is optically inactive because it has a plane of symmetry. Indicate the plane of symmetry in each case.

(a)
CH$_2$OH
\|
H—C—OH
\|
H—C—OH
\|
CH$_2$OH

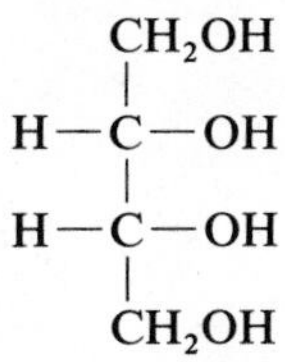

(b)

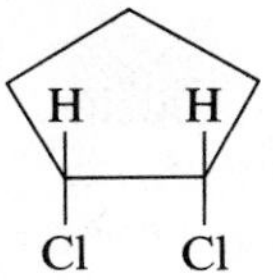

ANSWERS TO PROBLEM **8. (a)**
CH$_2$OH
\|
H—C—OH
------+------
H—C—OH
\|
CH$_2$OH

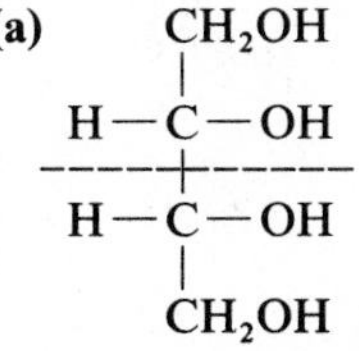

(b)

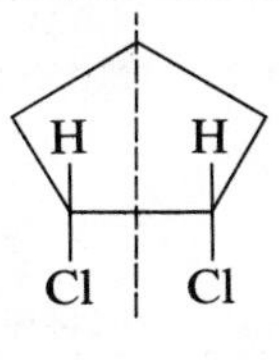

18-9 RACEMIC MIXTURE: *IN VITRO* AND *IN VIVO* LACTIC ACID

LEARNING OBJECTIVES

1. Define racemic mixture.
2. Explain the stereochemical difference between *in vivo* and *in vitro* syntheses of lactic acid.

Lactic acid exists as two enantiomers. The levorotatory isomer is found in yeast and the dextrorotatory isomer is isolated from muscle tissue.

$$
\begin{array}{cc}
\text{COOH} & \text{COOH} \\
| & | \\
\text{H}-\text{C}-\text{OH} & \text{HO}-\text{C}-\text{H} \\
| & | \\
\text{CH}_3 & \text{CH}_3 \\
(-)\text{-lactic acid} & (+)\text{-lactic acid}
\end{array}
$$

When lactic acid is synthesized in the laboratory from acetaldehyde (see Section 15-3.1), both the dextrorotatory and levorotatory forms of lactic acid are obtained. Furthermore, both forms of lactic acid are produced in equal amounts.

$$
\underset{\text{acetaldehyde}}{\overset{\text{O}}{\underset{\|}{\text{CH}_3\text{C}-\text{H}}}} \xrightarrow{\text{HCN}} \underset{\substack{\text{acetaldehyde} \\ \text{cyanohydrin}}}{\text{CH}_3-\overset{\text{H}}{\underset{\text{OH}}{\text{C}}}-\text{CN}} \xrightarrow{\text{H}_2\text{O}} \underset{\substack{(+)\text{- and }(-)\text{-lactic acid} \\ \text{formed in equal amounts}}}{\text{CH}_3-\overset{\text{H}}{\underset{\text{OH}}{\text{C}}}-\text{COOH}}
$$

A mixture that contains equal amounts of enantiomers is called a **racemic mixture**. A racemic mixture is optically inactive because it contains equal amounts of the isomers that rotate plane-polarized light to the right and to the left. The net rotation of the mixture is zero. Table 18.4 lists the physical properties of the three forms of lactic acid.

A **racemic mixture** is a mixture which contains equal amounts of enantiomers.

This situation seems to be quite common whenever chemical syntheses are carried out in the laboratory (*in vitro*). As a general rule, *the use of optically inactive reagents gives products that will be optically inactive.* Both acetaldehyde and hydrogen cyanide (HCN) are optically inactive. Even though the product, lactic acid, contains one chiral carbon, the product is a mixture of equal amounts of the (+)- and (−)- forms and, as explained above, is optically inactive.

In living systems (*in vivo*), the enantiomers of lactic acid are formed by reduction from pyruvic acid. However, the reduction is selective. If the reduction occurs in muscle tissue, (+)-lactic acid is formed; if the reduction occurs in yeast, (−)-lactic acid is formed.

$$
\underset{(-)\text{-lactic acid}}{\overset{\text{COOH}}{\underset{\text{CH}_3}{\text{H}-\text{C}-\text{OH}}}} \xleftarrow[\text{in yeast}]{\text{lactic reductase}} \underset{\text{pyruvic acid}}{\overset{\text{COOH}}{\underset{\text{CH}_3}{\text{C}=\text{O}}}} \xrightarrow[\text{in muscle}]{\text{lactic dehydrogenase}} \underset{(+)\text{-lactic acid}}{\overset{\text{COOH}}{\underset{\text{CH}_3}{\text{HO}-\text{C}-\text{H}}}}
$$

Why, in this situation, is one isomer selectively formed? The answer is that the reaction is controlled by enzymes. The enzyme lactic dehydrogenase (found in muscle tissue) selectively catalyzes the formation of (+)-lactic acid. On the other hand, lactic reductase (found in yeast) selectively catalyzes the formation of (−)-lactic acid.

Table 18.4 Physical Properties of Various Forms of Lactic Acid

Physical Property	Dextrorotatory Form	Levorotatory Form	Racemic Mixture
Melting point	52.8°	52.8°	18°
Specific rotation	+2.6°	−2.6°	0°

Enzymes are called stereoselective catalysts. This is because they can recognize subtle differences in the shapes and structures of biomolecules. Stereoselective catalysts are, in fact, themselves optically active stereoisomers. For this reason enzymes are capable of producing one stereoisomer to the exclusion of the other.

SUMMARY

Stereoisomers are isomers that are different from one another only in the way in which the atoms are oriented in space. Stereoisomers are best viewed in the Fischer projection. The Fischer projection is a planar representation of a tetrahedral model.

$$
\begin{array}{cc}
\text{COOH} & \text{COOH} \\
| & | \\
\text{H} \blacktriangleright \text{C} \blacktriangleleft \text{OH} & \text{H} - \text{C} - \text{OH} \\
| & | \\
\text{CH}_3 & \text{CH}_3
\end{array}
$$

tetrahedral model   Fischer projection

Stereoisomers that are not superimposable on their mirror image are called *enantiomers*. In modern terminology, two molecules that are enantiomers are said to be chiral. In many cases a chiral molecule can be identified as a molecule that contains a chiral carbon, a carbon that has four different groups attached to it.

$$
\begin{array}{cc}
\text{COOH} & \text{COOH} \\
| & | \\
\text{H} - \text{C} - \text{NH}_2 & \text{H}_2\text{N} - \text{C} - \text{H} \\
| & | \\
\text{CH}_3 & \text{CH}_3
\end{array}
$$

enantiomers

A stereoisomer that is optically active can rotate plane-polarized light. An optically active molecule is not superimposable on its mirror image.

Diastereomers are stereoisomers that are not superimposable mirror images.

$$
\begin{array}{cc}
\text{CHO} & \text{CHO} \\
| & | \\
\text{H} - \text{C} - \text{OH} & \text{HO} - \text{C} - \text{H} \\
| & | \\
\text{H} - \text{C} - \text{OH} & \text{H} - \text{C} - \text{OH} \\
| & | \\
\text{CH}_2\text{OH} & \text{CH}_2\text{OH}
\end{array}
$$

diastereomers

A *meso compound* contains a plane of symmetry and is optically inactive. The configuration of a molecule is specified as D- or L-, using D- and L-glyceraldehyde as the standard.

The synthesis of a molecule with a chiral center *in vitro* results in a racemic mixture. The synthesis of a molecule with a chiral center *in vivo* results in the exclusive formation of one enantiomer.

The molecular configurations of molecules (cis-trans isomers, enantiomers, diastereomers, and meso compounds) are of primary importance to life. Life is dependent on molecular configuration and in the following chapters we will see how the subtle differences in molecular configurations that we have discussed in this chapter can and do affect the chemistry of life.

$$
\begin{array}{c}
CO_2H \\
| \\
H-C-OH \\
| \\
H-C-OH \\
| \\
CO_2H
\end{array}
$$

meso compound

ADDITIONAL PROBLEMS

9. Define or explain each of the following terms.
   - (a) isomers
   - (b) tetrahedral carbon
   - (c) Fischer projection
   - (d) enantiomers
   - (e) diastereomers
   - (f) chiral carbon
   - (g) optical activity
   - (h) specific rotation
   - (i) meso compound
   - (j) configuration
   - (k) racemic mixture

10. Draw the structures of the isomers of heptane that have chiral carbon atoms.

11. *cis*-9-Tricosine is the active sex attractant in the common housefly. Draw the structure of the cis isomer of *cis*-9-tricosine, $CH_3(CH_2)_7CH=CH(CH_2)_{12}CH_3$.

12. Identify the chiral carbon(s) in each of the following molecules, if any.

   (a) H_2NCH_2COOH

   glycine
   an amino acid

   (b) $CH_3CHCH_2CH_2CH_2CH_3$
       $\quad\; |$
       $\quad\; CH_3$

   2-methylhexane
   a component of gasoline

   (c)
   $$
   \begin{array}{c}
   COOH \\
   | \\
   H_2N-CH \\
   | \\
   CH_2 \\
   | \\
   CH_2 \\
   | \\
   SCH_3
   \end{array}
   $$

   methionine
   an amino acid and a thioether

   (d)
   $$
   \begin{array}{c}
   CH_2OH \\
   | \\
   CHOH \\
   | \\
   CH_2OPO_3H_2
   \end{array}
   $$

   glycerol phosphate
   parent compound of certain lipids

   (e)
   $$
   \begin{array}{c}
   CHO \\
   | \\
   H-C-OH \\
   | \\
   HO-C-H \\
   | \\
   HO-C-H \\
   | \\
   H-C-OH \\
   | \\
   CH_2OH
   \end{array}
   $$

   D-galactose
   a component of milk sugar

13. Deoxyribose is a monosaccharide occurring in DNA. Draw Fischer formulas for all the stereoisomers. Identify which compounds are enantiomers, and diastereomers, and which are optically active.

   $$
   \begin{array}{c}
   CHO \\
   | \\
   CH_2 \\
   | \\
   CHOH \\
   | \\
   CHOH \\
   | \\
   CH_2OH
   \end{array}
   $$

   deoxyribose

14. Erythritol and threitol correspond to the following molecular structure. Erythritol is optically inactive

and threitol is optically active. Draw the configurational formulas of each.

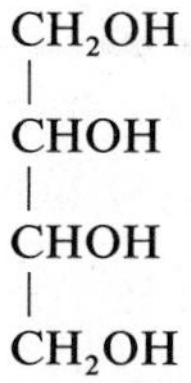

$$CH_2OH$$
$$|$$
$$CHOH$$
$$|$$
$$CHOH$$
$$|$$
$$CH_2OH$$

erythritol and threitol
molecular structure

15. Which of the following compounds is optically active?

(a)

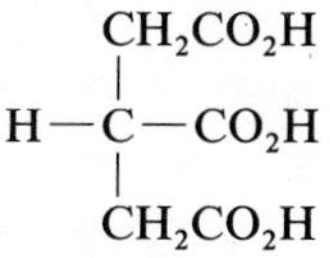

$$CH_2CO_2H$$
$$|$$
$$H-C-CO_2H$$
$$|$$
$$CH_2CO_2H$$

citric acid

(b)

H
H△H
H
Br   Br

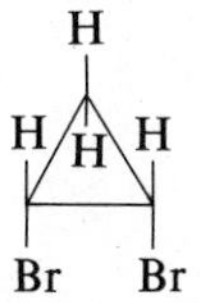

cis-1,2-dibromocyclopropane

(c)

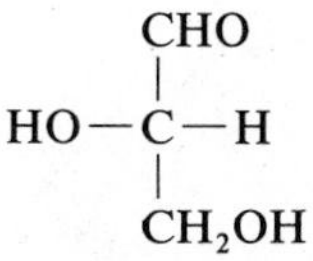

$$CHO$$
$$|$$
$$HO-C-H$$
$$|$$
$$CH_2OH$$

glyceraldehyde

16. Four stereoisomers are possible for chloramphenicol. Draw these four stereoisomers. Which are enantiomers? Which are diastereomers? (Only one of these four stereoisomers possesses antibacterial activity.)

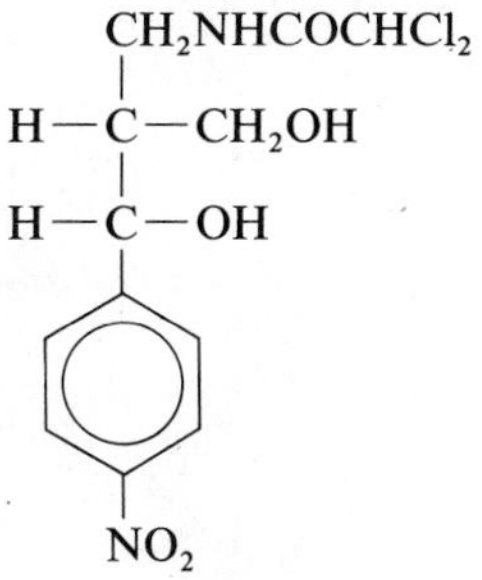

$$CH_2NHCOCHCl_2$$
$$|$$
$$H-C-CH_2OH$$
$$|$$
$$H-C-OH$$

NO_2

chloramphenicol

17. In Chapter 15 you learned that the reaction of an aldehyde with HCN formed a cyanohydrin. Glyceraldehyde forms the two cyanohydrins shown. What is the stereochemical relationship between them?

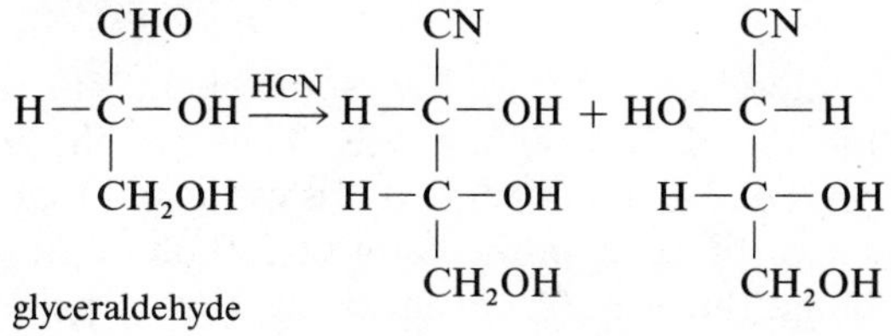

$$CHO \qquad\qquad CN \qquad\qquad CN$$
$$| \qquad\qquad\quad | \qquad\qquad\quad |$$
$$H-C-OH \xrightarrow{HCN} H-C-OH + HO-C-H$$
$$| \qquad\qquad\quad | \qquad\qquad\quad |$$
$$CH_2OH \qquad H-C-OH \qquad H-C-OH$$
$$\qquad\qquad\qquad | \qquad\qquad\quad |$$
$$\qquad\qquad\qquad CH_2OH \qquad CH_2OH$$

glyceraldehyde

CHAPTER 19

CARBOHYDRATES

Almost every day we hear someone talking about carbohydrates. Carbohydrates are found in all living cells. They are the most abundant of the organic compounds—so abundant that it is estimated that well over half of the organic carbon on earth exists in the form of carbohydrates. Carbohydrate molecules make up about three-fourths of the dry weight of plants. While fats give more calories per unit weight than carbohydrates, carbohydrates provide about half of the calories in the average human diet. Table 1.6 is a list of some representative values of the number of calories released by combustion of various foods.

The **dry weight** is what is left after the removal of water; water may comprise 90% or more or the actual weight of living organisms.

Carbohydrates were introduced in Chapter 15. In Chapter 19, we will examine, in a little more depth, some of the chemical properties of the naturally occurring carbohydrates in preparation for describing their metabolism, which will be done in Chapter 24.

19-1 CLASSIFICATION

LEARNING OBJECTIVES

1. Define carbohydrate, monosaccharide, aldose, ketose, triose, tetrose, pentose, hexose, heptose, oligosaccharide, disaccharide, and polysaccharide.
2. Classify monosaccharides with names

indicating both the number of carbon atoms and the kind of carbonyl group in the molecule.
3. Classify carbohydrates as mono-, oligo-, and polysaccharides.

In Chapter 15 we learned that the basic carbohydrate molecule possesses an aldehyde or ketone group and a hydroxyl group on every carbon atom except the one involved in the carbonyl group. As a result, *carbohydrates* are defined as aldehyde or ketone derivatives of polyhydroxy alcohols. (*Poly-* is a prefix signifying many.)

A look at the formula for glucose ($C_6H_{12}O_6$) shows that it contains hydrogen and oxygen atoms in the ratio in which they are found in water. The basic carbohydrate molecule has the formula $(CH_2O)_n$. The name carbohydrate (*hydrate of carbon*) is derived from this fact.

$$
\begin{array}{cc}
& CH_2OH \\
& | \\
HC{=}O & C{=}O \\
| & | \\
(CHOH)_n & (CHOH)_n \\
| & | \\
CH_2OH & CH_2OH \\
\end{array}
$$

polyhydroxy aldehyde   polyhydroxy ketone
aldose   *ketose*

Monosaccharides (commonly referred to as the simple sugars) are carbohydrates that cannot be broken down by hydrolysis. They are classified according to the kind of carbonyl group and the number of carbon atoms contained in the molecule (see Table 19.1). The two classification systems can be joined in a single-word description. For example, a three-carbon aldose is an aldotriose, a three-carbon ketose is a ketotriose, a five-carbon aldose is an aldopentose, a six-carbon aldose is an aldohexose, a six-carbon ketose is a ketohexose, and so on. Examples of aldoses and ketoses are given in Table 19.2. From the examples given in Table 19.2, we can see that, in general, the ending (suffix) for the names of aldehyde sugars is *-ose* and for ketones, *-ulose*. However, there are exceptions to this rule in both cases.

An **aldose** is a polyhydroxy aldehyde, an aldehyde that has a hydroxyl group on every carbon atom except the carbonyl carbon atom.

A **ketose** is a polyhydroxy ketone.

Table 19.1   Classification of Monosaccharides
I. Kind of carbonyl group
A. Aldoses (aldehyde group)
B. Ketoses (keto group)
II. Number of carbon atoms
A. Trioses (3 carbon atoms)
B. Tetroses (4 carbon atoms)
C. Pentoses (5 carbon atoms)
D. Hexoses (6 carbon atoms)
E. Heptoses (7 carbon atoms)

Most monosaccharides that occur in nature have the D configuration (see Section 18-6). An exception is arabinose which is found as L-arabinose. All possible structures of the aldotriose, aldotetroses, aldopentoses, and aldohexoses with the

Table 19.2   Aldoses and Ketoses

Kind of Carbonyl Group and Generic Name	Number of Carbon Atoms and Generic Name	Example
Aldehyde (aldose)	3 (aldotriose)	$HC\!=\!O$ | $HCOH$ | CH_2OH D-glyceraldehyde
Aldehyde (aldose)	4 (aldotetrose)	$HC\!=\!O$ | $HCOH$ | $HCOH$ | CH_2OH D-erythrose
Aldehyde (aldose)	5 (aldopentose)	$HC\!=\!O$ | $HCOH$ | $HOCH$ | $HCOH$ | CH_2OH D-xylose
Aldehyde (aldose)	6 (aldohexose)	$HC\!=\!O$ | $HCOH$ | $HOCH$ | $HCOH$ | $HCOH$ | CH_2OH D-glucose
Ketone (ketose)	3 (ketotriose or triulose)	CH_2OH | $C\!=\!O$ | CH_2OH dihydroxyacetone*

*Common names.

Table 19.2 Aldoses and Ketoses (*cont'd.*)

Kind of Carbonyl Group and Generic Name	Number of Carbon Atoms and Generic Name	Example
Ketone (ketose)	5 (ketopentose or pentulose)	CH_2OH | $C=O$ | HCOH | HCOH | CH_2OH D-ribulose*
Ketone (ketose)	6 (ketohexose or hexulose)	CH_2OH | $C=O$ | HOCH | HCOH | HCOH | CH_2OH D-fructose
Ketone (ketose)	7 (ketoheptose or heptulose)	CH_2OH | $C=O$ | HOCH | HCOH | HCOH | HCOH | CH_2OH D-sedoheptulose*

*Common names.

D configuration are given in Figure 19.1. The naturally occurring sugars are highlighted in this figure. A similar family tree could be drawn for the L sugars.

Monosaccharides may be joined together in chains, forming oligosaccharides and polysaccharides. Oligosaccharides are classified according to the number of monosaccharide units in them (see Table 19.3).

Oligosaccharides are carbohydrates that yield 2–10 monosaccharide molecules upon hydrolysis. *Oligo-* is a prefix signifying few.

Polysaccharides are carbohydrates that yield many monosaccharide molecules upon hydrolysis. *Poly-* is a prefix signifying many.

Figure 19.1 The D family of aldoses having three to six carbon atoms. The accepted three-letter abbreviation is given for those sugars most commonly found in nature. These structures are shown in color. The arabinose molecule most often found is L-arabinose not D-arabinose as shown here. No abbreviation is used for D-glyceraldehyde.

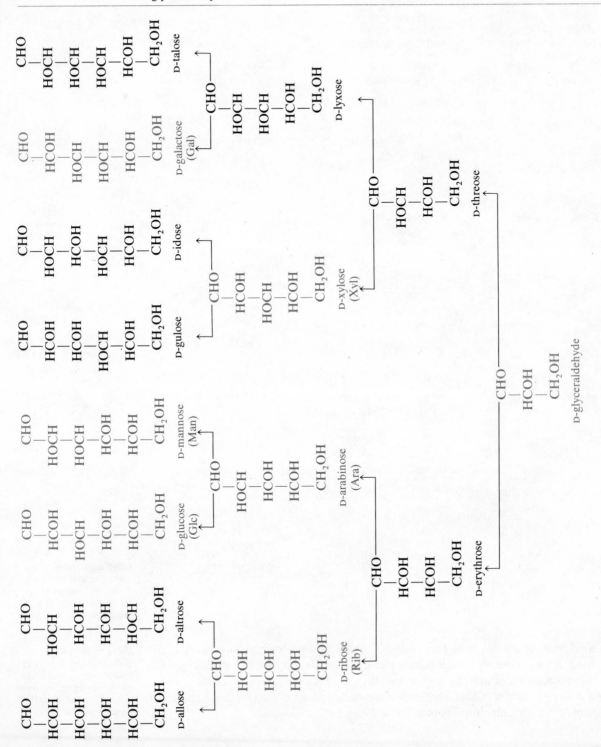

Table 19.3 Classification of Oligosaccharides

Number of Monosaccharide Units	Generic Name
2	Disaccharide
3	Trisaccharide
4	Tetrasaccharide
5	Pentasaccharide
6	Hexasaccharide
7	Heptasaccharide
8	Octasaccharide
9	Nonasaccharide
10	Decasaccharide

Polysaccharides may be linear (unbranched) or branched (see Figure 19.2). They may be made up of a single kind of monosaccharide residue (*homopolysaccharides*) or of two or more different monosaccharide residues (*heteropolysaccharides*).

Figure 19.2
Types of branched polysaccharides. R indicates the sugar residue at the reducing end.

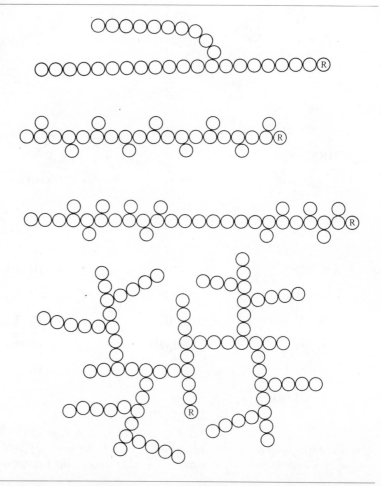

Most carbohydrates exist in the form of polysaccharides. Polysaccharides give structure to the cell walls of land plants (cellulose), seaweeds, and some micro-organisms, and store energy (starch in plants and glycogen in animals). We use the cellulose of woody plants to make paper. Cotton is almost pure cellulose. We use the polysaccharides from the cell walls of seaweeds to thicken chocolate milk and salad dressings and in many other food products. We use the starch of plants as the principal source of carbohydrate in our diets.

PROBLEM

1. Identify each of the following with the best one-word description; for example, aldopentose.

(a)
```
HC=O
 |
HCOH
 |
HCOH
 |
HOCH
 |
 CH₂OH
```

(b)
```
CH₂OH
 |
C=O
 |
HCOH
 |
HCOH
 |
HCOH
 |
CH₂OH
```

(c)
```
 CHO
  |
HOCH
  |
HCOH
  |
CH₂OH
```

(d)
```
HC=O
 |
HOCH
 |
 CH₂OH
```

(e)
```
CH₂OH
 |
C=O
 |
HOCH
 |
HCOH
 |
CH₂OH
```

(f)
```
CH₂OH
 |
C=O
 |
CH₂OH
```

(g)
```
 CHO
  |
HCOH
  |
HOCH
  |
HOCH
  |
HCOH
  |
CH₂OH
```

(h)
```
CH₂OH
 |
C=O
 |
HCOH
 |
CH₂OH
```

(i)
```
HC=O
 |
HOCH
 |
HOCH
 |
HCOH
 |
HCOH
 |
CH₂OH
```

ANSWERS TO PROBLEM **1. (a)** aldopentose **(b)** ketohexose **(c)** aldotetrose **(d)** aldo-triose **(e)** ketopentose **(f)** ketotriose **(g)** aldohexose **(h)** ketotetrose **(i)** aldohexose

19-2 REACTIONS OF MONOSACCHARIDES

LEARNING OBJECTIVES

1. Define pyranose ring, furanose ring, glycoside, reducing end, and nonreducing end.

2. Given the structure of an aldopentose, aldohexose, or ketohexose in the open-chain form, write the structure in an α- or β-D pyranose or furanose ring form and vice versa.

3. Write open-chain and pyranose ring structures of D-glucose.

4. Write the structures of the alcohol and the aldohexose or aldopentose (in the pyranose ring form) that are formed upon hydrolysis of an α- or β-D-pyranoside.

5. Write the structures of the monosaccharide(s) (in the pyranose ring form) that is (are) formed upon hydrolysis of an oligo- or polysaccharide.

6. Write the open-chain structure of D-fructose.

7. Indicate, on the structural formula, the reducing end and the nonreducing end of an oligo- or polysaccharide.

8. Write open-chain and pyranose ring structures of D-glucose 6-phosphate.

9. Write open-chain and furanose ring structures of D-ribose and 2-deoxy-D-ribose.

10. Write the structures of the aldose and ketose that can be formed from a given aldose by isomerization.

11. Write the structures of the two aldoses that can be formed from a given ketose by isomerization.

12. Write the structure of D-fructose 1,6-diphosphate in the open-chain form.

Reactions that occur in living organisms are typical chemical reactions of the kinds that we discussed in Chapters 11–17. Most carbohydrates have two kinds of reactive groups: the carbonyl (aldehyde or ketone) group and the hydroxyl (alcohol) groups. We discussed reactions of these functional groups in Sections 14-3 and 15-3 and list biologically important reactions of them in Table 19.4. We will discuss the reactions listed in Table 19.4 in the following sections.

Table 19.4 Some Biologically Important Reactions of Carbohydrate Molecules

Reactions of	Reaction
the carbonyl group (see Section 15-3)	1. Formation of hemiacetals: pyranose or furanose ring forms. This reaction also involves a hydroxyl group.
	2. Formation of acetals: pyranosides and furanosides (glycosides).
	3. Oxidation to a carboxylic acid.
hydroxyl groups (see Section 14-3)	1. Ester formation
	2. Oxidation to a carbonyl group
carbonyl and hydroxyl groups	1. Isomerization
	2. Aldol condensation

19-2.1 REACTIONS OF THE CARBONYL GROUP

We first looked at reactions of carbonyl groups in Section 15-3. Here, we will review them as preparation to an examination of the role of carbohydrates in living organisms.

Ring Forms

One reaction discussed in Section 15-3.1 is the reaction of carbonyl groups with hydroxyl groups. Aldehydes and ketones react with alcohols (compounds containing a hydroxyl group) to form first, hemiacetals, and then acetals. Because aldoses and ketoses have a carbonyl group and hydroxyl groups on the same carbon chain, they can form hemiacetals by reacting with themselves (*an intramolecular reaction*), rather than by reacting with another molecule (*an intermolecular reaction*). As we shall see later, an intramolecular reaction forms a ring, as when a dog bites his own tail. The most common rings are the six-membered ring (**pyranose ring**) and the five-membered ring (**furanose ring**). The structure of D-glucose illustrates the six-membered ring and that of D-ribose, the five-membered ring. D-Glucose is essential for human life and, if both free and combined forms are considered, is probably the most abundant organic molecule on earth. D-Ribose is a component of the nucleic acid RNA (see Chapter 20).

The **pyranose ring** is a cyclic structure composed of five carbon atoms and one oxygen atom.

The **furanose ring** is a cyclic structure composed of four carbon atoms and one oxygen atom.

$$HC={O} + R'OH \rightleftharpoons HCOH + R''OH \rightleftharpoons HCOR'' + HOH$$
(with OR' above the hemiacetal carbon, R below each)

aldehyde   alcohol   hemiacetal   alcohol   acetal   water

Let us first examine the basic structure of aldoses and ketones. D-Glucose is an aldohexose. It contains six carbon atoms and an aldehyde group. Four of the six carbon atoms are chiral carbon atoms. We learned in Section 18-4 that four different groups are attached to a *chiral carbon atom*. As a result, the chiral carbon atom can exist in two different arrangements, or configurations, of these groups in space. These two different configurations are mirror images of each other. In other words, one is the reflection of the other that we would see in a mirror, with everything that is on the right in one on the left in the other configuration and vice versa.

We learned in Sections 18-2, 18-3, and 18-4 that, to compare the arrangements of atoms in stereoisomers, we write structural formulas using the convention that the bonds connecting carbon atoms are projecting away from us and the ones to the other atoms are projecting toward us. When the structural formulas of sugars are written in this way, it is understood that all the vertical bonds connecting carbon atoms project away from us and the bond to the hydrogen atom and the hydroxyl group on each carbon atom projects toward us. This is shown for carbon atom number 5 (C-5) of D-glucose. Horizontal bonds are often omitted as in the formulas in Table 19.2 and Figure 19.1. The structural formula of D-glucose shown is known as the open-chain, or Fischer, formula. If the hydroxyl group on the most distant chiral carbon atom is on the right, when the carbon chain of an aldose or ketose is written using this convention, the sugar is said to have the D configuration; if the hydroxyl group is on the left, the sugar is said to have the L configuration.

Because a hexose contains four chiral carbon atoms, there are $2^4 = 16$ different possible arrangements of the hydroxyl groups in space. That is, there are 16 different stereoisomers. The structures of half of these, the 8 D isomers, are given in Figure 19.1. This figure highlights the 3 of the 16 stereoisomers that are commonly found in nature. D-Glucose and D-galactose are the most abundant of the aldohexoses.

$HC={O}$ (1), $HCOH$ (2), $HOCH$ (3), $HCOH$ (4), $HCOH$ (5), CH_2OH (6)

D-glucose in the open chain form

configuration of carbon atom 5

It is easy to picture the formation of a ring from an open-chain structure if we remember that the carbon chain is curving away from us. Then, if we lay this structure on its side, we find that it is naturally curved into almost the correct shape. However, in order to close the ring, that is, to form the hemiacetal between the hydroxyl group on C-5 and the aldehyde group, C-5 must be rotated to bring the hydroxyl group closer to the aldehyde group. The result of this rotation is that the $-CH_2OH$ group sticks up in the representation of the pyranose ring of the D sugars and the $-CH_2OH$ group hangs down in representations of the L sugars. Most, but not all, naturally occurring sugars are D sugars. For C-2, C-3, and C-4,

D-glucose
Fischer projection

α-D-glucopyranose
Haworth projection

the carbon atoms that are not involved in ring formation, the hydroxyl groups that are on the right in the Fischer projection hang down in the Haworth ring form and the hydroxyl groups that are on the left in the Fischer projection stick up in the Haworth ring form. Both structural formulas are formulas of D-glucose. One is D-glucose in the open-chain form and the other is D-glucose in the *pyranose ring* form. To distinguish between them, we call the one on the right D-*glucopyranose* to indicate that it is D-glucose in the pyranose ring form.

When the pyranose (six membered) ring is formed, a new chiral carbon atom is formed from C-1. Thus, there can be two forms of the pyranose ring. D Sugars with the hydroxyl group at C-1 sticking up are said to be in the beta (β) configuration; D sugars with the hydroxyl group at C-1 hanging down are said to be in the alpha (α) configuration. For this reason, the particular form of D-glucopyranose we have used as an example is α-D-glucopyranose.

The structure we have used for α-D-glucopyranose requires further explanation. The complete structure for this compound is shown in Figure 19.3(a). However, following the convention of organic chemistry used for writing ring structures such as those of benzene, the Cs used to designate carbon atoms can be omitted, giving the representation of Figure 19.3(b). Following the same convention, the Hs designating hydrogen atoms bonded to ring carbon atoms can be omitted, giving the representation of Figure 19.3(c). Although this representation is used, it can be confusing because a simple line is also used by organic chemists to indicate a methyl group. The shorthand representation preferred by most chemists is that of Figure 19.3(d) in which both the ring carbon atoms and the hydrogen atoms connected to them are omitted, so the structure is less cluttered and easier to analyze at a glance. We will continue to use this shorthand version in this book.

Figure 19.3 Four representations of the structure of α-D-glucopyranose. These two-dimensional representations of the pyranose ring are Haworth projections. Actually, the ring is not flat but puckered. (a) All C, H, and O atoms are shown, (b) C-2, C-3, C-4, C-5, and O-5 are represented by a hexagon. (c) Hydrogen atoms attached to C-1, C-2, C-3, C-4, and C-5 are indicated but not shown. (d) Hydrogen atoms attached to C-1, C-2, C-3, C-4, and C-5 are omitted.

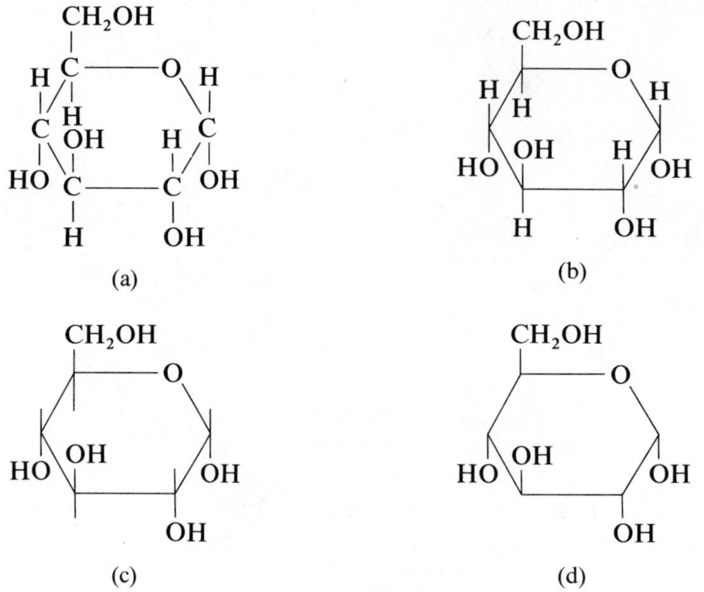

(a)

(b)

(c)

(d)

Monosaccharides also occur in a five-membered ring called the *furanose ring*. Most free pentoses, hexoses, and heptoses occur in pyranose rings. The furanose ring is less common than is the pyranose ring, but it is quite important. The furanose ring is formed in the same way as the pyranose ring and also occurs in α and β forms. We have demonstrated this with D-ribose. β-D-Ribofuranose is a component of RNA, a most important biomolecule that we will discuss in Chapter 20.

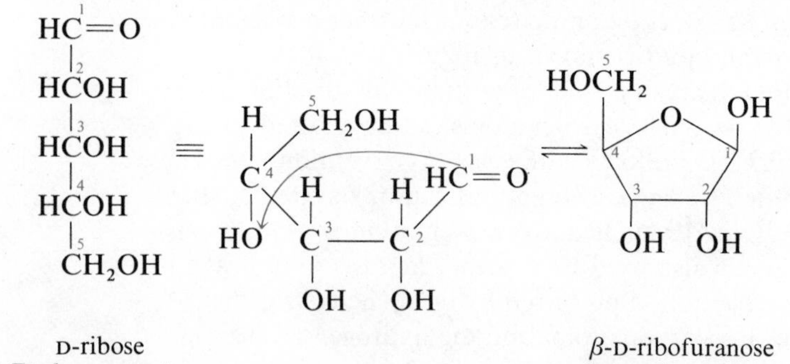

D-ribose
Fischer projection

β-D-ribofuranose
Haworth projection

While furanose rings are almost, but not quite, flat, pyranose rings are not. Therefore, the Haworth representations are not true representations of the actual molecular shape. Pyranose rings can assume several shapes. We have shown one in Figure 19.4. This is by far the most prevalent shape of the β-D-glucopyranose molecule. However, Haworth representations are widely used and are quite satisfactory for most purposes; we will use them throughout this book.

Figure 19.4 A conformational representation (chair form) of the β-D-glucopyranose molecule. The two bonds projecting straight up and the three bonds projecting straight down are known as *axial* bonds. The five bonds projecting out around the edge of the ring are known as *equatorial* bonds. In β-D-glucopyranose, the larger groups, the four hydroxyl groups and the hydromethyl ($-CH_2OH$) group, are equatorial where they can be the maximum distance from each other. Only the small hydrogen atoms are axial where they are crowded together. (a) A ball-and-stick model of the pyranose ring in a chair form. The oxygen atom is shown in color. (b) The same as that used in Figure 19.3(c) is used. (c) A space-filling molecular model of β-D-glucopyranose. Black = carbon atoms, grey = oxygen atoms, and white = hydrogen atoms. C-1 is at the upper right.

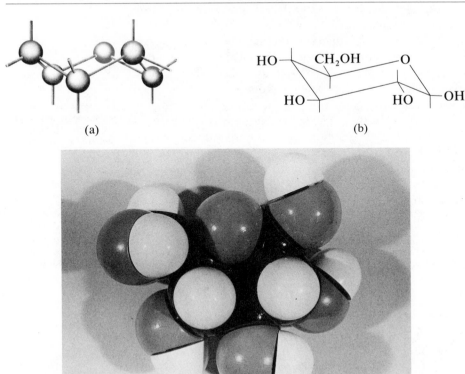

(a)

(b)

(c)

Glycosides, Oligosaccharides, and Polysaccharides

We have just reviewed the fact that the pyranose and furanose ring forms of carbohydrate molecules are the hemiacetal forms. Thus, each of these ring forms can then be reacted with an alcohol to form a **glycoside,** the acetal form of the sugar. Hydrolysis of a glycoside in an acidic solution releases the monosaccharide and the alcohol. This is illustrated by the equation for the formation and hydrolysis of a methyl glycoside of D-glucose in the pyranose ring form. Alcohols other than methanol would form the corresponding glycoside of that alcohol.

> **A glycoside** is the acetal form of a monosaccharide.

α-D-glucopyranose   methyl α-D-glucopyranoside

Every day, our bodies must rid themselves of a number of substances. Some are drugs that we have taken. Some are substances that we have taken into our bodies with the food we eat. Some are waste products of our metabolism. Some are harmless; some are poisonous. These substances are often transformed into other substances before excretion. For example, toxic (poisonous) substances are often converted into nontoxic compounds in a process known as detoxification, and then excreted. Detoxification usually involves converting the substance into something that is more soluble in water, that can then be excreted in the urine. The most common conversion reactions are hydroxylations, oxidations, reductions, and **conjugations.** A great many substances, including many common drugs, are hydroxylated then added to a polar substance to make them soluble in water. Acetaminophen, an analgesic used as an aspirin substitute (see Section 16-3.2), is a drug that does not need to be hydroxylated before detoxification by conjugation. This is because it already contains a hydroxyl group. Acetaminophen is added to a sugar known as D-**glucuronic acid** to form the water-soluble β-D-pyranoside. After deacetylation, aspirin (see Section 16-3.2) is conjugated with both D-glucuronic acid and the amino acid glycine (see Section 17-2.1).

> **Conjugation** refers to a joining together; specifically, the joining together of two compounds to produce a third compound.

> D-**Glucuronic acid** is a monosaccharide in which the primary hydroxyl group (C-6) of D-glucose is oxidized to a carboxyl group.

D-glucuronic acid glycoside of
acetaminophen (ionized form)

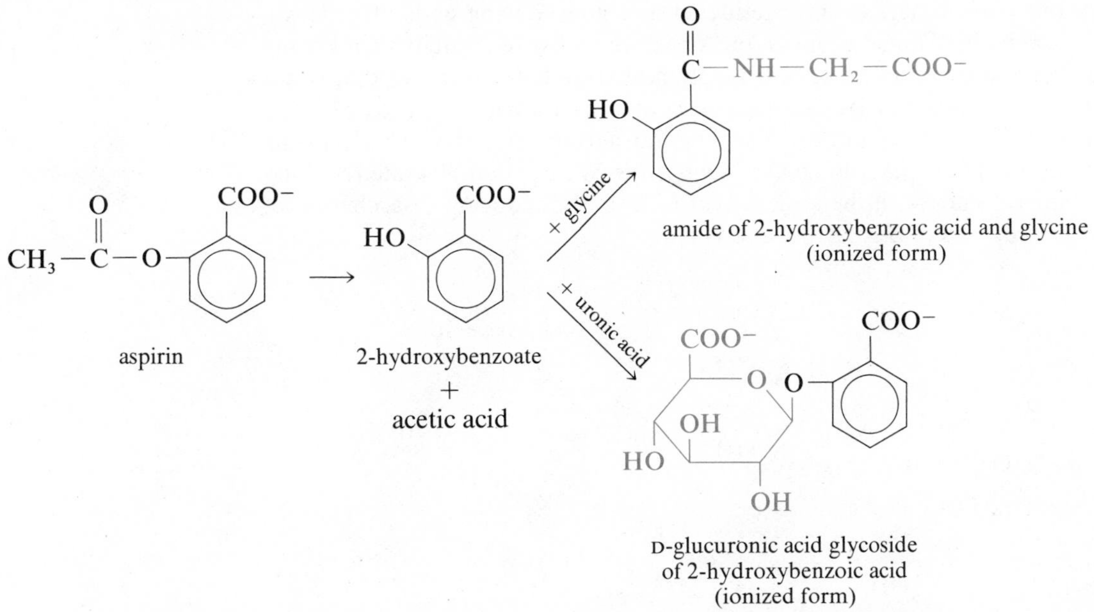

$$C-NH-CH_2-COO^-$$

amide of 2-hydroxybenzoic acid and glycine
(ionized form)

+ glycine

+ uronic acid

aspirin

2-hydroxybenzoate
+
acetic acid

D-glucuronic acid glycoside
of 2-hydroxybenzoic acid
(ionized form)

Frequently, the alcohol that forms a glycoside with a sugar is a hydroxyl group of another sugar. The formation of a glycoside between two sugar units joins them, forming a disaccharide (see Table 19.3). This joining together can be seen by looking at the hydrolysis of maltose, a disaccharide composed of two D-glucopyranosyl residues. The symbol $\diagdown\!\!\!\wedge\!\!\!\!\diagup$OH indicates a mixture of α and β forms.

A **glucopyranosyl residue** is a glucopyranose ring without the hemiacetal hydroxyl group, a hydrogen atom of an alcoholic hydroxyl group, or both, just as a methyl group is methane without one hydrogen atom.

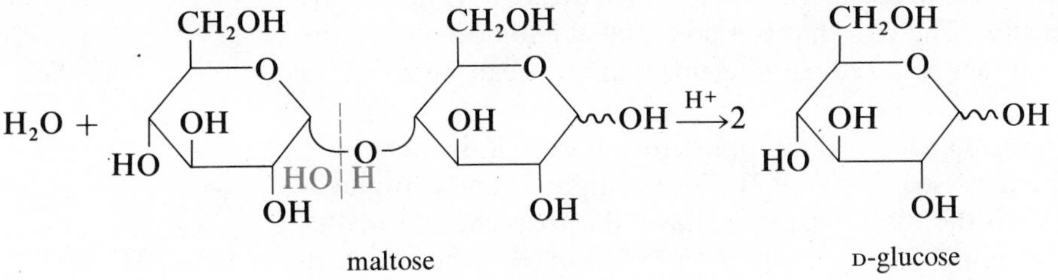

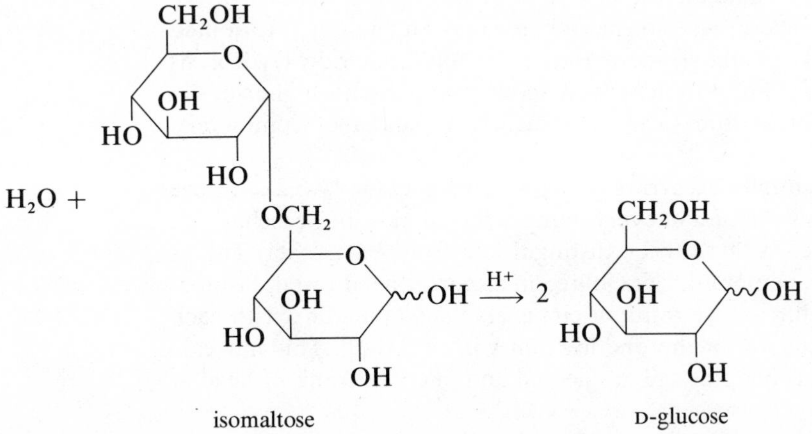

$$H_2O + \quad \xrightarrow{H^+} 2$$

maltose

D-glucose

$$H_2O + \quad \xrightarrow{H^+} 2$$

isomaltose

D-glucose

Oligo- and polysaccharides have reducing and nonreducing ends. A *reducing sugar* is a carbohydrate that contains a free, or a potentially free, aldehyde or ketone group. The *reducing end* of an oligo- or polysaccharide is the one end that is not involved in a glycosidic linkage and can, therefore, react as an aldehyde or ketone (see Figure 19.2). The sugar units of all other ends are attached through glycosidic (acetal) bonds and are, therefore, *nonreducing ends.* We can demonstrate reducing and nonreducing ends with the structure of lactose, the reducing disaccharide of milk that is composed of D-galactose and D-glucose.

D-galactose residue   D-glucose residue
(nonreducing end)   (reducing end)

β-lactose

In lactose, the reducing end is the D-glucopyranosyl residue on the right and the nonreducing end, the D-galactopyranosyl residue on the left. This is because the sugar residue on the right, a hemiacetal, can open to an aldehyde, while the one on the left, a glycoside, cannot. The end that is a potential aldehyde is called the reducing end because it can act as a reducing agent, that is, it can be oxidized to an acid.

Additional sugar units can be added to the nonreducing end of disaccharides to form higher oligosaccharides (see Table 19.3). For example, if one α-D-glucopyranosyl residue is added to the disaccharide maltose, the trisaccharide maltotriose (see Figure 19.5) is obtained. For maltotriose, $n = 1$. If $n = 2$, the structure is that of the tetrasaccharide maltotetraose; if $n = 3$, the structure is that of the pentasaccharide maltopentaose, and so on. When many sugar units are joined together by glycosidic linkages, the structure is that of a polysaccharide. Hydrolysis of the polysaccharides of starch to release D-glucose, maltose, and higher oligosaccharides of glucose forms corn syrup, the main component of commercial pancake syrup.

Polysaccharides are naturally occurring polymers (see Section 11-5.2). All the naturally occurring polymers found in every living cell (proteins, polysaccharides, and nucleic acids) have ends that can be distinguished from each other. This is because the individual monomer units are joined in a specific head-to-tail fashion. We can picture the structures if we think of circus elephants parading with each elephant holding on to the tail of the one in front with its trunk. This makes a string of elephants that has one free tail at one end and one free trunk or head at

Figure 19.5   The structure of amylose, a polysaccharide of starch, and of oligosaccharides formed by hydrolysis of starch. For amylose, $n = 400-1700$; for oligosaccharides, $n = 2-10$. The reducing end is on the right and the nonreducing end is on the left.

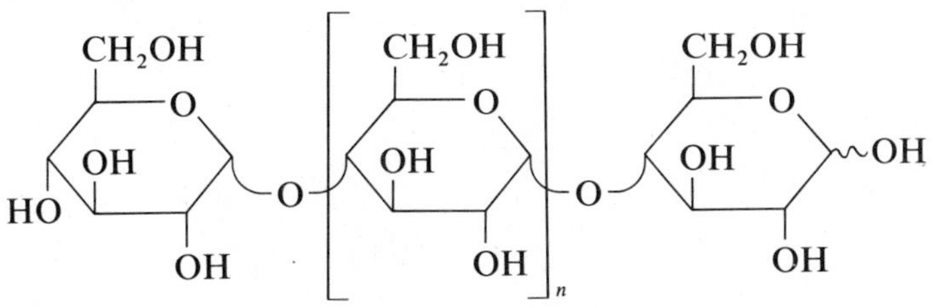

the other end. So it is with polymers such as polysaccharides that are joined together in a head-to-tail fashion. Polysaccharides have one reducing end (free aldehyde or potential aldehyde) and at least one nonreducing end (the tail end).

The polysaccharides of greatest interest to us are those of starch and glycogen. Because starch is the carbohydrate storage material of many plants and the principal component of corn, wheat, rice, potato, tapioca, arrowroot, sorghum, and other seeds and tubers, it is the principal source of carbohydrate in our diet. Starch is also widely used in the paper and textile industries and to make adhesives, D-glucose, and syrups such as pancake syrup. Starch occurs in the form of granules composed of two polysaccharides. Both contain only α-D-glucopyranosyl residues. One, amylose, is a linear polysaccharide of α-D-glucopyranosyl residues. Amylose has the structure given in Figure 19.5, where $n = 400$ to 1700 or more depending on the source. Most starches contain about 25% amylose. All linkages in a polysaccharide involve the aldehyde or ketone group of a monosaccharide unit. Consequently, as we see in Figure 19.5, there can be only one sugar residue in any polysaccharide that is not joined to another through a glycosidic bond. That is, each polysaccharide has only one reducing end, just as the string of circus elephants has only one trunk end. However, since a polysaccharide chain such as amylose has many hydroxyl groups that can be used to form other glycosidic linkages, branching can occur (see Figure 19.2). The second polysaccharide in starch granules, amylopectin, is a branched molecule. Its structure is similar to that of glycogen (see Figure 19.6). It contains 1,800 to 50,000 glucose units, depending upon the source. Most starches contain about 75% amylopectin.

Glycogen is the storage form of carbohydrate in humans and in other higher animals. The structure of the glycogen molecule is given in Figure 19.6, except that the structure as drawn has only four tiers (layers of branches) while the actual molecule has about 20 tiers. If one could see a glycogen molecule, it probably would have the feathery appearance of the head of a dandelion that has gone to seed. The molecular weight of glycogen varies from 1 million to 100 million daltons (6,200–620,000 glucose units), depending upon the tissue of origin, the state of

Figure 19.6 Structure of glycogen. Amylopectin has a similar structure. ϕ indicates the reducing end; all other ends are nonreducing ends.

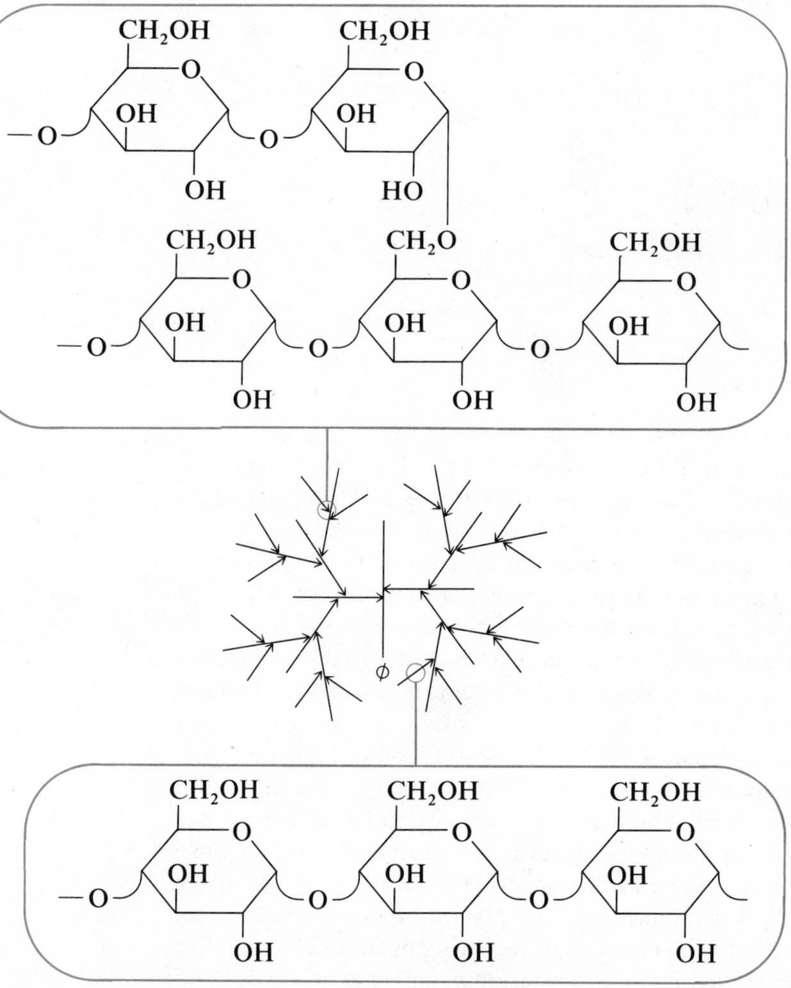

nutrition, and the presence or absence of disease. We will discuss glycogen as a storage form of carbohydrate and as an energy source in humans in Chapters 24 and 27.

Sucrose, ordinary table sugar from sugar cane or sugar beets, is a disaccharide. It contains two sugar residues, α-D-glucopyranose (the upper unit in the structure), and β-D-fructofuranose (the lower unit in the structure). The disaccharide is unique in that the two sugar residues are linked head-to-head rather than head-to-tail. This is like two elephants holding the trunk of the other with their trunk. Because the two reducing groups are joined together in an acetal bond, the molecule has no hemiacetal group and no reducing end, and is, therefore, classified as a nonreducing sugar. In sucrose, D-fructose, a ketohexose, is in a five-membered ring (fura-

sucrose

nose ring) formed between the carbonyl group at C-2 and the hydroxyl group on C-5.

D-fructose   β-D-fructofuranose

Oxidation

As we learned in Section 15-3.4, the aldehyde group can be oxidized to a carboxylic acid group. When the aldehyde group of an aldose is oxidized, the resulting compound is an **aldonic acid**. Some aldonic acids are products of carbohydrate metabolism.

An **aldonic acid** is a carboxylic acid formed from an aldose by oxidation of the aldehyde group to a carboxyl group.

D-glyceraldehyde   D-glycerate
(an aldose)   *(an aldonate)*

We can use the oxidation of the aldehyde group of an aldose to form a carboxylic acid or carboxylic acid anion to determine the amount of sugar in blood, urine, and foods. The Benedict test is one method for measuring the amount of sugar present in a fluid. In this reaction, the oxidant Cu^{2+}, is reduced to Cu^+. Cu^+ precipitates as Cu_2O, which can be measured in a variety of ways.

$$
\begin{array}{c}
HC\!=\!O \\
|\\
HCOH\\
|\\
HOCH\\
|\\
HCOH\\
|\\
HCOH\\
|\\
CH_2OH
\end{array}
\;+\;2\,Cu(OH)_2 + NaOH \longrightarrow
\begin{array}{c}
\overset{\displaystyle O}{\overset{\|}{C}}\!-\!O^-Na^+\\
|\\
HCOH\\
|\\
HOCH\\
|\\
HCOH\\
|\\
HCOH\\
|\\
CH_2OH
\end{array}
\;+\;Cu_2O + 3\,H_2O
$$

Monosaccharides in the pyranose or furanose ring forms, can also be oxidized. In this case, oxidation produces an internal ester, a *lactone,* that can subsequently open to the open-chain form. The amount of glucose in blood or urine is most often determined in the hospital laboratory by this kind of oxidation catalyzed by an enzyme, glucose oxidase. We will discuss the use of this reaction to determine the presence and amount of glucose in Section 23-4.

19-2.2 REACTIONS OF HYDROXYL (ALCOHOL) GROUPS

Ester Formation

The hydroxyl groups of sugars can react with organic and inorganic acids just as other alcohols do (see Section 14-3.3). The most common biological reaction of the hydroxyl groups of carbohydrates (other than glycosidic linkage formation) is the formation of phosphate esters. Recall that this is the result of the reaction of an alcohol (hydroxyl group) with phosphoric acid. In the interconversion of the open-chain and pyranose ring forms of D-glucose 6-phosphate shown, the phosphate ester group is represented in two ways.

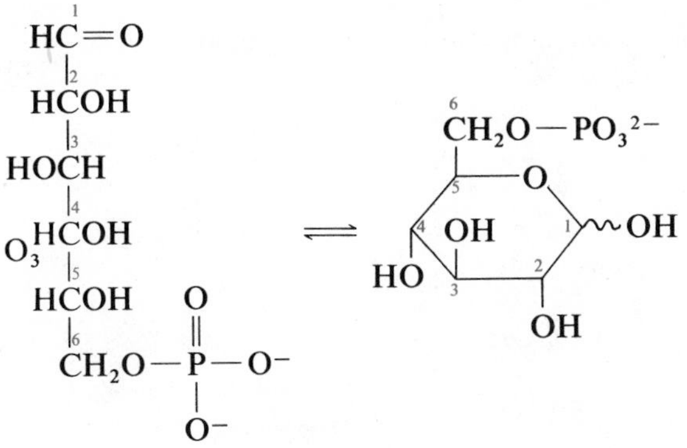

$$\underset{1}{HC}\!\!=\!\!O$$
$$\underset{2}{HCOH}$$
$$\underset{3}{HOCH}$$
$$\underset{4}{HCOH}$$
$$\underset{5}{HCOH}\quad O$$
$$\underset{6}{CH_2O}\!-\!\overset{\displaystyle O}{\underset{\displaystyle O^-}{P}}\!-\!O^-$$

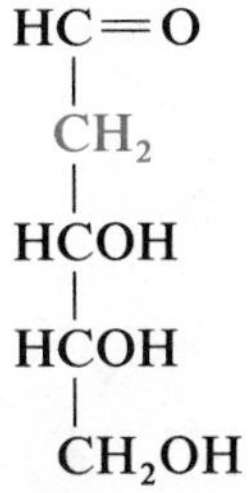

$\rightleftharpoons$

(ring structure of D-glucose 6-phosphate with $CH_2O-PO_3{}^{2-}$ at C-6, OH groups, HO)

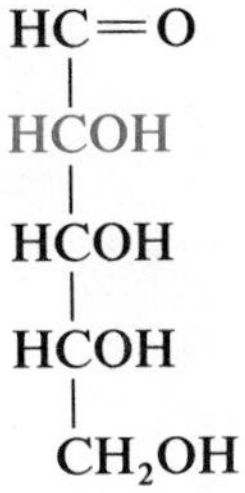

D-glucose 6-phosphate

Reduction and Oxidation

We learned in the chapters on organic chemistry that hydroxyl groups can be both oxidized to carbonyl groups and removed by reduction. Sugars that have the hydroxyl group missing from one or more of the carbon atoms are called *deoxy sugars*. The sugar known by the common name of 2-deoxy-D-ribose is an example. A comparison of the structure of 2-deoxy-D-ribose with its parent sugar, D-ribose, shows us that the hydroxyl group on C-2 of D-ribose has been replaced with a hydrogen atom in 2-deoxy-D-ribose. The latter is designated 2-deoxy because it is missing an oxygen atom from C-2.

Isomerization and Condensation

Isomerization is a reaction that gives a product that is an isomer of the starting compound. Isomerization of monosaccharides involves both the carbonyl group and the adjacent hydroxyl group and produces sugars that are isomers of the original one. This reaction can be catalyzed by either a base or an enzyme.

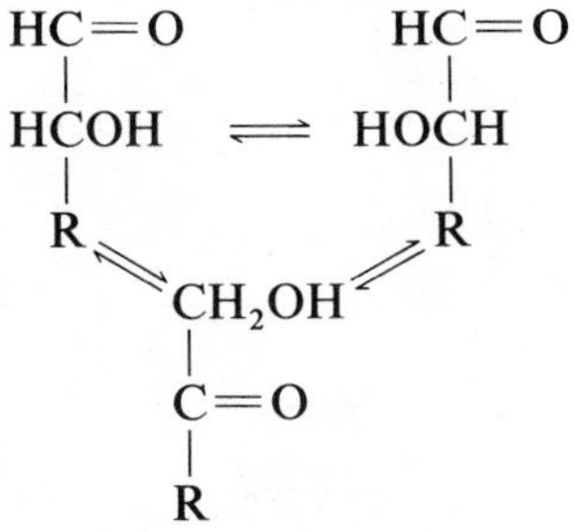

$$HC\!\!=\!\!O$$
$$HCOH$$
$$HCOH$$
$$HCOH$$
$$CH_2OH$$

D-ribose

$$HC\!\!=\!\!O$$
$$CH_2$$
$$HCOH$$
$$HCOH$$
$$CH_2OH$$

2-deoxy-D-ribose

$$HC\!\!=\!\!O \qquad\qquad HC\!\!=\!\!O$$
$$HCOH \;\rightleftharpoons\; HOCH$$
$$R \qquad\qquad\qquad R$$
$$CH_2OH$$
$$C\!\!=\!\!O$$
$$R$$

isomerization

We see from this reaction that an aldose can be converted into another aldose and a ketose and that a ketose can be converted into two aldoses. It is for this reason that ketoses are reducing sugars. They cannot act as reducing agents because they cannot be oxidized to acids, but they can be isomerized to aldoses that are reducing agents. By this isomerization reaction, D-glucose, D-fructose, and D-mannose can be interconverted.

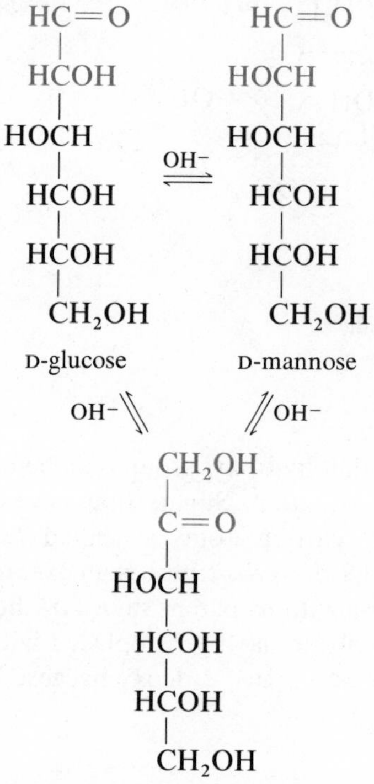

$$
\begin{array}{ccc}
\text{HC=O} & & \text{HC=O} \\
| & & | \\
\text{HCOH} & & \text{HOCH} \\
| & & | \\
\text{HOCH} & \overset{\text{OH}^-}{\rightleftharpoons} & \text{HOCH} \\
| & & | \\
\text{HCOH} & & \text{HCOH} \\
| & & | \\
\text{HCOH} & & \text{HCOH} \\
| & & | \\
\text{CH}_2\text{OH} & & \text{CH}_2\text{OH} \\
\text{D-glucose} & & \text{D-mannose}
\end{array}
$$

$$
\text{OH}^- \diagdown \qquad \diagup\!\diagup \text{OH}^-
$$

$$
\begin{array}{c}
\text{CH}_2\text{OH} \\
| \\
\text{C=O} \\
| \\
\text{HOCH} \\
| \\
\text{HCOH} \\
| \\
\text{HCOH} \\
| \\
\text{CH}_2\text{OH} \\
\text{D-fructose}
\end{array}
$$

Another important metabolic reaction is the aldol condensation (see Section 15-3.2). The most common aldol condensation is the reversible aldol condensation of the ketotriose dihydroxyacetone phosphate and the aldotriose D-glyceraldehyde 3-phosphate to form the ketohexose D-fructose 1,6-diphosphate.*

$$
\begin{array}{lcc}
 & \text{CH}_2\text{O}\,\textcircled{P} & \text{CH}_2\text{O}\,\textcircled{P} \\
 & | & | \\
\text{dihydroxyacetone phosphate} & \text{C=O} & \text{C=O} \\
 & | & | \\
 & \text{CH}_2\text{OH} & \text{HOCH} \\
 & + & | \\
 & & \text{HCOH} \\
 & \text{HC=O} & | \\
 & | & \text{HCOH} \\
\text{D-glyceraldehyde 3-phosphate} & \text{HCOH} & | \\
 & | & \text{CH}_2\text{O}\,\textcircled{P} \\
 & \text{CH}_3\text{O}\,\textcircled{P} & \\
\end{array}
$$

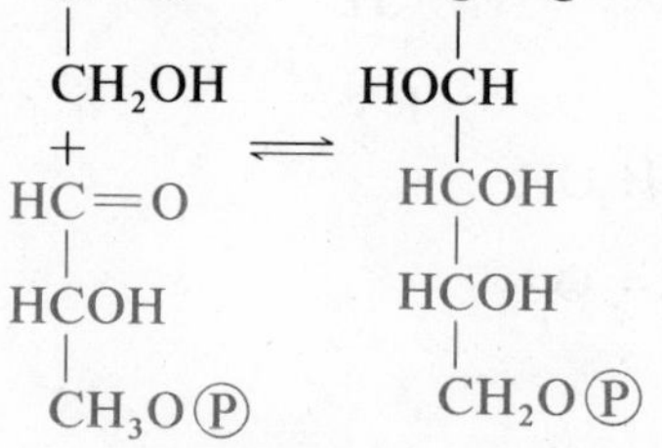

D-fructose 1,6-diphosphate

---

*In this representation, $\textcircled{P}$ is a shorthand symbol for a phosphate group. We used two other forms of designating the phosphate ester group in writing the structures of D-glucose 6-phosphate. At the pH of a living cell (pH 6.5–7.5), the phosphate ester group will be in the form of monoanions ($ROPO_3H^-$) and dianions ($ROPO_3^{2-}$). For convenience, the biochemist frequently uses the symbol $\textcircled{P}$ to signify a mixture of these two ionic forms.

PROBLEMS

2. Write the structures of the following molecules in the α-D-pyranose ring form (Haworth representation).

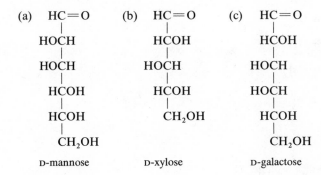

(a) HC=O
     |
     HOCH
     |
     HOCH
     |
     HCOH
     |
     HCOH
     |
     CH₂OH

     D-mannose

(b) HC=O
     |
     HCOH
     |
     HOCH
     |
     HCOH
     |
     CH₂OH

     D-xylose

(c) HC=O
     |
     HCOH
     |
     HOCH
     |
     HOCH
     |
     HCOH
     |
     CH₂OH

     D-galactose

3. Write structures of the following molecules in the β-D-furanose ring form (Haworth representation).

(a) HC=O
     |
     HCOH
     |
     HCOH
     |
     HCOH
     |
     CH₂OH

     D-ribose

(b) CH₂OH
     |
     C=O
     |
     HOCH
     |
     HCOH
     |
     HCOH
     |
     CH₂OH

     D-fructose

(c) HC=O
     |
     HCOH
     |
     HOCH
     |
     HOCH
     |
     HCOH
     |
     CH₂OH

     D-galactose

4. Write structures of the following molecules in the open-chain form (Fischer representation).

(a) β-D-idopyranose

(b) α-D-arabinofuranose

(c) α-D-fructopyranose

(d) β-D-psicofuranose

5. Write the structures of the other two sugars that can be formed by isomerization of the following molecules.

(a) 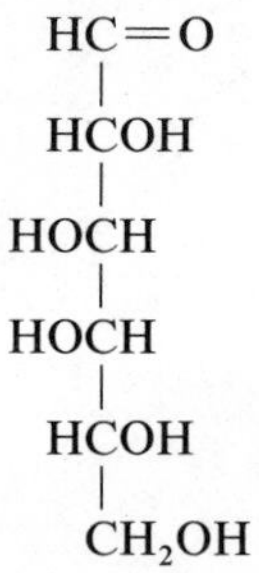
```
    HC=O
     |
    HCOH
     |
    HOCH
     |
    HOCH
     |
    HCOH
     |
    CH₂OH
```
D-galactose

(b) 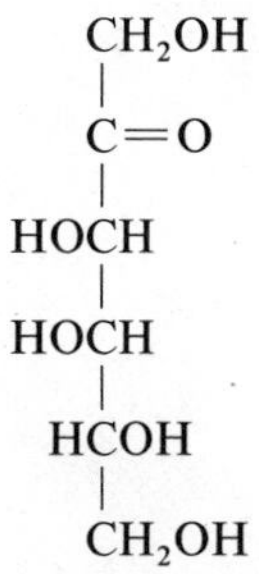
```
    CH₂OH
     |
    C=O
     |
    HOCH
     |
    HOCH
     |
    HCOH
     |
    CH₂OH
```
D-tagatose

(c) 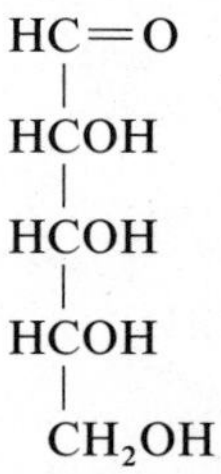
```
    HC=O
     |
    HCOH
     |
    HCOH
     |
    HCOH
     |
    CH₂OH
```
D-ribose

6. Write the open-chain structure of D-fructose 1,6-diphosphate.
7. Write the structures of the monosaccharides that are formed from the following trisaccharide upon hydrolysis. Use the Haworth representation of the pyranose ring form.

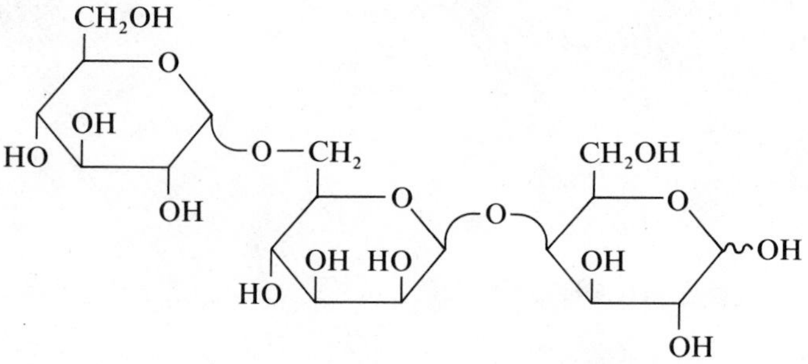

8. Indicate the reducing end and the nonreducing end of the structure of the molecule given in problem 7.
9. Write an equation that illustrates the hydrolysis of sucrose.
10. Write the structure of β-D-glucopyranose 6-phosphate.
11. Write the structure of 2-deoxy-β-D-ribofuranose.

ANSWERS TO PROBLEMS

2. (a)

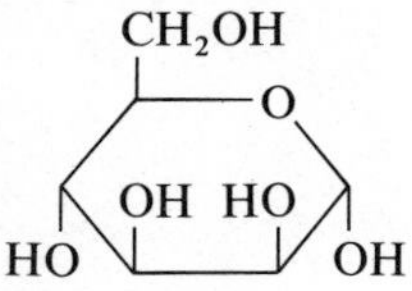

(b)

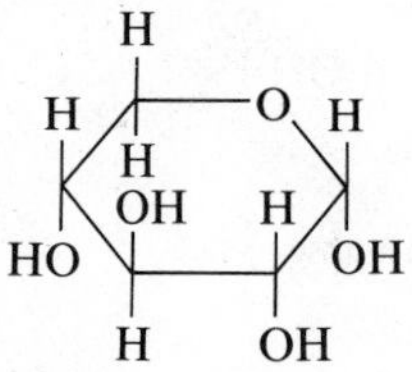

(c)

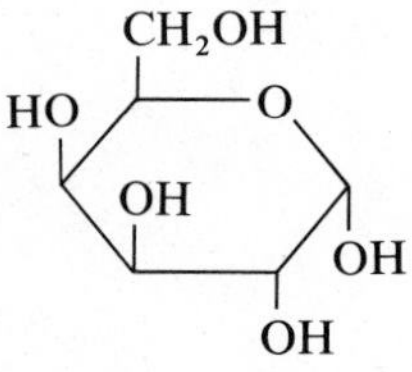

3. (a)

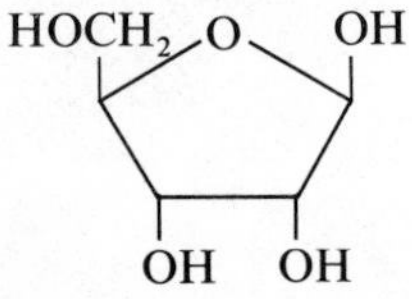

(b)

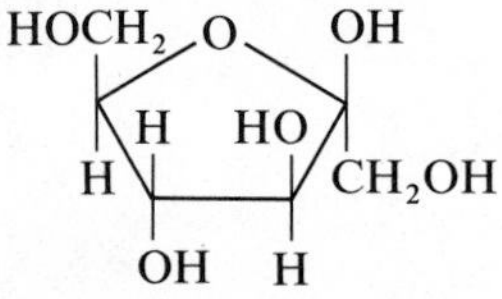

(c)

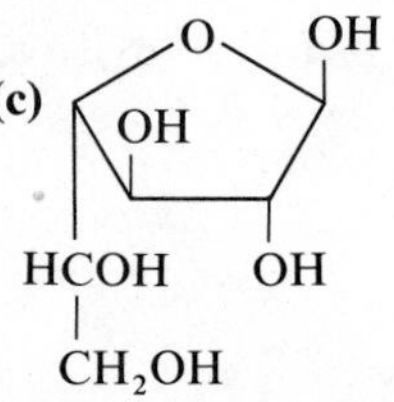

4. 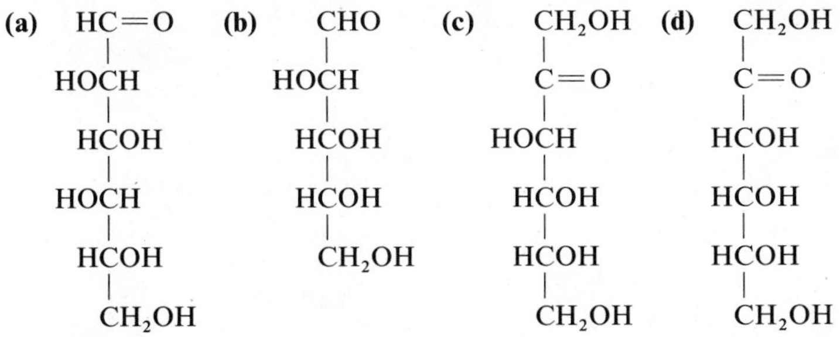 (a)

```
  HC=O
   |
 HOCH
   |
 HCOH
   |
 HOCH
   |
 HCOH
   |
 CH₂OH
```

(b)

```
  CHO
   |
 HOCH
   |
 HCOH
   |
 HCOH
   |
 CH₂OH
```

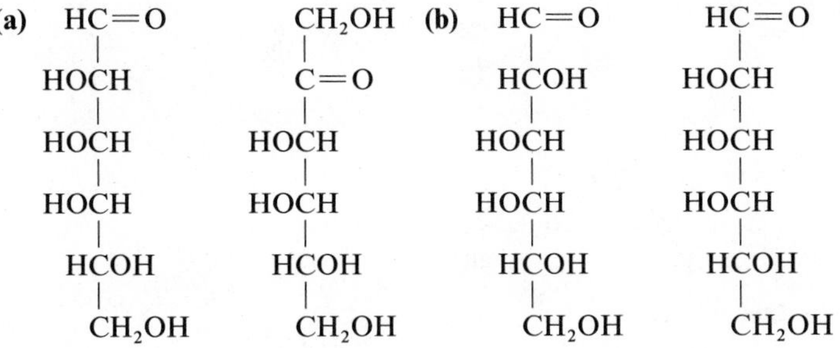

(c)

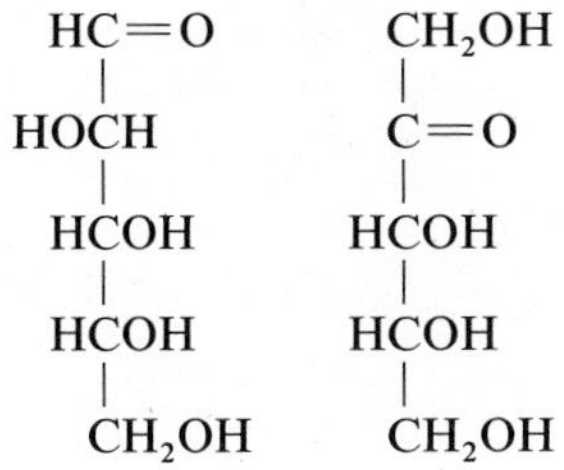

```
 CH₂OH
   |
  C=O
   |
 HOCH
   |
 HCOH
   |
 HCOH
   |
 CH₂OH
```

(d)

```
 CH₂OH
   |
  C=O
   |
 HCOH
   |
 HCOH
   |
 HCOH
   |
 CH₂OH
```

5. (a)

```
  HC=O
   |
 HOCH
   |
 HOCH
   |
 HOCH
   |
 HCOH
   |
 CH₂OH
```

```
 CH₂OH
   |
  C=O
   |
 HOCH
   |
 HOCH
   |
 HCOH
   |
 CH₂OH
```

(b)

```
  HC=O
   |
 HCOH
   |
 HOCH
   |
 HOCH
   |
 HCOH
   |
 CH₂OH
```

```
  HC=O
   |
 HOCH
   |
 HOCH
   |
 HOCH
   |
 HCOH
   |
 CH₂OH
```

(c)

```
  HC=O
   |
 HOCH
   |
 HCOH
   |
 HCOH
   |
 CH₂OH
```

```
 CH₂OH
   |
  C=O
   |
 HCOH
   |
 HCOH
   |
 CH₂OH
```

6.

```
          O
          ||
 CH₂O—P—O⁻
 |        |
O=C      O⁻
 |
 HOCH
 |
 HCOH
 |
 HCOH      O
 |         ||
 CH₂O—P—O⁻
           |
           O⁻
```

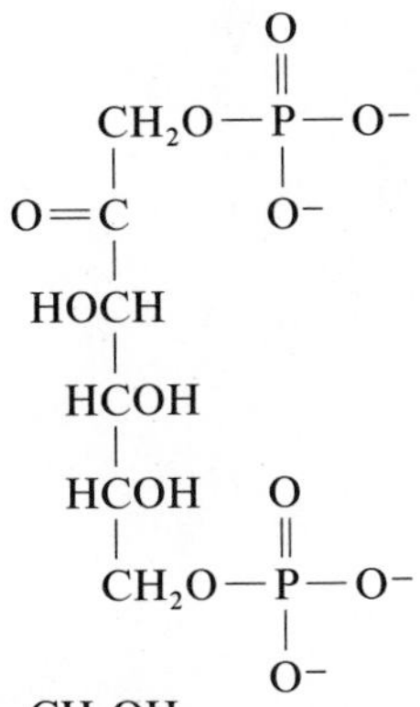

7.

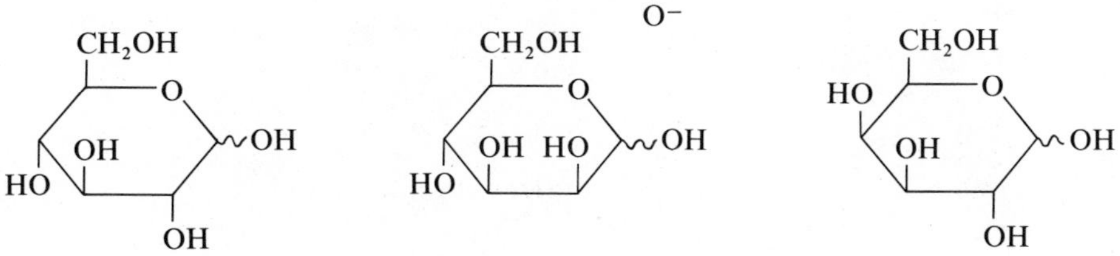

8.

nonreducing end

reducing end

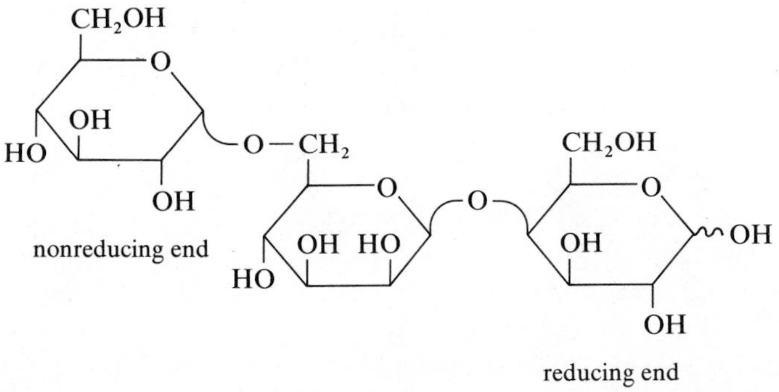

9.

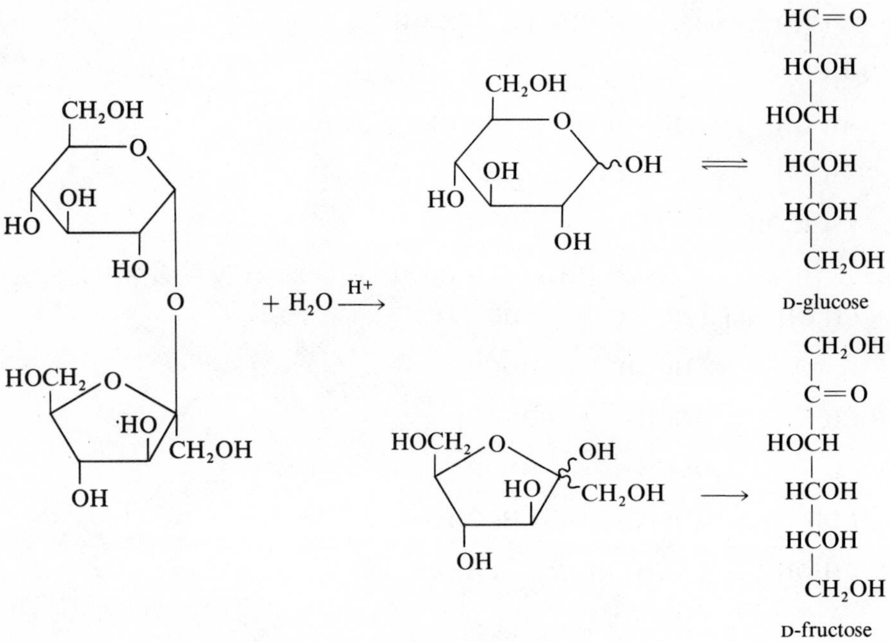

$$+ \; H_2O \xrightarrow{\text{H}^+}$$

D-glucose

D-fructose

10. $CH_2OPO_3{}^{2-}$

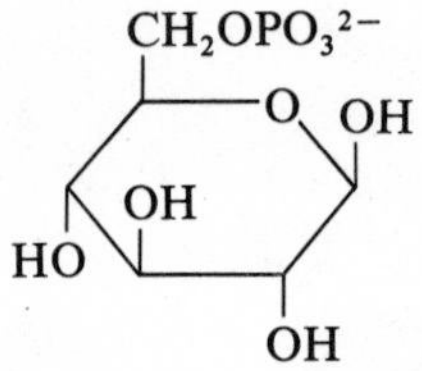

11.

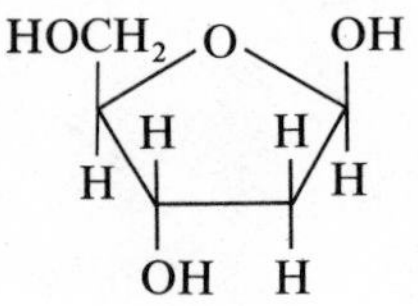

SUMMARY

In this chapter, we learned that carbohydrates occur in many different forms and that they are grouped into three classes: monosaccharides, oligosaccharides, and polysaccharides. Each of these classes is subdivided into subclasses. Monosaccharides are grouped according to the kind of carbonyl group (aldehyde or ketone) and according to the number of carbon atoms they contain. Oligosaccharides are grouped according to the number of monosaccharide units they contain. Polysaccharides are grouped according to whether they contain one or more than one kind of monosaccharide unit and according to whether the molecules are linear or branched.

We found that polysaccharides are the most abundant of the carbohydrates in living organisms. However, monosaccharides such as D-glucose and disaccharides such as sucrose and lactose also are common. Glucose is necessary for the maintenance of life. Because polysaccharides are polymers of monosaccharides, polysaccharide molecules can be converted into monosaccharide molecules by hydrolysis.

We also reviewed the fact that monosaccharide molecules contain both carbonyl groups and hydroxyl groups and, therefore, exhibit many of the chemical properties of aldehydes (or ketones) and alcohols. The most important reactions of carbohydrates are the formation of hemiacetal rings, the joining of monosaccharide units by glycosidic (acetal) linkages to form oligo- and polysaccharides, oxidation of aldehyde groups to carboxylic acids, isomerization to other sugars, the formation of phosphate esters, and aldol condensations.

In the next chapter, we will see how monosaccharide units are involved in the structures of the nucleic acids, DNA and RNA. In Chapters 22 and 23 we will discuss the nature of the enzyme molecules required for metabolism. Then we will examine the use of carbohydrates as a source of energy in Chapters 24, 25, and 27. In Chapter 26, we will learn how carbohydrate units play an important role in the structure and function of cell membranes. In Chapters 27 and 28, we will examine the relationship of carbohydrates to lipids and amino acids.

ADDITIONAL PROBLEMS

12. Using structural formulas in the Fischer projection, give examples of the following.
    (a) a D-ketohexose (D-hexulose)
    (b) an aldotetrose
    (c) an aldotriose
    (d) a D-aldopentose
    (e) a D-ketoheptose (D-heptulose)
13. Using the Haworth representation, write structural formulas for the α-D-pyranose ring form of the structures you have written for problem 12(d) and 12(e).
14. Using the Haworth representation, write structural formulas for the β-D-furanose ring form of the structures you have written for problems 12(a) and 12(d).
15. Complete the following equation giving the structures of the products in the Fischer projection.

16. Circle the reducing group in the structure in problem 15.

17. Draw structures of an aldotetrose, an aldopentose, and a ketohexose. Circle the chiral carbon atoms in each structure.
18. Draw the structure of β-D-glucopyranose in the Haworth representation.
19. Complete the following equations.

(a)
$$
\begin{array}{c}
\text{HC}{=}\text{O} \\
|\\
\text{HCH} \\
|\\
\text{HCOH} \\
|\\
\text{HCOH} \\
|\\
\text{CH}_2\text{OH}
\end{array}
\quad \xrightarrow{\text{oxidation with Benedict's reagent}}
$$

(b)
$$
\begin{array}{c}
\text{HC}{=}\text{O} \\
|\\
\text{HCOH} \\
|\\
\text{HCOH} \\
|\\
\text{HCOH} \\
|\\
\text{CH}_2\text{OH}
\end{array}
\quad \xrightarrow{\text{isomerization with OH}^-}
$$

CHAPTER 20

NUCLEIC ACIDS

It is the **proteins** (see Section 5-1.2, Section 17-4.1, and Chapter 21) of a cell, and primarily those proteins that function as catalysts, the enzymes (see Section 10-1 and Chapter 23), that are responsible for the differences between organisms and between cells of the same organism. Genes determine which enzymes are present in an organism. Genes are regions of molecules of DNA (deoxyribonucleic acid) that are found in the nucleus* of a cell (see Figure 20.1) in particles called **chromosomes**. The information contained in genes is passed on from generation to generation. In this chapter, we will examine this conservation of information.

The information contained in DNA molecules in the nucleus of a cell is transmitted to RNA (ribonucleic acid) molecules. RNA molecules leave the nucleus and pass into the **cytoplasm** of the cell where they are used to direct the synthesis of proteins (see Chapter 22). Proteins are, in turn, used to control all the activities of a cell and are ultimately responsible for observed differences. Proteins, the products of genes, determine inherited characteristics such as eye color, make the contraction of muscle possible, and allow us to use glucose for energy. We will study muscle contraction, the use of glucose, and other activities of cells in Chapters 23–28.

Before we consider the various chemical reactions in living cells, we should first examine the structures of the nucleic acids, DNA and RNA. In order to understand how genes control cellular activities and how genetic information is passed on from generation to generation, we must be familiar with the structures of DNA and RNA.

Proteins are naturally occurring polymers of amino acids joined together by peptide (amide) bonds in specific sequences.

Chromosomes are thread-like structures composed of nucleic acid and protein that contain the hereditary material of cells and viruses.

The **cytoplasm** is everything inside a cell with the exception of the nucleus.

---

*Not all cells have a nucleus. Cells of higher or multicellular organisms do. Organisms containing cells with a defined nucleus are called *eukaryotes*. Unicellular organisms without a defined nucleus are called *prokaryotes*.

Figure 20.1 The parts of an animal cell.

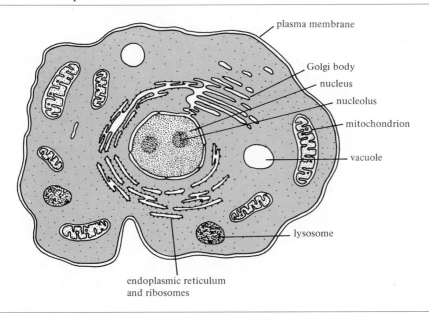

20-1 NUCLEOSIDES AND NUCLEOTIDES

LEARNING OBJECTIVES

1. Define ribonucleoside, ribonucleotide, deoxyribonucleoside, and deoxyribonucleotide.
2. Identify examples of, or give examples using structural formulas of, purines and pyrimidines.
3. Given the structure of a purine or pyrimidine, write the structure of the corresponding ribonucleoside, deoxyribonucleoside, ribonucleotide, and deoxyribonucleotide.
4. Write complete structures (using accepted condensed formulas) of adenine, AMP, ADP, and ATP.

A **nucleotide** is a phosphate ester of a nucleoside.

A **nucleoside** is a compound consisting of a purine or pyrimidine residue joined to a sugar residue.

Nucleic acids are polymers of nucleotides and can be broken down by hydrolysis into **nucleotides**. Nucleotides can be further hydrolyzed to **nucleosides** and the phosphate ion (see Figure 20.2). We will begin this chapter with a discussion of nucleosides.

Figure 20.2 General structures of nucleic acids (DNA and RNA), nucleotides, and nucleosides, and an indication of the way in which they are related through hydrolysis. Base = purine or pyrimidine. Sugar = D-ribofuranose (RNA) or 2-deoxy-D-ribofuranose (DNA).

```
                 base                base               base                base
                  |                   |                  |                   |
~~ phosphate—sugar—phosphate—sugar—phosphate—sugar—phosphate—sugar ~~
```

nucleic acids

↓ hydrolysis + HOH

phosphate—sugar—base

nucleotides

↓ hydrolysis + HOH

sugar—base

nucleosides

Figure 20.3 (a) General equation for the formation and hydrolysis of a nucleoside of D-ribofuranose. (b) Specific equation for the formation and hydrolysis of the nucleoside of uracil.

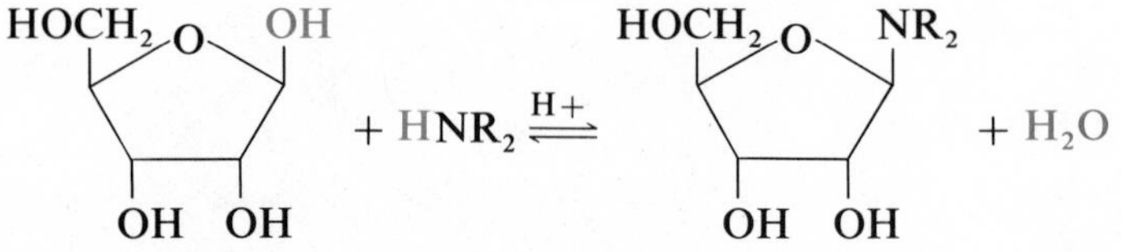

β-D-ribofuranose nucleoside

(a)

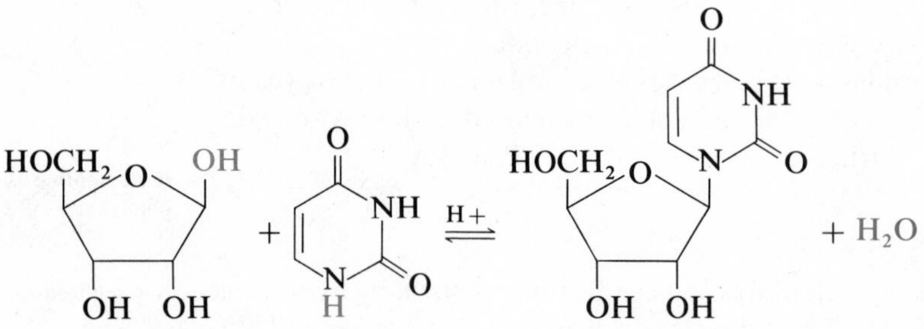

β-D-ribofuranose uracil uridine

(b)

20-1.1 RIBONUCLEOSIDES AND DEOXYRIBONUCLEOSIDES*

In Chapter 19, we learned that reaction of an alcohol with a pyranose or furanose ring form of a sugar yields a glycoside. An amine may be substituted for the alcohol in the reaction with the ring form of a sugar because reactions of alcohols and amines are very similar (see Sections 14-3 and 17-4). The product of the reaction of a sugar with an amine is a nucleoside (see Figure 20.3).

The amine that is joined to a sugar unit to form a nucleoside is usually one of the five heterocyclic bases given in Figures 20.4 and 20.5. We learned in Section 17-1 that the term *heterocyclic* refers to an organic ring structure that contains at least one atom other than carbon. We also learned earlier that a base is a proton acceptor and that an organic base is almost always a compound containing nitrogen. Heterocyclic bases are, therefore, organic ring structures containing both carbon and

Figure 20.4 Structures of purine and the two derivatives of purine found in nucleic acids. The numbering system for ring atoms is given for the purine molecule. Abbreviations used are given in parentheses under the names.

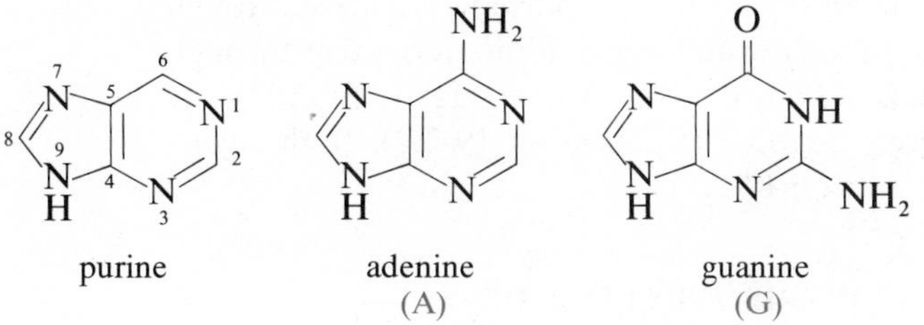

Figure 20.5 Structures of pyrimidine and the three derivatives of pyrimidine found in nucleic acids. The numbering system for ring atoms is given for the pyrimidine molecule. Abbreviations used are given in parentheses under the names.

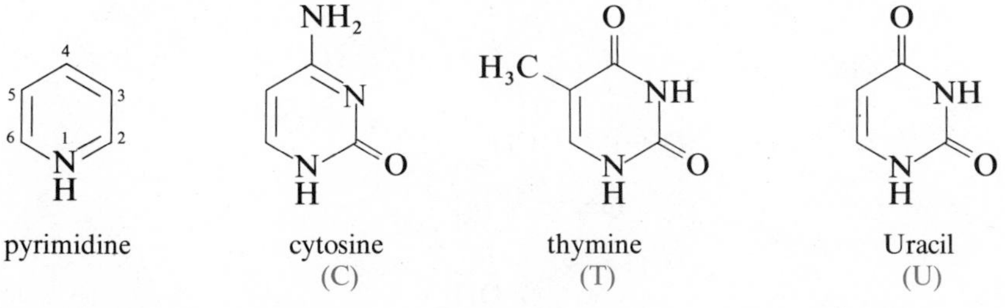

*Deoxyribonucleosides are often referred to simply as *deoxynucleosides.*

Figure 20.6 Examples of nucleosides. Deoxyguanosine is a deoxyribonucleoside of the purine guanine and 2-deoxy-D-ribose. (The numbering of the carbon atoms of the sugar unit is given for this molecule.) Uridine is a ribonucleoside of the pyrimidine uracil and D-ribose. The sugar unit is attached to the nitrogen atom at position 9 of purines and to the nitrogen atom at position 1 of pyrimidines.

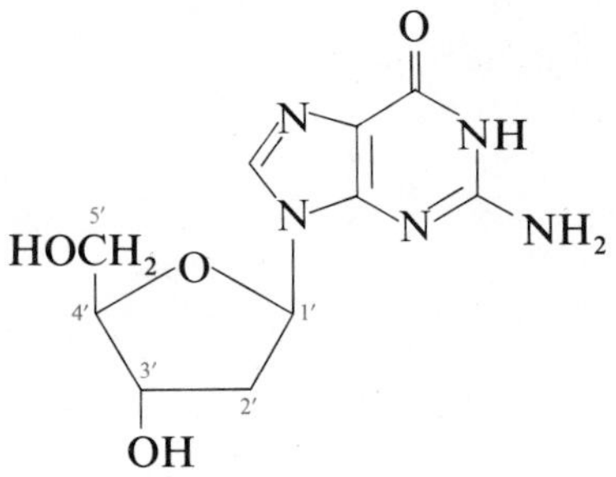

deoxyguanosine 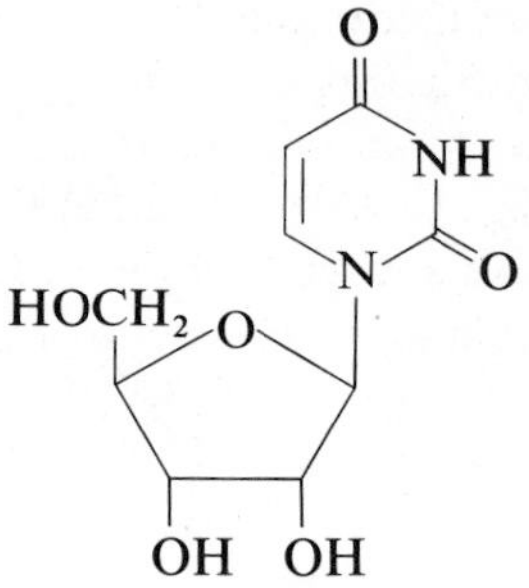   uridine

nitrogen atoms in the ring. Addition of a sugar unit to a purine derivative through bonding to nitrogen atom number 9 (N-9) or to a pyrimidine derivative through bonding to N-1 forms a nucleoside (see Figure 20.6). Two sugars are found in the nucleic acids — D-ribose and 2-deoxy-D-ribose (see Section 19-2.2). Both sugars exist in the furanose ring form (see Section 19-2.1).

20-1.2 RIBONUCLEOTIDES AND DEOXYRIBONUCLEOTIDES*

Addition of a phosphate ester group at the 5′ position of the sugar unit of a nucleoside forms a nucleotide (see Figure 20.7). A second and third phosphate group

Figure 20.7 Examples of nucleotides. Adenosine 5′-monophosphate is a ribonucleotide of the purine adenine and D-ribose. Deoxycytidine 5′-monophosphate is a deoxyribonucleotide of the pyrimidine cytosine and 2-deoxy-D-ribose.

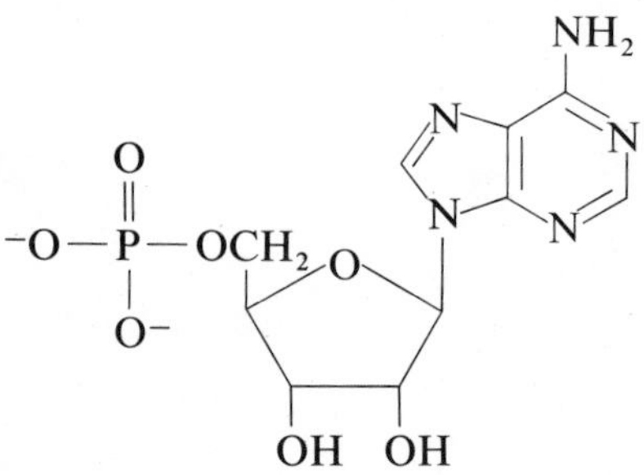

adenosine 5′-monophosphate   deoxycytidine 5′-monophosphate
(AMP)   (dCMP)

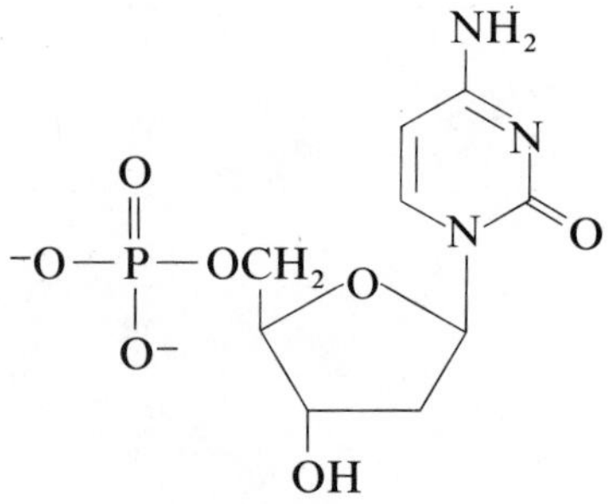

*Deoxyribonucleotides are often referred to simply as *deoxynucleotides*.

Figure 20.8 The hydrolysis of adenosine 5′-triphosphate (ATP) to adenosine 5′-diphosphate (ADP). Further hydrolysis of ADP would yield AMP and P_i. $P_i =$ inorganic phosphate in its two physiological forms, $H_2PO_4^-$ and $H_2PO_4^{2-}$.

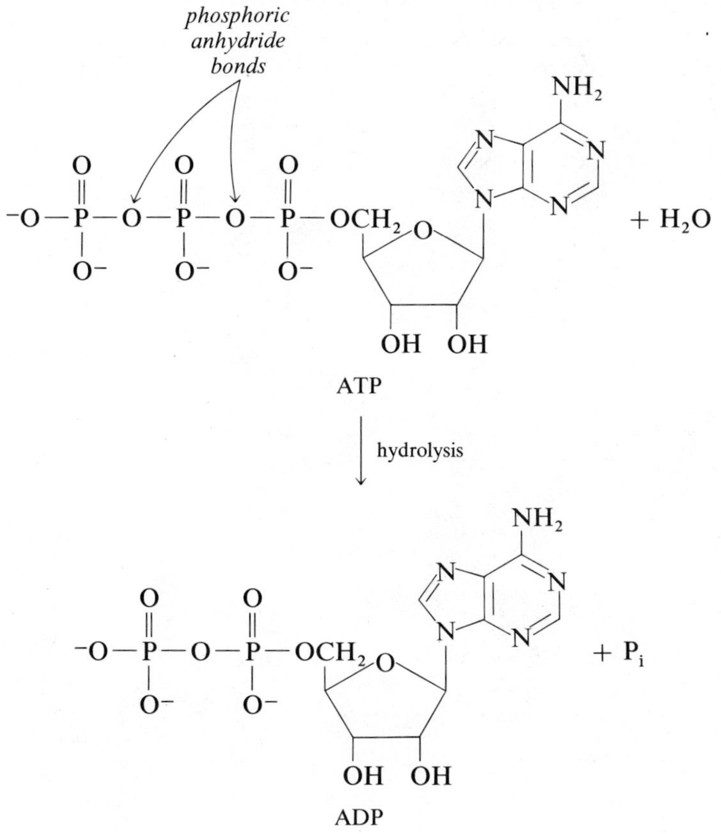

can be added to a nucleotide to form nucleoside di- and triphosphates, respectively (see Figure 20.8).

Thus far, we have learned the following:
1. The combination of a monosaccharide (D-ribofuranose or 2-deoxy-D-ribofuranose) with a heterocyclic base (a purine or a pyrimidine derivative) forms a nucleoside (see Table 20.1).
2. The conversion of a nucleoside into its phosphate ester forms a nucleotide (see Table 20.2).
3. Nucleic acids (DNA and RNA) are polymers of nucleotides.
4. Nucleosides may be substituted with a pyrophosphate (Ⓟ — Ⓟ) or a triphosphate (Ⓟ — Ⓟ — Ⓟ) group. The latter compounds are used to provide the chemical energy for cellular reactions and processes.

Table 20.1   Names of Nucleosides and Nucleotides of D-Ribose*

Base	Ribonucleoside (base + D-ribose)	Ribonucleotide (nucleoside phosphate)	Abbreviation of Ribonucleotide
Adenine (A)	Adenosine	Adenosine 5'-monophosphate	AMP
Guanine (G)	Guanosine	Guanosine 5'-monophosphate	GMP
Cytosine (C)	Cytidine	Cytidine 5'-monophosphate	CMP
Uracil (U)	Uridine	Uridine 5'-monophosphate	UMP

*The ribonucleoside and ribonucleotide of thymine are not components of nucleic acids.

Table 20.2   Names of Nucleosides and Nucleotides of 2-Deoxy-D-Ribose*

Base	Deoxynucleoside (base + 2-deoxy-D-ribose)	Deoxynucleotide (nucleoside phosphate)	Abbreviation of Deoxynucleotide
Adenine (A)	Deoxyadenosine	Deoxyadenosine 5'-monophosphate	dAMP
Guanine (G)	Deoxyguanosine	Deoxyguanosine 5'-monophosphate	dGMP
Cytosine (C)	Deoxycytidine	Deoxycytidine 5'-monophosphate	dCMP
Thymine (T)	Deoxythymidine	Deoxythymidine 5'-monophosphate	dTMP

*The deoxyribonucleoside and deoxyribonucleotide of uracil are not components of nucleic acids.

All life processes depend upon a continual supply of energy. This energy is needed for various activities in living cells, including mechanical work (for example, contraction of muscle), the transport of some molecules and ions through membranes (including the uptake of certain molecules and ions from the blood and the digestive tract), and the synthesis of the great number of carbohydrate, nucleic acid, protein, and smaller molecules required continually to maintain a living cell. Cells use the energy stored in the chemical bonds of molecules for these processes. The bonds used most for storing energy are **high-energy** phosphate **bonds**. The source of high-energy phosphate bonds used most is ATP (adenosine triphosphate).

A chemist faced with a synthesis involving the addition of an acyl group uses an activated form of the acid. For acetylations, reactions involving the transfer of acetyl groups, the chemist will use reagents such as acetic anhydride or acetyl chloride (see Section 16-4.2). For *phosphorylations,* reactions involving the transfer of phosphate groups, a living cell uses activated forms of phosphoric acid, high-energy phosphate bonds. ATP (see Figure 20.8) has two high-energy phosphate bonds: the two phosphoric acid anhydride bonds linking the three phosphate

A **high-energy bond** is one that releases energy of more than 5,000 cal/mol upon hydrolysis.

$$CH_3 - \overset{\displaystyle O}{\overset{\displaystyle \|}{C}} -$$

acetyl group

residues. They are the bonds that are used for phosphorylation reactions. Like the chemist, a living cell uses an anhydride of phosphoric acid (an activated form of phosphoric acid) to bring about phosphate transfer reactions. It will become apparent time and time again that reactions of living cells are not magical but are ordinary chemical reactions. The difference between the reactions of cells and the reactions we studied in the chapters on inorganic and organic chemistry is that the former are very efficiently catalyzed by enzymes (see Chapter 23).

$$
\underset{\text{acetic anhydride}}{CH_3 - \overset{\overset{\displaystyle O}{\|}}{C} - O - \overset{\overset{\displaystyle O}{\|}}{C} - CH_3}
\qquad
\underset{\text{acetyl chloride}}{CH_3 - \overset{\overset{\displaystyle O}{\|}}{C} - Cl}
$$

From an examination of the structures of high-energy phosphate compounds, we find that there are four general types of high-energy phosphate bonds. These four types and an example of a compound containing each are given in Table 20.3.

Table 20.3 Examples of High-Energy Phosphate Compounds

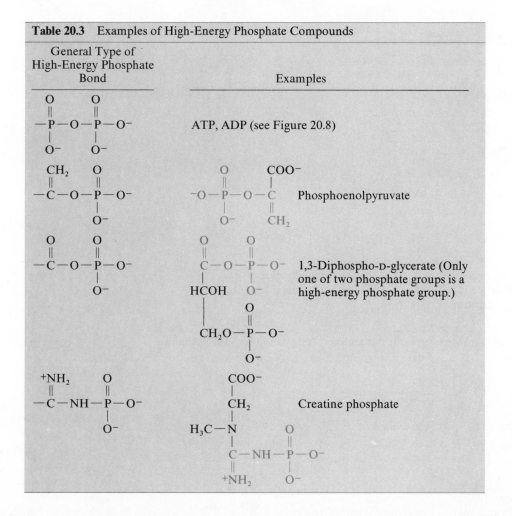

General Type of High-Energy Phosphate Bond	Examples
$-\overset{\overset{\displaystyle O}{\|}}{\underset{\underset{\displaystyle O^-}{\|}}{P}} - O - \overset{\overset{\displaystyle O}{\|}}{\underset{\underset{\displaystyle O^-}{\|}}{P}} - O^-$	ATP, ADP (see Figure 20.8)
$-\overset{\overset{\displaystyle CH_2}{\|}}{C} - O - \overset{\overset{\displaystyle O}{\|}}{\underset{\underset{\displaystyle O^-}{\|}}{P}} - O^-$	Phosphoenolpyruvate
$-\overset{\overset{\displaystyle O}{\|}}{C} - O - \overset{\overset{\displaystyle O}{\|}}{\underset{\underset{\displaystyle O^-}{\|}}{P}} - O^-$	1,3-Diphospho-D-glycerate (Only one of two phosphate groups is a high-energy phosphate group.)
$-\overset{\overset{\displaystyle ^+NH_2}{\|}}{C} - NH - \overset{\overset{\displaystyle O}{\|}}{\underset{\underset{\displaystyle O^-}{\|}}{P}} - O^-$	Creatine phosphate

Although living cells contain other high-energy compounds, ATP is by far the most prevalent, so prevalent that it is involved in all the life processes that we will discuss in the remainder of this book. In Chapters 24, 25, and 27, we shall see how carbohydrates and fats are oxidized to release energy that is stored in the high-energy phosphate bonds of ATP. ATP molecules are stable. However, when its high-energy phosphate bonds are cleaved by an enzyme, energy in the amount of about 13,000 cal/mole is released. Together, the two properties of stability and the ability to release large amounts of energy when cleaved allow ATP to be transported from one part of a cell to another and then cleaved to provide the energy for useful work. If high-energy phosphate compounds were organized in a list beginning with the one that releases the greatest amount of energy upon hydrolysis and ending with the one that releases the least amount of energy upon hydrolysis, ATP would fall in the middle. Because ATP occupies this intermediate place, the energy released by the degradation of carbohydrates and fats can be used to make ATP and can then be transferred to other compounds for use in various cellular activities.

PROBLEMS

1. Identify the following structures as pyrimidines or purines.

(a)

(b)

(c)

(d)

(e)

2. Write the structural formula of each of the following.
   (a) the deoxyribonucleoside of 1(d)   (b) the ribonucleotide of 1(c)
   (c) the deoxyribonucleotide of 1(e)   (d) the ribonucleotide of 1(a)
3. Write the complete structure of ADP.

ANSWERS TO PROBLEMS   **1. (a)** purine **(b)** pyrimidine   **(c)** pyrimidine   **(d)** pyrimidine
(e) purine   **2. (a)**   **(b)**

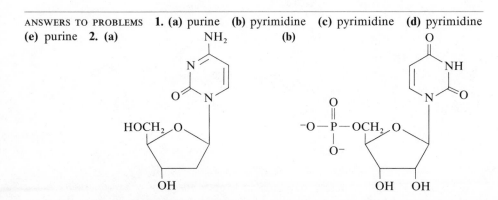

(c)

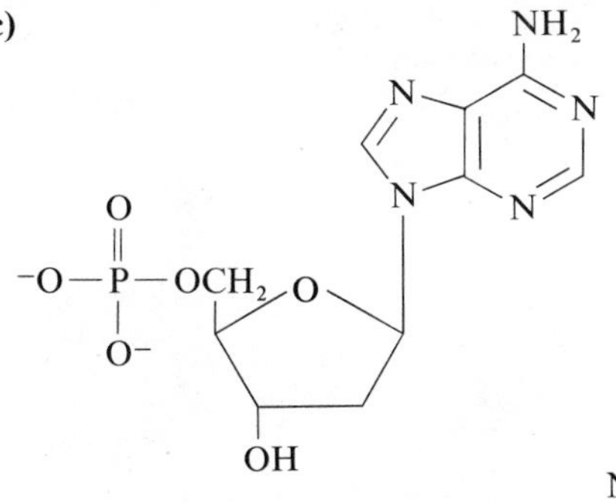

(d)

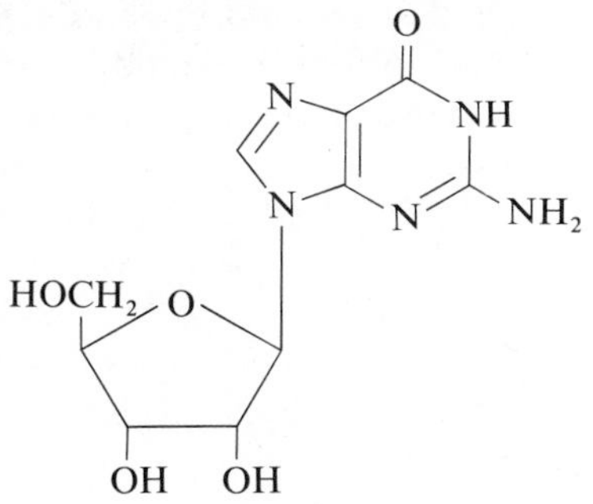

3.

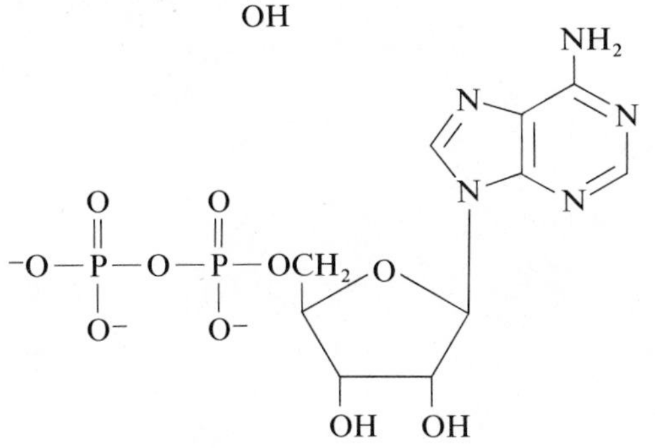

20-2 DNA (DEOXYRIBONUCLEIC ACID)

LEARNING OBJECTIVES

1. Define base pair, double helix, complementary bases, and replication.
2. Describe the relationship of a nucleotide to a nucleic acid.
3. Identify the 3′ and 5′ ends of an oligonucleotide.
4. Given a shorthand notation for a sequence of a DNA molecule and the structures of the heterocyclic bases, write the complete structure of that segment.
5. Using the abbreviations A, C, T, and G, write the structure of a trideoxyribonucleotide.
6. Describe the function of DNA.
7. Describe how the two antiparallel chains of DNA are held together.
8. Given the sequence of nucleotides in one strand of DNA and the 5′ end, give the complementary sequence of nucleotides in the other strand and identify its 5′ and 3′ ends.
9. Illustrate replication with a diagram.

DNA is one of the two kinds of polymers of nucleotides. It is an extremely high-molecular-weight compound that is found in the nucleus of a cell and carries the genetic information for the organism. In this section, we will first examine its structure, and finally its packaging.

20-2.1 STRUCTURE OF DNA

DNA is a polymer of four deoxynucleotide units: dAMP, dCMP, dGMP, and dTMP (see Table 20.2). It has the general structure depicted in Figure 20.2. In

DNA, the sugar is 2-deoxy-D-ribose and the bases are adenine (A), guanine (G), cytosine (C), and thymine (T). A four-nucleotide segment typical of those found in DNA molecules can be written as shown.

Possible segment of a DNA molecule

From the structure we can see that, as with polysaccharides, the monomer units of nucleic acids are joined together in a head-to-tail fashion and the two ends of nucleic acid molecules are different and can be distinguished from each other. The end with a free hydroxyl group at C-3 of the 2-deoxy-D-ribofuranose residue is called the 3′ end (read the *3 prime end*); the other end is called the 5′ end.

As we have already pointed out, genetic information is carried in DNA molecules.* Because the backbone sequence of 2-deoxy-D-ribofuranose units connected by phosphate diester groups is the same for all DNA molecules, and because a gene is a segment of a DNA molecule, we can conclude that the differences between DNA molecules that allow them to carry different genes must reside in differences

---

*The chromosomes of some viruses are composed not of DNA but of RNA, a chemically similar compound that also has the ability to carry genetic information.

Figure 20.9 (a) A trideoxynucleotide composed of deoxyadenosine monophosphate (dAMP), deoxythymidine monophosphate (dTMP), and deoxyguanosine monophosphate (dGMP), reading from the 5' end to the 3' end. (b) Three shorthand designations of the molecule.

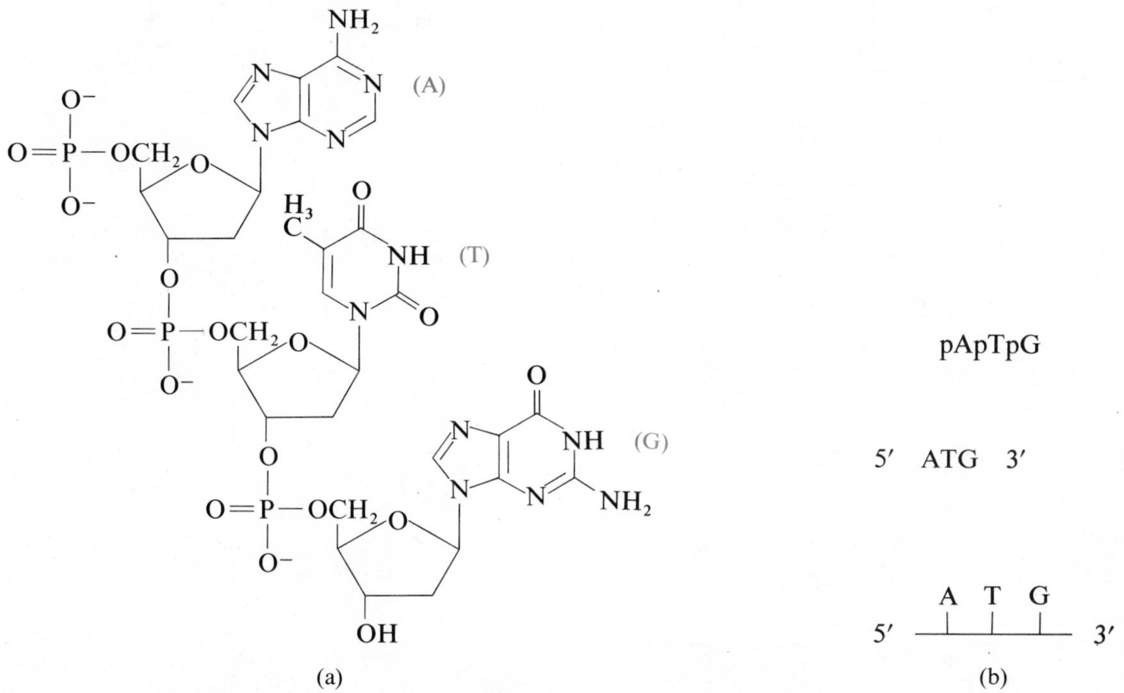

pApTpG

5' ATG 3'

A T G
5' ——┴——┴——┴—— 3'

(a)   (b)

in the sequence of the purine and pyrimidine bases. Because each DNA molecule has the same backbone and the sequence of bases is the important distinguishing difference between molecules, biochemists have developed several shorthand notations for DNA. Figure 20.9 shows three of these shorthand notations for an **oligonucleotide** (a trideoxynucleotide).

An **oligonucleotide** is a chain of 2–10 nucleotide residues.

Of course, if DNA is to serve as the carrier of genetic information, each cell must have the same genes. As cells divide, beginning with a single fertilized egg, to form a complete organism, each cell must have the genetic information required to allow it to carry out its function. In addition, all these genes must be passed on from parents to children. To meet these requirements, DNA molecules must be able to direct synthesis of exact copies of themselves. This property is related to an important structural feature of DNA, namely that DNA consists of two polynucleotide chains.

The two chains of DNA are joined together by hydrogen bonds (see Section 5-1.2) between pairs of bases (see Figure 20.10). The purine adenine is always paired with the pyrimidine thymine, and the purine guanine is always paired with the pyrimidine cytosine. Other pairs do not occur. The two bases that make up a

Figure 20.10
Hydrogen bonding between the adenine–thymine and guanine–cytosine base pairs in DNA. The A=T pair has two hydrogen bonds and the G≡C pair, three. Hydrogen bonds are indicated by dashed lines.

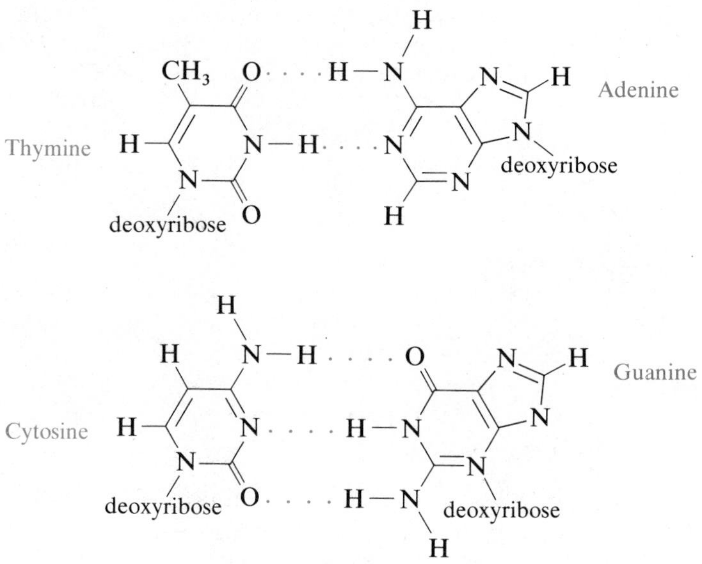

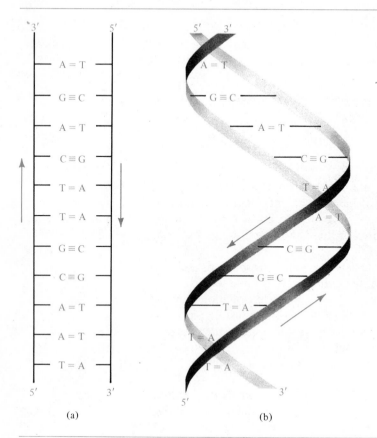

Figure 20.11
Diagram of the two polynucleotide chains of a DNA molecule showing (a) the pairing of the bases and the reversed directions of the chains and (b) the twisting of the two chains into a double helix. Note that there are three hydrogen bonds holding the guanosine–cytosine base pair together and two hydrogen bonds holding the adenine–thymine pair together.

Figure 20.12
A space-filling model of a DNA double helix. The phosphate diester groups point to the outside of the helix; the base pairs are in the interior. The size of the units representing the various atoms reflects the actual relative size of the atoms. (Photograph courtesy of Professor M. H. F. Wilkins, Biophysics Department, King's College, London.)

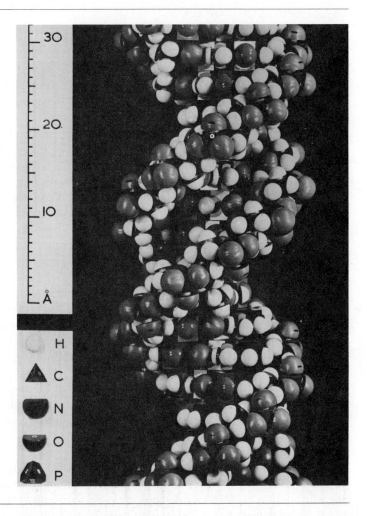

pair of bases are called **complementary bases**. Because the larger purine residue is always paired with the smaller pyrimidine residue, in DNA the number of adenine residues always equals the number of thymine residues (A = T) and the number of guanine residues always equals the number of cytosine residues (G = C). As a consequence, the sum of the purine residues equals the sum of the pyrimidine residues (A + G = T + C). Another result of this pairing rule is a complementary relationship between the sequences of bases on the two polynucleotide chains. For example, if one chain has the sequence GTACT, the opposite chain must have the sequence CATGA.

The two chains run in opposite directions and are twisted together in a *double helix* (see Figures 20.11 and 20.12), the structure obtained when two strands are coiled around a common axis. The double helix of DNA is what we would get if we had a ladder with the two side pieces made out of something flexible like rubber, held it at both ends, and twisted it in opposite directions.

Complementary bases are the specific pairs of heterocyclic bases formed by hydrogen bonds between a purine residue on one nucleic acid strand and a pyrimidine residue on another strand or another segment of the same strand.

20-2.2 REPLICATION OF DNA

A cell divides into two daughter cells by a process called *mitosis.* During mitosis, all DNA molecules are duplicated in such a way that each of the two new cells has all the same genetic information as the parent cell. Two complete sets of chromosomes must be formed before the cell divides so that each new cell gets an identical set. As we have already mentioned, the DNA double helix has the property of directing synthesis of an exact copy of itself.

Replication of DNA molecules and, hence, replication of genes, occurs during cell division. Replication is a rather simple process. When a cell divides, the two DNA strands separate, as shown in Figure 20.13. Then a new polynucleotide chain builds up on each of the two DNA strands. Each heterocyclic base on a single strand of DNA pairs with only that heterocyclic base that is complementary to it: guanine pairs with cytosine and adenine, with thymine. Deoxynucleoside triphosphate molecules similar in structure to the ribonucleoside triphosphate molecule ATP are used for the synthesis of DNA. As shown in Figure 20.13, base pairing orders these deoxynucleoside triphosphate molecules one at a time in a specific sequence along the polymer strand. The enzyme DNA polymerase* then connects the deoxynucleotides together in a stepwise manner using the energy stored in the high-energy phosphate bond to form a chemical bond joining the nucleotides. Because only a complementary base fits at a given position, each DNA strand directs the formation of a new strand that is complementary to itself (see Figure 20.13). Only one nucleotide can fit at a given position, so the duplication is highly accurate.

> **Replication** is the production of an identical copy of the double-stranded DNA molecule.

We can see how replication works to pass along the proper information when cells divide if we compare it to photography. If we have a negative of a picture, a print of the picture can be made. Also, another negative can be made from the print by photographing it. As long as either the negative or the positive image is available, both can be generated and the picture can be preserved. One DNA strand can be compared to the negative and the other to the print. Just as a new photographic positive image can be made from a negative image and vice versa, the information in DNA can be copied by separating the two strands and synthesizing a new complementary strand on each.

Figure 20.14 illustrates how the high-energy phosphate bond is used to form a new phosphate ester bond, splitting off a pyrophosphate group. This group is immediately cleaved by hydrolysis, ensuring that the reaction is not reversible.

The DNA strands do not separate completely before synthesis begins. Rather, new strands are formed even as the double helix unwinds. Nucleotides are added only to the 3′ end of the growing chain and the two strands of a double helix lie in opposite directions (that is, are *antiparallel*). Therefore, the two new complementary daughter strands will grow in opposite directions (see Figure 20.13). This would seem to require complete separation of the two strands before synthesis. For example, suppose a planter is suspended by cables made of two smaller cables wound around each other. Now suppose we decided that it would look nice if one

---

*Other enzymes and proteins also play important roles in DNA replication, but DNA polymerase catalyzes the reaction joining the nucleotides.

Figure 20.13
Diagram of replication of a short segment of double-stranded DNA showing
strand separation, the addition of deoxynucleotides to the growing ends (3′ end)
of the new complementary chains, and
the final product. Note that the two new
double-stranded segments are identical
to the original one. The original strands
are black and the new strands are in
color. The enzyme DNA polymerase is
present at the growing point at all times.

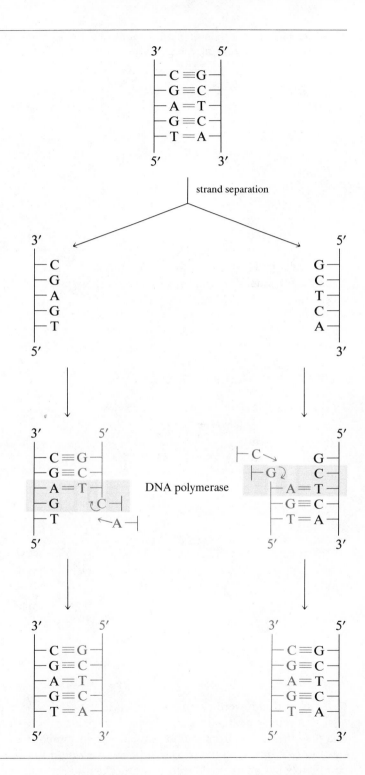

Figure 20.14   The addition of a dAMP unit to a single strand of DNA that has dCMP at the growing end (3′ end).

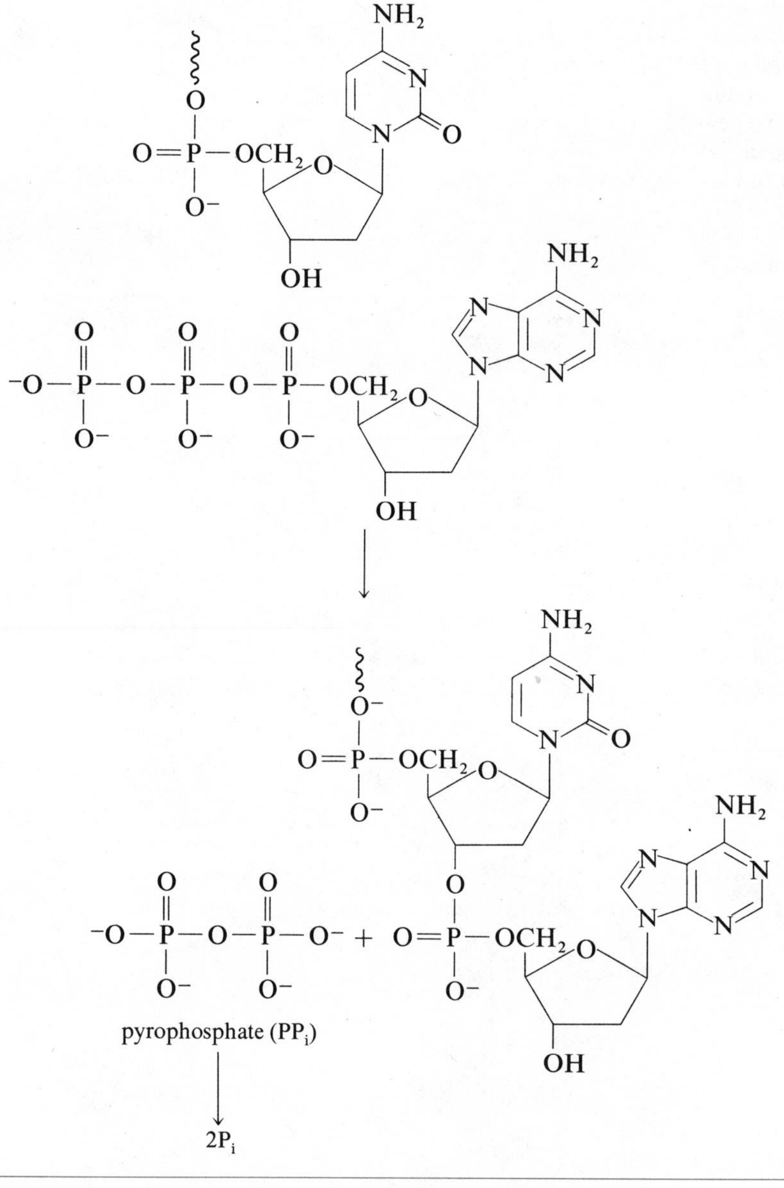

of the two strands were painted white while the other was painted blue. Suppose, also, that for some reason we could only start painting one strand from the top end and the other strand from the bottom end. Then someone could start painting one strand as we unwound the two strands, but we would have to separate the strands completely before we could start painting the other strand because we would have

Figure 20.15

Growing point in a DNA molecule. Black lines represent parent DNA polynucleotide strands, and color lines represent daughter DNA strands. Solid circles represent 5′ ends. Arrowheads represent the 3′ ends and, hence, indicate the direction of growth. The upper arm of the fork shows Okazaki fragments.

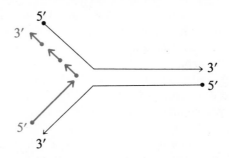

to wait until we got to the other end. So, because the two DNA strands lie in opposite directions and grow in only one direction, the two strands would have to be completely separated before new DNA could be synthesized. Biochemists now know that some simple organisms get around this problem by making short pieces called Okazaki fragments, named after the scientist who first proposed and discovered them. These fragments are then joined together to make a complete strand (see Figure 20.15).

The DNA molecule is very large and its replication is relatively rapid, so many origins are involved. These origins are called *initiation points.* As an example, one of the 46 human chromosomes contains approximately 270 million pairs of bases. The DNA double helix in this chromosome is about 10 cm (4 in.) long when it is stretched out. This gives some idea of the compactness of DNA in a chromosome, because the nucleus of a cell is so small that it can be seen only with a good microscope. The average diameter of a nucleus in a human cell is usually less than 2 micrometers (2×10^{-4} cm). It would take about 2 months to replicate this chromosome if there were only one origin. However, it probably takes only minutes for the chromosome to replicate because of the cooperative action of thousands of replication forks (Figure 20.15). We do know that the actual replication time of a large chromosome of the fruit fly (*Drosophila*) is less than 3 minutes because of the cooperative action of more than 6000 replication forks.

20-2.3 PACKAGING OF DNA

Of the many marvels of nature, one is the manner in which genetic information is inherited as discussed in the previous section. Another is how DNA is used to control cellular functions, which we will examine in Chapter 21. A third is the way in which DNA is packaged. In human cells, about 4 m (12 ft) of DNA are packed into 46 paired chromosomes with a total length of only about 0.2 mm (0.0008 in.), a ratio of about 20,000 to 1. The DNA molecules must be packaged in such a way that segments are independently accessible for information transfer because not all genes are expressed in any cell. The *expression* of genetic information refers to the use of genes to direct protein synthesis, that is, putting genes into action just as you express thoughts by putting them into words. All our cells contain the same genetic information (DNA); but just as we do not express all our thoughts, so any one cell does not express all its genes. For example, the genes responsible for synthesis of the pigments that determine eye color are not expressed in skin cells; the genes responsible for the formation of gastric juice are expressed only in certain cells of the stomach; and so on.

Like replication, the packaging of DNA is a fairly simple and logical process. A single long DNA molecule is complexed with positively charged protein molecules to form a fiber called *chromatin.* The chromatin fiber is then coiled and folded into a chromosome.

The positively charged proteins are *histones.* Eight histone molecules form a roughly spherical complex around which DNA is wound, forming a superhelix and giving a bead-like structure. Two more histone molecules on the outside of this bead-like structure, a *nucleosome,* further stabilize it. These nucleosomes are separated by linear segments of the DNA molecule and are evenly spaced so that the chromatin fiber looks like beads on a string. The nucleosome particles contain segments of double-helical DNA about 140 base pairs long; the segments separating the nucleosomes are 20–30 base pairs long. The beads on a string structure is then organized into an even larger structure, probably an even larger helix. This

A **superhelix** is a helix of the double helix of DNA.

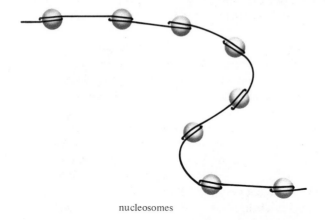

nucleosomes

organization of DNA molecules reduces the linear length about 20,000 times. It allows the DNA molecule to fit into the nucleus of a cell and yet be accessible to the enzymes required to transfer the genetic information to RNA molecules.

20-2.4 RECOMBINANT DNA

We have heard much recently about recombinant DNA. Recombinant DNA is made in the following way. DNA fragments containing the gene of interest are isolated from cells of a eukaryotic organism, for example cells of a human hypothalamus gland. The DNA fragment is obtained by isolating the cells' DNA and cleaving it at specific points with an enzyme known as a *restriction enzyme.* At the same time, a particular kind of DNA is isolated from a microorganism such as *Escherichia coli* (*E. coli*), a common intestinal bacterium. The particular DNA molecule used, *plasmid DNA,* is a relatively small, circular molecule. The plasmid

Eukaryotic (sometimes spelled eucaryotic) cells are those cells with a defined nucleus.

DNA is cleaved at one specific point by the same restriction enzyme used to cut the eukaryotic DNA. Because both the eukaryotic DNA and the plasmid DNA have been cleaved by the same enzyme, the ends of each fragment are very similar. As a result, these "sticky" ends recognize each other and reform the circle. Using another enzyme, the eukaryotic gene can be spliced to the bacterial DNA (Figure 20.16). The plasmid DNA molecule now contains the eukaryotic gene and can be put back into the bacterium, which can be cultured. The bacterium can be grown in culture to produce any number of cells, so the gene can be replicated many times to produce any number of cloned copies.

The bacterial cells then use this genetic information to make the protein for which it codes. The hormones insulin and somatostatin have been made in this way. Somatostatin is secreted by the hypothalamus gland and regulates the function of the pituitary gland. It functions by inhibiting the production of pituitary hormones that regulate growth and the production of two other hormones, insulin and glucagon, by the pancreas. At present, doctors still use insulin from cattle and pigs to treat diabetes mellitus (see Sections 10-1 and 24-1.8), but before long, they will be using human insulin made by *E. coli* cells. Other biomolecules will also be produced in bacteria cultures to solve problems in medicine and agriculture.

Figure 20.16 Technique of making recombinant DNA. By this technique, nucleotide sequences obtained from one organism or by chemical synthesis in the laboratory can be inserted into the DNA of another organism. This figure shows DNA being inserted into the circular DNA (plasmid DNA) of a bacterium such as *E. coli*. The bacterial cell will then make the protein coded for by the inserted DNA fragment. (The relative sizes of the cells and the DNA molecules shown are inaccurate; the DNA has been enlarged for clarity.)

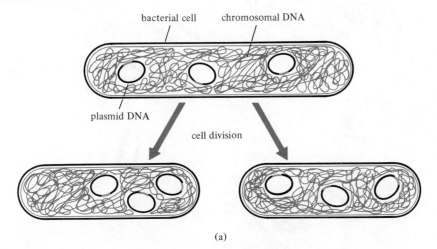

(a)

Figure 20.16 continues

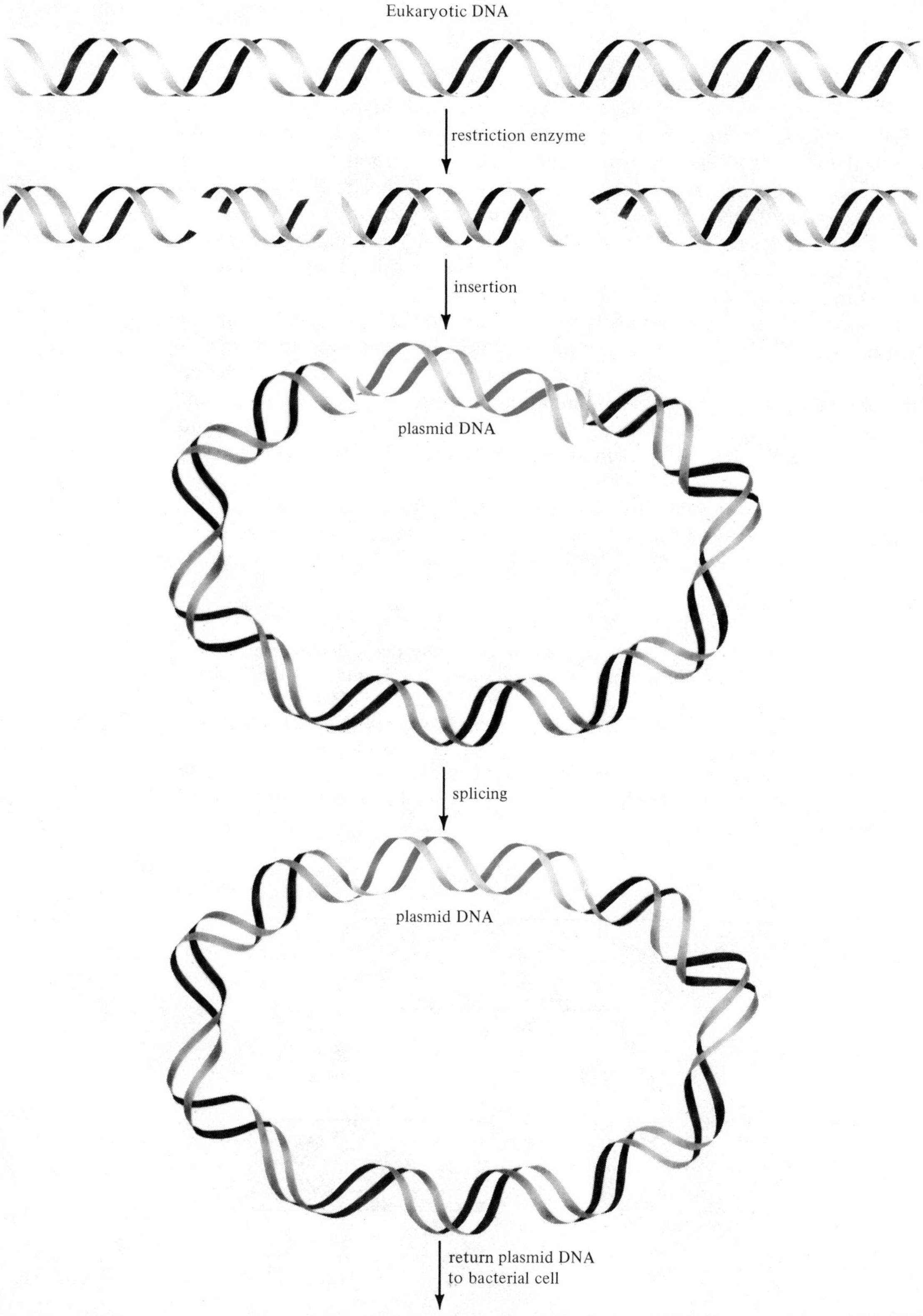

Eukaryotic DNA

restriction enzyme

insertion

plasmid DNA

splicing

plasmid DNA

return plasmid DNA
to bacterial cell

PROBLEMS

4. Using the structures of the purine and pyrimidine bases given in Figures 20.4 and 20.5, write the complete structures of the following segments of DNA molecules.

(a) 5′ pCpTpA 3′

(b) 5′ TGGC 3′

(c) 5′ ⎯⎯⎯⎯ 3′   (with A A T above)

(d) 5′ ⎯⎯⎯⎯ 3′   (with C A G T above)

5. Using the same shorthand notations given in problem 7, write the sequences of nucleotides that are complementary to those given in problem 7 and identify the 3′ and 5′ ends.
6. Using the one-letter abbreviations for the purine and pyrimidine bases, write the structure of the trideoxynucleotide 5′ GTA 3′.
7. Diagram the replication of the following DNA segment.

$$3' \quad \underbrace{\quad\quad\quad\quad}_{} \quad 5'$$

A G T C

T C A G

$$5' \quad \overline{\quad\quad\quad\quad} \quad 3'$$

ANSWERS TO PROBLEMS

4. (a)

(b)

(c)

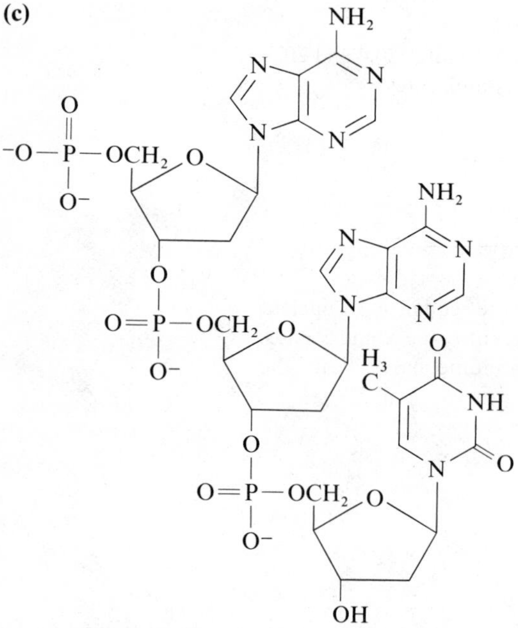

(d)

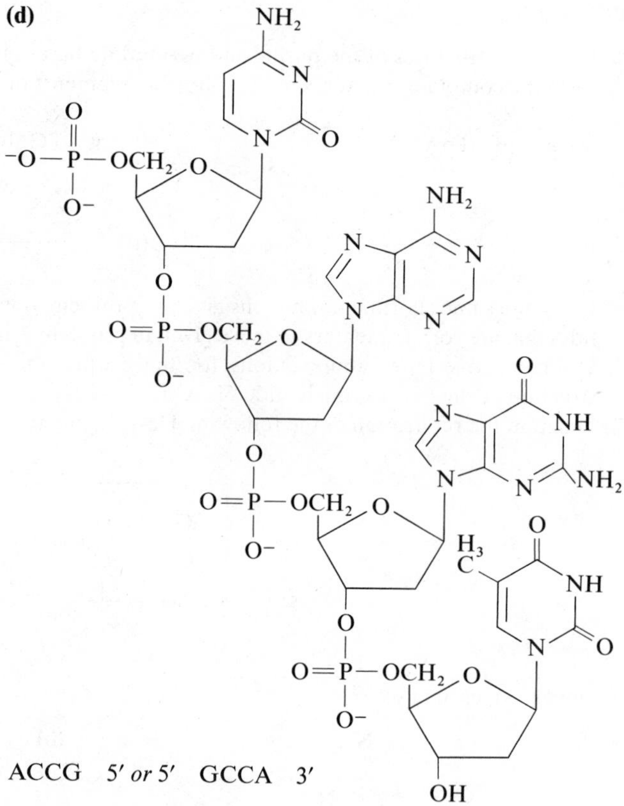

5. (a) 3′ GpApTp 5′ *or* 5′ pTpApG 3′   **(b)** 3′ ACCG 5′ *or* 5′ GCCA 3′

(c) 3′ ─┬──┬──┬── 5′ *or*
                T   T   A

        A   T   T
   5′ ──┴──┴──┴── 3′

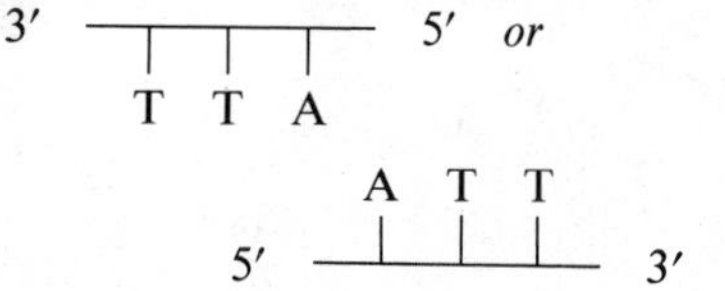

(d) 3′ ─┬──┬──┬──┬── 5′ *or*
               G   T   C   A

         A   C   T   G
   5′ ──┴──┴──┴──┴── 3′

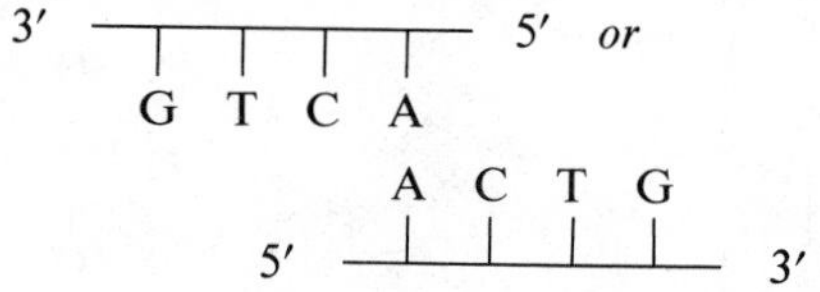

6.

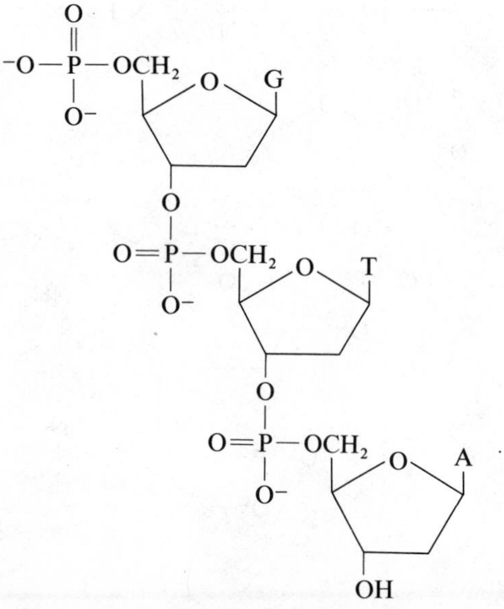

7.

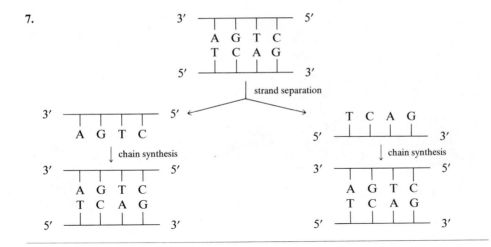

20-3 RNA (RIBONUCLEIC ACID)

LEARNING OBJECTIVES

1. List the structural differences between DNA and RNA.
2. Given any one of the several shorthand notations for a sequence of an RNA molecule and the structures of the heterocyclic bases, write the complete structure of that segment.
3. Using the abbreviations A, C, U, and G, draw a triribonucleotide and identify its 3′ and 5′ ends.

4. Define transcription.
5. Illustrate transcription with a diagram.
6. Given a double-stranded DNA sequence, write the RNA base sequence obtained by transcribing either one of the DNA chains.

RNA molecules, like DNA molecules, are polymers of nucleotides. They are made in the nucleus, where their synthesis is directed by DNA molecules. The RNA molecules then pass from the nucleus to the cytoplasm where they direct protein synthesis. Thus, RNA molecules are the link between genes and their expression.

20-3.1 STRUCTURE OF RNA

RNA is chemically very similar to DNA. Like DNA, RNA is a long, unbranched molecule containing four types of nucleotides linked together by phosphate diester bonds. There are two chemical differences between RNA and DNA.

1. RNA contains ribonucleotides rather than deoxyribonucleotides; that is, RNA contains D-ribofuranose rather than 2-deoxy-D-ribofuranose.
2. RNA contains not thymine but uracil, a closely related pyrimidine that differs from thymine in the absence of a methyl group (see Figure 20.5).

Table 20.4   Comparison of DNA and RNA Molecules

Feature	DNA	RNA
Sugar	2-Deoxy-D-ribofuranose	D-Ribofuranose
Purines	Adenine	Adenine
	Guanine	Guanine
Pyrimidines	Cytosine	Cytosine
	Thymine	Uracil
Organization	Usually double stranded	Usually single stranded
	A=T and G=C	

These differences between RNA and DNA are summarized in Table 20.4.

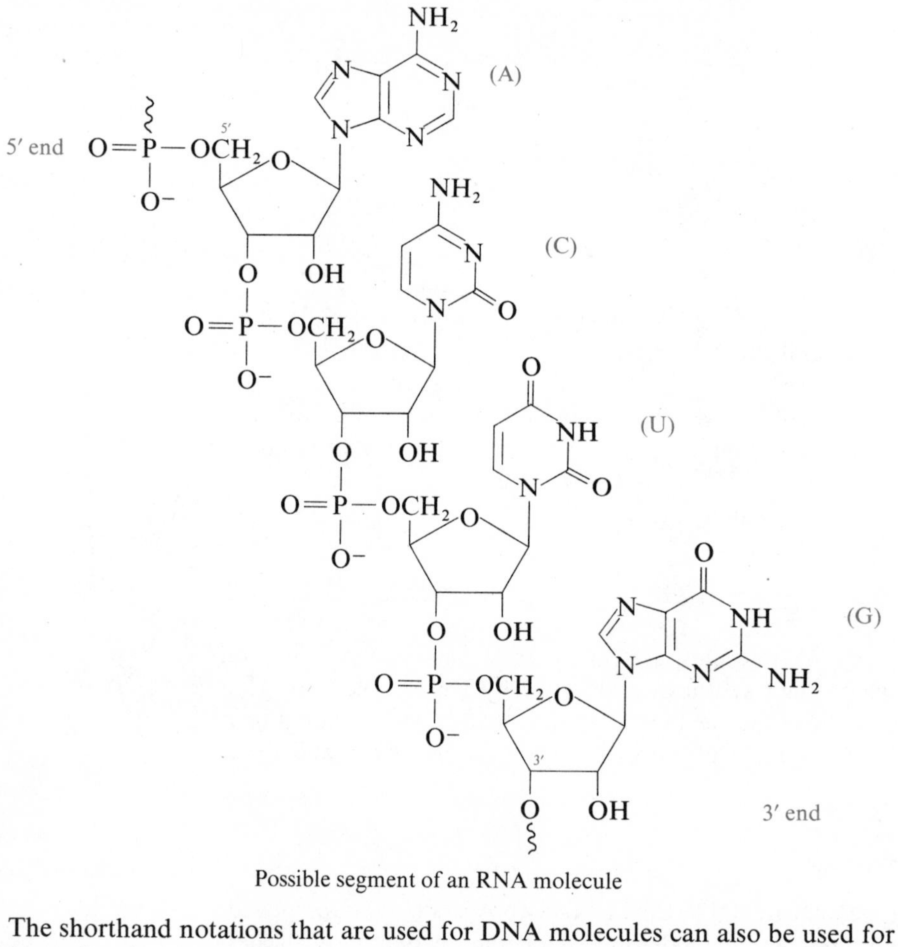

Possible segment of an RNA molecule

The shorthand notations that are used for DNA molecules can also be used for RNA molecules. Figure 20.17 gives these shorthand notations for a trinucleotide from RNA. A shorthand notation for the segment shown above is

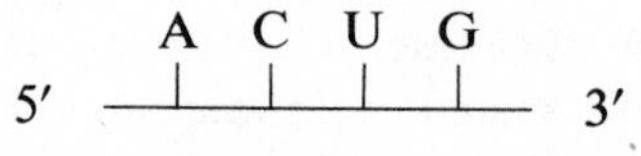

$$5' \quad \underset{|}{\overset{A}{\rule{0pt}{0pt}}} \; \underset{|}{\overset{C}{\rule{0pt}{0pt}}} \; \underset{|}{\overset{U}{\rule{0pt}{0pt}}} \; \underset{|}{\overset{G}{\rule{0pt}{0pt}}} \quad 3'$$

Figure 20.17   (a) A trinucleotide composed of cytidine monophosphate (CMP), adenosine monophosphate (AMP), and uridine monophosphate (UMP), reading from the 5′ end to the 3′ end. (b) Three shorthand designations of the molecule.

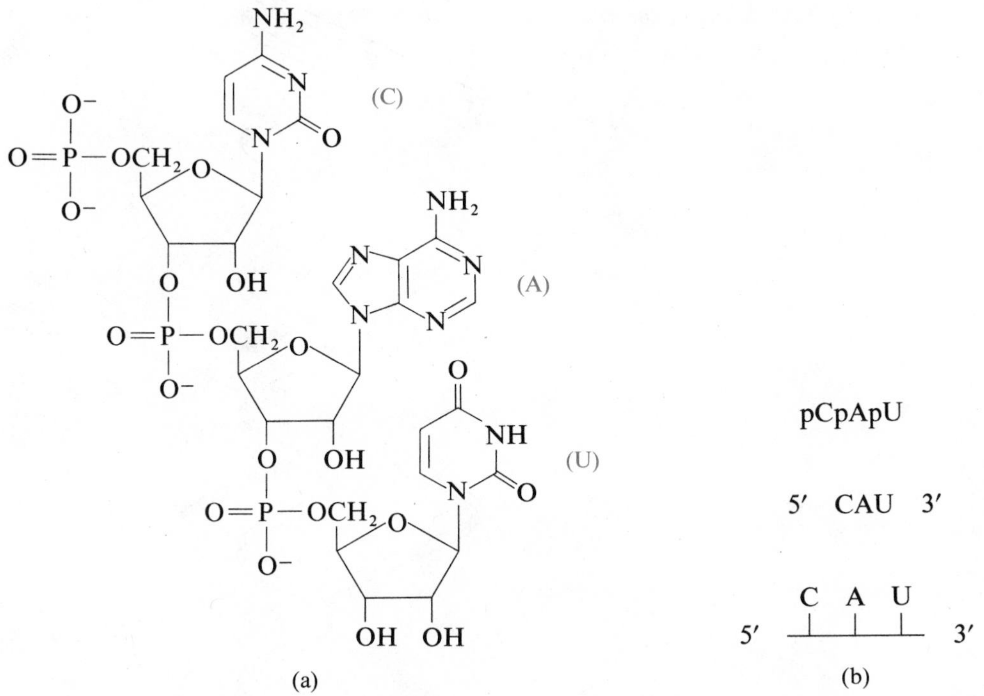

pCpApU

5′ CAU 3′

5′ C A U 3′

(a)   (b)

20-3.2 SYNTHESIS OF RNA (TRANSCRIPTION)

As has already been mentioned, DNA must not only replicate so that a cell can transmit its information to its two daughter cells when it divides, but it must also pass its genetic information to RNA molecules so that protein synthesis can occur. The process of passing the genetic information carried by DNA molecules to RNA molecules is called **transcription**. Replication refers to making an exact copy, just as you might make a carbon copy of a letter or run a document through a copying machine to make an exact copy. Transcription refers to putting the same words in another form. Just as you might transcribe spoken words into shorthand notes or shorthand notes into a typewritten record, the information contained in a sequence of deoxynucleotides (heterocyclic bases) in a DNA molecule is transcribed into a sequence of nucleotides (bases) on an RNA molecule, a molecule that contains the same language in a different form.

Transcription is the production of an RNA molecule that is the complement of a segment of a DNA molecule.

The process of transcription is quite similar to the process of replication. Most DNA molecules carry several genes, but the cell may only need to express one at a given time. As a result, RNA polymerase, the enzyme that will catalyze the

Figure 20.18 The opening up of a DNA double helix following attachment of the enzyme RNA polymerase and synthesis of RNA on one of the two DNA strands.

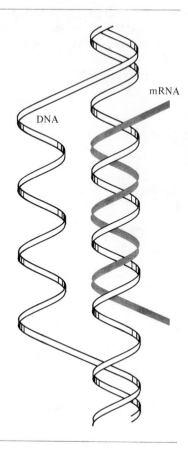

mRNA

DNA

synthesis of RNA on a DNA **template,** must locate the proper place on the DNA molecule to begin the process of transcription. It does this by recognizing a specific region (specific nucleotide sequence) on the DNA molecule. This region of a DNA molecule containing the start signal is called a *promoter,* which is subdivided into an initial recognition site and a binding site.

Because DNA molecules serve as templates for RNA molecules, the two strands of DNA must separate. Separation of the two DNA strands as shown in Figure 20.18 occurs only in the specific region where the RNA polymerase molecule has attached. Only one of the two polynucleotide strands of DNA acts as a template for the synthesis of an RNA molecule. This strand is called the *sense strand.* The other, complementary strand is called the *antisense strand.*

The process of the synthesis of RNA is quite similar to that of the synthesis of DNA. Propagation of the RNA chain also occurs by addition of nucleotides to the 3′ end of the growing polynucleotide chain (see Figure 20.19). We see from Figure 20.19 that nucleotides are added by means of reaction with nucleoside triphosphates. The energy stored in a high-energy phosphate bond is used to form a new phosphate ester bond. Inorganic pyrophosphate splits off in the process. We also

A **template** is a pattern or mold. In the case of transcription, a strand of DNA provides the pattern for the synthesis of an RNA molecule.

see that, just as in the synthesis of DNA, the pyrophosphate ion (PP_i) is immediately cleaved by hydrolysis, ensuring that the reaction is irreversible.

Also, as in the synthesis of DNA, the nucleotides are positioned along the DNA template strand by base pairing. The pairing of complementary bases is the same as in DNA with purines pairing with pyrimidines, but with uracil replacing thymine. Therefore, when adenine is found on the DNA template, uracil will be

Figure 20.19 The addition of a UMP unit to a strand of RNA that has GMP at the growing end (3' end).

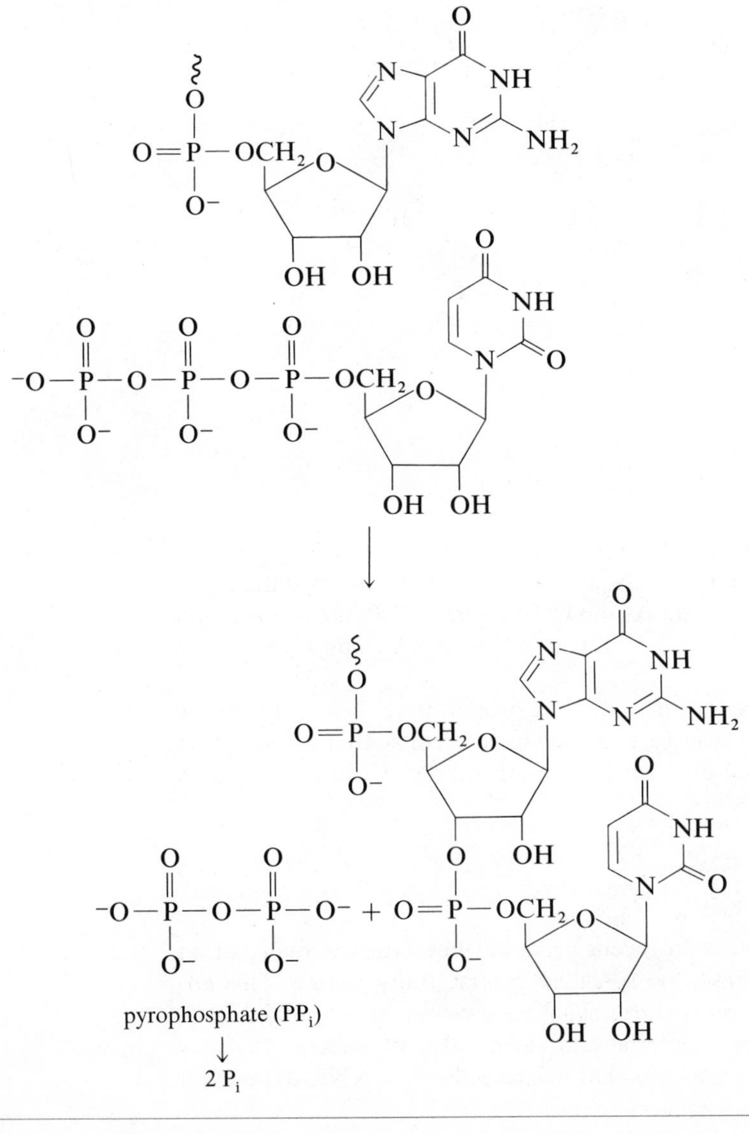

Figure 20.20   Diagram of transcription of a short segment of a single strand of DNA showing the addition of nucleotides to the growing end (3′ end) of the RNA molecule. The DNA molecule is indicated by a straight line and the RNA molecule by a wavy line.

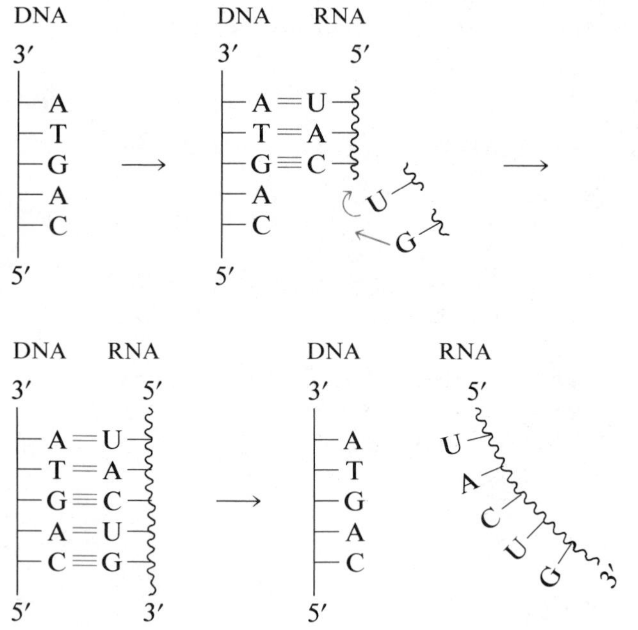

inserted into the RNA molecule. The pairing of guanine with cytosine is the same as in replication. The result is that the RNA product is complementary to the DNA template as we can see in Figure 20.20. As shown in Figure 20.20, the information on the DNA chain is read in the 3′ ⟶ 5′ direction. The RNA chain grows in the 5′ ⟶ 3′ direction.

Finally, transcription must be terminated at the proper place, just as it must be started (initiated) at the proper place. Termination must occur so that transcription is not interrupted in the middle of a gene and so it does not run over into an adjacent gene.

20-3.3 REVERSE TRANSCRIPTASE

One of the more exciting advances in recent years has been the discovery of an RNA-dependent DNA polymerase, the so-called reverse transcriptase. This enzyme is associated with *RNA tumor viruses*. These are viruses that contain RNA rather than DNA as their genetic material and are capable of infecting various eukaryotic cells, transforming them into cancer-like cells. The RNA-dependent

DNA polymerase is capable of catalyzing the synthesis of single-stranded DNA using viral RNA as a template. The DNA molecules so formed can direct the synthesis of a complementary strand and form a double helix. These DNA molecules then direct the synthesis of more RNA for the assembly of more virus particles.

20-3.4 BLOCKING OF REPLICATION AND TRANSCRIPTION

As stated in Table 17.1, *antibiotic* is a general term that refers to a substance produced by one organism that has an inhibitory effect on another organism. Antibiotics have a variety of chemical structures and act in a variety of ways. Some prevent replication of DNA, thus blocking cell division. Preventing cells from dividing stops the growth of microorganisms (bacteria) and cancer cells. Others prevent transcription, thus blocking the expression of the genes required for cellular activities. Still other antibiotics act in other ways, for example, by blocking synthesis of bacterial cell wall materials.

Chemists have made synthetic compounds that block replication, transcription, and other cellular activities. Some of these were made intentionally, some unintentionally. As might be expected, many of these antibiotics and synthetic drugs are quite toxic to animals and humans. Those that are less toxic can be used to treat bacterial infections and cancers.

A *tumor* is a mass of new tissue that grows independently. A tumor results from uncontrolled division of cells that should not be dividing. A substance that will block division of these cells is needed to stop the growth of a tumor. Infections are treated with substances that will destroy or stop the division of bacterial and fungal cells. Some of the compounds used are listed in Table 20.5. The search for better chemotherapeutic agents is continuing.

Chemotherapeutic refers to the treatment of disease with chemicals that have a specific toxic effect.

Certain antibiotics, such as adriamycin, daunomycin, and rifampicin, are inhibitors of reverse transcriptase. These antibiotics are being investigated for the treatment of cancers caused by RNA tumor viruses.

PROBLEMS

8. Using the one-letter abbreviations for the purine and pyrimidine bases, write the structure 5′ GUG 3′
9. Write the RNA base sequence obtained by transcribing the upper chain of the following DNA segment.

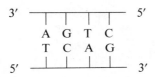

10. List the structural differences between DNA and RNA.
11. Define transcription.

Table 20.5   Selected Inhibitors of Replication and Transcription

Agent	Inhibition	Site of Action	Use
Alkylating Agents			
Nitrogen mustards (chlorambucil, cyclophosphamide, and uracil mustards) BCNU Busulfan	Replication and transcription	Alkylate purines bringing about depurination and cross-linking of DNA*	Cancer chemotherapy
Anthracycline antibiotics			
Actinomycin	Transcription	Binds to DNA (inhibits RNA polymerase)	Laboratory research
Adriamycin	Replication and transcription	Binds to DNA (inhibits DNA polymerase and RNA polymerase)	Cancer chemotherapy
Daunomycin (daunorubicin)	Replication and transcription	Binds to DNA (inhibits DNA polymerase and RNA polymerase)	Cancer chemotherapy
Colchicine	Cell division	Mitotic spindle proteins	Laboratory research
Folic acid analogs			
Aminopterin	Synthesis of dTMP and purine nucleotides (required for DNA synthesis)	Folic acid reductase	Cancer chemotherapy†
Methotrexate	Synthesis of dTMP and purine nucleotides (required for DNA synthesis)	Folic acid reductase	Cancer chemotherapy
Pyrimidine derivatives			
Arabinosylcytosine (AraC)	DNA synthesis	DNA polymerase	Cancer chemotherapy
5-Fluorouracil	Synthesis of dTMP (required for DNA synthesis)	Thymidylate synthetase‡	Cancer chemotherapy
Rifamycin, rifampicin	RNA synthesis	Bacterial RNA polymerase	Treatment of bacterial infections
Vincristine, vinblastine	Cell division	Mitotic spindle proteins	Cancer chemotherapy

*Depurination refers to the removal of adenine and guanine residues. Cross linking refers to the covalent joining of DNA strands.
†Leukemia chemotherapy.
‡Thymidylic acid (thymidylate) is an alternative name for deoxythymidine monophosphate (dTMP). In this case, the common name of the enzyme uses this common name for the nucleotide.

ANSWERS TO PROBLEMS

8.

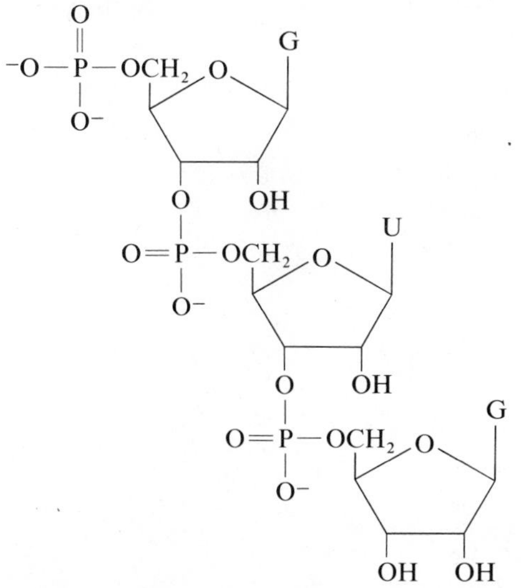

$$\overset{\displaystyle O}{\underset{\displaystyle O^-}{\overset{\displaystyle \|}{-O-P}}}-OCH_2 \quad O \quad G$$

OH

$$O=\overset{\displaystyle }{\underset{\displaystyle O^-}{P}}-OCH_2 \quad O \quad U$$

OH

$$O=\overset{\displaystyle }{\underset{\displaystyle O^-}{P}}-OCH_2 \quad O \quad G$$

OH   OH

9.  5′   U C A G   3′

---

10. (a) DNA contains 2-deoxy-D-ribofuranose and RNA contains D-ribofuranose. **(b)** RNA contains uracil in place of thymine, which is found in DNA. **(c)** DNA is double stranded and RNA is single stranded. **11.** Transcription is the biosynthesis of an RNA molecule using the sequence of bases in a DNA molecule to direct the sequence of bases in the RNA molecule.

SUMMARY

In this chapter, we have found that nucleic acids are polymers of nucleotides and that there are two nucleic acids, DNA and RNA. By hydrolysis, DNA (deoxyribonucleic acid) can be converted into four nucleotides: deoxyadenosine monophosphate (dAMP), deoxyguanosine monophosphate (dGMP), deoxycytidine monophosphate (dCMP), and deoxythymidine monophosphate (dTMP). All four of these nucleotides (called deoxyribonucleotides) contain 2-deoxy-D-ribofuranose as the sugar unit. RNA (ribonucleic acid) can also be converted into four nucleotides by hydrolysis, but these four nucleotides are different from those obtained from DNA. They are the ribonucleotides adenosine monophosphate (AMP), guanosine monophosphate (GMP), cytidine monophosphate (CMP), and uridine monophosphate (UMP). All these nucleotides contain D-ribofuranose as the sugar unit. DNA usually consists of two complementary polynucleotide chains twisted together into a double helix. RNA molecules usually occur as single strands.

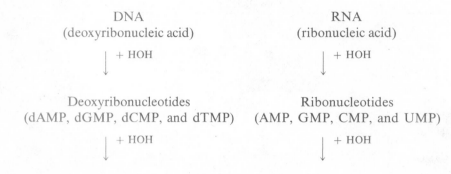

We have learned that both RNA and DNA molecules are synthesized by the reaction of a nucleoside triphosphate (a ribonucleoside triphosphate for RNA and a deoxynucleoside triphosphate for DNA) with the 3′ end of a polynucleotide chain. The energy in a high-energy phosphate bond is used to form a phosphate ester bond, and, thus, to add a nucleotide to the 3′ end. For both DNA and RNA, the incoming nucleotide is positioned by pairing with a complementary purine or pyrimidine base on a DNA molecule. Adenine pairs with thymine (in DNA synthesis) or uracil (in RNA synthesis), and guanine pairs with cytosine in both. Substances that block the synthesis of DNA or RNA will block cell division and can be used to treat bacterial or fungal infections and in cancer chemotherapy.

We also learned that DNA molecules contain the information for the primary structure of proteins in segments called genes. When a cell divides, an exact copy of this genetic information is made so that each daughter cell contains the same genes that were in the original cell. The process by which exact copies of DNA molecules are made is called replication. The information contained in the genes (in segments of DNA molecules) is transferred to RNA molecules in a process known as transcription. In higher organisms (those organisms whose cells have a nucleus), the DNA molecules are in the nucleus, and replication and transcription take place in the nucleus. The RNA molecules made by transcription then pass from the nucleus into the cytoplasm of the cell where they direct the synthesis of protein molecules.

We will examine how RNA molecules direct the synthesis of protein molecules in Chapter 22 after we have looked at the structures of proteins in Chapter 21.

ADDITIONAL PROBLEMS

12. Using the structures of the purine and pyrimidine bases given in Figures 20.4 and 20.5, write the complete structures of the following segments of RNA molecules.

(a) 5′   C A U A   3′

(b) 5′ GUU 3′

(c) 5′ pApGpU 3′

13. Classify each of the following molecules as oligo-ribonucleotides or oligodeoxyribonucleotides.
    (a) GUG (b) CTA
    (c) TAG (d) UAA
    (e) CAU

14. Use the segment of a DNA molecule shown to answer the following questions.

    5′ G–T–C–A–A–T–G–G–A 3′

    (a) What would be the product of replication?
    (b) What would be the product of transcription?

15. Complete the following table.

	DNA	RNA
Sugar unit		
Purine bases		
Pyrimidine bases		

16. Define nucleoside, complementary bases, double helix, and base pair.

17. Describe the relationship of a nucleotide to a nucleic acid.

CHAPTER 21

AMINO ACIDS AND PROTEINS

Biochemistry is the study of the molecular basis of life. In earlier chapters, we discussed a number of the molecules that play roles in life processes. In Chapters 19 and 20, we examined in some detail two of the four main classes of biologically important molecules (the carbohydrates and the nucleic acids). In this chapter, we will examine a third main class of molecules essential to life, the proteins (see Sections 5-1.2 and 17-4.1).

Proteins, like nucleic acids and polysaccharides, are polymers. These naturally occurring polymers are often referred to as macromolecules.* In Section 17-4.1, we learned that proteins are naturally occurring polymers of amino acids joined together by amide (peptide) bonds. Proteins are required for virtually all biological processes. They can have a variety of roles because they can have a variety of structures. In this chapter, we will look at some differences in the structures of protein molecules. We will study how the structure of a protein molecule controls its shape and how the shape of the protein allows it to carry out its specific function. We will begin by looking at the basic building blocks of protein molecules, the amino acids.

---

*Macro- is a prefix meaning *large*. Hence, a *macromolecule* is a *large molecule*. The opposite of macro- is *micro-*, meaning *small*, as in *microorganism*. These two prefixes are widely used in biology and biochemistry.

21-1 AMINO ACIDS

Amino acids are important constituents of living cells. Quantities of free amino acid molecules are not usually found in living systems; rather, most amino acids are found as constituents of protein molecules. The peptide bonds of proteins can be broken to release the amino acids making up the protein molecules by acid-, base-, or enzyme-catalyzed hydrolysis (see Section 16-4.1). We already know that amino acids are both acids and bases and, therefore, ionic organic molecules.

21-1.1 STRUCTURES OF AMINO ACIDS

Most protein molecules are composed of 20 amino acids. Amino acids that are constituents of proteins are L-α-amino acids. They are called α-amino acids because they have an amino group on the carbon atom that is adjacent to the carboxyl group (the alpha carbon atom*). Thus, α-amino acids have both an amino group and a carboxyl group on the same carbon atom. This puts four different groups on this alpha carbon atom (in all amino acids with the exception of glycine which has two hydrogen atoms on the alpha carbon atom) and makes it a chiral carbon atom (see Section 18-4). The configuration of this chiral atom is L. Recall that this means that the amino group is on the left when the carboxyl group is on the top and the alpha carbon atom is oriented in a Fischer projection so that the vertical bonds project away from you and the horizontal bonds project toward you (see Section 18-6). The general structure of an L-α-amino acid in its ionized form is shown.

$$\begin{array}{c} O \\ \parallel \\ C-O^- \\ | \\ H_3\overset{+}{N}-C-H \\ | \\ R \end{array}$$

L-α-amino acid

The structures of the 20 amino acids most commonly found as constituents (monomer units) of proteins are given in Table 21.1. In addition, there are many more L-amino acids that do not occur in proteins. We will discuss some of these in Chapter 28.

Only L-amino acids are found in proteins. However, there are more than 20 known D-amino acids that occur naturally, including the D-alanine and D-glutamic acid of certain bacterial cell walls and a variety of D-amino acids in antibiotics.

---

*Organic chemists frequently number carbon atoms according to their distance from a functional group with letters of the Greek alphabet. The carbon atom adjacent to the functional group is designated with the letter alpha. For example,

$$\overset{\epsilon}{C}-\overset{\delta}{C}-\overset{\gamma}{C}-\overset{\beta}{C}-\overset{\alpha}{C}-COOH$$

Table 21.1   L-α-Amino Acids Found As Constituents of Most Proteins

I. Amino Acids with Nonpolar (Hydrophobic) Groups
   (a) Amino acids with aliphatic groups

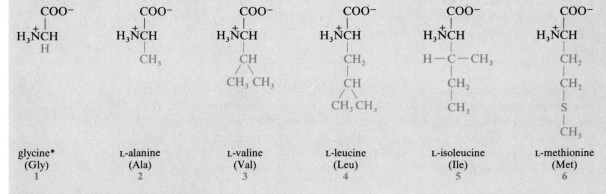

| glycine*
(Gly)
1 | L-alanine
(Ala)
2 | L-valine
(Val)
3 | L-leucine
(Leu)
4 | L-isoleucine
(Ile)
5 | L-methionine
(Met)
6 |

   (b) Amino acids with aromatic groups

L-phenylalanine
(Phe)
7

L-tryptophan
(Trp)
8

II. Amino Acids with Polar (Hydrophilic) Groups
   (a) Amino acids with uncharged polar groups at pH 6–7

| L-serine
(Ser)
9 | L-threonine
(Thr)
10 | L-cysteine†
(Cys)
11 | L-tyrosine‡
(Tyr)
12 | L-asparagine
(Asn)
13 | L-glutamine
(Gln)
14 |

*Glycine is the only amino acid without a chiral carbon atom.
†The amino acid cystine (Cys—Cys) consists of two cysteine residues linked by a disulfide bond. It is sometimes
listed as a separate amino acid.
‡Tyrosine could also be classified as an aromatic amino acid (group I–b).

Table 21.1 L-α-Amino Acids Found as Constituents of Most Proteins (*cont'd*)

(b) Amino acids with carboxyl-containing groups (groups that are negatively charged at pH 6–7)

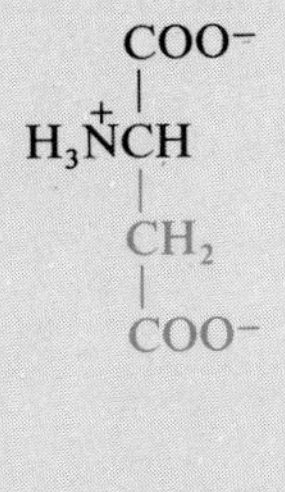

$$COO^-$$
$$H_3\overset{+}{N}CH$$
$$CH_2$$
$$COO^-$$

L-aspartic acid
(Asp)
15

$$COO^-$$
$$H_3\overset{+}{N}CH$$
$$CH_2$$
$$CH_2$$
$$COO^-$$

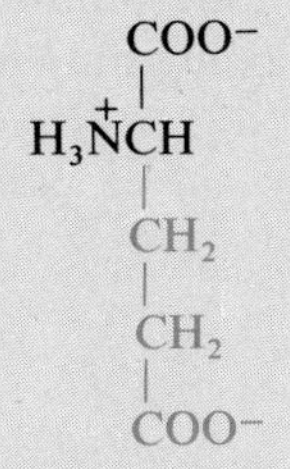

L-glutamic acid
(Glu)
16

(c) Amino acids with basic groups (groups that are positively charged at pH 6–7)

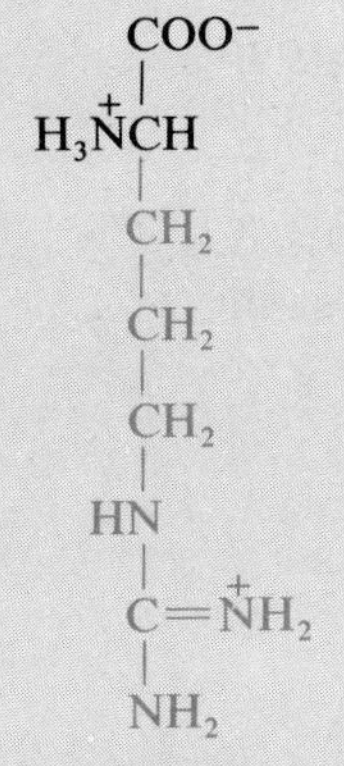

$$COO^-$$
$$H_3\overset{+}{N}CH$$
$$CH_2$$
$$CH_2$$
$$CH_2$$
$$HN$$
$$C\!=\!\overset{+}{N}H_2$$
$$NH_2$$

L-arginine
(Arg)
17

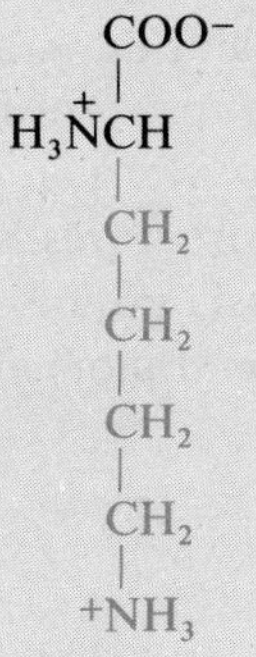

$$COO^-$$
$$H_3\overset{+}{N}CH$$
$$CH_2$$
$$CH_2$$
$$CH_2$$
$$CH_2$$
$$^+NH_3$$

L-lysine
(Lys)
18

$$COO^-$$
$$H_3\overset{+}{N}CH$$
$$CH_2$$
$$HN \quad NH$$

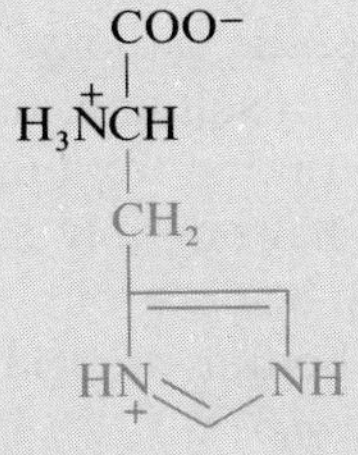

L-histidine
(His)
19

(d) Imino acids*

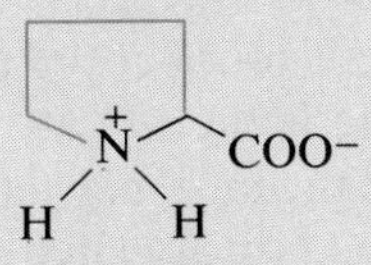

$$N\!-\!COO^-$$
$$H \quad H$$

L-proline†
(Pro)
20

---

*An imino group is $=$NH (see Section 17-2.1).
†Proline (and hydroxyproline) are the only constituents of proteins that are imino acids rather than amino acids. Proline is nonpolar and could be placed in group I–a.

Amino acids with nonpolar (hydrocarbon) R groups are less soluble in water than those with polar R groups. Amino acids with R groups that tend to be ionic at the pH of most living cells (pH about 7) are the most polar and most soluble. One polar amino acid is glutamic acid. We have all heard of monosodium glutamate, or MSG. Monosodium glutamate is the monosodium salt of glutamic acid. It is used to enhance the flavor of food products containing meats.

$$
\begin{array}{c}
COO^- \\
\overset{+}{H_3}NCH \\
CH_2 \\
CH_2 \\
COO^-Na^+
\end{array}
$$

monosodium glutamate

21-1.2 ISOELECTRIC pH AND ZWITTERION

The general structural formula of an amino acid clearly shows that all amino acids contain at least two ionizable groups, the carboxyl group and the α-amino group (or, for proline, the α-imino group). In solution, these groups form internal salts through the movement of the acidic proton from the carboxylic acid group to the amino group (a Brønsted base) to form the doubly charged structures we have used. Amino acids, therefore, contain both acidic groups and basic groups on the same molecule. Molecules that can serve both as an acid (proton donor) and as a base (proton acceptor) are said to be *amphoteric*.

$$
\begin{array}{cc}
\overset{O}{\underset{\parallel}{R-C-OH}} \rightleftharpoons \overset{O}{\underset{\parallel}{R-C-O^-}} + H^+ \\
R-NH_3^+ \rightleftharpoons R-NH_2 + H^+
\end{array}
$$

protonated or   *unprotonated or*
acid forms   *base forms*

We learned in Section 8-3.2 that the pK is the negative logarithm of the ionization constant. The pK of a carboxylic acid is also the pH at which, in a population of molecules, one-half of the molecules are in the un-ionized (protonated) state and one-half of the molecules are in the ionized (unprotonated) state. As the pH value drops below the pK, the equilibrium shifts so that there is more of the protonated form than the unprotonated form. The pK of an amine is the pH at which, in a population of molecules, one-half of the molecules are in the un-ionized (unprotonated) state and one-half of the molecules are in the ionized (protonated) state.

As the pH value rises above the pK, the equilibrium shifts to the other side. As the pH value drops to 1 pH unit below the pK, the group in question is 90% in the protonated form; at 2 pH units below the pK, the group is 99% in the protonated form. At 1 and 2 pH units above the pK, the group in question is 90% and 99%, respectively, in the base form.

The pK of the carboxyl group of amino acids is about 2.1; therefore, at pH 2.1, one-half of the carboxyl groups exist as R—COOH and one-half exist as R—COO⁻. The pK for the amino group of amino acids is about 9.8; therefore, at pH 9.8, the amino group on one-half of the molecules will be in the form R—NH₂ and the amino group on the other half of the molecules will be in the form R—NH₃⁺.

We know that the pH of a living cell is characteristic of that cell and that the pH value of most human cells is close to 7.0. At this pH, the carboxyl group is almost entirely in the form of the carboxylate ion (R—COO⁻) and the amino group is also almost entirely ionized (R—NH₃⁺). Therefore, at the pH of most cells, the general structure of L-α-amino acids is that given in Section 21-1.1. This structure is called a **zwitterion**. The un-ionized form of an amino acid, as shown, cannot exist at any pH value.

A **zwitterion** is an amino acid molecule that has one positive and one negative charge.

$$\underset{\displaystyle\text{R}}{\overset{\displaystyle\text{COOH}}{\vert\ \ \ \ \vert}}$$

COOH
|
H₂NCH
|
R

un-ionized amino acid

We have shown the various ionic forms of L-alanine. At the pH half way between the two pK values, both the carboxyl group and the amino group are predominantly ionized, and at any instant, there will be equal numbers of minus and plus charges in the population of molecules. At this pH, called the **isoelectric pH (pI)**, the net charge on the population of molecules is zero. For amino acids with uncharged R groups, the isoelectric pH is the pH at which there is a maximum concentration of the zwitterion form.

The **isoelectric pH** is the pH of a solution at which an amino acid or protein molecule has an equal number of positively and negatively charged groups so that the molecule has zero net electric charge overall.

COOH   COO⁻   COO⁻
⁺| ⁺| |
H₃NCH   ⇌   H₃NCH   ⇌   H₂NCH
| −H⁺ | −H⁺ |
CH₃   +H⁺   CH₃   +H⁺   CH₃

pK = 2.1   pK = 9.8

in strong acid   at pH 5.95,   in strong base
(below pH 1),   net charge = 0   (above pH 11),
net charge = +1   net charge = −1

L-alanine

Recall from Table 21.1 that many amino acids also have other ionizable groups in the R group. The various ionic forms of one of these amino acids can be indicated using L-aspartic acid as a model as shown. pH 3.0 is the isoelectric pH of L-aspartic acid. Because pH 3.0 is half way between pK 2.1 and pK 3.9 (the pKs for the ionization of the two carboxyl groups), it is the pH at which the net charge on the population of molecules is zero.

$$pI = \frac{2.1 + 3.9}{2} = 3.0$$

$$\begin{array}{ccccccc}
\text{COOH} & & \text{COO}^- & & \text{COO}^- & & \text{COO}^- \\
| & & | & & | & & | \\
\overset{+}{H_3}\text{NCH} & & \overset{+}{H_3}\text{NCH} & & \overset{+}{H_3}\text{NCH} & & H_2\text{NCH} \\
| & \underset{+H^+}{\overset{-H^+}{\rightleftharpoons}} & | & \underset{+H^+}{\overset{-H^+}{\rightleftharpoons}} & | & \underset{+H^+}{\overset{-H^+}{\rightleftharpoons}} & | \\
\text{CH}_2 & & \text{CH}_2 & & \text{CH}_2 & & \text{CH}_2 \\
| & & | & & | & & | \\
\text{COOH} & & \text{COOH} & & \text{COO}^- & & \text{COO}^-
\end{array}$$

pK = 2.1　　　　　pK = 3.9　　　　　pK = 9.8

in strong acid　　at pH = 3.0,　　at pH 6.85,　　in strong base
(below pH 1),　　net charge = 0　　net charge = −1　　(above pH 11),
net charge = +1　　　　　　　　　　　　　　　net charge = −2

L-aspartic acid

The concept of isoelectric pH is more important to the understanding of the behavior of proteins than of the reactions of amino acids, but amino acids serve as a convenient model to describe this property of amphoteric molecules. We will look at isoelectric pHs of proteins later. We will see then that proteins, like amino acids, are least soluble at their isoelectric pH. This is true because, at the isoelectric pH, the positively charged group on one molecule is attracted to the negatively charged group on another molecule and aggregates are formed. At a pH value either above or below the isoelectric pH, the molecules have either a net negative charge or a net positive charge and repel each other. The molecules have less tendency to aggregate at these pH values and so, they are more soluble.

PROBLEM

1. Write structures of amino acids with the following R groups in the zwitterionic form.

(a)
$$\begin{array}{c} | \\ \text{CH}_2 \\ | \\ \text{CH}_2 \\ | \\ \text{C}-\text{NH}_2 \\ \| \\ \text{O} \end{array}$$

(b)
$$\begin{array}{c} | \\ \text{CH}_2 \\ \end{array}$$
(benzene ring)

(c)
$$\begin{array}{c} | \\ \text{H}-\text{C}-\text{CH}_3 \\ | \\ \text{CH}_2 \\ | \\ \text{CH}_3 \end{array}$$

(d)
$$\begin{array}{c} | \\ \text{CH}_2\text{OH} \end{array}$$

(e)
$$\begin{array}{c} | \\ \text{CH}_2\text{SH} \end{array}$$

(f)
$$\underset{+}{\text{HN}} \diagdown \diagup \text{NH}$$

ANSWERS TO PROBLEM   **1. (a)**
$$\begin{array}{c} \text{COO}^- \\ | \\ \overset{+}{H_3}\text{NCH} \\ | \\ \text{CH}_2 \\ | \\ \text{CH}_2 \\ | \\ \text{C}-\text{NH}_2 \\ \| \\ \text{O} \end{array}$$
(b)
$$\begin{array}{c} \text{COO}^- \\ | \\ \overset{+}{H_3}\text{NCH} \\ | \\ \text{CH}_2 \end{array}$$
(benzene ring)
(c)
$$\begin{array}{c} \text{COO}^- \\ | \\ \overset{+}{H_3}\text{NCH} \\ | \\ \text{H}-\text{C}-\text{CH}_3 \\ | \\ \text{CH}_2 \\ | \\ \text{CH}_3 \end{array}$$

(d) 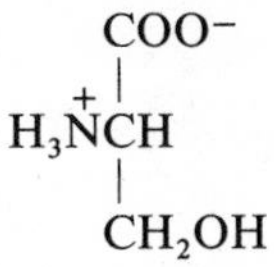 $\overset{\displaystyle COO^-}{\underset{\displaystyle CH_2OH}{\overset{|}{\underset{|}{H_3\overset{+}{N}CH}}}}$
    (e) 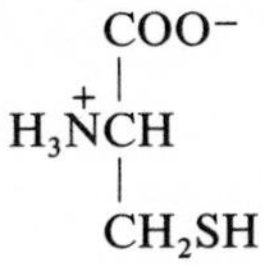 $\overset{\displaystyle COO^-}{\underset{\displaystyle CH_2SH}{\overset{|}{\underset{|}{H_3\overset{+}{N}CH}}}}$
    (f) 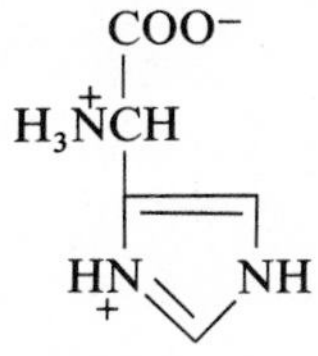 $\overset{\displaystyle COO^-}{\overset{|}{H_3\overset{+}{N}CH}}$

(f) R group:

HN   NH
$\overset{+}{}$

21-2 STRUCTURES AND FUNCTIONS OF PROTEINS

LEARNING OBJECTIVES

1. Define peptide, peptide bond, protein, polypeptide, N-terminal end, C-terminal end, primary structure, secondary structure, α-helix, β-pleated sheet, tertiary structure, quaternary structure, globular protein, fibrous protein, denaturation, and glycoprotein.
2. Write the complete structure of a peptide, given the sequence of the peptide and the structures of the R groups.

3. Determine the structure of a peptide, given the structures of fragments from it.
4. Describe four kinds of bonding that contribute to the tertiary structure of a protein.
5. Using hemoglobin as an example, explain how the sequence of a polypeptide influences its biological activity.

We already know that proteins are long chains of amino acids, that the amino acids are joined together by amide bonds (Section 17-4.1) between the amino and carboxyl groups (see Figure 21.1), and that the amide bonds linking amino acids

Figure 21.1
General formula for a segment of a protein molecule. The atoms involved in the peptide bond are shown in color.

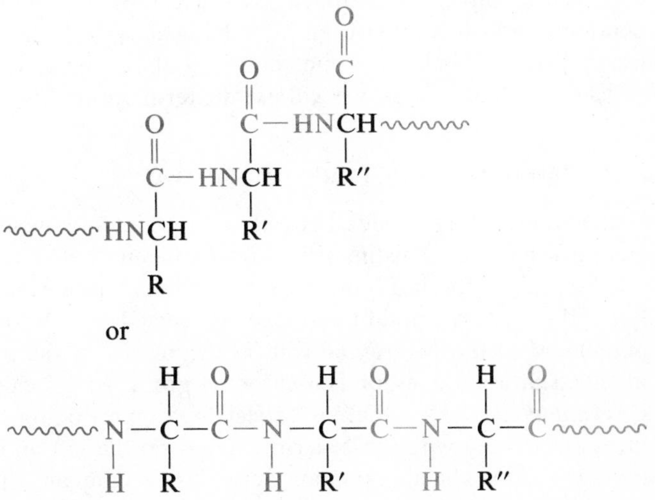

Table 21.2 Oligopeptides

Oligopeptides	Number of Amino Acids
Dipeptides	2
Tripeptides	3
Tetrapeptides	4
Pentapeptides	5
Hexapeptides	6
Heptapeptides	7
Octapeptides	8
Nonapeptides	9
Decapeptides	10

together are called *peptide bonds.* We will now consider the structures of protein molecules in a little more detail. We will discuss the relationship between the structures of proteins and their functions.

Chains of amino acids can be various lengths. When a few (usually 2–10) amino acids are joined together by peptide bonds, the molecule is called an *oligopeptide* just as a molecule of 2–10 sugar units joined together is an oligosaccharide and a molecule of a few nucleotide units joined together is an oligonucleotide. Oligopeptides are named according to the number of amino acids contained in them (Table 21.2). The same prefixes that are used in naming oligosaccharides are used in naming oligopeptides.

The term *polypeptide* is not always used in the same way by biochemists. It most often is used to describe (1) any naturally occurring chain of 10–100 amino acids, (2) any chain of more than 10 amino acids produced by breakdown of a **native protein** molecule, and (3) any one chain of a native protein that is made up of two or more chains of amino acids joined by disulfide bonds. For example, each of the two chains making up the insulin molecule (Figure 21.2) is called a polypeptide. The term *protein* is used only to describe a naturally occurring polymer of amino acids and is sometimes limited to those naturally occurring chains of more than 100 amino acids. However, because the word *polypeptide* refers to many peptide bonds, it is often used interchangeably with the term *protein.* In general usage, therefore, there is some confusion about the actual definition of the term *polypeptide.* In this book, we will use the term in the three ways listed above.

A **native protein** is a protein as it occurs in a living cell without any change in its chemical structure or shape.

21-2.1 PRIMARY STRUCTURES OF PROTEINS

Proteins and other peptides, like polysaccharides and nucleic acids, have two different ends that can be distinguished from one another. The two ends of the tripeptide Ser—Tyr—Lys in Figure 21.3 are labeled. Upon examination, we see that the left end of the tripeptide has a free (unreacted) α-amino group and is called the **amino end,** or the **N-terminal end.** The right end of the tripeptide has a free (not involved in a peptide bond) α-carboxyl group and is called the **carboxyl end** or **C-terminal end.** Unless there is a special reason for doing otherwise, peptide structures are written with the N-terminal end on the left and the C-terminal end on the right. Thus, the abbreviation Ser—Tyr—Lys means that L-serine is the amino acid with a free amino group and L-lysine is the amino acid at the C-terminal end.

The **N-terminal end (amino end)** is the end of a peptide having the only amino acid with the α-amino group not involved in a peptide bond.

The **C-terminal end (carboxyl end)** is that end of a peptide having the only amino acid with an α-carboxyl group not involved in a peptide bond.

Figure 21.2 Amino acid sequence of human insulin. The A chain contains 21 amino acids. The B chain contains 30 amino acids.

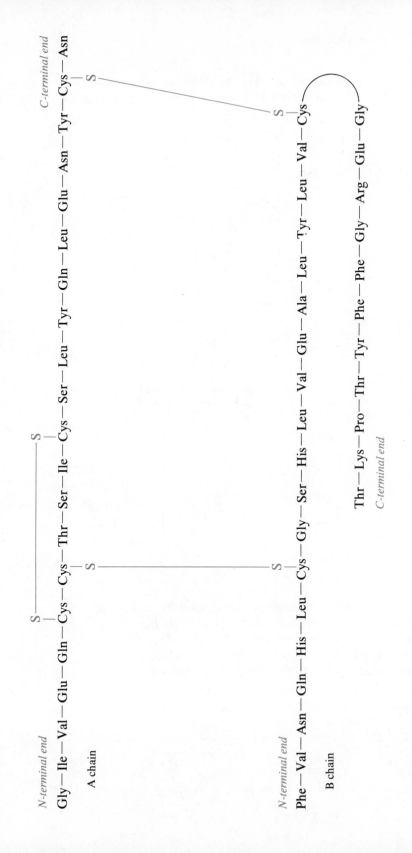

Figure 21.3   Complete structure of the tripeptide Ser—Tyr—Lys, showing the amino and carboxyl ends, and its complete hydrolysis. The ionic forms of the peptide and amino acids are those that would predominate at pH 7.

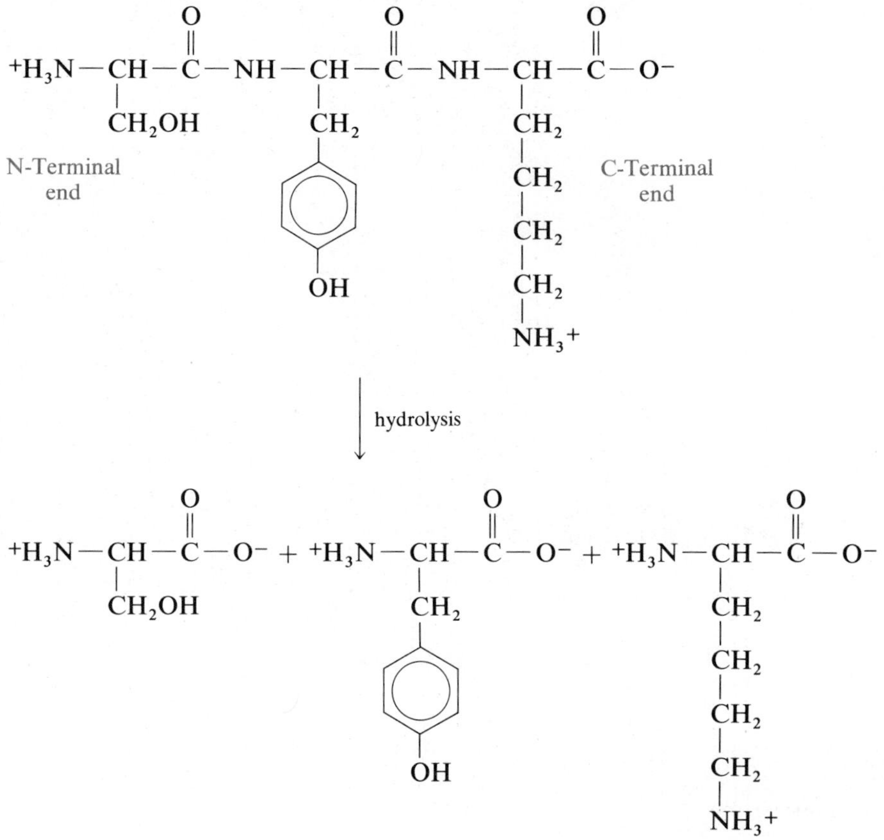

In Figure 21.2 we have also shown the release of amino acids by acid-, base-, or enzyme-catalyzed hydrolysis of the peptide bonds. Hydrolysis (see Section 16-4.1) breaks the carbon-to-nitrogen bond between $\diagup C{=}O$ and $-NH-$ by adding the elements of water: OH to $\diagup C{=}O$ to form $-COOH$ (a carboxyl group), which then ionizes, and H to $-NH-$ to form $-NH_2$ (an amino group), which then is protonated.

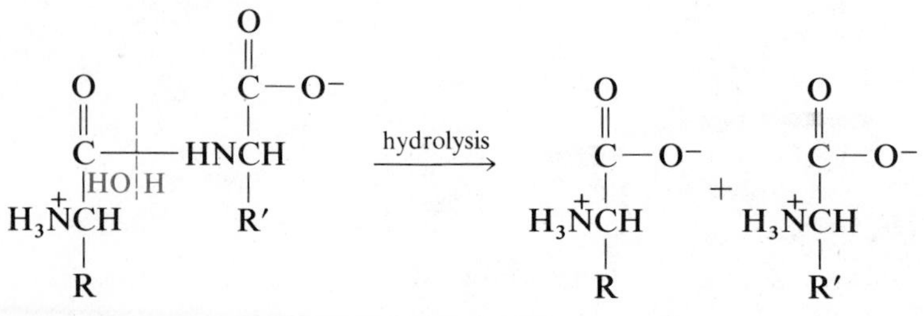

A formula such as that in Figure 21.2 that shows the number, kind, and sequence (order) of amino acids is the primary structure of a protein or peptide. The primary structure is the structure that describes which atoms are connected. It is the same sort of chemical structure that we have used to describe other organic molecules. The primary structures of peptides can be determined by biochemists in two different ways. One way is to use reagents or enzymes that remove one amino acid at a time from one of the ends. In this way, the order in which the amino acids are joined together can be determined. Sequential removal of amino acids from an end of a peptide works well with small peptides. For larger peptides, the method of sequential removal of amino acids is coupled with the method of determining overlapping sequences. By this method, a large peptide is cleaved into smaller peptides by a reagent or an enzyme that is specific for certain peptide bonds (peptide bonds involving a specific amino acid). The structure of the resulting peptides is then determined. The original peptide is then treated with a second reagent or enzyme that is specific for peptide bonds involving a different amino acid, and the structures of the resulting peptides are again determined. Finally, the biochemist looks for amino acid sequences common to peptides from the first reaction and peptides from the second reaction to determine how the peptides originally related to each other.

The **primary structure** of a protein molecule is the specific sequence of amino acids within it, including the location of disulfide bridges (see Section 14-3.6).

The determination of the amino acid sequence in an unknown peptide Z illustrates how the method of overlapping sequences works. Peptide Z is treated with the proteolytic enzyme trypsin. *Trypsin* is a digestive enzyme that catalyzes the hydrolysis of peptide bonds involving the carboxyl group of lysine and arginine. It is made in the pancreas and works in the small intestine in the digestion of proteins. We will take a closer look at trypsin in Section 28-1.3. Proteolytic enzymes are **proteases**. Three fragments are obtained when peptide Z is treated with trypsin. When the amino acid sequences in these three fragments are determined, it is found that they have the following structures:*

A **proteolytic enzyme** is an enzyme that catalyzes the breakdown of proteins (-*lytic* refers to a breaking down).

A **protease** is an enzyme that catalyzes the hydrolysis of peptide bonds of protein molecules.

Peptide	Structure
A	Ala—Thr
B	Gly—Lys
C	Leu—Phe—Glu—Arg

This information tells us something about the structure of the original peptide. What we do not know is the order of the fragments in the original peptide. There are six ways to combine three different amino acids, as shown. We know that the enzyme trypsin cleaves peptides at points involving the carboxyl groups of lysine and arginine, creating peptides such as peptides B and C with lysine or arginine at the carboxyl end. Thus, peptide A is the only one that could have been at the original C-terminal end (see final structure). Therefore, we know that only the two peptides shown in color are possible combinations.

A—B—C
A—C—B
B—A—C
B—C—A
C—A—B
C—B—A

peptide combinations

Now, the original peptide Z is treated with the proteolytic enzyme chymotrypsin. *Chymotrypsin* is another digestive enzyme that is made in the pancreas and acts in the small intestine. It catalyzes the hydrolysis of peptide bonds involving the carboxyl group of the three aromatic amino acids (phenylalanine, tyrosine, and

---

*As usual, the amino end is on the left.

tryptophan). When chymotrypsin is used to catalyze the hydrolysis of peptide Z, two fragments are obtained. When the amino acid sequences in the two fragments are determined, it is found that they have the following structures:

Peptide	Structure
D	Glu—Arg—Ala—Thr
E	Gly—Lys—Leu—Phe

The sequence in peptide D tells us that peptide C precedes peptide A because Glu—Arg of peptide C is in front of Ala—Thr of peptide A. The sequence in peptide E tells us that peptide B precedes peptide C because Gly—Lys of peptide B is in front of Leu—Phe of peptide C. Therefore, we now know that the sequence of the octapeptide Z is B—C—A. The structure of the complete peptide and the fragments formed from it are shown. T indicates the point of cleavage by trypsin and X, the point of cleavage by chymotrypsin.

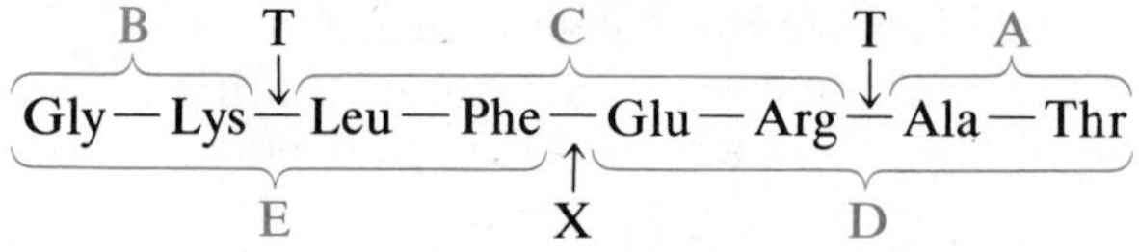

21-2.2 FUNCTIONS OF PROTEINS AND POLYPEPTIDES

Proteins have a wide variety of functions and are crucial to virtually all life processes. These functions can be categorized as follows:

1. *Structure*

Proteins provide the high tensile strength of tendons, bones, and skin. Cartilage, hair, wool, fingernails, and claws are largely composed of protein. Viruses have an outer layer of protein around a nucleic acid core. Proteins called histones are bound tightly to DNA in the cells of higher organisms helping to fold it and wrap it in an orderly fashion in chromosomes (see Section 20-2.3).

2. *Movement*

Proteins are the major components of muscles and are directly responsible for the ability of muscles to contract. The swimming of sperm is the result of contraction of protein filaments in their tails. The same is true of movements of particles within cells such as the movement of chromosomes during cell division.

3. *Catalysis*

Almost all chemical reactions in living organisms are catalyzed by enzymes. All enzymes are proteins. Enzymes will be discussed in Chapter 23.

4. *Transport*

Many molecules and ions are transported from one place to another by specific proteins. Oxygen is carried from the lungs to tissues by the protein hemoglobin in red blood cells (erythrocytes). Iron is transported in blood plasma by the protein transferrin, from the intestines where it is absorbed, to the spleen where it is stored, to the liver and bone marrow where it is used for synthesis. Proteins in the membranes of cells allow the passage of various molecules and ions through the membrane. Membranes and transport will be discussed in Section 26-2.1.

5. *Storage*

The protein ferritin in the liver, spleen, and bone marrow stores iron.

6. *Energy transformers, conductors, and trappers*

Myosin, a protein of muscle, transforms chemical energy of ATP into useful work, the contraction of the *actomyosin filament* (a complex of two proteins, actin and myosin). Rhodopsin, a protein of the retina of the eye, traps light energy and, with other membrane components, converts it into the electrical energy of a nerve impulse (see Section 15-3.1). Receptor protein molecules that combine with specific small molecules such as acetylcholine (see Section 17-3) are responsible for the transmission of nerve impulses at **synapses.**

A **synapse** is the region of contact between adjacent nerve cells where a nerve impulse is transmitted from one cell to another.

7. *Protection*

Antibodies (immunoglobulins) make up the gamma-globulin fraction of the plasma. They are special proteins that are synthesized in response to foreign substances and cells, such as bacterial cells, and then bind to those substances or cells. The foreign material is called an *antigen* (antibody-generating substance). Antibodies provide us with immunity to various diseases such as measles (rubeola virus), mumps, chicken pox, tetanus, and diphtheria. We acquire antibodies either from having had the disease or by receiving inactivated viruses, such as in the Sabin polio vaccine, and bacteria, such as in tetanus toxoid. Hay fever and food allergies are also caused by the interaction of antibodies with antigen.

Interferon is a small protein made and released by cells exposed to a virus. It protects other cells against viral infection.

Blood-clotting proteins protect against bleeding (hemorrhage).

Some antibiotics are polypeptides.

8. *Control*

We learned in Section 10-1 that hormones are chemical substances produced in the body that have specific effects on the activity of certain organs. Some hormones are proteins. For example, human growth hormone is a protein. Some hormones are smaller molecules that are sometimes classified as polypeptides. Examples are insulin and glucagon, which are made in the pancreas and control carbohydrate metabolism. Some hormones are oligopeptides (for example, vasopressin and oxytocin) which we will discuss later in this chapter. Some are even smaller peptides.

Expression of genetic information is under the control of proteins.

9. *Buffering*

Proteins can neutralize both acids and bases because they contain both acidic and basic groups on the R groups. Therefore, they provide some buffering (see Section 8-4.3) for blood and tissues.

The wide variety of functions exhibited by protein molecules indicates a wide variety of structures. There are, in fact, a great number of different proteins. We know that a single cell may contain as many as 3000 different proteins and that there are more than 200 different proteins in human blood plasma.

The importance of differences in the sequences of amino acids can be seen in the activity of two nonapeptides that are made in the pituitary gland. The two peptides, vasopressin and oxytocin, differ in only two amino acids, yet have quite different activities. We can determine from Table 21.1 what the differences in side

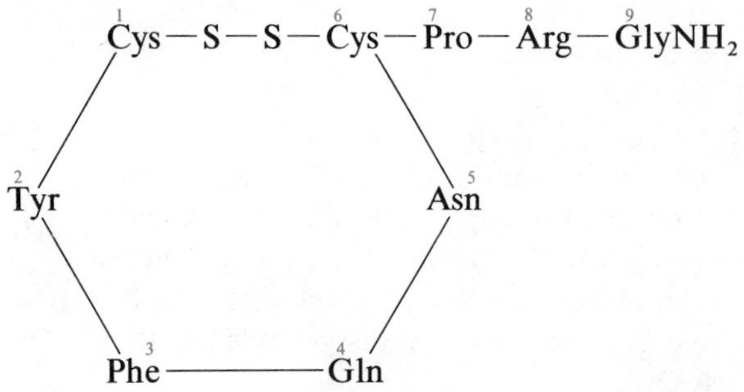

$$\overset{1}{Cys}-S-S-\overset{6}{Cys}-\overset{7}{Pro}-\overset{8}{Arg}-\overset{9}{Gly}NH_2$$

Tyr $\overset{2}{}$ Asn $\overset{5}{}$

Phe $\overset{3}{}$ —————— Gln $\overset{4}{}$

vasopressin

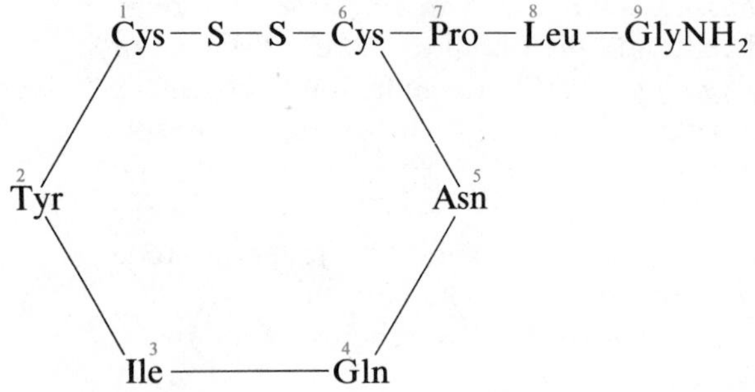

$$\overset{1}{Cys}-S-S-\overset{6}{Cys}-\overset{7}{Pro}-\overset{8}{Leu}-\overset{9}{Gly}NH_2$$

Tyr $\overset{2}{}$ Asn $\overset{5}{}$

Ile $\overset{3}{}$ —————— Gln $\overset{4}{}$

oxytocin

chains (R groups) would be. In both peptides, the cysteine residue at the N-terminal end and the cysteine residue at position 6* are joined by a disulfide bond (see Section 14-3.6). $GlyNH_2$ is glycinamide at the C-terminal end.

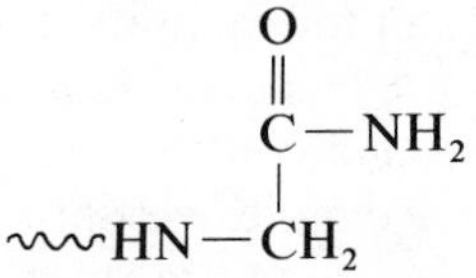

$$\overset{\displaystyle O}{\overset{\displaystyle \|}{\underset{\displaystyle |}{C}}}-NH_2$$
$$\sim\!\!\sim\!\!\sim HN-CH_2$$

glycinamide residue

Vasopressin is an **antidiuretic** hormone. It also raises blood pressure. Vasopressin has been used with other substances to elevate blood pressure when surgical shock occurs, in the management of hemorrhage occurring after childbirth, and during labor to overcome sluggishness of uterine contraction. Oxytocin is a uterus-contracting hormone and is employed when induction of uterus contraction is desired. It also causes contraction of the smooth muscles in the mammary glands, resulting in milk secretion.

Protein molecules have not only a variety of primary structures, but also a variety of shapes described by their secondary, tertiary, and quaternary structures. Ultimately, the biological function of proteins is dependent upon the shape of the molecules and their distribution of charges. In the following sections, we will examine the shapes of protein molecules.

Antidiuretic refers to the suppression of the secretion of urine.

---

*The amino acids in a peptide are numbered beginning at the N-terminal end.

21-2.3 SECONDARY STRUCTURES OF PROTEINS

Proteins can be grouped in two broad classes: fibrous proteins and globular proteins.

Fibrous Proteins

Fibrous proteins have peptide chains that interact with each other to form fibers or sheets. Fibrous proteins are physically tough, and are insoluble in water. These proteins are the basic structural elements of a number of tissues including connective tissue, the tissue that binds together and supports the various structures of the body. The most familiar of the fibrous proteins are the keratins which form such structures as skin, hair, fur, wool, nails, claws, hooves, horns, beaks, and scales. α-Keratin molecules have a secondary structure called an *alpha-helix*. As we have already seen in the double helix of DNA, a helix is a coil like an ordinary spring. The helix found in proteins is designated alpha, which refers to a particular pitch (the tightness of the coil or the length of one complete turn; see Figures 21.4 and 21.5). It is a right-handed helix which means that, if you could look down the center of the coil, the coil would be turning in a clockwise direction. This spiral arrangement of keratin molecules makes them flexible and elastic, like a spring. Extra elasticity and strength are achieved by winding the individual springs into larger coils.

The **secondary structure** of a protein describes the relationship of amino acids that are close to one another in the primary structure.

Silk represents another class of fibrous proteins. Silk has a secondary structure called a *beta-pleated sheet* (see Figure 21.6). This structure makes silk quite flexible, very strong, and resistant to stretching. The hydrogen bonds forming the α-helix are *intramolecular* (within the same molecule; see Figure 21.4) and the hydrogen bonds forming the β-pleated sheet are *intermolecular* (between molecules; see Figure 21.5).

Figure 21-4   (a) α-Helix. (b) β-pleated sheet. (c) Collagen triple helix (see also Figure 15.2). (d) A diagram showing the folding of a representative globular protein.

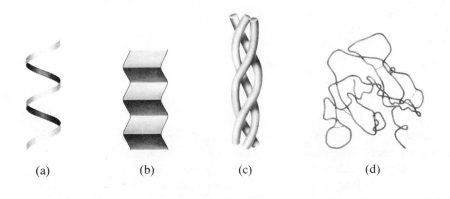

(a)   (b)   (c)   (d)

Figure 21.5
Schematic representation of the hydrogen bonding stabilizing a right-handed α-helix of a peptide chain.

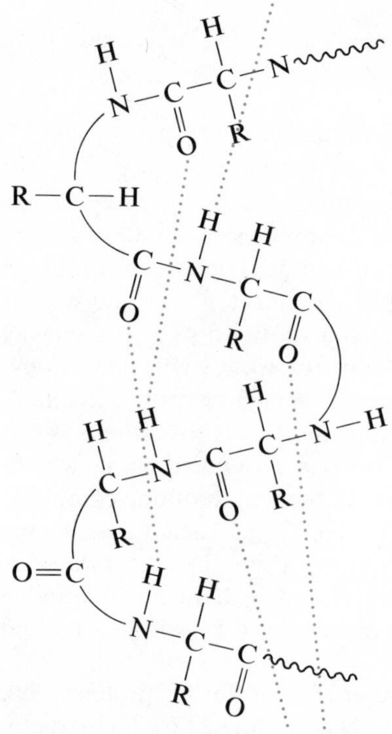

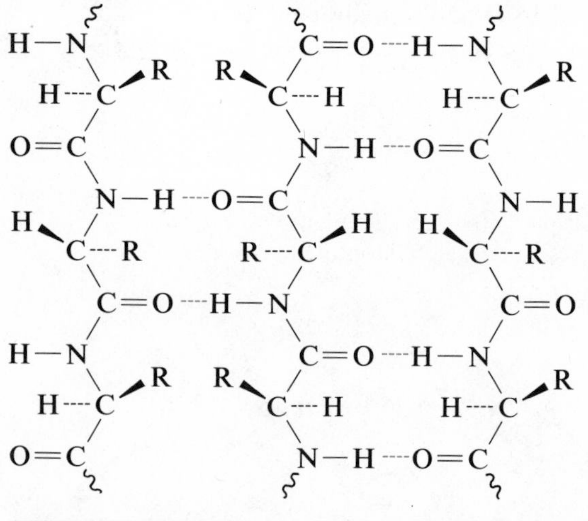

Figure 21.6
Hydrogen bonding in a β-pleated sheet. The R groups are perpendicular to the pleated sheet. Those in a single chain are alternately above and below the sheet. Silk has a high content of amino acids with small side chains: glycine (R = —H), L-alanine (R = —CH$_3$), and L-serine (R = —CH$_2$OH). Adjacent chains lie in opposite directions.

Collagens of tendons that form connective ligaments are in a third class of fibrous proteins. Collagens, introduced in Section 15-3.2, consist of a triple helix. In the triple helix, three individual molecules with left-handed helices are given a right-handed twist to form a super helix (see Figure 21.4). The triple helix structure is quite strong, resistant to stretching, and relatively rigid.

The secondary structure of a protein is stabilized by hydrogen bonding.

Globular Proteins

Proteins that have functions other than providing mechanical support are largely globular proteins. Globular proteins usually have some degree of solubility in water. They are defined primarily by their tertiary structure, although there may be regions in the protein where the peptide chain is in an α-helix or where the shape is maintained by two portions of the chain lying close together forming a β-pleated-sheet segment. Structures of globular proteins will be discussed in more detail in the next section.

Globular proteins have a generally globe-like (spherical) shape formed by the peptide chain folding back on itself.

21-2.4 TERTIARY STRUCTURES AND DENATURATION OF PROTEINS

Globular proteins, in addition to having a primary structure and perhaps some secondary structure, have a tertiary structure. The tertiary structure describes the relationship of amino acids that are far apart in the primary structure.

The tertiary structure of a protein is the folding that gives it its native, three-dimensional shape.

We can illustrate the primary structure of proteins by holding a piece of wire in a straight line. If we wind the wire around a pencil to form a coil, we put it in the shape of a helix (a secondary structure). If we crumple the wire in our hand, much as we would crumple a piece of string, we form a tertiary structure that may or may not retain some helical segments. A particular globular protein, however, always folds into the same unique, three-dimensional structure (see Figure 21.7). The specific way a protein chain folds into a compact structure is determined by its primary structure. Because we know that the peptide-bond backbones of all

Figure 21.7
Tertiary structure of the β-chain of hemoglobin. The dots in the protein chain represent α-carbon atoms of individual amino acids. The heme group is represented by a disk in the protein structure. For the structure of the heme group, see Figure 12.2.

β-chain of hemoglobin

Figure 21.8 A diagram of the types of bonds and interactions holding a protein in its tertiary structure. (a) Ionic or electrostatic interactions. (b) Hydrogen bonding. (c) Hydrophobic or nonpolar interactions. (d) Disulfide bond. (e) A β-pleated sheet region (hydrogen bonding).

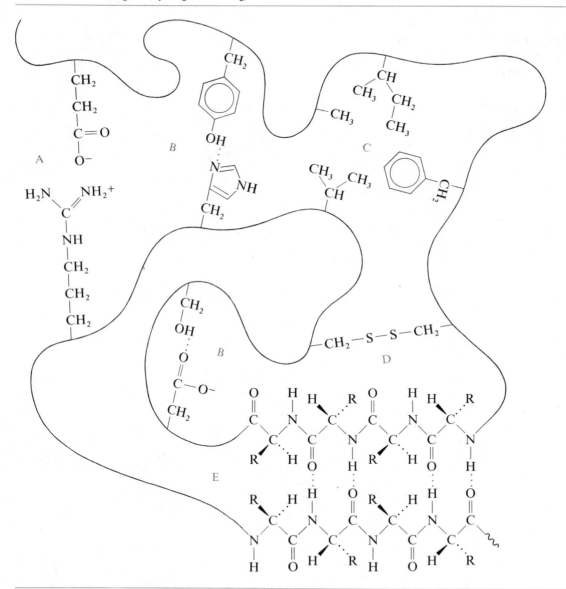

proteins are identical, as shown in Figure 21.1, we can assume that the folding is determined by the sequence of side chains.

Globular proteins are held in their tertiary structure by disulfide bonds (covalent bonds) and weak noncovalent interactions, as diagrammed in Figure 21.8. Of particular importance is the so-called *hydrophobic bonding*. This is not bonding in the sense described elsewhere in this book, but bonding that involves an association between nonpolar, hydrocarbon side chains. Most proteins (except some associated with membranes) are surrounded by water (a very polar environment). The hydrocarbon groups prefer not to associate with water and to get as far away from it as they can. Thus, whenever possible, they cluster together and associate with each other. These hydrophobic regions on the inside of a globular protein not only help determine the tertiary structure, but often are involved in biological activity, such as enzymatic activity.

We have learned that a protein is defined by its primary, secondary, and tertiary structures. Together, these three types of structures give the protein its properties; for example, its biological activity, size, and solubility. Together, they give the protein its native structure. Any change in this native structure that does not involve the breaking of peptide bonds is called **denaturation** (see Section 5-1.2). Denaturation can be caused by heat, strong acids, strong bases, alcohols, heavy metal ions, X-rays, microwaves, and other agents that break the weak, noncovalent bonds. During denaturation, the globular protein unfolds. More linear chains can now associate with each other to form aggregates (sheets) more like those of a fibrous protein, as diagrammed in Figure 21.9. As a result, the protein becomes less soluble.

We denature protein when we cook an egg. Both the white and the yolk of an egg are largely protein. Egg white is primarily a protein called ovalbumin (egg albumin) and the yolk contains the protein vitellin. Egg white is quite soluble in water.

Denaturation of protein is any change in its chemical or physical properties as a result of disruption of the weak physical forces holding it in its tertiary structure.

Figure 21.9 A diagrammatic representation of heat denaturation of a protein showing the unfolding of globular protein molecules and their association into aggregates.

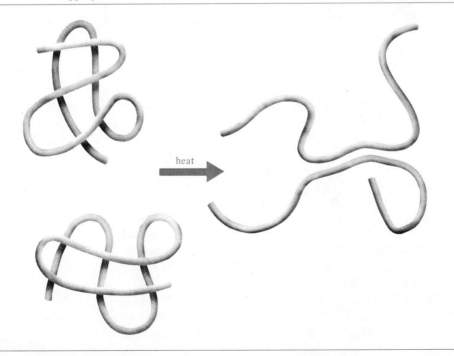

However, when an egg is cooked, the protein is denatured. It looks different, feels different, and is no longer soluble in water. We sometimes say that is has been *coagulated.*

A protein molecule, like an amino acid molecule, is amphoteric because it has both side chains containing acidic groups and side chains containing basic groups. (Remember the α-amino and α-carboxyl groups, except those at the ends, are tied up in the peptide bonds.) Therefore, like amino acids, proteins have isolectric pHs. With proteins, as with amino acids, the pI is the pH at which there are equal numbers of positive and negative charges so that the net charge on the molecule is zero.

As solution of protein becomes more acidic, the concentration of H_3O^+ increases and $R—COO^-$ is changed to $R—COOH$. As a result, the charge on the protein becomes less negative (more positive). The positive charges repel each other, and the molecule unfolds so that the like charges are as far apart as possible. Thus, addition of a strong acid denatures (unfolds) a protein. Likewise, as the pH rises,

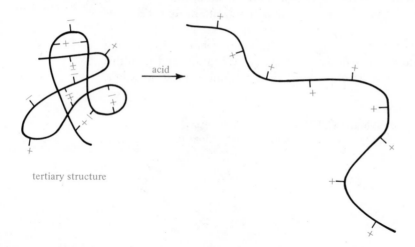

tertiary structure

the concentration of H_3O^+ decreases and the charge on the protein becomes more negative. Again, the like charges repel each other, and the molecule unfolds. Thus, addition of a strong base also denatures a protein.

Proteins, like amino acids (see Section 21-1.2), tend to be least soluble at their isoelectric pHs. As the pH value of a solution deviates a little from the isoelectric pH, the molecules take on a predominantly positive or negative charge. Having like charges, they tend to repel each other and, therefore, not to aggregate. As we know, most protein molecules are very large. Increasing the particle size by aggregation of two or more molecules pushes the size beyond the range of solubility and causes precipitation. As the tendency to aggregate decreases, solubility increases. Various ions that combine with either the positively charged or the negatively charged groups on protein molecules, and reagents that react with thiol (—SH) groups, also denature proteins. In addition to decreased solubility, denatured protein is characterized by increased digestibility (one reason we cook foods), loss of biological activity, and increased chemical reactivity.

If it is suspected that someone has swallowed lead, mercury, or silver salts, the best first aid treatment is to give the victim a concentrated protein solution and to induce vomiting. A raw egg provides a concentrated solution of protein. The heavy metal ions in the stomach will bind to the egg proteins. This will prevent their absorption. Then, of course, the protein and heavy metal ions must be removed by vomiting before the protein is digested.

21-2.5 QUATERNARY STRUCTURES OF PROTEINS

Some proteins are composed of several peptide chains that are not covalently linked to one another (see Figure 21.10). An example of this kind of protein is hemoglobin, the oxygen-carrying protein of red blood cells (erythrocytes) that contains four peptide chains (four subunits) in association with one another. In hemoglobin, there are two chains called β chains and two chains called α chains. The structures of the α and β chains are similar. Each subunit in the quaternary structure has its own primary, secondary, and tertiary structure. Many, but not all, proteins with molecular weights greater than 50,000 daltons are composed of subunits; in other words, they have a quaternary structure. When the subunit chains of some proteins are separated from one another, they retain their biological

The **quaternary structure** of a protein is the association of two or more polypeptide chains (called *subunits*) to form a stable, larger unit.

Figure 21.10   A diagrammatic representation of a quaternary structure that contains four subunits, two polypeptide chains of one kind and two slightly different polypeptide chains, as in the hemoglobin molecule.

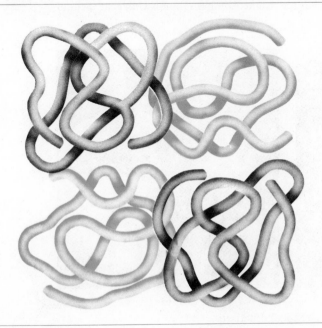

activity. In most cases, however, separation destroys biological activity because it results in a change in tertiary structure.

Sickle-cell anemia provides us with an example of how the primary structure determines the tertiary structure and how the tertiary structure affects biological activity. The hemoglobin of persons with sickle-cell anemia contains normal α chains composed of 141 amino acids. However, the β chains of sickle-cell hemoglobin, which contain 146 amino acids, contain the amino acid valine in place of glutamic acid at position 6. This substitution of valine for glutamic acid at position 6 is caused by a change in the gene coding for the β chain. This change, known as a *mutation,* occurred centuries ago. This mutation will be discussed further in Chapter 22.

Source of β chain	Position							
	1	2	3	4	5	6	7	8
normal hemoglobin	Val—	His—	Leu—	Thr—	Pro—	Glu—	Glu—	Lys—
sickle-cell hemoglobin	Val—	His—	Leu—	Thr—	Pro—	Val—	Glu—	Lys—

Sickle-cell hemoglobin differs from normal hemoglobin in a change of only 1 out of 287 amino acids. This one small alteration changes the properties of hemoglobin. Deoxygenated sickle-cell hemoglobin (sickle-cell hemoglobin without oxygen bound to it) forms a fibrous precipitate. The precipitate deforms the red blood cells (erythrocytes or RBCs) and gives them their sickle shape (see Figure 21.11). The sickled cells then become trapped in the small blood vessels. Blockage of the

Figure 21.11 Photographs, taken with a scanning electron microscope, of (a) a normal appearing red blood cell among sickle cells and (b) a deoxygenated red blood cell from a patient with sickle-cell anemia. (Photographs courtesy of Donald Caspery and Ruth Blumershine, Southern Illinois University School of Medicine).

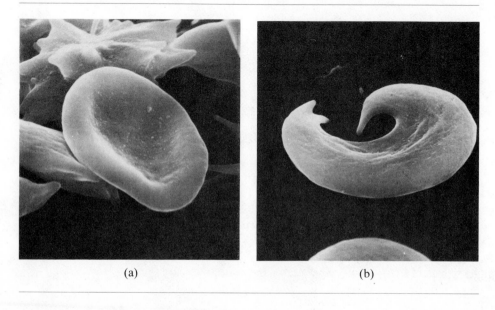

(a)   (b)

vessel creates a local region of low oxygen concentration. As a result, more hemoglobin becomes deoxygenated, and more sickling occurs. Circulation is impaired, resulting in damage, particularly to bone tissue and the kidneys. More than 100 other inherited abnormalities of hemoglobin have been found.

21-2.6 CONJUGATED PROTEINS

Hemoglobin is also an example of a protein that contains a group that is not composed of amino acids. The part that is not composed of amino acids is called the **prosthetic group**. A protein containing a prosthetic group is a *conjugated protein*. In the conjugated protein hemoglobin, the prosthetic group is the *heme* group (see Figure 21.7). The protein moiety is *globin*. Heme is a large heterocyclic, organic ring with an iron atom in its center (see Figure 12.2). Oxygen binds to the iron atom.

A **prosthetic group** is a component, not composed of amino acids, that is attached or strongly bound to a peptide chain of a protein molecule.

Examples of conjugated proteins are noted in later chapters. Some proteins contain a metal ion, such as a Zn^{2+} or an Mo^{2+} ion (as discussed in Chapter 10) as an integral part of their structure. Casein (from milk) and vitellin (from egg yolk) are phosphoproteins; that is, they contain a phosphate ester group on the hydroxyl group of a serine or a threonine residue. Many proteins have carbohydrate side chains. These proteins are called *glycoproteins*. Most of the proteins that are found outside of cells, such as the proteins of blood plasma and saliva, are glycoproteins. Mucin is a glycoprotein of saliva. Mucin thickens saliva. It also lubricates, making it easy for us to swallow.

Membrane proteins have very interesting structures. Glycophorin is a protein of the **plasma membrane** of the red blood cell. The glycophorin molecule has three regions or domains (see Figure 21.12): a hydrophilic section (the N-terminal end) that contains many oligosaccharide units and is on the outside of the cell, a hydrophobic region that is embedded in the lipid matrix of the membrane (to be discussed in Section 26-2.1), and another hydrophilic region (the C-terminal end) that projects from the lipid bilayer in toward the cytoplasm of the cell. *Hydrophilic* means *water-loving.* As we have already learned, hydrophilic groups are polar in nature and prefer to associate with water. We have also learned that the opposite of hydrophilic is *hydrophobic,* which means *water-hating.* This term refers to groups that are insoluble in water and nonpolar in nature.

The **plasma membrane** is the membrane surrounding a cell.

The oligosaccharide units of glycophorin are terminated at the nonreducing end with a sugar called sialic acid. Like other acids, this sugar acid residue is ionized at the pH of most tissues and fluids, giving the glycophorin molecule a negative charge. Maintaining a negative charge on the surface of red blood cells seems to be important. Because of the charge, the cells repel each other instead of grouping together and blocking capillaries. The presence of sialic acid is also required for keeping red blood cells in circulation. The sialic acid residues are removed from the ends of the oligosaccharide chains of glycophorin molecules of old cells. As soon as this happens, the cells are removed by the spleen. In every cell, molecules such as glycophorin that extend across the plasma membrane serve as part of the system that allows the cell to communicate with and react to its environment. Such molecules can bind hormones and transmit messages across the membrane and are involved in various interactions between cells.

Figure 21.12   The orientation of a molecule of glycophorin in the plasma membrane of an erythrocyte. We will examine the composition of the lipid bilayer in Section 26-3.1.

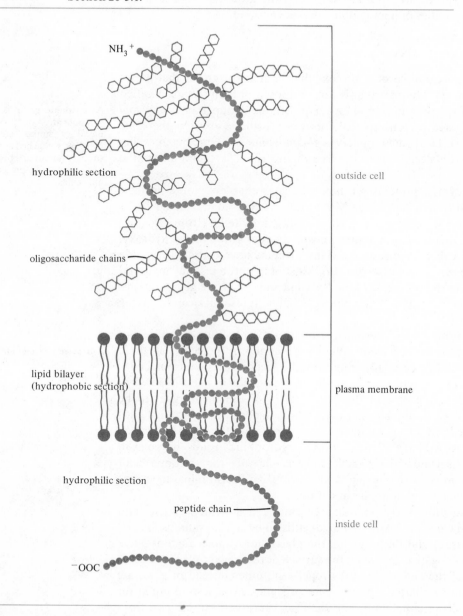

PROBLEMS

2. Using the structures of the amino acids in Table 21.1, write the structure of the tripeptide Glu—Ser—Val.

3. The following fragments were obtained by digestion of a peptide. What was the original structure of the peptide?

Using Enzyme 1	Using Enzyme 2
Ser—Gly	Arg—Thr—Phe
Thr—Phe—Lys	Val—Try
Val—Tyr—Arg	Lys—Ser—Gly

4. The specific sequence of amino acids in a protein is called its _____.

5. The α-helix and β-pleated sheet are examples of _____.

6. The three-dimensional shape of a native protein is its _____.

7. The process of unfolding the three-dimensional shape of a native protein is called _____.

8. The relationship of subunit polypeptide chains to form the total protein is called its _____.

9. A protein that contains covalently bound carbohydrate residues is a(n) _____.

ANSWERS TO PROBLEMS   **2.**

$$\underset{+}{H_3N}-CH-\overset{\overset{\displaystyle O}{\|}}{C}-NH-CH-\overset{\overset{\displaystyle O}{\|}}{C}-NH-CH-\overset{\overset{\displaystyle O}{\|}}{C}-O^-$$

with side chains:
- CH$_2$, CH$_2$, COO$^-$ (first residue)
- CH$_2$OH (second residue)
- HC—CH$_3$, CH$_3$ (third residue)

3. Val—Tyr—Arg—Thr—Phe—Lys—Ser—Gly   **4.** primary structure
5. secondary structures   **6.** tertiary structure   **7.** denaturation   **8.** quaternary structure
9. glycoprotein

SUMMARY

In this chapter, we learned that proteins are polymers of L-amino acids joined by peptide (amide) bonds. These peptide bonds join the carboxyl group on one amino acid residue to the amino group on the alpha carbon atom of another amino acid residue. There are 20 different amino acids that are the building units of protein molecules. A great number of different protein molecules actually exist in nature, and proteins have a wide variety of functions.

We also learned that protein molecules are defined by their primary, secondary, tertiary, and in some cases, quaternary structures. The primary structure of protein molecules determines the secondary, tertiary, and if present, quaternary structures. The shape (conformation) of protein molecules, as determined by these structures, enables them to serve their particular biological function. The conformation of a protein molecule is a rather fragile structure held together by weak bonds that can be easily disrupted, resulting in denaturation.

In the following chapter, we will see how the genetic information in RNA molecules, having been transferred to RNA molecules from segments of DNA molecules by the process of transcription (Section 20-3.2), controls the order of amino acid residues in protein molecules. In the remaining chapters of this text, we will see how the genetic information expressed as the sequences of amino acids in protein molecules ultimately is expressed as the biological activity of these protein molecules.

ADDITIONAL PROBLEMS

10. Determine the structure of the peptide that gives the following fragments.

    Cleavage with reagent 1   *Cleavage with reagent 2*
      Ser   Ser—Phe—Lys
      Leu—Glu   Ala—Leu—Glu
      Phe—Lys—Ala

11. Determining the structure of the peptide that gives the following fragments.

    Cleavage with reagent 2   *Cleavage with reagent 3*
      Asp   Tyr—Pro—Lys—Asp
      Thr—Arg   Thr—Arg—Val—Met
      Val—Met—Tyr—Pro—Lys

12. Using the formulas for the amino acids in Table 21.1, write the complete structures of the following peptides with the N-terminal end on the left.
    (a) Gly—Lys—Val
    (b) His—Ile

13. Using Table 21.1, write the structure of L-lysine at its pI. (*Hint:* pK of α-COOH = 2.18, pK of α-NH$_2$ = 8.95, and pK of ϵ-NH$_2$ = 10.53)

14. Consider the following peptide.

    (a) Classify the peptide on the basis of the number of amino acids.
    (b) Draw a box around the amino end.
    (c) Circle the peptide bonds.
    (d) Write structural formulas for the amino acids that would be released upon complete hydrolysis.

15. List the four types of bonding that contribute to the tertiary structure of a protein.

CHAPTER 22

PROTEIN SYNTHESIS

In Chapter 21 we learned that it is the proteins of a cell that determine the cell's activities and equip it to do its job. The genetic information in the DNA molecules, and whether the genes in question are used to make RNA molecules, determines which proteins can be made by the cells of an organism.

Just as the sequence of deoxyribonucleotides in DNA molecules determines the sequence of ribonucleotides in RNA molecules (see Chapter 20), the sequence of ribonucleotides in RNA molecules determines the sequence of amino acids in protein molecules. The concept that the genetic information in DNA molecules is transferred to RNA molecules, which in turn serve as the templates ordering the amino acids in proteins, is a key element of the central dogma of molecular biology (see Figure 22.1). The definition of molecular biology is expanding continually. Classically, molecular biology has been considered the science of molecular descriptions of the structure, organization, and control of expression of genes. Therefore, this definition includes the broad areas of replication, transcription, and translation as described in this chapter and in Chapter 20. However, the field of molecular biology is now considered by some also to include molecular descriptions of other cellular activities.

The **central dogma of molecular biology** describes the basic relationship between DNA, RNA, and protein. The central dogma is a statement of the principle that DNA serves as a template for both its own duplication and the synthesis of RNA and that RNA, in turn, is the template in protein synthesis.

Figure 22.1 A diagrammatic representation of the central dogma. Arrows indicate the direction of transfer of genetic information.

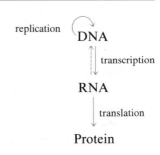

In this chapter, we will discuss the synthesis of protein molecules. The process of synthesis of a protein molecule using the information contained in the sequence of nucleotides in an RNA molecule to determine and direct the order of amino acids in the protein molecule is called *translation.* Translation refers to putting the same information in another language, just as one might translate the first sentence of this paragraph into German, French, or Russian. When biochemists and molecular biologists use the term *translation,* they are referring to changing the information in a sequence of heterocyclic bases in an RNA molecule into a sequence of amino acids in a protein molecule.

LEARNING OBJECTIVES

1. Define central dogma, translation, messenger RNA, mRNA, genetic code, codon, degenerate code, transfer RNA, tRNA, anticodon, aminoacyl-tRNA synthetase, ribosome, ribosomal RNA, rRNA, polyribosome, polysome, initiation complex, translocation, mutation, and mutagen.
2. Describe the functions of mRNA, tRNA, ribosomes, aminoacyl-tRNA synthetase, fMet, and termination codons.

3. Describe the structures of eukaryotic mRNA and tRNA (include the cloverleaf model, modified bases, and the overall three-dimensional structure).
4. Given an mRNA sequence and a table of codons, write the sequence of the peptide formed by translation.
5. List and describe the three phases of protein synthesis.
6. List and describe the three steps of the elongation phase of protein synthesis.

There are three classes of cellular RNA molecules: *messenger RNA (mRNA), transfer RNA (tRNA),* and *ribosomal RNA (rRNA)* molecules. All of these molecules participate in the process of translation (protein synthesis) shown in Figure 22.2, and all are made on DNA templates by transcription. We will begin this chapter by describing these species of RNA and how they participate in the translation process.

22-1 MESSENGER RNA AND THE GENETIC CODE

The RNA molecules that determine the sequence of amino acids in a protein molecule are called **messenger RNA (mRNA)** molecules. mRNA molecules come in many sizes. Chain lengths of mRNA molecules vary greatly because, as we learned in Chapter 21, the lengths of protein chains vary greatly. Chain lengths of mRNA molecules also vary because some mRNA molecules contain codes for more than one protein molecule, and because mRNA molecules contain varying amounts of a region at the 5′ end that does not code for a protein molecule. When an mRNA molecule contains codes for more than one protein molecule, the protein products frequently have related functions.

Sequences of groups of three bases (three nucleotides) on mRNA molecules are used to determine the specific orders of amino acids in peptide chains. Because

Messenger RNA (mRNA) molecules serve as templates for protein synthesis.

Figure 22.2   An overview of protein synthesis. All three species of RNA molecules are made in the nucleus of a cell on DNA templates. mRNA molecules code for the specific protein that is being made. rRNA molecules and proteins make up ribosomes, particles containing the protein synthesizing machinery. tRNA molecules are adapter molecules that, by attaching to the amino acids, allow the amino acids to recognize codons on the mRNA molecules.

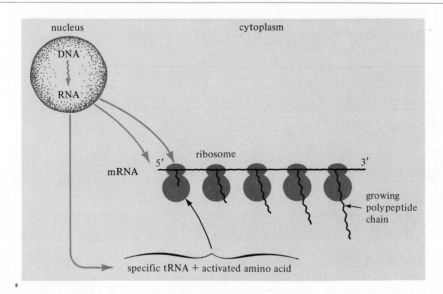

there are four bases (adenine, guanine, cytosine, and uracil) in an mRNA molecule (see Section 20-3.1), there are 64 possible three-base sequences that can be made from the four nucleotides making up an mRNA molecule.* All 64 three-base sequences, codons, are used in protein synthesis. Sixty-one of these codons code for specific amino acids; the other three codons code for chain termination. The specific codon assignments are given in Figure 22.3.

The genetic code is degenerate. Degeneracy of the code refers to the fact that some amino acids are positioned along the mRNA chain by more than one codon. For example, as we can see by examination of Figure 22.3, there are four different codons for glycine: GGU, GGC, GGA, and GGG. We also can see from Figure 22.3 that, in some cases, such as in the codons for glycine, the first two nucleotides determine the amino acid. In other words, to specify glycine there must be a sequence GGN, where N stands for any of the four nucleotides of RNA.

A **codon** is a sequence of three bases of an mRNA molecule that codes for an amino acid (or chain termination).

The **genetic code** is the sequence of bases in an mRNA molecule (or a DNA molecule) that determines the sequence of amino acids in a protein molecule.

In a **degenerate code** there are two or more codons that code for the same amino acid.

22-2 TRANSFER RNA

Before an amino acid molecule can be positioned at the proper places along the mRNA template molecules, it must acquire an identity that allows it to recognize

---

*The number of ways that four different items can be joined in groups of three is given by the formula $4^3 = 4 \times 4 \times 4 = 64$

Figure 22.3 The 64 codons of the genetic code of mRNA molecules. The base of the nucleotide in the first position (5′ end) is given in the inner circle. The middle circle gives the base of the nucleotide in the second or middle position. The outer circle specifies the base of the nucleotide at the third position (3′ end). For example, 5′ UUU 3′ and 5′ UUC 3′ on an mRNA molecule specify phenylalanine, 5′ UGG 3′ specifies tryptophan, and so on. 5′ AUG 3′ specifies methionine and, in addition, is the initiation signal. 5′ UAA 3′, 5′ UAG 3′, and 5′ UGA 3′ are termination (stop) signals indicating that the end of the message has been reached.

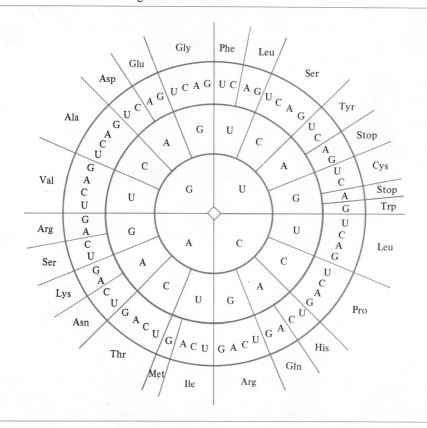

its specific codon and vice versa. The amino acid obtains this identity by chemical modification. Specifically, the amino acid is attached to a **transfer RNA (tRNA)** molecule that has a three-base sequence (the **anticodon**) that recognizes and forms hydrogen bonds with the three bases in the codon in the same way that complementary bases pair up during replication and transcription (see Figure 22.4).

Just as the monomer units of polysaccharides and nucleic acids must be activated before they can react to form polymers, amino acid molecules must be activated before they can be joined together in peptide bonds to form a protein molecule. The energy required for activation comes from high-energy bonds of ATP molecules (see Figure 20.8). Amino acids are activated by combining them with AMP

Transfer RNA (tRNA) molecules are able to combine covalently with a specific amino acid and to form hydrogen bonds with at least one mRNA codon.

The **anticodon** is the three-base sequence on a tRNA molecule that recognizes and pairs through hydrogen bonding to a three-base codon on mRNA molecules.

(adenosine monophosphate) through a high-energy bond (see Figure 22.5). The reaction of amino acid molecules with ATP molecules is catalyzed by enzymes called aminoacyl-tRNA synthetases. There is a specific aminoacyl-tRNA synthetase for each of the 20 amino acids found in proteins.

Figure 22.4 Transfer RNA. (a) The two-dimensional cloverleaf structure of a tRNA molecule. (b) The folding of a tRNA molecule into its three-dimensional shape. O = nucleotide, ——— = hydrogen bonding, Y = pyrimidine nucleotide, R = purine nucleotide, H = modified purine nucleotide, Ψ = a specific modified pyrimidine nucleotide known as a pseudouridine nucleotide, and . . . = variable number of nucleotides. All tRNA molecules except initiator tRNA molecules have this sequence.

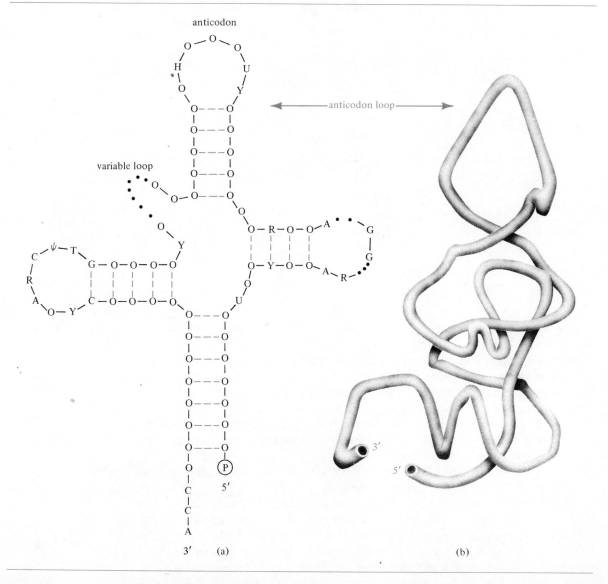

Figure 22.5   Reaction of an amino acid with ATP to form aminoacyl adenosine monophosphate, a molecule in which the amino acid residue is attached to the AMP
residue through a high-energy bond.

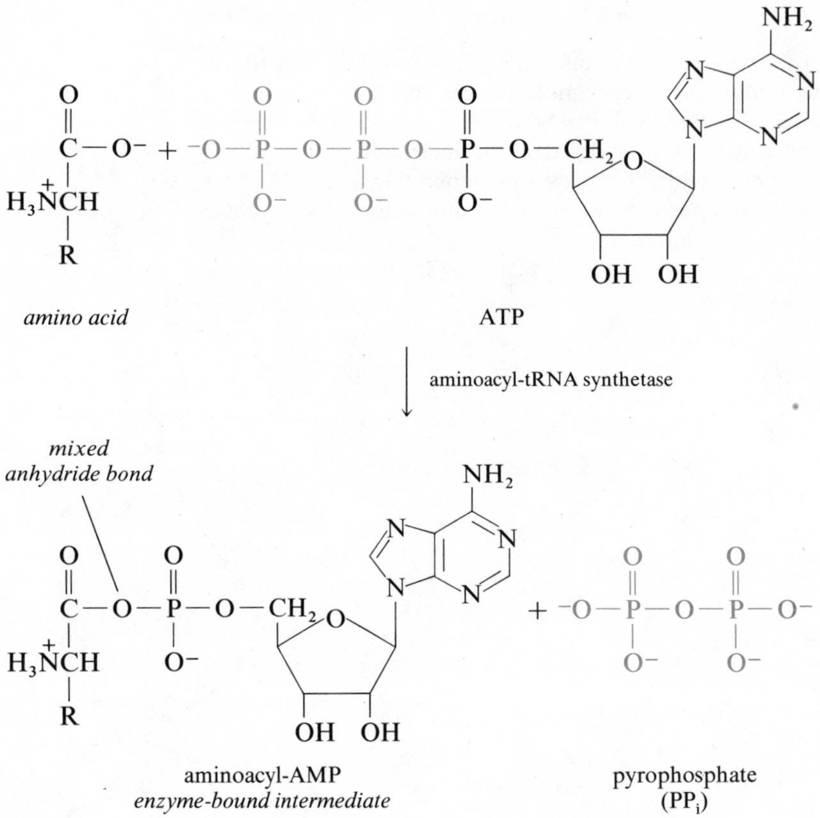

The activated amino acids are always tightly bound to the enzyme molecules
that produce them (see Figure 22.5). The same enzyme molecule that catalyzes
the activation of an amino acid by forming an aminoacyl-AMP compound then
transfers the aminoacyl residue to an appropriate tRNA molecule (see Figure 22.6).
During transfer, the amino acid residue becomes attached to the ribofuranose
residue at the 3′ end of the tRNA molecule by formation of an ester bond. The
amino acid remains activated after transfer. Thus, the activating enzymes, the
aminoacyl-tRNA synthetases, are able to recognize both a specific amino acid and
the specific tRNA molecules of that amino acid.

As we learned in Chapter 21, every cell has 20 different amino acids that are
used to make proteins. Every cell, therefore, needs 20 different aminoacyl-tRNA
synthetase enzymes, each one specific for an amino acid. Also, every cell needs at
least 20 different tRNA molecules because there must be at least one tRNA molecule for each amino acid. In general, more than one species of tRNA can act as
an acceptor for each amino acid. These acceptors are called *isoacceptor tRNA
molecules.* However, there is only one aminoacylating enzyme for each amino acid.

The process of attaching an activated amino acid to a tRNA molecule is highly specific. For example, the aminoacyl-tRNA synthetase for leucine only activates leucine, but can aminoacylate all the tRNA molecules for leucine (the entire iosacceptor family). There are five isoacceptor tRNA molecules for leucine in *E. coli*. The enzymes rarely, if ever, make mistakes. The leucine aminoacyl-tRNA synthetase, for example, does not make the mistake of attaching leucine to a tRNA for any other amino acid.

tRNA molecules are relatively small molecules. They contain 73 to 93 nucleotides in a single polynucleotide chain. They also contain heterocyclic bases other than the four (A, U, G, and C) found in other RNA molecules. Many of these different bases, of which there are typically 7 to 15 per molecule, are methylated derivatives of adenine, uracil, guanine, and cytosine. The polynucleotide chain folds back on itself and is held in a cloverleaf structure by hydrogen bonding between complementary bases similar to the bonding that holds together the two chains of a DNA molecule (see Section 20-2.1). In fact, the hydrogen-bonded regions seen in Figure 22.4(a) are DNA-like double-helical arrangements with adenine hydrogen-bonded to uracil (A $=$ U) and guanine hydrogen-bonded to cytosine (G $\equiv$ C). The two-dimensional cloverleaf structure in Figure 22.4(a) shows the three large loops and the smaller, variable loop. One of the three large loops

Figure 22.6   Attachment of an amino acid to a tRNA molecule by reaction of the activated amino acid (the aminoacyl adenosine monophosphate molecule of Figure 22.5) with the tRNA molecule specific for it.

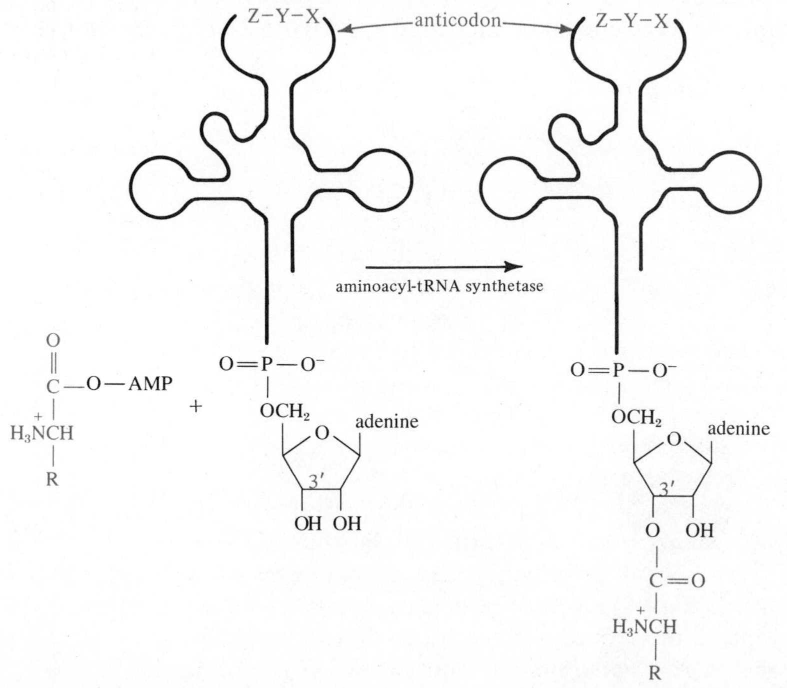

is the anticodon loop, the function of which will be described later. As shown in Figure 22.4(b), the three-dimensional structure of a tRNA molecule is actually L-shaped.

Each tRNA molecule must have at least two specific recognition sites—one for its activating enzyme and one for specific codons. The 3′ end of the tRNA molecule always consists of the sequence CCA. A specific aminoacyl-tRNA synthetase enzyme molecule with its attached amino acid recognizes a tRNA molecule specific for that amino acid. Then, the enzyme attaches the amino acid to the ribofuranose residue of the AMP residue at the 3′ end of the tRNA molecule (see Figure 22.6). tRNA molecules such as these, with attached amino acid residues, are called *charged tRNA* molecules. Finally, the three-base sequence of the anticodon of the charged tRNA molecule recognizes the codon for that specific amino acid on an mRNA molecule and positions the amino acid residue in the correct position along the mRNA template.

Thus, adding an adaptor molecule (a tRNA molecule) to each amino acid gives an identity to the amino acid molecule that allows it to be positioned at the correct place along the mRNA molecule. The tRNA molecules for the different amino acids all have different structures; each is uniquely adapted for lining up with a specific nucleotide sequence on the template. The differences in tRNA structures that allow for this specific positioning are in a three-nucleotide sequence (the anticodon) that hydrogen bonds to the three-nucleotide sequence of a codon for the particular amino acid that is attached to the tRNA molecule.

22-3 RIBOSOMES

Protein synthesis always takes place on the surface of organelles called ribosomes. Ribosomes consist of two subunits—a large one and a small one (see Figure 22.7).

Organelles are membrane-bound particles within a cell that are made up of molecules organized in a specific way. Organelles contain enzymes for a specialized function.

Ribosomes are organelles, composed of rRNA and protein, that are the site of protein synthesis.

Figure 22.7 A ribosome and its constituent parts.

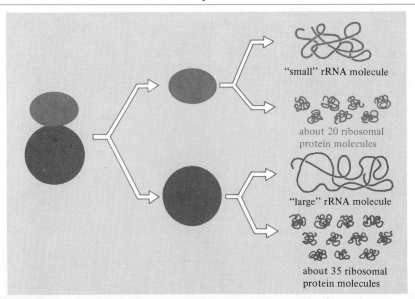

"small" rRNA molecule

about 20 ribosomal protein molecules

"large" rRNA molecule

about 35 ribosomal protein molecules

The large subunit consists of a large **ribosomal RNA (rRNA)** molecule and about 35 protein molecules. The small subunit contains a smaller rRNA molecule and about 20 protein molecules. About half of the mass of human ribosomes is rRNA.

Ribosomal RNA (rRNA) is the nucleic acid component of ribosomes.

One function of ribosomes is to position the charged tRNA molecules and the mRNA molecules so that the genetic code can be read accurately. Therefore, ribosomes contain specific surfaces that bind the mRNA molecule, the charged tRNA molecules, and the growing polypeptide chain (the protein that is being made). An mRNA molecule attaches to the surface of the smaller ribosomal subunit. The section of an mRNA molecule that is in contact with a single ribosome is relatively short. As a result, an mRNA molecule may bind to more than one ribosome. The complex formed by attachment of several ribosomes to a single mRNA molecule is called a **polyribosome** or, more simply, a **polysome**.

A **polyribosome** or **polysome** is a complex of an mRNA molecule and two or more ribosomes actively engaged in protein synthesis.

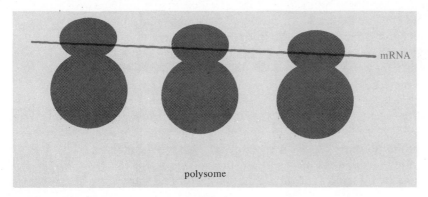

mRNA

polysome

22-4 PEPTIDE-BOND FORMATION

Protein synthesis consists of three phases: (1) *initiation,* (2) *elongation,* and (3) *termination.* We will consider these three phases in order, beginning with initiation.

All bacterial (prokaryotic) proteins are made with the amino acid N-formylmethionine at the C-terminal end. A protein chain is made by adding one amino acid at a time, beginning with the amino end and ending with the carboxyl end. In other words, the polypeptide chain grows by adding amino acids one by one to the C-terminal end. N-Formylmethionine is a molecule of methionine (see Table 21.1) that is modified by the addition of a formyl group to the α-amino group.

$$
\begin{array}{ccc}
& & \overset{\displaystyle O}{\underset{\displaystyle \|}{}} \\
& & C-O-tRNA \\
\overset{\displaystyle O}{\underset{\displaystyle \|}{}} & \overset{\displaystyle O}{\underset{\displaystyle \|}{}} & | \\
-C-H & H-C-HNCH \\
\text{formyl group} & & | \\
& & CH_2 \\
& & | \\
& & CH_2 \\
& & | \\
& & S \\
& & | \\
& & CH_3
\end{array}
$$

N-formylmethionine attached to a tRNA molecule

N-Formylmethionine (fMet) is, and can only be, used to start a protein chain in bacteria.* The formyl group is added after methionine is attached to a tRNA molecule. Not all methionine-tRNA (Met-tRNA) molecules can be formylated. There are two species of Met-tRNA, only one of which permits formylation of the methionine residue. However, they contain the same anticodon sequence — an anticodon that recognizes AUG, the initiating codon.

Protein synthesis starts with the formation of an *initiation complex*. The initiation complex is formed from the smaller ribosomal subunit, an mRNA molecule, a charged methionyl-tRNA molecule in eukaryotes or a charged *N*-formylmethionyl-tRNA molecule in prokaryotes, a molecule of GTP,† and several protein molecules called *initiation factors* (see Figure 22.8). A larger ribosomal subunit then binds to the first initiation complex to form an entire ribosome and a second initiation complex. The energy required for this binding of the larger ribosomal subunit comes from the hydrolysis of GTP (see Figure 22.8). Next, the Met-tRNA molecule is positioned so that its anticodon is paired with the initiating AUG codon. The fMet-tRNA molecule occupies what is described as the P, or peptidyl, site on the ribosome. Another site on the ribosome described as the A, or aminoacyl, site is empty. Initiation is now complete.

Elongation, the second phase of protein synthesis, consists of three steps: (1) binding of an aminoacyl-tRNA to the next codon, (2) peptide-bond formation, and (3) translocation (see Figure 22.9). In step 1, an aminoacylated tRNA molecule binds to the empty A site on the ribosome. As we learned from Section 22-2, the aminoacylated tRNA molecule that binds is the one with the anticodon for the codon on the mRNA molecule. A protein called an elongation factor and energy obtained from the hydrolysis of a GTP molecule are required for this binding.

At this point, Met-tRNA occupies the P site and a different aminoacylated tRNA molecule occupies the A site. To form a peptide bond (step 2), the methionine residue on Met-tRNA in the P site is transferred to the amino group of the amino acid residue on the aminoacyl-tRNA in the A site (see Figure 22.10). The transfer is catalyzed by the enzyme peptidyl transferase, and the product is a dipeptidyl-tRNA molecule. Peptidyl transferase is part of the larger ribosome subunit.

Translocation (step 3) consists of three steps. First, the tRNA molecule in the P site, which now is without an amino acid, leaves the ribosome. Then, the peptidyl-tRNA molecule moves from the A site to the P site (see Figure 22.9), and the mRNA molecule moves a distance of three nucleotides (one codon) across the

Translocation is the movement of the mRNA molecule the distance of one codon across a ribosome.

N-Formylmethionine is used to start synthesis of all protein molecules in prokaryotic organisms. Initiation of protein synthesis in eukaryotic organisms is less well understood. We do know that protein synthesis in eukaryotes also begins with methionine, although the methionine residue does not have an added formyl group. In this chapter we will assume that the general process of protein synthesis is the same in humans as it is in bacteria, with the exception that eukaryotes such as humans have larger ribosomes and begin protein synthesis with methionine rather than with *N*-formylmethionine.

†Nucleoside triphosphates other than ATP can be used as an energy source in certain cellular reactions. Here, GTP is cleaved, forming GDP and phosphate ion. This reaction provides the energy to form the second initiation complex (see Figure 22.8), just as ATP is cleaved to form ADP and phosphate ion in order to provide energy for various reactions. Compounds other than nucleoside triphosphates also serve as cellular energy sources.

Figure 22.8 The initiation phase of protein synthesis, showing formation of the first and second initiation complexes.

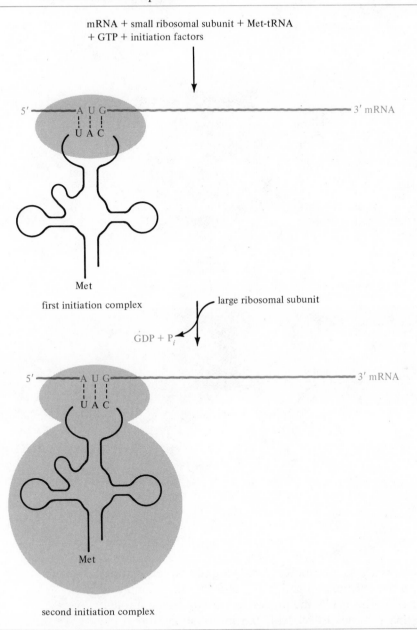

mRNA + small ribosomal subunit + Met-tRNA
+ GTP + initiation factors

5′ ———— A U G ———————————————————— 3′ mRNA
          U A C

Met

first initiation complex

large ribosomal subunit

GDP + P$_i$

5′ ———— A U G ———————————————————— 3′ mRNA
          U A C

Met

second initiation complex

ribosome. This translocation requires another protein elongation factor and hydrolysis of a second molecule of GTP. Now, the next codon is positioned for reading and the A site is empty and ready to bind the aminoacyl-tRNA molecule with the proper anticodon to start another elongation sequence to add the third amino acid.

Figure 22.9   The elongation phase of protein synthesis: binding of an aminoacyl-tRNA molecule, peptide-bond formation, and translocation. A = A site and P = P site.

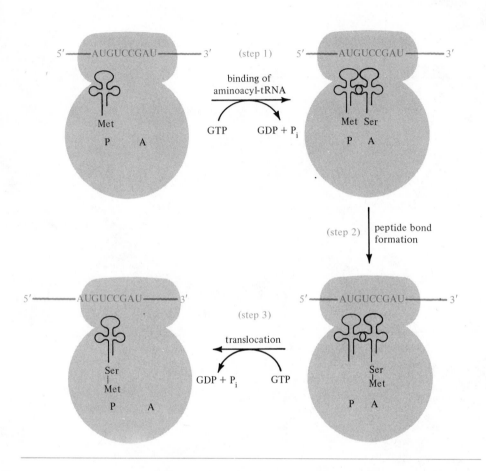

The mRNA molecule moves across the ribosome allowing the protein synthesizing machinery to read off the information and translate it into the amino acid sequence of a protein molecule. This is like the process of playing a tape through a tape recorder, allowing the recorder to read off the information stored and translate it into sounds. As shown in Figure 22.11, the direction of movement of the mRNA molecule with respect to the ribosomes is such that the ribosomes move in the 5' $\longrightarrow$ 3' direction across the mRNA molecule because mRNA is translated in the 5' $\longrightarrow$ 3' direction.

As shown in Figure 22.3, the codons UAA, UGA, and UAG are the stop, or termination, signals. No aminoacyl-tRNA molecules can bind to these codons

Figure 22.10   Transfer of a methionine residue to the amino group of another amino acid residue with the formation of a peptide bond.

P site   *A site*

tRNA   tRNA   tRNA   tRNA
  | | | |
  O   O   OH   O
  | | |
  C=O   C=O   O   C=O
  | | peptidyl transferase   ‖ |
H₃N⁺CH → H₃N⁺CH ────────────────→   C—HN—CH
  | | | |
  CH₂   R   H₃N⁺CH   R
  | |
  CH₂   CH₂
  | |
  S   CH₂
  | |
  CH₃   S
                                                |
                                                CH₃

because normal cells do not contain tRNA molecules with anticodons complementary to these three-nucleotide sequences. Rather than coding for amino acids, these codons are recognized by *protein-release factors*. A release factor binds to the termination codon. The bond connecting the **nascent** protein chain to the tRNA molecule in the P site is hydrolyzed, releasing the newly synthesized protein molecule. After the new protein molecule has left the ribosome, the ribosome dissociates into its two subunits (see Figure 22.11).

Although there are several steps involved in the addition of each amino acid, and although most proteins are rather long chains of amino acids, the synthesis

Nascent means *beginning to exist.* In this case, it refers to a polypeptide chain that is ready to be set free.

Figure 22.11   A polysome during protein synthesis. The mRNA molecule is moving from right to left; the ribosomes are moving from left to right. The ribosomes function independently of each other.

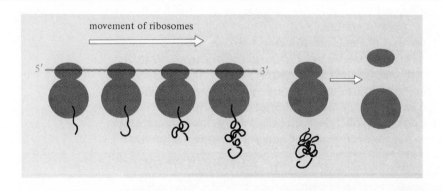

of a protein molecule, like most reactions in a living cell, is quite fast. In mammalian cells, elongation occurs at a rate of about one amino acid each second. It takes 2 to 3 minutes to synthesize the α chain of hemoglobin that contains 141 amino acids. Elongation is about 20 times faster in bacterial cells than in eukaryotic cells. As the chain elongates, it assumes much of its final tertiary structure through formation of secondary bonds. The final protein is obtained by removal of one or more amino-terminal amino acid residues. The formyl group is removed enzymically from the N-formylmethionine residue at the amino end of bacterial proteins. Then, in both prokaryotes and eukaryotes, the terminal methionine residue is removed from many, but not all, proteins. These reactions may occur before all amino acids are added to the growing chain. Additional amino acid residues may also be removed from the amino end. Disulfide bonds may be formed by oxidation of two cysteine residues to complete the tertiary structure.

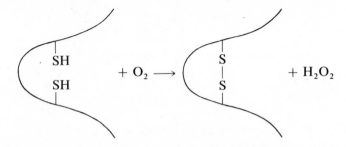

22-5 DISCONTINUOUS GENES OF EUKARYOTES

The coding sequences of eukaryotic genes, in contrast to the coding sequences of prokaryotic genes, exist as blocks of DNA separated by DNA regions of unknown function. For example, the gene coding for ovalbumin, the major egg-white protein, is split into eight pieces. Figure 22.12(a) shows the general organization of the ovalbumin gene. Figure 22.12(b) shows the mRNA for ovalbumin. The relation between the two is indicated by the dotted lines connecting them. As can be seen from Figures 22.12(a) and 22.12(b) there are eight sequences of the DNA molecule whose complementary RNA sequences do not appear in the final mRNA moelcule. Such sequences have been called *intervening sequences* and *introns*. The DNA sequences whose complementary RNA sequences appear in the final mRNA molecule and are, therefore, expressed are called *exons*.

The ovalbumin gene is about 7700 nucleotides long.* It consists of a leader (47 nucleotides long), 7 exons [1 to 7; see Figure 22.12(a)] and 7 introns (A to G, Figure 22.12(a)]. As a result, the transcribed RNA also is about 7700 nucleotides long. The final mRNA is about one-fourth that size, about 1875 nucleotides long. After transcription, the RNA molecule is processed. Processing involves the addition of sequences to both ends and splicing. All eukaryotic mRNA molecules have a special unit known as a *cap* added to the 5′ end and a sequence of AMP

---

*Double helical DNA made up of two strands each 1000 nucleotides long is said to contain one kilo base pairs (1 Kbp). The double helical DNA containing the ovalbumin gene is 7.7 Kbp in length.

Figure 22.12   The ovalbumin gene and the processing of the precursor mRNA. (a) The ovalbumin gene. (b) Ovalbumin mRNA. (c) A possible mechanism for excision of the transcribed intervening sequences to form the ovalbumin mRNA molecule (not drawn to scale). (Adapted from F. Gannon *et al., Nature,* 278 (1979), pp. 428–434.)

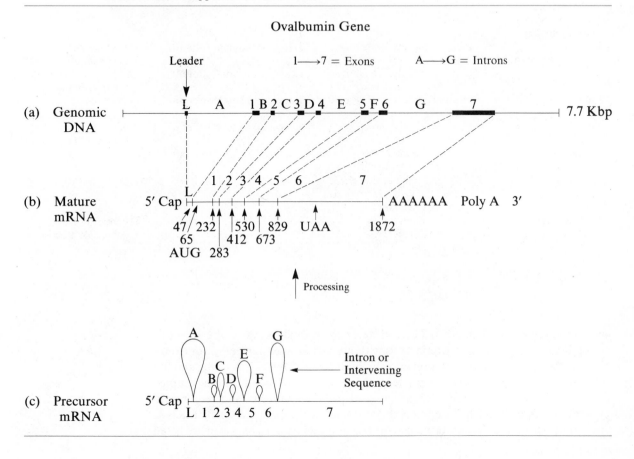

residues, called *poly A,* added to the 3′ end. The mechanism of the splicing is not known. However, in Figure 22.12(c), we have shown a looping out of the transcribed introns to bring together the noncontiguous exons. Enzymes may then perform the cutting and splicing (*ligation*) to join the coding sequences of the mRNA in a very precise manner.

22-6 INHIBITION OF PROTEIN SYNTHESIS BY ANTIBIOTICS

Prokaryotes and eukaryotes have different ribosomes. The ribosomes of eukaryotes are larger than those of prokaryotes. In addition, prokaryotes and eukaryotes begin protein synthesis with a different unit (*N*-formylmethionine in prokaryotes

Table 22.1   Selected Inhibitors of Translation

Agent	Inhibition	Site of Action	Use
Aminoglycosides (gentamycin, kanamycin, neomycin, and streptomycin)	Inhibit initiation and cause codon misreading	Smaller ribosomal subunit	Treatment of bacterial infections
Chloramphenicol (chloromycetin)	Inhibits peptidyl transferase activity in prokaryotes	Larger ribosomal subunit	Treatment of bacterial infections
Cycloheximide	Inhibits peptidyl transferase activity in eukaryotes	Larger ribosomal subunit	Laboratory research, potent fungicide
Lincomycin	Inhibits peptidyl transferase activity	Larger ribosomal subunit	Treatment of bacterial infections
Erythromycin	Inhibits translocation	Larger ribosomal subunit	Treatment of bacterial infections
Puromycin	Replaces aminoacyl-tRNA as an acceptor of the nascent polypeptide chain from the P site, leading to premature chain release	Larger ribosomal subunit	Laboratory research
Tetracyclines (tetracycline, chlortetracycline, doxycycline, methacycline, and oxytetracycline)	Inhibit binding of aminoacyl-tRNA molecules	Smaller ribosomal subunit	Treatment of bacterial infections

and methionine in eukaryotes). These differences provide keys to the selective disruption of the growth of bacterial cells without harming human cells. For example, the antibiotic streptomycin prevents the binding of fMet-tRNA to the smaller ribosomal subunit and, therefore, prevents the formation of the initiation complex in bacteria. This, in turn, prevents initiation of protein synthesis, and, as a result, prevents bacterial growth. Streptomycin also causes the misreading of mRNA. Some other drugs that inhibit protein synthesis and their modes of action are listed in Table 22.1.

22-7 MUTATIONS

Mutations are differences from the normal that are inheritable. We know that inheritable characteristics are determined by the genes in chromosomes, so we could guess that mutations are changes in the sequence of base pairs in DNA molecules that are passed on to succeeding generations. We know that transcription of a different sequence of bases in a DNA molecule results in a different sequence of bases in the RNA molecule. We know that translation of a new mRNA molecule results in an altered sequence of amino acids in the corresponding protein molecule. And we know that any change in the primary structure of a protein has the potential for changing its secondary, tertiary, and quaternary structures and, as a result, its biological activity. This was dramatically presented in Section 21-2.5

where we learned that a change in a single amino acid in a polypeptide subunit of hemoglobin results in the condition known as sickle-cell anemia.

Three classes of mutations bring about the introduction of defects into the sequence of bases in a DNA molecule: (1) one or more nucleotides are simply inserted in place of others, (2) one or more nucleotides are added, and (3) one or more nucleotides are deleted (see Figure 22.13). Mutagens can cause mutations either by converting one base to another or by being incorporated into DNA without destroying the capacity of DNA for replication. Nitrous acid (HNO_2) is a powerful mutagen because it can convert one heterocyclic base to another. Physical agents, such as radiation, can also be mutagens.

A **mutagen** is a physical or chemical agent that increases the frequency of mutations.

In Section 21-2.5, we learned that sickle-cell anemia is a genetic (inherited) disease. In sickle-cell anemia, the amino acid at position 6 in the β chain of hemoglobin is L-valine. In the β chain of normal hemoglobin, this amino acid is L-glutamic acid. From Figure 22.3, we can determine that the codons for valine and glutamic acid are

L-glutamic acid:	GAA	GAG		
L-valine:	GUA	GUG	GUU	GUC

It is obvious that the substitution of uracil for adenine in the codon in the mRNA molecule for the β chain of hemoglobin results in the substitution of valine for glutamic acid at position 6 of the protein. This means that there was a substitution of adenine for thymine in the appropriate position of the DNA molecule

Figure 22.13 Demonstration of how a message consisting of three-letter code words can be changed by changing a letter, adding a letter, or deleting a letter. In the first example, the letter R is changed to the letter H. In the second example, the letter P is inserted. In the third example, the letter T in the third word is deleted. Note that since the code is read in three-letter words, all the words after an insertion or deletion are changed. Of course, only four letters are used in the genetic code. (The different colors are used to make it easier to see how insertions and deletions change the codons.)

Normal Message:
THE CAT ATE THE RAT AND WAS SAD ALL DAY ＿ ＿ ＿

Changing one letter:
THE CAT ATE THE HAT AND WAS SAD ALL DAY ＿ ＿ ＿

Inserting one letter:
THE CAT PAT ETH ERA TAN DWA SSA DAL LDA Y＿ ＿

Deleting one letter:
THE CAT AET HER ATA NDW ASS ADA LLD AY＿ ＿ ＿ ＿

at some time in the past and that this change in the gene for the β chain of hemo-globin has been carried through successive generations. This kind of mutation, a single-base change, is known as a *point mutation.*

22-8 VIRUSES

Viruses exist in a variety of shapes and sizes, but they are always smaller than bacteria. They are made up of a core of either DNA or RNA that is surrounded by a protective coat composed primarily of protein molecules. Each virus has the ability to infect certain specific cells, causing diseases such as measles, influenza, Rocky Mountain spotted fever, and cold sores.

Viruses differ from cells in two important respects. Viruses contain only one type of nucleic acid, either DNA or RNA, and cells contain both. And viruses, unlike cells, cannot divide by themselves but always require a host cell for replication.

The nucleic acid molecules of most viruses are like the nucleic acid molecules we studied in Chapter 20. For example, the smallpox virus and other viruses contain double-helical DNA. The influenza virus, the poliomyelitis virus, and other viruses contain single-stranded RNA. However, there are viruses that contain single-stranded DNA and viruses that contain double-stranded RNA. As we have already discussed, a single-stranded DNA molecule can be used to form a comple-mentary DNA molecule. Replication gives new DNA double helices. We also learned from Section 20-3.3 that the enzyme reverse transcriptase catalyzes the syn-thesis of DNA using viral RNA as a template. Therefore, we can see that the genetic information present as a sequence of bases in an RNA molecule can be used to make a DNA moelcule that can replicate, making copies of itself. The new DNA molecules can then be transcribed to make more RNA molecules.

In order for a virus to reproduce, its nucleic acid must enter a specific host cell and use the cell's ribosomes, enzymes, and other molecules to make additional nucleic acid and protein molecules. New virus particles can then be constructed from these nucleic acid and protein molecules. The nucleic acid of the virus must code for the protein molecules that make up the virus' coat. It also codes for one or more enzymes involved in virus multiplication. One process of virus replication is shown in Figure 22.14.

When the DNA or RNA of a virus enters a cell, it interferes with the normal functions of the cell and causes it to concentrate on making the virus' nucleic acid and protein molecules. The DNA or RNA molecules and the molecules of the protein coat then spontaneously associate to form new virus particles. The newly formed virus particles are then released from the cell. After release, they are free to infect other cells, and the process can be repeated. The infected cells often die as a result of the virus infection.

Not all viruses cause cells to die. Some viruses change infected cells so that they begin to divide rather than die. Most cells in an adult person no longer divide because the tissue in which they are found is no longer growing. A cell, surrounded by other cells of the same type, ceases to divide or move about. We refer to this phenomenon as *contact inhibition of growth and movement.* When a tumor virus

Figure 22.14 Life cycle of a double-stranded DNA virus that infects bacterial cells.

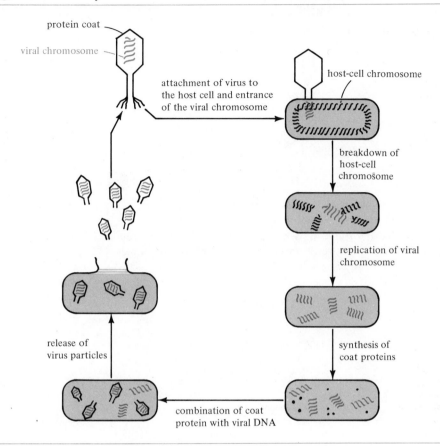

protein coat

viral chromosome

attachment of virus to
the host cell and entrance
of the viral chromosome

host-cell chromosome

breakdown of
host-cell
chromosome

replication of viral
chromosome

synthesis of
coat proteins

release of
virus particles

combination of coat
protein with viral DNA

infects a cell, the cell loses its contact inhibition of growth and begins to divide in
an uncontrolled manner, causing a tumor. Such a cell is called a *transformed cell.*
Transformed cells have cell surfaces and contain enzymes and other proteins that
are different from those of normal cells.

Cells apparently become transformed when viral DNA or DNA made from
viral RNA by reverse transcriptase (see Section 20-3.3) becomes incorporated
within the cells' own DNA. The tumor may become progressively worse and result
in death. In this case, the tumor is called a *malignant tumor* or a *cancer.* Transformed cells in malignant tumors may lose contact inhibition of movement, in
addition to losing contact inhibition of growth, and move about, transferring the
cancer to other tissues. The transfer of a cancer from one organ to another is
called *metastasis.*

PROBLEMS

1. What is the amino acid sequence resulting from the following segments of mRNA molecules? (Use Figure 22.3.) Be sure that you know the three-letter abbreviations for the amino acids (see Table 21.1).

   (a) 5′ ∿∿ CGAGAAGUC ∿∿
   (b) 5′ ∿∿ AACUCCAUG ∿∿
   (c) 5′ ∿∿ UGGGGUAAG ∿∿
   (d) 5′ ∿∿ CCACUAAGU ∿∿
   (e) 5′ ∿∿ GACCACACA ∿∿

2. What are the anticodons for the following codons?
   (a) Ser: 5′   UCU   3′
   (b) Asp: 5′   GAC   3′
   (c) Leu: 5′   CUA   3′
   (d) Met: 5′   AUG   3′
   (e) Arg: 5′   AGG   3′

3. A segment of a DNA molecule has the following sequence of bases. (a) What is the base sequence on an mRNA molecule produced from this segment of the DNA molecule? (b) What would be the order of amino acids produced from this portion of the mRNA molecule?

   DNA: 5′ ∿∿AGGTACTGA ∿∿ 3′

4. Answer the questions in problem 3 for the following DNA segment.

   DNA: 3′ ∿∿ GTCAGTCGG ∿∿ 5′

5. List the three phases of protein synthesis.
6. List the three steps of the elongation phase of protein synthesis.

ANSWERS TO PROBLEMS   1. (a)   ∿∿ Arg—Glu—Val ∿∿
(b) ∿∿ Asn—Ser—Met ∿∿   (c) ∿∿ Trp—Gly—Lys ∿∿
(d) ∿∿ Pro—Leu—Ser ∿∿   (e) ∿∿ Asp—His—Thr ∿∿   2. (a) 3′   AGA   5′
(b) 3′   CUG   5′   (c) 3′   GAU   5′   (d) 3′   UAC   5′   (e) 3′   UCC   5′
3. (a) mRNA: 3′ ∿∿UCCAUGACU ∿∿ 5′
   (b) protein: C-terminal Pro—Val—Ser   N-terminal
4. (a) mRNA: 5′ ∿∿ CAGUCAGCC ∿∿ 3′
   (b) protein: N-terminal Gln—Ser—Ala   C-terminal
5. initiation, elongation, and termination   6. binding of aminoacyl-tRNA, peptide-bond formation, and translocation

SUMMARY

In Chapter 21, we learned that proteins are polymers of amino acids and that the molecules of each specific protein have a fixed sequence of amino acid residues in the polymer chain. In Chapter 20, we learned that the information specifying this sequence of amino acids (the primary structure) is carried in a molecule of

DNA (the gene). We also learned that this information, in the form of a linear sequence of heterocyclic bases, is transferred to an mRNA molecule by transcription. In this chapter, we learned how the resulting linear sequence of heterocyclic bases in the RNA molecule serves as a template to order the amino acids in the correct sequence to manufacture a particular protein molecule. We also learned that a mutation is caused by a change in the sequence of bases in a DNA molecule that results in a change in the sequence of bases in an mRNA molecule and shows up as a change in the sequence of amino acids in a protein.

We learned that amino acids do not interact directly with mRNA molecules, but first combine with tRNA molecules that serve as adaptor molecules. A given tRNA molecule is specific for a given amino acid and contains a sequence of three bases that serves as the anticodon for that particular amino acid. These three bases are complementary to, and pair through hydrogen bonding with, a sequence of three bases on the mRNA molecule known as the codon.

Before the amino acids combine with their own tRNA molecules, they are activated by reaction of their carboxyl groups with ATP molecules. These activated amino acid residues are then attached through their carboxyl groups to the 3′ end of the tRNA molecule. In becoming attached to tRNA molecules, the amino acids remain activated. There is a specific activating enzyme for each amino acid that is the same enzyme that attaches the amino acid residue to its specific tRNA molecule. The tRNA molecule with its attached amino acid then makes its way to a ribosome. An mRNA molecule also attaches to the ribosome and moves across it, bringing successive codons into position to interact with the correct tRNA molecule charged with an amino acid residue. On the ribosome, the three bases of the anticodon of the tRNA molecule interact with the three complementary bases of the codon of the mRNA molecule.

We discussed the fact that protein synthesis consists of initiation, elongation, and termination. Bacterial proteins always begin with the amino acid derivative N-formylmethionine because formylmethionyl-tRNA recognizes a special initiation codon on the mRNA molecule. Initiation of protein synthesis in eukaryotes begins with the binding of methionyl-tRNA.

Elongation consists of the binding of another aminoacyl-tRNA molecule, peptide-bond formation, and translocation. Each ribosome has two special regions on its surface: the P (peptidyl) site and the A (aminoacyl) site. The P site contains a tRNA molecule with an attached growing peptide chain. When an aminoacyl-tRNA molecule enters the A site, the partially synthesized peptide chain on the tRNA molecule in the P site is transferred to the amino group of the amino acid residue attached to the tRNA molecule on the A site. Thus, the peptide chain grows by stepwise addition of amino acids to its carboxyl end. Following transfer of the growing peptide chain to the amino acid residue to lengthen the chain by one amino acid residue, the peptidyl-tRNA molecule, now in the A site, is moved to the P site and the mRNA is moved across the ribosome to orient the next codon in the A site. This is the process of translocation.

Termination occurs when the translocation process reaches a termination codon. Protein release factors bind to the termination codons and effect hydrolysis of the bond between the protein and the tRNA molecule, releasing the completed protein chain.

Figure 22.15   Diagrammatic summary of replication, transcription, and translation beginning with a portion of a sense strand of a DNA molecule coding for phenylalanine, alanine, and glycine. Arrows under the new strands of DNA and
RNA indicate the direction of synthesis. Other arrows indicate the transfer
of genetic information.

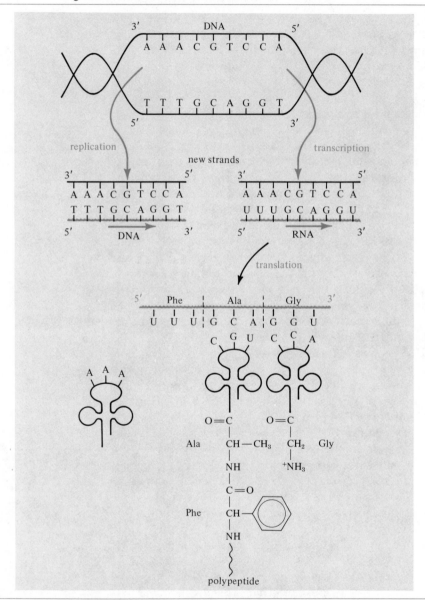

A summary of replication, transcription, and translation as discussed in Chapters
20 and 22 is presented in Figure 22.15. Now that we have learned how protein
molecules are made, in the next chapter we will learn about the properties of the
proteins that act as enzymes.

ADDITIONAL PROBLEMS

7. Using Figure 22.3, write the sequence of the peptides formed by translation of the following theoretical mRNA molecules from humans. (The 5′ end is on the left.)

 (a) AUGAACUACGGAUGCGUCCCCAUUGACUUCUGA
 (b) AUGACAGUCUCAAUGCACUGGUUAGAAAAGUAG
 (c) AUGUGUACGACGCGAAUCACUUGCGCUGCGUAA
 (d) AUGUGCCAGGAAAGCAUGACCACGUACUCGUGA

8. Consider the following sequence of codons on an mRNA molecule.

      5′ ∿ UUUGGCAUGC ∿ 3′

 (a) What is the amino acid sequence formed by translation?
 (b) What would be the amino acid sequence formed if the guanidine nucleotide residue (in color) were changed to an adenosine nucleotide residue by a mutation?
 (c) What would be the amino acid sequence formed if the guanidine nucleotide residue (in color) were deleted by a mutation?
 (d) What would be the amino acid sequence formed if a cytidine nucleotide residue were inserted just before the guanidine nucleotide residue (in color) by a mutation?
 (e) What would be the amino acid sequence formed if the underlined cytidine nucleotide residue were changed to a uridine nucleotide residue?

CHAPTER 23

ENZYMES

In Chapter 21, we examined the structures and functions of proteins. Then in Chapter 22, we discussed how protein molecules are made, using the cells' genetic information. We know from Chapter 21 that certain protein molecules function as *enzymes* and that enzymes are proteins that are catalysts. In this chapter, we will learn more about enzymes and their properties.

Enzymes are necessary for the normal functioning of cells. No one knows the exact number of enzymes in any cell, but there are certainly thousands in most living cells. Disease states may be caused by the absence of an enzyme (a genetic disease), the introduction of an enzyme inhibitor (for example, a bacterial toxin or a poison), the overproduction of an enzyme, or the introduction of a foreign enzyme (for example, from an infection). Some pharmaceuticals are inhibitors that prevent the action of specific enzymes. Enzyme assays are important in medical diagnosis. A decrease or increase in enzyme activities in tissues and fluids is indicative of the various causes of disease. Injury may release tissue enzymes to the blood; therefore, the presence of enzymes in the blood is often an indication of the site of injury. For all these reasons and others, it is important that everyone interested in health sciences, or in any other field based on the life sciences, have an understanding of how enzymes act and how their activity is controlled. In this chapter, we will discuss the properties of enzymes. In the remaining chapters, we will examine the reactions catalyzed by enzymes.

An **assay** is a quantitative measurement of the amount of enzyme activity.

Chemical reactions that occur in living cells are the same chemical reactions that we have studied in previous chapters and run in the laboratory. What makes the chemical reactions of living cells unique is that they are very rapid at body temperature, which is relatively low. These reactions are so fast at low temperatures because they are catalyzed by enzymes, and enzymes are very efficient catalysts (see Section 7-1.2). Because enzymes are proteins that are catalysts, the study of enzymes focuses on how catalysts act and how the tertiary and quaternary structures of enzymes are related to their efficiency in catalyzing reactions.

1. Define enzyme, catalyst, substrate, enzyme-substrate complex, enzyme assay, maximum velocity (V_{max} or V_m), Michaelis-Menten constant (K_m), cofactor, activator, coenzyme, active site, competitive inhibition, noncompetitive inhibition, allosteric enzyme, negative allosteric effector, positive allosteric effector, regulatory site, international unit (IU) of enzyme activity, and isoenzyme (isozyme).

2. List the seven factors that affect enzyme activity and explain how each affects activity.

3. Describe competitive inhibition and its use in medicine.

4. Explain the concept of the active site of an enzyme.

23-1 NAMING ENZYMES

In several earlier chapters, specific enzymes were identified by name. It will be important in this chapter, and each of the remaining chapters, for us to know the names of enzymes and to determine the action of enzymes from their names. You will have to memorize many enzyme names. An understanding of how enzymes are named will help in this memorization task.

There are five ways of naming enzymes. Four of these five ways give the enzyme a trivial or common name (see Table 23.1). The first enzymes discovered were given names that have no relationship to their activities. These are primarily digestive enzymes. Examples of enzymes named in this way are pepsin, trypsin, and chymotrypsin (all of which catalyze the hydrolysis of proteins) and catalase (which catalyzes the decomposition of hydrogen peroxide into water and oxygen).

$$2 \, H_2O_2 \xrightarrow{\text{catalase}} 2 \, H_2O + O_2$$

Table 23.1 Methods for Naming Enzymes

Method for Giving Enzymes Common Names	Examples of Common Names	Some Systematic Names
1. No relationship to activity	Pepsin Trypsin Chymotrypsin Catalase	Peptide peptidohydrolases
2. Substrate name plus -*ase*	Peroxidase Fumarase Glucose 6-phosphatase Alkaline phosphatase Lactase Ribonuclease Urease	Donor: H_2O_2 oxidoreductase Fumarate hydratase D-Glucose 6-phosphate phosphohydrolase
3. Reaction type plus -*ase*	Invertase	β-D-Fructofuranoside fructohydrolase
4. Substrate name and reaction catalyzed plus -*ase*	Lactate dehydrogenase Glucose oxidase Glucokinase	L-Lactate: NAD oxidoreductase β-D-Glucose: O_2 oxidoreductase ATP: D-Glucose 6-phosphotransferase

Other enzymes have been named by adding the suffix *-ase* to the substrate name. Examples of enzymes named in this way are peroxidase (which catalyzes the oxidation of an organic compound using hydrogen peroxide as the oxidant); alkaline phosphatase (which catalyzes the hydrolysis of phosphate esters at an alkaline pH); fumarase [which catalyzes the conversion of fumarate into L-malate (see Section 25-1.2)]; and maltase, glucose 6-phosphatase, ribonuclease, and urease (which catalyze the hydrolysis of maltose, D-glucose 6-phosphate, RNA, and urea, respectively).

The **substrate** is the reactant in an enzyme-catalyzed reaction.

A third method for naming enzymes is to add the suffix *-ase* to the name for the reaction that is catalzyed. An example of an enzyme named in this way is invertase, which catalyzes the hydrolysis of sucrose into D-glucose and D-fructose, a reaction called an inversion.

A fourth method names the substrate and the reaction catalyzed and adds the suffix *-ase*. Examples of enzymes named in this way are lactate dehydrogenase (see Sections 23-5 and 24-2.2) and glucose oxidase (see Section 23-4).

Biochemists also use an international system of enzyme classification that has specific rules for naming enzymes. This classification method assigns to each enzyme a name that describes the activity of the enzyme and a number that is part of an indexing system. The six main groupings in this international system are as follows:

1. oxidoreductases (catalyze oxidation-reduction reactions)
2. transferases (catalyze group transfers)
3. hydrolases (catalyze the cleavage of bonds by the addition of water)
4. lyases (catalyze the cleavage of carbon-to-carbon bonds and certain other bonds)
5. isomerases (catalyze isomerization reactions)
6. ligases (catalyze the formation of bonds)

The international, systematic name is the most descriptive of all names for enzymes because it includes the substrates and the reaction catalyzed (see Table 23.1). In most cases, however, the systematic name is rather long, and biochemists, biologists, and clinical personnel tend to use common names.

There are no official abbreviations for enzyme names, although the use of abbreviations is expanding, particularly in computer-controlled data-processing systems. In some cases, two or more abbreviations are used for the same enzyme. The names and abbreviations used in this book are ones that are in general usage, and are primarily common names. However, biochemists, medical technologists, and pathologists may use different names for the same enzyme, and names and abbreviations may vary from one laboratory to another.

23-2 HOW CATALYSTS FUNCTION

As catalysts (see Section 7-1.2), enzymes have the following characteristics.

1. They are unchanged in the reaction. In other words, they are not used up in the reaction.

2. They are effective in small amounts.

3. They do not affect the equilibrium of a reversible chemical reaction. The function of enzymes, like other catalysts, is to speed up the process in either direction. Therefore, the same chemical equilibrium will be reached with or without the enzyme, although it may take days, months, or even years to reach equilibrium without the enzyme. For example, consider the reaction A + B ⇌ C + D, with equilibrium concentrations of 30% (A + B) and 70% (C + D). The enzyme will shorten the time it takes to reach these equilibrium concentrations, whether 100% (A + B) or 100% (C + D) is the starting point.

4. They exhibit specificity in their ability to accelerate chemical reactions, although the degree of specificity varies. Most enzymes act only with a single substrate or with a very limited number of chemically similar compounds. Some enzymes are less specific, however. For example, as we learned in Section 21-2.1, trypsin and chymotrypsin are proteases that catalyze the hydrolysis of peptide bonds involving only two or three specific amino acids. Other proteases are less discriminating about the nature of the amino acids that form the peptide bond.

Consider the following general reaction.

$$\text{substrate (S)} \rightleftharpoons \text{product (P)}$$

In any given population of molecules, there is a distribution of energies (see Figure 23.1). Because molecules must have a certain energy before they can react, substrate molecules must be activated before they can be converted into product molecules.

Figure 23.1 Energy distributions for a population of molecules (black curve) and the same molecules after they have been heated (color curve). The activation energy (E_a) for the reaction in question is indicated by x on the abscissa. Therefore, the entire population of molecules to the right of the dashed vertical line has sufficient energy to react (shaded area). Note that there are many more molecules that can react in the population of molecules at the higher temperature. Note also that lowering the activation energy to, for example, the point indicated by • on the abscissa also increases the population of molecules with sufficient energy to react without raising the temperature (average energy of the molecules).

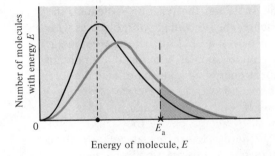

Energy of molecule, E

Figure 23.2   Diagram of the average energies of the populations of substrate molecules, product molecules, and activated substrate molecules and the activation energy (E_a) required for the reaction S $\longrightarrow$ P. The path of the enzyme-catalyzed reaction is given in color. The enzyme-catalyzed reaction goes by a different route, forming an intermediate enzyme-substrate complex (ES). The energies of activation for the formation of (E + S)*, ES*, and (E + P)* are smaller than that for the uncatalyzed reaction. (* = activated state) Note that the energy of ES is not necessarily lower than that of E + S.

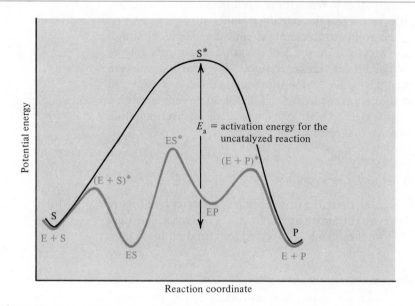

As we learned in Section 7-1.3, the amount of energy that must be acquired by molecules before they can react is called the *activation energy* (see Figure 23.2, E_a, and Figure 23.1, dashed vertical line).

There are two ways that a chemical reaction can be accelerated: (1) by raising the temperature and (2) by adding a catalyst. When the reaction is heated, the energy of the molecules is increased and, hence, the percentage of molecules with the required activation energy is increased (see Figure 23.1, color curve versus the black curve). However, living cells of animals, such as humans, that maintain a constant body temperature cannot use heat energy as a means of speeding up chemical reactions. Catalysts provide the second way to accelerate a chemical reaction: they lower the activation energy. In other words, catalysts lower the energy barrier that substrate molecules must overcome before they can be converted into product molecules (see Figure 23.1, vertical line on the left versus vertical line on the right). Enzymes lower the activation energy of a reaction by changing the path of the reaction. The enzyme (E) reacts with the substrate molecule (S) to form an **enzyme-substrate complex (ES).*** In a second step, the substrate molecule is converted into a product molecule. The result is that the enzyme-substrate complex is converted into an enzyme-product complex (EP). EP then dissociates to regenerate the enzyme and release the product (P).

An **enzyme-substrate complex (ES)** is the combination of an enzyme with its substrate.

$$E + S \underset{\text{release}}{\overset{\text{binding}}{\rightleftharpoons}} ES \overset{\text{catalysis}}{\rightleftharpoons} EP \underset{\text{binding}}{\overset{\text{release}}{\rightleftharpoons}} E + P$$

Each of these reactions has its own activation energy, which is much lower than that for the uncatalyzed reaction (see Figure 23.2).

The result of catalysis by an enzyme is like that of taking a road that goes around a mountain over several smaller passes as opposed to one that goes directly over the mountain. Taking the road around the mountain requires less fuel (less energy) because less climbing is involved. Lowering of the activation energy by an enzyme has the effect of increasing the proportion of molecules having sufficient energy for reaction. As a result, more molecules are reacting at a given time, just as reducing the amount of fuel required to get cars around a mountain allows more cars to make the trip. As a result, the reaction is more rapid.

23-3 FACTORS THAT AFFECT ENZYME-CATALYZED REACTIONS

There are several factors that affect enzyme-catalyzed reactions:

1. concentration of substrates and products
2. concentration of enzyme
3. concentration of cofactors (activators and coenzymes)
4. concentration of inhibitors
5. temperature
6. pH
7. concentration of allosteric effectors

Each of these factors is discussed in the sections that follow.

23-3.1 EFFECT OF SUBSTRATE CONCENTRATION

Enzyme assays are the determination of enzyme activity. When doing an enzyme assay (for example, when determining the amount of an enzyme in the blood for diagnostic purposes), a solution containing the enzyme activity to be measured is added to a substrate solution and the initial rate of conversion of substrate to product is measured. Under these conditions, the concentration of product (P) is zero and the equation becomes

$$E + S \rightleftharpoons ES \rightleftharpoons EP \longrightarrow E + P$$

---

*The combination of an enzyme molecule with a substrate molecule is called a *complex* because the interaction often involves weak bonding such as hydrogen bonding, hydrophobic bonding, or ionic bonding. However, some substrate molecules actually react chemically with their enzyme molecules, giving an intermediate that is covalently bonded to the enzyme molecule.

Figure 23.3   Effect of the concentration of substrate on the initial velocity (rate) of an enzyme-catalyzed reaction.

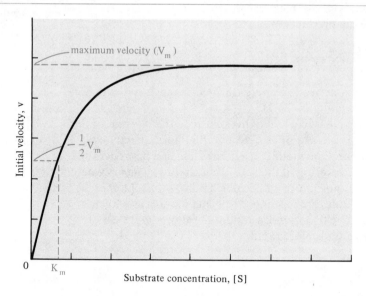

As the ratio of substrate molecules to enzyme molecules increases, the initial rate of reaction* increases in proportion to the substrate concentration ([S])†. The reaction rate continues to increase as the substrate concentration increases until essentially all enzyme molecules are tied up in an enzyme-substrate complex. At this point there are, for all practical purposes, no free enzyme molecules that can combine with more substrate molecules, so the velocity cannot increase, no matter how many more substrate molecules are added. The curve showing the effect of substrate concentration on reaction rate is given in Figure 23.3.

We can compare the effect of substrate concentration on reaction rate to a person picking apples. If there are only a few apples (substrate) on a tree, the picker (enzyme) must walk around to get them and can pick only slowly. The more apples there are on the tree, the faster they can be picked. If the tree is loaded so that the picker can stand in one spot and pick apples as fast as possible, then having more apples on the tree doesn't mean that the apples can be picked faster, because now the limiting factor is the speed at which the picker can work, not the availability of apples.

The maximum rate is the rate that results when the enzyme is saturated with substrate. As a result, this value is called the *maximum velocity* (V_m or V_{max}) or *saturation velocity*. The concentration of substrate that gives one-half V_m is the *Michaelis-Menten constant* (K_m) (see Figure 23.3). The K_m is roughly a measure of the affinity (attraction) of the enzyme for the substrate. The higher the K_m, the

---

*Biochemists refer to the rate of an enzyme-catalyzed reaction as the *velocity* of the reaction.

†As we learned in Section 6-2.2, the symbol [ ] is used to designate molar concentration. Therefore [S] is read *concentration of substrate*, [E] is read *concentration of enzyme*, and so on.

lower the affinity because, if a high concentration of substrate is needed to form an enzyme-substrate complex with one-half of the enzyme molecules at any instant, then the enzyme and substrate must have a low attraction for each other. If the K_m is low, it means that only a low concentration of substrate is needed to form an enzyme-substrate complex with one-half of the enzyme molecules at any instant. Therefore, the enzyme and the substrate must have a high affinity for each other. We can use the analogy of apple pickers again to illustrate this concept. If picker 1 picks apples from a tree with only a few apples on it at the same rate as picker 2 picks apples from a tree loaded with apples, then we might conclude that picker 1 has a stronger interest in (or attraction for) picking apples than does picker 2. The Michaelis-Menten constant for an enzyme and its specific substrate under specified conditions is characteristic of that enzyme, just as the melting point is characteristic of a simpler organic compound.

23-3.2 EFFECT OF ENZYME CONCENTRATION

The enzyme concentration also affects the rate of enzyme-catalyzed reactions. Consider a solution with substrate concentration in saturating amounts (large excess). If the enzyme concentration is doubled, the rate of the reaction is doubled; if the enzyme concentration is tripled, the rate of the reaction is tripled; and so forth (see Figure 23.4). This fact is the basis for enzyme assays.

The importance of K_m can be seen in the following examples. Brain hexokinase and liver glucokinase are two enzymes that catalyze the transfer of a phosphate

A **kinase** is an enzyme that catalyzes the transfer of a phosphate group from ATP to an acceptor molecule, forming a phosphate ester (from an alcohol), a phosphate amide (from an amine), or a phosphoric anhydride (from an acid).

Figure 23.4 Effect of the concentration of the enzyme on the velocity (rate) of an enzyme-catalyzed reaction in the presence of an excess of substrate [graph (a)]. Note that, for any given time of reaction, there is a linear relationship between the amount of product formed and the concentration of enzyme present. Therefore, a plot of the amount of product formed in a given time of reaction versus enzyme concentration gives a straight line [graph (b)]. [E] = concentration of enzyme.

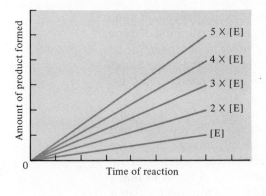

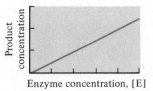

(a)

(b)

Figure 23.5 D-Glucose + ATP $\longrightarrow$ D-glucose 6-phosphate + ADP.

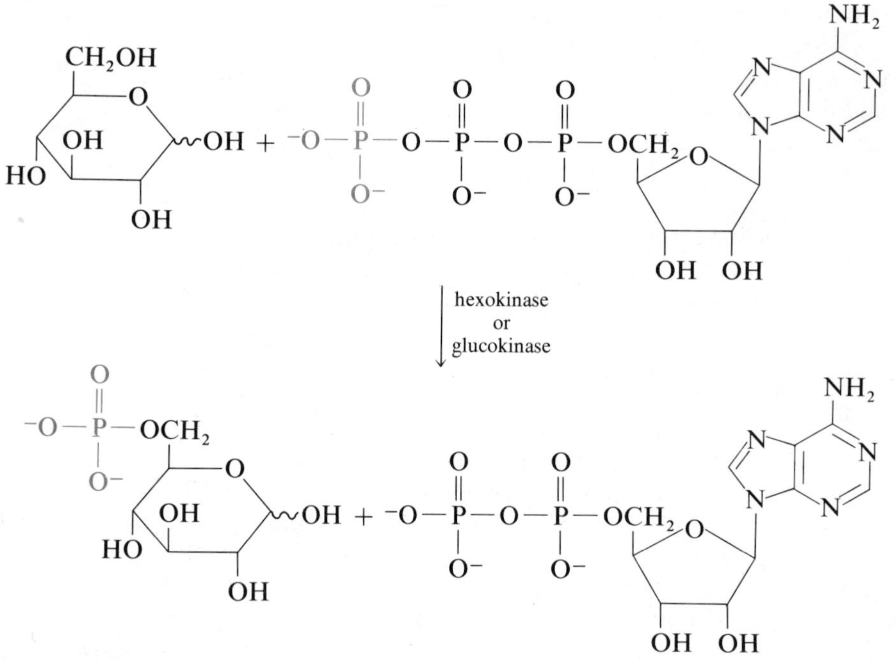

group from ATP to D-glucose to yield D-glucose 6-phosphate (see Figure 23.5). This reaction is involved in the trapping and use of glucose in cell. Brain hexokinase has a K_m for glucose of 0.05 **mM** and liver glucokinase has a K_m for glucose of 20 mM. Because the K_m is the substrate concentration that gives $\frac{1}{2} V_m$ and the normal concentration of glucose in the blood is about 4.5 mM (80 mg of glucose in 100 mL of blood plasma), the normal concentration in the blood is about 90 times the K_m for brain hexokinase (4.5 mM ÷ 0.05 mM = 90). Thus, this enzyme is completely saturated at all normal blood glucose concentrations (see the far right-hand part of the curve in Figure 23.3). The brain is almost totally dependent upon blood glucose for its energy supply. It is very efficient in glucose utilization because hexokinase catalyzes the conversion of glucose into glucose 6-phosphate at an essentially constant rate, no matter what the concentration of glucose in the blood. This allows the brain to capture sufficient glucose for its energy needs even if the concentration of glucose in the blood falls to low levels.

On the other hand, the concentration of glucose in the blood is only about one-fifth the K_m for liver glucokinase. Therefore, glucokinase acts in the left-hand part of the curve in Figure 23.3 where the rate changes with any change in glucose concentration. This property of glucokinase allows the liver to carry out its physiological role in controlling blood-glucose levels (see Chapter 24). Another example of the physiological use of K_m was mentioned in Section 14-3.5. Drinking methanol can be fatal. Methanol itself is not toxic, but it is converted by the liver enzyme alcohol dehydrogenase into formaldehyde, which is toxic. Alcohol dehydrogenase normally oxidizes ethanol to acetaldehyde.

mM is the abbreviation for millimolar, which is $10^{-3} M = 0.001 M$.

$$CH_3-CH_2OH + NAD^+ \xrightarrow{\text{alcohol dehydrogenase}} CH_3-CHO + NADH + H^+$$

ethanol   coenzyme   acetaldehyde

$$CH_3OH + NAD^+ \xrightarrow{\text{alcohol dehydrogenase}} HCHO + NADH + H^+$$

methanol   coenzyme   formaldehyde

One treatment for methanol poisoning is to give the patient large amounts of ethanol. You might ask, "Why get the patient drunk?" The answer is that the K_m of alcohol dehydrogenase for ethanol is lower than its K_m for methanol. Flooding the patient's system with ethanol ties up much of the alcohol dehydrogenase in an enzyme-ethanol complex. This prevents the enzyme from forming an enzyme-methanol complex and catalyzing the oxidation of methanol. The treatment must continue until most of the methanol has been excreted.

23-3.3 EFFECT OF THE CONCENTRATION OF COENZYMES AND ACTIVATORS

Many enzymes require inorganic ions such as Mg^{2+} and Zn^{2+} as **activators**. Activators put the enzyme molecule in the proper state to combine with the substrate or remove inhibitors. They are often metal ions whose presence is required for full enzymic activity. Many enzymes (but not all) require organic molecules called **coenzymes** in order to act as catalysts. The concentration of **cofactors**, activators and coenzymes, affects the rate of enzyme-catalyzed reactions. Hence, in any determination of enzyme activity, required cofactors must be present in amounts that are in excess of the minimum amounts required for maximum activity.

Enzyme-catalyzed reactions often involve direct reaction of the substrate and a coenzyme. The enzyme brings the two molecules together. Some coenzymes, like enzymes, are unchanged in the reaction; others participate in the reaction as acceptor or donor molecules and, therefore, are used up in the reaction. The latter type are regenerated in other reactions. The best way to learn how coenzymes function is through the study of specific examples. As we encounter enzymes and their coenzymes in the upcoming chapters, we will study their specific mode of action.

With some enzymes, the molecules functioning as coenzymes are attached to the enzyme molecules. In these cases, the group functioning as a coenzyme is usually referred to as a coenzyme rather than as a prosthetic group (see Section 21-2.6).

protein molecule + organic molecule = biologically active enzyme

    apoenzyme   *prosthetic group*   *holoenzyme*
                   or coenzyme

23-3.4 EFFECT OF ENZYME INHIBITORS

Much of **pharmacology** and chemotherapy is based on the selective inhibition of enzymes by natural or synthetic compounds. Enzyme inhibition, therefore, is a significant factor in the treatment of disease. Before we discuss inhibitors and their action, we must first discuss in more detail the mechanism of enzyme action.

Activators are inorganic ions required for enzymatic activity.

Coenzymes are low-molecular-weight, non-protein, organic compounds that are required for enzymatic activity.

Cofactors are any and all inorganic ions and organic molecules required for the activity of an enzyme.

An **apoenzyme** is the inactive protein that remains after a cofactor or prosthetic group is removed from an enzyme.

Pharmacology is the science of drugs, including their preparation, uses, and effects.

The Active Site

Each enzyme molecule has an active site. This active site is the part of the enzyme molecule that interacts directly with the substrate, determining both the specificity (high affinity for specific compounds) and the catalytic function of the active protein molecule. The active site is frequently described in terms of binding sites that hold the substrate in the proper place and shape, and the catalytic site where reaction takes place.

As far as we know, there are two ways in which active sites are formed. The folding of the peptide chain into a tertiary structure forms a crack or crevice that is complementary in shape to that of the substrate. This is called the *lock-and-key model*. Just as only one key has the specific shape that will open a lock, only specific substrate molecules have the shape that allows them to fit into the crack or crevice of the active site of the enzyme so that reaction can take place. An active site can

The **active site** is the specific region of the enzyme molecule where the substrate binds and reacts to form product.

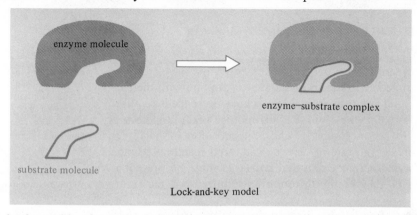

Lock-and-key model

also be formed by the enzyme changing shape after binding with the substrate. In this model, the active site has the shape that allows catalysis of a reaction involving the substrate only after the substrate is bound. This is called the *induced-fit model*.

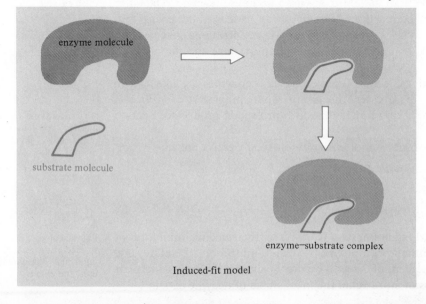

Induced-fit model

The active site contains the side chains of at least three amino acid residues brought together in a specific, active, geometrical pattern by the folding of the peptide chain. Therefore, the nature of the active site is determined by weak bonds (the secondary and tertiary structure) as well as by the primary structure (amino acid sequence).

Enzymes are often much larger than their substrates; at least, they are much larger than that part of the substrate with which they interact. Therefore, only a small portion of the enzyme can be in contact with the substrate. The rest of the enzyme is usually required to put the reactive groups of the active site into the proper arrangement in space.

Competitive Inhibition

There are two general types of enzyme inhibition: competitive and noncompetitive. In *competitive inhibition,* the inhibitor has a shape similar to that of the substrate. This often means that the substrate and the inhibitor have similar structures (see below). A *competitive inhibitor* binds with the enzyme at the active site, filling the active site and preventing the substrate from binding. Therefore, no reaction takes place. We call this kind of inhibitor a competitive inhibitor because there is competition between the substrate and the inhibitor for the active site of the enzyme. A diagrammatic representation of competitive inhibition is shown. Using

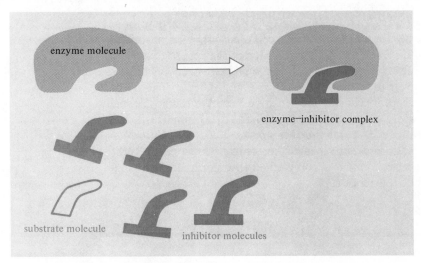

the lock and key analogy again, we can explain competitive inhibition in the following way. More than one key may fit in a lock. However, only the key with the correct notches will open the lock. When the other keys are in the key hole, the correct key cannot be inserted and, therefore, is prevented (inhibited) from opening the lock.

Bacteria can synthesize folic acid; humans cannot. Because folic acid is necessary for the normal functioning of the human body but cannot be made by our bodies, it is a vitamin. Bacteria use *p*-aminobenzoic acid (PABA) to make folic acid. The structures of PABA and the sulfa drugs (see Section 17-4.2) are similar.

Therefore, the sulfa drugs bind to the bacterial enzyme that uses PABA as a substrate to make folic acid and are competitive inhibitors of the synthesis of folic acid. By inhibiting the synthesis of folic acid, sulfa drugs prevent the growth of bacteria.

Noncompetitive Inhibition

Noncompetitive inhibitors destroy the condition of any part of the enzyme necessary for catalytic action. General protein denaturants such as heavy metal ions (for example, mercury and lead ions) that react with thiol ($-SH$) groups fall into this category. Another noncompetitive inhibitor is the cyanide ion (CN^-). This ion inhibits enzymes with a heme group similar to that of hemoglobin. These heme-containing enzymes catalyze oxidation-reduction reactions. Inhibition of these reactions, which we will discuss in Section 25-2.4, brings about cell death. Therefore, the cyanide ion and hydrogen cyanide gas are extremely poisonous.

$$\begin{array}{c} H_3C \\ \diagdown \\ \qquad CH-O-\overset{\overset{\displaystyle O}{\|}}{\underset{\displaystyle F}{P}}-O-CH \\ H_3C \diagup \qquad\qquad \diagdown CH_3 \\ \qquad\qquad\qquad CH_3 \end{array}$$

diisopropyl fluorophosphate (DFP)

Another noncompetitive inhibitor of enzymes is diisopropyl fluorophosphate (DFP). This ester inhibits the enzyme acetylcholinesterase, which is involved in the transmission of many nerve impulses. A nerve transmits an impulse from one point in the body to another through a traveling wave of changes in concentrations of ions. When the wave reaches the end of the nerve cell (*neuron*), it is transmitted to the next nerve cell, a muscle fiber, or another receptor cell in one of two ways. In many nerve cells, the impulse is transmitted through the release of acetylcholine, (see Section 17-3) which travels to the receptor cell and stimulates it. The nerve cells, muscle cells, or other receptor cells that are sensitive to acetylcholine are said to be *cholinergic*. Once the acetylcholine has brought about the desired response, it must be removed to permit recovery of the receptor so that the receptor can receive the next stimulus, and to prevent repeated, uncontrolled responses to the first stimulus. Removal of acetylcholine is accomplished through hydrolysis catalyzed by the enzyme acetylcholinesterase.

$$CH_3-\overset{\overset{\displaystyle CH_3}{|}}{\underset{\displaystyle CH_3}{N^+}}-CH_2-CH_2-O-\overset{\overset{\displaystyle O}{\|}}{C}-CH_3 + H_2O \xrightarrow{\text{acetylcholinesterase}}$$

acetylcholine

$$CH_3-\overset{\overset{\displaystyle CH_3}{|}}{\underset{\displaystyle CH_3}{N^+}}-CH_2-CH_2OH + HO-\overset{\overset{\displaystyle O}{\|}}{C}-CH_3$$

choline   acetic acid

As we can see, acetylcholinesterase activity is essential. Sometimes, however, this activity must be modulated. Pharmaceutical chemists and pharmacologists have developed drugs that alter acetylcholinesterase and other drugs that affect cholinergic nerves by enhancing or inhibiting the action of acetylcholine on the receptor. For example, drugs that inhibit the effect of acetylcholine on muscle fibers are used by anesthesiologists to relax muscles. Stronger reagents completely and irreversibly inhibit the action of acetylcholinesterase, allowing the accumulation of acetylcholine. This results in the uncontrolled contraction of muscles, an effective means of exterminating insects and people.

The German chemists who were first affected by esters such as DFP during their preparation suggested that the esters might be useful as insecticides. Their suggestion was ignored. However, just before World War II, others realized that these phosphate esters could be used in warfare as poisonous gases. The development of such nerve gases was rapid and continued for many years. Following World War II, the original suggestion that the esters could be used as insecticides was investigated. The problem was to reduce the volatility of the esters and, thus, their danger to humans, while retaining their deadly effect on insects. This has proved to be possible, and several effective phosphate-ester insecticides that are safe enough for careful use have been developed.

Interest in the development of nerve gases and insecticides stimulated research into their mechanism of action. It was discovered that DFP reacts with the hydroxyl group of a residue of the amino acid serine in the active site of acetylcholinesterase. In this way the enzyme is irreversibly blocked from acting as a catalyst for the hydrolysis of acetylcholine.

diisopropyl fluorophosphate

23-3.5 EFFECT OF TEMPERATURE ON ENZYME-CATALYZED REACTIONS

In Section 7-1.2, we explained that, like other chemical reactions, the rates of enzyme-catalyzed reactions are increased by increases in temperature. However, because enzymes are proteins, all the factors that affect the tertiary structures of proteins will also affect the activity of enzymes. For example, denaturation of the enzyme as it is heated, as discussed in Section 21-2.4, destroys the active site and, therefore, reduces the effective concentration of the enzyme. The general effect of temperature on an enzyme-catalyzed reaction is depicted in Figure 23.6. The upper limit for most human enzymes is 40°–50°C. (Normal body temperature is 37°C.)

23-3.6 EFFECT OF pH ON ENZYME-CATALYZED REACTIONS

Changes in pH alter the degree of ionization of various ionic groups on enzymes and other proteins (see Section 21-2.4). If the substrate is ionic, its ionization also changes with changes in pH. Thus, as the ionic nature of either the substrate or the enzyme is changed from the form that combines with the other, the reaction is slowed and a curve such as that in Figure 23.7 results.

Figure 23.6   Effect of temperature (heat) on an enzyme-catalyzed reaction. The colored dashed line indicates the increasing rate due to increasing temperature. The colored dotted line indicates the decreasing activity due to thermal denaturation. The solid line indicates the actual observed effect which is the composite effect of increasing rate and thermal denaturation. The point x on the abscissa indicates the optimum temperature. Velocity = reaction rate. Note that, once the enzyme begins to become denatured by heat, its activity drops rapidly.

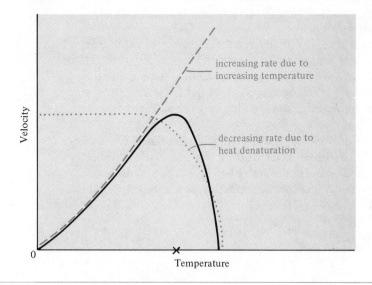

Figure 23.7
One general effect of pH on the velocity (rate) of an enzyme-catalyzed reaction. The point x on the abscissa indicates the optimum pH. Other types of pH dependence are also observed.

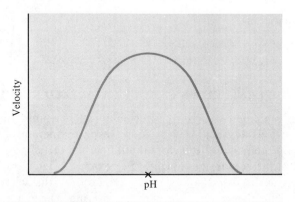

The pH at which an enzyme is most efficient is called the pH optimum for that enzyme. The range of values for the pH optimum may be either narrow or broad. As the pH increases or decreases above or below the optimum pH, the activity of the enzyme (the rate of the enzyme-catalyzed reaction) decreases for one or more of the following reasons.

1. There is a decrease in enzyme-to-substrate binding due to repulsion of molecules with the same charge or loss of a charge on either the substrate or the enzyme.

2. The catalytic functional groups in the active site (that is, the R groups on the amino acid residues in the active site) are not in the proper ionization state.

3. The enzyme unfolds (becomes denatured) (see Section 21-2.4).

23-3.7 EFFECT OF ALLOSTERIC EFFECTORS ON ENZYME-CATALYZED REACTIONS

Allosteric, or *regulatory, enzymes* are enzymes whose activity is altered by regulatory molecules. With allosteric enzymes, a regulatory molecule interacts at some position on the enzyme molecule other than the active site. A regulatory molecule that decreases the activity of an allosteric enzyme is a *negative effector* and a regulatory molecule that increases the activity of an allosteric enzyme is a *positive effector.* These effector molecules have shapes and structures that are quite different from the shape and structure of the substrate molecule. The place on the enzyme molecule at which an effector molecule binds is called a *regulatory,* or *allosteric, site.*

Allosteric control is dependent upon binding sites for the effectors and the existence of at least two conformational states* (two tertiary structures) of the enzyme. In one of these states, the binding sites have a high affinity for the substrate, and in the other, they bind weakly, if at all, with the substrate. One model of the action of allosteric enzymes is that binding of the effector molecules stabilizes the enzyme in one of its possible conformations. Binding of a negative effector stabilizes the enzyme in a less active, three-dimensional shape, and binding of a positive effector stabilizes the enzyme in a more active shape. This idea is represented in Figure 23.8. All allosteric enzymes have a quaternary structure, that is, they are composed of more than one subunit (see Section 21-2.5).

This kind of control is of great physiological significance because of its importance in the regulation of biochemical reactions. Effector molecules can be hormones, products resulting from the action of a hormone, intermediate products, or the end-product of processes involved in the synthesis or breakdown of various molecules in cells. We will study examples of this kind of control as we go along.

23-4 ENZYME ASSAYS

Often a physician must know the concentration of an enzyme in a patient's blood because this information is important in diagnosis. In some cases, a low concentration of enzyme is significant. For example, the activity of glucose 6-phosphate dehydrogenase (see Section 27-3.2) in red blood cells is decreased in one kind of anemia. In other cases, a high concentration of enzyme is significant. For example, the activity of certain enzymes in blood serum increases following a heart attack (*myocardial infarction†*). A myocardial infarction is the formation of an area of dead cells in the myocardium (the middle and thickest layer of the heart wall, which is composed of cardiac muscle) as a result of interruption of the blood supply

Blood serum is the clear, pale yellow liquid that remains after removal of cells and the clotting components from blood.

---

*As we learned in Section 11-4.3, the *conformations* of a molecule are the arrangements of atoms not determined by covalent bonding. *Conformational states* are the various shapes a molecule can assume due to rotation about single bonds. In this case, the term refers to the tertiary structures of a peptide chain.

†*Myo-* refers to muscle. *Cardi-* or *cardio-* refers to the heart. An *infarct* is an area of dead tissue.

Figure 23.8 One model for the action of allosteric effectors. □ = positive allosteric effector, ◯ = negative allosteric effector, and △ = substrate. The enzyme is a tetrameric enzyme with identical subunits.

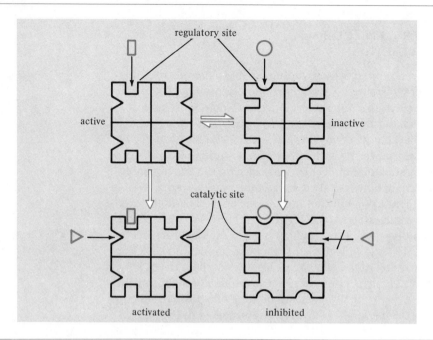

to the area by the formation of a clot in the coronary artery. One of the enzymes that leaks out of heart muscle cells when they die is creatine phosphokinase. In Section 27-2.2, we will discuss how determination of the activity of creatine phosphokinase in serum is used in diagnosis of a myocardial infarction.

Enzyme concentration is not reported in units such as grams per liter or moles per liter; rather, it is determined by the amount of substrate converted to product in a given time under specified conditions. This is called the *activity of the enzyme.* Enzyme assay results are reported in activity units. Because there are several methods for assaying most enzymes, the measure of the activity of an enzyme depends upon the method chosen by a particular laboratory. Thus, results from one laboratory cannot necessarily be compared with those from another unless identical methods are used. In an attempt to achieve uniformity, international units have been established. One *international unit (IU)* of enzyme activity is the amount of enzyme that will catalyze the conversion of one micromole of substrate to product in one minute. The pH, temperature, and other factors affecting enzyme activity must be specified and carefully controlled in determining units of enzyme activity.

Enzymes in Blood Serum

The concentration of an enzyme in serum is a result of (1) its rate of release from the the tissue and (2) the rate at which it is removed from circulation. There is some leakage of enzymes into serum from normal cells, but these small amounts are rapidly removed. When a cell is damaged, enzymes are released from it into the serum at a rate faster than they can be removed. Therefore, an increase in the serum concentration results.

Increases in the permeability of the plasma membrane of muscle cells, such as is caused by progressive muscular dystrophy, results in a partial loss of soluble enzymes from the cytoplasm of muscle cells to the serum. This, in turn results in a moderate increase in the activity of the enzymes in the serum.

Cell death, such as is caused by a blockage of blood flow to heart muscle (a myocardial infarction), acute hepatitis, and acute pancreatitis, results in a complete loss of all cellular enzymes from the dead cells. As a result, there are high levels of enzymes from the affected tissues in the serum. (The prefix *hepat-* or *hepato-* indicates a relationship to the liver and the suffix *-itis* denotes an inflammation. Therefore, hepatitis is an inflammation of the liver. Pancreatitis is an inflammation of the pancreas.) Also increases in the number of cells in a tissue with an accompanying increased production of enzymes give increased levels of enzymes in serum. Therefore, enzyme assays can be used to determine both the presence, and the location, of areas of cell death and areas of cell proliferation (tumors).

> **Acute** refers to having a short and relatively severe course.

Diagnostic Use of Enzyme Assays

Some enzymes are fairly localized; that is, only specialized tissues have a high concentration of them. Amylase (see Sections 24-1.2 and 24-1.4) is an example of this type of enzyme. It is found in high concentration only in the pancreas, and therefore, a high level of amylase activity in the serum reflects damage of the pancreas. Other enzymes, such as lactate dehydrogenase (LDH) (see Section 24-2.2), are present in almost all tissues. Thus, determination of total LDH activity provides less information about specific tissue damage. However, as we will see in Section 23-5, determination of LDH activity can, nevertheless, be an important diagnostic tool.

> An **amylase** is an enzyme that catalyzes the breakdown of starch.

Enzyme assays may also be performed on fluids and tissues other than blood serum. One application that is not as common, but is important, is the determination of certain enzymes in *amniotic fluid.* Amniotic fluid is the fluid that is produced by the cells that make up the membrane surrounding a fetus at very early stages of development within the uterus. Many inherited diseases are caused by the absence of an enzyme. When there is a history of a disease in the families of the parents, it is often desirable to know if the developing fetus is afflicted with the disorder. Whether a fetus is afflicted with a particular inherited disease can be determined by examining the cells in a sample of amniotic fluid to find out if the particular enzyme in question is present.

Clinical assay procedures make use of synthetic substrates. For example, the Bessey-Lowry method for the determination of the enzyme alkaline phosphatase in human serum uses *p*-nitrophenyl phosphate. This substrate, in the presence of alkaline phosphatase, undergoes hydrolysis to yield the *p*-nitrophenolate anion and inorganic phosphate (abbreviated P_i to indicate unesterified phosphate in any of its ionic forms). The rate of reaction is determined by measuring the formation of the product, *p*-nitrophenolate anion, which has a distinctive color. In enzyme assays, products are always measured because an excess of substrate is used. Small changes in concentration can be determined with much less error when the initial concentration is zero than when the initial concentration is some large value as it would be if disappearance of substrate were to be measured. Also, measurement of product formation is more desirable than measurement of substrate disappearance because the fluids assayed contain many enzymes and, hence, substrate disappearance may occur by means of other reactions.

$$O_2N-\langle\bigcirc\rangle-O-\overset{\overset{\textstyle O}{\|}}{\underset{\underset{\textstyle O^-}{|}}{P}}-O^- + H_2O \xrightarrow{\text{alkaline phosphatase}} O_2N-\langle\bigcirc\rangle-O^- + P_i$$

p-nitrophenyl phosphate   p-nitrophenolate

The use of glucose oxidase to determine the presence of glucose (see Section 19-2.1) is an example of the use of an enzyme as a reagent and the use of coupled enzyme reactions.

β-D-glucopyranose   D-gluconolactone   D-gluconic acid

$$H_2O_2 + \text{reduced dye} \xrightarrow{\text{peroxidase}} \text{oxidized dye} + 2\,H_2O$$
$$\text{(colorless)} \qquad\qquad\qquad \text{(green)}$$

The intensity of the green color produced is proportional to the original glucose concentration because there is a direct relationship between the amount of glucose oxidized and the amount of hydrogen peroxide (H_2O_2) produced and between the amount of hydrogen peroxide produced and the formation of the green color by oxidation of the dye. Therefore, a simple color measurement provides a determination of the amount of glucose present in a sample of blood, urine, or a food.

23-5 ISOENZYMES

Some proteins are associations of several polypeptide chains. When a protein is made up of two or more polypeptide chains, we call each individual chain a *subunit* (see Section 21-2.5). The arrangement of the subunits is called the quaternary structure of a protein (see Section 21-2.5). Allosteric enzymes have such a structure. We call enzymes such as allosteric enzymes that consist of two or more peptide chains in a quaternary structure *oligomeric enzymes.* Those enzymes consisting of a single peptide chain we call *monomeric enzymes.*

Some enzymes occur in multiple forms that catalyze the same reaction in the same organism. These forms are known as isoenzymes or, more simply, isozymes. The most thoroughly studied of these enzymes is lactate dehydrogenase (LDH). LDH can occur in five hybrid forms, each of which catalyzes the reaction shown. In this reversible reaction, the carbonyl group is reduced to a secondary alcohol group. The reducing agent is the coenzyme NADH, a coenzyme that will be discussed in the next chapter (see Section 24-2.1 and Figures 12.3 and 24.9). The activity of lactate dehydrogenase will be described further in Section 24-2.2.

Isoenzymes (isozymes) are different forms of the same enzyme found in the same cell or tissue. That is, isoenzymes are forms of an enzyme with different quaternary structures that catalyze the same reaction.

$$
\begin{array}{ccc}
\underset{\substack{|\\ \text{C}=\text{O} \\ |\\ \text{CH}_3}}{\text{COO}^-} + \text{NADH} + \text{H}^+ \underset{\text{LDH}}{\rightleftharpoons} \underset{\substack{|\\ \text{HOCH} \\ |\\ \text{CH}_3}}{\text{COO}^-} + \text{NAD}^+ \\
\text{pyruvate} \quad \underset{\text{coenzyme}}{\text{reduced}} \qquad \underset{\text{L-lactate}}{} \quad \underset{\text{coenzyme}}{\text{oxidized}}
\end{array}
$$

Using a technique called electrophoresis, five different LDH molecules can be found in blood serum. Each of these five LDH isoenzymes is composed of four polypeptide chains. There are two types of polypeptide chains in lactate dehydrogenase: H subunits and M subunits. Each molecule of LDH_1 contains four identical H (heart) subunits and each molecule of LDH_5 contains four identical M (muscle) subunits. The other LDH molecules are hybrid isoenzymes containing both types of subunits (see Table 23.2). Lactate dehydrogenase (LDH) is found in essentially all tissues, but the isoenzyme pattern is different for each tissue. As we have already learned, because enzymes get into the blood serum by leaking from injured or diseased cells, elevation of the activity of an enzyme such as LDH in blood serum is an indication of injury or disease. Because increases in specific serum LDH isoenzymes reflect the isoenzyme content of the injured or diseased tissue, LDH isoenzyme analysis is used in medical diagnosis.

Electrophoresis is the movement of molecules or particles under the influence of an electric field.

When a person suddenly experiences severe chest pain, the immediate problem is to decide whether or not this person has suffered a myocardial infarction. When heart muscle has been damaged, marked increases in concentrations of LDH_1 and LDH_2 (especially LDH_1) may persist for up to 14 days after the episode. Therefore, we can use the separation of LDH isoenzymes in confirming the diagnosis of suspected heart attacks.

Table 23.2 Lactate Dehydrogenase Isoenzymes

Isoenzyme	Subunits	Tissues in Which It Is the Principal Isoenzyme
LDH_1	HHHH	Heart muscle and kidney
LDH_2	HHHM	Brain and erythrocytes
LDH_3	HHMM	Thyroid gland, adrenal glands, lymph nodes, and pancreas
LDH_4	HMMM	Thymus, spleen, and leukocytes
LDH_5	MMMM	Liver and skeletal muscle

Another application of the LDH isoenzyme assay is in the diagnosis of liver diseases. LDH_5 is the predominant lactate dehydrogenase isoenzyme in the liver. Therefore, the presence of LDH_5 in serum is characteristic of liver diseases, including hepatitis, infectious mononucleosis, and cirrhosis. The presence of increased amounts of LDH_5 in the serum is most apparent in hepatitis.

PROBLEMS

1. The substrate concentration that gives one-half of the maximum velocity is the _____.
2. The cleft or crevice in the tertiary structure of an enzyme that binds the substrate and positions it for reaction is the _____.
3. A _____ that is a catalyst for biological reactions is an enzyme.
4. The amount of an enzyme that will convert 1 μmole of product per minute under defined conditions is 1 _____.
5. LDH_1 and LDH_5 are examples of _____.
6. Draw a representative curve for the activity of an enzyme with a temperature optimum of 33°C as a function of temperature.
7. Plot concentration of enzyme (x axis) against the amount of product formed in five minutes (y axis).

ANSWERS TO PROBLEMS   **1.** K_m **2.** active site   **3.** protein   **4.** international unit
5. isoenzymes   **6.**

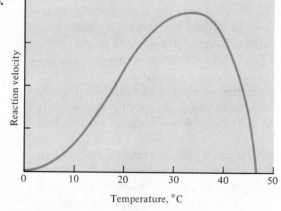

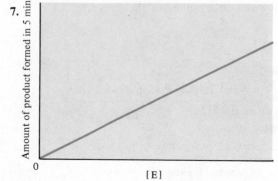

SUMMARY

In this chapter, we learned that most reactions of living organisms are catalyzed by certain proteins called enzymes. Enzymes have properties that are the same as those of other catalysts, but are especially characterized by their specificity and their high efficiency. They also have properties that are the same as those of other proteins: they exhibit thermal denaturation and changes in shape and activity with changes in pH.

We learned that the enzyme reacts with the substrate to form an enzyme-substrate complex, with the substrate situated at the active site of the enzyme. As a result, the rates of enzyme-catalyzed reactions are affected by the concentration of the substrate(s) and the concentration of the enzyme. Some enzymes require metal ions as activators and some require coenzymes. All are inhibited by one or more inhibitors of various classes. Therefore, the rates of enzyme-catalyzed reactions are also affected by the concentrations of cofactors and inhibitors. Competitive inhibition of enzymes is important in the treatment of disease. The activity of some enzymes is controlled by the binding of a molecule other than the substrate at a site (allosteric site) on the enzyme other than the active site. Such regulatory (effector) molecules may either activate or inhibit the enzyme.

Finally, we learned that some enzymes exist in multiple molecular forms called isoenzymes or isozymes. Determination (assay) of enzymes and isoenzymes, especially those in blood serum, is an important diagnostic tool for physicians.

With this background, we are ready to study how enzymes catalyze the reactions that are essential for life. In the remaining chapters, we will examine the reactions essential for life and the specific enzymes that catalyze them, beginning with the energy-producing reactions.

ADDITIONAL PROBLEMS

8. Are the following statements true or false?
   (a) All enzymes are proteins.
   (b) All proteins are enzymes.
   (c) All enzymes are catalysts.
   (d) All enzymes require coenzymes.
   (e) All enzymes are made as proenzymes.
   (f) All enzymes have an optimum temperature.
   (g) All enzymes can be denatured.
   (h) All enzymes have a V_{max}.
   (i) All enzymes have a K_m.
   (j) All enzymes form enzyme-substrate complexes.

9. The substance on which an enzyme acts is called the _____.

10. The location within the tertiary structure of an enzyme molecule where a reactant molecule binds is called the _____.

11. The reactant in an enzyme-catalyzed reaction is called the _____.

12. A substance that interacts with an enzyme at a site other than the one at which the reactant binds and reduces the activity of the enzyme by changing its shape is called a(n) _____.

CHAPTER 24

DIGESTION AND METABOLISM OF CARBOHYDRATES

Now that we have examined the structure, function, synthesis, and mode of action of enzymes, we are ready to discuss the reactions they catalyze. All life is dependent on a source of energy. In this chapter and chapters 25 and 27, we will study energy-producing reactions. Much of the energy for most living organisms comes from carbohydrate molecules. In muscles of mammals, the main store of energy is the polysaccharide glycogen. D-Glucose is an extremely important source of energy for the human brain.

It is important for us to be acquainted with the series of chemical reactions involved in the use of carbohydrates for energy. This series of reactions is found in the great majority of all tissues; in many, it is essential for life. We will study the reactions in this chapter, but first we will see how the digestible carbohydrates of our diets are made available for use as a source of energy.

24-1 DIGESTION AND ABSORPTION OF CARBOHYDRATES

LEARNING OBJECTIVES

1. Define amylase, hydrolase, gluconeogenesis, and glycogenolysis.
2. Trace the digestion of starch, sucrose, and lactose through the mouth, the stomach, and the small intestine.

3. Describe and diagram the action of alpha-amylase.

Carbohydrates supply 40–60% of the calories in most diets. Food preferences and economics may raise the percentage of diets having inadequate amounts of protein in segments of the population. Included in these groups are older persons who have trouble chewing, alcoholics, growing children who, because of food preferences, may consume a diet consisting largely of cereal foods, and those with incomes inadequate for the purchase of high-protein foods.

The digestible carbohydrates in an average human diet in the United States include the following groups.

Digestible carbohydrates in diet	Percentage
Starch (polysaccharide)	About 60%
Sucrose (disaccharide)	About 30%
Lactose (disaccharide)	About 5%
Glucose and other monosaccharides	About 5%

The food we eat also contains considerable amounts of indigestible polysaccharides. Cellulose is the indigestible polysaccharide consumed in the largest amounts; lesser amounts of polysaccharides known as pectin and hemicelluloses are also present in plant cell walls. These indigestible polysaccharides, which add fiber or roughage to our diets, are found principally in vegetables. Various indigestible polysaccharides are added to prepared foods to improve their physical properties. Of course, these indigestible polysaccharides add no calories to the diet.

24-1.1 THE ROLE OF SALIVA IN THE DIGESTION OF CARBOHYDRATES

The food we eat is first chewed and mixed with saliva in our mouths. Saliva contains an *alpha-amylase (α-amylase)*. Amylases are enzymes that catalyze the hydrolysis of the glycosidic bonds of the two polysaccharides of starch. α-Amylases are amylases that catalyze the cleavage of internal bonds of the starch polysaccharides rather than the cleavage of glycosidic bonds at the ends of the molecules. As a result, the α-amylases give a rapid reduction in molecular weight.

The functions of saliva are primarily mechanical. Saliva assists in chewing and swallowing and aids in speech. Saliva also solubilizes dry foods, aiding in the stimulation of taste nerves that play a role in the secretion of **gastric juice.** The α-amylase of saliva is relatively unimportant in starch digestion because food is in contact with the active enzyme for only a short time. Food does not stay in the mouth long, and the enzyme is inactivated by the acid and the enzymes of the stomach that digest proteins.

Gastric juice is the digestive juice of the stomach.

24-1.2 ACTION OF α-AMYLASES

Amylases act on the polysaccharides of starch (amylose and amylopectin). Hydrolytic enzymes acting on polymers may either remove residues starting at one end of the chain or catalyze the hydrolysis of linkages within the chain. As we stated in the previous section, α-amylases act by splitting the polysaccharide molecules internally rather than by removing units from the ends of chains.

We know that α-amylases hydrolyze amylose and amylopectin to oligosaccharides (primarily to oligosaccharides having six or seven glucose units). The

Figure 24.1 Action of α-amylase on amylose. Glc = D-glucose or D-glucosyl residue.

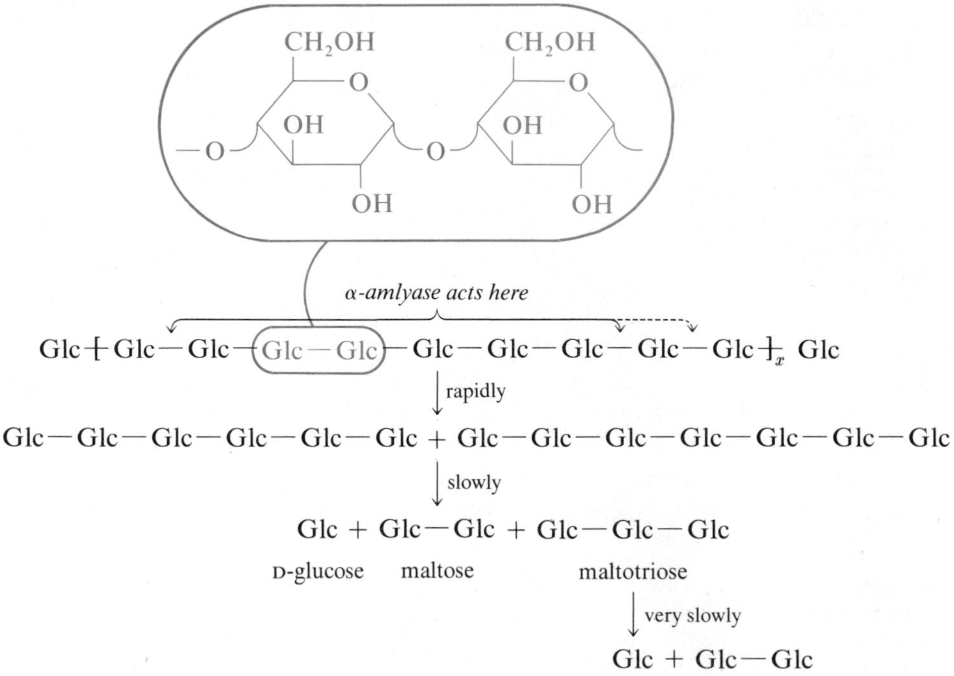

α-amlyase acts here

Glc ┼ Glc — Glc ┤Glc — Glc├ Glc — Glc — Glc — Glc — Glc ┤$_x$ Glc

↓ rapidly

Glc — Glc — Glc — Glc — Glc — Glc + Glc — Glc — Glc — Glc — Glc — Glc — Glc

↓ slowly

Glc + Glc — Glc + Glc — Glc — Glc

D-glucose   maltose   maltotriose

↓ very slowly

Glc + Glc — Glc

α-amylases then act more slowly on these oligosaccharides to break them down into smaller fragments (maltose and maltotriose). The α-amylases act even more slowly on maltotriose (a trisaccharide), but eventually can convert a chain that contains only α-D-(1 → 4)-linked glucopyranosyl units into D-glucose and maltose (see Figure 24.1). They do not catalyze the hydrolysis of the glucosidic bond of the disaccharide maltose (see Section 19-2.1). Additionally, α-amylases only catalyze the hydrolysis of α-D-(1 → 4) linkages, never α-D-(1 → 6) linkages. Therefore, they, by themselves, cannot convert starch completely into D-glucose and maltose.

The action of α-amylase on amylose, the unbranched fraction of starch, is shown in Figure 24.1. The action of α-amylase on amylopectin, the branched fraction, forms oligosaccharides containing the α-D-(1 → 6) linkage, such as isomaltose (see Section 19-2.1), in addition to glucose, maltose, and maltotriose. However, the α-amylase of saliva brings about little of this digestion because it is not in contact with the starch polysaccharides long enough, as we have just learned. Another α-amylase acts in the small intestine as we will learn in Section 24-1.4.

24-1.3 ROLE OF THE STOMACH IN THE DIGESTION OF CARBOHYDRATES

Almost no digestion of carbohydrates takes place in the stomach. Some sucrose is hydrolyzed by the acid present; very little, if any, other carbohydrate digestion occurs. The only carbohydrase present is the α-amylase of saliva, and that is soon inactivated by the acid and the protease of gastric juice. The pH of saliva is 6.0–7.0, and the pH of gastric juice is 1.0–3.5. Thus, gastric juice is, on the average, at least 10,000 times more acidic than saliva.

A **carbohydrase** is an enzyme that catalyzes the hydrolysis of an oligo- and/or polysaccharide.

24-1.4 ROLE OF THE PANCREAS AND THE SMALL INTESTINE IN THE DIGESTION OF CARBOHYDRATES

Most carbohydrate digestion and absorption takes place in the small intestine. Most digestive enzymes that are present in the small intestine were secreted into it from the pancreas. Pancreatic juice contains an α-amylase. This α-amylase is the primary digester of starch, acting in the same manner as the α-amylase of saliva. In Section 23-4, we learned that the presence of α-amylase in blood serum is an indicator of inflammation of the pancreas that allows the enzyme to leak out into the blood. As a result, assays for serum α-amylase are used as a diagnostic tool for pancreatitis.

Absorption is the taking up of substances. Here, it refers to the passage of the products of digestion into the blood through the cells lining the small intestine.

Other enzymes are needed to catalyze the hydrolysis of the disaccharides produced from starch (maltose and isomaltose) and dietary disaccharides (sucrose and lactose; see Table 24.1) because complete hydrolysis to monosaccharides is required before absorption from the small intestine can take place. The enzymes that catalyze the hydrolysis of disaccharides are called *dissaccharidases* and are located on the surface of cells lining the inner surface of the small intestine.

Table 24.1   Enzymes of the Intestines That Catalyze the Hydrolysis of Disaccharides

Substrate	Enzyme	Products
Maltose*	Maltase	D-Glucose
Isomaltose*	Isomaltase	D-Glucose
Sucrose*	Sucrase (invertase)	D-Glucose + D-fructose
Lactose	Lactase	D-Glucose + D-galactose

*These enzymes act slowly on higher oligosaccharides.

24-1.5 NEED FOR CONTROL OF BLOOD SUGAR LEVEL

After absorption, the monosaccharides (at least 75% of which is D-glucose) are transported via the blood to various tissues. Glucose is about the only energy source that can be used by the brain, the retina, the germinal epithelium (cells from which sperm are produced), mature erythrocytes (red blood cells), and the innermost portion of the adrenal gland. As a result, sufficient glucose in the blood is necessary to prevent damage to these tissues.

The concentration of glucose (or of anything else) in the blood is determined by the rate at which it enters the blood and the rate at which it leaves. The key organ in regulating blood glucose is the liver. After absorption, most of the glucose is carried in the blood directly to the liver. The liver is very sensitive to changes in blood glucose concentrations and responds accordingly. When the concentration of glucose in the blood is high, the liver takes up glucose and stores it as glycogen (see Figure 19.6). When the concentration is low, there is a net loss of glucose from the liver to the blood stream.

$$\text{glycogen in liver} \xrightarrow[\text{when blood sugar is high}]{\text{when blood sugar is low}} \text{D-glucose in the blood}$$

24-1.6 REMOVAL OF GLUCOSE FROM THE BLOOD

Nerve tissue (primarily the brain) uses 125–150 g of glucose per day. An additional 50 g of glucose is extracted by the other tissues that rely solely on glucose as an energy source. No more than 50 g of glucose is converted into glycogen by muscle tissue at rest (see Section 27-2.2). Liver and fat cells become important in the removal of glucose from the blood primarily after carbohydrate ingestion.

Ingestion is the act of taking food or medicines into the body by mouth.

The ways that glucose is removed from the blood can be summarized as follows.

1. *Intake by cells* and then
   (a) oxidation for energy (ATP) production (see this chapter and Chapter 25)
   (b) use for synthesis of glycogen (storage) (see this chapter)
   (c) conversion into fat (see Chapter 27)
   (d) use for synthesis of other tissue components, such as other sugars, amino acids, nucleotides, and cholesterol.
   (e) oxidation in the hexose monophosphate pathway to generate reducing power (NADPH) for synthetic reactions (see Chapter 27)
2. *Excretion in the urine.* Glucosuria* occurs only in severe hyperglycemia† such as in uncontrolled diabetes mellitus. *Diabetes mellitus* is a metabolic disorder in which much of the ability to utilize glucose as a source of energy is lost due to disturbances in the normal functioning of insulin. It is characterized by an inability of glucose to pass from the blood into cells, resulting in high blood sugar concentrations. Diabetes mellitus is caused by either a lack of insulin or an ineffectiveness of its action. We will discuss diabetes in several places later in this book.

Insulin is the most important factor tending to lower blood glucose concentrations. Insulin facilitates the passage of glucose into skeletal muscle, diaphragm, heart, and adipose tissue cells.

Adipose tissue is fatty tissue.

24-1.7 ADDITION OF GLUCOSE TO THE BLOOD

Glucose is released by the liver into the blood at a rate of 2–3.5 mg/kg/min or 200–350 g/day. Glycogen breakdown accounts for 75–80% of hepatic glucose production; an additional 15–20% is formed from lactate, and the remaining 5–10% is produced from amino acids (see Section 28-2.4), particularly alanine, by a process known as gluconeogenesis. *Gluconeogenesis‡* refers to the formation of molecules of D-glucose from molecules that are not themselves carbohydrate. With

Hepatic is an adjective referring to the liver.

---

*-Uria is a suffix denoting a characteristic or constituent of urine, so *glucosuria* is the presence of glucose in the urine.

†-Emia is a suffix denoting a characteristic or constituent of blood; *hyper-* is a prefix denoting greater than normal, or excessive, and *glycose* is a general term for sugar. Therefore, *hyperglycemia* is an abnormally high concentration of glucose in the blood.

‡*Neo-* is a prefix meaning new and *genesis* refers to production or creation, so *gluconeogenesis* is the production of new glucose molecules.

prolonged fasting, glycogen stores are depleted and the contribution from gluco-neogenesis becomes increasingly important. Gluconeogenesis will be discussed in Section 28-2.4.

The ways that glucose is added to the blood to increase blood glucose levels can be summarized as follows.

1. Ingestion of carbohydrate
*2. Glycogenolysis**

$$\text{glycogen} \xrightarrow{\text{glycogenolysis}} \text{D-glucose}$$

3. Gluconeogenesis

$$\text{lactate and certain amino acids} \xrightarrow{\text{gluconeogenesis}} \text{D-glucose}$$

A low blood sugar level is called *hypoglycemia.*† The normal concentration of glucose in whole blood is 70–100 mg/100 mL = 3.9–5.6 mM. A low blood sugar level causes us to be hungry. As a result, we eat, digest starch, and absorb glucose from the small intestine to raise the level of sugar in our blood. The first response to low blood sugar, however, is breakdown of glycogen in the liver, which occurs within minutes. Glycogenolysis is promoted by two hormones that are rapidly secreted in response to a drop in the blood glucose level. These two hormones are adrenaline (epinephrine)‡ and glucagon. Gluconeogenesis is a much slower response taking hours.

It is interesting that the physiological state that is found when diabetes mellitus is uncontrolled is much like that found during starvation. If insulin is not present, glucose cannot enter the cells. This results in high blood sugar (hyperglycemia). However, although there is a high concentration of glucose in the blood, the cells are actually starved for glucose.

PROBLEMS

1. An enzyme that catalyzes the hydrolysis of starch is a(n) _____.
2. The synthesis of glucose from noncarbohydrate molecules is called _____.
3. An enzyme that catalyzes bond cleavage by the addition of water is called a(n) _____.
4. The conversion of glycogen into glucose is a process known as _____.
5. The digestion of starch occurs primarily in the _____.
6. The digestion of lactose takes place in the _____.
7. The hydrolysis of sucrose occurs in the _____.

ANSWERS TO PROBLEMS **1.** amylase **2.** gluconeogenesis **3.** hydrolase **4.** glycogenolysis **5.** small intestine **6.** small intestine **7.** small intestine

---

*-*Lysis* refers to a breaking down, so *glycogenolysis* refers to the breaking down of glycogen; that is, the conversion of glycogen into glucose.

†*Hypo-* is a prefix denoting less than normal, so *hypoglycemia* is an abnormally low concentration of glucose in the blood.

‡American physiologists and pharmacologists changed the name of adrenaline to epinephrine when a pharmaceutical company marketed the compound under the trade name Adrenalin. However, the name adrenaline is used in other countries and by chemists everywhere.

24-2 METABOLISM OF CARBOHYDRATES

LEARNING OBJECTIVES

1. Define anabolism, catabolism, metabolism, glycogenesis, kinase, and mutase.
2. Write, in the form $A + B \longrightarrow C + D$, reactions such as

$$\text{D-glucose} \xrightarrow[\quad]{\text{ATP} \quad \text{ADP}} \text{D-glucose 6-phosphate}$$

3. Describe the functions of phosphorylase, adrenaline (epinephrine), glucagon, cAMP, the conversion of pyruvate to lactate, glucose 6-phosphatase, and the Cori cycle.
4. Write the structure of cAMP.
5. Describe the relationship between nicotinic acid (niacin, nicotinamide) and NAD and NADP. Describe the types of reactions in which NAD and NADP function.
6. Summarize glycolysis as in Figure 24.5.
7. Given the substrates, coenzymes, and enzyme required for any of the reactions of glycolysis, write a balanced equation showing the products of the reaction.
8. Explain how analysis of LDH isozymes in serum can be used in diagnosis.
9. Diagram the Cori cycle in its simplest form.
10. Describe the cause and results of lactose intolerance.

A living organism such as the human body is made up of living cells that take in nutrients (foodstuffs) and use these for energy, as materials to make specific products such as hemoglobin, and when necessary, as materials to make the components of a new cell so that cell division can take place. Some cells, such as photosynthetic cells of green plants, use the energy of sunlight and simple raw materials (carbon dioxide and water) to make carbohydrates and other molecules in which energy is stored.

Each of the activities of a cell requires energy. Cells constantly use energy for synthesis of chemical substances and for transport of materials across membranes. Chemical energy is also transformed into mechanical work (for example, muscle contraction) and into the electrical energy of nerve impulses. In addition, cells use energy to maintain themselves. A living cell is complex and unstable. It maintains its fragile structure by constantly rebuilding itself. Each cell component, with the exception of DNA, is in a dynamic state. Cell components are not permanent, static structures but are constantly being used up or broken down and replaced by newly made molecules. Even DNA is repaired as required. Cells are also replaced; a human body makes about three billion new cells each minute.

All the chemical reactions that take place within a living organism are called, collectively, *metabolism*. Metabolism is the sum of anabolism and catabolism. *Anabolism* refers to the synthetic activities of a cell. Anabolic reactions use energy and form more complex molecules from simpler ones. *Catabolism* refers to the degradative reactions in a cell. Catabolic reactions produce energy. This energy is most commonly stored as ATP (see Section 20-1.2).

As explained in Chapter 23, reactions of a living cell are catalyzed by enzymes. These chemical reactions are not independent, but occur in sequence. For example, substance A might be converted into substance E in a sequence of reactions catalyzed by four enzymes.

$$A \xrightleftharpoons{\text{enzyme}_1} B \xrightleftharpoons{\text{enzyme}_2} C \xrightleftharpoons{\text{enzyme}_3} D \xrightarrow{\text{enzyme}_4} E$$

Such a sequence of biochemical reactions is called a *pathway,* and each pathway is given a name. We have already studied the synthesis of DNA, RNA, and protein. In this section and in the following chapters, we will study the pathways for the synthesis and utilization of the common foodstuffs—carbohydrates, fats, and proteins—beginning with carbohydrates.

Life is characterized by growth, movement, and reproduction. Life depends upon controlled chemical reactions for the synthesis of organic compounds (anabolism) and for the production of usable energy (from catabolism). These controlled chemical reactions are the enzyme-catalyzed reactions of metabolic pathways.

24-2.1 THE CATABOLISM OF GLUCOSE VIA THE GLYCOLYTIC PATHWAY

The catabolism of glucose ordinarily involves its complete oxidation to carbon dioxide and water.

$$C_6H_{12}O_6 + 6\,O_2 \longrightarrow 6\,CO_2 + 6\,H_2O + 686\text{ kcal/mol}$$

(180 g of glucose)

Much of the released energy is stored as ATP. However, as we will see, not all the energy released by oxidation of glucose in cells can be converted to useful work.

Total oxidation of glucose is accomplished in a pathway that, for convenience, is often discussed in two parts. The first, the pathway of *glycolysis,* or the *Embden-Myerhof pathway,* occurs in the soluble cytoplasm in either the presence or absence of oxygen and produces little energy. The second, known as the citric acid, the tricarboxylic acid (TCA), or the Krebs cycle, takes place in organelles known as mitochondria. The second pathway requires oxygen indirectly and, when coupled with the respiratory chain, and oxidative phosphorylation, produces most of the cells' energy (see Chapter 25).

Glycose is a general term for carbohydrate and *-lysis* refers to a breaking down, so glycolysis literally means the breaking down of carbohydrate. *Glycolysis* is the breaking down of a glucose molecule (or a glucopyranosyl residue of glycogen) to a salt of the three-carbon acid pyruvic acid, yielding energy that is stored in ATP molecules (see Figure 24.2). Pyruvate may subsequently be converted into lactate, another salt of a three-carbon acid (see Section 24-2.2), or into ethanol and carbon dioxide. When glycolysis begins with glycogen, the first reaction in the pathway is catalyzed by the enzyme phosphorylase.

Figure 24.2   The pathway of glycolysis.

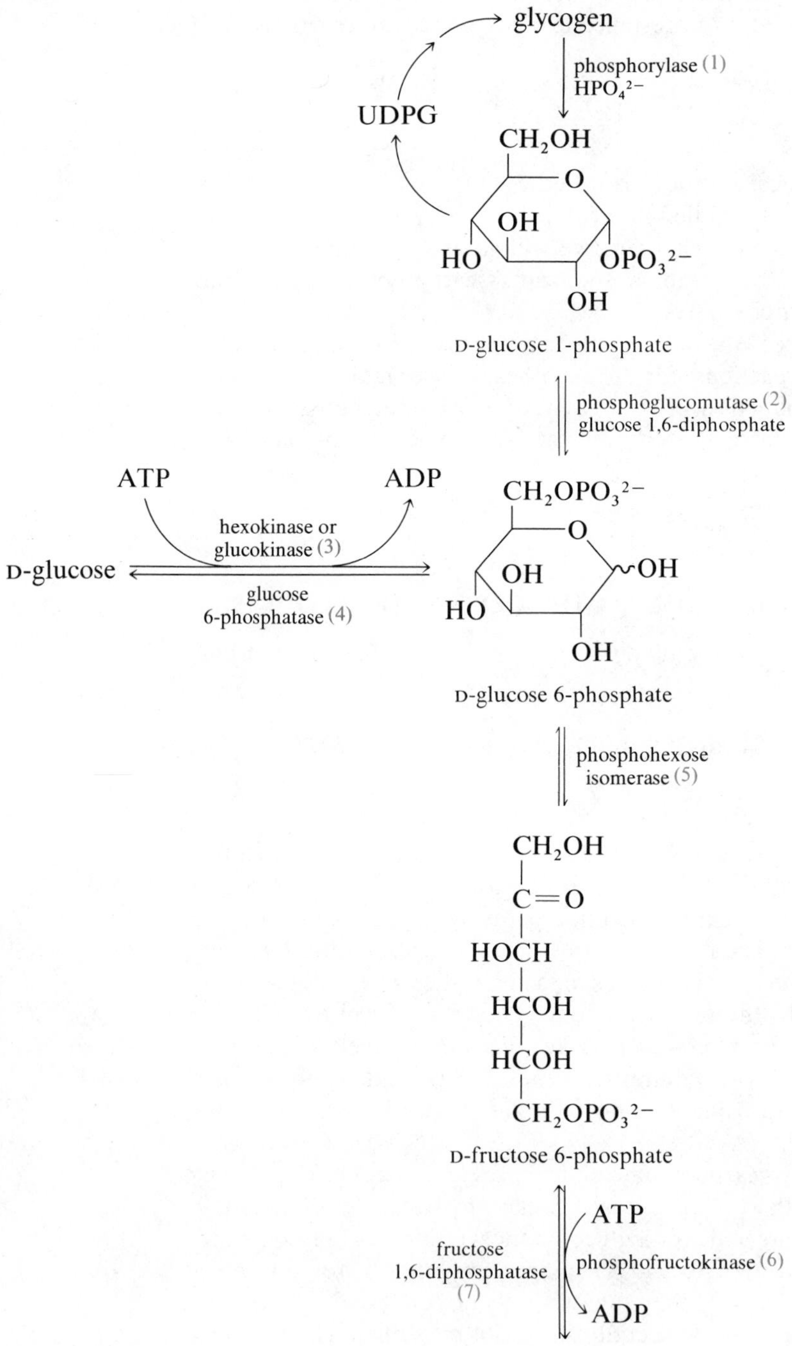

Figure 24.2   The pathway of glycolysis (*cont'd*).

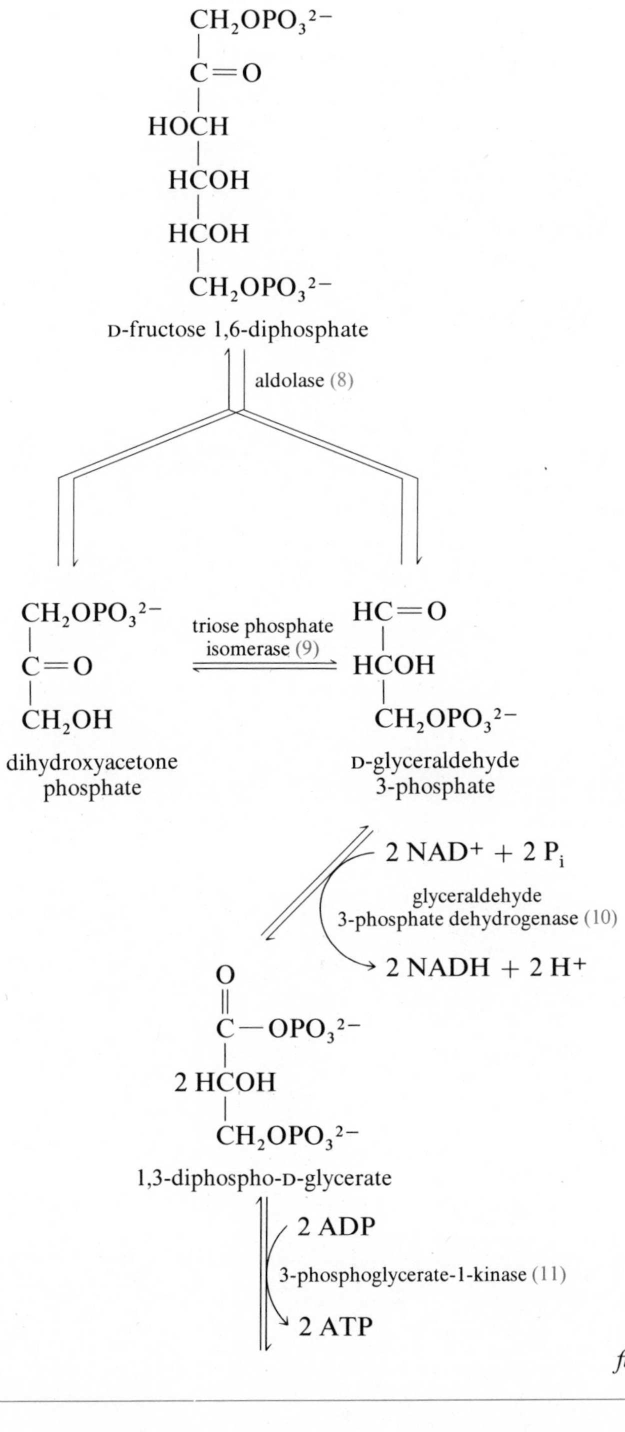

$$CH_2OPO_3{}^{2-}$$
$$|$$
$$C{=}O$$
$$|$$
$$HOCH$$
$$|$$
$$HCOH$$
$$|$$
$$HCOH$$
$$|$$
$$CH_2OPO_3{}^{2-}$$

D-fructose 1,6-diphosphate

aldolase (8)

$$CH_2OPO_3{}^{2-}$$
$$|$$
$$C{=}O$$
$$|$$
$$CH_2OH$$

dihydroxyacetone
phosphate

triose phosphate
isomerase (9)

$$HC{=}O$$
$$|$$
$$HCOH$$
$$|$$
$$CH_2OPO_3{}^{2-}$$

D-glyceraldehyde
3-phosphate

2 NAD$^+$ + 2 P$_i$

glyceraldehyde
3-phosphate dehydrogenase (10)

2 NADH + 2 H$^+$

$$O$$
$$\|$$
$$C{-}OPO_3{}^{2-}$$
$$|$$
$$2\ HCOH$$
$$|$$
$$CH_2OPO_3{}^{2-}$$

1,3-diphospho-D-glycerate

2 ADP

3-phosphoglycerate-1-kinase (11)

2 ATP

figure continues

Figure 24.2 The pathway of glycolysis (*cont'd*).

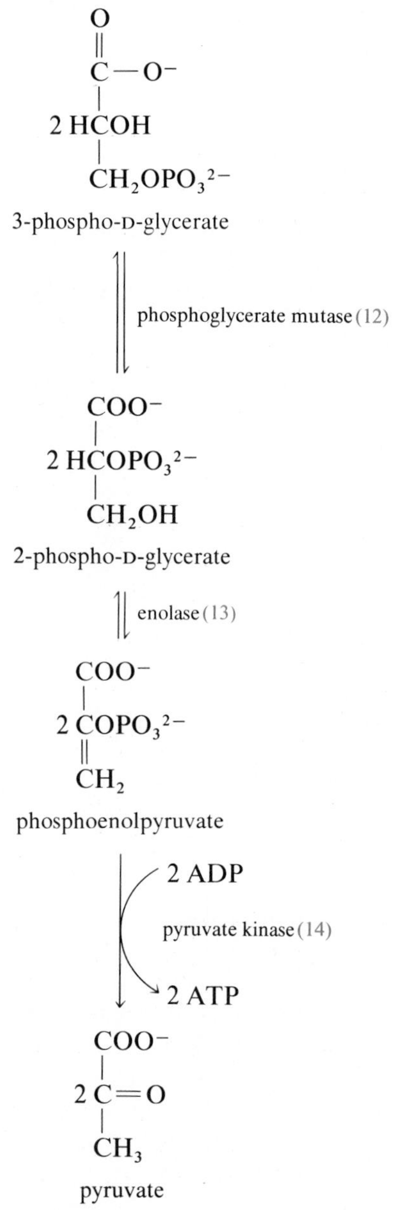

$$
\begin{array}{c}
\text{O} \\
\parallel \\
\text{C}-\text{O}^- \\
| \\
2\ \text{HCOH} \\
| \\
\text{CH}_2\text{OPO}_3{}^{2-}
\end{array}
$$

3-phospho-D-glycerate

phosphoglycerate mutase (12)

$$
\begin{array}{c}
\text{COO}^- \\
| \\
2\ \text{HCOPO}_3{}^{2-} \\
| \\
\text{CH}_2\text{OH}
\end{array}
$$

2-phospho-D-glycerate

enolase (13)

$$
\begin{array}{c}
\text{COO}^- \\
| \\
2\ \text{COPO}_3{}^{2-} \\
\parallel \\
\text{CH}_2
\end{array}
$$

phosphoenolpyruvate

2 ADP

pyruvate kinase (14)

2 ATP

$$
\begin{array}{c}
\text{COO}^- \\
| \\
2\ \text{C}=\text{O} \\
| \\
\text{CH}_3
\end{array}
$$

pyruvate

Phosphorylase (1)

The first reaction in the pathway of glycolysis when stored glycogen is the starting material is the removal of units of D-glucose from the nonreducing ends of glycogen (see Figure 24.3). This removal is catalyzed by the enzyme phosphorylase, which catalyzes a phosphorolysis. Recall that hydrolysis is the splitting of a chemical bond with the concomitant addition of water. Similarly, *phosphorolysis* is the splitting of a chemical bond with the addition of phosphoric acid.

When we encounter a situation that calls for a fight, flight, or fright response, adrenaline is released from the adrenal glands. Adrenaline binds to the surface (plasma membrane) of skeletal muscle and heart cells, activating the enzyme adenyl cyclase which catalyzes the conversion of ATP into cyclic adenosine 3',5'-monophosphate, or *cyclic AMP (cAMP)*. cAMP is called the *second messenger* of hormonal control. It is produced by the action of adrenaline, noradrenaline (see Section 28-2.5), thyroxine, and all polypeptide hormones. These hormones bind with the plasma membrane and bring about the production of cAMP molecules without entering the cells. It is the cAMP molecules, then, rather than the hormone molecules themselves, that are the actual mediators of metabolic change inside the cell.

cAMP

cAMP is an activator of protein kinase, an enzyme that catalyzes the formation of phosphate esters (phosphorylation) of proteins at the expense of ATP. Here, the protein that is phosphorylated, phosphorylase *b* kinase, in turn, catalyzes the conversion of phosphorylase *b* to phosphorylase *a* (see Figure 24.4). Phosphorylase *b* has little activity; phosphorylase *a* is a much more active enzyme. Phosphorylase catalyzes the conversion of glycogen into D-glucose 1-phosphate, the first step in glycolysis.

The arrangement for the breakdown of glycogen, that is, the conversion of glycosyl units of glycogen into D-glucose 1-phosphate, is called a *cascade*. In a cascade, the product of each step is a catalyst for the next step (see Figure 24.4). The result is that a minute amount of adrenaline has a very large effect—so large that 500 micrograms (0.0005 g or 0.000003 mol) is a potent dose for humans. The blood concentration of adrenaline during normal function is in the nanomolar (0.000000001, or 10^{-9}, mol/L) range. Adrenaline is, therefore, a very powerful hormone. The same cascade occurs in the liver, except that *glucagon,* a polypeptide hormone made by the pancreas in response to low blood sugar (hypoglycemia),

Figure 24.3   Action of phosphorylase. (For the structure of glycogen, see Figure 19.10.)

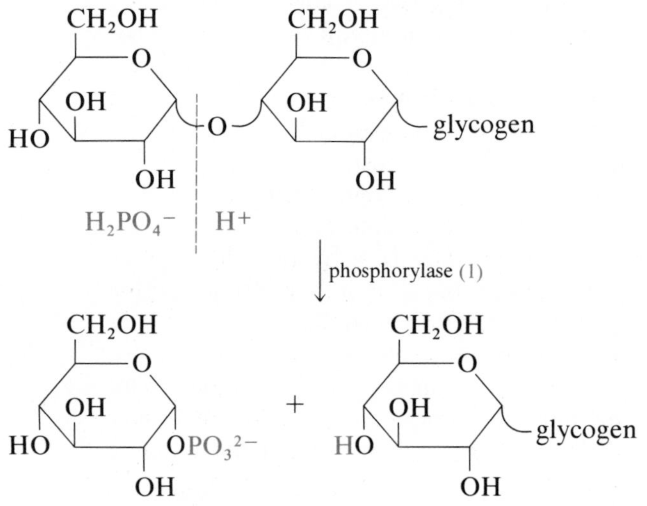

α-D-glucopyranose 1-phosphate

initiates the cascade and liver phosphorylase *a* contains only two polypeptide sub-units rather than four. Adenyl cyclase, protein kinase, and phosphorylase *b* are all good examples of allosteric enzymes (see Section 23-3.7).

Phosphoglucomutase (2)

Phosphoglucomutase is an enzyme that requires a coenzyme and provides a good example of one way in which a coenzyme works. Phosphoglucomutase requires α-D-glucopyranose 1,6-diphosphate, usually referred to simply as D-glucose 1,6-diphosphate (G-1,6-diP) as a coenzyme. A *mutase* is an enzyme that catalyzes the apparent migration of a phosphate group from one hydroxyl group to another of the same molecule. *Phosphoglucomutase* catalyzes the apparent migration of a phosphate group from the hydroxyl group on C-1 of α-D-glucopyranose to the hydroxyl group on C-6. The enzyme first transfers a phosphate group from G-1, 6-diP to the hydroxyl group of one of its serine residues to form a phosphate ester of the enzyme. Then the phosphate group is transferred to α-D-glucopyranose 1-phosphate, usually referred to simply as D-glucose 1-phosphate (G-1-P),* form-ing G-1,6-diP from G-1-P and D-glucose 6-phosphate (G-6-P) from the original G-1,6-diP.

$$\text{enzyme} + \text{G-1,6-diP} \rightleftharpoons \text{enzyme-P} + \text{G-6-P}$$
$$\text{enzyme-P} + \text{G-1-P} \rightleftharpoons \text{enzyme} + \text{G-1,6-diP}$$

---

*The official and proper three-letter abbreviation for glucose is Glc (because Glu is the abbreviation for glutamic acid). However, biochemists frequently use G for glucose in compound abbreviations such as G-1-P, G-6-P, and UDPG.

Figure 24.4 The cascade (arrangement) responsible for the activation of the enzyme phosphorylase.

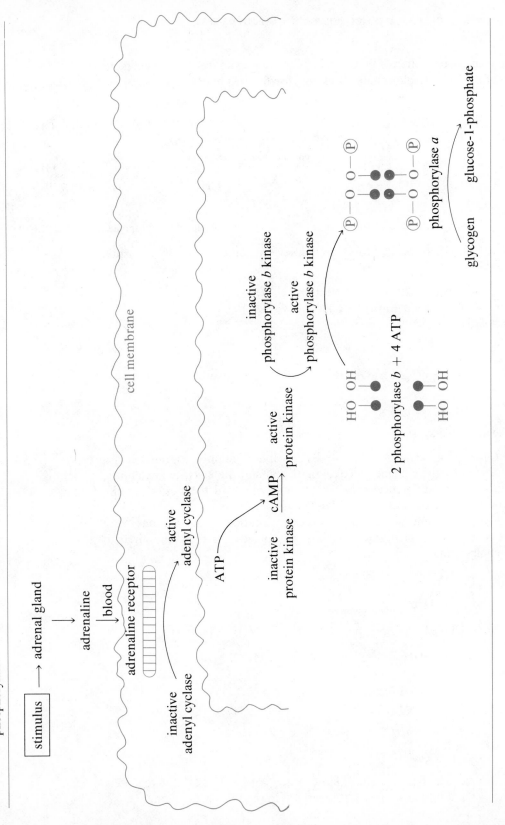

The overall result of the action of the enzyme phosphoglucomutase is the reversible conversion of D-glucose 1-phosphate into D-glucose 6-phosphate as shown.

D-glucose 1,6-diphosphate   D-glucose 1-phosphate   D-glucose 6-phosphate   D-glucose 1,6-diphosphate

D-Glucose 6-phosphate is a key compound in several pathways. We will consider three of these in this book:

1. glycolytic pathway (described in this section)
2. hexose monophosphate pathway (see Section 27-3.2)
3. synthesis of glycogen (**glycogenesis**)

> **Glycogenesis** is the synthesis (genesis) of glycogen.

$$\text{glycogen} \underset{\text{glycogenesis}}{\overset{\text{glycogenolysis}}{\rightleftarrows}} \text{G-1-P} \rightleftharpoons \text{G-6-P} \rightleftharpoons \text{glucose}$$

Glucokinase and Hexokinase (3)

D-Glucose 6-phosphate can also be made by phosphorylation of D-glucose in a reaction catalyzed by the enzyme glucokinase (liver) and the enzyme hexokinase (liver and brain). The differences in these two enzymes and their effect on the utilization of blood glucose was discussed in Section 23-3.2.

Although D-glucose and other hexoses occur in solution in the pyranose ring form, it is easier for us to picture the reactions by considering the compounds to be in open-chain forms.

D-glucose   D-glucose 6-phosphate

Because the reaction catalyzed by glucokinase or hexokinase is one of the reactions in a pathway, we find it more convenient to write it in the following manner.

$$
\begin{array}{ccc}
\text{HC}=\text{O} & & \text{HC}=\text{O} \\
| & & | \\
\text{HCOH} & & \text{HCOH} \\
| & \text{ATP} \quad \text{ADP} & | \\
\text{HOCH} & \xrightarrow{} & \text{HOCH} \\
| & \text{glucokinase } or & | \\
\text{HCOH} & \text{hexokinase} & \text{HCOH} \\
| & & | \\
\text{HCOH} & & \text{HCOH} \\
| & & | \\
\text{CH}_2\text{OH} & & \text{CH}_2\text{O}\,\textcircled{P}
\end{array}
$$

The value of writing a reaction in this way is apparent in Figure 24.2. It is important to note that when blood glucose is the source of G-6-P, one mole of ATP is used at this point.

Phosphohexose Isomerase (5)

The conversion of D-glucose 6-phosphate into D-fructose 6-phosphate is an aldose $\rightleftharpoons$ ketose isomerization of the kind we discussed in Section 19-2.2. It is catalyzed by the enzyme phosphohexose isomerase, also called phosphoglucoisomerase.

$$
\begin{array}{ccc}
\text{HC}=\text{O} & & \text{CH}_2\text{OH} \\
| & & | \\
\text{HCOH} & & \text{C}=\text{O} \\
| & \text{phosphohexose} & | \\
\text{HOCH} & \underset{\xrightarrow{}}{\text{isomerase}} & \text{HOCH} \\
| & & | \\
\text{HCOH} & & \text{HCOH} \\
| & & | \\
\text{HCOH} & & \text{HCOH} \\
| & & | \\
\text{CH}_2\text{O}\,\textcircled{P} & & \text{CH}_2\text{O}\,\textcircled{P}
\end{array}
$$

D-glucose 6-phosphate   D-fructose 6-phosphate

Phosphofructokinase (6)

The conversion of D-fructose 6-phosphate into D-fructose 1,6-diphosphate also uses up a high-energy phosphate bond.

$$
\begin{array}{ccc}
& \text{CH}_2\text{OH} & \\
& | & \\
& \text{C}=\text{O} & \\
& | & \text{phosphofructokinase} \\
\text{ATP} + & \text{HOCH} & \xrightarrow{} \\
& | & \\
& \text{HCOH} & \\
& | & \\
& \text{HCOH} & \\
& | & \\
& \text{CH}_2\text{O}\,\textcircled{P} &
\end{array}
\qquad
\begin{array}{c}
\overset{\displaystyle \text{O}}{\underset{\displaystyle \parallel}{}} \\
\text{CH}_2\text{O}-\text{P}-\text{O}^- \\
| \qquad | \\
\text{C}=\text{O} \quad \text{O}^- \\
| \\
\text{ADP} + \text{HOCH} \\
| \\
\text{HCOH} \\
| \\
\text{HCOH} \\
| \\
\text{CH}_2\text{O}\,\textcircled{P}
\end{array}
$$

D-fructose 6-phosphate   D-fructose 1,6-diphosphate

Again, because this reaction is one of the reactions in a pathway, we write it as shown.

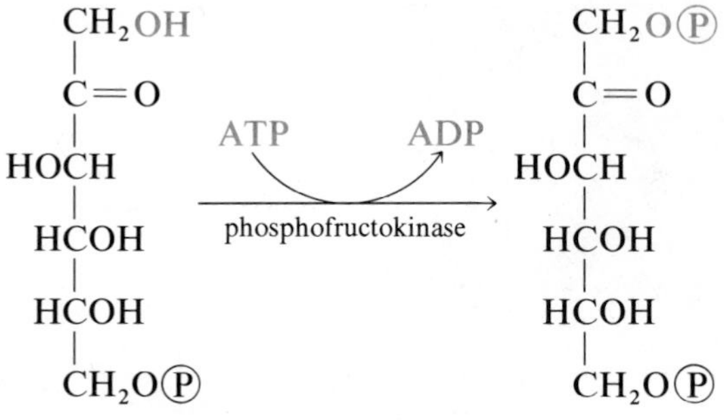

$$
\begin{array}{ccc}
\text{CH}_2\text{OH} & & \text{CH}_2\text{O}\textcircled{P} \\
| & & | \\
\text{C}=\text{O} & & \text{C}=\text{O} \\
| & \overset{\text{ATP} \qquad \text{ADP}}{\underset{\text{phosphofructokinase}}{\longrightarrow}} & | \\
\text{HOCH} & & \text{HOCH} \\
| & & | \\
\text{HCOH} & & \text{HCOH} \\
| & & | \\
\text{HCOH} & & \text{HCOH} \\
| & & | \\
\text{CH}_2\text{O}\textcircled{P} & & \text{CH}_2\text{O}\textcircled{P}
\end{array}
$$

Aldolase (8)

Aldolase catalyzes the following reversible aldol condensation reaction (see Section 15-3.2 and 19-2.2). In this reaction, a hexose (six-carbon sugar) has been split into two trioses (three-carbon sugars).

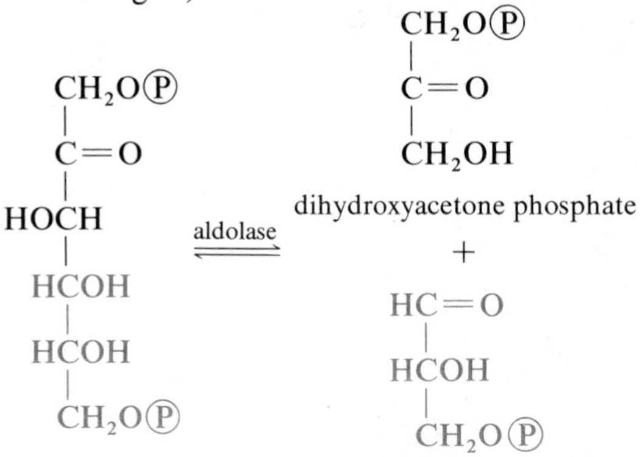

$$
\begin{array}{ccc}
& & \text{CH}_2\text{O}\textcircled{P} \\
& & | \\
\text{CH}_2\text{O}\textcircled{P} & & \text{C}=\text{O} \\
| & & | \\
\text{C}=\text{O} & & \text{CH}_2\text{OH} \\
| & \overset{\text{aldolase}}{\rightleftharpoons} & \text{dihydroxyacetone phosphate} \\
\text{HOCH} & & \\
| & & + \\
\text{HCOH} & & \text{HC}=\text{O} \\
| & & | \\
\text{HCOH} & & \text{HCOH} \\
| & & | \\
\text{CH}_2\text{O}\textcircled{P} & & \text{CH}_2\text{O}\textcircled{P}
\end{array}
$$

D-fructose 1,6-diphosphate   D-glyceraldehyde 3-phosphate

Triose Phosphate Isomerase (9)

The reaction catalyzed by triose phosphate isomerase is another aldose $\rightleftharpoons$ ketose isomerization. This reaction, coupled with the reaction catalyzed by aldolase, brings about the conversion of one mole of hexose into two moles of D-glyceraldehyde 3-phosphate because, as soon as any D-glyceraldehyde 3-phosphate is used up in the next reaction, more dihydroxyacetone phosphate is converted into D-glyceraldehyde 3-phosphate. The remainder of the pathway reflects this two-from-one conversion.

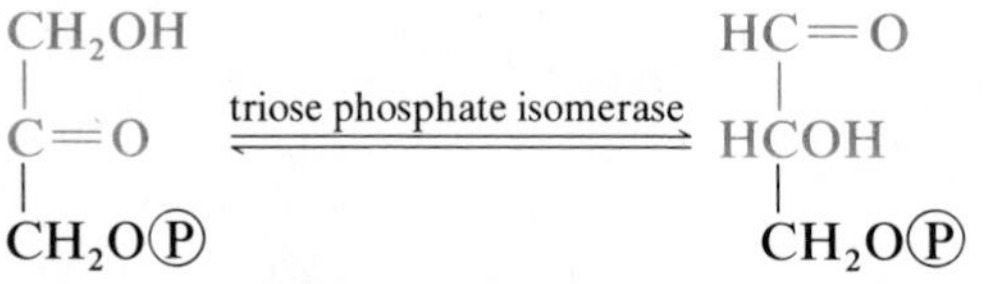

$$
\begin{array}{ccc}
\text{CH}_2\text{OH} & & \text{HC}=\text{O} \\
| & \overset{\text{triose phosphate isomerase}}{\rightleftharpoons} & | \\
\text{C}=\text{O} & & \text{HCOH} \\
| & & | \\
\text{CH}_2\text{O}\textcircled{P} & & \text{CH}_2\text{O}\textcircled{P}
\end{array}
$$

dihydroxyacetone   D-glyceraldehyde
phosphate   3-phosphate

Glyceraldehyde 3-Phosphate Dehydrogenase (10)

The only oxidative step in the glycolytic pathway is that catalyzed by the enzyme glyceraldehyde 3-phosphate dehydrogenase. In this step, D-glyceraldehyde 3-phosphate is oxidized to 1,3-diphospho-D-glycerate, a compound containing a high-energy phosphate bond (see Section 20-1.2). As in all other areas of chemistry, when one compound is oxidized, another must be reduced. In this case, it is the coenzyme NAD⁺ that is reduced. NAD stands for nicotinamide adenine dinucleotide. We learned in Section 20-1 that a nucleotide is a compound of the following structure.

$$\text{nitrogen base} - \text{(ribose or deoxyribose)} - \text{phosphate}$$

The structure of NAD, a dinucleotide, is shown below and in Figure 12.3.

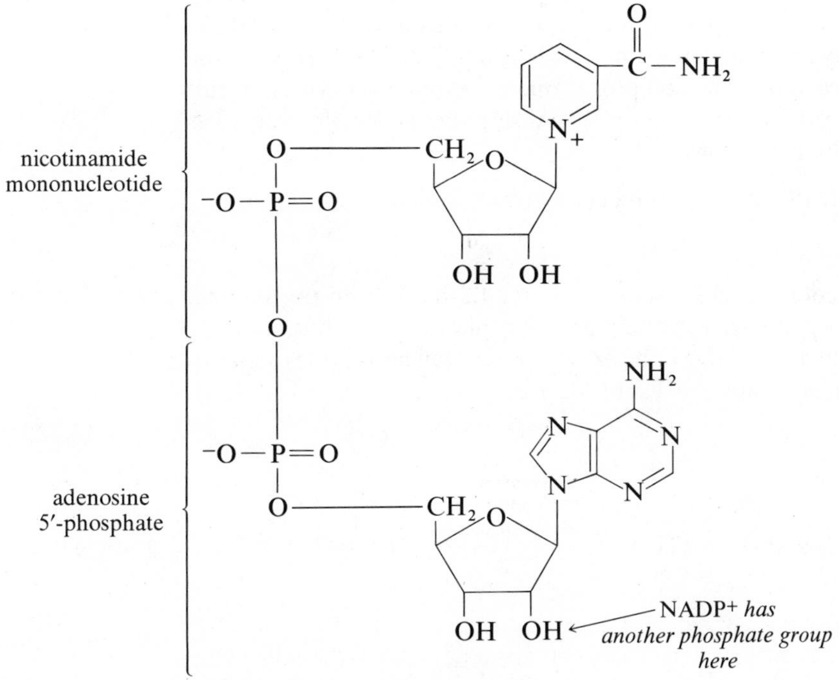

NAD⁺ (nicotinamide adenine dinucleotide)

In other words, this dinucleotide has the structure

$$\text{nitrogen base} - \text{ribose} - \text{phosphate} - \text{phosphate} - \text{ribose} - \text{nitrogen base}$$

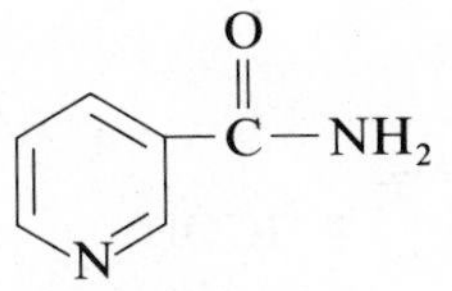

nicotinamide

In NAD, the nitrogen bases are adenine (see Figure 20.4) and nicotinamide. *Nicotinamide* is the amide of the B vitamin *nicotinic acid* or *niacin*. NAD and NADP are coenzymes in oxidation–reduction reactions of the type shown.

$$\text{substrate (SH}_2) + \text{NAD}^+ \rightleftharpoons \text{product (P)} + \text{NADH} + \text{H}^+$$

reduced   oxidized   oxidized   reduced

The general term nicotinamide adenine dinucleotide (NAD) can mean either the oxidized or reduced form, so NAD can be described simply as a coenzyme for redox reactions.

In Section 9-1, we learned that oxidation of an atom, ion, or compound involves (1) the loss of electrons, (2) the loss of hydrogen atoms, or (3) the addition of oxygen atoms. Biological oxidations most frequently involve the loss of hydrogen atoms (that is, the loss of electrons and protons). The enzymes that catalyze the removal of hydrogen atoms from substrate molecules are called *dehydrogenases*. An acceptor for the electrons and one or both of the protons must be present because, as we have already learned, when a molecule is oxidized, another molecule must be reduced and vice versa. The acceptor is one of several coenzymes, in this case NAD⁺. In the reduced form (for example, NADH), these coenzymes serve as electron donors in reduction reactions.

$$\text{substrate (S H}_2) \longrightarrow \text{product (P)} + 2\,e^- + 2\,H^+$$

$$\text{NAD}^+ + 2\,e^- + 2\,H^+ \longrightarrow \text{NADH} + H^+$$

The product of the reaction we are discussing here, 1,3-diphospho-D-glycerate, is an anhydride of 3-phospho-D-glyceric acid and phosphoric acid. Thus, an aldehyde has been oxidized to a derivative of a carboxylic acid (an anhydride) containing a high-energy phosphate bond (see Table 20.3).

D-glyceraldehyde 3-phosphate   1,3-diphospho-D-glycerate

As a reaction in a pathway, this reaction is written as shown.

3-Phosphoglycerate 1-Kinase (11)

In Section 23-3.2, we learned that a *kinase* is an enzyme that can phosphorylate a substrate. In other words, a kinase is an enzyme that can make the phosphate ester of a substrate using the high-energy phosphate of ATP as the source of the phosphate group. The reaction catalyzed by the enzyme phosphoglycerate 1-kinase is reversible and generates ATP by transfer of the high-energy phosphate group of 1,3-diphospho-D-glycerate to ADP. The enzyme is named for the reaction from right to left, although it goes from left to right in glycolysis.

$$
\begin{array}{ccc}
& \begin{array}{c} O \\ \parallel \\ C-O\,\text{P} \\ | \\ 2\ ADP + 2\ HCOH \\ | \\ CH_2O\,\text{P} \end{array} & \xrightleftharpoons[\text{1-kinase}]{\text{phosphoglycerate}} & \begin{array}{c} O \\ \parallel \\ C-O^- \\ | \\ 2\ ATP + 2\ HCOH \\ | \\ CH_2O\,\text{P} \end{array} \\
& \text{1,3-diphospho-D-glycerate} & & \text{3-phospho-D-glycerate}
\end{array}
$$

As a reaction in a pathway, this reaction is written as shown.

$$
\begin{array}{c} O \\ \parallel \\ C-O\,\text{P} \\ | \\ 2\ HCOH \\ | \\ CH_2O\,\text{P} \end{array}
\quad
\begin{array}{c} 2\ ADP \quad 2\ ATP \\ \\ \xrightleftharpoons{} \\ \text{phosphoglycerate} \\ \text{1-kinase} \end{array}
\quad
\begin{array}{c} O \\ \parallel \\ C-O^- \\ | \\ 2\ HCOH \\ | \\ CH_2O\,\text{P} \end{array}
$$

Here, two high-energy phosphate bonds (two moles of ATP) are made.

Phosphoglycerate Mutase (12)

Earlier in this section, we learned that mutase is a common name for an enzyme that catalyzes the apparent migration of a phosphate group from one to another hydroxyl group of the same molecule. The action of phosphoglucomutase has already been described. Phosphoglycerate mutase catalyzes the conversion of 3-phospho-D-glycerate to 2-phospho-D-glycerate in a manner analogous to the conversion of D-glucose 1-phosphate to D-glucose 6-phosphate, using 2,3-diphospho-D-glycerate as a coenzyme.

$$
\begin{array}{ccc}
\begin{array}{c} O \\ \parallel \\ C-O^- \\ | \\ 2\ HCOH \\ | \\ CH_2O\,\text{P} \end{array}
& \xrightleftharpoons{\text{phosphoglycerate mutase}} &
\begin{array}{c} O \\ \parallel \\ C-O^- \\ | \\ 2\ HCO\,\text{P} \\ | \\ CH_2OH \end{array} \\
\text{3-phospho-D-glycerate} & & \text{2-phospho-D-glycerate}
\end{array}
$$

Enolase (13)

Enolase catalyzes the dehydration of 2-phospho-D-glycerate to phosphoenol-pyruvate. Note that phosphoenolpyruvate is the phosphate ester of the enol form of ionized pyruvic acid. (Recall that these organic acids are ionized at the pH of a living cell, which is 6.8–7.2.)

$$
\begin{array}{ccc}
& \text{O} & & & \text{O} \\
& \parallel & & & \parallel \\
& \text{C}-\text{O}^- & & & \text{C}-\text{O}^- \\
& | & \xrightarrow{\text{enolase}} & & | \\
2\ \text{HCO}\,\text{P} & & & 2\ \text{CO}\,\text{P} & & +\ 2\ \text{H}_2\text{O} \\
& | & & & \parallel \\
& \text{CH}_2\text{OH} & & & \text{CH}_2
\end{array}
$$

2-phospho-D-glycerate   phosphoenolpyruvate

$$
\begin{array}{cc}
\text{COO}^- & \text{COO}^- \\
| & | \\
\text{C}=\text{O} & \text{COH} \\
| & \parallel \\
\text{CH}_3 & \text{CH}_2
\end{array}
$$

ionized pyruvic acid   ionized pyruvic acid
enol form

Pyruvate Kinase (14)

Phosphoenolpyruvate is a high-energy phosphate compound (see Table 20.3) and can be used to generate ATP by transfer of its high-energy bond to ADP. The reaction is catalyzed by the enzyme pyruvic kinase, which is named for the reverse reaction. The overall reaction can be written as shown. Two more moles of ATP are made in this reaction.

$$
\begin{array}{ccc}
& \text{O} & & & \text{O} \\
& \parallel & & & \parallel \\
& \text{C}-\text{O}^- & & & \text{C}-\text{O}^- \\
& | & \xrightarrow{\substack{\text{pyruvate} \\ \text{kinase}}} & & | \\
2\ \text{CO}\,\text{P}\ +\ 2\ \text{ADP} & & & 2\ \text{C}=\text{O}\ +\ 2\ \text{ATP} \\
& \parallel & & & | \\
& \text{CH}_2 & & & \text{CH}_3
\end{array}
$$

phosphoenolpyruvate   pyruvate

Summary of Glycolysis

Glycolysis releases energy. This released energy is, in part, stored as chemical energy in the high-energy phosphate bonds of ATP; the remainder is released as heat. Overall, glycolysis is about 30% efficient in converting the energy of carbohydrate into the energy of ATP. We will discuss the concept of biochemical efficiency in the next chapter. When glycolysis begins with glycogen, the summary reaction is

glucosyl unit of glycogen $+$ 3 HPO_4^{2-} $+$ 3 ADP $+$ 2 NAD^+ $\longrightarrow$ 2 pyruvate $+$ 3 ATP* $+$ 2 NADH

---

*This ATP value is a net value. It is the difference between the moles of ATP produced and the moles of ATP used up in the oxidation of one mole of glucosyl units of glycogen or D-glucose in glycolysis.

When glycolysis starts with D-glucose (blood sugar), the summary reaction is

$$\text{D-glucose} + 2\,HPO_4{}^{2-} + 2\,ADP + 2\,NAD^+ \longrightarrow 2\,\text{pyruvate} + 2\,ATP^* + 2\,NADH$$

Actually the process never stops at pyruvate. Usually, pyruvate is further oxidized to CO_2 in the tricarboxylic acid cycle (see Section 25-1) and NADH is reoxidized to NAD^+ by oxygen (see Section 25-2). These are efficient pathways and most of the ATP generated from carbohydrate is made by these processes.

The two moles of NADH produced require two moles of oxygen atoms for reoxidation[†] as described in Section 25-2. These two moles of oxygen atoms would yield two moles of water. Therefore, we can write the equation

$$\text{D-glucose}\,(C_6H_{12}O_6) + O_2 \longrightarrow 2\,\text{pyruvate} + 2\,H_2O$$

As we will learn in the next chapter, the oxidation of pyruvate can be summarized as follows.

$$2\,\text{pyruvate} + 5\,O_2 \longrightarrow 6\,CO_2 + 4\,H_2O$$

The complete catabolism of D-glucose satisfies the stoichiometry for oxidation because the sum of the two equations is

$$C_6H_{12}O_6 + 6\,O_2 \longrightarrow 6\,CO_2 + 6\,H_2O$$

24-2.2 FORMATION OF LACTATE

During the early stages of intense exercise, oxygen may be in short supply in muscle tissue (see Section 27-2.2). When oxygen is in short supply, pyruvate may be reduced to lactate using NADH to regenerate NAD^+. This is depicted in Figure 24.5.

Lactate Dehydrogenase (15)

As we learned in Section 23-5, there are five different lactate dehydrogenase (LDH) enzymes in the human body. These LDH isoenzymes contain four polypeptide chains of two different types—H (heart) subunits and M (muscle) subunits (see Table 23.1). Each subunit combines with one NAD coenzyme molecule, so there are four NAD coenzyme molecules with each LDH isoenzyme molecule.

---

*See footnote on p. 610.

[†]$2\,NADH + 2\,H^+ + 2\,O \longrightarrow 2\,H_2O + 2\,NAD^+$

Figure 24.5 Summary of glycolysis showing the recycling of the NAD coenzyme to allow glycolysis to continue in the absence of oxygen.

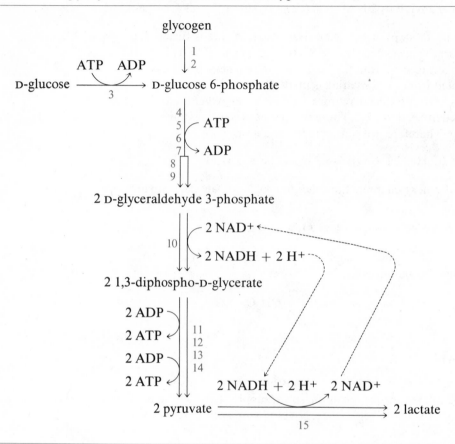

As was discussed in Section 23-4, lactate dehydrogenase isoenzymes are not ordinarily found to any appreciable extent in blood serum. Lactate dehydrogenase is, however, found in all tissues, with the isoenzyme pattern being somewhat unique to each tissue. We also learned that when a tissue is damaged by injury or disease, enzymes leak out of cells into the blood. The serum level of LDH is elevated in a variety of diseases, and changes in the serum isoenzyme pattern reflect the isoenzyme complement of the tissues affected. Therefore, LDH isoenzyme analysis is used in diagnosis as we have already discussed in detail in Section 23-5.

24-2.3 THE CORI CYCLE

Some of the glucose needed for energy by muscle cells is converted into lactate to provide ATP without requiring the use of oxygen. This lactate is transported by the blood to the liver where it can be converted back into glucose by a process

Figure 24.6
The Cori cycle in its simplest form.

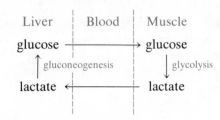

known as gluconeogenesis. *Gluconeogenesis* refers to the making of glucose from noncarbohydrate sources, such as lactate. We will discuss the pathway of gluconeogenesis later in Section 28-2.4.

The complete process by which glucose is converted into lactate to produce ATP in a working muscle and lactate is converted back to glucose in the liver is called the Cori cycle, which is diagrammed in its simplest form in Figure 24.6. The resynthesis of glucose from lactate is a relatively slow process.

In the liver, 15% of the lactate is oxidized to CO_2 and water by oxidation to pyruvate and then oxidation of pyruvate in the tricarboxylic acid cycle (see Section 25-1) to provide the ATP required to convert the remaining 85% of lactate into glucose by gluconeogenesis. Glycolysis and the Cori cycle combined produce 37% as much ATP as glycolysis and the tricarboxylic acid cycle combined. The efficiency of glycolysis plus the tricarboxylic acid cycle is discussed in Chapter 25. The complete Cori cycle and its relationship to other pathways is diagrammed in Figure 24.7.

Figure 24.7   A detailed representation of the Cori cycle. TCA cycle = tricarboxylic acid cycle. The synthesis of glycogen involves the sequence

$$\text{G-6-P} \longrightarrow \text{G-1-P} \xrightarrow{\text{several steps}} \text{glycogen}$$

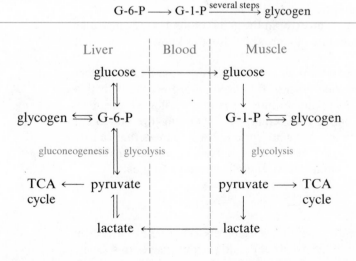

24-2.4 CONVERSION OF GLYCOGEN INTO GLUCOSE

Glycogen is the storage form of glucose. It is found primarily in the liver with smaller amounts in muscles and the brain. When there is an excess of glucose in the blood after a meal, some is converted into glycogen and/or fat. When the blood sugar level drops, liver glycogen is converted into glucose via the following sequence.

$$\text{glycogen}$$
$$\downarrow (1)$$
$$\text{D-glucose 1-phosphate}$$
$$\downarrow (2)$$
$$\text{D-glucose} \xleftarrow[\substack{\text{glucose}\\\text{6-phosphatase}\\(4)}]{} \text{D-glucose 6-phosphate}$$

The liver and the kidneys are the only organs that contain the enzyme glucose 6-phosphatase. Therefore, they are the only organs that can convert glycogen into glucose. In all others, all carbohydrate in cells must be catabolized there. It cannot leave. The liver is the principal organ involved in the regulation of blood glucose levels.

Muscle glycogen is a direct source of energy. The structure of glycogen is shown in Figure 19.6. Because glycogen is a very large, only slightly soluble molecule containing hundreds of thousands of glucose residues, cells can store glycogen without increasing the osmotic pressure (see Section 6-3.3). A very high osmotic pressure that would cause the cell to swell and burst would result if a cell stored the glucose in glycogen molecules as free glucose. In other words, a low concentration of glycogen is as effective as a store of energy as a high concentration of glucose without giving a high osmotic pressure.

There are a group of diseases known as *glycogen storage diseases* in which there is some abnormality in glycogen metabolism, usually in glycogen breakdown, but occasionally in glycogen synthesis. These diseases result from enzyme deficiencies. Each disease can be traced to an error in the gene coding for the particular enzyme missing, and each disease is one of a group of diseases known as inherited or genetic diseases.

One glycogen storage disease is known as Von Gierke's disease and is characterized by a deficiency of the enzyme glucose 6-phosphatase. Because with this deficiency the conversion of glycogen into D-glucose is greatly slowed, the liver becomes stuffed with glycogen and greatly enlarged, distending the abdomen. Development is slowed, and there are frequent episodes of weakness that can be eliminated by eating. The patient easily becomes hypoglycemic because glycogen is not converted into glucose to maintain blood sugar levels between meals.

24-2.5 LACTOSE INTOLERANCE

Lactose (see Section 19-2.1), or milk sugar, is a specific disaccharide produced by the mammary gland. The amount of lactose in milk varies from species to species. For example, cow and goat milks contain 4.5% lactose and human milk has approximately 7.0%. We take in lactose in milk and other dairy products, such as ice cream, that are not fermented. The fermented dairy products, such as most yogurt and cheese, contain less lactose because during fermentation, some of the lactose is converted into lactate.

Lactose is not digested until it reaches the small intestine, where the hydrolytic enzyme lactase is located (see Section 24-1.4). Lactase catalyzes the hydrolysis of lactose into its monosaccharide constituents, D-glucose and D-galactose. Both are rapidly absorbed and enter the blood stream.

$$\text{lactose} \xrightarrow{\text{lactase}} \text{D-glucose} + \text{D-galactose}$$

If for some reason the ingested lactose is only partially hydrolyzed or is not hydrolyzed at all and subsequently passes into the large intestine, a clinical **syndrome** results. The **symptoms** of this syndrome are abdominal distention, cramps, flatulence (gas), and diarrhea. If there is a deficiency of lactase, some lactose is not absorbed from the small intestine and remains in the **lumen**. The presence of lactose tends to draw fluid into the intestinal lumen by osmosis. It is this fluid that leads to the abdominal distention, cramps, and diarrhea. From the small intestine, the lactose passes into the large intestine (colon) where bacteria cause it to undergo fermentation to lactic acid (present as the lactate anion; see

A **syndrome** is the pattern of symptoms in a disease.

Symptoms are evidence of a disease.

A **lumen** is a cavity or channel within a tube or a tubular organ such as the intestine.

$$\text{lactose} \xrightarrow[\text{of bacteria}]{\beta\text{-galactosidase}} \text{D-glucose} + \text{D-galactose}$$

fermentation in bacteria ↓

$$\begin{array}{c} COO^- \\ | \\ HOCH \\ | \\ CH_3 \end{array}$$

L-lactate

Section 24-2.2) and other short-chain acids. Production of more molecules results in still greater retention of fluid. The acidic products of fermentation lower the pH and irritate the lining of the colon, leading to an increased movement of the contents. Diarrhea is caused by the retention of fluid and the increased movement of the intestinal contents. The gaseous products of fermentation cause bloating. The major gases of the colon, carbon dioxide, hydrogen, methane, and oxygen, are odorless. The components having disagreeable odors make up much less than 1% of the total gases. These are ammonia, hydrogen sulfide, short-chain fatty acids (see Chapter 26), and amines, particularly skatole and indole.

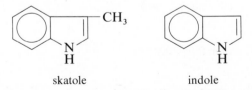

skatole   indole

This whole syndrome that we have described is called *lactose intolerance.* It results from the absence or deficiency of the enzyme lactase. Lactose intolerance is not usually seen in children until after about 6 years of age. At this point, the incidence of lactose-intolerant individuals begins to rise, but no one knows why.

There are varying degrees of lactose intolerance. The difference in incidence among whites of western European descent and blacks is large. By 12 years of age, 45% of blacks develop the symptoms of lactose intolerance; among teenage blacks, the incidence climbs to 70%; and by adulthood, 90% of the black population in America shows symptoms of lactose intolerance. Lactose intolerance is also high among Orientals. Among the whites of Western European ancestry, the peak incidence in adulthood is about 15%. This information indicates that the presence or absence of lactase is under genetic control.

24-2.6 INCOMPLETE DIGESTION AND/OR ABSORPTION OF OTHER CARBOHYDRATES

Other carbohydrates that are not completely broken down into monosaccharides by intestinal enzymes and are not absorbed also pass into the colon. There they are metabolized by microorganisms producing lactate and gas. Again diarrhea and bloating result. This problem occurs from eating beans because beans contain a trisaccharide and a tetrasaccharide that are not hydrolyzed to monosaccharides by intestinal enzymes and, thus, pass into the colon where they are fermented.

PROBLEMS

8. Write, in the form $A + B \longrightarrow C + D$, the following equations.

(a)

$$
\begin{array}{c}
\text{COO}^- \\
| \\
\text{C}-\text{O}\textcircled{P} \\
\| \\
\text{CH}_2
\end{array}
\quad
\xrightarrow[]{\text{ADP} \quad \text{ATP}}
\quad
\begin{array}{c}
\text{COO}^- \\
| \\
\text{C}=\text{O} \\
| \\
\text{CH}_3
\end{array}
$$

      phosphoenolpyruvate   pyruvate

(b)

$$
\begin{array}{c}
\text{CH}_2\text{OH} \\
| \\
\text{C}=\text{O} \\
| \\
\text{HOCH} \\
| \\
\text{HCOH} \\
| \\
\text{HCOH} \\
| \\
\text{CH}_2\text{O}\textcircled{P}
\end{array}
\quad
\xrightarrow[]{\text{ATP} \quad \text{ADP}}
\quad
\begin{array}{c}
\text{CH}_2\text{O}\textcircled{P} \\
| \\
\text{C}=\text{O} \\
| \\
\text{HOCH} \\
| \\
\text{HCOH} \\
| \\
\text{HCOH} \\
| \\
\text{CH}_2\text{O}\textcircled{P}
\end{array}
$$

     D-fructose 6-phosphate   D-fructose 1,6-diphosphate

9. Describe the function of the Cori cycle.
10. How much ATP is produced in glycolysis from the following?
    (a) glycogen
    (b) glucose
11. How many moles of ATP are produced from each mole of glucose?
12. Write the overall equation for the catabolism of glucose in glycolysis.
13. Describe the function of NAD^+.

14. The enzyme that catalyzes the conversion of D-glucose 6-phosphate to D-glucose, allowing the liver to convert glycogen into glucose, is _____.

ANSWERS TO PROBLEMS   **8. (a)**

$$
\begin{array}{ccc}
\text{COO}^- & & \text{COO}^- \\
| & & | \\
\text{C}-\text{O}\,\textcircled{P} + \text{ADP} \longrightarrow & & \text{C}=\text{O} + \text{ATP} \\
\| & & | \\
\text{CH}_2 & & \text{CH}_3
\end{array}
$$

(b)

$$
\begin{array}{ccc}
\text{CH}_2\text{OH} & & \text{CH}_2\text{O}\,\textcircled{P} \\
| & & | \\
\text{C}=\text{O} & & \text{C}=\text{O} \\
| & & | \\
\text{HOCH} + \text{ATP} \longrightarrow & \text{HOCH} & + \text{ADP} \\
| & & | \\
\text{HCOH} & & \text{HCOH} \\
| & & | \\
\text{HCOH} & & \text{HCOH} \\
| & & | \\
\text{CH}_2\text{O}\,\textcircled{P} & & \text{CH}_2\text{O}\,\textcircled{P}
\end{array}
$$

9. to increase the efficiency of the oxidation of glucose to produce ATP when oxygen is in short supply in muscle tissue   **10. (a)** 3 mol from each mole of glucosyl units   **(b)** 2 mol/ mol   **11.** 2   **12.** D-glucose $+ 2\,\text{HPO}_4{}^{2-} + 2\,\text{ADP} + 2\,\text{NAD}^+ \longrightarrow 2$ pyruvate $+ 2\,\text{ATP} + 2\,\text{NADH}$ **13.** It is a coenzyme that functions as an oxidizing agent   **14.** glucose 6-phosphatase

---

24-3 REGULATION OF CARBOHYDRATE METABOLISM

LEARNING OBJECTIVES

1. Describe the effects of insulin and glucagon on the metabolism of glucose.
2. Diagram the concurrent regulation of glycogenesis and glycogenolysis.
3. Discuss feedback inhibition.
4. Describe the characteristics that identify feedback inhibition.

Living organisms are not wasteful. Reactions proceed and molecules are used as they are needed. Some kind of control is necessary. The end products of a pathway must be present in sufficient supply for the process for which they are needed. Yet when sufficient end product for a given purpose is present, the pathway producing it must be slowed down or stopped to save the starting materials. Energy needs to be supplied by catabolic processes such as glycolysis. However, these catabolic processes should not go on at full speed when sufficient usable energy (ATP) is present to prevent wasteful using up of stores of chemical energy. There are regulatory mechanisms for every aspect of carbohydrate metabolism. Some of these are gross controls. Others provide fine tuning. We will now look at a few, selected controls.

24-3.1 TRANSPORT

Transport can be very important in the control of metabolism because anabolic and catabolic pathways frequently begin with transport through membranes into or out of cells or organelles. In muscle and adipose tissue, intake of glucose by cells is the rate-limiting step for all subsequent glucose metabolism. This transport is regulated by the hormone insulin.

24-3.2 HORMONAL CONTROL OF CARBOHYDRATE METABOLISM

Most of the factors that tend to influence the blood glucose level, with the exception of physical and mental activity, are under hormonal control.

Insulin

Insulin is a polypeptide hormone synthesized in special cells of the pancreas called beta cells, first as a proinsulin polypeptide which is subsequently converted into insulin by removal of a connecting peptide (see Figure 24.8) by the proper enzyme. Insulin affects carbohydrate metabolism in the following ways.

1. Insulin increases intake and utilization of glucose by skeletal muscle, heart muscle, diaphragm, adipose, and lactating mammary gland tissue.
2. Insulin promotes glycogen synthesis in the liver and muscle.
3. Insulin retards hepatic glycogenolysis.
4. Insulin decreases hepatic gluconeogenesis.

$$\text{glycogen} \underset{\text{glycogenesis}}{\overset{\text{glycogenolysis}}{\rightleftarrows}} \text{glucose} \underset{\text{gluconeogenesis}}{\overset{\text{glycolysis}}{\rightleftarrows}} \text{pyruvate}$$

The secretion of insulin in response to a rise in blood glucose concentration demonstrates its importance as a regulator of carbohydrate metabolism. Insulin lowers blood glucose mainly by promoting transport of glucose into cells. It also activates enzymes that are responsible for catabolism of glucose inside cells and antagonizes the metabolic processes that tend to raise the blood glucose level.

Liver cells do not require insulin for the intake of glucose. However, even though glucose can freely enter liver cells without insulin, insulin does affect metabolism in the liver. Again, insulin stimulates glycolysis, glycogenesis, and the synthesis of protein. Insulin inhibits glycogenolysis and gluconeogenesis. Insulin increases the amount of glucokinase. The level of glucokinase activity is important because the conversion of glucose into glucose 6-phosphate is the first step in the use of glucose.

Insulin does not enter cells but becomes attached to a specific receptor on the plasma membrane. It acts in opposition to hormones such as adrenaline and glucagon by lowering the concentration of cAMP (see Section 24-2.1).

Diabetes mellitus is a condition in which insulin is missing, is in a low concentration, or is ineffective for some reason. Without insulin, glucose cannot enter many cells such as those of skeletal muscle and adipose tissue. Therefore, in diabetes mellitus, although the concentration of glucose in the blood is so high that glucose spills over into the urine because its entry into cells is prohibited or greatly reduced, the cells are actually starved for glucose.

Figure 24.8 Conversion of **porcine** proinsulin into insulin. The A chain contains 21 amino acids, the B chain contains 30 amino acids, and the connecting peptide (shown in color) contains 33 amino acids. Porcine insulin is commonly used in the treatment of diabetes mellitus (see also Figure 21.2).

Porcine refers to swine (pigs).

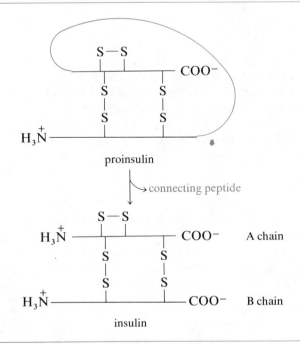

Glucagon

Glucagon is another polypeptide hormone. It is made by cells of the pancreas called alpha cells and contains 29 amino acid residues. In contrast to insulin, glucagon is released when the concentration of blood sugar is low. Glucagon raises the insulin level of the blood in its initial action. Its longer-term effect, however, is to increase the blood glucose level (see below). Other hormones that tend to raise the blood glucose level are adrenaline, glucocorticoids, and somatotropic hormone.

Glucagon acts in the liver and affects carbohydrate metabolism in the following ways.

1. Glucagon increases glycogenolysis via activation of adenyl cyclase and the resulting cascade (see Section 24-2.1).
2. Glucagon increases protein catabolism and gluconeogenesis.

However, as mentioned above, the initial and outstanding effect of glucagon is to raise rapidly the concentration of insulin in blood plasma. The effect of glucagon on raising the insulin concentration is greater than that of glucose and is not dependent on a rise in the blood glucose level. In fact, low blood sugar stimulates glucagon release. At first, these effects might seem to be in opposition to each other. However, they are not. During periods of hypoglycemia, the body would want to do two things: (1) raise blood sugar levels, and (2) allow the glucose present to get into cells. (Insulin concentration would be low when the blood sugar concentration is low.) Both events are activated by glucagon. Glucagon effects the release of insulin so that the glucose of the blood can pass into cells. Then glucagon promotes the conversion of glycogen stored in the liver into glucose to raise the blood sugar level.

Adrenaline

Adrenaline increases glycogenolysis in nonliver cells via activation of adenyl cyclase. Noradrenaline acts in a similar manner. The activation of adenyl cyclase results in the production of cAMP and initiates the cascade activating phosphorylase (see Section 24-2.1). Adrenaline also raises blood sugar indirectly by inhibiting the secretion of insulin. Emotional stimuli bring about a release of adrenaline. For the structures of adrenaline and noradrenaline see Sections 17-5.2 and 28-2.5.

Glucocorticoids

Glucocorticoids, such as cortisol, are steroid hormones (see Section 26-2.1) produced in the adrenal gland. They are released in response to a great variety of "alarming" stimuli that include pain and practically all unpleasant emotions, such as anger, fear, and anxiety, as well as other types of stress, such as exposure to cold and physical and mental tension. Glucocorticoids act in the following ways.

1. They reduce the intake of glucose by tissues.
2. They promote gluconeogenesis; that is, they have a catabolic effect on protein metabolism.
3. They increase glucose 6-phosphatase activity; that is, they promote the conversion of glucose 6-phosphate, and therefore glycogen, into glucose.

However, cortisol also promotes the deposition of glycogen (glycogenesis) in the liver.

cortisol

Somatotropic (Growth) Hormone

Somatotropic or growth hormone, a protein, acts in the following ways.

1. It inhibits the oxidation of carbohydrates.
2. It increases the activity of glucose 6-phosphatase.

3. It inhibits the action of insulin.

All the anti-insulin hormones (adrenaline, glucagon, cortisol, and somatotropic hormone) help to maintain the blood glucose level, partly by increasing the release of glucose into the blood and partly by reducing its intake by tissues.

Combined Actions

The processes of glucose formation in the liver and utilization of glucose in other tissues are sensitive to relatively slight deviations from the normal blood sugar concentration. For example, as the blood sugar concentration rises, glycogen synthesis is accelerated and utilization is increased, with a consequent fall in blood sugar. The reverse occurs as the blood sugar concentration falls. The response is determined to a considerable extent by the balance between insulin, on the one hand, and glucagon and hormones of the adrenal gland (adrenaline, noradrenaline, and glucocorticoids) and the pituitary gland (growth hormone) on the other hand. The overall effect of insulin is to lower the blood sugar and that of glucagon and the glucocorticoids and growth hormone is to raise it. It is the ratio between them rather than their absolute amounts that is of primary importance in this connection.

The processes of glycogenesis and glycogenolysis in the liver, the various processes of glucose utilization, and the blood sugar concentration are continually exposed to disturbing influences under physiological conditions. These disturbing influences include absorption of glucose from the intestine, physical and mental activity, and emotional states. The primary effect of the majority of these is a rise in blood sugar. This automatically results in a net decrease in the outflow of glucose from the liver and an acceleration of its utilization by the tissues. There is a simultaneous increase in the secretion of insulin, stimulated by the elevated blood sugar concentration, with a consequent increase in the ratio of insulin to the glucocorticoid and growth hormones. This change in hormonal balance results in increased synthesis of glycogen in the liver, decreased gluconeogenesis, decreased output of glucose from the liver, and increased utilization of glucose. As a result, the blood sugar concentration falls. Overshoot sometimes occurs when a starchy food such as a baked potato is digested and the products of digestion are absorbed.

A drop in blood sugar below the normal resting level causes a decreased secretion of insulin, an increased secretion of glucagon, a decrease in the ratio of insulin to the antagonistic adrenocortical and growth hormones, increased production of blood sugar, and decreased glucose utilization. Accordingly, the blood sugar rises. If the blood sugar falls to hypoglycemic levels, an additional mechanism comes into operation. This emergency mechanism works by stimulation of the secretion of adrenaline, resulting in acceleration of glycogenolysis in the liver and a rise in blood sugar.

Thus, acting in concert, these hormones maintain blood glucose levels in the normal person. If the control seems complicated, it is. The control of hormone levels and their action is determined by an elaborate system involving several endocrine cells, each of which produces and secretes a specific hormone, that may be regulated by the product of their action on a distant target cell. This system allows for a rapid response and close control of blood sugar levels.

24-3.3 REGULATION OF PHOSPHORYLASE AND GLYCOGEN SYNTHETASE ACTIVITIES

We have already discussed hormonal regulation of levels of cAMP, which controls the activity of protein kinase. Protein kinase controls the conversion of phosphorylase *b* to phosphorylase *a*, which in turn controls the conversion of glycogen into glucose 1-phosphate (see Section 24-2.1). Phosphorylase exists in two molecular forms and its activity is regulated by allosteric effectors and reversible phosphorylation, with modifiers acting on both processes. The activity of glycogen synthetase, which catalyzes the synthesis of glycogen using an activated form of glucose, is regulated in a similar way. The more active form (glycogen synthetase *I*) can be phosphorylated in the presence of ATP and protein kinase to the less active form (glycogen synthetase *D*), which requires glucose 6-phosphate as a positive allosteric effector. As a result, glycogen synthesis is also controlled by a cascade effect. As the concentration of cAMP increases, phosphorolysis increases and glycogen synthesis decreases, and vice versa. The overall control is depicted by the diagram in Figure 24.9. cAMP, through activation of protein kinase, affects many other cellular processes.

Figure 24.9   Diagram of the concurrent control of glycogen synthetase activity (glycogen synthesis) and phosphorylase activity (glycogen breakdown) by hormones that regulate the concentration of cAMP. The colored arrows mean that that form of the enzyme has some, but low, activity and then only in the presence of a positive allosteric effector. The black arrows indicate that that form of the enzyme is very active as is.

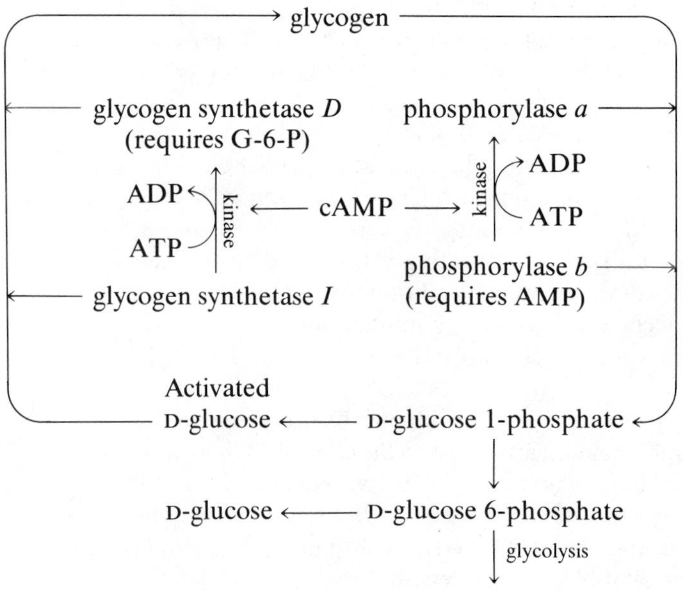

In addition to requiring glucose 6-phosphate, whose binding stabilizes the catalytically active conformation, the *D* (dependent) form is markedly more sensitive to negative allosteric control by cellular metabolites than is the *I* (independent) form, suggesting that the *D* form is normally inactive in the cell while the *I* form is active. Thus, as with phosphorylase, covalent modification provides the "on-off" switch and allosteric controls add the "fine tuning."

24-3.4 PHOSPHOFRUCTOKINASE AND FRUCTOSE DIPHOSPHATASE IN THE REGULATION OF GLYCOLYSIS

$$\text{D-fructose 6-phosphate} \xrightleftharpoons[\text{fructose diphosphatase}]{\text{phosphofructokinase}} \text{D-fructose 1,6-diphosphate}$$
$$\text{(F-6-P)} \qquad\qquad\qquad\qquad\qquad\qquad \text{(F-1,6-diP)}$$

Phosphofructokinase (PFK) is the enzyme of glycolysis with the lowest activity in the liver. Hence, it is the rate-limiting enzyme in liver glycolysis. Its activity appears to be geared to the demand for energy. It is an allosteric enzyme and catalyzes the most important control point in glycolysis.

Reduce Phosphofructokinase Activity	Increase Phosphofructokinase Activity
ATP	AMP
citrate	ADP
2,3-diphosphoglycerate (liver and red blood cell enzymes)	
free fatty acids (liver enzyme)	

These effects can be explained easily in terms of feedback control.

Feedback Inhibition

In many pathways, the first enzyme in a pathway is inhibited by the end product, or a late intermediate, of that pathway acting as a negative allosteric effector (see Section 23-3.7). The first step is the most strategic point of control because regulation of the first step governs the rate at which molecules enter the pathway. Inhibition of the first step, for example, decreases the rate of end-product formation but also avoids unnecessary accumulation of intermediates in the pathway. The molecule that inhibits the enzyme generally has no structural relationship to the normal substrate for the enzyme. A diagrammatic representation of feedback inhibition showing the end product (F) of a pathway as an inhibitor of the enzyme (e_1) catalyzing the first reaction is shown.

$$A \xrightleftharpoons{e_1} B \xrightleftharpoons{e_2} C \xrightleftharpoons{e_3} D \xrightleftharpoons{e_4} E \xrightarrow{e_5} F$$

In this kind of inhibition, the end product binds to a specific site (the regulatory site, not the catalytic site or active site) on the enzyme. The binding results in the induction or stabilization of a molecular shape of the enzyme with a lower specific activity. We have already seen in Section 23-3 that such enzymes are called allosteric or regulatory enzymes, and the end products that do the regulating are called allosteric effectors, or simply effectors. When there is sufficient ATP in a cell, there is no need to produce more and, therefore, ATP inhibits glycolysis. Likewise, if there is a high concentration of AMP, ADP, or both, it means that there is a low concentration of ATP. Therefore, carbohydrate catabolism needs to be increased in order to increase ATP production. If there is a high level of 2,3-diphospho-D-glycerate, an intermediate of glycolysis farther down the pathway, there is no need to make more, so the cell slows down the earlier conversion of F-6-P into F-1,6-diP to conserve carbohydrate. Also, if there is a high level of free fatty acids that also can be catabolized to produce ATP, or if there is a high level of citrate, an intermediate of the tricarboxylic acid cycle (see Section 25-1) that utilizes the end product of glycolysis, pyruvate, carbohydrate catabolism is slowed. In short, when the energy level (stored ATP) in a cell is low, glycolysis is increased by increasing the activity of PFK, and when the energy level is high, glycolysis is decreased by decreasing PFK activity.

Fructose diphosphatase is an important enzme in gluconeogenesis. It reverses the conversion of D-fructose 6-phosphate into D-fructose 1,6-diphosphate. When AMP concentration is high, it means that ATP concentration is low. Therefore, it is necessary to have glycolysis go forward for ATP production (conversion of AMP into ATP). As a result, fructose diphosphatase, and therefore gluconeogenesis, is inhibited by AMP. There are a number of other feedback controls on glycolysis.

PROBLEMS

15. List the four ways that insulin affects carbohydrate metabolism.
16. List the two ways that glucagon affects carbohydrate metabolism.
17. Describe the characteristics that identify feedback inhibition.
18. Diagram the concurrent regulation of glycogenesis and glycogenolysis.

ANSWERS TO PROBLEMS   **15.** (1) Insulin increases intake and utilization of glucose by many tissues, especially skeletal and heart muscle tissue.   (2) Insulin promotes glycogen synthesis in the liver and muscle.   (3) Insulin retards glycogenolysis in the liver.   (4) Insulin decreases gluconeogenesis.   **16.** (1) Glucagon increases glycogenolysis in the liver.   (2) Glucagon increases gluconeogenesis.   **17.** Feedback inhibition occurs when the end product, or a late intermediate, of a pathway is a negative allosteric effector for an enzyme catalyzing a reaction near the beginning of the pathway.   **18.** See Figure 24.9.

SUMMARY

In this chapter, we studied the digestion of carbohydrates and the first pathway (the pathway of glycolysis) involved in the use of glucose as a source of energy. We learned that digestible carbohydrate in our diets exists largely in the form of starch. Starch consists of two polysaccharides, amylose and amylopectin, both of which are polymers of D-glucopyranosyl units.

When cooked starch is eaten, it is first mixed with saliva. Saliva contains an α-amylase. α-Amylases readily cleave starch molecules to oligosaccharides of six or seven glucose units by catalyzing the hydrolysis of internal bonds of the polysaccharide chain. α-Amylases act more slowly on these oligosaccharides, breaking them into smaller fragments. However, food does not stay in the mouth long enough for the α-amylase to catalyze the breakdown of starch molecules significantly.

The digestion of starch is completed in the small intestine by the action of pancreatic α-amylase and other enzymes that catalyze the hydrolysis of the resulting oligosaccharides into D-glucose. Other digestible carbohydrates, such as sucrose and lactose, are also broken down into monosaccharides in the small intestine through the action of specific enzymes.

Monosaccharides, primarily glucose from the digestion of starch, are absorbed. Glucose is transported in the blood and taken up by cells. In muscle and liver, the glucose molecules are used for the synthesis of glycogen, a molecule that stores glucose units until they are needed for energy. Glycogen is a branched polymer of D-glucose that is very similar in structure to amylopectin.

Carbohydrates that are not digested and absorbed in the small intestine pass into the large intestine where bacteria convert them into lactate, carbon dioxide, hydrogen, and other small molecules. The result is gas and diarrhea.

When it is necessary to catabolize glucose to make ATP or other molecules, glucose or glucopyranosyl units of glycogen are oxidized to pyruvate in a pathway called glycolysis. This pathway can proceed in either the presence or the absence of oxygen. The overall reaction for the oxidation of a molecule of glucose in the glycolytic pathway is D-glucose $+ 2 HPO_4^{2-} + 2 ADP + 2 NAD^+ \longrightarrow 2$ pyruvate $+ 2 ATP + 2 NADH$.

When sufficient oxygen is present, the two pyruvate molecules are further oxidized in the tricarboxylic acid cycle and the two molecules of the reduced coenzyme NADH are reoxidized to two molecules of the oxidized coenzyme NAD^+ in a pathway that requires oxygen and produces three molecules of ATP. Both pathways are presented in the next chapter.

When there is insufficient oxygen present in a tissue or cell, some of the pyruvate molecules are reduced to lactate molecules. This reaction uses NADH as the reducing agent, so the other product of the reaction is NAD^+. The purpose of this reaction is to reoxidize the reduced coenzyme NADH to the oxidized form of the coenzyme (NAD^+) so that glycolysis can continue and ATP can be made even though there is not enough oxygen to effect the oxidation of NADH to NAD^+. Lactate made by the reduction of pyruvate is transported to the liver where most of it is converted back into D-glucose via the pathway of gluconeogenesis.

Hormones are important in the control of carbohydrate metabolism. Insulin is required for the transport of glucose into most cells. Insulin also promotes glycogen synthesis and glycolysis. The effect of insulin is to lower the blood sugar level. Glucagon has the opposite effect. It promotes the conversion of glycogen into glucose and raises the blood sugar level. Adrenaline activates adenyl cyclase and brings about the formation of cAMP. cAMP initiates two cascades. One cascade activates phosphorylase. Phosphorylase then catalyzes the formation of D-glucose 1-phosphate from glycogen, the reaction that initiates glycogenolysis and glycolysis. The other cascade results in a reduction of glycogen synthesis. The overall effect of adrenaline is to increase the use of glycogen to make ATP.

Feedback control is also important in the regulation of glycolysis. In feedback inhibition, the end product, or a late intermediate, of a pathway is a negative allosteric effector for an enzyme catalyzing a reaction near the beginning of the pathway. In this way, high concentrations of substances such as ATP and 2,3-diphospho-D-glycerate slow the pathway that produces them (glycolysis) so that larger amounts do not accumulate. This prevents waste. Other substances, such as AMP and ADP, are positive allosteric effectors that activate key enzymes in the pathway. High concentrations of AMP and ADP reflect low concentrations of ATP, so they activate the pathway converting them to ATP.

Balance is maintained through the combined actions of various stimuli. Reactions are speeded when there is a need for them and slowed when they are not needed.

ADDITIONAL PROBLEMS

19. Describe the differences between glycogenesis, glycogenolysis, and gluconeogenesis.
20. Complete the following equations.

(a)
$$\begin{array}{c} COO^- \\ | \\ HCO\,\textcircled{P} \\ \| \\ CH_2 \end{array} + ADP \xrightarrow{\text{pyruvate kinase}}$$

(b)
$$\begin{array}{c} HC=O \\ | \\ HCOH \\ | \\ CH_2O\,\textcircled{P} \end{array} \xrightleftharpoons{\text{triose phosphate isomerase}}$$

(c)
$$\begin{array}{c} CH_2O\,\textcircled{P} \\ | \\ C=O \\ | \\ HOCH \\ | \\ HCOH \\ | \\ HCOH \\ | \\ CH_2O\,\textcircled{P} \end{array} \xrightleftharpoons{\text{aldolase}}$$

21. Summarize glycolysis with a diagram showing the reactions that generate ATP, those that use ATP, and those that generate NADH.
22. Describe the function of the conversion of pyruvate to lactate.
23. Write the structure of cAMP.

CHAPTER 25

REACTIONS OF MITOCHONDRIA

In the preceding chapter, we learned that carbohydrate molecules such as D-glucose are oxidized to a three-carbon compound, pyruvate, in the pathway known as glycolysis. In this chapter, we will examine the oxidation of pyruvate in organelles called mitochondria and the production of ATP molecules as a result of this oxidation.

The major role of mitochondria is the conversion of ADP and inorganic phosphate into ATP. Because this reaction results in the formation of a high-energy phosphate bond, it requires energy. The energy is provided by oxidation of foodstuffs. The food we eat consists mainly of carbohydrates, fats, and proteins. Each of these substances is broken down into less complex molecules. Ultimately, each can be oxidized to a common intermediate (acetyl coenzyme A), the acetyl group of which is completely oxidized to carbon dioxide. We learned in Section 9-1 that oxidation involves the loss of electrons. The electrons removed from carbohydrates, fats, proteins, acetyl coenzyme A, and other molecules as they are oxidized are used to reduce oxygen molecules. The reaction of electrons with oxygen molecules in the presence of protons produces water molecules and a large amount of energy. In mitochondria, all these reactions are coupled to produce ATP molecules.

Oxidation of acetyl coenzyme A to carbon dioxide occurs in a metabolic pathway known as the *tricarboxylic acid cycle*. The reduction of oxygen molecules occurs at the end of a pathway known as the *respiratory chain*. The production of ATP molecules occurs by a process known as *oxidative phosphorylation* that occurs as electrons are passed through the respiratory chain. Each of these pathways will be discussed in this chapter.

25-1 THE TRICARBOXYLIC ACID (TCA) CYCLE

LEARNING OBJECTIVES

1. Describe the function of the tricarboxylic acid cycle.
2. Describe "pyruvate dehydrogenase" and "α-ketoglutarate dehydrogenase" as multienzyme complexes.

3. List the name of the coenzyme and the general type of reaction in which the coenzyme functions for the following B vitamins that are coenzymes or that make up part of a coenzyme structure: thiamine, lipoic acid, riboflavin, and pantothenic acid.

4. Define α-keto acid, decarboxylation, and flavoprotein.

5. Given the substrate, coenzyme(s), and the enzyme required for any of the reactions of the tricarboxylic acid cycle, write a balanced equation showing the products of the reaction.

The function of the *tricarboxylic acid, Krebs,* or *citric acid cycle* is to oxidize acetyl coenzyme A and produce the reduced coenzymes NADH and $FADH_2$. The production of NADH and $FADH_2$ is accomplished by coupling the reduction of NAD^+ (see Section 24-2.1) and FAD with the oxidation of intermediates of the TCA cycle (see Figure 25.1 on pp. 630 and 631). The structures of riboflavin, FAD, and another riboflavin-containing coenzyme, FMN, are shown. See Sections 20-1.1 and 20-1.2 for the structure of adenosine.

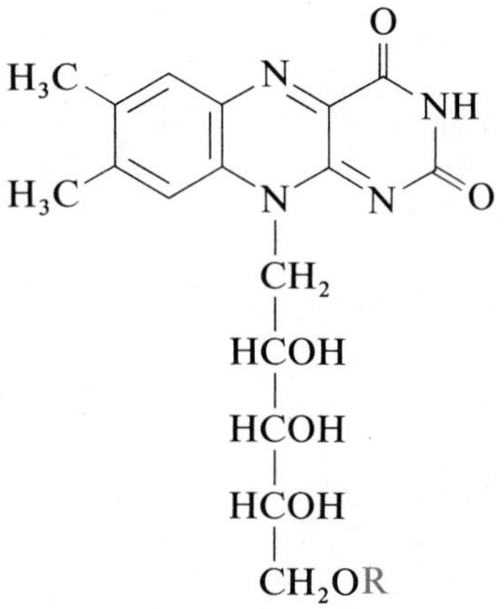

$R = H$

riboflavin (Vitamin B_2)

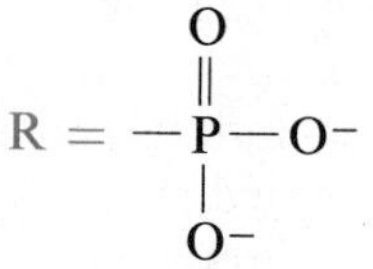

$$R = -\overset{\overset{\displaystyle O}{\|}}{\underset{\underset{\displaystyle O^-}{|}}{P}}-O^-$$

flavin mononucleotide (FMN)

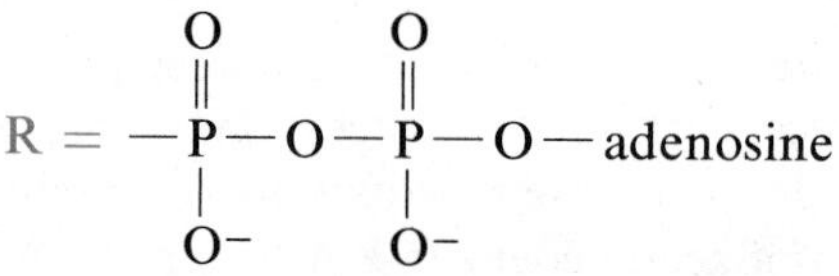

$$R = -\overset{\overset{\displaystyle O}{\|}}{\underset{\underset{\displaystyle O^-}{|}}{P}}-O-\overset{\overset{\displaystyle O}{\|}}{\underset{\underset{\displaystyle O^-}{|}}{P}}-O-\text{adenosine}$$

flavin adenine dinucleotide (FAD)

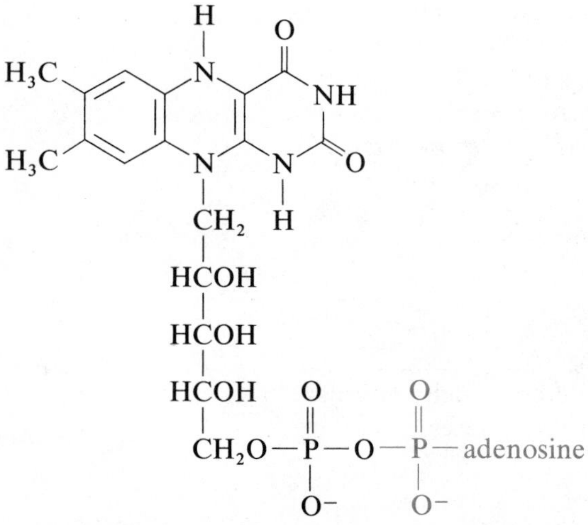

reduced flavin adenine dinucleotide ($FADH_2$)

25-1.1 JOINING GLYCOLYSIS WITH THE TRICARBOXYLIC ACID CYCLE

(1) The "enzyme" connecting glycolysis with the tricarboxylic acid (TCA) cycle is the pyruvate dehydrogenase enzyme complex. This **multienzyme complex** is made up of three enzymes and five coenzymes, one of which is the actual pyruvate dehydrogenase. The five coenzymes are NAD, FAD, thiamine pyrophosphate, lipoic acid, and coenzyme A.

Thiamine pyrophosphate (TPP) is formed from thiamine (vitamin B_1) and is the coenzyme for most oxidative decarboxylations (see Section 16-3.3).

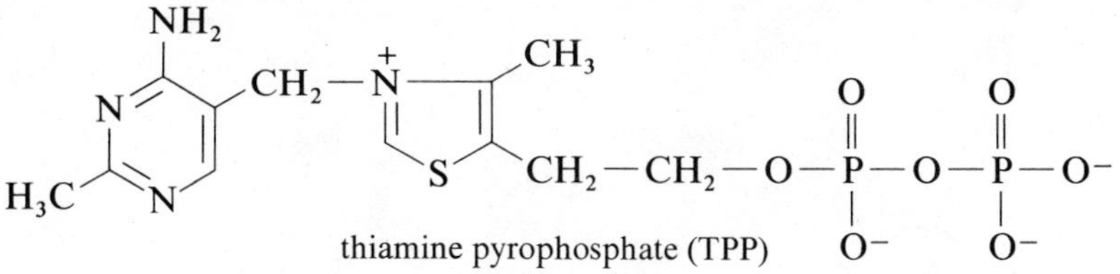

A **multienzyme complex** is an organized association of several different enzymes in one integrated unit in which individual enzymes act cooperatively to catalyze an overall reaction.

thiamine pyrophosphate (TPP)

Lipoic acid is another *B vitamin.* Recall that the B vitamins are the water-soluble vitamins. Like NAD and FAD, lipoic acid functions in oxidation-reduction (redox) reactions. It carries out its function while attached to an enzyme by an amide linkage to the epsilon amino group* of a lysine residue. The structure of lipoic acid as attached to the lysine residue of a protein molecule is given.

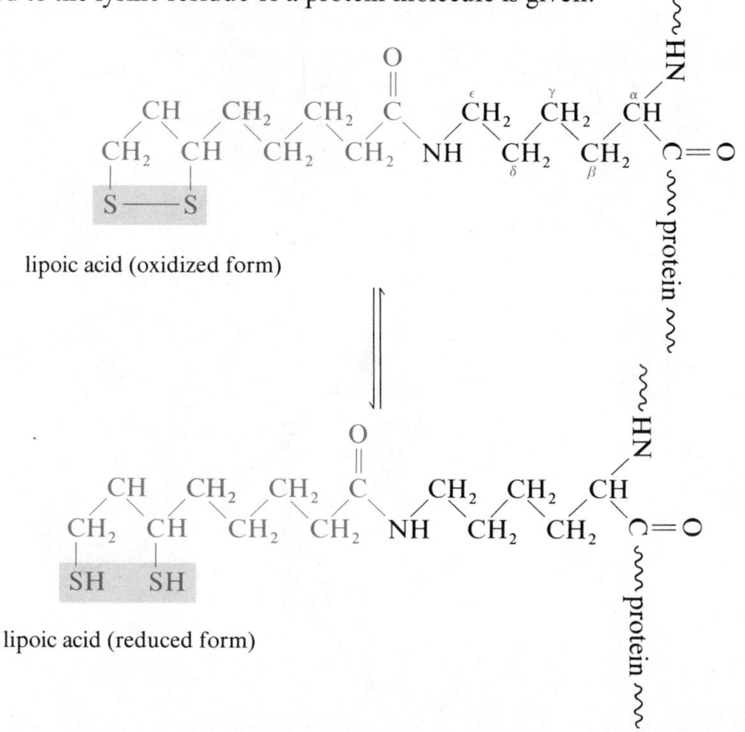

lipoic acid (oxidized form)

lipoic acid (reduced form)

---

*Remember that carbon atoms of an organic compound are numbered. They are also designated by their distance from a functional group with the letters of the Greek alphabet. Because epsilon is the fifth letter of the Greek alphabet, the epsilon amino group is the amino group on the fifth carbon atom from the carboxyl group of the amino acid lysine. $\epsilon \quad \delta \quad \gamma \quad \beta \quad \alpha$
C—C—C—C—C—COOH

Figure 25.1 The tricarboxylic acid, Krebs, or citric acid cycle, and the reaction connecting glycolysis to the cycle.

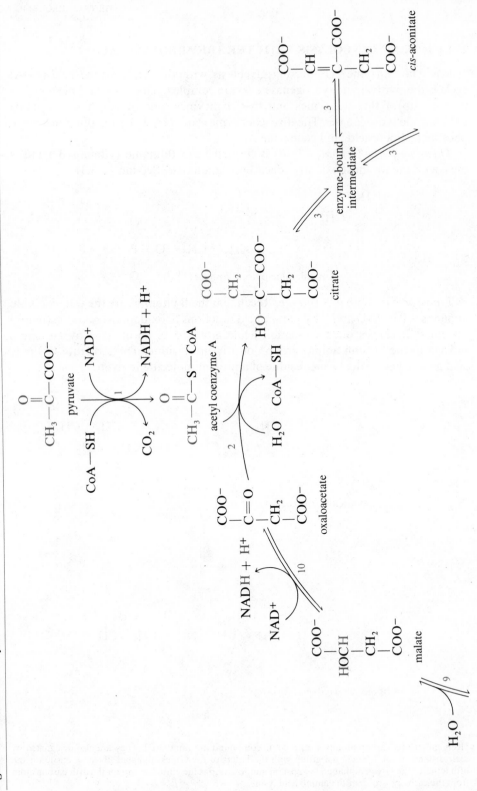

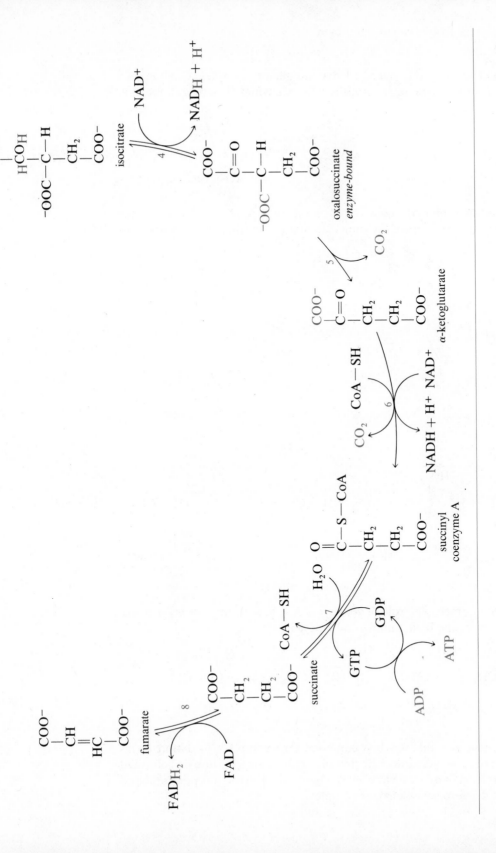

The three enzymes of the pyruvate dehydrogenase complex catalyze five separate reactions, each involving one of the five coenzymes. The overall reaction can be written as shown.

$$CH_3-\overset{\overset{\displaystyle O}{\|}}{C}-COO^- + CoA-SH + NAD^+ \xrightarrow[\text{dehydrogenase}]{\text{pyruvate}} CH_3-\overset{\overset{\displaystyle O}{\|}}{C}-S-CoA + CO_2 + NADH + H^+$$

pyruvate   coenzyme A   acetyl coenzyme A
                                                           (acetyl-CoA)

The abbreviation commonly used for *coenzyme A* is CoA—SH because the thiol (—SH) group is the part of the molecule that reacts when such compounds

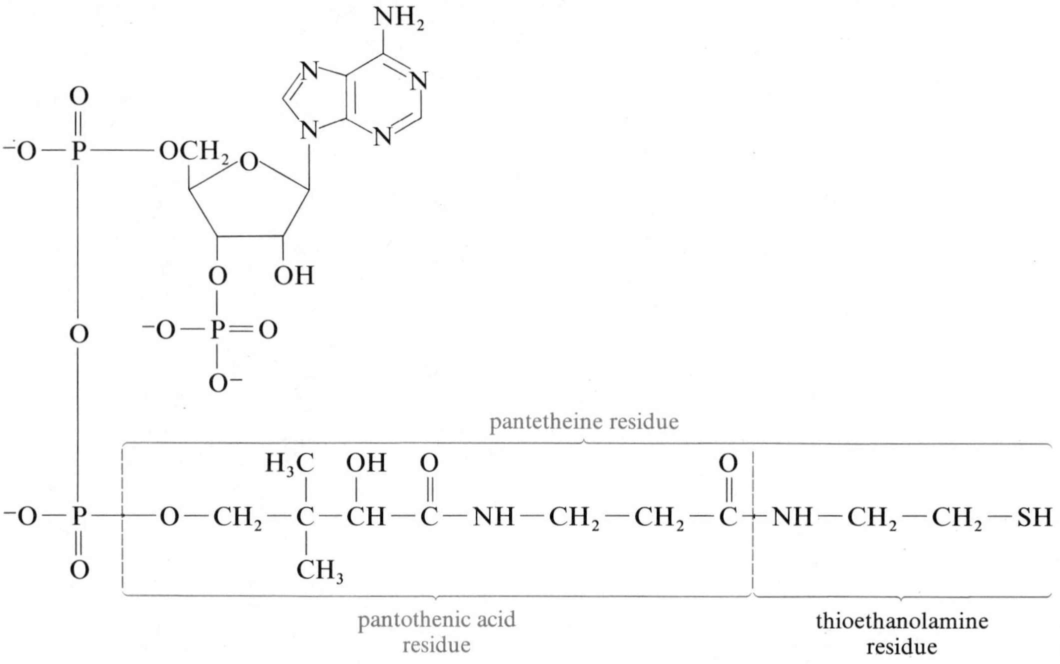

coenzyme A (CoA—SH)

as acetyl-CoA are formed. The role of coenzyme A is to activate carboxylic acids. Coenzyme A contains the B vitamin pantothenic acid.

$$HOCH_2-\overset{\overset{\displaystyle CH_3}{|}}{\underset{\underset{\displaystyle CH_3}{|}}{C}}-CHOH-\overset{\overset{\displaystyle O}{\|}}{C}-NH-CH_2-CH_2-COOH$$

pantothenic acid

Because pyruvate is a three-carbon compound, its complete oxidation forms three molecules of carbon dioxide. The first molecule of carbon dioxide is formed during the conversion of pyruvate into acetyl coenzyme A in the reaction catalyzed by the pyruvate dehydrogenase enzyme complex.

25-1.2 THE TRICARBOXYLIC ACID CYCLE

(2)   *Acetyl coenzyme A* is an activated form of acetic acid (see Section 20-1.2). The two-carbon acetyl group adds to the four-carbon anion of oxaloacetic acid to form citrate, the anion of the six-carbon tricarboxylic acid citric acid, in a reaction catalyzed by the enzyme citrate synthetase. This reaction begins the TCA cycle (see Figure 25.1).

$$
\begin{array}{c}
COO^- \\
| \\
C{=}O \\
| \\
CH_2 \\
| \\
COO^-
\end{array}
+ CH_3{-}\overset{O}{\overset{\|}{C}}{-}S{-}CoA + H_2O
\xrightarrow[2]{\text{citrate synthetase}}
\begin{array}{c}
COO^- \\
| \\
CH_2 \\
| \\
HO{-}C{-}COO^- \\
| \\
CH_2 \\
| \\
COO^-
\end{array}
+ CoA{-}SH
$$

oxaloacetate   citrate

(3)   Citrate then undergoes isomerization to isocitrate, the anion of the six-carbon, tricarboxylic acid, isocitric acid. This reaction involves an enzyme-bound intermediate. The enzyme aconitase catalyzes this reaction.

$$
\begin{array}{c}
COO^- \\
| \\
CH_2 \\
| \\
HO{-}C{-}COO^- \\
| \\
CH_2 \\
| \\
COO^-
\end{array}
\underset{3}{\overset{\text{aconitase}}{\rightleftharpoons}}
\text{intermediate}
\underset{3}{\overset{\text{aconitase}}{\rightleftharpoons}}
\begin{array}{c}
COO^- \\
| \\
HCOH \\
| \\
{}^-OOC{-}C{-}H \\
| \\
CH_2 \\
| \\
COO^-
\end{array}
$$

citrate   isocitrate

$$
\begin{array}{c}
COO^- \\
| \\
HC \\
\| \\
C{-}COO^- \\
| \\
CH_2 \\
| \\
COO^-
\end{array}
$$

aconitate

(4, 5)   Oxidation of isocitrate by NAD^+ in a reaction catalyzed by the enzyme isocitrate dehydrogenase yields oxalosuccinate and the reduced coenzyme NADH. Oxalosuccinate remains bound to the enzyme. Decarboxylation of oxalosuccinate in a reaction catalyzed by the same enzyme, isocitrate dehydrogenase, gives α-ketoglutarate and the second molecule of carbon dioxide. α-Ketoglutarate is the anion of a five-carbon, dicarboxylic acid.

$$\begin{array}{ccc}
\text{COO}^- & \text{COO}^- & \text{COO}^- \\
| & | & | \\
\text{H COH} & \text{C}=\text{O} & \text{C}=\text{O} \\
| & | & | \\
^-\text{OOC}-\text{C}-\text{H} & \text{H}-\text{C}-\text{COO}^- & \text{CH}_2 \\
| & | & | \\
\text{CH}_2 & \text{CH}_2 & \text{CH}_2 \\
| & | & | \\
\text{COO}^- & \text{COO}^- & \text{COO}^- \\
\text{isocitrate} & \text{oxalosuccinate} & \alpha\text{-ketoglutarate}
\end{array}$$

4 isocitrate dehydrogenase — NAD⁺ → NADH + H⁺

5 isocitrate dehydrogenase — CO₂

(Reaction 4: isocitrate dehydrogenase, $NAD^+ \to NADH + H^+$; Reaction 5: isocitrate dehydrogenase, CO_2)

(6) Next, α-ketoglutarate is converted into succinyl coenzyme A and the third molecule of carbon dioxide. If we compare the structures of pyruvate and acetyl

$$\text{NAD}^+ + \begin{array}{c}
\text{COO}^- \\
| \\
\text{C}=\text{O} \\
| \\
\text{CH}_2 \\
| \\
\text{CH}_2 \\
| \\
\text{COO}^-
\end{array} + \text{CoA}-\text{SH} \xrightarrow[\substack{\text{FAD,} \\ \text{lipoic acid,} \\ \text{TPP,} \\ \text{Mg}^{2+}}]{\substack{6 \\ \alpha\text{-ketoglutarate} \\ \text{dehydrogenase}}} \text{NADH} + \text{H}^+ + \begin{array}{c}
\text{O} \\
\| \\
\text{C}-\text{S}-\text{CoA} \\
| \\
\text{CH}_2 \\
| \\
\text{CH}_2 \\
| \\
\text{COO}^-
\end{array} + \text{CO}_2$$

α-ketoglutarate · · · · · · · · · · · succinyl coenzyme A

coenzyme A to the structures of α-ketoglutarate and succinyl coenzyme A, we see that the latter compounds are like the former. The only difference between them is that the latter two have two additional carbon atoms (the $-\text{CH}_2-\text{COO}^-$ group). Not only are the structures of α-ketoglutarate and succinyl coenzyme A similar to the structures of the substrate and products of the reaction catalyzed by pyruvate dehydrogenase, the entire reaction is similar. In both reactions, α-ketoacids are oxidized and decarboxylated and the products are joined to coenzyme A. This reaction is also catalyzed by a multienzyme complex (α-ketoglutarate dehydrogenase) of three enzymes and the same five coenzymes. The product is the ionized coenzyme A derivative of the four-carbon, dicarboxylic acid, succinic acid. This derivative, like acetyl coenzyme A, is a thioester. These thioesters are high-energy compounds.

$$\begin{array}{cc}
\text{O} & \text{O} \\
\| & \| \\
\text{C}-\text{O}-\text{R} & \text{C}-\text{S}-\text{R} \\
| & | \\
\text{CH}_2 & \text{CH}_2 \\
| & | \\
\text{CH}_2 & \text{CH}_2 \\
| & | \\
\text{COO}^- & \text{COO}^-
\end{array}$$

ester of ionized · · · · · thioester of ionized
succinic acid · · · · · · · · · · succinic acid

(7)   In this case, the high-energy compound is not used to make a new carbon-to-carbon bond as in the reaction of acetyl coenzyme A with oxaloacetate, but it is used to make a high-energy phosphate bond in a molecule of guanosine triphosphate. This high-energy phosphate bond is then used to generate a molecule of ATP.

$$
\begin{array}{c}
\text{O} \\
\parallel \\
\text{C—S—CoA} + \text{H}_2\text{O} \\
\mid \\
\text{CH}_2 \\
\mid \\
\text{CH}_2 \\
\mid \\
\text{COO}^-
\end{array}
\underset{\text{synthetase}}{\overset{\substack{7 \\ \text{succinyl-CoA}}}{\rightleftharpoons}}
\begin{array}{c}
\\
\text{COO}^- + \text{CoA—SH} \\
\mid \\
\text{CH}_2 \\
\mid \\
\text{CH}_2 \\
\mid \\
\text{COO}^-
\end{array}
$$

GDP   GTP

ATP   ADP

succinyl coenzyme A   succinate

(8)   Succinate is then oxidized to fumarate by a flavoprotein enzyme containing FAD, which is reduced to FADH$_2$ in the process. This enzyme is succinate dehydrogenase.

$$
\begin{array}{c}
\text{COO}^- \\
\mid \\
\text{HCH} \\
\mid \\
\text{HCH} \\
\mid \\
\text{COO}^-
\end{array}
+ \text{FAD}
\underset{8}{\overset{\substack{\text{succinate} \\ \text{dehydrogenase}}}{\rightleftharpoons}}
\begin{array}{c}
\text{COO}^- \\
\mid \\
\text{CH} \\
\parallel \\
\text{HC} \\
\mid \\
\text{COO}^-
\end{array}
+ \text{FADH}_2
$$

succinate   fumarate

A *flavoprotein* is an enzyme whose cofactor (see Section 23-3.3) is either flavin adenine dinucleotide (FAD) or flavin mononucleotide (FMN). FAD and FMN are usually tightly bound to the enzyme; they also may be loosely bound as coenzymes. When FAD is reduced to FADH$_2$, the conversion involves the net addition of two hydrogen atoms. These two hydrogen atoms can be thought of as either ($2 \text{ H}^+ + 2 \text{ e}^-$) or ($\text{H}^- + \text{H}^+$).

(9)   A water molecule is added to fumarate to convert it to L-malate in a reaction catalyzed by the enzyme fumarase.

$$
\begin{array}{c}
\text{COO}^- \\
\mid \\
\text{CH} \\
\parallel \\
\text{HC} \\
\mid \\
\text{COO}^-
\end{array}
+ \text{HOH}
\underset{9}{\overset{\text{fumarase}}{\rightleftharpoons}}
\begin{array}{c}
\text{COO}^- \\
\mid \\
\text{HOCH} \\
\mid \\
\text{HCH} \\
\mid \\
\text{COO}^-
\end{array}
$$

fumarate   L-malate

(10)   The cycle is completed by the oxidation of L-malate to oxaloacetate in a reaction catalyzed by malate dehydrogenase, an enzyme utilizing NAD as the redox coenzyme.

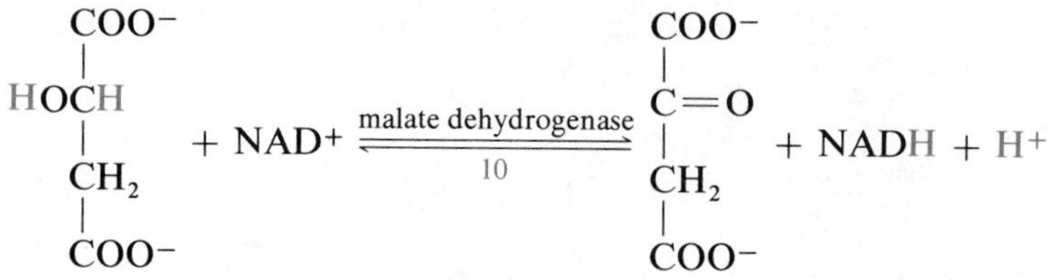

$$
\begin{array}{c}
COO^- \\
| \\
HOCH \\
| \\
CH_2 \\
| \\
COO^-
\end{array}
\;+\; NAD^+ \;\xrightleftharpoons[10]{\text{malate dehydrogenase}}\;
\begin{array}{c}
COO^- \\
| \\
C{=}O \\
| \\
CH_2 \\
| \\
COO^-
\end{array}
\;+\; NADH \;+\; H^+
$$

malate   oxaloacetate

An examination of the complete TCA cycle now reveals that, for each mole of acetyl-CoA oxidized, two moles of carbon dioxide, three moles of NADH, one mole of $FADH_2$, and one mole of ATP are produced.

$$
\begin{array}{c}
O \\
\| \\
CH_3-C-S-CoA
\end{array}
+ 3\,NAD^+ + FAD + ADP + P_i + 3\,H_2O \longrightarrow
$$

acetyl coenzyme A

$$2\,CO_2 + 3\,NADH + 3\,H^+ + FADH_2 + ATP + CoA-SH$$

Another mole of NADH and another mole of CO_2 are produced in the conversion of pyruvate into acetyl coenzyme A so that the equation for the complete oxidation of pyruvate is as follows.

$$
\begin{array}{c}
O \\
\| \\
CH_3-C-COO^-
\end{array}
+ 4\,NAD^+ + FAD + ADP + P_i + 3\,H_2O \longrightarrow
$$

pyruvate

$$3\,CO_2 + 4\,NADH + 4\,H^+ + FADH_2 + ATP$$

A summary of the TCA cycle showing the products produced by the oxidation of the two-carbon acetyl group of acetyl-CoA is given in Figure 25.2.

The ultimate reason for the complete oxidation of carbohydrate to carbon dioxide and water via glycolysis and the TCA cycle and the oxidation of fat by β-oxidation (see Chapter 28) and the TCA cycle is to change their chemical energy into the high-energy phosphate bonds of adenosine triphosphate (ATP). Most of the ATP in the body is made in mitochondria in other pathways (the respiratory chain and oxidative phosphorylation) as described in the sections that follow. ADP molecules enter mitochondria and are converted into ATP molecules. ATP molecules leave the mitochondria for use elsewhere in the cell.

Figure 25.2   Summary of the tricarboxylic acid cycle.

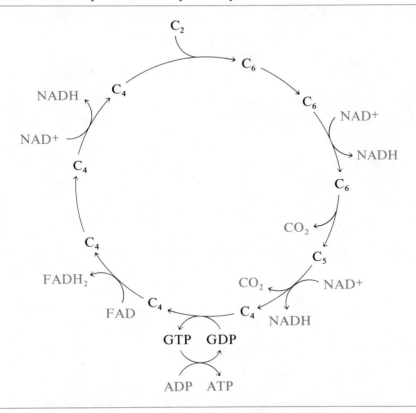

PROBLEMS

1. Complete the following table.

	B Vitamin	Coenzyme	Reaction in Which the Enzyme Functions
(a)		FAD	
(b)	Lipoic acid		
(c)			Activation of acids
(d)		TPP	

2. A protein containing covalently bound FAD is a(n) _____.

3. The process of removing a molecule of carbon dioxide from a carboxylic acid is called _____.

4. Complete the following equations.

(a)

$$
\begin{array}{l}
\text{COO}^- \\
| \\
\text{C}{=}\text{O} \\
| \\
\text{CH}_2 \\
| \\
\text{CH}_2 \\
| \\
\text{COO}^-
\end{array}
+ \text{CoA}{-}\text{SH} + \text{NAD}^+ \xrightarrow[\text{FAD}]{\substack{\alpha\text{-ketoglutarate} \\ \text{dehydrogenase}}}
$$

(b)
$$
\begin{array}{l}
\text{COO}^- \\
| \\
\text{HOCH} \\
| \\
\text{CH}_2 \\
| \\
\text{COO}^-
\end{array}
+ \text{NAD}^+ \xrightarrow{\text{malate dehydrogenase}}
$$

ANSWERS TO PROBLEMS

1. B Vitamin	Coenzyme	Reaction in Which the Enzyme Functions
(a) Riboflavin	FAD	Oxidation-reduction
(b) Lipoic acid	Itself	Oxidation-reduction
(c) Pantothenic acid	Coenzyme A	Activation of acids
(d) Thiamine	TPP	Decarboxylations

2. flavoprotein   **3.** decarboxylation

4. (a)
$$
\begin{array}{l}
\text{COO}^- \\
| \\
\text{C}{=}\text{O} \\
| \\
\text{CH}_2 \\
| \\
\text{CH}_2 \\
| \\
\text{COO}^-
\end{array}
+ \text{CoA}{-}\text{SH} + \text{NAD}^+ \xrightarrow[\text{FAD}]{\alpha\text{-ketoglutarate dehydrogenase}}
\begin{array}{l}
\text{O} \\
\| \\
\text{C}{-}\text{S}{-}\text{CoA} \\
| \\
\text{CH}_2 \\
| \\
\text{CH}_2 \\
| \\
\text{COO}^-
\end{array}
+ \text{CO}_2 + \text{NADH} + \text{H}^+
$$

(b)
$$
\begin{array}{l}
\text{COO}^- \\
| \\
\text{HOCH} \\
| \\
\text{CH}_2 \\
| \\
\text{COO}^-
\end{array}
+ \text{NAD}^+ \xrightarrow{\text{malate dehydrogenase}}
\begin{array}{l}
\text{COO}^- \\
| \\
\text{C}{=}\text{O} \\
| \\
\text{CH}_2 \\
| \\
\text{COO}^-
\end{array}
+ \text{NADH} + \text{H}^+
$$

25-2 THE RESPIRATORY CHAIN

LEARNING OBJECTIVES

1. Define reducing equivalents, respiratory chain, coenzyme Q, cytochrome, and dehydrogenase.
2. Using symbols and a diagram, describe the respiratory chain.

3. Describe the function of the respiratory chain.
4. Describe the action of dehydrogenases.

The end product of glycolysis (see Section 24-2.1) is a three-carbon unit, pyruvate, that has the ability (unlike the phosphorylated intermediates of glycolysis) to penetrate the mitochondrial membrane. Pyruvate releases energy during its oxidation to CO_2. The reducing equivalents released by its oxidation are transferred to coenzymes.

The major source of energy in a eukaryotic cell is ultimately derived from an oxidation-reduction reaction involving oxygen. The *respiratory,* or *electron trans-*

Reducing equivalents are electrons and/or hydrogen atoms participating in oxidation-reduction reactions.

port, chain is a series of reactions that transfers the electrons removed from a substrate, with or without accompanying protons, through a group of enzymes and other substances to oxygen (see Section 9-4.1).

25-2.1 ELECTRON TRANSPORT

The respiratory or electron transport chain is a series of reactions that begins with the oxidation of substrates by the extraction of electrons. The electrons are passed through the molecules making up the chain, ultimately to the last enzyme in the chain, which transfers them to oxygen. This electron flow provides the energy for converting ADP to ATP. Two or three moles of ATP are produced from the passage of one electron pair removed from the molecules of one mole of substrate oxidized. This means that a minimum of 39 kcal of energy per mole of substrate (see Section 25-3.2) is stored in the high-energy bonds of ATP as a result of certain oxidative steps and 26 kcal, as the result of other oxidative steps of a pathway.

A simplified symbolic representation of the respiratory chain is given in Figure 25.3. In Figure 25.3, electrons flow from left to right. A pair of electrons and two protons are removed from a substrate by a dehydrogenase, resulting in oxidation of the substrate. The two electrons and one proton are passed together through the first part of the chain to coenzyme Q. Then the electrons alone are passed through a series of proteins called cytochromes. Finally, oxygen is reduced. Two electrons are added to each atom in a molecule of oxygen. The reduction of oxygen results in the formation of water according to the following simplified representation.

$$O_2 + 4\,e^- + 4\,H^+ \longrightarrow 2\,H_2O$$

Figure 25.3 A symbolic representation of the sequence of events in the respiratory chain. Fp = flavoprotein, CoQ = coenzyme Q, and Cyt = cytochrome.

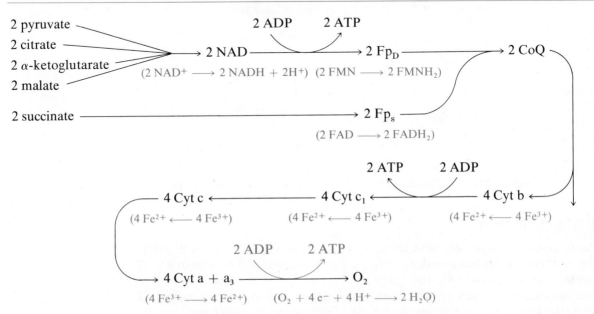

We can see in Figure 25.3 that oxygen oxidizes cytochrome a + a$_3$ by removing electrons from it, oxidized cytochrome a + a$_3$ oxidizes cytochrome c, and so on, with each intermediate oxidizing the one to the left of it until the substrate is oxidized. Recall that oxidation is the loss of electrons.

25-2.2 DEHYDROGENASES

Most enzymes that use NAD$^+$ as a coenzyme are dehydrogenases. The enzyme alcohol dehydrogenase uses NAD$^+$ as a coenzyme and catalyzes the reaction in which ethanol is converted to acetaldehyde. In this reaction, ethanol has lost two

$$
\begin{array}{ccc}
& \overset{\displaystyle H}{\underset{\displaystyle |}{}} & \\
\overset{\displaystyle H}{\underset{\displaystyle |}{}}\;\; \overset{\displaystyle O}{\underset{\displaystyle |}{}} & & \overset{\displaystyle H}{\underset{\displaystyle |}{}}\;\; \overset{\displaystyle O}{\underset{\displaystyle \|}{}} \\
H-\overset{|}{\underset{|}{C}}-\overset{|}{\underset{|}{C}}-H & \longrightarrow & H-\overset{|}{\underset{|}{C}}-C-H \\
H\;\; H & & H \\
\text{ethanol} & & \text{acetaldehyde}
\end{array}
$$

hydrogen atoms. The one in color is lost as H$^-$ (that is, H$^+$ plus 2 electrons); the other is lost as H$^+$. Note that H$^-$ plus H$^+$ adds up to two hydrogen atoms. The H$^-$ is picked up by NAD$^+$ as it goes to NADH and H$^+$ is left over as shown. Thus,

$$
\text{CH}_3-\overset{\displaystyle H}{\underset{\displaystyle H}{\overset{|}{\underset{|}{C}}}}-O-H \;\xrightarrow[\text{NAD}^+\quad \text{NADH}+\text{H}^+]{\overset{\text{alcohol}}{\text{dehydrogenase}}}\; \text{CH}_3-\overset{\displaystyle H}{\overset{|}{C}}=O
$$

$$
\begin{array}{ccc}
\text{ethanol} & & \text{acetaldehyde}
\end{array}
$$

the mechanism for the loss of hydrogen from ethanol is

$$
\text{CH}_3-\overset{\displaystyle H}{\underset{\displaystyle \underset{\textstyle \text{H}^-}{\underset{|}{\textcircled{H}}}}{\overset{|}{\underset{|}{C}}}}-O-\textcircled{H}\;\nearrow\;\text{H}^+
$$

where H$^-$ = H$^+$ + 2 e$^-$.

In flavoproteins that are dehydrogenases, FAD functions as the coenzyme.

25-2.3 COENZYME Q

All coenzyme Q molecules have the same central structure. This central ring is the part that participates in redox reactions. The side chain of the coenzyme Q of humans has 10 isoprene units (see Section 11-2.2); therefore, it is known as CoQ$_{10}$. The net change between oxidized coenzyme Q (CoQ) and reduced coenzyme Q (CoQH$_2$) is two hydrogen atoms. As with the flavin cofactors, the change may be thought of as the addition of (2 H$^+$ + 2 e$^-$).

CoQ
oxidized form (quinoid form)

CoQH$_2$
reduced form (hydroquinone form)

where

human coenzyme Q

25-2.4 CYTOCHROMES

Cytochrome is a protein of the respiratory chain that transfers electrons. It has a prosthetic group that is related to the heme group in hemoglobin (see Figure 12.2) and undergoes cyclic reduction and oxidation during electron transport.

There are several cytochromes in the respiratory chain. As indicated in Figure 25.3, they are designated b_1, c_1, c, a, and a_3. The oxidized form is represented as $Cyt(Fe^{3+})$ and the reduced form is represented as $Cyt(Fe^{2+})$. The cytochromes transfer electrons only, while the other components of the chain (for example, NAD^+, flavoproteins, and CoQ) transfer hydrogen in some form plus electrons. The ultimate electron acceptor is oxygen (O_2). The individual reactions of the respiratory chain may be represented as shown.*

$$2 \text{ NAD}^+ + 2 \text{ R}_2\text{C} \overset{\text{OH}}{\underset{\text{H}}{\diagdown}} \longrightarrow 2 \text{ R}_2\text{C}{=}\text{O} + 2 \text{ NADH} + 2 \text{ H}^+$$

$$2 \text{ NADH} + 2 \text{ H}^+ + 2 \text{ Fp}_D \longrightarrow 2 \text{ NAD}^+ + 2 \text{ Fp}_D\text{H}_2$$

$$2 \text{ Fp}_D\text{H}_2 + 2 \text{ CoQ} \longrightarrow 2 \text{ Fp}_D + 2 \text{ CoQH}_2$$

$$2 \text{ CoQH}_2 + 4 \text{ Cyt b(Fe}^{3+}) \longrightarrow 2 \text{ CoQ} + 4 \text{ Cyt b(Fe}^{2+}) + 4 \text{ H}^+$$

$$4 \text{ Cyt b(Fe}^{2+}) + 4 \text{ Cyt c}_1(\text{Fe}^{3+}) \longrightarrow 4 \text{ Cyt b(Fe}^{3+}) + 4 \text{ Cyt c}_1(\text{Fe}^{2+})$$

$$4 \text{ Cyt c}_1(\text{Fe}^{2+}) + 4 \text{ Cyt c(Fe}^{3+}) \longrightarrow 4 \text{ Cyt c}_1(\text{Fe}^{3+}) + 4 \text{ Cyt c(Fe}^{2+})$$

$$4 \text{ Cyt c(Fe}^{2+}) + 4 \text{ Cyt a} + a_3(\text{Fe}^{3+}) \longrightarrow 4 \text{ Cyt c(Fe}^{3+}) + 4 \text{ Cyt a} + a_3(\text{Fe}^{2+})$$

$$4 \text{ Cyt a} + a_3(\text{Fe}^{2+}) + O_2 + 2 \text{ HOH} \longrightarrow 4 \text{ Cyt a} + a_3(\text{Fe}^{3+}) + 4 \text{ OH}^-$$

Net reaction: 4 hydrogen atoms ($4 \text{ H}^+ + 4 \text{ e}^-$) $+ O_2 + 2 \text{ HOH} \longrightarrow 4 \text{ H}^+ + 4 \text{ OH}^-$

or $\quad\quad\quad\quad 4 \text{ H}^+ + 4 \text{ e}^- + O_2 \longrightarrow 2 \text{ HOH}$

*In these reactions, F_P = flavoprotein and F_{P_S} = succinate dehydrogenase. F_{P_D} is an enzyme that oxidizes the NADH from NAD-linked dehydrogenases.

Figure 25.4
The respiratory chain, showing the continuous flow of electrons from substrates (SH_2) to oxygen.
S = oxidized substrate.
SH_2 = reduced substrate.

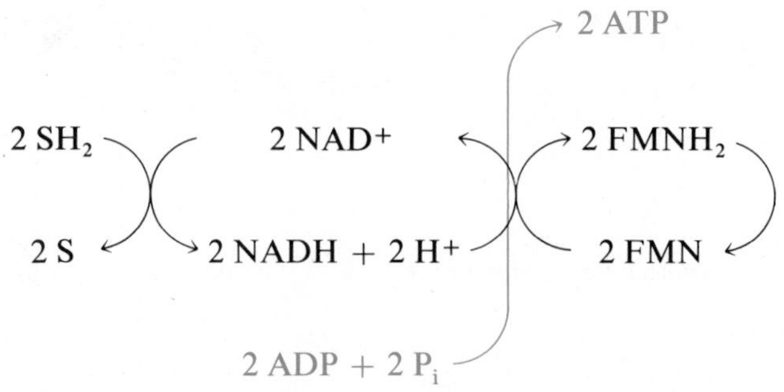

Cytochromes a and a_3 occur together as a complex. This complex is sometimes called *cytochrome oxidase* or *cytochrome c oxidase*. Cytochrome c oxidase consists of two heme-containing subunits (like hemoglobin), two copper-containing subunits, and two other subunits.

Another representation of the reactions of the respiratory chain showing successive redox reactions is given in Figure 25.4.

25-2.5 STOICHIOMETRY OF THE TCA CYCLE

The tricarboxylic acid cycle, coupled with the respiratory chain, satisfies stoichiometric requirements for the oxidation of pyruvate by molecular oxygen to carbon dioxide and water. The reduction of five atoms of oxygen for each molecule of pyruvate oxidized corresponds to the removal of five pairs of electrons during operation of the TCA cycle. Each oxygen atom that is reduced forms one molecule of water. As shown in Figure 25.1, three molecules of water are taken up during operation of the cycle, so the net change corresponds to the formation of two molecules of water per molecule of pyruvate oxidized.

$$2\ CH_3-\overset{\displaystyle O}{\overset{\|}{C}}-COO^- + 5\ O_2 + 2\ H^+ \longrightarrow 6\ CO_2 + 4\ H_2O$$
oxidation of pyruvate

or

$$CH_3-\overset{\displaystyle O}{\overset{\|}{C}}-COO^- + 5\ O + H^+ \longrightarrow 3\ CO_2 + H_2O$$
oxidation of pyruvate

PROBLEMS

5. Electrons and/or hydrogen atoms participating in oxidation-reduction reactions are called _____.

6. Describe the function of the respiratory chain.

7. Describe the action of dehydrogenases.
8. The molecule that can exist in either a quinone or hydroquinone form, has a poly-isoprene side chain, and undergoes oxidation-reduction in the respiratory chain is _____.

9. _____ are proteins of the respiratory chain that have a heme prosthetic group and are alternately oxidized and reduced as they transfer electrons.

ANSWERS TO PROBLEMS   **5.** reducing equivalents   **6.** The energy needed by cells comes from oxidation of substrates, such as carbohydrates and fats, by oxygen. The oxidation of substrates involves the transfer of electrons from the substrates to oxygen. The chemical energy of substrate molecules must be converted into the chemical energy of ATP molecules so that it can be used by cells. The respiratory chain transfers electrons from substrate molecules to oxygen molecules in a series of steps so that three (or sometimes two) molecules of ATP can be made for each pair of electrons transferred. This stepwise transfer of electrons increases the efficiency of ATP production because only one molecule of ATP could be made from the oxidation of a molecule of substrate if the electrons were transferred directly to oxygen.   **7.** Dehydrogenases catalyze the oxidation of substrates by removing two hydrogen atoms from them.   **8.** coenzyme Q   **9.** cytochromes

25-3 OXIDATIVE PHOSPHORYLATION AND EFFICIENCY OF MITOCHONDRIAL REACTIONS

LEARNING OBJECTIVES

1. Define oxidative phosphorylation, coupling site, uncoupling agent, and respiratory inhibitor.
2. Give the number of ATP molecules formed by complete oxidation of each of the following: one molecule of D-glucose, one molecule of pyruvate, and one molecule of acetyl coenzyme A.
3. Explain why mitochondria are called the "powerhouses of the cell."
4. Give the approximate overall efficiency of the formation of ATP by the complete oxidation of D-glucose.
5. Using a diagram, describe the structure of a mitochondrion.

We learned in the preceding section that ATP molecules are made from ADP molecules as the electrons or hydrogen atoms removed from substrates when they are oxidized move through the respiratory chain. This conversion of ADP into ATP is another mitochondrial process known as oxidative phosphorylation. We will examine this synthesis of ATP from ADP and inorganic phosphate in Section 25-3.1. Then we will take an overall look at the combination of glycolysis, the tricarboxylic acid cycle, the respiratory chain, and oxidative phosphorylation in Section 25-3.2. In Section 25-3.4, we will calculate the efficiency of converting the energy in glucose molecules into the energy of ATP molecules.

25-3.1 OXIDATIVE PHOSPHORYLATION

Oxidative phosphorylation is the process by which high-energy phosphate (ATP) is generated during the flow of reducing equivalents through the respiratory chain. The oxidation of NADH by the respiratory chain releases energy.

$$2 \, NADH + 2 \, H^+ + O_2 \longrightarrow 2 \, NAD^+ + 2 \, H_2O + \text{free energy}$$

This energy is used to phosphorylate ADP.

$$\text{free energy} + P_i + ADP \longrightarrow ATP$$

In the process of oxidative phosphorylation, the oxidation of NADH and the phosphorylation of ADP are coupled together in some ratio. The respiratory chain has three **coupling sites.** These three sites are the three places where ATP is made from ADP as shown in Figures 25.3 and 25.4. When all three coupling sites are utilized, a maximum of six molecules of ATP can be made per oxygen molecule used.

$$2 \, NADH + 2 \, H^+ + O_2 + 6 \, ADP + 6 \, P_i \longrightarrow 2 \, NAD^+ + 6 \, ATP + 2 \, H_2O$$

A **coupling site** is a reaction of the respiratory chain that releases enough energy to allow an ATP molecule to be made from an ADP molecule and an inorganic phosphate ion.

In addition to entering the respiratory chain through NADH, reducing equivalents can also enter through a flavoprotein called flavoprotein S (Fp$_S$). Flavoprotein S is the enzyme succinate dehydrogenase. Flavoprotein S is reduced during the oxidation of succinate (see Figures 25.1 and 25.3). In this case, the first coupling site is by-passed and only four ATP molecules are formed per molecule of oxygen reduced.

$$2 \, FADH_2 + O_2 + 4 \, ADP + 4 \, P_i \longrightarrow 2 \, FAD + 4 \, ATP + 2 \, H_2O$$

The passing of reducing equivalents through a coupling site causes inorganic phosphate (P$_i$) to combine with ADP to form ATP. In the presence of certain inhibitors, the transfer of electrons from NADH or FADH$_2$ to oxygen is no longer coupled to the production of ATP, and no ATP is made. These chemical agents are called uncoupling agents. An *uncoupling agent* is a compound that prevents the formation of ATP at a coupling site in the respiratory chain. Uncoupling agents do not block the transfer of electrons through the respiratory chain; they simply prevent the coupling of the transfer of electrons with the formation of ATP molecules. 2,4-Dinitrophenol (DNP) is an uncoupling agent.

As we have seen, the purpose of the mitochondrial reactions is to produce ATP molecules that can be used as a source of energy in various cellular reactions

OH

NO$_2$

NO$_2$

2,4-dinitrophenol

and processes. If the oxidation of acetyl coenzyme A in the TCA cycle no longer results in the production of ATP molecules due to the action of an uncoupling agent such as 2,4-dinitrophenol, the rate of substrate oxidation will increase as the cell tries to make ATP to meet its needs. Such oxidation is nonproductive, however, because it does not result in the production of ATP molecules. When uncoupling agents block the production of ATP, cells catabolize more carbohydrate and fat in an attempt to produce more ATP molecules. However, heat energy rather than the chemical energy of high-energy phosphate bonds is produced. The result is a rise in body temperature (fever). The increased catabolism causes increased oxygen utilization and carbon dioxide production.

Salicylates probably act as uncoupling agents also. The most common cause of salicylate poisoning is an overdose of aspirin (see Section 16-3.2). The most common symptom in acute poisonings is fever. Salicylates cause up to a 100% increase in oxygen utilization and in carbon dioxide production. The increased carbon dioxide production results in hyperventilation as the body tries to rid itself of carbon dioxide. Other symptoms are mild intoxication, headache, dizziness, ringing in the ears, difficulty in hearing, impaired vision, general mental confusion, drowsiness, heavy sweating, thirst, nausea, vomiting, and occasionally diarrhea.

Acute means brief and relatively severe.

Hyperventilation is abnormally prolonged, rapid, and deep breathing.

acetylsalicylic acid
(aspirin)

In addition to uncoupling agents, there are also respiratory inhibitors and inhibitors of oxidative phosphorylation. Respiratory inhibitors are compounds that prevent the flow of electrons through the chain. An example is cyanide, which binds to cyt a + a_3 and prevents the flow of electrons from cyt a + a_3 to oxygen. Another is the antibiotic antimycin, which blocks the flux of reducing equivalents from cyt b to cyt c_1. The antibiotics oligomycin and rutamycin are examples of inhibitors of oxidative phosphorylation.

25-3.2 EFFICIENCY OF MITOCHONDRIAL REACTIONS

The tricarboxylic acid cycle is combined with electron transport and oxidative phosphorylation to provide most of the energy used by a cell. Hence, the mitochondria are sometimes called the "powerhouses of the cell." Evidence of this is their high concentration in muscle and around the base of the tail in sperm cells, where they supply the tail with energy in the form of ATP so that it can propel the cell.

In mitochondria, there is a close association of enzymes of the TCA cycle with enzymes involved in reoxidation of reduced coenzymes with enzymes of the respiratory chain that transfer electrons from NADH and $FADH_2$ to oxygen. The energy released by reoxidation of the coenzymes is transferred to ATP.

In the course of the removal and transfer of electrons during oxidation of a substrate, energy is liberated. This energy either is stored temporarily in a high-energy bond (ATP) or is directly dissipated as heat. A living cell does not use heat energy to do work. Rather, it uses high-energy bonds, primarily high-energy phosphate bonds, as its source of energy for work. The storage of energy in the form of a high-energy phosphate bond is limited to one bond for each reaction step. Thus, at least 13,000 cal/mol of substrate oxidized can be stored as a result of a one-step electron transfer.* The transfer of electrons from substrate to oxygen liberates much more energy than this, so electrons are transferred in a series of steps, rather than in one step, so that more than one step might result in the formation of a high-energy phosphate bond. By this means, a large proportion of the liberated energy of the substrate is stored and relatively little is "wasted" as heat. This stepwise system, which is characteristic of biological processes, allows energy production to occur with a high level of efficiency. In other words, the stepwise system allows a high percentage of the energy to be recovered as useful work.

Energy in the form of ATP is constantly used by the cell for its biosynthetic and physiological activities. As mentioned in Section 20-1.2, ATP is the primary energy source for a wide variety of cellular reactions including synthesis of various components of the living cell, contraction of muscle fibers, and transport of materials across membranes. The rate at which the high-energy bond of ATP is used (ATP $\longrightarrow$ ADP + P_i) and the rate at which ATP is resynthesized (ADP + P_i $\longrightarrow$ ATP) is indicated by the high rate of **turnover** of ATP. For example, the half-life of ATP in a kidney cell is less than 1 minute. This means that, in 1 minute, more than half of the ATP molecules in a kidney cell are used, largely in transport processes, and that the ADP molecules formed when the high-energy bond is used are immediately reconverted into ATP molecules.

The **turnover** of a compound is its use or breakdown and concurrent synthesis.

25-3.3 STOICHIOMETRY OF THE FORMATION OF ATP FROM PYRUVATE AND ACETYL COENZYME A

One molecule of NADH is formed from each molecule of pyruvate that is oxidized to acetyl coenzyme A. This molecule of NADH can form three molecules of ATP by oxidative phosphorylation during passage of the reducing equivalents through the respiratory chain.

Twelve ATP molecules are produced from each acetyl coenzyme A molecule that enters the TCA cycle. The origins of these ATP molecules are shown in the table at the top of the next page. In Figure 25.5, the origin of the 12 molecules of ATP that result from oxidation of one molecule of acetyl coenzyme A in the TCA cycle and the origin of the 15 molecules of ATP that result from oxidation of one molecule of pyruvate are indicated.

As we will discover in Chapter 27, catabolism of fatty acids from fats and oils also takes place in mitochondria. Their catabolism also produces acetyl-CoA that

---

*It is known that hydrolysis of ATP releases about 7300 cal/mol of energy when ATP is at a concentration of 1 M. However, at the much lower concentrations of ATP and ADP found in cells, at least 13,300 cal/mol of energy are released upon hydrolysis.

Reaction	Coenzyme Reduced	Yield of ATP Molecules/Mole of Acetyl-CoA
Directly from the TCA cycle		
Succinyl-CoA $\longrightarrow$ succinate	None	1
From electron transport through the respiratory chain and oxidative phosphorylation		
Isocitrate $\longrightarrow$ α-ketoglutarate	NAD	3
α-Ketoglutarate $\longrightarrow$ succinyl-CoA	NAD	3
Succinate $\longrightarrow$ fumarate	FAD	2
Malate $\longrightarrow$ oxaloacetate	NAD	3
		12

Figure 25.5 Origin of ATP molecules from the oxidation of pyruvate and acetyl-CoA in mitochondria through the combined action of the TCA cycle, the respiratory chain, and oxidative phosphorylation. The coupling of the respiratory chain and oxidative phosphorylation is given in color.

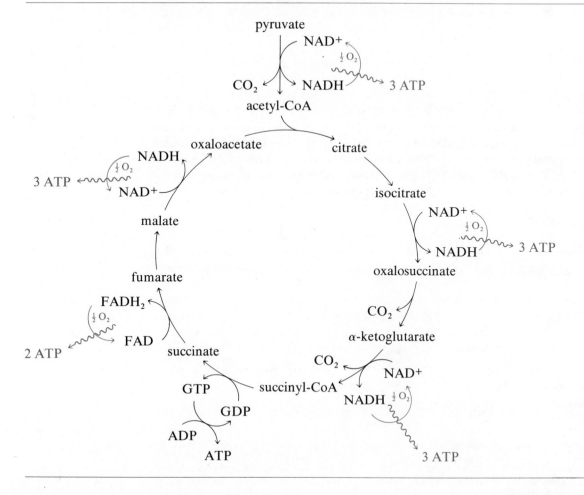

can enter the TCA cycle. As we learned in Chapter 24, catabolism of carbohydrate produces pyruvate: two molecules of pyruvate per molecule of glucose entering the pathway of glycolysis. The conversion of these two molecules of pyruvate into two molecules of acetyl-CoA is accompanied by the reduction of two molecules of NAD to NADH (see Section 25-1.1 and Figure 25.3). Reoxidation of these molecules of NADH results in the production of 6 (2×3) molecules of ATP. Thus, for every two molecules of pyruvate (produced from one molecule of glucose) that enter mitochondria, 30 molecules of ATP are produced.

The total number of ATP molecules that can be formed by the complete oxidation of a molecule of glucose to carbon dioxide and water in glycolysis and the TCA cycle is 38. The origins of these 38 ATP molecules are given in the following tabulation and in Figure 25.6.

	ATP Formation (net yield)	
	Directly	Oxidative Phosphorylation
Glycolysis	2	
2 NADH		6*
Citric acid cycle	2	
8 NADH		24
2 FADH$_2$		4
	4	34

A total of 38 moles of ATP can be produced by the total oxidation of 1 mole of glucose. The high-energy phosphate bonds of 1 mole of ATP molecules release upon hydrolysis at least 13 kcal of energy. Therefore, the 38 moles of ATP produced by the oxidation of 1 mole of glucose can store 38 mol $\times$ 13 kcal/mol = 494 kcal of energy. Burning 1 mole of glucose releases about 686 kcal of energy, so the efficiency of the oxidation of glucose in glycolysis and the TCA cycle and the storage of the released energy in chemical bonds is about $\frac{494}{686} \times 100 = 72\%$. The overall reaction can be written

$$C_6H_{12}O_6 + 6\,O_2 + 38\,ADP + 38\,P_i \longrightarrow 6\,CO_2 + 6\,H_2O + 36\,ATP + heat$$

This reaction can be said to be at least 70% efficient in terms of generating ATP.

25-4 REGULATION OF THE TCA CYCLE

LEARNING OBJECTIVES

1. Describe the interrelationship of glycolysis and the TCA cycle.
2. Describe the feedback control of the TCA cycle by positive and negative allosteric effectors.

---

*In muscle and nerve cells, the NADH molecules produced in glycolysis, which occurs outside the mitochondria in the soluble part of the cytoplasm, yield only two molecules of ATP each rather than three. This reduction in the yield of ATP molecules results because the reducing equivalents must enter the mitochondria by a special process, called the glycerol phosphate shuttle, that uses energy. In liver and heart cells, another type of shuttle called the malate-aspartate shuttle allows six ATP molecules to be formed from reoxidation of the two NADH molecules formed in glycolysis.

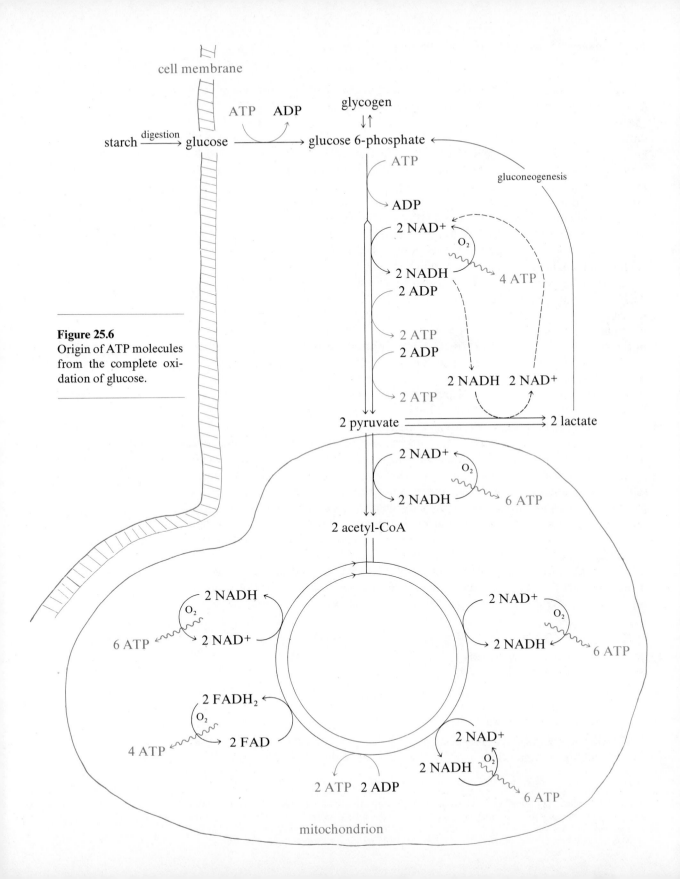

Figure 25.6
Origin of ATP molecules from the complete oxidation of glucose.

We learned in Section 24-3 that carbohydrate metabolism is regulated by hormones and feedback. We learned that citrate, an intermediate of the TCA cycle, and ATP, the ultimate product of mitochondrial reactions, inhibit phosphofructokinase (PFK), the enzyme catalyzing a key reaction of glycolysis. Citrate and ATP are negative allosteric effectors of PFK (see Section 24-3.4). By inhibiting PFK, they slow glycolysis. Recall that high concentration of citrate* and ATP means that the cell has enough of these substances for its current needs and so, can slow the production of citrate and ATP and prevent a wasteful depletion of its energy stores. On the other hand, AMP and ADP are positive allosteric effectors for PFK and increase its activity. A high concentration of AMP and ADP means a low concentration of ATP. In this case, the pathways converting AMP and ADP into ATP need to be made more active, and they are.

All reactions controlling glycolysis also control the TCA cycle. The TCA cycle itself is regulated by both positive and negative feedback. Succinyl-CoA, a late intermediate of the TCA cycle, NADH, a product of the cycle, and ATP, the ultimate product of mitochondrial reactions, inhibit enzymes catalyzing early steps in the cycle. AMP, ADP, and D-fructose 1,6-diphosphate (FDP), an intermediate of glycolysis, activate enzymes catalyzing early steps in the cycle. A high concentration of FDP means that there is a buildup of intermediates of glycolysis because of a sluggishness in the TCA cycle. Some of the actions of allosteric effectors on regulatory enzymes of the TCA cycle that control the cycle are shown in Figure 25.7.

The reactions of mitochondria are separated physically from glycolysis and other reactions of the cytosol. Therefore, the reactions of mitochondria are also regulated through control of proteins that transport substances into and out of them. Some advantages of compartmentalization of enzymes and substrates in organelles such as mitochondria are (1) an increased concentration of intermediates, (2) the possibility of an assembly-line process by keeping enzymes close together in membranes, (3) separation of substances that might otherwise be reactants, and (4) localized pH differences. Mitochondria have a highly organized structure (see Figure 25.8) that allows them to carry out their varied reactions and to maintain a high degree of efficiency. The location of all mitochondrial enzymes has not been determined, but more than 50 have been located. The location of some of those that we have discussed are given in the table below.

Inner Membrane	Matrix
Succinate dehydrogenase	Citrate synthetase
Cytochromes b, c, c_1, and a + a_3	Aconitase
	Isocitrate dehydrogenase
	Fumarase
	Malate dehydrogenase

The enzymes of the respiratory chain and oxidative phosphorylation are located on the inner membrane, while many of the enzymes of the TCA cycle, but not all,

*Citrate can come out of mitochondria into the soluble cytoplasm (the cytosol) where glycolysis takes place. Citrate comes out of mitochondria to transport acetyl-CoA units to the cytosol for the synthesis of fatty acids (see Section 27-3.1).

Figure 25.7 Some of the actions of allosteric effectors on regulatory enzymes of the TCA cycle. FDP = D-fructose 1,6-diphosphate.

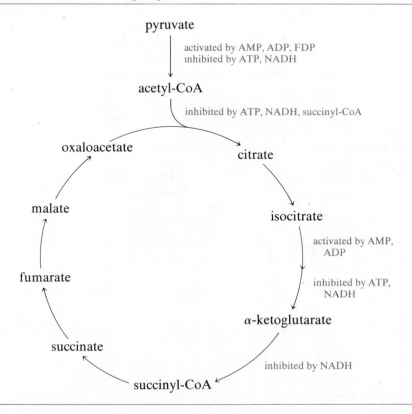

pyruvate

activated by AMP, ADP, FDP
inhibited by ATP, NADH

acetyl-CoA

inhibited by ATP, NADH, succinyl-CoA

oxaloacetate

citrate

malate

isocitrate

activated by AMP,
ADP

inhibited by ATP,
NADH

fumarate

α-ketoglutarate

succinate

inhibited by NADH

succinyl-CoA

Figure 25.8 Diagram of the cross-section of a mitochondrion.

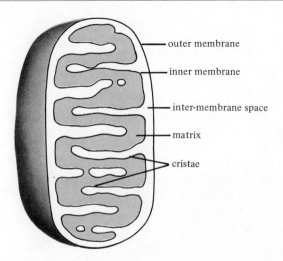

outer membrane

inner membrane

inter-membrane space

matrix

cristae

are located in the matrix. None of the enzymes of the TCA cycle, the respiratory chain, or oxidative phosphorylation have been found in the outer membrane or in the intermembrane space (outer compartment).

PROBLEMS

10. Give the number of ATP molecules formed by complete oxidation of one molecule of each of the following.
    (a) D-glucose   (b) pyruvate   (c) acetyl-CoA
11. A compound that prevents the formation of ATP as electrons are transferred through a coupling site in the respiratory chain is known as a(n) _____ _____.
12. The process by which ATP molecules are generated during the flow of reducing equivalents through the respiratory chain is known as _____ _____.
13. Indicate which of the following is the approximate efficiency of converting the chemical energy of glucose into the chemical energy of high-energy phosphate bonds of ATP.
    (a) 10%   (b) 30%   (c) 50%
    (d) 70%   (e) 90%

ANSWERS TO PROBLEMS **10. (a)** 38   **(b)** 15   **(c)** 12   **11.** uncoupling agent   **12.** oxidative phosphorylation   **13. (d)** 70%

SUMMARY

In Chapter 24, we learned that D-glucose molecules are broken down to pyruvate molecules in a metabolic pathway called glycolysis. In this chapter, we learned that the product of glycolysis (pyruvate) is converted into acetyl coenzyme A. Acetyl coenzyme A molecules are then oxidized to carbon dioxide in the tricarboxylic acid (TCA) cycle. This oxidation releases energy, and the coupling of oxidation in the TCA cycle with electron transport in the respiratory chain and oxidative phosphorylation produces ATP molecules.

We also have learned that enzymes of the TCA cycle, the respiratory chain, and oxidative phosphorylation are located inside organelles called mitochondria. Glycolysis takes place in the soluble cytoplasm outside the mitochondria.

Oxidation of pyruvate to acetyl coenzyme A and oxidation of acetyl coenzyme A to carbon dioxide uses the coenzymes NAD^+ and FAD as oxidizing agents and produces the reduced coenzymes NADH and $FADH_2$. Reducing equivalents carried by NADH and $FADH_2$ enter the respiratory chain where the energy released by their reoxidation to NAD^+ and FAD using molecular oxygen as the ultimate oxidizing agent is used to make ATP from ADP and inorganic phosphate. The process by which ATP is made is called oxidative phosphorylation.

Oxidation of each mole of pyruvate produces 15 moles of ATP. Oxidation of each mole of acetyl coenzyme A produces 12 moles of ATP. Oxidation of each mole of glucose by the combined action of the glycolytic pathway and the TCA cycle produces approximately 38 moles of ATP, making the process at least 70% efficient in terms of storing the energy released by the oxidation of glucose in the high-energy phosphate bonds of ATP where it can be used for various cellular processes.

Fatty acids are completely oxidized in mitochondria, first to acetyl coenzyme A molecules in a pathway that we will study in Chapter 27. Then, of course, acetyl coenzyme A molecules are oxidized to carbon dioxide in the TCA cycle.

Proteins are broken down into amino acids, which are then converted into α-keto acids. α-Keto acids eventually yield acetyl coenzyme A molecules that can enter the TCA cycle. We will discuss the metabolism of amino acids in Chapter 28.

ADDITIONAL PROBLEMS

14. Using Figure 25.1, write equations for the two steps responsible for the production of carbon dioxide from the acetyl group of acetyl-CoA.

15. Using Figure 25.1, write equations to illustrate the following.
    (a) an isomerization
    (b) a reaction that forms an α-keto acid
    (c) an oxidation using a flavoprotein as the oxidant
    (d) the addition of water to a double bond
    (e) a reaction in which NAD^+ is reduced

    (f) an oxidation of a hydroxyl group

16. Oxygen is the ultimate acceptor of reducing equivalents in the respiratory chain. What is produced as a result?

17. If reducing equivalents enter the respiratory chain as NADH, how many molecules of ATP are formed?

18. If reducing equivalents enter the respiratory chain as reduced flavoprotein, how many molecules of ATP are formed?

CHAPTER 26

LIPIDS

When we think of substances in living systems, we usually think first of water-soluble, or at least hydrophilic, substances because living cells are about 90% water. Indeed, most carbohydrates, most proteins, and all nucleotides and nucleic acids are water-soluble. **Lipids** are quite different. Lipid molecules are hydrophobic, have a low degree of solubility in water, and have a high degree of solubility in organic solvents such as pentane, hexane, and chloroform.

Lipids are compounds of living organisms that are soluble in nonpolar solvents such as chloroform, ether, and ethanol.

Many lipids are derived from *fatty acids.* As we learned in Section 11-5.2, fatty acids are aliphatic, carboxylic acids. Most have a straight hydrocarbon chain with an even number of carbon atoms (between 4 and 24) and 0 to 6 double bonds. The lipids that are esters or amides of fatty acids have two main biological roles: they serve as a source of energy and as components of membranes. The lipids that are not related to fatty acids serve as hormones and vitamins in humans.

In the next chapter, we will discuss the digestion and metabolism of lipids. In this chapter, we will examine the structures and chemical and physical properties of lipids in a little more detail than we did in previous chapters. Lipids can be classified as shown in Table 26.1. Each class will be discussed in this chapter.

26-1 FATTY ACIDS AND NEUTRAL GLYCERIDES

LEARNING OBJECTIVES

1. Define lipid, fatty acid, fat, oil, saturated fatty acid, unsaturated fatty acid, essential fatty acid, monoglyceride, diglyceride, triglyceride, mixed glyceride, polyunsaturated oil, hydrogenated oils, emulsifying agent, and soap.
2. Given a shorthand notation for a fatty acid, write the structure of the fatty acid and vice versa.
3. Describe how the properties of glycerides are determined by the fatty acid composition.
4. Given a triglyceride structure, show the products of saponification.

Table 26.1 Classification of Lipids

A. Fatty Acids
   1. Saturated fatty acids
   2. Unsaturated fatty acids
   3. Other fatty acids
B. Glycerides (esters of fatty acids with glycerol)
   1. Neutral glycerides
      (a) Monoacylglycerols (monoglycerides)
      (b) Diacylglycerols (diglycerides)
      (c) Triacylglycerols (triglycerides)
      (d) Other neutral glycerides
   2. Phosphoglycerides
      (a) Phosphatidylethanolamines (cephalins)
      (b) Phosphatidylcholines (lecithins)
      (c) Other phosphoglycerides
C. Sphingolipids (ceramides, amides of fatty acids with sphingosine)
   1. Sphingomyelins
   2. Cerebrosides
   3. Gangliosides
D. Waxes (esters of fatty acids with long-chain alcohols)
E. Nonsaponifiable Lipids (lipid-soluble substances that do not contain fatty acid residues)
   1. β-Carotene, retinol, and vitamin A family
   2. Cholesterol, other steroids, and vitamin D family
   3. Vitamin E family
   4. Vitamin K family

26-1.1 FATTY ACIDS

Fatty acids occur primarily as esters. The most common fatty acids in nature contain an even number of carbon atoms because they are synthesized from two-carbon units, as we will see in the next chapter.

The saturated fatty acids of greatest metabolic importance are presented in Table 26.2. Butyric (C_4), caproic (C_6), caprylic (C_8), and capric (C_{10}) acids are

Saturated fatty acids are fatty acids that do not have double bonds.

Table 26.2 Saturated Fatty Acids

Common Name	Systematic Name	Structure	Shorthand Designation
Butyric acid	Butanoic acid	$CH_3-(CH_2)_2-COOH$	C_4
Caproic acid	Hexanoic acid	$CH_3-(CH_2)_4-COOH$	C_6
Caprylic acid	Octanoic acid	$CH_3-(CH_2)_6-COOH$	C_8
Capric acid	Decanoic acid	$CH_3-(CH_2)_8-COOH$	C_{10}
Lauric acid	Dodecanoic acid	$CH_3-(CH_2)_{10}-COOH$	C_{12}
Myristic acid	Tetradecanoic acid	$CH_3-(CH_2)_{12}-COOH$	C_{14}
Palmitic acid	Hexadecanoic acid	$CH_3-(CH_2)_{14}-COOH$	C_{16}
Stearic acid	Octadecanoic acid	$CH_3-(CH_2)_{16}-COOH$	C_{18}
Arachidic acid	Eicosanoic acid	$CH_3-(CH_2)_{18}-COOH$	C_{20}
Lignoceric acid	Tetracosanoic acid	$CH_3-(CH_2)_{22}-COOH$	C_{24}

found in only small amounts in certain **fats** and **oils,** especially butter. Caprylic and capric acids are also present as esters in certain plant oils. Lauric acid (C_{12}) is peculiar to certain plant oils, particularly those from palm trees. The source of the lauric acid that is sold commercially is coconut oil. Myristic acid (C_{14}) is a constituent of nutmeg and coconut oil. Palmitic (C_{16}) and stearic (C_{18}) acid esters are common to all plant and animal fats and oils and are, therefore, the most abundant of the saturated fatty acids. Arachidic acid (C_{20}) is present in small amounts in peanut oil. Lignoceric acid (C_{24}) is a component of sphingolipids (see Section 26-2.1). Longer chain, saturated fatty acids are found in **waxes.** Naturally occurring waxes are produced by plants

The **unsaturated fatty acids** of greatest metabolic importance are listed in Table 26.3. Oleic acid, a monounsaturated fatty acid, is found in most fats and oils and is probably the most abundant fatty acid in nature. Linoleic acid, a fatty acid that contains two double bonds, occurs in many plant (vegetable) oils such as corn, peanut, cottonseed, and soybean oils. Linolenic acid contains three double bonds and, again, is a component of many vegetable oils, especially linseed oil. Fatty acids with two or more double bonds are called *polyunsaturated fatty acids.* The configuration of the double bonds in unsaturated fatty acids is usually cis.

There are several shorthand notations used to describe unsaturated fatty acids. Two styles are given in Table 26.3. Both notations describe the number of carbon atoms in the carbon chain, the number of double bonds, and the position of the double bonds by giving the number of the first atom involved in the double bond.

Two unsaturated fatty acids, linoleic acid and linolenic acid, are **essential fatty acids** in our diets. They are required in our diets because mammals such as humans lack the enzymes required to introduce double bonds at carbon atoms beyond C-9 in the fatty acid chain. Linoleic acid and linolenic acid are important constituents of phosphoglycerides (see Section 26-2.1). They are also the starting materials for the synthesis of other unsaturated fatty acids, one of which is used

Fats are lipids that are solid or semi-solid at room temperature.

Oils are lipids that are liquid at room temperature.

Waxes are esters of fatty acids and alcohols with a long hydrocarbon chain.

Unsaturated fatty acids are fatty acids containing one or more double bonds.

Essential fatty acids are unsaturated fatty acids that cannot be synthesized in the body and, therefore, must be included in the diet.

Table 26.3.   Unsaturated Fatty Acids

$CH_3-CH_2-CH_2-CH_2-CH_2-CH_2-CH=CH-CH_2-CH_2-CH_2-CH_2-CH_2-CH_2-CH_2-COOH$
palmitoleic acid
$C_{16}{}^{\Delta 9}$ $16{:}1^{\Delta 9}$

$CH_3-CH_2-CH_2-CH_2-CH_2-CH_2-CH_2-CH_2-CH=CH-CH_2-CH_2-CH_2-CH_2-CH_2-CH_2-CH_2-COOH$
oleic acid
$C_{18}{}^{\Delta 9}$ $18{:}1^{\Delta 9}$

$CH_3-CH_2-CH_2-CH_2-CH_2-CH=CH-CH_2-CH=CH-CH_2-CH_2-CH_2-CH_2-CH_2-CH_2-CH_2-COOH$
linoleic acid
$C_{18}{}^{\Delta 9,12}$ $18{:}2^{\Delta 9,12}$

$CH_3-CH_2-CH=CH-CH_2-CH=CH-CH_2-CH=CH-CH_2-CH_2-CH_2-CH_2-CH_2-CH_2-CH_2-COOH$
linolenic acid
$C_{18}{}^{\Delta 9,12,15}$ $18{:}3^{\Delta 9,12,15}$

$CH_3-(CH_2)_4-CH=CH-CH_2-CH=CH-CH_2-CH=CH-CH_2-CH=CH-(CH_2)_3-COOH$
arachidonic acid
$C_{20}{}^{\Delta 5,8,11,14}$ $20{:}4^{\Delta 5,8,11,14}$

for the biosynthesis of prostaglandins. Infants lacking essential fatty acids in their diets will lose weight and develop *eczema*. Eczema is an inflammatory skin disease characterized by the development of scales and crusts.

Prostaglandins (see Section 16-3.2) are twenty-carbon fatty acids that are hormones. There are a variety of molecules that make up the prostaglandin family. Fourteen naturally occurring prostaglandins have been identified; six of them are widely distributed in the human body. Each brings about a physiologic response. Prostaglandins are synthesized from arachidonic acid (see Table 26.3). Arachidonic acid is cyclized and modified to form the various prostaglandins. The structure of one of the primary prostaglandins in the human body, prostaglandin E_2, is given. The need for prostaglandins is undoubtedly an important reason that the polyunsaturated fatty acids are required in the human diet.

arachidonic acid   prostaglandin E_2

Different prostaglandin molecules have different effects, indicating that prostaglandins have a variety of functions. In general, they are potent, smooth-muscle relaxants. They also influence the breakdown of fats, secretion, permeability, and blood coagulation. Prostaglandins may be used in the treatment of *hypertension* (high blood pressure), prevention of conception, induction of abortions, relief of bronchial asthma, relief of nasal congestion, healing of peptic ulcers, treatment of inflammation, and prevention of blood clots. Recent information suggests that one class of prostaglandins regulates cAMP (see Section 24-2.1) concentrations and another regulates *cGMP* concentrations. cGMP is 3′,5′-cyclic guanosine monophosphate, a molecule with the same structure as cAMP except that the purine guanine is substituted for the purine adenine. Both molecules are very active in the regulation of a number of cellular activities, and it is by controlling the ratio of the two that different prostaglandins can have such different effects. Aspirin (see Section 16-3.2) decreases the synthesis of prostaglandins, which is probably the reason aspirin acts as an anti-inflammatory drug and reduces aggregation of blood platelets.

In addition to saturated and unsaturated fatty acids, there are branched-chain fatty acids, fatty acids containing rings of carbon atoms, fatty acids containing triple bonds, fatty acids containing hydroxyl groups, and others. An example of a fatty acid containing a hydroxyl group is ricinoleic acid. Ricinoleic acid accounts for about 87% of the fatty acids esterified to glycerol in castor oil.

Asthma is a disease marked by recurrent attacks of difficult breathing due to spasmodic contraction of the larger air passages within the lung.

$$CH_3-(CH_2)_5-\overset{\overset{\textstyle OH}{\textstyle |}}{CH}-CH_2-CH=CH-(CH_2)_7-COOH$$

ricinoleic acid

26-1.2 NEUTRAL GLYCERIDES

Glycerides are esters of fatty acids and glycerol; that is, they are acylglycerols in which a fatty acid moiety is the acyl group (see Table 26.1). Glycerides are subdivided according to the number of hydroxyl groups of glycerol that are esterified with a fatty acid and according to whether any hydroxyl groups of the glycerol moiety are substituted with groups other than fatty acid residues [for example, the phosphate group in phosphoglycerides (see Section 26-2.1)].

Glycerides that have no charge are commonly called *neutral glycerides.* The neutral glycerides that are fatty acid esters of glycerol without other groups are subdivided into those with a single fatty acid acyl group on the glycerol residue (called *monoacylglycerols* or *monoglycerides*), those with two fatty acid acyl groups (called *diacylglycerols* or *diglycerides*), and those with all the hydroxyl groups of glycerol esterified (called *triacylglycerols* or *triglycerides*) (see Figure 26.1).

Lipids that are stored as an energy source, primarily in the fatty (adipose) tissue of animals and the seeds of plants, are triglycerides. Triglycerides are also the major transport form of fatty acids and, hence, are the most abundant of the acylglycerols. Seldom do these triglyceride molecules contain the same fatty acid residue in all three positions of the glycerol moiety. Rather, triglyceride molecules are almost

Figure 26.1
General structures of monoglyceride, diglyceride, and triglyceride molecules. $R-\overset{\overset{\displaystyle O}{\|}}{C}-$, $R'-\overset{\overset{\displaystyle O}{\|}}{C}-$, and $R''-\overset{\overset{\displaystyle O}{\|}}{C}-$ are fatty acyl groups.

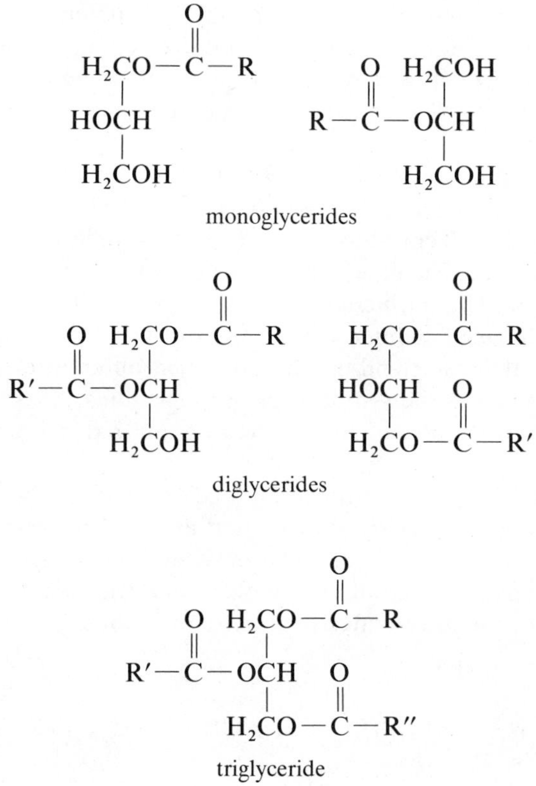

monoglycerides

diglycerides

triglyceride

always **mixed acylglycerols** or **mixed glycerides.** In addition, usually many different kinds of individual triglyceride molecules are present in a fat or an oil. In other words, any two molecules are likely to differ from each other in the kinds of fatty acids esterified to the glycerol residue and in their location. Hence, we speak of *mixtures of mixed glycerides* meaning that individual triglyceride molecules differ from each other and that each glycerol molecule has two or more different fatty acids esterified to it. The result is that most triglyceride stores do not contain just one, two, or three fatty acids, although one, two, or three may predominate (see Table 26.4). For example, four triglyceride molecules of an animal fat might contain the following fatty acid residues: one stearic + two oleic, two palmitic + one oleic, two palmitic + one linoleic, and one stearic + one oleic + one linoleic.

> **Mixed acylglycerols (glycerides)** are triglyceride molecules in which there are two or three different fatty acid residues esterified to a glycerol residue.

Lipids that are liquid at room temperature are called oils. Examples of oils are corn oil, cottonseed oil, soybean oil, safflower oil, and peanut oil. Lipids that are solid or semi-solid at room temperature are called fats. Examples of fats are lard, beef tallow, and butter. Oils contain primarily unsaturated fatty acids; fats contain a higher concentration of saturated fatty acids. This relationship can be seen in Table 26.4. Oils are found in plants (vegetable oils) and fish (fish oils). Oils with especially high amounts of polyunsaturated fatty acids are called *polyunsaturated oils.* (A method for determining the amount of unsaturation in an oil was described in Section 11-5.2.) Fatty acids with fewer double bonds and triglycerides containing more saturated fatty acids have higher melting points than unsaturated fatty acids and triglycerides containing unsaturated fatty acids because the flexible, extended hydrocarbon chains of saturated fatty acids and triglycerides can pack together well. Unsaturated fatty acid molecules have rigid bends in their hydrocarbon chains produced by the double bonds that prohibit rotation; therefore, they do not pack as compactly and are easier to separate by heating. As a result, they have lower melting points.

Table 26.4 Fatty Acid Composition of Some Common Fats and Oils*

Fat or Oil	Saturated Fatty Acids				Unsaturated Fatty Acids				
	Myristic	Palmitic	Stearic	Other	Palmitoleic	Oleic	Linoleic	Linolenic	Other
Land Animals									
Butterfat	11	29	9	10	5	27	4	0	6
Human depot fat	3	24	8	0	5	47	10	0	3
Beef tallow	6	27	14	0		50	2	0	0
Marine Animals									
Cod-liver oil	6	8	1	0	20	←— 29 —→		0	35
Sardine oil	5	15	3	0	12	←— 18 —→		0	48
Plants									
Coconut oil	18	11	2	60	0	8	0	0	0
Corn oil	1	10	3	0	2	50	34	0	0
Cottonseed oil	1	23	1	1	2	23	48	0	0
Linseed oil	0	6	3	1	0	19	24	47	0
Peanut oil	0	8	3	2	0	56	26	0	4
Safflower oil	←— 7 —→				0	19	70	3	0
Soybean oil	0	10	2	1	0	29	51	7	0

*The unit of measure for these values is g/100 g of total fatty acids. Numbers have been rounded off to the nearest whole number, so the total is not always 100.

Oils can be converted into fats by hydrogenation (see Section 11-5.2) according to the following equation.

$$R-\underset{\underset{\displaystyle H}{|}}{C}=\underset{\underset{\displaystyle H}{|}}{C}-R' + H_2 \xrightarrow{\text{catalyst}} R-\underset{\underset{\displaystyle |}{\overset{\displaystyle H}{|}}{}}{\overset{\displaystyle H}{C}}-\underset{\underset{\displaystyle H}{\overset{\displaystyle H}{|}}}{C}-R'$$

Hydrogenation converts some polyunsaturated fatty acids to less unsaturated fatty acids and some unsaturated fatty acids to saturated fatty acids. By hydrogenation, linolenic, linoleic, and oleic acids can be converted into stearic acid. A representation of the hydrogenation of linolenic acid is given.

$$CH_3-CH_2-CH=CH-CH_2-CH=CH-CH_2-CH=CH-(CH_2)_7-\overset{\displaystyle O}{\overset{\displaystyle \|}{C}}-OH + 3\,H_2 \longrightarrow$$

linolenic acid

$$CH_3-(CH_2)_{16}-\overset{\displaystyle O}{\overset{\displaystyle \|}{C}}-OH$$

stearic acid

The melting points of these four 18-carbon acids are given in Table 26.5. We can see that, at room temperature (about 19–24°C), the three unsaturated fatty acids will be melted (liquid). Therefore, these three fatty acids are oils. Stearic acid (the saturated fatty acid), however, is a waxy solid that melts high above room temperature and, in fact, is used to make high quality candles. Just as hydrogenation of linolenic acid (an oil) converts it into stearic acid (a fat), so hydrogenation of a vegetable oil (triglyceride) converts it into a fat. The degree of hardness of the fat is controlled by the amount of unsaturation remaining. This reaction is used to convert vegetable oils into **hydrogenated oils** to make such foods as margarine and vegetable shortening. By reducing the degree of unsaturation, an oil is converted into a semi-solid material. In Section 11-5.2, we learned that the degree of unsaturation can be determined by measuring the amount of halogen taken up by a fat or oil. Nutritionists recommend the use of polyunsaturated oils because the amount of cholesterol in the blood seems to be related, in part, to the amount of saturated fatty acids in the diet. In turn, the amount of cholesterol in the blood is related to **atherosclerosis**.

A **hydrogenated oil** is an oil that has been converted into a fat-like material by the addition of hydrogen atoms to the double bonds of the fatty acids.

Atherosclerosis is the plugging of large and medium-sized arteries with deposits of plaques containing cholesterol, other lipids, and cells.

Table 26.5	Melting Points of Four 18-Carbon Fatty Acids	
		Melting Point, °C
Linolenic acid	$C_{18}{}^{\Delta 9,12,15}$	−10
Linoleic acid	$C_{18}{}^{\Delta 9,12}$	− 5
Oleic acid	$C_{18}{}^{\Delta 9}$	13
Stearic acid	C_{18}	70

26-1.3 SAPONIFICATION OF TRIGLYCERIDES

In Section 16-4.1 we learned that triglycerides, as esters, undergo acid- and base-catalyzed hydrolysis. In the latter reaction, salts of the constituent fatty acids are formed. Salts of fatty acids are called *soaps*. This reaction, called *saponification,* yields diglycerides as one fatty acid is removed, then monoglycerides as a second fatty acyl ester bond is broken, and, finally, glycerol in addition to the salts of the fatty acids. Complete saponification of a triglyceride molecule is given. Enzyme-catalyzed hydrolysis of triglycerides occurs during digestion (see Section 27-1.2) and during the release into the blood of fatty acids from triglycerides stored in fatty tissues.

triglyceride   glycerol   soaps

Mono- and diglycerides are metabolic intermediates that are important in digestion as **emulsifying agents**. Monoglycerides and their derivatives are added to potato products, peanut butter, animal and vegetable shortenings, baked goods, whipped toppings, ice cream, coffee whiteners, coatings, candy, margarine, salad dressings, and cereal products such as macaroni to improve their texture, appearance, and stability.

An **emulsifying agent** is a substance used to make and stabilize an emulsion.

PROBLEMS

1. Write complete structures for the following fatty acids.
   (a) C_{12} (b) $C_{18}^{\Delta9,12}$ (c) $18:1^{\Delta9}$
2. Complete the following equation.

3. Is the triglyceride molecule in problem 2 that of a fat or an oil?
4. Using Tables 26.2 and 26.3, write the structure of a mixed triglyceride that contains esters of lauric, linoleic, and palmitoleic acids.
5. Write the structure of the triglyceride in problem 4 after complete hydrogenation.

ANSWERS TO PROBLEMS   **1. (a)** $CH_3-(CH_2)_{10}-COOH$

(b) $CH_3-(CH_2)_4-CH=CH-CH_2-CH=CH-(CH_2)_7-COOH$

(c) $CH_3-(CH_2)_7-CH=CH-(CH_2)_7-COOH$

2.

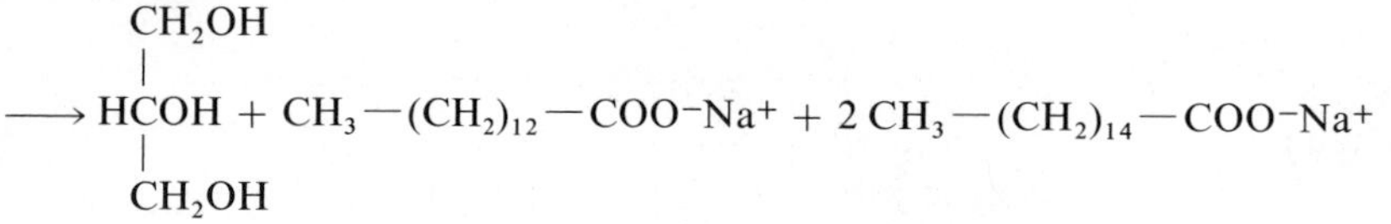

$$\longrightarrow \begin{matrix} CH_2OH \\ | \\ HCOH \\ | \\ CH_2OH \end{matrix} + CH_3-(CH_2)_{12}-COO^-Na^+ + 2\ CH_3-(CH_2)_{14}-COO^-Na^+$$

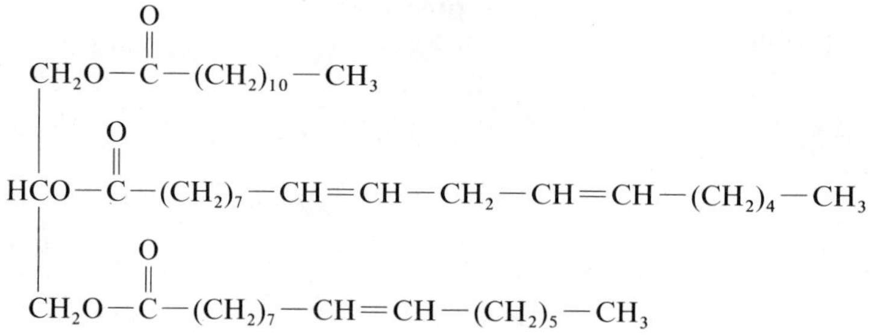

3. a fat

$$\begin{matrix} & & O \\ & & \| \\ CH_2O-C-(CH_2)_{10}-CH_3 \\ | \\ & & O \\ & & \| \\ \end{matrix}$$

4. one possibility is
$$\begin{matrix} & & O \\ & & \| \\ HCO-C-(CH_2)_7-CH=CH-CH_2-CH=CH-(CH_2)_4-CH_3 \\ | \\ & & O \\ & & \| \\ CH_2O-C-(CH_2)_7-CH=CH-(CH_2)_5-CH_3 \end{matrix}$$

5.
$$\begin{matrix} & & & & O \\ & & & & \| \\ & & & CH_2O-C-(CH_2)_{10}-CH_3 \\ & & & | \\ & O & & \\ & \| & & \\ CH_3-(CH_2)_{16}-C-OCH \\ & & | \\ & & & O \\ & & & \| \\ & & CH_2O-C-(CH_2)_{14}-CH_3 \end{matrix}$$

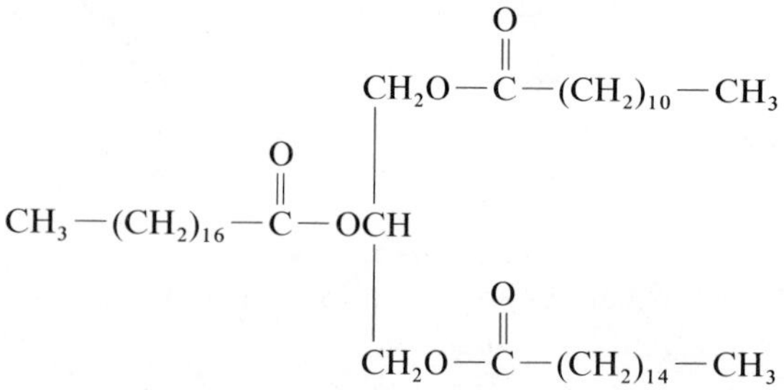

26-2 MEMBRANE COMPONENTS AND OTHER LIPIDS

LEARNING OBJECTIVES

1. Define phosphoglyceride, phosphatidylethanolamine, phosphatidylcholine, ceramide, sphingomyelin, ganglioside, and cholesterol.

2. Given the structure of a lipid molecule, identify the lipid as a member of one of the following classes: triglyceride, phosphoglyceride, sphingolipid, or steroid.

3. Define bilayer, extrinsic proteins, intrinsic proteins, passive transport, facilitated diffusion, active transport, exocytosis, endocytosis, and contact inhibition.

4. Diagram and describe the fluid mosaic model of membrane structure.

5. Diagram the structure of glycophorin, locating and describing its three domains (see Figure 21.12).

Membranes of cells are composed and structured to perform their particular roles. Generally, the fewer the functions that are required of the membrane, the simpler is its structure. Sarcoplasmic reticulum (the network of tubules between filaments of muscles), myelin (the sheath around certain nerve fibers), and the disk membranes of retinal rod cells (the light-sensitive cells of the retina of the eye) are all membranes with very specialized functions and unique and relatively simple compositions. On the other hand, mitochondrial and red blood cell membranes have a variety of functions and are more complex. In this section, the general features of the architecture of the normal, mammalian plasma membrane and its behavior will be discussed.

The **plasma membrane**, or **plasmalemma**, is the membrane surrounding a cell.

26-2.1 PLASMA MEMBRANE COMPONENTS AND FUNCTIONS

Membranes are composed of lipids and proteins (see Figure 26.2). The lipids are primarily of three types: the phosphoglycerides, the sphingolipids, and cholesterol. There are two types of plasma membrane proteins: extrinsic, or peripheral, proteins and intrinsic, or integral, proteins.

The lipid molecules are arranged in a bilayer. The molecules in the bilayer arrange themselves with the hydrophobic hydrocarbon chains (represented by lines in Figure 26.2) extending toward the interior of the bilayer and the hydrophilic (polar) groups (also called head groups; represented by circles in Figure 26.2) on the outside. In this arrangement, the polar groups face the polar, aqueous environments, both inside and outside the cell, and the hydrocarbon chains associate with each other away from the aqueous environment. The lipid bilayer is not a rigid structure but a flexible one in which the components can move sideways. This picture of a membrane is called the *fluid mosaic model.*

Extrinsic, or peripheral, proteins are only loosely attached to the membrane surface and can be removed in soluble form by mild extraction procedures.

Intrinsic, or integral, proteins are very tightly bound to the membrane and can be removed only be destruction of the membrane.

A bilayer (double layer) is the special arrangement of lipid molecules in a biological membrane as diagrammed in Figure 26.2.

Phosphoglycerides

Phosphoglycerides are derivatives of *phosphatidic acid,* the phosphate ester of a diglyceride. The structures of phosphatidic acid and the two most abundant types of phosphoglycerides (phosphatidylethanolamine and phosphatidylcholine) are given. The circles in Figure 26.2 represent the polar end of the phosphoglyceride

Figure 26.2   Diagrammatic representation of the arrangement of lipid and protein molecules in a plasma membrane. EP = extrinsic protein and IP = intrinsic protein.

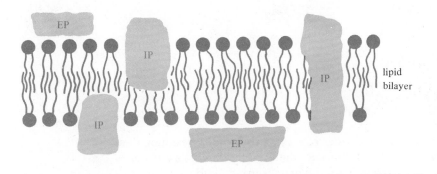

molecules containing the two ions, and the lines represent the hydrocarbon chains of the two fatty acids.

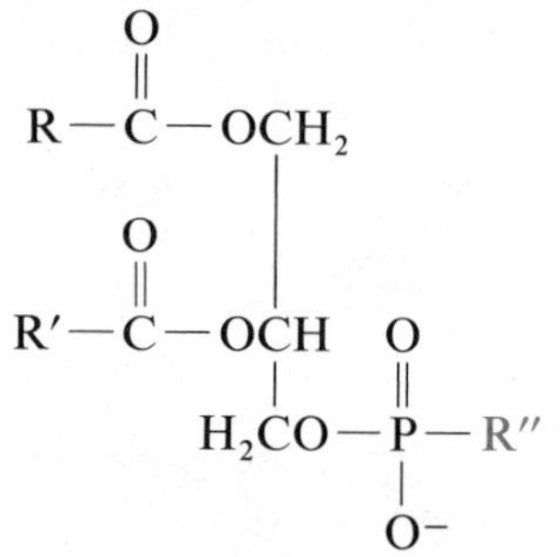

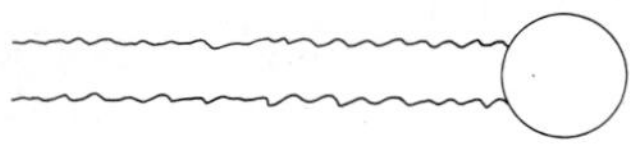

$$R-\overset{\overset{\displaystyle O}{\|}}{C}-OCH_2$$

$$R'-\overset{\overset{\displaystyle O}{\|}}{C}-OCH \quad \overset{\overset{\displaystyle O}{\|}}{\underset{\underset{\displaystyle O^-}{|}}{H_2CO-P-R''}}$$

$$R'' = -O^- \quad \text{phosphatidic acid}$$

$$= -O-CH_2-CH_2-NH_3^+$$

phosphatidylethanolamine (cephalin)

$$= -O-CH_2-CH_2-\overset{+}{N}(CH_3)_3$$

phosphatidylcholine (lecithin)

The phosphoglyceride composition of membranes varies, and any one type of phosphoglyceride represents a variety of molecules containing fatty acid chains of differing lengths and degrees of unsaturation. In general, the higher the degree of unsaturation in the fatty acid chains, the more fluid the membrane.

Sphingolipids

The **sphingolipids** are of two types: **sphingomyelins** and **glycosphingolipids.** The structures of these molecules are given in Figure 26.3. Recall that the circles in Figure 26.3 represent the hydrophilic portion of the molecule and the lines represent the hydrocarbon chains. In sphingolipids, one of the hydrocarbon chains is part of the sphingosine molecule and the other is provided by the fatty acid attached to sphingosine in an amide linkage. The two principal fatty acids of sphingolipids are lignoceric acid, the 24-carbon saturated fatty acid, and cerebronic acid, the 24-carbon saturated fatty acid with a hydroxyl group on C-2.

A **sphingolipid** is a fatty acid amide of the long-chain amino alcohol sphingosine.

A **sphingomyelin** contains residues of sphingosine, a fatty acid, phosphate, and choline.

A **glycosphingolipid** is a substance containing residues of sphingosine, a fatty acid, and carbohydrate.

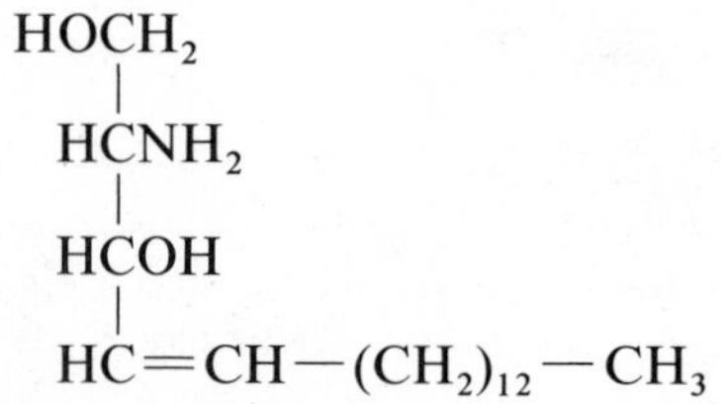

$$HOCH_2$$
$$|$$
$$HCNH_2$$
$$|$$
$$HCOH$$
$$|$$
$$HC=CH-(CH_2)_{12}-CH_3$$

sphinogosine

Glycosphingolipids are sphingolipids that contain carbohydrate residues. The gangliosides are a class of glycosphingolipids containing oligosaccharide residues attached to ceramide. The oligosaccharide units of glycosphingolipids contain up to

Figure 26.3 Shorthand notations for sphingomyelin, ceramide, glucosyl ceramide, and lactosyl ceramide. The fatty acid shown connected to sphingosine through an amide linkage is lignoceric acid (C_{24}). Individual $-CH_2-$ groups in both the fatty acid and sphingosinyl moieties have been omitted.

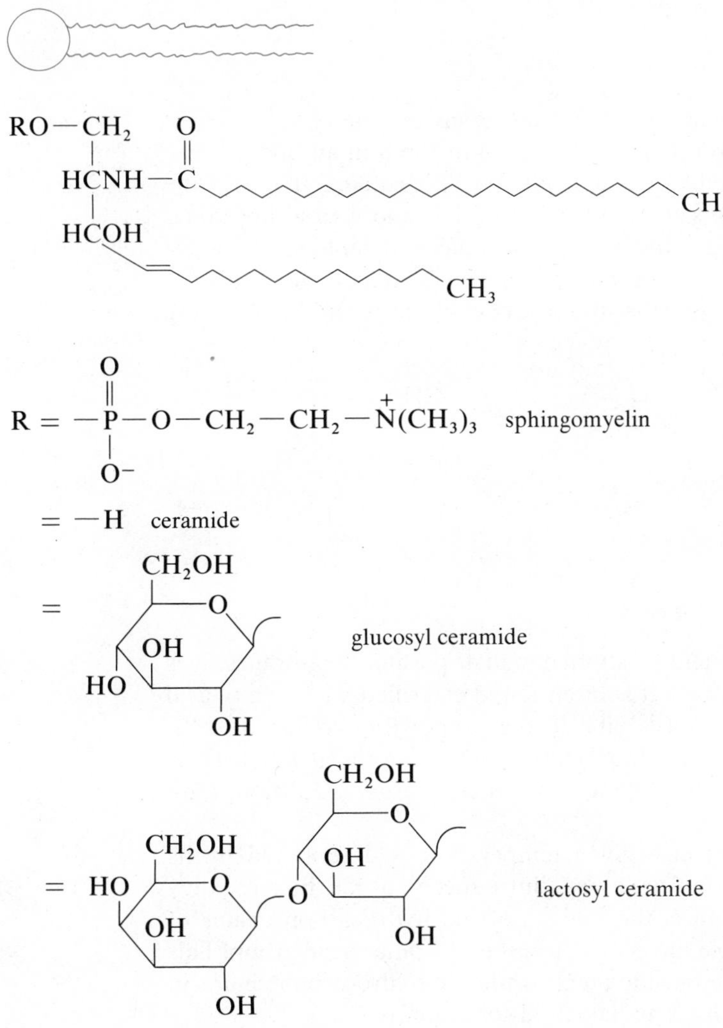

seven sugar residues. The simplest glycosphingolipid is glucosyl ceramide; the next simplest one is lactosyl ceramide. The carbohydrate portion of glycosphingolipids plays an important role in the functions of the plasma membrane, as we will learn.

Steroids

Steroids are not composed of fatty acids; rather, they are derivatives of the following ring system.

Sterols are steroids containing a hydroxyl group at C-3 and a chain of at least eight carbon atoms at C-17. Cholesterol is the major sterol in the human body. It is a structural component of cell membranes and plasma **lipoproteins**. It is also the starting material from which bile salts (see Section 27-1.2) and steroid hormones (see Figure 26.4) are synthesized. Most of the differences in biologically active steroids involve carbon atoms 3, 10, 13, and 17 (shown in color), which form a straight line through the molecule (see the structure of cholesterol and Figure 26.4).

Lipoproteins are complexes of lipids (triglycerides, phosphoglycerides, cholesterol, and cholesterol esters) and lipid-transport proteins that transport lipids in an aqueous medium (the blood) in which they are insoluble.

cholesterol

Cholesterol is a major component of all mammalian plasma membranes. It is synthesized from acetyl coenzyme A. It is often found esterified with a fatty acid in lipoproteins in blood. Cholesterol is vital to the proper functioning of cells. However, excessive amounts due to abnormalities in cholesterol metabolism or transport may be related to the development of atherosclerosis. Gallstones are made up primarily of cholesterol.

The fluidity of a lipid bilayer at a given temperature is determined by its composition. When cholesterol molecules are introduced into a bilayer, they interact with, and partly immobilize, the regions of the hydrocarbon chains of phosphoglyceride and sphingolipid molecules closest to the polar head group. This arrangement gives stiffness to the outside areas, while the hydrocarbon chains in the interior remain flexible in an intermediate fluid condition.

We will now spend a little time examining the action of steroid hormones. Steroid hormones control gene expression. Their effect is slow as compared to the effect of adrenaline and insulin. It is hours before the full effect of steroid hormones is expressed because they work by stimulating the synthesis of new mRNA and, subsequently, new protein molecules. Steroid hormones pass through the plasma membrane, unlike other hormones we have studied (adrenaline, insulin, and glucagon) that act at the cell surface. Once inside the cell, they bind to specific receptor protein molecules. This binding to one or more subunits of a protein brings about a conformational change in the protein molecule, allowing it to enter the nucleus by passing through the nuclear membrane. In the nucleus, the entire

Figure 26.4   Some steroid hormones.

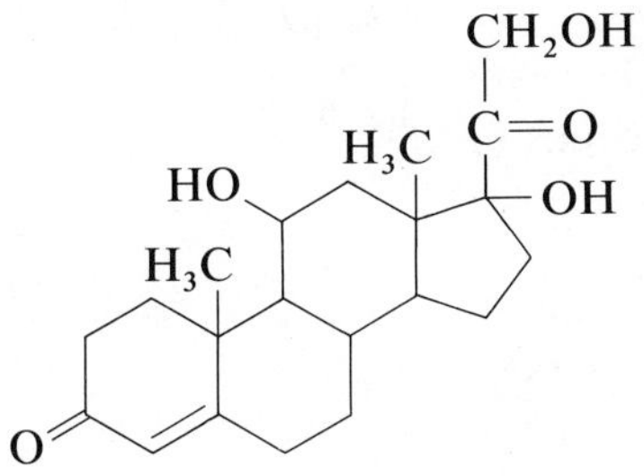

cortisol
(hydrocortisone)
(Controls carbohydrate
metabolism, particularly
gluconeogenesis.)

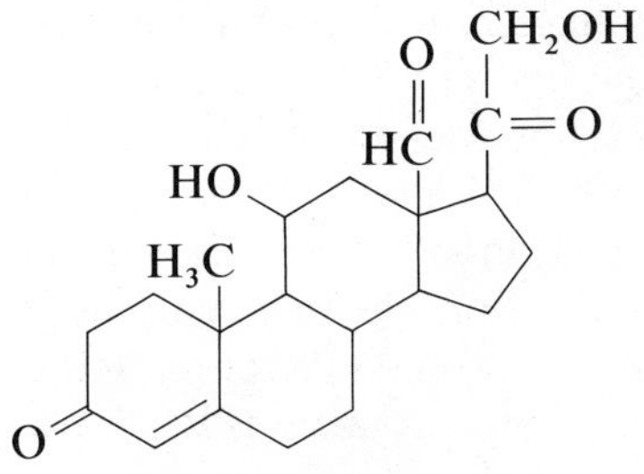

aldosterone
(Regulates ion
concentrations,
i.e., electrolyte balance.)

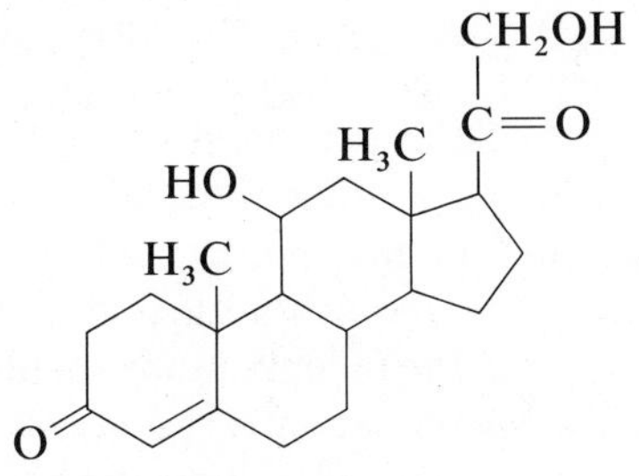

corticosterone
(Like cortisol and
aldosterone, formed in the
adrenal gland. Like cortisol,
raises blood sugar by
stimulating gluconeogenesis.)

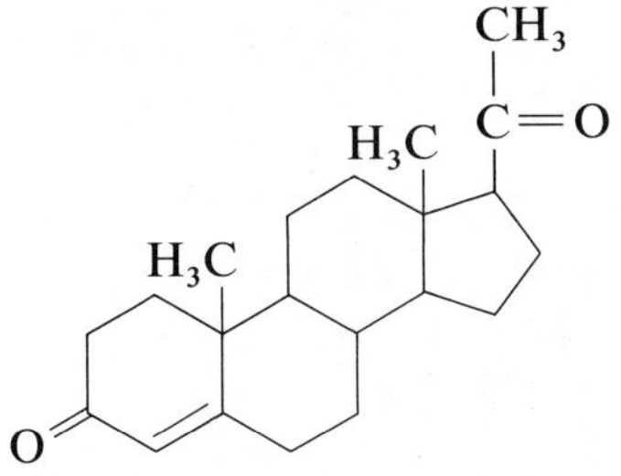

progesterone
(Prohibits the release
of an ovum [an egg] from
the ovaries.)

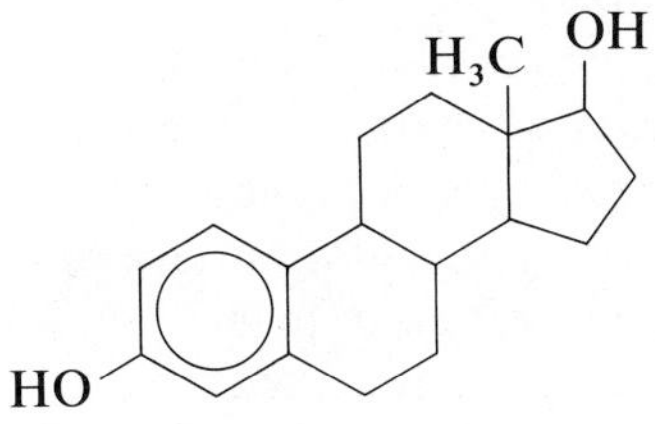

17-β-estradiol

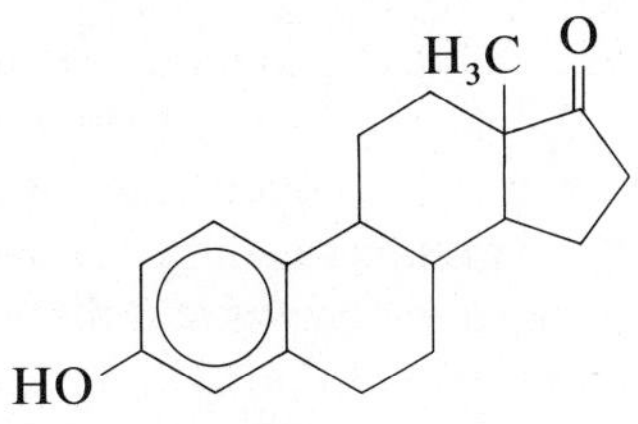

estrone

estrogens
(Prepares the uterus for reception and
development of a fertilized ovum.)

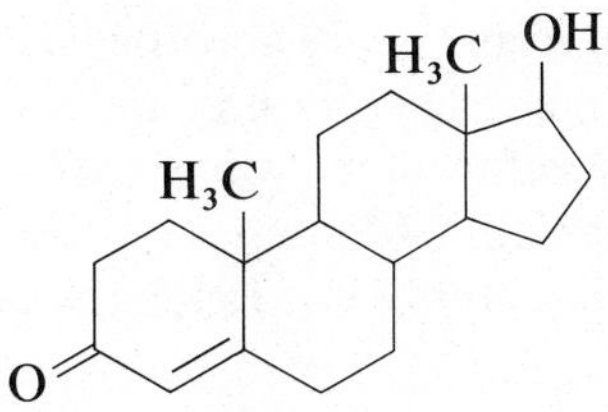

testosterone
(Produces and maintains male secondary
sex characters)

complex interacts with DNA and somehow stimulates synthesis of specific mRNA molecules.

Synthetic derivatives of the female sex hormones are widely taken as contraceptives known as "the pill." In order to understand how these oral contraceptives work, we must examine the female menstrual cycle. The cycle begins when the pituitary gland (a small **endocrine gland** located at the base of the brain) releases

An **endocrine gland** releases substances directly into the blood.

follicle-stimulating hormone (FSH) into the blood. FSH is a glycoprotein that stimulates the growth of one of the eggs (an ovum) in an ovary. Under the influence of FSH, the follicle surrounding the ovum secretes estrogens (see Figure 26.4). The estrogens then prepare the lining of the uterus to receive a fertilized ovum. The output of estrogens increases with time. The higher concentration of estrogens in the blood causes the pituitary gland to decrease the secretion of FSH and to begin secreting another hormone, luteinizing hormone (LH). LH, a glycoprotein, performs several functions including bringing about the release of the ovum from the ovary into the fallopian tube (ovulation) and stimulating progesterone production. Following ovulation, a third protein hormone, prolactin (PRL), is secreted by the pituitary gland. PRL influences the formation of a glandular mass from a follicle that has matured and discharged its ovum. This glandular tissue in the ovary is called the *corpus luteum*. As the corpus luteum grows, it releases progesterone (see Figure 26.4), a hormone that prohibits the release of another ovum from an ovary by bringing about a decrease in LH secretion. Prolactin also stimulates milk production in the mammary gland.

The **follicle** is a small secretory sac surrounding the egg.

If the ovum is fertilized, the corpus luteum increases in size and persists for several months. If the ovum is not fertilized, the corpus luteum shrinks and disintegrates. As it degenerates, small blood vessels rupture. Fragments of the gland, blood, and mucus form the menstrual discharge. At the same time, there is a sharp drop in the concentrations of estrogens and progesterone in the blood. As a result, the pituitary gland can now release FSH, and the cycle starts over again.

When fertilization occurs and the fertilized ovum is implanted in the uterus, the corpus luteum remains, and the pituitary gland is prevented from releasing FSH and LH by the estrogens and progesterone the corpus luteum produces. In the last six months of pregnancy, the estrogens and progesterone are produced by the placenta.

Oral contraceptives contain both a derivative of estrogen and a derivative of progesterone (see Figure 26.4). Estrogen and progesterone themselves are not absorbed and are, therefore, not effective if taken orally. The slightly modified derivatives used in oral contraceptives are absorbed and are active even when taken orally. However, these synthetic compounds are enough like the real ones that they prevent the pituitary gland from releasing FSH and LH. In other words, they fool the pituitary gland, making it respond as if fertilization has occurred and a pregnancy has resulted so that no FSH or LH is released, no follicle is developed, and no ovulation occurs.

ethyinyl estradiol

mestranol

Examples of Derivatives of Estrogen Used
in Oral Contraceptives

norethindrone

norethynodrel

Examples of Derivatives of Progesterone Used
in Oral Contraceptives

Membrane Proteins

More than half the mass (about 60%) of most plasma membranes is protein. An exception is the myelin membrane surrounding certain nerve cells (neurons). This membrane may contain up to 75% lipid. The composition of the myelin membrane reflects its special function. It has little enzymatic activity; rather, its principal function is as an insulator of the neuron.

Proteins are associated with the lipid bilayer. The ease with which protein molecules can be removed from a membrane varies. Some are loosely attached to the membrane surface and can be easily removed. Those are the extrinsic or peripheral proteins. The intrinsic or integral proteins are embedded in the lipid bilayer and interact with the lipids through hydrophobic bonding. Usually, at least 70% of the membrane proteins are intrinsic proteins. Some protein molecules extend completely across the lipid bilayer. An example is the glycophorin molecule of the erythrocyte plasma membrane (see Figure 21.12).

Many membrane proteins, like glycophorin, are glycoproteins (see Section 21-2.6). In both glycoproteins and glycosphingolipids, the carbohydrate (oligosaccharide) residues are located at the external surface (on the outside of the cell), rather than on the cytoplasmic side of the plasma membrane. These oligosaccharide units play an important role in several of the functions of membranes.

The intrinsic membrane protein molecules have been described as icebergs floating in a sea of lipid (see Figure 26.5). Not only are the protein molecules partially embedded in the lipid bilayer, but they are also mobile. Membrane protein molecules, like membrane lipid molecules, can move sideways through the membrane. This description of a plasma membrane as mobile, globular protein molecules floating in a fluid, lipid bilayer is called the *fluid-mosaic model* (Figure 26.5). The lipid bilayer consists primarily of phosphoglyceride molecules with some sphingolipid molecules and cholesterol dispersed throughout.

Functions of Plasma Membranes

The animal cell surface plays a critical role in interactions between cells and in the response of cells to the external environment, which may contain biologically active substances such as nutrients, drugs, hormones, toxins, and infectious agents. Some functions of plasma membranes are outlined in Table 26.6.

Figure 26.5 The fluid-mosaic model of membrane structure. Extrinsic proteins have been removed and are not seen in this diagram. Sphingolipid molecules are depicted in color and cholesterol molecules in black. (Modified from S. J. Singer and G. L. Nicolson, *Science,* 175 (1972), pp. 720–731.)

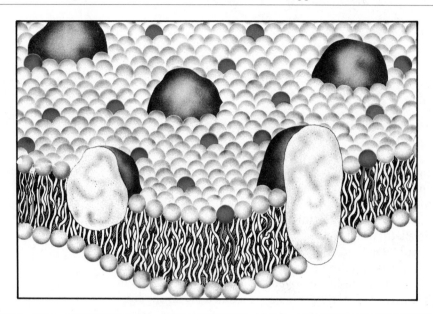

Table 26.6 Functions of Plasma Membranes

Compartmentalization

Transport
A. Passive transport
   1. Diffusion
   2. Facilitated diffusion
B. Active transport
C. Exocytosis*
D. Endocytosis†

Receiving and transducing‡ of various extracellular signals

Organization of enzymes

Locomotion

Exocytosis is the process by which materials packaged in vesicles are secreted outside cells.

†*Endocytosis* is the process by which substances are engulfed by a portion of the plasma membrane to form a vesicle inside the cell.

‡*Transducing* refers to the transformation of energy from one system to another such as the changing of light energy into the electrical energy of a nerve impulse.

A principal function of any membrane is to act as a mechanical boundary separating environments. The plasma membrane separates the intracellular environment from the extracellular environment. The membranes of organelles such as the nucleus separate their contents from the cytoplasm. Although membranes act as boundaries separating environments, they are not inert barriers. Among other activities, the plasma membrane controls the passage of nutrient substances into the cell and the passage of products out of the cell. Organelle membranes control the passage of substrates and products into and out of the organelle.

There are two types of transport of small molecules through membranes: passive transport and active transport. In one kind of **passive transport,** molecules move through membranes by simple **diffusion.** Simple diffusion does not require energy and does not require any special transporting apparatus. It probably involves solubility in the membrane or passage through channels (pores) formed by intrinsic proteins. Water, urea, ammonia, oxygen, carbon dioxide, chloride ion, and many drugs pass in and out of cells by simple diffusion. The second kind of passive transport is *facilitated diffusion.* Like simple diffusion, facilitated diffusion proceeds only from a region of higher concentration to a region of lower concentration and does not require energy. It does require specific transporting molecules (intrinsic proteins that form channels). This carrier-mediated transport is used to transport larger, lipid-insoluble substances such as glucose through membranes. Facilitated diffusion has the characteristics of enzyme-substrate interactions because it involves the interaction of specific molecules with specific transporting proteins. Facilitated diffusion shows specificity, competition (competitive inhibition), and saturation kinetics (a V_m).

Active transport requires energy and is capable of transporting molecules and ions against a concentration gradient. One active transport system is the Na^+/K^+ pump. Using ATP as an energy source, this transport system pumps sodium ions out of, and potassium ions into, cells. In a living system, the extracellular concentration of Na^+ is much higher than the intracellular concentration (see Table 26.7). Na^+ is continually pumped out of cells to maintain this difference. Similarly, the intracellular concentration of K^+ is much higher than the extracellular concentration (see Table 26.7) and so K^+ must be pumped in the other direction. Similar active transport systems are the H^+ pump in the kidney and the Ca^{2+} pump in muscle.

Plasma membranes contain recognition and binding sites (called **receptors**) for a variety of molecules. There are receptor sites for certain hormones (such as insulin,

Passive transport refers to the passage of molecules through a membrane without the expenditure of energy.

Diffusion is a scattering of molecules or ions until the concentration is equal in all parts of the system. It is the movement of molecules or ions from regions of high concentration to regions of low concentration until they are evenly distributed.

Receptors are protein molecules in a membrane that specifically recognize and bind other molecules.

Table 26.7	Distribution of Sodium and Potassium Ions in a Typical Cell	
	Concentration Inside Cell	Concentration Outside Cell
Na^+	10 mEq/L	142 mEq/L
K^+	141 mEq/L	5 mEq/L

glucagon, and adrenaline), the molecules (neurotransmitters) involved in the transmission of an impulse from one nerve cell to another and from a nerve cell to a muscle cell, **antigens,** groups on the surfaces of other cells, and viruses. The binding (adhesion) of many of these molecules to the plasma membrane is a signal that triggers the cell to respond in a specific way.

Cell-cell recognition and adhesion is a type of intercellular reaction that is important for several reasons. It is the primary process in the development of embryonic organs. Some cells migrate to predetermined areas of the embryo where they attach to other cells and/or multiply. Intercellular interactions are also involved in the normal growth of tissues and in malignancy.

Most cells in the adult human body have ceased to divide and move about. The cessation of cell division and movement that occurs when freely growing cells from a multicellular organism come into physical contact with each other is called *contact inhibition.* Contact inhibition of growth and movement undoubtedly involves the plasma membrane and the transfer of information. Malignancy, in part, involves a loss of this contact inhibition so that normally nondividing cells become dividing cells and sometimes move about. Changes occur in the plasma membrane when a normal cell becomes a cancer cell (a transformed cell).

The carbohydrate-containing substances, such as the glycoproteins and glycosphingolipids, are good candidates for participating in intercellular recognition and in the binding of regulatory molecules because (1) they are present only on the outside surface, (2) they are present on the outside surface in relatively large amounts, and (3) carbohydrate-protein interactions are very specific and very strong.

An **antigen** is a foreign molecule which, when introduced into the body, stimulates the formation of a specific antibody and which binds specifically to that antibody.

26-2.2 FAT-SOLUBLE VITAMINS

Nonpolar substances will dissolve in lipids. Among such substances are the fat-soluble vitamins (A, D, E, and K) that, like the steroids, are classified as non-saponifiable lipids (see Table 26.1).

Vitamin A is found in liver. Fish liver oils are especially high in vitamin A. Vitamin A is made from the carotenes in plants. β-Carotene, for example, is oxidatively cleaved in the cells of the small intestine to give two molecules of the aldehyde retinal, which is reduced to vitamin A_1 (retinol). Retinol is involved in the formation and maintenance of the tissues covering both the internal and external surfaces of the body (for example, skin). Certain structural proteins appear to be stabilized by combination with retinol or its derivatives. As we learned in Section 15-3.1, retinal, the aldehyde form of retinol, is a constituent of the conjugated protein rhodopsin, the light-absorbing protein involved in night vision.

retinol (vitamin A_1)

Vitamin D (D_2 and D_3) regulates calcium and phosphorus metabolism (see Section 10-2) as follows:

1. It causes an increased absorption of calcium ion from the intestinal tract by inducing the synthesis of a transport protein for Ca^{2+}.
2. It brings about the release of Ca^{2+} from bone to blood serum when there is insufficient calcium in the diet. Here again, vitamin D may affect transport and the biosynthesis of a transport protein.
3. It influences intestinal and renal absorption of phosphate.

Vitamin D deficiency results in the disease rickets. Sunlight helps to prevent rickets because it brings about the photooxidation of a derivative of cholesterol to produce a form of vitamin D, cholecalciferol. This reaction occurs in the skin.

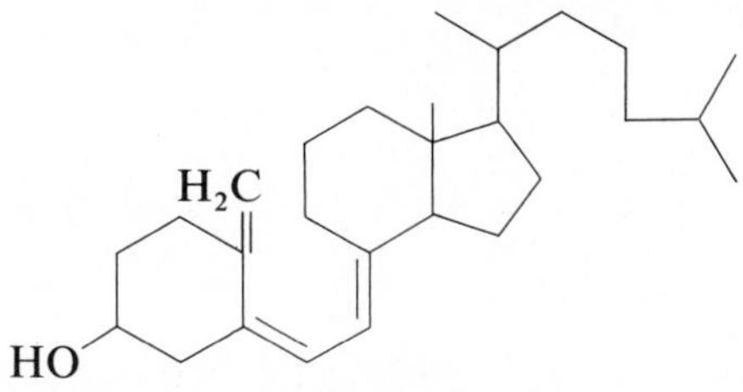

cholecalciferol
(vitamin D)

Vitamin E ocurs in several forms, the most active of which is α-tocopherol. The biological function of vitamin E is unknown. It may serve as an antioxidant for preserving cell membranes, perhaps by protecting the unsaturated fatty acids.

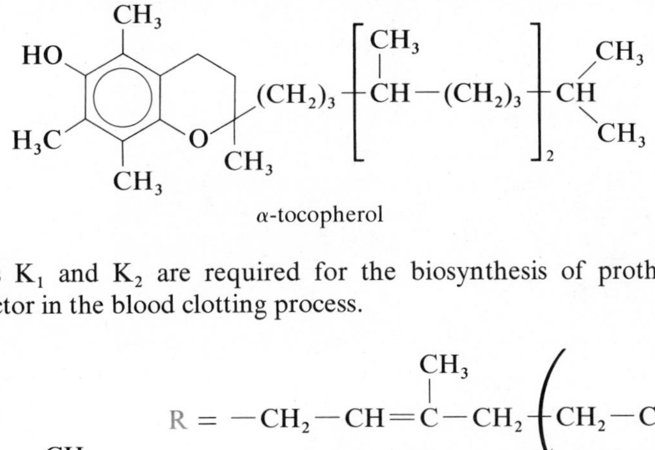

α-tocopherol

Vitamins K_1 and K_2 are required for the biosynthesis of prothrombin, an essential factor in the blood clotting process.

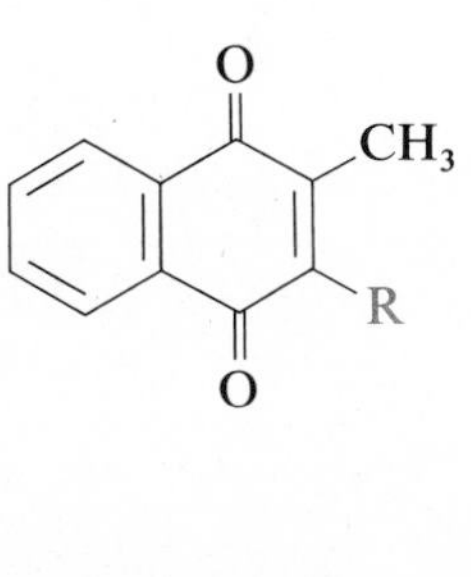

$$R = -CH_2-CH=\overset{\displaystyle CH_3}{\overset{|}{C}}-CH_2\left(CH_2-CH_2-\overset{\displaystyle CH_3}{\overset{|}{C}H}-CH_2\right)_3H \quad \text{Vitamin } K_1$$

$$= \left(CH_2-CH=\overset{\displaystyle CH_3}{\overset{|}{C}}-CH_2\right)_{6-8}H \quad \text{Vitamin } K_2$$

PROBLEM

6. To what class of lipids do each of the following compounds belong?

(a)

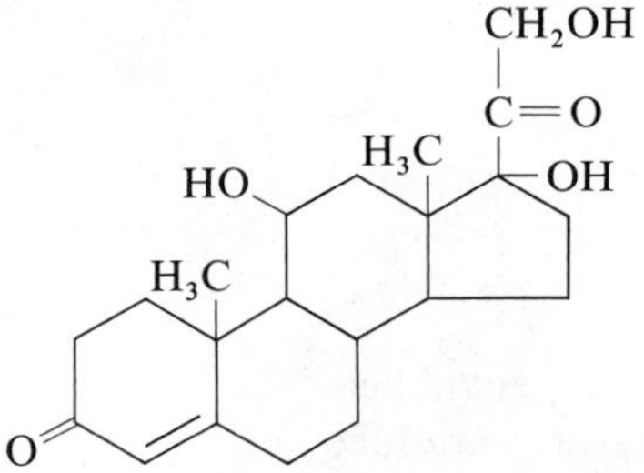

(b)

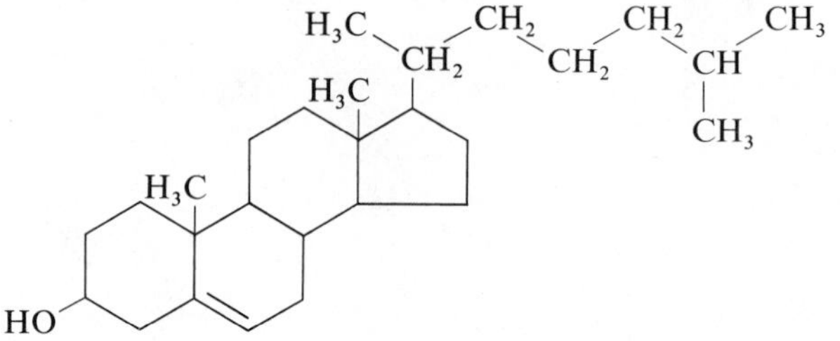

(c)

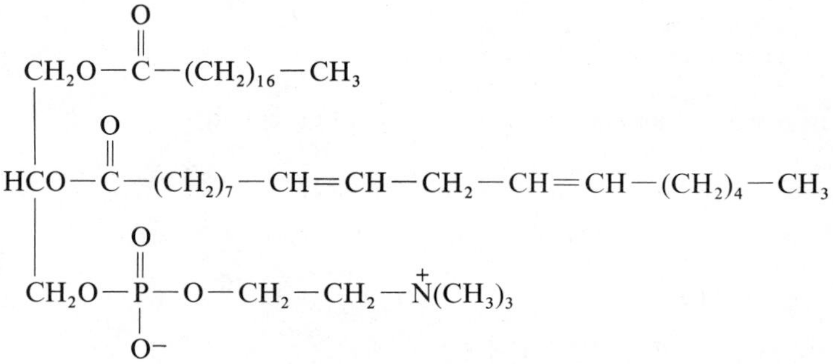

$$CH_2O-\overset{\overset{\displaystyle O}{\|}}{C}-(CH_2)_{16}-CH_3$$

$$HCO-\overset{\overset{\displaystyle O}{\|}}{C}-(CH_2)_7-CH=CH-CH_2-CH=CH-(CH_2)_4-CH_3$$

$$CH_2O-\overset{\overset{\displaystyle O}{\|}}{\underset{\underset{\displaystyle O^-}{|}}{P}}-O-CH_2-CH_2-\overset{+}{N}(CH_3)_3$$

(d)

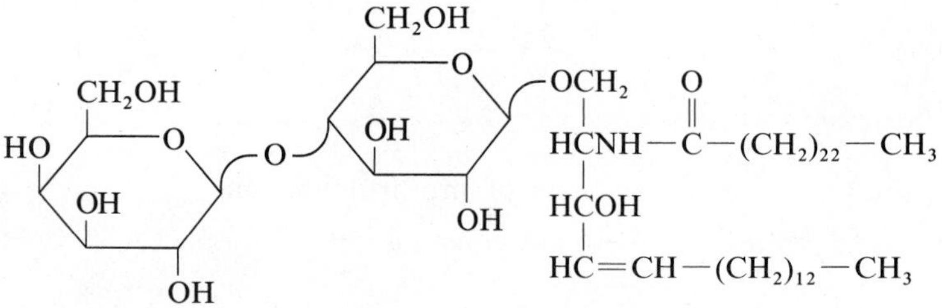

(e) $CH_3-(CH_2)_{12}-CH=CH$

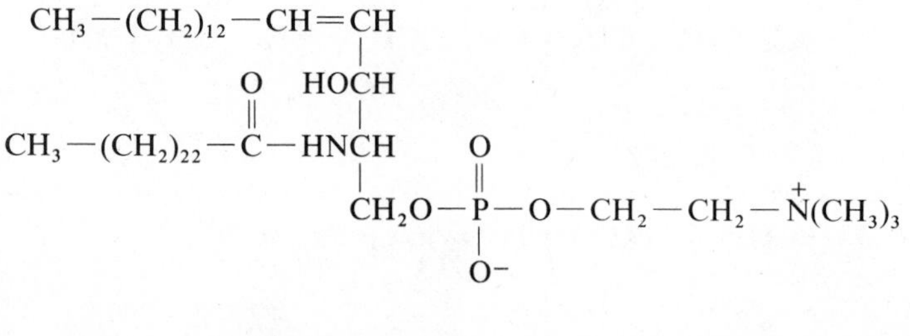

(f)

$$CH_3-CH_2-CH=CH-CH_2-CH=CH-CH_2-CH=CH-(CH_2)_7-\overset{\displaystyle O}{\overset{\displaystyle \|}{C}}-OCH$$

$$\begin{array}{c} \overset{\displaystyle O}{\overset{\displaystyle \|}{CH_2O-C}}-(CH_2)_{12}-CH_3 \\ | \\ \\ \overset{\displaystyle O}{\overset{\displaystyle \|}{CH_2O-C}}-(CH_2)_{14}-CH_3 \end{array}$$

(g) $HOCH_2$

$$\begin{array}{c} HCNH-\overset{\displaystyle O}{\overset{\displaystyle \|}{C}}-(CH_2)_{22}-CH_3 \\ | \\ HCOH \\ | \\ CH \\ \| \\ HC \\ | \\ (CH_2)_{12} \\ | \\ CH_3 \end{array}$$

ANSWERS TO PROBLEM **6. (a)** steroid   **(b)** steroid   **(c)** phosphoglyceride (a phosphatidylcholine)   **(d)** sphingolipid (a ganglioside)   **(e)** sphingolipid (a sphingomyelin)
(f) triglyceride   **(g)** sphingolipid (ceramide)

SUMMARY

In this chapter, we learned that lipids are organic compounds that are only slightly
soluble in water but quite soluble in organic solvents. All but the nonsaponifiable
lipids are esters or amides of fatty acids.

The two main classes of fatty acids are the saturated and the unsaturated fatty
acids. Until they are needed as a source of energy, fatty acids are stored in the
form of triglycerides (triacyglycerols), which are molecules of glycerol esterified
with three fatty acid molecules. Lipids, including triglycerides, that are solid or
semi-solid at room temperature are called fats; those that are liquid at room temperature are called oils.

Cell membranes are composed of lipids (about 40% by weight) and proteins
(about 60% by weight). Carbohydrate residues are also present on glycosphingolipid and glycoprotein molecules. Carbohydrate residues make up only a few
percent by weight of the plasma membrane, but they are concentrated on the
external surface and impart important biological properties to the cells.

The most abundant class of lipids in a membrane is the phosphoglycerides.
Phosphoglycerides are arranged in the membrane bilayer with the glycosphingolipids. Cholesterol molecules are present in this bilayer in different amounts in
different membranes. Globular proteins, some of which are glycoproteins, partially
or completely penetrate this bilayer. The composition of membranes is related to
their functions, which are many and varied.

The nonsaponifiable lipids are lipid-soluble substances that are not esters or amides of fatty acids. These include steroids and the fat-soluble vitamins: vitamins A, D, E, and K.

In Chapter 27, we will examine the digestion of fats and oils, the transport of lipids in the blood, and then the pathway by which fatty acids are oxidized to produce ATP. We will also learn how carbohydrates can be used to make fatty acids that can be stored in triglyceride molecules until needed for energy.

ADDITIONAL PROBLEMS

7. A lipid that is solid or semi-solid at room temperature is called a(n) _____.
8. A fatty acid containing one or more double bonds is a(n) _____.
9. The addition of hydrogen to the double bonds of a vegetable oil is called _____.
10. A mixture of triglyceride molecules that is liquid at room temperature is called a(n) _____.
11. Which of the following will not release a fatty acid upon hydrolysis?
    (a) triglyceride   (b) phosphatidylcholine
    (c) monoglyceride   (d) phosphoglyceride
    (e) diglyceride   (f) ceramide
    (g) steroid

(h) phosphatidylethanolamine
(i) cholesterol
(j) ganglioside
(k) sphingomyelin

12. Which of the compounds listed in problem 11 will release carbohydrate molecules upon hydrolysis?
13. Which of the compounds listed in problem 11 have a glycerol residue as part of their structure?
14. Which of the compounds listed in problem 11 have a sphingosine residue as part of their structure?
15. Diagram the fluid mosaic model of membrane structure.
16. What three classes of lipids are found in plasma membranes?

CHAPTER 27

DIGESTION AND METABOLISM OF LIPIDS

In Chapters 24 and 25, we learned how carbohydrate is stored and subsequently used for energy in a living cell. In mammals, including humans, however, the principal storage form of energy is triglyceride (fat), which we studied in Chapter 26. Fatty acids are stored in triglyceride molecules. Fatty acids are a better source of chemical energy than are carbohydrates because the release of energy is due to oxidation and carbohydrates are already partially oxidized. Carbohydrates contain $-CHOH-$ groups, while fatty acids contain primarily $-CH_2-$ groups. More energy is released upon oxidation of fatty acids because more oxidation is required to convert $-CH_2-$ to CO_2 than to convert $-CHOH-$ to CO_2. Fatty acids contain as much as 90% carbon and hydrogen atoms and about 9 kcal/g. Carbohydrates contain only about 46% carbon and hydrogen atoms and about 4 kcal/g. The energy density difference is even greater because glycogen is hydrated and fat is not. This hydration reduces the energy of glycogen to about 1 kcal/g. We learned in Section 1-3.3 that a calorie is the amount of heat required to raise the temperature of one milliliter of water one degree Celsius. The *calorie* used to describe the energy content of foods as used by nutritionists and persons on diets is a kilocalorie (1 kilocalorie = 1000 calories = 1 Calorie).

The triglyceride molecules of stored fat can be broken down to yield fatty acid molecules that can be oxidized to acetyl coenzyme A molecules. These acetyl coenzyme A molecules can then be oxidized in the tricarboxylic acid cycle. Coupling of oxidation in the TCA cycle with the respiratory chain and oxidative phosphorylation as described in Chapter 25 produces the ATP molecules needed to provide the energy for cellular processes. This catabolism of fatty acid molecules will be discussed in this chapter. We will also learn about the digestion of lipids that we take in with our food and the biosynthesis of fatty acids from acetyl-CoA.

27-1 DIGESTION, ABSORPTION, AND TRANSPORT OF TRIGLYCERIDES

LEARNING OBJECTIVES

1. Define bile salts, lipase, detergent, lipoprotein, and chylomicron.
2. Describe the actions of bile salts and lipases.
3. Describe the functions of plasma albumins, chylomicrons, and lipoproteins.

4. Trace the digestion of triglycerides.
5. Write an equation illustrating the digestion of a triglyceride.
6. Explain how the absence of bile would adversely affect digestion.

Forty to forty-two percent of the energy of the average adult in the United States is obtained from lipids, 12–14% from protein, and the rest (about 46%) from the carbohydrate in the diet. Dietary lipids are primarily triglycerides of 16-carbon and 18-carbon fatty acids. Dairy products are the chief source of the shorter chain fatty acids.

As discussed in Section 26-1.1, triglycerides that are liquid at room temperature are called oils, usually contain high amounts of esterified unsaturated fatty acids, and are found primarily in plant seeds (vegetable oils) and in fish. Triglycerides that are semi-solid or solid at room temperature are known as fats, are more saturated than the oils, and are found primarily in animals (meat products). Fats and oils often have lipid-soluble materials dissolved in them. Vegetable oils may contain vitamins A, E, and K, depending on the degree of refinement. Fish oils may contain vitamin D. Animal fats may contain all four fat-soluble vitamins and dissolved cholesterol.

The major site of lipid digestion is the small intestine, but before hydrolysis of triglycerides can occur effectively, the fat or oil must be emulsified. **Emulsification** is effected by (1) bile salts, (2) monoglycerides and soaps, and (3) mechanical action. Bile salts are salts of amides of cholic (25–60%), deoxycholic (5–25%), and cheno-deoxycholic (30–50%) acids and glycine or taurine. They are made from cholesterol as indicated in Figure 27.1.

Bile salts are detergents and, thus, promote emulsification of lipids. Because hydrolysis of fatty acid esters of glycerol and other alcohols involves water and water-soluble enzymes, and because lipids are essentially insoluble in water, hydrolysis can occur only at the surface of a lipid droplet where it is in contact with the aqueous phase. Hence, the rate of reaction is determined in part by the surface area of the droplets. Therefore, as the degree of emulsification increases, the size of the individual lipid droplet decreases, the total available surface area increases, the rate of hydrolysis increases, and the hydrolysis becomes more complete in a given time. The function of bile salts is to facilitate digestion of fats and oils through their emulsifying action, thus greatly increasing the surface area of the substrate exposed to lipases.

Emulsification is the process of suspending tiny particles of fat or oil in an aqueous fluid.

A **detergent** is a molecule with a nonpolar, hydrocarbon group and a polar (usually ionic) group that can emulsify fats, oils, and greases and, thus, has a cleansing action in water.

A **lipase** is an enzyme that catalyzes the hydrolysis of the ester linkages between glycerol and the fatty acids of glycerides.

Figure 27.1 Biosynthesis of the primary bile salts from cholesterol.

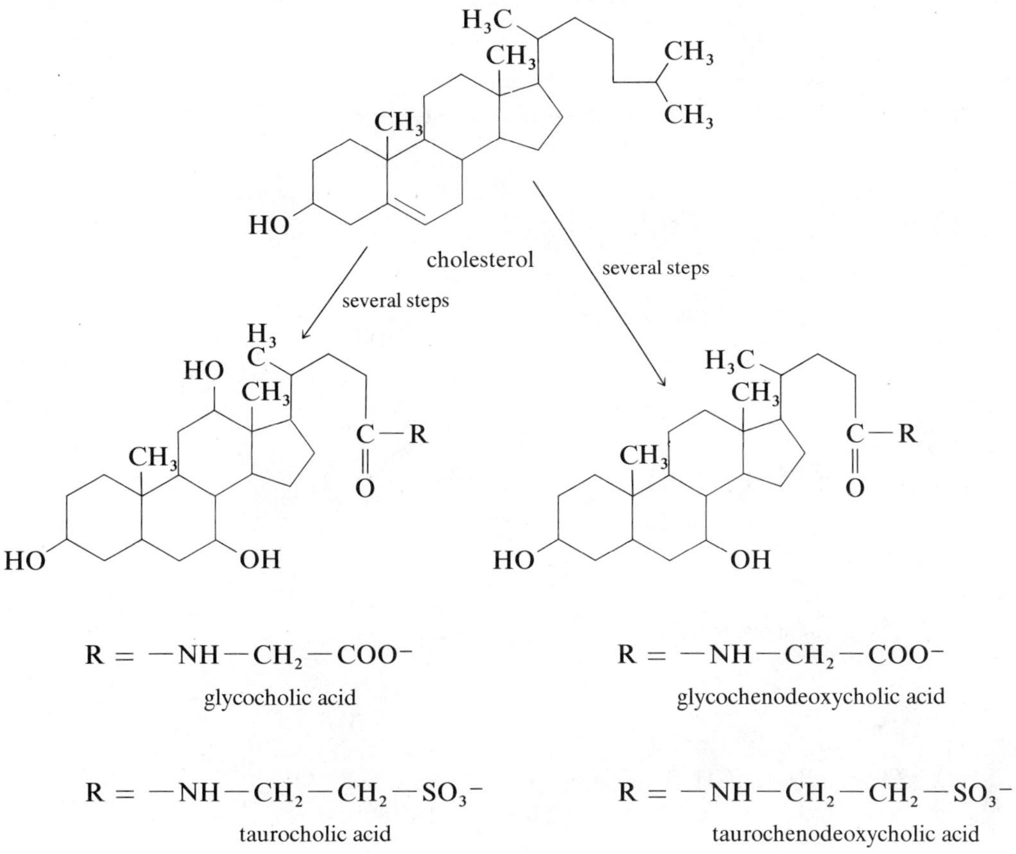
cholesterol
several steps
several steps
R = —NH—CH₂—COO⁻
glycocholic acid
R = —NH—CH₂—COO⁻
glycochenodeoxycholic acid
R = —NH—CH₂—CH₂—SO₃⁻
taurocholic acid
R = —NH—CH₂—CH₂—SO₃⁻
taurochenodeoxycholic acid

Several enzymes are involved in lipid digestion. Pancreatic lipase is secreted by the pancreas. It requires a small protein cofactor (colipase) that is also secreted by the pancreas. Lipase acts on triglycerides emulsified with bile salts. The pH of the small intestine is not optimal for lipase activity; therefore, hydrolysis is incomplete. Pancreatic lipase most effectively catalyzes the hydrolysis of triglycerides (as opposed to diglycerides), esters of the higher-molecular-weight fatty acids (for example, C_{18}), and esters of the more unsaturated fatty acids. Pancreatic lipase specifically attacks the ester linkages at the primary (1 and 3) positions of triglycerides, leaving a monoglyceride. Another esterase catalyzes the hydrolysis of the third fatty acid ester bond. The products of hydrolysis, the fatty acid salts (*soaps*) and monoglycerides (formed by partial hydrolysis of triglycerides), are also detergents and aid in emulsification. The process of hydrolysis of triglycerides during digestion is shown.

$$
\begin{array}{l}
\overset{\displaystyle O}{\underset{\displaystyle \|}{}} \\
H_2CO-C-(CH_2)_{14}-CH_3 \\[4pt]
\overset{\displaystyle O}{\underset{\displaystyle \|}{}} \\
HCO-C-(CH_2)_7-CH{=}CH-CH_2-CH{=}CH-(CH_2)_4-CH_3 \\[4pt]
\overset{\displaystyle O}{\underset{\displaystyle \|}{}} \\
H_2CO-C-(CH_2)_7-CH{=}CH-(CH_2)_7-CH_3
\end{array}
$$

triglyceride

↓ pancreatic lipase

$$
\begin{array}{l}
H_2COH \\[4pt]
\overset{\displaystyle O}{\underset{\displaystyle \|}{}} \\
HCO-C-(CH_2)_7-CH{=}CH-CH_2-CH{=}CH-(CH_2)_4-CH_3 \\[8pt]
H_2COH
\end{array}
$$

monoglyceride

+

$$
CH_3-(CH_2)_{14}-\overset{\displaystyle O}{\overset{\displaystyle \|}{C}}-O^- \quad + \quad CH_3-(CH_2)_7-CH{=}CH-(CH_2)_7-\overset{\displaystyle O}{\overset{\displaystyle \|}{C}}-O^-
$$

palmitate   oleate

Absorption of digestion products, primarily free fatty acids (70%) and mono-glycerides (25%), occurs in the small intestine. Digestion products enter the cells lining the inner surface of the small intestine by simple diffusion. Absorbed products are used by these cells for synthesis of triglycerides and phosphoglycerides.

Within the intestinal mucosal cell, triglyceride-phosphoglyceride particles acquire a coating of protein, forming specific lipoprotein particles. These particles, about 1 micron in diameter, are called chylomicrons. *Chylomicrons* are small droplets of lipid synthesized in cells lining the small intestine from products of lipid digestion. They are composed primarily of triglyceride with small amounts of phosphoglyceride, cholesterol, and protein. Chylomicrons are released into lymphatic vessels and eventually end up in the blood. Their fate is discussed later in this chapter. Neither the hydrolysis of triglycerides nor their absorption in the small intestine is complete. Lipids make up 5–25% of the dry weight of normal feces. (Normally, less than 5 g of fecal fat is excreted daily. One-third of this is cholesterol or its derivatives.)

An increase in fecal fat may result from defective emulsification, digestion, or absorption. Defective emulsification and digestion may occur because of an absence of bile (for example, in cases of obstructive jaundice). Defective digestion may occur because of an absence, or defective formation, of pancreatic juice. Defective absorption may occur because of a loss of extensive areas of absorptive surface of the small intestine.

After an average meal containing fat, there is a characteristic hyperlipemia (hyperlipidemia). The concentration of plasma lipids usually begins to rise within 2 hours, reaches a peak in 4–6 hours, and then drops rather rapidly to the resting level. Fat transported in the blood is found in at least three forms:

1. chylomicrons
2. lipoproteins
3. unesterified (free) fatty acids bound to plasma albumin (a protein of blood plasma)

There are several types of lipoproteins. They differ from each other in density, electric charge, particle size, and absorption characteristics. Lipoproteins form stable dispersions. They transport water-insoluble lipids to and from the liver, where triglyceride molecules are synthesized, and to and from adipose tissue, where triglyceride molecules are stored.

Two-thirds of the unesterified fatty acids are bound to plasma albumin molecules; one-third are found as a part of lipoproteins. The half-life of unesterified fatty acids in the plasma is 1–3 min. Their concentration in blood plasma decreases with increases in the concentration of glucose or insulin and with exercise. The concentration of unesterified fatty acids increases as the concentration of either adrenaline, noradrenaline, or growth hormone increases; during fasting; with diabetes mellitus; and when exercise is stopped. Fatty acids are an important energy source for muscle contraction as we will discuss in Section 27-2.2. Therefore, for example, as adrenaline is secreted to prepare you to fight or to run, fatty acids are also released to provide some of the energy required by your muscles.

> Hyperlipemia (hyperlipidemia) is a condition in which there is an abnormally high concentration of lipids in the blood.

PROBLEMS

1. Describe the process and site of digestion of triglyceride.
2. An association of triglyceride, phosphoglyceride, cholesterol, and protein molecules is called a(n) _____.
3. Describe the function of plasma albumins.
4. Describe the function of lipoproteins.
5. Explain how the absence of bile would adversely affect digestion.

ANSWERS TO PROBLEMS   **1.** triglyceride $\xrightarrow{\text{pancreatic lipase}}$ monoglyceride + fatty acids   This reaction occurs in the small intestine.   **2.** lipoprotein   **3.** to transport free fatty acids in the blood   **4.** to transport insoluble lipids, primarily triglycerides, in the blood   **5.** Without bile, there would be no bile salts. Without bile salts, the triglycerides (fats and oils) cannot be sufficiently emulsified, the action of lipase on them would be inefficient, and digestion of the fats and oils would be incomplete.

27-2 CATABOLISM OF FATTY ACIDS

LEARNING OBJECTIVES

1. Define lipolysis.
2. Describe the function of β-oxidation and L-carnitine.
3. Without using equations, describe the process of β-oxidation of fatty acids, including its purpose, the products obtained, and the ultimate destination of the products.
4. Given the substrates, coenzymes, and the enzyme required for any of the reactions of fatty acid oxidation, write a balanced equation showing the products of the reaction.
5. Describe the fates of the NADH and FADH$_2$ molecules produced by the β-oxidation of fatty acids.
6. Given a fatty acid with an even number of carbon atoms, calculate the number of ATP molecules generated by its complete oxidation.
7. Describe the interrelationships of β-oxidation, the TCA cycle, and the respiratory chain.
8. Describe the three phases of the production of ATP during strenuous exercise.
9. Describe the roles of creatine phosphate, myokinase, ATP, glycogen, fatty acids, and oxygen in skeletal muscle metabolism.
10. Explain what determines whether fatty acids or carbohydrate is used as an energy source in working muscle.
11. Describe the control of phosphorylase in skeletal muscle and its role in skeletal muscle energy production (see Section 24-2.1).
12. Describe the function of lactate formation in skeletal muscle metabolism (see Section 24-2.2).
13. Describe and state the function of the Cori cycle (see Section 24-2.3).

Fat (triglyceride) is the major form of energy storage in humans. Fat can be stored in much larger quantities than can carbohydrate (glycogen) because the human body contains special cells (adipose cells) that can store triglyceride. To be used for energy, stored triglyceride molecules first must be hydrolyzed to release their constituent fatty acids. Hydrolysis (**lipolysis**) is effected by a hormone-activated lipase. The free fatty acids are released to the blood where they are transported by physical combination with plasma albumin. After uptake by tissues, fatty acids are oxidized in a pathway known as β-oxidation (see Figure 27.2), which is the subject of our next discussion.

Lipolysis is the hydrolysis of triglyceride molecules into free fatty acids.

27-2.1 β-OXIDATION

β-Oxidation occurs in mitochondria. In this pathway, two-carbon fragments in the form of acetyl-coenzyme A molecules are removed, one at a time, from a fatty acid molecule. The acetyl-CoA molecules then enter the TCA cycle where they are catabolized to CO$_2$ and H$_2$O. Before fatty acid molecules can be catabolized, however, they must first be converted to an activated form. This activation is accomplished by reacting the fatty acid molecule with coenzyme A in the presence of the enzyme acyl-CoA synthetase (thiokinase) (1), using ATP as the source of

Figure 27.2 β-oxidation of fatty acids.

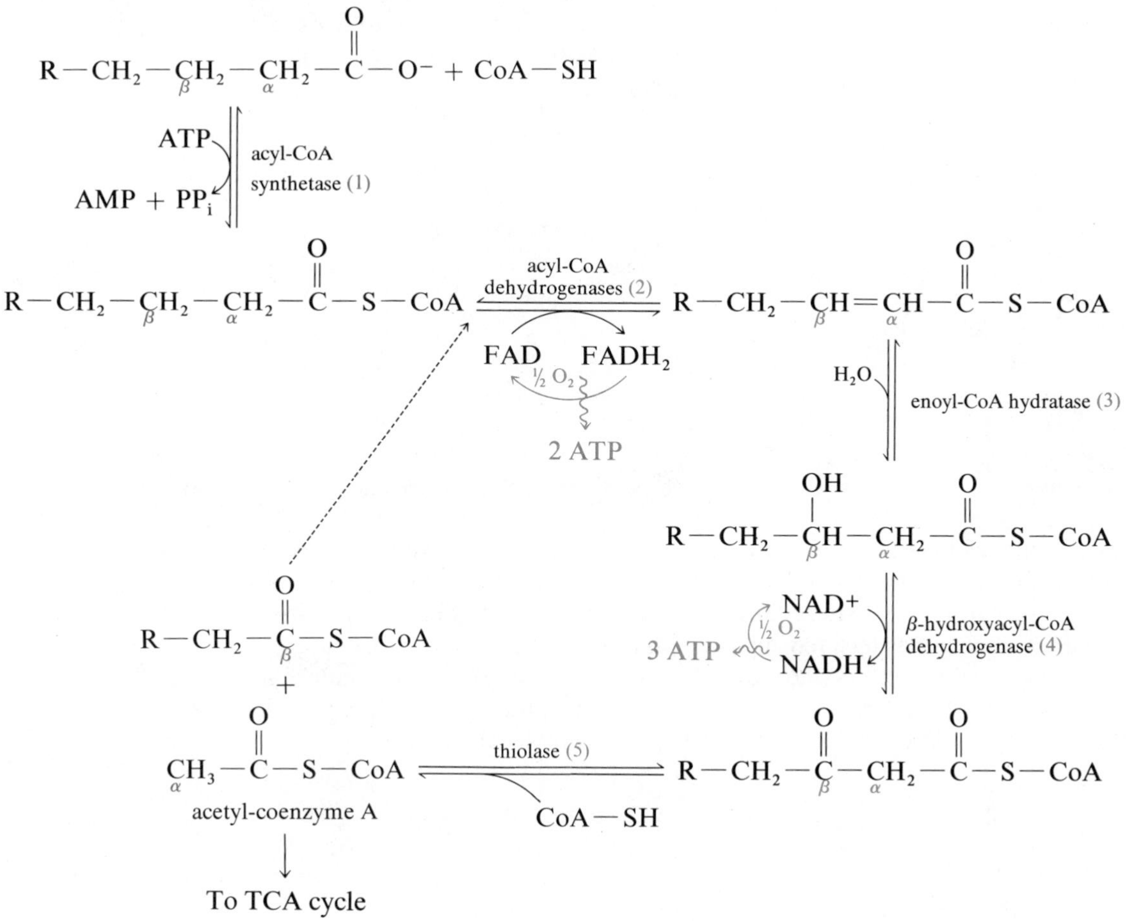

energy. The product is the fatty acyl-CoA thioester. The two steps in this reaction are shown. The intermediate, enzyme-bound fatty acyl-AMP molecule has the same type of structure as that of the activated amino acids used in protein synthesis (see Figure 22.5).

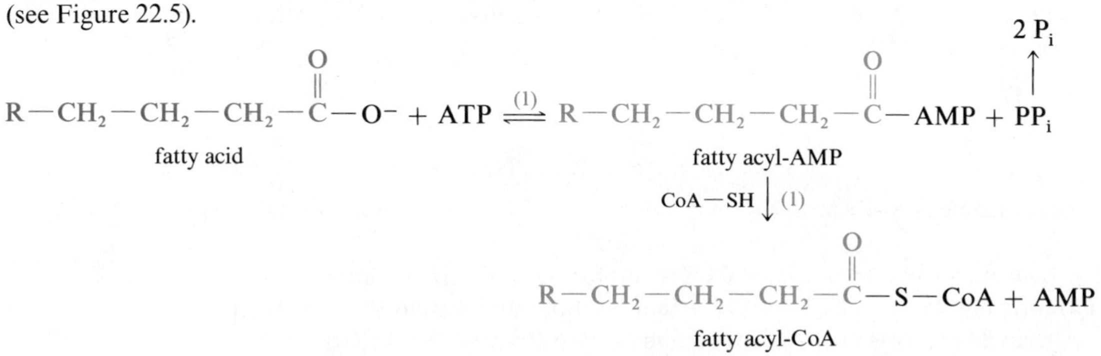

Fatty acyl-CoA synthesis occurs in the outer mitochondrial membrane. However, the fatty acyl-CoA molecules cannot penetrate the inner mitochondrial membrane to reach the site of the β-oxidation enzyme system. In order to cross this membrane, fatty acyl molecules are transferred to carrier molecules, L-carnitine, in a reaction catalyzed by L-carnitine acyltransferase.

$$
\begin{array}{c}
\text{COO}^- \\
|\\
\text{CH}_2 \\
|\\
\text{HOCH} \\
|\\
\text{CH}_2 \\
|\\
\text{CH}_3-\text{N}^+-\text{CH}_3 \\
|\\
\text{CH}_3
\end{array}
\;+\; R-\overset{\overset{\text{O}}{\|}}{\text{C}}-S-\text{CoA}
\;\underset{}{\overset{\text{L-carnitine acyltransferase}}{\rightleftharpoons}}\;
R-\overset{\overset{\text{O}}{\|}}{\text{C}}-O
\begin{array}{c}
\text{COO}^- \\
|\\
\text{CH}_2 \\
|\\
\text{CH} \\
|\\
\text{CH}_2 \\
|\\
\text{CH}_3-\text{N}^+-\text{CH}_3 \\
|\\
\text{CH}_3
\end{array}
\;+\; \text{CoA}-\text{SH}
$$

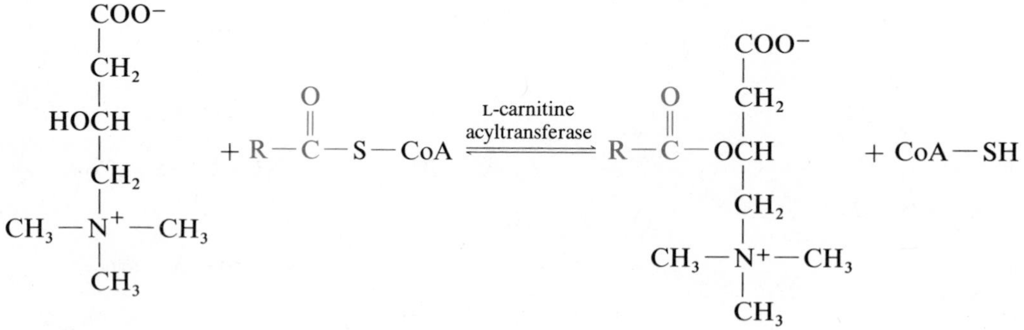

| carnitine | fatty acyl coenzyme A | fatty acyl carnitine | coenzyme A |

Once a fatty acid is transported through the mitochondrial membrane in combination with L-carnitine, it is transferred back to coenzyme A. Then the fatty acyl-CoA molecule can be oxidized. First, two hydrogen atoms are removed from the α and β carbon atoms by a flavoprotein enzyme, acyl-CoA dehydrogenase (2). The products of this reaction are an α,β-unsaturated acyl-CoA molecule and reduced FAD (FADH$_2$). FADH$_2$ is reoxidized in the respiratory chain (see Section 25-2.1), forming two ATP molecules for each FADH$_2$ molecule oxidized.

$$
R-\text{CH}_2-\text{CH}_2-\text{CH}_2-\overset{\overset{\text{O}}{\|}}{\text{C}}-S-\text{CoA} + \text{FAD} \overset{(2)}{\rightleftharpoons}
$$

fatty acyl-CoA

$$
R-\text{CH}_2-\text{CH}=\text{CH}-\overset{\overset{\text{O}}{\|}}{\text{C}}-S-\text{CoA} + \text{FADH}_2
$$

α,β-unsaturated acyl-CoA

Hydration is the next reaction. Water is added to the double bond to form β-hydroxyacyl-CoA in a reaction catalyzed by enoyl-CoA hydratase (3).

$$
R-\text{CH}_2-\text{CH}=\text{CH}-\overset{\overset{\text{O}}{\|}}{\text{C}}-S-\text{CoA} + \text{H}_2\text{O} \overset{(3)}{\rightleftharpoons} R-\text{CH}_2-\overset{\overset{\text{OH}}{|}}{\text{CH}}-\text{CH}_2-\overset{\overset{\text{O}}{\|}}{\text{C}}-S-\text{CoA}
$$

α,β-unsaturated acyl-CoA $\qquad\qquad$ β-hydroxyacyl-CoA

The β-hydroxyacyl-CoA molecule undergoes further dehydrogenation (oxidation) to form a β-ketoacyl-CoA molecule in a reaction catalyzed by β-hydroxyacyl-CoA dehydrogenase (4). In this reaction, the oxidant is NAD$^+$, which is reduced to NADH. Oxidation of NADH back to NAD$^+$ in the respiratory chain forms three ATP molecules for each molecule of NADH oxidized (see Section 25-2.1).

$$R-CH_2-\underset{\underset{\text{OH}}{|}}{CH}-CH_2-\underset{\underset{\text{O}}{\|}}{C}-S-CoA + NAD^+ \underset{}{\overset{(4)}{\rightleftharpoons}}$$

β-hydroxyacyl-CoA

$$R-CH_2-\underset{\underset{\text{O}}{\|}}{C}-CH_2-\underset{\underset{\text{O}}{\|}}{C}-S-CoA + NADH + H^+$$

β-ketoacyl-CoA

The final reaction is a cleavage catalyzed by the enzyme thiolase (5) that forms acetyl-CoA and a fatty acyl-CoA molecule that is two carbon atoms shorter than the original fatty acyl-CoA molecule that entered the β-oxidation sequence. The new, shorter fatty acyl-CoA molecule then reenters the β-oxidation sequence. Eventually, the fatty acid is completely converted into acetyl-CoA molecules.

$$R-CH_2-\underset{\underset{\text{O}}{\|}}{C}-CH_2-\underset{\underset{\text{O}}{\|}}{C}-S-CoA + CoA-SH \overset{(5)}{\longrightarrow}$$

β-ketoacyl-CoA

$$R-CH_2-\underset{\underset{\text{O}}{\|}}{C}-S-CoA + CH_3-\underset{\underset{\text{O}}{\|}}{C}-S-CoA$$

fatty acyl-CoA   acetyl-CoA

Conversion of a molecule of palmitic acid (the saturated fatty acid with 16 carbon atoms) into 8 acetyl-CoA molecules requires 7 cleavages or passages through the β-oxidation sequence. These cleavages are indicated in Figure 27.3. In fact, for any fatty acid with an even number of carbon atoms, the number of cleavages is the number of acetyl-CoA molecules formed minus one, because the last cleavage forms two acetyl-CoA molecules.

A single cleavage here after β-oxidation forms two acetyl-CoA molecules.

$$CH_3-CH_2-CH_2-\underset{\underset{\text{O}}{\|}}{C}-SCoA + FAD + NAD^+ + CoA-SH \longrightarrow$$

$$CH_3-\underset{\underset{\text{O}}{\|}}{C}-SCoA + CH_3-\underset{\underset{\text{O}}{\|}}{C}-SCoA + FADH_2 + NADH + H^+$$

Each of the acetyl coenzyme A molecules formed from a fatty acid molecule can then enter the TCA cycle. Through the combination of oxidation of the acetyl-CoA molecules to CO_2 and H_2O in the TCA cycle, oxidation in the respiratory chain of the reduced coenzyme molecules formed by oxidation of the acetyl-CoA

Figure 27.3   Conversion of a molecule of palmitic acid into eight acetyl-CoA molecules. The acetyl-CoA molecules are released sequentially, beginning at the carboxyl end. Ac-CoA = $CH_3 - \overset{\overset{O}{\|}}{C} - S - CoA$.

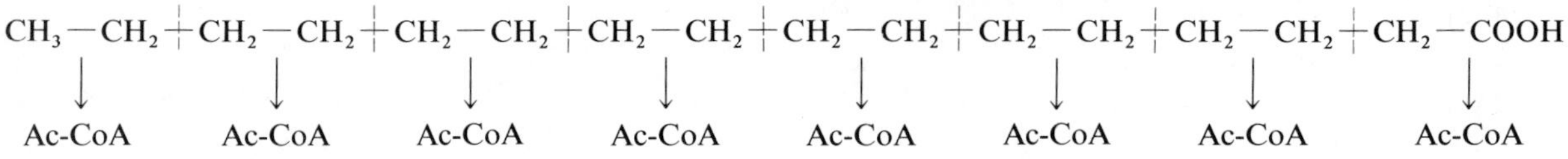

$$CH_3-CH_2 + CH_2-CH_2 + CH_2-CH_2 + CH_2-CH_2 + CH_2-CH_2 + CH_2-CH_2 + CH_2-CH_2 + CH_2-COOH$$

| Ac-CoA | Ac-CoA | Ac-CoA | Ac-CoA | Ac-CoA | Ac-CoA | Ac-CoA | Ac-CoA |

molecules, and oxidative phosphorylation (see Chapter 25), 12 ATP molecules are formed from each acetyl-CoA molecule. Therefore, the total number of high-energy phosphate bonds (ATP molecules) formed by oxidation of one saturated fatty acid molecule is determined as follows.

1. Multiply the number of acetyl-CoA molecules formed (the number of carbon atoms in the fatty acid ÷ 2) by 12.
2. Add 5 ATP molecules (3 from reoxidation of NADH and 2 from reoxidation of $FADH_2$) for each sequence of β-oxidation required to convert the fatty acid molecule completely into acetyl-CoA molecules. The number of β-oxidation sequences is (the number of carbon atoms in the fatty acid ÷ 2) − 1.
3. Subtract the two high-energy phosphate bonds of the ATP molecule used in the initial activation of the fatty acid molecule. (Two high-energy phosphate bonds are lost because AMP is a reaction product; see Figure 27.4).

Figure 27.4   Formation of ATP molecules by catabolism of a fatty acid. Numbers shown in color relate to the complete oxidation of palmitic acid.

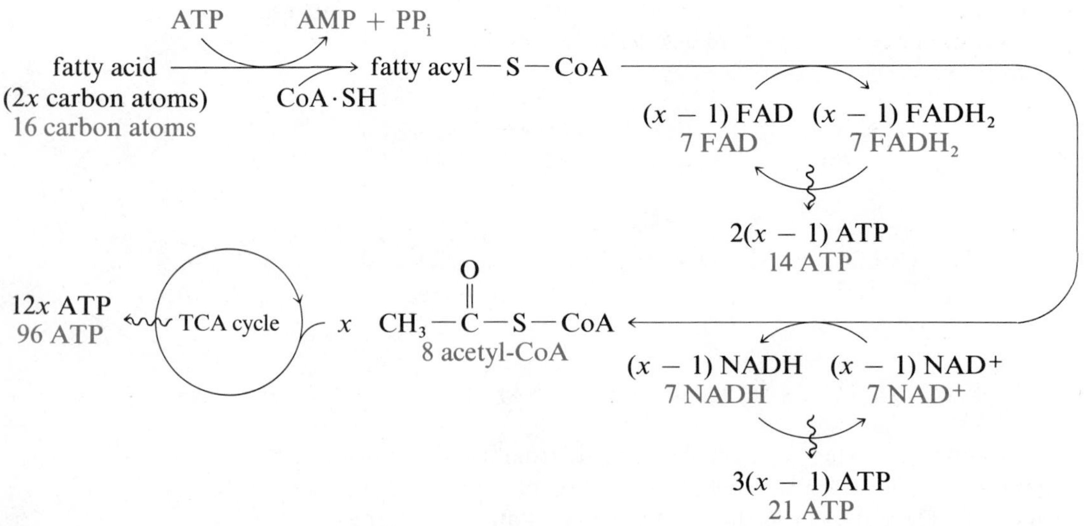

Table 27.1 High-Energy Phosphate Bonds (ATP Molecules) Formed by Complete Oxidation of Palmitic Acid

Source	ATP Molecules
Used in initial activation of palmitic acid	−2
From oxidation of acetyl-CoA molecules in the TCA cycle ($\frac{16}{2}$ Ac-CoA $\times$ 12 = 96)	96
From oxidation of FADH$_2$ ([$\frac{16}{2}$ − 1] $\times$ 2 = 14)	14
From oxidation of NADH [($\frac{16}{2}$ − 1) $\times$ 3 = 21]	21
	129

This stoichiometry is given in Table 27.1 for palmitic acid, the 16-carbon, saturated fatty acid.

The interaction of glycolysis, the anabolism and the catabolism of fatty acids, and the TCA cycle is given in Figure 27.5. The glycerol formed by lipolysis can enter glycolysis as dihydroxyacetone phosphate after phosphorylation and oxidation.

Insulin affects the total energy supply because it affects lipid metabolism as well as carbohydrate metabolism. Insulin increases fat storage in adipose tissue by stimulating lipogenesis and by inhibiting lipolysis.

$$\text{triglyceride} \underset{\text{lipogenesis}}{\overset{\text{lipolysis}}{\rightleftarrows}} \text{fatty acids} + \text{glycerol}$$

$$\uparrow \text{lipogenesis}$$

$$\text{acetyl-CoA}$$

Figure 27.5 The sources of high-energy phosphate bonds during strenuous exercise.

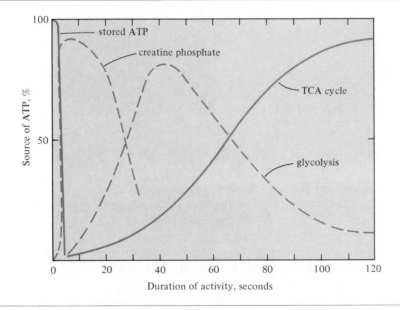

27-2.2 ENERGY METABOLISM IN SKELETAL MUSCLE

Hydrolysis of ATP into ADP and an inorganic phosphate ion provides the energy for the contraction of a muscle fiber. When working maximally, mammalian muscle uses approximately one millimole of ATP per gram per minute. There are, however, only about five micromoles of ATP present per gram of resting muscle. This amount of ATP can meet the demands for energy for less than one-half second of intense activity. As a result, the rate at which a muscle can work depends upon the rate at which the ADP produced in the work can be converted back into ATP. There are three ways that muscle forms ATP from ADP. In the order in which they are employed, they are as follows.

1. by phosphate transfer from creatine phosphate (enough for only a few seconds of work, certainly less than a minute)
2. through the action of myokinase
3. through the oxidation of substrates
   (a) carbohydrate (primarily glycogen)
      (i) **anaerobically** (by glycolysis alone)
      (ii) **aerobically** (by glycolysis coupled with the TCA cycle)
   (b) lipids (oxidation of fatty acids)

Anaerobic reactions and pathways are biological reactions and pathways that may proceed in the absence of oxygen.

Aerobic reactions and pathways are biological reactions and pathways that require oxygen.

Phosphate Transfer from Creatine Phosphate

Stored ATP is the most immediate source of high-energy phosphate bonds. Creatine phosphate is the next most immediate source. Skeletal muscles contain the enzyme creatine kinase that catalyzes the following reversible reaction.

$$H_2N-\underset{\underset{+NH_2}{\overset{\|}{C}}}{\overset{\overset{CH_3}{|}}{}}-N-CH_2-COO^- + ATP \xrightleftharpoons{\text{creatine kinase}}$$

creatine

$$^-O-\underset{\underset{O^-}{\overset{|}{}}}{\overset{\overset{O}{\|}}{P}}-NH-\underset{\underset{+NH_2}{\overset{\|}{C}}}{\overset{\overset{CH_3}{|}}{}}-N-CH_2-COO^- + ADP$$

creatine phosphate

Creatine kinase (CK), also called creatine phosphokinase (CPK), occurs in three forms (see Section 23-5), designated MM, MB, and BB on the basis of the subunits contained in them. A determination of the amounts of these isoenzymes in the blood is used in the diagnosis of disease. The only human tissue containing an appreciable amount of the MB isoenzyme is heart muscle. Thus, an increase in the MB-CK activity in the blood plasma is a sensitive and specific indicator of myocardial injury (heart muscle damage).

The skeletal muscle isoenzyme is almost exclusively the MM form. In **hypothy-roidism** and skeletal muscle diseases, such as muscular dystrophy, serum creatine phosphokinase activity increases and the isoenzyme found is almost entirely of the MM form.

Hypothyroidism is a lower than normal activity of the thyroid gland.

An interesting disorder in pigs can also be followed using a CPK assay. Three to five percent of swine on their way to the market die of a condition called porcine stress syndrome. This condition results in a large annual loss to producers. Porcine stress syndrome causes swine to die under stressful situations such as transportation, fighting, mixing, weaning, and castration. Serum CPK levels are higher in affected animals following stress than they are in resistant animals following stress. An assay for serum CPK following stress allows swine breeders to identify susceptible and resistant pigs. Thus, the assay provides a means of improving their herds.

Creatine can be phosphorylated in a reversible reaction using ATP as the phosphate donor as shown. The high-energy phosphate group can be transferred back to ADP to make ATP, which can be used in muscle contraction. This energy source is important because, while there is a relatively high content of ATP in skeletal muscles, the supply is limited. For sustained muscular activity, a rapid resynthesis of ATP is necessary.

Muscle cells store creatine phosphate because they have an immediate need for high-energy phosphate bonds. When muscles contract quickly, high-energy phosphate is needed more rapidly than it can be provided by the catabolism of glycogen and fatty acids. If ATP itself were stored, the increased concentration of ATP would affect the rate of all the processes utilizing ATP and upset metabolism in general. Therefore, muscle cells use another compound as a high-energy phosphate source. In human muscle, this compound is creatine phosphate, which can quickly provide a high-energy phosphate bond for the synthesis of ATP.

Although both ATP and creatine phosphate are stored in skeletal muscles, their importance as an energy source is due primarily to a high rate of use and replacement, and the available energy from them at any instant is quite limited. During a single day, a human may expend 2000–5000 kcal of energy. During maximal exercise, the energy demand may exceed 50 kcal/min. Under these conditions, the total stored ATP would cover the effort of less than one-half second and regeneration of ATP from stored creatine phosphate would cover only a few more seconds of work. Therefore, maximum effort can be maintained for only a few seconds without additional sources of energy.

Action of Myokinase

Skeletal muscle cells contain the enzyme myokinase. This enzyme catalyzes the transfer of a high-energy phosphate group from one ADP molecule to another. During this process, a molecule of AMP is formed from the ADP molecule losing the high-energy phosphate bond and a molecule of ATP is formed from the ADP molecule gaining the high-energy phosphate bond.

$$2 \text{ ADP} \xrightarrow{\text{myokinase}} \text{ATP} + \text{AMP}$$

As a result of the action of myokinase, both of the high-energy bonds of ATP can be used for muscle contraction.

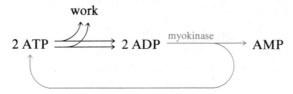

This coupling can be demonstrated in another way.

Reaction catalyzed by myokinase: $ADP + ADP \longrightarrow ATP + AMP$

In muscle contraction: $ATP \longrightarrow ADP + P_i$

Sum: $ADP \longrightarrow AMP + P_i$

The overall result of these coupled reactions is the use of the high-energy bond of ADP in contraction. The reaction catalyzed by myokinase occurs only after the disappearance of creatine phosphate.

Oxidation of Substrates

There are three phases in the provision of energy for the operation of muscle during strenuous exercise (see Table 27.2 and Figure 27.5). During the first phase, lasting only a few seconds, the replenishment of the ATP supply is provided solely by the transfer of the high-energy phosphate group from creatine phosphate to ADP and by the action of myokinase. The second phase involves the conversion of glycogen to lactate (glycolysis). In the third phase, oxygen is used in mitochondria to oxidize pyruvate from glycogen and acetyl-CoA from fatty acids to CO_2 and H_2O, producing ATP in the process.

Stores of ATP and creatine phosphate will sustain intense muscle activity for less than one minute. Then, glycogen catabolism begins to replenish ATP via glycolysis. It takes a while longer for the respiratory system (lungs) and the cardiovascular system (heart and blood vessels) to respond so, initially, the catabolism is partly anaerobic. As the respiratory and cardiovascular systems begin to respond and more oxygen reaches tissues, the percentage of carbohydrate (glucose units of glycogen) that is converted into lactate declines and the percentage of pyruvate that is completely oxidized to carbon dioxide in mitochondria increases. During this third phase, oxidation of fatty acids becomes increasingly important as a source of energy.

Glycogen stored in muscle cells is more readily available for energy production than is glucose from the blood. The content of glycogen in skeletal muscle controls the ability to perform both intense short-term activity and heavy, prolonged work. When the glycogen content approaches zero, the person becomes exhausted.

With a sugar content of approximately 1 g/L, the total glucose content of the blood is 5–6 g. Because as much as 3 g of carbohydrate may be utilized during every minute of heavy exercise, the available glucose can sustain only about 2 minutes of heavy work. If half of the blood sugar were used, leaving 0.5 g/L, severe symptoms of hypoglycemia would develop. However, skeletal muscles continue to use carbohydrate even when blood sugar (glucose) falls to low levels.

Table 27.2 The Three Phases of ATP Production in Skeletal Muscle During Strenuous Exercise

PHASE 1

Phosphate Transfer from Creatine Phosphate (CP)	Myokinase
$CP + ADP \rightleftharpoons Creatine + ATP$	After disappearance of CP, the high-energy bond of ADP is used in muscle contraction.
Only small amounts of creatine phosphate are stored in muscle cells, so other sources of energy are required.	Myokinase catalyzes the reaction $ADP + ADP \rightarrow AMP + ATP$
	For prolonged muscular activity, substrate oxidation is required.

PHASE 2 (Anaerobic)

Carbohydrate Catabolism

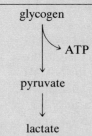

glycogen

→ ATP

pyruvate

lactate

This process occurs until the heart and lungs respond allowing catabolism to be aerobic.

Anaerobic catabolism is inefficient.

PHASE 3 (Aerobic)

Carbohydrate Catabolism	Lipid Catabolism
glycogen	fatty acid
↓	↓ β-oxidation
pyruvate	acetyl-CoA
↓	↓
acetyl-CoA	TCA cycle → CO_2, H_2O
↓	→ ATP
TCA cycle → CO_2, H_2O → ATP	

Then, the oxidation of glycogen declines, and the oxidation of fatty acids increases.

Fat is readily available and is an important energy source for working muscle. Fats stored in adipose tissues are broken down by a lipase to free fatty acids (FFA). These free fatty acids are then transported in the blood, primarily as a complex with albumin, as mentioned in Section 27-1.

At the beginning of exercise, there is a drop in the FFA of plasma due to an increased uptake in active muscles. After some time, there is a rise in the plasma FFA concentration if the work is not extremely heavy. This rise is due to an enhanced mobilization of FFA from adipose tissues.

We learned in Chapters 24 and 25 that glucose can be oxidized to pyruvate in the glycolytic pathway without the use of oxygen and this oxidation produces small amounts of ATP. However, oxygen is required for the oxidation of pyruvate and fatty acids in mitochondria, where most of the ATP is produced. It is the availability of oxygen in the cell that determines the extent to which the various metabolic processes producing ATP are used. Normally the oxygen supply to muscle cells is sufficient to reoxidize the NADH formed in glycolysis as quickly as it is formed and to oxidize pyruvate in mitochondria as quickly as it is formed.

As the severity of the exercise or work increases, the breakdown of glycogen and the catabolism of glucose speeds up. As a result, the reduction of NAD^+ is speeded up. Under these conditions, sufficient oxygen may not be supplied to the muscle cells to oxidize NADH back to NAD^+. If this is the case, some of the NADH must be oxidized to NAD^+ in another way because, if NAD^+ is not available, the catabolism of carbohydrate stops. In muscle cells, NADH is oxidized to NAD^+ during the reduction of pyruvate to lactate (see Section 24-2.2). The reformed NAD^+ can participate in glycolysis. Thus, more ATP can be formed, even when the supply of oxygen is insufficient. The lactate so produced enters the blood stream. It undergoes gluconeogenesis to form glucose (to complete the Cori cycle; see Section 24-2.3) in the liver, or it is completely oxidized to CO_2 and H_2O in heart muscle.

We learned in Section 24-2.2 that, without oxygen, glycogen can be catabolized as follows.

$$\text{glucose unit of glycogen} + 3\ \text{ADP} + 3\ P_i \longrightarrow 2\ \text{lactate} + 3\ \text{ATP}$$
$$(C_6H_{10}O_5)$$

When oxygen is present, glycolysis is linked to mitochondrial processes (see Section 25-3.3) and the complete equation becomes

$$\text{glucose unit of glycogen} + 36\ \text{ADP} + 36\ P_i + 6\ O_2 \longrightarrow 6\ CO_2 + 5\ H_2O + 36\ \text{ATP*}$$

It is clear that glycogen provides about 12 times more energy aerobically than anaerobically (see Table 27.3).

During moderate exercise, muscles do not produce lactate. Lactate is produced only when there is insufficient oxygen, because of a limit to the rate of delivery of oxygenated blood, to (1) allow all the pyruvate to be oxidized in the TCA cycle and (2) reoxidize the NADH produced in glycolysis.

During very strenuous exercise, the production of lactate begins after about 15 seconds and reaches its maximum rate in about 35 seconds. The concentration of

---

*See footnote, p. 648.

Table 27.3 Moles of ATP Formed from Each Gram of Carbohydrate and Fatty Acid Oxidized and from Each Mole of Oxygen Used

Fatty Acid (Palmitic Acid)

$$CH_3-(CH_2)_{14}-COOH + 129\ ADP + 129\ P_i + 23\ O_2 \longrightarrow$$
$$16\ CO_2 + 16\ H_2O + 129\ ATP$$

$$\text{moles of ATP/gram of palmitic acid} = \frac{129}{256} = 0.50$$

$$\text{moles of ATP/mole of } O_2 = \frac{129}{23} = 5.61$$

Carbohydrate (Glucosyl Residue of Glycogen)

$$C_6H_{10}O_5 + 6\ O_2 \longrightarrow 6\ CO_2 + 5\ H_2O + 36\ ATP$$

$$\text{moles of ATP/gram of glucosyl residue} = \frac{36}{162} = 0.22$$

$$\text{moles of ATP/mole of } O_2 = \frac{36}{6} = 6.0$$

Carbohydrate can also produce ATP anaerobically
$$C_6H_{10}O_5 \longrightarrow 2\ \text{lactate} + 3\ ATP$$

$$\text{moles of ATP/gram of glucosyl unit} = \frac{3}{162} = 0.019$$

lactate then increases linearly with the duration of the exercise. The formation of lactate begins because the muscle has been called upon to contract more rapidly and strongly than can be sustained by the ATP generated through oxidative phosphorylation, which requires oxygen. Skeletal muscle cells have a greater concentration of the enzymes of glycolysis (in the soluble cytoplasm) than of the enzymes of mitochondria. However, muscles are aerobic organs, and lactate production is only a small part of muscle metabolism.

The third phase of ATP generation in active muscles is the highly aerobic phase. In this phase, about 90% of the ATP is supplied initially by oxidation of glycogen. However, the oxidation of glycogen steadily declines, and the oxidation of fat increases until it accounts for 80% or more of the total ATP produced. This shift accounts for the increased oxygen consumption because fatty acids can only be oxidized aerobically (via beta-oxidation and the TCA cycle) and because oxidation of fatty acids inherently requires more oxygen. (The consumption of oxygen may increase sixteen-fold or greater in the first five minutes of intense activity.)

The equations in Table 27.3 indicate that, when carbohydrate is oxidized, each mole of oxygen results in the formation of about 6.2 mol of ATP, but that only 5.7 mol of ATP are formed for each mole of oxygen used in the oxidation of fat. Therefore, when the oxygen supply to a muscle becomes limited during heavy work or exercise, glycogen contributes relatively more to the energy yield than does fat. This allows the most effective utilization of oxygen.

During light or moderate work or exercise, energy is supplied by fat and carbohydrate (mainly glycogen) in about equal amounts. As the muscular work increases in intensity so that catabolism becomes more and more aerobic, fat is used in increasing amounts as an energy source. Fat probably plays a somewhat greater role in the metabolism of trained athletes because of their greater aerobic power. So, the more inadequate the oxygen supply, the greater the utilization of glycogen. Assuming 13 kcal of usable chemical energy for muscle contraction per mole of ATP, in the catabolism of glycogen to carbon dioxide and water, 1 mol of oxygen (22.4 L) provides about 78 kcal of usable energy as ATP (see Table 27.1). In the complete catabolism of a fatty acid (for example, palmitic acid), 1 mol of oxygen provides about 73 kcal of usable energy as ATP. In other words, the energy yield in terms of ATP production per liter of oxygen used when carbohydrate is catabolized is about 7% higher than when fat is catabolized.

Factors that affect the participation of fat versus carbohydrates in muscle metabolism are as follows.

1. intensity of work in relation to the person's total work capacity (physical conditioning) or, stated another way, the person's ability to accomplish the work or exercise under aerobic conditions
2. duration of the work or exercise
3. diet

A low-carbohydrate diet tends to favor the utilization of fat as an energy source, but will reduce the capacity for prolonged heavy work. A carbohydrate-rich diet enhances the storage of glycogen, which improves endurance.

PROBLEMS

6. Circle the β-carbon atom in the following molecule.

$$CH_3-CH_2-CH_2-CH_2-CH_2-CH_2-CH_2-CH_2-CH_2-COOH$$

7. Calculate the number of ATP molecules formed by complete oxidation of the molecule in problem 6.
8. Without the use of equations, describe the process of β-oxidation of fatty acids.
9. Describe the three phases of the production of ATP during strenuous exercise.
10. Describe the fates of the NADH and $FADH_2$ produced by the β-oxidation of fatty acids.

ANSWERS TO PROBLEMS
6. $CH_3-CH_2-CH_2-CH_2-CH_2-CH_2-CH_2-\textcircled{C}H_2-CH_2-COOH$
7. 78 **8.** In β-oxidation, the β-carbon atoms of fatty acid molecules are oxidized to carbonyl groups by the following process. The fatty acid is converted to a coenzyme A thioester. Then, the CoA ester is dehydrogenated to an α,β-unsaturated compound. Next, a water molecule is added to form a β-hydroxy compound. This reaction is followed by a second dehydrogenation, by which a β-keto compound is formed. Finally, the oxidized molecule is cleaved to form acetyl-CoA and a fatty acyl-CoA ester that has two fewer carbon atoms. **9.** Phase 1. ATP is formed by transfer of a high-energy phosphate group from creatine phosphate to ADP and by the action of myokinase. Phase 2. ATP is formed without sufficient oxygen. In this phase, lactate is formed. Phase 3. ATP is formed in mitochondrial reactions by oxidation of acetyl-CoA from fatty acids and carbohydrate in the presence of oxygen. **10.** These reduced coenzymes are reoxidized by transferring electrons and protons to oxygen molecules in the respiratory chain that, when coupled with oxidative phosphorylation, produces ATP.

27-3 BIOSYNTHESIS OF FATTY ACIDS AND KETONE BODIES

LEARNING OBJECTIVES

1. Define and identify structures of ketone bodies.

2. Describe the function of biotin, the HMP pathway, NADPH, acetyl-CoA, and malonyl-CoA in fatty acid synthesis.

3. Given the substrates, the coenzymes, and the enzyme required for any of the reactions of fatty acid biosynthesis, write a balanced equation showing the products of the reaction.

4. Explain why most fatty acids have an even number of carbon atoms.

5. Describe how ingested carbohydrate is converted into stored fat.

6. Explain why acetyl-CoA is a central intermediate in carbohydrate and fatty acid metabolism.

7. Outline the hexose monophosphate pathway showing the oxidative and regenerative phases as in Figure 27.9.

8. Describe or diagram the utilization of β-hydroxybutyrate and acetoacetate as sources of energy.

As we see in Figure 27.6, fatty acids are not only broken down to acetyl-CoA molecules, but also synthesized from acetyl-CoA molecules. This synthesis of fatty acid molecules occurs in the liver and can occur at the same time that fatty acids are being broken down because synthesis occurs by a slightly different pathway that we will discuss in the next section. This simultaneous catabolism and anabolism of fatty acids is necessary to convert the animal fats and vegetable oils ingested in our diets into human fat. It is because fatty acids are made from the two-carbon acetyl group of acetyl-coenzyme A that they have an even number of carbon atoms.

27-3.1 BIOSYNTHESIS OF FATTY ACIDS

Fatty acids are synthesized from acetyl-CoA. This reaction occurs in the soluble cytoplasm (*cytosol*) of a cell. However, the acetyl-CoA molecules are generated within mitochondria. Therefore, a transport mechanism is necessary to get acetyl-CoA molecules out of the mitochondria into the cytosol.

Acetyl-CoA leaves mitochondria in the form of citrate, the initial intermediate of the TCA cycle. In mitochondria, acetyl-CoA combines with oxaloacetate to form citrate (see Section 25-1.2). In the cytosol, the reaction is reversed in an energy-requiring reaction catalyzed by the enzyme citrate lyase.

$$H_2O + \begin{array}{c} COO^- \\ | \\ C{=}O \\ | \\ CH_2 \\ | \\ COO^- \end{array} + CH_3 - \overset{\overset{\displaystyle O}{\|}}{C} - S - CoA \longrightarrow HO - \overset{\overset{\displaystyle COO^-}{|}}{\underset{\underset{\displaystyle COO^-}{|}}{\overset{\overset{\displaystyle CH_2}{|}}{\underset{\underset{\displaystyle CH_2}{|}}{C}}}} - COO^- + CoA - SH$$

oxaloacetate   acetyl-CoA   citrate

Figure 27.6 The interrelationships of glycolysis, the anabolism and catabolism of fatty acids, and the TCA cycle.

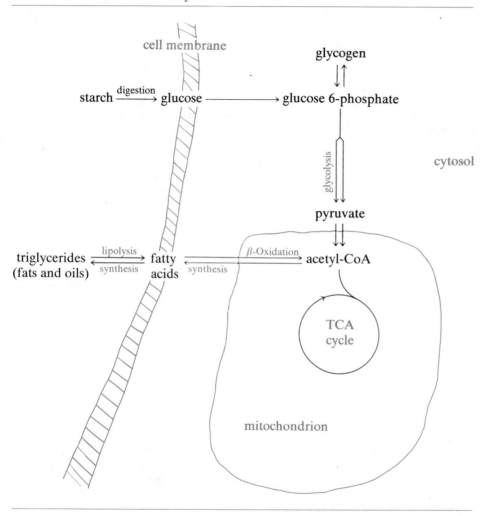

$$\text{citrate} + \text{CoA}\!-\!\text{SH} \xrightarrow[\text{citrate lyase}]{\overset{\displaystyle \text{ATP} \quad \text{ADP} + \text{P}_i}{\curvearrowright}} \text{oxaloacetate} + \text{acetyl}\!-\!\text{CoA}$$

One acetyl-coenzyme A molecule enters the synthetic pathway directly. The remaining acetyl-CoA molecules must first be converted to malonyl-coenzyme A molecules before they can be used for the synthesis of a fatty acid. This reaction is catalyzed by the acetyl-CoA carboxylase (6, see Figure 27.7) enzyme complex, an enzyme complex that contains the B vitamin biotin as a prosthetic group and utilizes bicarbonate ion (from CO_2). The formation of malonyl-CoA from acetyl-

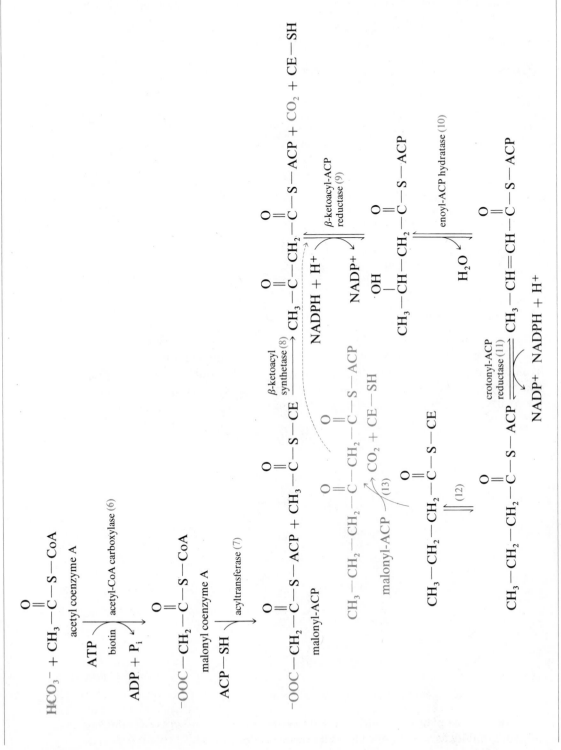

Figure 27.7 Details of the fatty acid biosynthesis pathway. ACP—SH = acyl carrier protein, CE—SH = condensing enzyme, and $HCO_3^- = CO_2 + OH^- \rightleftharpoons HCO_3^-$.

CoA activates the methyl group of acetyl-CoA so that condensation of units can occur.

$$CH_3-\overset{\overset{\textstyle O}{\|}}{C}-S-CoA + HCO_3^- \xrightarrow[\quad(6)\quad]{\quad ATP \quad ADP + P_i \quad} {}^-O-\overset{\overset{\textstyle O}{\|}}{C}-CH_2-\overset{\overset{\textstyle O}{\|}}{C}-S-CoA$$

acetyl-CoA   malonyl-CoA

biotin

Biotin is covalently attached by means of an amide bond to a lysine residue of one of the enzymes of the acetyl-CoA carboxylase enzyme complex. The bicarbonate ion reacts with the biotin residue to form a carboxybiotin residue containing a carboxyl group (shown in color) that can be transferred to other molecules. Biotin is a coenzyme for several carboxylation reactions of this type.

carboxybiotin attached to a lysine residue of acetyl-CoA carboxylase

In bacteria and plants, a soluble protein called acyl carrier protein (ACP) holds the growing fatty acyl chain during synthesis of a fatty acid. This protein contains a pantetheinyl group identical to that in coenzyme A (see Section 25-1.1). In addition to ACP, six enzymes are needed for fatty acid synthesis. The acyl carrier protein and the six enzymes form the fatty acid synthetase complex. In mammals, the fatty acid synthetase multienzyme complex cannot be separated into individual components without loss of activity, so less is known about it. However, there is reason to believe that the system in humans is very similar to that in bacteria and plants (see Figures 27.7 and 27.8).

In the first step, the acetyl group of acetyl-CoA is transferred to the thiol group of the pantetheinyl group of ACP in a reaction catalyzed by acetyl-CoA:ACP transacylase (also called acyltransferase).

$$\text{acetyl} - \text{S} - \text{CoA} + \text{ACP} - \text{SH} \rightleftharpoons \text{acetyl} - \text{S} - \text{ACP} + \text{CoA} - \text{SH}$$

Next, the acetyl group is temporarily transferred to a thiol group of another enzyme in the fatty acid synthetase complex called the condensing enzyme (CE—SH).

$$\text{acetyl} - \text{S} - \text{ACP} + \text{CE} - \text{SH} \rightleftharpoons \text{acetyl} - \text{S} - \text{CE} + \text{ACP} - \text{SH}$$

This reaction frees the thiol group of ACP to accept the next incoming molecule, malonyl-CoA. The addition of the malonyl group to the thiol group of ACP is catalyzed by malonyl CoA:ACP transacylase (another acyltransferase) (7).

$$\text{malonyl} - \text{S} - \text{CoA} + \text{ACP} - \text{SH} \stackrel{(7)}{\rightleftharpoons} \text{malonyl} - \text{S} - \text{ACP} + \text{CoA} - \text{SH}$$

Figure 27.8 Summary of the biosynthesis of the eight-carbon fatty acid octanoic acid.

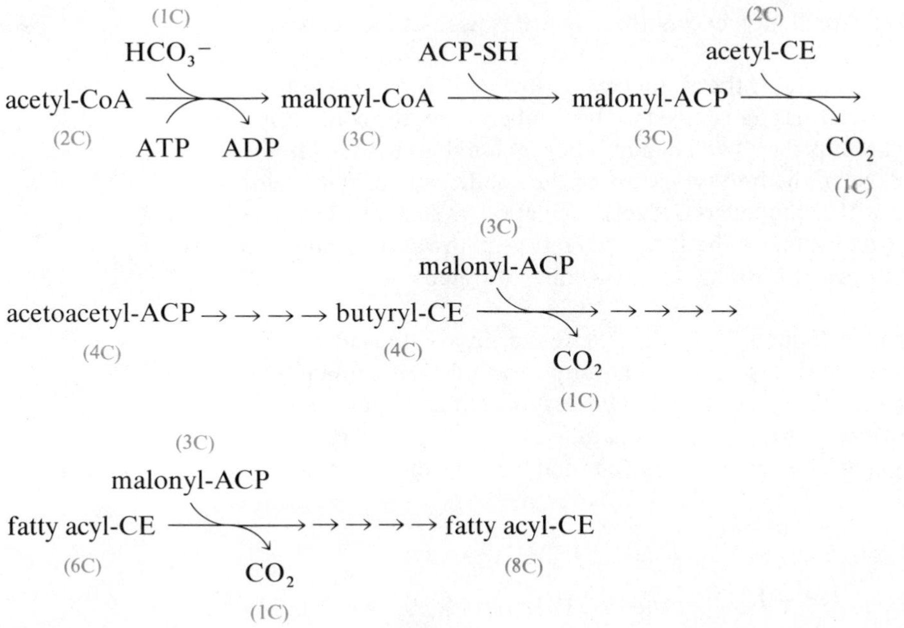

When the enzyme complex contains both an acetyl group (on the condensing enzyme) and a malonyl group (on the acyl carrier protein), the acetyl group is transferred to the malonyl residue in a reaction catalyzed by the condensing enzyme, β-ketoacyl synthetase (8). In this reaction, the free carboxyl group of the malonyl residue is released as CO_2 (HCO_3^-). It is this decarboxylation that provides the energy for the reaction. This energy originally came from an ATP molecule during the formation of malonyl-CoA in the reaction catalyzed by acetyl-CoA carboxylase.

$$CH_3-\overset{\overset{\displaystyle O}{\|}}{C}-S-CE$$

acetyl-CE

CE—SH

(8)

$$CH_3-\overset{\overset{\displaystyle O}{\|}}{C}-CH_2-\overset{\overset{\displaystyle O}{\|}}{C}-S-ACP$$

acetoacetyl-ACP

CO_2

$$^-O-\overset{\overset{\displaystyle O}{\|}}{C}-CH_2-\overset{\overset{\displaystyle O}{\|}}{C}-S-ACP$$

malonyl-ACP

In the next step, acetoacetyl-ACP is reduced by NADPH (see Section 24-2.1) to β-hydroxybutyryl-ACP in a reaction catalyzed by β-ketoacyl-ACP reductase (9). β-Hydroxybutyryl-ACP is dehydrated to an α,β-unsaturated thioester in a reaction catalyzed by enoyl-ACP hydratase (10). Finally, the α,β-unsaturated thioester is reduced to butyryl-ACP in a reaction utilizing NADPH as the reductant and catalyzed by crotonyl-ACP reductase (11). These three reactions are essentially a reverse of the reactions of β-oxidation except that NADP is used as the coenzyme in place of NAD and FAD.

Following the second reduction, the sequence is repeated. First the butyryl group (a four-carbon acyl group) is transferred to the thiol group of the condensing enzyme where the acetyl group was (12). Then another malonyl group adds to the free thiol group of ACP. Next, the butyryl group on the condensing enzyme adds to the malonyl residue with simultaneous decarboxylation as before (13). The product now has six carbon atoms. Then, the reduction pathway utilizing two NADPH molecules takes place, forming a six-carbon, saturated fatty acyl-ACP intermediate.

This pathway, depicted in Figures 27.7 and 27.8, results in a fatty acid with an even number of carbon atoms because the acetyl group of acetyl-CoA contains two carbon atoms and each malonyl residue adds two carbon atoms to the growing carbon chain. Synthesis usually stops after the chain reaches a length of 16 carbon atoms. In mammalian systems, free palmitate is liberated from the enzyme complex by hydrolysis.

$$CH_3-(CH_2)_{14}-\overset{\overset{\displaystyle O}{\|}}{C}-S-ACP + H_2O \longrightarrow CH_3-(CH_2)_{14}-COO^- + ACP-SH$$

palmityl acyl carrier protein   palmitate   acyl carrier
                                                                                     protein

In mitochondria, additional two-carbon units can be added to the carboxyl end of a fatty acid. In this chain-elongation pathway, acetyl-CoA is added to a fatty acyl-CoA molecule. Reduction of the β-keto acid so formed requires both NADPH and NADH.

Double bonds can be introduced by an enzyme system requiring O_2 and NAD. The human enzyme system utilizes fatty acids containing 16 or more carbon atoms and inserts the double bond in position 9,10. Thus, palmitic acid (C_{16}) can be converted into palmitoleic acid ($C_{16}{}^{\Delta 9}$) and oleic acid ($C_{18}{}^{\Delta 9}$) can be made from stearic acid (C_{18}). However, human cells cannot introduce a second double bond between the position 9,10 and the end of the molecule that does not have the carboxyl group. Therefore, human cells cannot make linoleic and linolenic acid. As a result, these polyunsaturated fatty acids are essential fatty acids required in our diets (see Section 26-1.1).

27-3.2 THE HEXOSE MONOPHOSPHATE PATHWAY

The overall reaction of the hexose monophosphate (HMP) pathway is as follows.

D-glucose 6-phosphate + 12 NADP$^+$ $\longrightarrow$
$$6\,CO_2 + 12\,NADPH + 12\,H^+ + HPO_4{}^{2-}$$

The primary function of the HMP pathway is to produce the reduced coenzyme NADPH. NADPH is the coenzyme for anabolic (synthetic) reactions. Most of the reducing power in the form of NADPH for the biosynthesis of fatty acids and steroid hormones is produced by the hexose monophosphate pathway. The pathway is most active in those tissues in which there is greatest fatty acid synthesis, that is, tissues with the greatest demand for NADPH. Examples of such tissues are adipose tissues and lactating mammary glands. In these tissues, the HMP pathway may be more significant for the oxidation of glucose than the glycolytic pathway and the TCA cycle.

A secondary function of the HMP pathway is the production of D-ribose 5-phosphate for the synthesis of nucleotides. The overall pathway is given in Figure 27.9. There are two oxidative steps that produce all the NADPH. The pathway can be balanced most easily by starting with 6 moles of D-glucose 6-phosphate. Thus, in the two oxidative steps, 6 moles of hexose phosphate ($6 \times C_6 = 36$ carbon atoms) are converted into 6 moles of pentose phosphate ($6 \times C_5 = 30$ carbon atoms) and 6 moles of carbon dioxide. In the regenerative phase, the 6 moles of pentose phosphate (30 carbon atoms) are converted into 5 moles of hexose phosphate (30 carbon atoms). Thus, the net reaction is the conversion of 1 mole of hexose phosphate into 6 moles of CO_2.

27-3.3 KETONE BODIES

Under certain metabolic conditions associated with a high rate of β-oxidation of fatty acids—for example, during starvation and in uncontrolled diabetes mellitus— the liver produces large amounts of *ketone bodies*. The ketone bodies are aceto-

Figure 27.9   The hexose monophosphate pathway showing the two oxidative steps that generate NADPH.

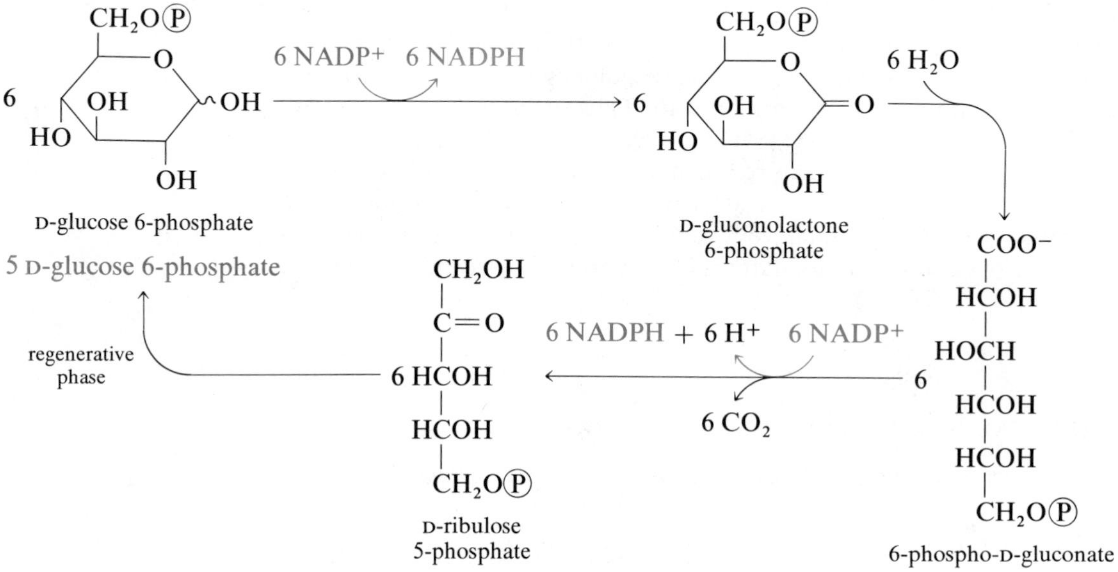

$CH_2O\textcircled{P}$

6 NADP$^+$ 6 NADPH

D-glucose 6-phosphate

D-gluconolactone 6-phosphate

$6 H_2O$

5 D-glucose 6-phosphate

CH_2OH
$C{=}O$
6 HCOH
HCOH
$CH_2O\textcircled{P}$

D-ribulose 5-phosphate

regenerative phase

6 NADPH + 6 H$^+$ 6 NADP$^+$

$6 CO_2$

COO^-
HCOH
HOCH
HCOH
HCOH
$CH_2O\textcircled{P}$

6-phospho-D-gluconate

acetate, β-hydroxybutyrate, and acetone. These three metabolic products are formed from acetyl-CoA by the pathway given in Figure 27.10. Acetoacetate and β-hydroxybutyrate can be used as an energy source by skeletal muscle, heart muscle, and the brain via the pathway given in Figure 27.11.

When the concentration of ketone bodies in the blood becomes too high (*ketonemia*), they are excreted in the urine, a condition known as *ketonuria*. Acetone is also removed through the lungs. When the ketone body concentration in the blood is high, the odor of acetone can be easily detected in the breath. Acetoacetic and β-hydroxybutyric acids are both moderately strong acids and are ionized when present in blood. Their continual excretion is accompanied by the loss of cations, which progressively causes a lowering of the pH of the blood (*ketoacidosis*). This acidosis may be fatal in uncontrolled diabetes mellitus. The overall condition of an abnormally high level of ketone bodies in tissues and fluids is known as *ketosis*.

The normal pH of blood plasma is approximately 7.4 (pH 7.35–7.45 is the normal range). Any drop in pH below 7.35 results in *acidosis*. Any rise in pH above 7.45 results in *alkalosis*. As we learned in Section 8-4.3, acidosis can result from hypoventilation, which results in a build-up of CO_2, and alkalosis can result from abnormal metabolism such as occurs when there is a deficient activity of insulin resulting in decreased utilization of glucose and increased catabolism of fatty acids. When this happens, ketone bodies are produced from the excess acetyl-CoA. As a result, the pH, the concentration of HCO_3^- in the blood, and

Figure 27.10 Synthesis of the three ketone bodies: acetoacetate, β-hydroxybutyrate, and acetone.

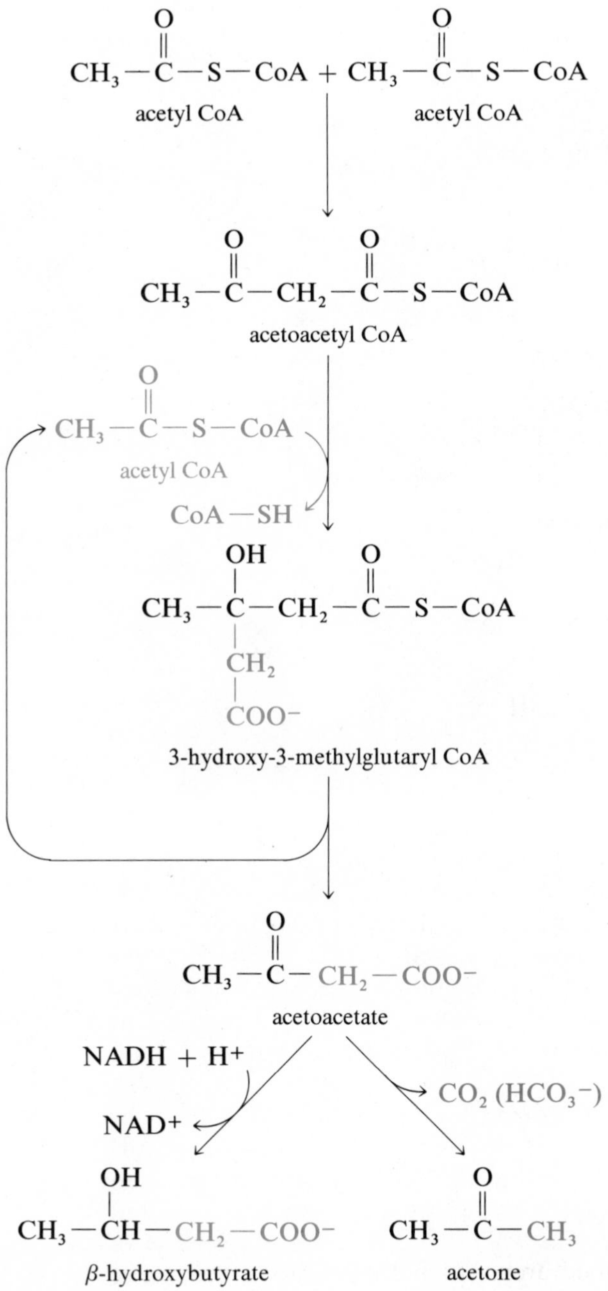

Figure 27.11   Utilization of ketone bodies as an energy source in skeletal and heart muscle. Both β-hydroxybutyrate and acetoacetate may be present in blood. The ATP molecules shown in color are formed by oxidative phosphorylation as reducing equivalents pass through the respiratory chain.

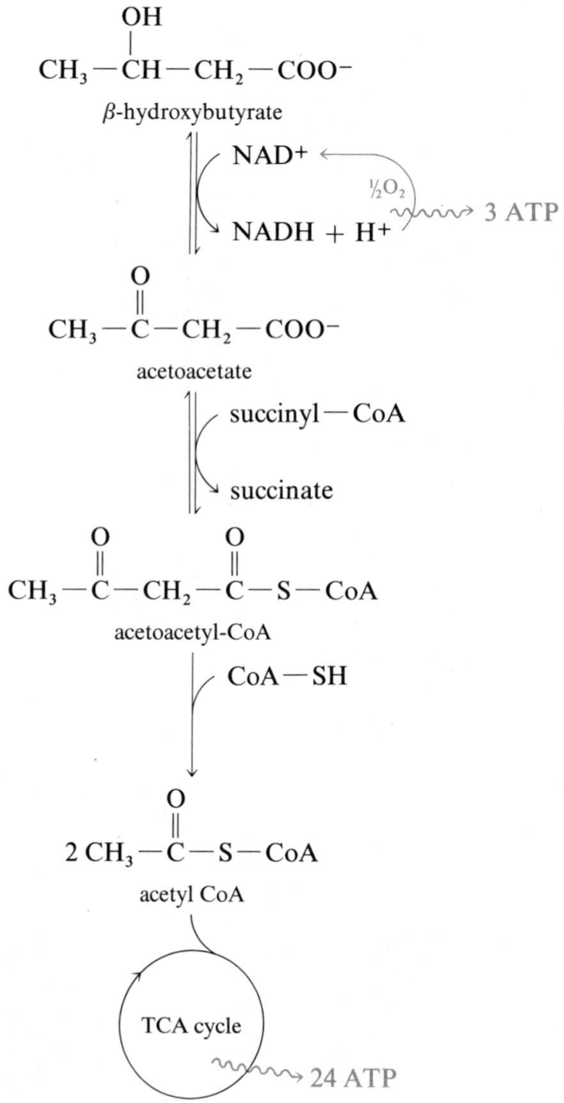

$$OH$$
$$|$$
$$CH_3-CH-CH_2-COO^-$$

β-hydroxybutyrate

$$NAD^+ \leftarrow$$
$$\tfrac{1}{2}O_2$$
$$NADH + H^+ \rightsquigarrow 3\ ATP$$

$$O$$
$$||$$
$$CH_3-C-CH_2-COO^-$$

acetoacetate

$$succinyl-CoA$$
$$succinate$$

$$O \qquad O$$
$$|| \qquad ||$$
$$CH_3-C-CH_2-C-S-CoA$$

acetoacetyl-CoA

$$CoA-SH$$

$$O$$
$$||$$
$$2\ CH_3-C-S-CoA$$

acetyl CoA

TCA cycle

$$\rightsquigarrow 24\ ATP$$

the ratio of HCO_3^- to CO_2 in the blood drop. When the pH drops, the patient characteristically develops rapid and deep breathing as the body tries to raise the pH by getting rid of CO_2. If the pH drops as low as 7.0, a coma may result. Control of blood pH was discussed earlier in Section 8-4.3.

PROBLEMS

11. Explain why most fatty acids have an even number of carbon atoms.
12. Describe how the carbohydrate we eat is converted into fat.
13. Describe the function of biotin.
14. Describe the function of the HMP pathway.

ANSWERS TO PROBLEMS   **11.** Most fatty acids have an even number of carbon atoms because they are made from two-carbon units.   **12.** Ingested carbohydrate, largely in the form of starch, is digested to form D-glucose. Glucose is catabolized to acetyl-CoA through pyruvate formed by glycolysis. Acetyl-coenzyme A is used for the synthesis of fatty acid thioesters, which are then used to make the triglyceride molecules of stored fat.   **13.** Biotin is a coenzyme for carboxylation reactions.   **14.** The HMP pathway makes reducing power in the form of NADPH for anabolic reactions such as the synthesis of fatty acids.

SUMMARY

Most of the lipid in our diets is triglyceride. The triglyceride molecules are acted upon by lipases in the small intestine. These lipases convert the triglyceride molecules into a mixture of fatty acids, monoglyceride, and lesser amounts of diglyceride molecules. However, before the lipases can act effectively, the fat or oil must be emulsified to increase the surface area exposed to the enzyme. Emulsification is effected first by bile salts; products of partial hydrolysis, especially monoglycerides and salts of fatty acids (soaps), also help.

The products of digestion are absorbed and used in the intestinal lining to remake triglyceride molecules. These triglyceride molecules are transported to the liver as chylomicrons. Human fat is synthesized in the liver and transported as lipoprotein particles containing molecules of triglyceride, phosphoglyceride, protein, and cholesterol to adipose tissue, where it is stored. When needed for energy, the triglyceride molecules of adipose tissue are acted upon by lipases, releasing fatty acids that are transported in the blood as complexes with plasma albumin.

Fatty acids are oxidized in mitochondria in a pathway called β-oxidation. In this pathway, the beta carbon atom of an activated fatty acid molecule (a fatty acyl-CoA molecule) is oxidized to a carbonyl group forming a β-ketoacyl-CoA molecule. The two carbon atoms at the carboxyl end are then cleaved off, forming a molecule of acetyl-CoA and a fatty acyl-CoA molecule two carbon atoms shorter than before. The reduced coenzyme molecules formed during β-oxidation enter the respiratory chain, and the acetyl-CoA molecules enter the TCA cycle. The combination of oxidation of acetyl-CoA in the TCA cycle, oxidation of reduced coenzyme molecules in the respiratory chain, and oxidative phosphorylation, all of which occur in the mitochondria in close proximity to the enzymes involved in the pathway of β-oxidation, produces ATP molecules.

Oxidation of fatty acids requires oxygen. Resting muscle cells use fatty acids as an energy source. Working muscle cells in a highly aerobic state also use fatty acids to a large extent, but not exclusively. Fatty acids are preferred as a source of energy because their oxidation yields more ATP per mole and per gram than does oxidation of carbohydrate. However, their oxidation yields less ATP per liter of oxygen required than does oxidation of carbohydrate. Therefore, during early states of strenuous exercise, before the heart and lungs have had a chance to respond, there is insufficient oxygen present and muscle cells get some of the needed ATP by conversion of glycogen to lactate (glycolysis).

Fatty acids are synthesized in the cytoplasm from acetyl-CoA. To get acetyl-CoA outside the mitochondria, citrate is transported to the cytoplasm where it is split into acetyl-CoA and oxaloacetate. Some acetyl-CoA molecules react with CO_2 (HCO_3^-) to form malonyl-CoA. Synthesis of a fatty acid molecule begins by reaction of a malonyl thioester with an acetyl thioester. During this reaction, the free carboxyl group of the malonyl thioester comes off as CO_2 (HCO_3^-) giving a four-carbon β-ketoacyl thioester. Reduction yields butyryl thioester. This process is then repeated in reaction of the butyryl thioester with another malonyl thioester, with the simultaneous loss of CO_2 to add two more carbon atoms to the growing chain. This process continues, adding two carbon units at a time, until a chain length of 16–18 carbon atoms is reached.

The entire synthetic pathway takes place on a multienzyme complex containing a carrier protein with a pantetheinyl group identical to that in coenzyme A. The acetyl group of acetyl-CoA is transferred to the pantetheinyl group of this carrier protein. This acyl group remains attached to the carrier protein in a thioester linkage during its growth, except that it is temporarily transferred to a thiol group of an enzyme while a malonyl group is transferred from malonyl-CoA to the carrier protein. Then reaction takes place. The result of the addition of an acyl group to the malonyl group, accompanied by decarboxylation, is a fatty acyl group, two carbon atoms longer, that remains on the carrier protein while reduction takes place.

The reducing power for the synthesis of fatty acids comes from NADPH. This reduced coenzyme is made in the hexose monophosphate pathway during the oxidation of D-glucose 6-phosphate into CO_2 and H_2O.

We have learned that acetyl-CoA can be made from carbohydrate, acetyl-CoA can be used for the synthesis of fatty acids, acetyl-CoA can also be made from fatty acids, acetyl-CoA can be used for the synthesis of cholesterol and other steroids, and acetyl-CoA can be used for the synthesis of ketone bodies. In the next chapter, we will discover that acetyl-CoA can be made from certain amino acids and that ketone bodies can be made from certain others. Figure 27.12 shows acetyl-CoA as a central intermediate in carbohydrate and fatty acid metabolism.

ADDITIONAL PROBLEMS

15. What is the function of bile salts?
16. Describe the action of lipases.
17. What two oxidizing coenzymes are involved in β-oxidation? How are they regenerated?
18. What determines whether fat or carbohydrate is used as an energy source in working muscle?
19. Describe the function of lactate formation in skeletal muscle metabolism.
20. Triglycerides are transported in the blood primarily as _____ complexes.
21. Fatty acids are transported in the blood as complexes with _____.

Figure 27.12   Acetyl-coenzyme A is a central intermediate in carbohydrate and fatty acid metabolism.

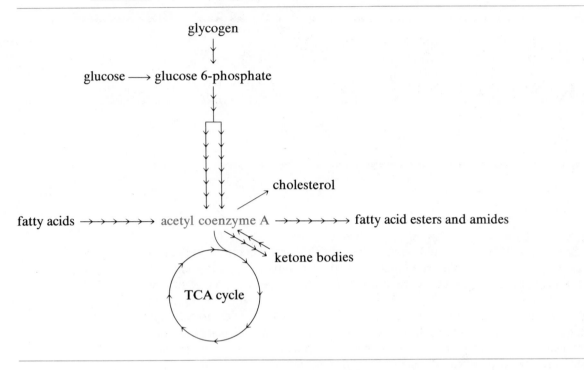

CHAPTER 28

PROTEIN DIGESTION
AND AMINO ACID METABOLISM

In Chapters 24, 25, and 27, we discussed the digestion and metabolism of carbohydrates and fats, two of the three main foodstuffs in our diets. In this chapter, we will examine the digestion of the third, the proteins, and the anabolism and catabolism of amino acids.

The proteins in our bodies are continually being replaced. Old molecules of protein are broken down and replaced by newly synthesized molecules. In Chapter 22, we learned how protein molecules are made according to instructions contained in genes. In this chapter, we will learn of the synthesis and disposal of amino acids. We also will explore further the interrelationships in metabolism, particularly the relationships between amino acid metabolism, carbohydrate metabolism, and the TCA cycle.

28-1 PROTEIN DIGESTION

LEARNING OBJECTIVES

1. Define proenzyme, zymogen, protease, proteolytic enzyme, endopeptidase, and exopeptidase.
2. Describe the actions of pepsin, enterokinase, trypsin, chymotrypsin, and carboxypeptidase.

3. Trace the complete digestion of a protein.
4. Given the structure of a peptide, show the products of digestion by trypsin, chymotrypsin, and carboxypeptidase.

The proteins we eat are broken down into amino acids in the digestive tract, or the gastrointestinal tract, as nurses and physicians call it. The amino acids can then be absorbed and used to make body proteins and other nitrogen-containing compounds.

28-1.1 PROENZYMES

The terms *proenzyme* and *zymogen* are synonymous. A proenzyme, or a zymogen, is an inactive precursor form of an enzyme that requires covalent modification by cleavage of the peptide chain to convert it into an active catalyst. *Proteases* belong to the family of enzymes that catalyze the hydrolysis of proteins to break them down into smaller peptides. In other words, proteases are *proteolytic enzymes.* All digestive proteases are synthesized as proenzymes that are converted into active enzymes in the gastrointestinal tract. If active enzymes were made by direct synthesis, they would digest the cells that made them. A mucous lining protects the gastrointestinal tract from digestion by proteases. Proenzymes of several of the digestive enzymes, among the first enzymes discovered, were given unsystematic names. These proenzymes were named by adding the suffix *-ogen* to the name of the enzyme. The modern way of naming proenzymes is to add the prefix *pro-*. For example, pepsinogen, the proenzyme of pepsin, is named by the old system; and procarboxypeptidase, the proenzyme of carboxypeptidase, is named by the new system.

28-1.2 ROLE OF THE STOMACH AND PEPSIN IN THE DIGESTION OF PROTEIN

Pepsinogen, the proenzyme of the protease pepsin, is produced in secretory cells (chief cells) located in the stomach wall. Pepsinogen is converted into pepsin both by the acidity of the gastric juice and by pepsin itself. Therefore, the process is said to be **autocatalytic.** In other words, pepsin catalyzes the formation of more pepsin. To convert pepsinogen into pepsin, several peptides totaling 42 amino acid residues are removed from the N-terminal end (see Section 21-2.1).

Autocatalytic describes a reaction that accelerates in rate because the products of the reaction are themselves catalysts in the reaction.

$$\text{pepsinogen} \quad \xrightarrow[\text{pH 2}]{\text{pepsin}} \quad \text{pepsin} \quad + \quad \text{inactive peptides}$$
$$\text{(M.W. 40,400)} \qquad\qquad \text{(M.W. 32,700)}$$

The optimum pH of pepsin is 1.5–2.5; therefore, it is most active at the very acid pH of the stomach. *Pepsin* catalyzes the hydrolysis of peptide bonds, breaking down proteins into peptides of various sizes. It is less specific in its action than the other proteases of the digestive tract (see Section 21-2.1). Hydrolysis brought about by pepsin is favored by hydrophobic side chains on the amino acids forming the peptide bond. Pepsin catalyzes the hydrolysis of peptide bonds formed by either the amino or carboxyl groups of the aromatic amino acids (phenylalanine, tyrosine, tryptophan), other amino acids with nonpolar R groups (leucine, isoleucine, alanine, valine, methionine), glutamic acid, cysteine, and cystine. Pepsin is, therefore, a protease that is an **endopeptidase** that breaks down protein in the food we eat to a mixture of peptides. The acid of the stomach does a good job of denaturing dietary protein, making it more susceptible to attack by this proteolytic enzyme (see Section 21-2.4).

An **endopeptidase** is an enzyme that catalyzes the hydrolysis of peptide bonds in the interior of a protein molecule rather than specifically at one end.

28-1.3 ROLE OF THE PANCREAS AND THE SMALL INTESTINE IN THE DIGESTION OF PROTEIN

Most digestion of proteins occurs in the small intestine. Two hormones stimulate the pancreas to release its digestive juice. The two hormones are released when the contents of the stomach enter the small intestine. Pancreatic juice contains the proenzymes of several proteases: trypsinogen, two chymotrypsinogens, two procarboxypeptidases, and others.

The enzyme *enterokinase* is made in, and released from, the cells lining the small intestine. It specifically and rapidly cleaves a hexapeptide from the N-terminal end of trypsinogen, forming trypsin.

$$\text{trypsinogen} \xrightarrow[\textit{or trypsin}]{\text{enterokinase}} \text{trypsin} + \text{a hexapeptide}$$

Trypsin then converts all the other proenzymes to active enzymes. In fact, it catalyzes its own formation by converting trypsinogen into trypsin, another autocatalytic reaction. Two additional transformations are

$$\text{chymotrypsinogens} \xrightarrow{\text{trypsin}} \text{chymotrypsins} + 2 \text{ dipeptides}$$

$$\text{procarboxypeptidases} \xrightarrow{\text{trypsin}} \text{carboxypeptidases}$$

The actions of trypsin and chymotrypsin are discussed in Section 21-2.1, in connection with the determination of structures of peptides, and in Table 28.1. Carboxypeptidases contain a zinc ion in their active site. They catalyze the removal of the C-terminal amino acid. As a result, carboxypeptidases convert peptides into their constituent amino acids by a stepwise peeling process from the C-terminal end.

Cells of the intestinal wall contain aminopeptidases that liberate amino acids from the N-terminal end of polypeptides and dipeptides. Aminopeptidases and carboxypeptidases are **exopeptidases**. Dipeptidases are also present in cells of the intestinal wall.

An **exopeptidase** is an enzyme that catalyzes the hydrolysis of peptide bonds at one end of a peptide chain rather than in the middle.

Thus, the proenzyme forms of proteolytic enzymes are made in the pancreas. In the intestine, they are converted into active enzymes. These proteases, with enzymes of cells of the intestinal wall, convert proteins into amino acids. The digestion of proteins is usually 95% complete due to the combined action of these enzymes. Digestion of proteins, lipids, and carbohydrates is summarized in Table 28.2.

Table 28.1 Optimum pH and Specificity of the Three Primary Proteases of the Small Intestine

Enzyme	Optimum pH	Site of Action
Trypsin	7.5–8.5	Peptide bonds involving the carboxyl group of arginine and lysine; works best on denatured, partially digested proteins
Chymotrypsins	7.5–8.5	Peptide bonds involving the carboxyl group of the aromatic amino acids (phenylalanine, tryptophan, tyrosine)
Carboxypeptidases	7.5–8.5	Peptide bond through which the C-terminal amino acid is attached

Table 28.2 Summary of the Sites and Sequences of Steps in the Digestion of the Principal Foodstuffs

Foodstuff	Mouth	Stomach	Small Intestine*
Starch	$\xrightarrow{\alpha\text{-amylase}\dagger}$		$\xrightarrow{\alpha\text{-amylase}}$ oligosaccharides $\xrightarrow{\alpha\text{-amylase}}$ maltose + isomaltose + D-glucose $\downarrow$ $\downarrow$ D-glucose D-glucose
Lactose			$\xrightarrow{\text{lactase}}$ D-glucose + D-galactose
Sucrose			$\xrightarrow{\text{sucrase}}$ D-glucose + D-fructose
Triglycerides			$\xrightarrow{\text{lipases}}$ fatty acids + monoglycerides + diglycerides
Proteins	$\xrightarrow{\text{pepsin}}$ peptides		$\xrightarrow[\text{chymotrypsins}]{\text{trypsin}}$ oligopeptides $\xrightarrow[\text{aminopeptidases}]{\text{carboxypeptidases}}$ amino acids

*There are three sources of enzymes in the small intestine: (1) enzymes secreted by the pancreas (proteolytic enzymes are secreted as proenzymes), (2) enzymes secreted by intestinal cells themselves, and (3) enzymes located in or on intestinal epithelial cells.

†Only a small degree of hydrolysis is effected by salivary α-amylase.

Released amino acids are then absorbed and enter the blood. The liver has the greatest capacity to remove circulating amino acids from the blood. The liver is the principal site of amino acid (nitrogen) metabolism and is responsible for a number of functions, including those listed below.

1. synthesis of its own proteins and of several protein components of blood plasma as well as the removal and breakdown of most plasma proteins
2. synthesis of various nitrogen-containing compounds such as purines and pyrimidines (see Section 20-1.1) and creatine (see Section 27-2.2)
3. synthesis of nonessential amino acids (see Sections 28-2.1 and 28-3.3) to the extent that this is necessary
4. delivery to other organs, via the blood, of a balanced mixture of amino acids
5. disposal of surplus amounts of both the carbon skeletons of ingested amino acids and nitrogen (synthesis of urea) (see Sections 28-2.2, 28-2.3, and 28-2.4).

PROBLEMS

1. An inactive percursor form of an enzyme that is converted into an active enzyme by cleavage of one or more peptide bonds is a(n) _____.
2. Categorize the following enzymes as an endopeptidase or an exopeptidase.
   (a) pepsin (b) trypsin (c) chymotrypsin (d) carboxypeptidase

ANSWERS TO PROBLEMS **1.** proenzyme or zymogen **2. (a)** endopeptidase **(b)** endopeptidase **(c)** endopeptidase **(d)** exopeptidase

28-2 AMINO ACID METABOLSIM

LEARNING OBJECTIVES

1. Define transamination, deamination, glycogenic amino acid, ketogenic amino acid, gluconeogenesis, and α-keto acid.

2. Given an α-keto acid, write the equation describing transamination involving it.

3. Given an amino acid, write the equation describing transamination involving it.

4. Describe the functions of pyridoxal phosphate, the urea cycle, urea, transamination, and L-glutamine.

5. Describe the interrelationships of glycolysis (carbohydrate metabolism), the TCA cycle, and amino acid metabolism.

6. Describe the cause of PKU (phenylketonuria).

7. Outline the biosynthesis of noradrenaline, adrenaline, and histamine.

8. Construct a table listing the nine coenzymes that are vitamins or contain vitamins as part of their structure (listed in Table 28.4), giving for each (1) the name of the vitamin, (2) the name of the coenzyme, and (3) the type of reaction in which the coenzyme functions.

9. Describe the key roles in metabolism played by D-glucose 6-phosphate and acetyl-coenzyme A.

The synthesis of nonessential amino acids, the breakdown of amino acids, the synthesis of urea, and the use of amino acids for the synthesis of other nitrogen-containing compounds are all important to our physical and mental health. In this section, we will look at these metabolic processes and some disorders that may occur in them. We will also summarize what we have learned so far about the B vitamins and their role as coenzymes.

28-2.1 TRANSAMINATION

The last step in the synthesis, and the first step in the breakdown, of most amino acids is *transamination*. This is the reversible transfer of an amino group from an amino acid to an α-keto acid that results in the synthesis of a new amino acid and the conversion of the original amino acid into an α-keto acid (see Figure 28.1). *Transaminases*, the enzymes that catalyze transaminations, are found in both the soluble cytoplasm and the mitochondria of eukaryotic cells.

L-Glutamate is the most active amino acid in transamination. The ionized α-keto acid related to glutamate is α-ketoglutarate, an intermediate of the TCA cycle (see Section 25-1.2). Other α-keto acid anions also are intermediates of carbohydrate metabolism (see Section 24-2.1). All amino acids are related to α-keto acids. The general transamination reaction involving L-glutamate and α-ketoglutarate is given.

$$\text{original amino acid} + \text{original } \alpha\text{-ketoacid} \xrightleftharpoons{\text{transamination}} \text{new } \alpha\text{-ketoacid} + \text{new amino acid}$$

amino acid zwitterion α-ketoglutarate α-keto acid anion   L-glutamate

Figure 28.1 Metabolic relationships between glycolysis, the TCA cycle, and amino acid metabolism.

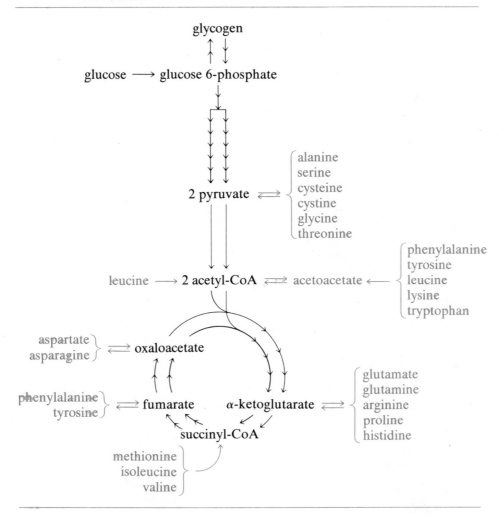

Transaminases require pyridoxal phosphate as a coenzyme. Pyridoxal, pyridoxamine, and pyridoxine all function as vitamin B_6. In the presence of the enzyme, the coenzyme pyridoxal phosphate reacts with an amino acid to form an imine (see Sections 15-3.1 and 17-2.1). The resulting imine then rearranges to form the imine of pyridoxamine phosphate and the corresponding α-keto acid, which subsequently adds water to form the latter two compounds. As an example of this

$$
\begin{array}{ccc}
\text{COO}^- & & \text{H} \\
| & & | \\
\text{H}_3\overset{+}{\text{N}}-\text{C}-\text{H} + \text{(P)}-\text{Pyr}-\text{C}=\text{O} & \rightleftharpoons & \text{(P)}-\text{Pyr}-\text{C}=\text{N}-\text{C}-\text{H} \\
| & & | \\
\text{R} & & \text{R}
\end{array}
$$

pyridoxal phosphate　　H$_2$O

$$
\begin{array}{c}
\text{COO}^- \\
| \\
\text{C}=\text{O} + \text{(P)}-\text{Pyr}-\text{CH}_2-\overset{+}{\text{NH}}_3 \rightleftharpoons \text{(P)}-\text{Pyr}-\text{CH}_2-\text{N}=\text{C} \\
| \\
\text{R}
\end{array}
$$

pyridoxamine phosphate　　H$_2$O

type of reaction, we will consider how L-alanine can be used in the synthesis of L-aspartate. The reaction involved is a transamination involving the enzymes glutamate-pyruvate transaminase (GPT) and glutamate-oxaloacetate transaminase (GOT). These two reactions are given.

$$
\begin{array}{ccc}
\text{COO}^- \quad \text{COO}^- & \xrightarrow{\text{GPT}} & \text{COO}^- \quad \text{COO}^- \\
\text{+H}_3\text{N}-\text{C}-\text{H} + \text{C}=\text{O} & \rightleftharpoons & \text{C}=\text{O} + \text{+H}_3\text{N}-\text{C}-\text{H} \\
\text{CH}_3 \quad \text{CH}_2 & & \text{CH}_3 \quad \text{CH}_2 \\
\text{CH}_2 & & \text{CH}_2 \\
\text{COO}^- & & \text{COO}^-
\end{array}
$$

L-alanine　　α-ketoglutarate　pyruvate　　L-glutamate

$$
\begin{array}{ccc}
\text{COO}^- \quad \text{COO}^- & \xrightarrow{\text{GOT}} & \text{COO}^- \quad \text{COO}^- \\
\text{+H}_3\text{N}-\text{C}-\text{H} + \text{C}=\text{O} & \rightleftharpoons & \text{C}=\text{O} + \text{+H}_3\text{N}-\text{C}-\text{H} \\
\text{CH}_2 \quad \text{CH}_2 & & \text{CH}_2 \quad \text{CH}_2 \\
\text{CH}_2 \quad \text{COO}^- & & \text{CH}_2 \quad \text{COO}^- \\
\text{COO}^- & & \text{COO}^-
\end{array}
$$

L-glutamate　　oxaloacetate　α-ketoglutarate　　aspartate

The liver is the organ containing the greatest amounts of GPT and GOT. Heart and skeletal muscle contain medium amounts. Only small amounts of GPT and GOT are present in the kidneys, pancreas, red blood cells, and lungs. As we discussed in Sections 23-4 and 24-1.4, when cells in any tissue are injured, enzyme is released and increased serum levels result. (SGPT = serum glutamate-pyruvate transaminase and SGOT = serum glutamate-oxaloacetate transaminase.) Elevated transaminase levels are most often associated with heart and liver disease. SGOT activity rises within the first 18 hours following an acute myocardial infarction. The degree of elevation is related to the number of myocardial muscle cells that have died. In medical terminology, the degree of elevation is related to the amount of muscle tissue that has become *necrotic* (*necrosis* refers to death). Because the liver contains the highest levels of both transaminases, any damage to the cells of the liver will also result in elevated levels of both SGOT and SGPT. The degree of elevation ususally reflects the severity of the damage. SGOT is also elevated after severe muscle trauma, and many drugs can give elevations.

There are other transaminases. In fact, most amino acids can be converted into the corresponding α-keto acids by transamination and many can be made from the corresponding α-keto acids by transamination.

28-2.2 REMOVAL OF NITROGEN FROM AMINO ACIDS

Most nitrogen is removed from amino acids by transfer to α-ketoglutarate to make L-glutamate; in other words, it is removed by transamination with α-ketoglutarate. L-Glutamate, being specifically permeable to the inner mitochondrial membrane, passes from mitochondria to the cytosol, and vice versa. Ammonia is removed from L-glutamate by oxidative **deamination**. This is a reaction catalyzed by glutamate dehydrogenase, an enzyme requiring NAD^+ or $NADP^+$ as a coenzyme. *Oxidative deamination* is the removal of ammonia (or the ammonium ion) from an amine accompanied by the removal of hydrogen atoms (an oxidation). A carbonyl compound is thereby formed.

Deamination is the removal of ammonia (or the ammonium ion) from an amine.

$$NAD^+ + {}^+H_3NCH \longrightarrow NADH + H^+ + {}^+H_2N{=}C \xrightarrow[H_2O]{} O{=}C + NH_4{}^+$$

L-glutamate ⇌ (glutamate dehydrogenase) α-ketoglutarate

$$H_2O + \begin{matrix} COO^- \\ | \\ {}^+H_3NCH \\ | \\ CH_2 \\ | \\ CH_2 \\ | \\ COO^- \end{matrix} \quad \underset{\text{glutamate dehydrogenase}}{\overset{\text{(NADP}^+\text{)} \; \text{(NADPH + H}^+\text{)}}{\underset{\text{NAD}^+ \quad \text{NADH + H}^+}{\rightleftharpoons}}} \quad \begin{matrix} COO^- \\ | \\ C{=}O \\ | \\ CH_2 \\ | \\ CH_2 \\ | \\ COO^- \end{matrix} + NH_4{}^+$$

The coupling of glutamate dehydrogenase and transamination can either make amino acids or deaminate them. The interconversions of pyruvate and L-alanine, α-ketoglutarate and L-glutamate, and oxaloacetate and L-aspartate demonstrate the interrelationship of amino acid metabolism with glycolysis (carbohydrate metabolism) and the TCA cycle. Other amino acids have similar relationships. These are outlined in Figure 28.1.

The reaction catalyzed by glutamate dehydrogenase is reversible, so nitrogen can also enter amino acid synthesis by reductive amination of α-ketoglutarate. *Reductive amination* is the addition of ammonia (or the ammonium ion) to a carbonyl group, accompanied by the addition of hydrogen atoms to form an amine. However, this reaction is of much less importance in the total nitrogen pool than is transamination. Reductive amination specifically requires NADPH for reversal.

There are several kinds of *nonoxidative deamination*. Two examples are given. Each involves the conversion of an amine salt to an imine salt by a reaction other than dehydrogenation, followed by hydrolysis to yield an ammonium ion and a carbonyl compound.

28-2.3 SYNTHESIS OF UREA

The ammonia (ammonium ion) removed from amino acids is very toxic to the brain, even in relatively low concentrations. As a result, all animals must get rid of ammonia. Humans convert ammonia into urea, a neutral, nontoxic, water-soluble molecule. Urea is then excreted in urine.

An overall view of the pathway for the formation of urea is given in Figure 28.2. In this cycle, one molecule of ammonia derived from an amino acid by transamination and deamination, and an amino group obtained from an amino acid by transamination, are combined to make urea.

Reactions of the urea cycle are primarily found in the liver, the principal site of urea formation. Some urea synthesis can also occur in the brain and the kidneys. Ammonia is stored and transported in the blood in the form of L-glutamine, which is formed from L-glutamate. L-Glutamine diffuses through the plasma membrane of cells into the blood and is carried to the liver and kidneys. The liver and kidneys are rich in the enzyme glutaminase, which catalyzes the hydrolysis of L-glutamine

Figure 28.2 An overall view of the formation of urea, showing the starting materials and final product in color.

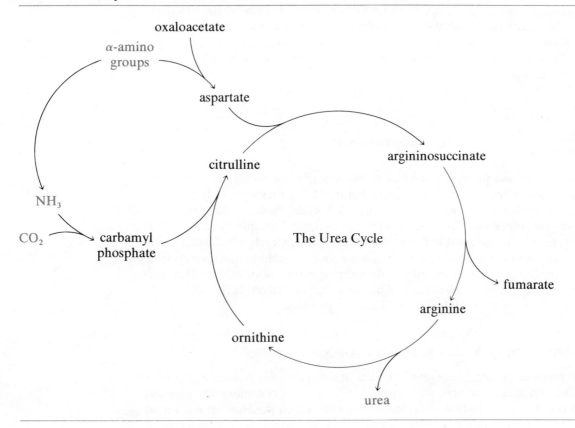

into L-glutamate and ammonium ion. In the liver, ammonium ions are converted into urea. Urea is excreted through the kidneys.

Carbamyl phosphate is important in several biosynthetic reactions. In urea formation, carbamyl phosphate is formed in mitochondria from ammonia and carbon dioxide using the energy of two ATP molecules. It then reacts with ornithine to form citrulline and, thereby, donates the carbon atom and one nitrogen atom of urea. The second nitrogen atom comes from glutamate via aspartate by a process that converts citrulline to arginine. The enzyme arginase then acts on the guanidino group

$$\overset{\displaystyle \overset{+NH_2}{\|}}{-NH-C-NH_2}$$

guanidino group

to release urea and to replace ornithine, thus completing the cycle. A detailed representation of the urea cycle is given in Figure 28.3. Also shown in Figure 28.3 is the relationship of the urea cycle to the TCA cycle. Note, at the top of the diagram, that fumarate, a product of the urea cycle and an intermediate of the TCA cycle, can be converted into oxaloacetate in the TCA cycle. Oxaloacetate can then pick up another amino group by transamination to feed into the urea cycle.

The urea cycle involves, as intermediates, two amino acids (L-citrulline and L-ornithine) that are not constituents of proteins. As pointed out in Section 21-1.1, there are other amino acids that are not found in proteins.

28-2.4 USE OF THE CARBON SKELETONS OF AMINO ACIDS

In the previous section, we learned that the amino groups removed from amino acids are converted into urea for excretion. Now we will consider what happens to the rest of the amino acid structure, that is, the carbon skeletons of the amino acids. After deamination by transamination, the carbon skeleton that is left is an α-keto acid anion (see Section 28-2.1).

Amino acids can be classified as **glycogenic** or **ketogenic** amino acids, depending on what happens to the α-keto acids formed from them. The glycogenic amino acids are those that give rise to pyruvate or an intermediate of the TCA cycle, and hence, can be converted to glucose. However, these glycogenic amino acids usually are oxidized to carbon dioxide and water via the TCA cycle.

The process by which D-glucose is made from noncarbohydrate molecules, *gluconeogenesis*, was mentioned in Section 24-2.3 in connection with the synthesis of D-glucose from L-lactate in the Cori cycle. Basically, gluconeogenesis involves a reversal of the pathway of glycolysis except for three steps.

A **glycogenic amino acid** is an amino acid whose carbon atoms can be used to form glucose or glycogen.

A **ketogenic amino acid** is an amino acid whose carbon atoms are used to form ketone bodies (see Section 27-3.3).

$$\begin{array}{ccc} COO^- & & COO^- \\ | & & | \\ C-O(P) + ADP \longrightarrow & C=O + ATP \\ \| & & | \\ CH_2 & & CH_3 \\ \\ \text{PEP} & & \text{pyruvate} \end{array}$$

Figure 28.3 Details of the urea cycle and its relationship to the TCA cycle.

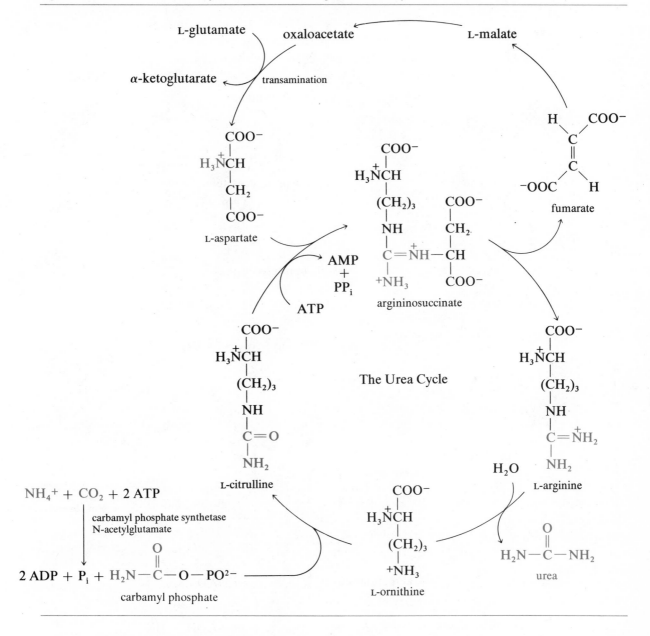

The reaction by which phosphoenolpyruvate (PEP) is converted into pyruvate is not reversible. Therefore, cells have to make PEP from pyruvate in a roundabout way that bypasses this reaction. The route used by cells in the livers of mammals is that depicted in Figure 28.4.

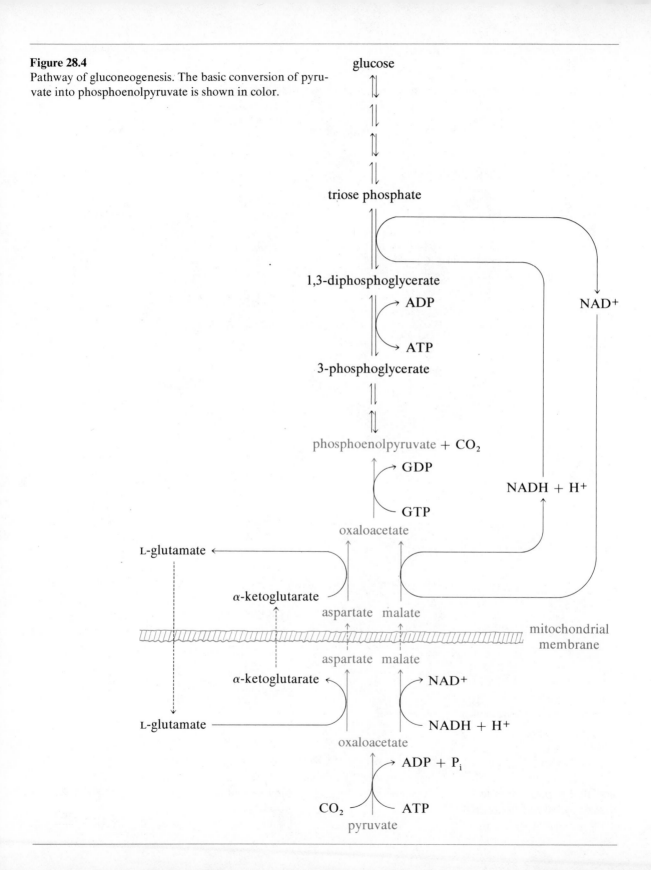

Figure 28.4
Pathway of gluconeogenesis. The basic conversion of pyruvate into phosphoenolpyruvate is shown in color.

Two other steps in the glycolytic pathway that require special enzymes for reversal are the following.

$$\text{D-glucose 6-phosphate} \xrightarrow{\text{glucose 6-phosphatase}} \text{D-glucose} + P_i$$

$$\text{D-fructose 1,6-diphosphate} \xrightarrow{\text{fructose 1,6-diphosphatase}} \text{D-fructose 6-phosphate} + P_i$$

Comparison with Figure 24.2 reveals that these two reactions are the reverse of reactions catalyzed, respectively, by glucokinase or hexokinase and by phosphofructokinase as far as the carbohydrate substrates and products are concerned. However, the formation of the phosphate esters in glycolysis requires ATP. Hydrolysis of the phosphate esters does not regenerate the high-energy phosphate bond.

Ketogenic amino acids are those which give rise to acetoacetate, which in turn can be reduced to 3-hydroxybutyrate and decarboxylated to give acetone. In other words, ketogenic amino acids give rise to ketone bodies. As pointed out in Section 27-3.3, ketone bodies can be cleaved to acetyl-CoA and oxidized in the TCA cycle. Ketone bodies are made in the liver and pass into the blood. Other organs may use them; for example, they are actively oxidized by muscles and sometimes by the brain to produce energy. Amino acids that produce acetoacetate are called ketogenic although acetoacetate eventually ends up as acetyl-CoA, an intermediate of the TCA cycle. The difference is that acetoacetate produced from amino acids in the liver passes into the blood, thus increasing serum levels of acetoacetate.

L-Phenylalanine and L-tyrosine are both glycogenic and ketogenic because part of their carbon skeleton is converted to fumarate and part to acetoacetate. In Table 28.3, the ultimate fates of the α-keto acids formed from the amino acids are listed. These relationships are diagrammed in Figure 28.1.

Table 28.3 The Products of Catabolism of Glycogenic and Ketogenic Amino Acids

Amino Acids	Product
Glycogenic	
Alanine, serine, cysteine, cystine, glycine, and threonine	Pyruvate
Leucine (2 of the 6 carbon atoms)	Acetyl-CoA
Glutamic acid, glutamine, arginine, proline, and histidine	α-Ketoglutarate
Methionine, isoleucine, and valine	Succinyl-coenzyme A
Phenylalanine and tyrosine	Fumarate
Aspartic acid and asparagine	Oxaloacetate (fumarate)
Ketogenic	
Phenylalanine, tyrosine, leucine, lysine, tryptophan	Acetoacetate

28-2.5 DECARBOXYLATION OF AMINO ACIDS

In addition to deamination, amino acids can undergo decarboxylation (see Section 17-2.2). Decarboxylation also requires pyridoxal phosphate as a coenzyme. Many of the amines formed in these reactions have potent physiological effects. For example, as we learned in Section 17-2.2, histamine is made by decarboxylation of the amino acid histidine. Histamine is produced in excessive amounts in allergic reactions. Antihistamines are drugs taken to counteract the action of histamine.

$$
\begin{array}{ccc}
\text{COO}^- & & ^+\text{NH}_3 \\
| & & | \\
\text{H}_3^+\text{NCH} & & \text{CH}_2 \\
| & \xrightarrow{\text{histidine decarboxylase}} & | \\
\text{CH}_2 & & \text{CH}_2 \qquad +\ \text{CO}_2 \\
\end{array}
$$

histidine   histamine

Figure 28.5 shows the synthesis of adrenaline from phenylalanine and/or tyrosine by a pathway involving amino acid decarboxylation. Adrenaline and noradrenaline* are synthesized and stored in specific cells of the adrenal gland. Upon stimulation by the central nervous system (the brain and spinal cord), they are released into the blood. Adrenaline and noradrenaline are also synthesized in, and released from, sympathetic nerve cells where they serve as neurotransmitters. Cells of adipose tissue receive these molecules from cells of the sympathetic nervous system. The molecules tell the cells to mobilize fatty acids rapidly.

As we discussed in Sections 24-2.1, adrenaline increases energy-producing reactions in carbohydrate metabolism. Adrenaline and noradrenaline do this by binding with a receptor on the outer surface of cells and activating adenyl cyclase, which causes an increased formation of cyclic AMP. This sets off a cascade of events that results in activation of phosphorylase, a rise in the concentration of blood glucose, and mobilization of fatty acids from adipose tissue. These hormones also effect an increase in the rate and depth of contraction of the heart and suppress contraction of many smooth muscles.

The enzymes catalyzing the conversion of tyrosine to noradrenaline are also contained in organelles near synapses of the sympathetic nervous system. Noradrenaline is bound and stored in particles associated with these organelles. The active structure of the synapse apparently contains a small amount of the amine that is replenished from these stores after discharge during transmission.

3,4-Dihydroxyphenylethylamine (dopamine) is an intermediate in the pathway for the formation of adrenaline (see Figure 28.5). Dopamine increases blood pressure, primarily by enhancing contraction of the heart. In Parkinson's disease, there is a marked decrease in the content of dopamine in the brain. Therefore, an increase in the concentration of dopamine is sought in the treatment of Parkinson's disease.

The **sympathetic nervous system** (or **autonomic nervous system**) is that part of the nervous system concerned with the regulation of the activity of the heart (cardiac) muscle, smooth muscle, and glands.

A **synapse,** the region of contact between nerve cells, is the place where an impulse is transmitted from one nerve cell (*neuron*) to another.

---

*In organic nomenclature, the prefix *nor-* indicates either a change from the trivial name of a branched-chain compound into that of a straight-chain compound or a lower homolog. A *homolog* is a member of a series of compounds whose structures differ regularly by some radical (for example, $-\text{CH}_2-$) from those of its adjacent neighbors in the series. *Homo-* is a prefix meaning similarity. In organic nomenclature, *homo-* indicates a difference of $-\text{CH}_2-$ among otherwise identical structures (see Figure 28.5). Noradrenaline differs from adrenaline in having $-\overset{+}{\text{N}}\text{H}_2-\text{CH}_3$ ($-\overset{+}{\text{N}}\text{H}_2-\text{CH}_2-\text{H}$) in place of $-\overset{+}{\text{N}}\text{H}_3$ ($-\text{NH}_2-\text{H}$).

Figure 28.5   The abnormal catabolism of L-phenylalanine (in color) that occurs when the enzyme phenylalanine hydroxylase is lacking, and the synthesis of adrenaline from L-phenylalanine.

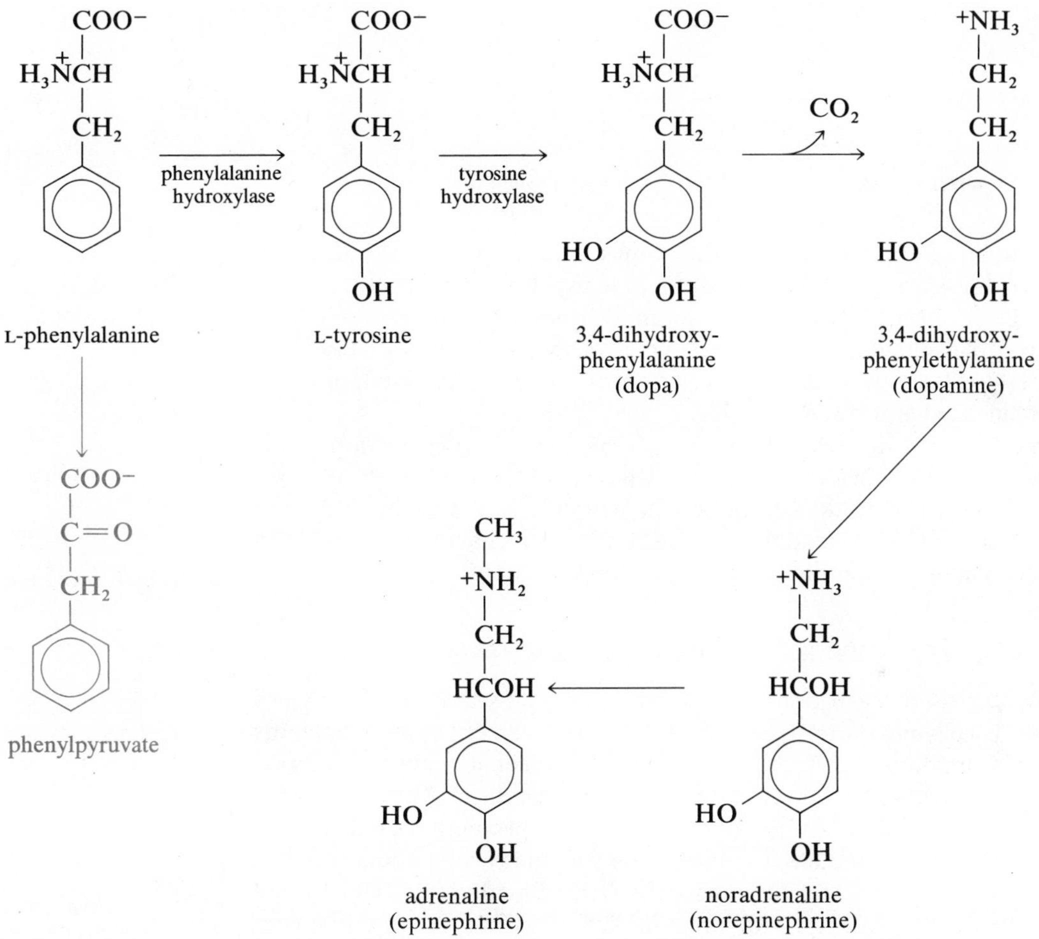

However, as already mentioned in Section 17-2.2, dopamine does not readily penetrate the blood-brain barrier, so L-dopa (3,4-dihydroxyphenylalanine) is administered rather than dopamine. L-Dopa, also called levo-dopa, presumably increases the concentration of dopamine in the brain.

There are a number of other amines with potent physiological activities (called *biogenic amines*) that are made by decarboxylation. Among them are 5-hydroxytryptamine (also called *serotonin*), which is formed from tryptophan via 5-hydroxytryptophan. 5-Hydroxytryptamine has a strong effect on the metabolism of the brain, is a potent constrictor of blood vessels, and is a stimulator of smooth muscle.

$$
\begin{array}{ccc}
COO^- & & NH_3^+ \\
| & & | \\
{}^+H_3NCH & & CH_2 \\
| & \xrightarrow{\text{decarboxylation}} & | \\
CH_2 & & CH_2 \\
\end{array}
\quad + \; CO_2
$$

L-tryptophan   5-hydroxytryptamine
(serotonin)

Decarboxylation of amino acids to form amines requires pyridoxal phosphate as a coenzyme. We learned in Section 28-2.1 that pyridoxal phosphate, formed from vitamin B_6, is the coenzyme for transaminations and deaminations of amino acids. Pyridoxal phosphate is also the coenzyme for racemization and aldol-like condensations (see Section 15-3.2) of amino acids. In addition, vitamin B_6 is involved in amino acid absorption.

An enzyme involved in breaking down biogenic amines, such as dopamine, noradrenaline, and serotonin, is monoamine oxidase (MAO). Low levels of MAO have been linked with chronic schizophrenia, particularly a specific subgroup known as chronic paranoid schizophrenia. This condition is characterized by delusions of persecution or grandeur and, frequently, hallucinations.

28-2.6 VITAMINS AS COENZYMES AND PARTS OF COENZYMES

We have now discussed seven water-soluble, or B, vitamins that are coenzymes or form parts of coenzymes or prosthetic groups. The involvement of these vitamins in metabolism is summarized in Table 28.4. Folic acid and cobalamine are vitamins and coenzymes involved in reactions of one-carbon metabolism. After its absorption, the vitamin folic acid (folate) is reduced through specific liver enzymes to tetrahydrofolate. Tetrahydrofolate is a coenzyme for one-carbon transfers. The one-carbon units on tetrahydrofolate can take the form of methyl ($-CH_3$), formyl ($-CHO$), methylene ($-CH_2-$), hydroxymethyl ($-CH_2OH$), methylidyne ($=CH-$), and formimino ($-CH_2=NH$) groups. With the exception of the hydroxymethyl group, all these groups are directly involved in one-carbon transfers.

One-carbon transfers are involved in the synthesis of many compounds in living cells. Among the reactions involving one-carbon transfers are (1) the synthesis of one amino acid from another, (2) the biosynthesis of pyrimidines and purines (see Section 20-1.1), (3) the biosynthesis of creatine (see Section 27-2.2), and (4) the biosynthesis of choline. Choline is used for the synthesis of acetylcholine (see Section 17-3), a compound used to transmit impulses from certain nerve cells to others or to muscle cells, and phosphatidylcholine, a phosphoglyceride (see Section 26-2.1).

Vitamin B_{12} (cobalamine) is a complicated molecule containing cobalt (see Section 10-1.5). It is used to form cobamide coenzyme, which is also involved in the transfer of methyl groups.

Table 28.4 Involvement of the B Vitamins As Coenzymes in Metabolic Processes

Vitamin	Coenzyme or Prosthetic Group	Process	Section of Discussion	Deficiency Disorders and Major Symptoms
Thiamine (B$_1$)	TPP (Cocarboxylase)	Decarboxylation (for example, in pyruvate $\longrightarrow$ acetyl-CoA + CO$_2$ and α-ketoglutarate $\longrightarrow$ succinyl-CoA + CO$_2$)	25-1	Improper carbohydrate metabolism with an accumulation of pyruvate and lactate results in insufficient energy for muscle and nerve cells. Also results in two syndromes: beriberi, in which there is partial paralysis of the smooth muscle of the gastro-intestinal tract causing digestive disturbances and skeletal muscle paralysis; and polyneuritis, in which there is impairment of reflexes related to muscle movement, impairment of sense of touch, decreased intestinal motility, stunted growth in children, and poor appetite.
Nicotinic acid and nicotinamide (niacin)	NAD$^+$ and NADP$^+$	Transfer of 2 e$^-$ + H$^+$ (oxidation-reduction)	24-2.2	Pellagra, characterized by inflamation of the skin, diarrhea, and psychological disturbances
Riboflavin (B$_2$)	FAD* and FMN*	Transfer of 2 e$^-$ + 2 H$^+$ (oxidation-reduction)	25-1	Improper utilization of oxygen resulting in blurred vision and cataracts, inflam-mation and cracking of skin, and breaks in the lining of the intestines
Pantothenic acid (B$_3$)	Coenzyme A	Acylation (for example, acetyl-CoA + oxaloacetate $\longrightarrow$ citrate)	25-1	Fatigue, muscle spasms, neuromuscular degenera-tion, and insufficient pro-duction of adrenal steroid hormones
Lipoic acid	Itself*	Transfer of 2 e$^-$ + 2 H$^+$ (oxidation-reduction)	25-1	(Unknown, definite requirement not established.)
Biotin	Itself*	Carboxylation (for example, acetyl-CoA + CO$_2$ $\longrightarrow$ malonyl-CoA in the biosynthesis of fatty acids)	27-3.1	Mental depression, muscular pain, fatigue, nausea, and inflammation of the skin
Pyridoxal, pyridoxine, and pyridoxamine (B$_6$)	Pyridoxal phosphate and pyridoxamine phosphate (Cotransaminase)	Transamination, decar-boxylation, and other reactions of amino acids	28-2.1	Inflammation of the tissues lining the eyes, nose, and mouth

*These are tightly (usually covalently) bound to enzymes, but function as coenzymes.

Table 28.4   Involvement of the B Vitamins As Coenzymes in Metabolic Processes (*cont'd*)

Vitamin	Coenzyme or Prosthetic Group	Process	Section of Discussion	Deficiency Disorders and Major Symptoms
Folic acid	Itself	One-carbon metabolism	28-2.6	Megaloblastic anemia, characterized by the presence of megaloblasts in the bone marrow (*Megaloblasts* are primitive red blood cells of large size. *Anemia* is a condition characterized by a reduced concentration of normal red blood cells.)
Cobalamine (B_{12})	Itself	Carbon–carbon bond cleavages, carbon–oxygen bond cleavages, carbon–nitrogen bond cleavages, and methyl group activation	28-2.6	Pernicious anemia, a megaloblastic anemia, and dysfunction of the nervous system, particularly the spinal cord

28-2.7 RESULTS OF IMPROPER AMINO ACID METABOLISM

Abnormalities of amino acid metabolism usually result in mental retardation and death. Almost all such disorders are inherited; thus they are called *"inborn errors" of metabolism.*

In metabolic defects, there is a block in a metabolic pathway. A block can result because (1) the enzyme may be completely missing, (2) there may be a mutation where some enzyme activity remains, but at a reduced level, or (3) there may be an error in the binding of a cofactor. In the presence of a block, the products of that pathway will be absent or reduced in quantity. For products several steps away from the block, there may be an alternative pathway that may provide the product; but it is likely that the immediate product of the blocked reaction will be missing or very low in concentration. Metabolites that are derived uniquely from that pathway or product will be missing. For example, below we have shown the same generalized metabolic pathway as that given in Section 24-2, but with the activity of enzyme 3 absent or reduced in quantity.

$$A \underset{}{\overset{enzyme_1}{\rightleftharpoons}} B \underset{}{\overset{enzyme_2}{\rightleftharpoons}} C \underset{}{\overset{}{\rightleftharpoons}}\!\!\!\times\!\!\!\rightleftharpoons D \overset{enzyme_4}{\longrightarrow} E$$

Because of the block, there may be an accumulation of starting compound A and intermediates B and C or there may be excessive production of substances derived from these precursors through alternative pathways. If a metabolite is soluble, its concentration in body fluids is raised and its excretion in the urine is increased. If a metabolite is poorly soluble, it may be stored, resulting in a *storage disease*. Gout is an example of this type of disease. Gout is caused by an abnormality in purine (see Section 20-1.1) catabolism. It is characterized by attacks of acute arthritis. The clinical symptoms resulting from a block may also arise from a lack of the product of the pathway.

In diagnosing errors of metabolism, the usual procedure is to check body fluids, particularly the urine and the blood, and sometimes cerebrospinal fluid (see Problem 17 in Chapter 1), for abnormal metabolites. If any are found, then parents, brothers, and sisters, and possibly other relatives of the patient, are checked to see if the same metabolites are found in the same or lesser amounts. Treatment may consist in providing missing products, eliminating excess precursors from the diet, or perhaps providing cofactors or vitamins that may increase the activity of the enzyme, if it is present at all.

Recall that the lack of a metabolic step leads to increased concentrations of precursors in cells. This, in turn, may lead to increases in the concentration of precursors in the blood. These high concentrations may affect pathways of other metabolites as well. For example, a high concentration of one amino acid will often affect transport and the distribution of other amino acids. Protein synthesis may also be affected. The immature brain in particular may be exposed to an abnormal and unbalanced supply of amino acids from which it cannot properly construct its essential, permanent proteins.

When an infant is unable to metabolize phenylalanine in the normal manner, phenylalanine and phenylpyruvate (the deamination product; see Figure 28.5) accumulate and result in *phenylketonuria (PKU)*. This inherited disorder is caused by an absence of the enzyme phenylalanine hydroxylase (which converts phenylalanine to tyrosine) in the liver (see Figure 28.5). PKU can result in mental retardation if it is not treated. PKU can be treated by giving the infant a special diet consisting of a mixture of amino acids with a regulated amount of phenylalanine. This diet must be continued until the child is at least 5 years old when the brain is developed and damage no longer results. An impairment of serotonin (5-hydroxytryptamine; see Section 28-2.5) formation accompanies PKU. Because serotonin has a potent effect on brain metabolism, a lower concentration of serotonin may be the reason that the brain does not develop properly and mental retardation results.

28-2.8 INTEGRATION OF METABOLISM

We have studied the metabolism of carbohydrates (see Chapter 24), lipids (see Chapter 27), proteins, and amino acids. We have studied the relationship of glycolysis to the TCA cycle (see Section 25-1.1), β-oxidation to the TCA cycle (see Section 27-2.1), β-oxidation to the respiratory chain (see Section 27-2.1), carbohydrate catabolism to the synthesis of fat (see Sections 24-2.1, and 25-1.1, and 27-3.1), and carbohydrate and amino acid metabolism (see Section 28-2.4). We can now look at the big picture of how metabolic pathways relate to each other.

Insulin has an effect on carbohydrate (see Section 24-3.2) and lipid (see Section 27-2.1) metabolism. Insulin also stimulates the synthesis of proteins. We learned in Sections 24-2.1, 27-3.2, and 28-2.4 that D-glucose 6-phosphate is a key intermediate in more than one pathway. Special mention has been made of the fact that molecules of acetyl-CoA can arise from carbohydrate, fat (fatty acids), and some amino acids, and that 12 ATP molecules are generated via oxidation of acetyl-CoA in the TCA cycle, the respiratory chain, and oxidative phosphorylation. Hence, carbohydrate, fatty acid, and amino acid molecules can all be funneled into energy

728

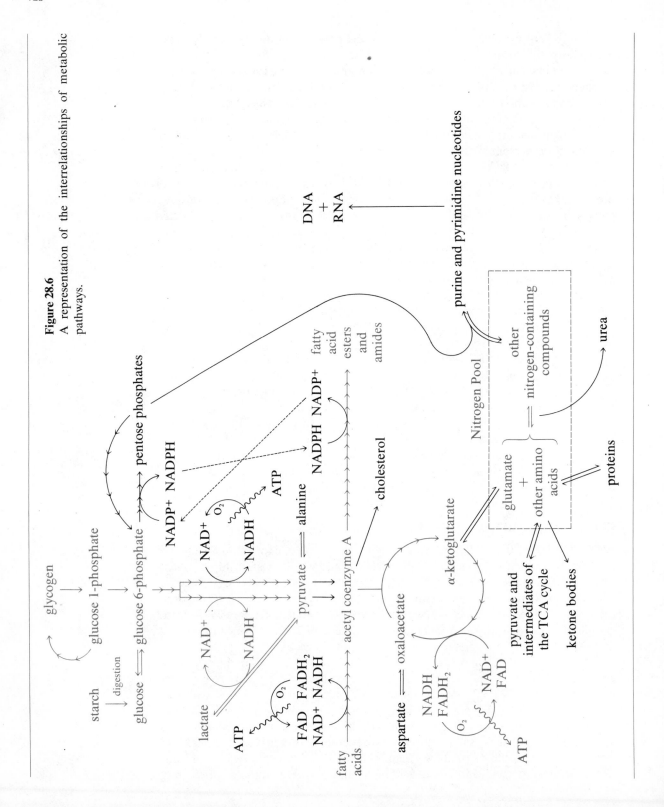

Figure 28.6
A representation of the interrelationships of metabolic pathways.

(ATP) production. Acetyl-CoA is also used for synthesis of fatty acids and cholesterol and for other anabolic reactions. Intermediates of the glycolytic pathway and the TCA cycle furnish the carbon skeletons for the glycogenic amino acids. These amino acids, along with others, make up part of the nitrogen pool. There is also a dependence of the urea cycle on the TCA cycle. Thus, there are interrelationships between carbohydrate, fatty acid, and amino acid metabolism. An abbreviated diagram showing some of these interrelationships is given in Figure 28.6. There is not room on such a diagram for all pathways. For example, ketone bodies (see Section 27-3.3) are also made from fats. The pathways shown in Figure 28.6 are made up of many different chemical reactions, each catalyzed by an enzyme, most of which are specific for a single reaction. The pathways proceed in a synchronized, controlled manner. Some of the controls were described in Sections 24-3 and 25-4.

PROBLEMS

3. Write an equation for the transamination of L-valine (see Table 21.1).
4. Noradrenaline is made from what amino acids?
5. Adrenaline is made from what amino acids?
6. Describe the function of pyridoxal phosphate.
7. What is the cause of PKU?
8. What is the function of the urea cycle?
9. What is a ketogenic amino acid?

ANSWERS TO PROBLEMS   **3.**

$$
\begin{array}{c}
\text{COO}^- \\
| \\
\text{C}=\text{O} \\
| \\
\text{CH}_2 \\
| \\
\text{CH}_2 \\
| \\
\text{COO}^-
\end{array}
\;+\;
\begin{array}{c}
\text{COO}^- \\
| \\
{}^+\text{H}_3\text{NCH} \\
| \\
\text{CH} \\
\diagup \; \diagdown \\
\text{CH}_3 \quad \text{CH}_3
\end{array}
\;\rightleftharpoons\;
\begin{array}{c}
\text{COO}^- \\
| \\
{}^+\text{H}_3\text{NCH} \\
| \\
\text{CH}_2 \\
| \\
\text{CH}_2 \\
| \\
\text{COO}^-
\end{array}
\;+\;
\begin{array}{c}
\text{COO}^- \\
| \\
\text{C}=\text{O} \\
| \\
\text{CH} \\
\diagup \; \diagdown \\
\text{CH}_3 \quad \text{CH}_3
\end{array}
$$

4. L-phenylalanine and L-tyrosine **5.** L-phenylalanine and L-tyrosine **6.** Pyridoxal phosphate is a coenzyme for transaminations, decarboxylations, and other reactions of amino acids. **7.** Phenylketonuria is caused by a reduction or absence of phenylalanine hydroxylase activity and a resulting inability to convert phenylalanine to tyrosine. **8.** The urea cycle converts the toxic ammonium ion into nontoxic urea for excretion. **9.** A ketogenic amino acid is one from which acetoacetate (a ketone body) is made.

28-3 PROTEIN NUTRITION

LEARNING OBJECTIVES

1. Define turnover rate, half-life, nitrogen pool, essential amino acid, nonessential amino acid, positive nitrogen balance, negative nitrogen balance, biologic value, and incomplete protein.
2. Explain what is meant by the phrase *dynamic state of tissue protein.*
3. Describe the cause of kwashiorkor.

Protein nutrition is very important to our well-being. Proteins are constantly being broken down and replaced. Amino acids are constantly being broken down. We need essential amino acids in our diets to make new protein molecules.

28-3.1 THE DYNAMIC STATE OF TISSUE PROTEIN

Amino acids are required for synthesis of tissue proteins. They are required even when a person is not growing because they are needed for replacement of proteins. The adult body makes about three billion new cells each minute. The growing body makes many more.

As we have discussed several times before, proteins, like all other tissue components, are in what is called a dynamic state. Proteins are constantly being broken down and replaced by newly synthesized protein, a process that uses some of the energy (ATP) generated in mitochondria. Turnover of proteins occurs in plant tissues, but it is less dynamic and less dramatic than in animals. The **turnover rate** of tissue protein is measured as its *half-life*. *Half-life* is the time required for one-half of the protein molecules to be replaced by new molecules. Typical half-lives are as follows:

> The **turnover rate** is the rate at which molecules are broken down or used up and replaced by new molecules.

Tissue Protein	Half-Life
Plasma	6 days
Liver	10 days
Muscle	180 days
Collagen	3 years

Plants use inorganic nitrogen to synthesize first amino acids and then proteins, sometimes getting assistance from nitrogen-fixing bacteria in this process. Plants do not excrete nitrogen.

28-3.2 NITROGEN POOL

A *metabolic pool* is a group of compounds that can provide a chemical entity for synthesis or can be metabolized to end products. A metabolic pool can be likened to a car pool, a group of cars that can be called upon for transportation when needed. In this case, the **nitrogen pool** consists of those compounds that can donate nitrogen for various syntheses or can be catabolized to excretory products. The components of the nitrogen pool vary from tissue to tissue, but are primarily amino acids.

> The **nitrogen pool** is the available and reactive compounds that are involved in the reactions of nitrogen metabolism.

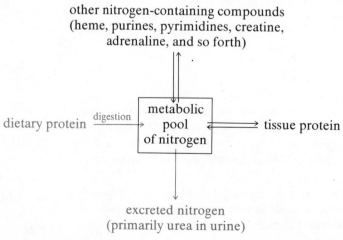

other nitrogen-containing compounds
(heme, purines, pyrimidines, creatine,
adrenaline, and so forth)

dietary protein —digestion→ metabolic pool of nitrogen ⇒ tissue protein

excreted nitrogen
(primarily urea in urine)

28-3.3 ESSENTIAL AMINO ACIDS

To synthesize proteins, both essential (or indispensable) and nonessential (or dispensable) amino acids are required. An amino acid is referred to as essential when the animal (in our case, the human) lacks the ability to make its corresponding α-keto acid. As we learned in Section 28-2.1, if the carbon skeleton is available, the amino acid can be formed by transamination. That is, the amino acid can be formed by the transfer of an amino group from another amino acid to its carbon skeleton (α-keto acid) to form the new amino acid.

The number of essential amino acids varies both from organism to organism and with the age of the organism. Those that are essential for humans are given in Table 28.5. All nonessential amino acids and some essential amino acids are glycogenic (compare Table 28.3 with Table 28.5).

Essential (indispensable) amino acids are the amino acids that are not synthesized at all or are not synthesized rapidly enough for normal growth and maintenance and, therefore, must be supplied in the diet.

Nonessential (dispensable) amino acids are the amino acids that may be synthesized in the body rapidly enough to meet the needs for protein synthesis.

Table 28.5 Essential and Nonessential Amino Acids (for Humans)

Essential*	Nonessential
Tryptophan	Alanine
Threonine	Aspartic acid
Histidine†	Asparagine
Arginine†	Cysteine
Lysine‡	Cystine
Leucine	Glutamic acid
Isoleucine	Glutamine
Methionine‡	Glycine
Valine	Proline
Phenylalanine	Serine
	Tyrosine

*Can be remembered by remembering the initials T. T. Hallim V. P. *or* Millpath T. V.

†Not necessary for nitrogen balance in the adult because some can be made, but necessary for growth.

‡Older people need twice as much as younger people.

28-3.4 NITROGEN BALANCE

An exact *nitrogen balance* occurs when the daily intake of nitrogen equals that lost in feces, urine, and sweat. However, the term *nitrogen balance* does not necessarily refer to an exact balance. Nutritionists speak of *positive nitrogen balance,* in which the amount of nitrogen ingested exceeds that excreted, and *negative nitrogen balance,* in which the intake is less than the loss.

A positive nitrogen balance is found in growing children (including a growing fetus) and in persons recovering from a disease in which there has been a loss of tissue mass. A negative nitrogen balance is found under the following conditions.

1. inadequate intake of protein
2. dietary deficiency of one or more essential amino acids
3. incomplete digestion of proteins or absorption of amino acids (high fecal nitrogen)
4. excessive catabolism of proteins in tissues (starvation)
5. loss of protein (for example, hemorrhage)

A dietary deficiency of one or more essential amino acids results in a negative nitrogen balance because cells cannot use other amino acids for protein synthesis if one is absent. For a cell to try to make protein molecules when an amino acid is missing would be like a typesetter trying to set the type for this sentence if the letter m were not available. The first 15 letters could be set, but the sentence could not be completed until a letter m was obtained. So protein molecules cannot be made if one of the amino acids is unavailable. Fragments could be started, but synthesis would stop until the missing amino acid was provided.

28-3.5 BIOLOGICAL VALUE OF PROTEINS, DEFICIENT PROTEINS, AND PROTEIN DEFICIENCY DISEASE

The *biologic value* of protein is a measure of the value of the protein in supplying amino acids for tissue-protein synthesis. To make protein, a cell needs a balance of essential and nonessential amino acids. Essential amino acids must be part of the dietary protein since they cannot be synthesized at a rate necessary for normal growth or repair. Also, there must be an adequate supply of nonessential amino acids to supply the amino groups that can be used to make other nonessential amino acids needed for protein synthesis. The biologic value of a protein takes into account the ratio of essential amino acids in the protein, and its ease of digestion. In order for dietary proteins to have biologic value, they must provide all essential amino acids. Animal proteins contain amino acids in about the same ratio as is found in human proteins. Therefore, animal proteins are, in general, well balanced and have a high biologic value. An exception is gelatin, which lacks the essential amino acid tryptophan. Gelatin is 100% deficient in tryptophan.

Plant proteins, in general, have a low biologic value because most plant proteins have a low level of one or more essential amino acids. That is, most plant proteins are deficient in one or more essential amino acids. Such proteins are called *incomplete proteins.* Zein, the predominant protein in corn, is 100% deficient in lysine and tryptophan. Corn contains other proteins in addition to zein. Whole grains (corn, wheat, rice, and sorghum) are 50–60% deficient in lysine. Soybean protein has the highest biologic value of the common plant proteins.

Two-thirds of the world's population subsists almost entirely on vegetable diets. Plant proteins in these diets should be mixed. However, a proper mixture of vegetables and grains can provide all the necessary protein. Unfortunately, the diet is not varied in most cases, and therefore, the proteins are not mixed. Most of the people in the world who live on vegetable diets live almost entirely on rice, corn, wheat, sorghum, or other cereal grains. In some areas of the world, 40% of all children die before reaching the age of four. One factor behind this statistic is the lack of sufficient protein of good quality in their diets. The protein-defficiency disease of children is called *kwashiorkor.*

Even in this country, there are population segments that subsist on suboptimal protein intake for economic reasons or because of poor dietary practices. These groups may include old people; poor people; people on reducing diets of poor quality; alcoholics; and growing children who, because of food preferences, consume a low-protein diet consisting largely of cereal foods. Wheat is the primary cereal grain in the diet of most Americans.

In general, animal proteins are more rapidly and completely hydrolyzed than are plant proteins; and, in general, cooked (denatured) proteins are more rapidly and completely hydrolyzed than are uncooked proteins. This is one reason we eat cooked meat rather than raw meat. In addition, raw soybeans and raw egg white contain rather powerful inhibitors of the proteolytic enzyme trypsin (see Section 28-1.3). Raw egg white also contains a basic protein, avidin, that combines with and inactivates the B vitamin biotin (see Section 27-3.1).

Biologic values have been placed on a relative scale; whole-egg protein or egg albumin is scored at 100. The recommended daily requirement for protein is calculated for a biologic value of 70, which is the average biologic value of protein in the average North American diet. Corrections are necessary for dietary proteins of different average values. An adult has a recommended daily requirement of 65 g of protein. If the dietary protein is of plant origin, with a biologic value of 40 instead of the 70 assumed, then the daily intake must be increased to $(65 \times 70/40)$ g or 91 g.

PROBLEMS

10. The time it takes for half of the molecules of a substance to be destroyed or used up and replaced by new molecules of a substance is called the _____ of the substance.

11. Pyridoxamine, glutamate, and glutamine are examples of compounds making up the
    _____.

12. When a person takes in more nitrogen than is excreted, the person is said to be in a(n)
    _____.

13. Amino acids that can be synthesized in the human body in amounts sufficient to meet the body's needs are called _____.

ANSWERS TO PROBLEMS   **10.** half-life   **11.** nitrogen pool   **12.** positive nitrogen balance   **13.** nonessential amino acids

---

SUMMARY

Nitrogen is ingested by humans largely in the form of protein. Protein is hydrolyzed in the stomach and the small intestine by proteases. The enzymes that digest protein are made in inactive forms called proenzymes, which are converted into active enzymes in the digestive tract itself.

In the stomach, pepsin cleaves proteins into peptides of various sizes. These peptides are cleaved to form smaller peptides in the small intestine by the action of trypsin and chymotrypsin. The resulting oligopeptides are converted into amino acids by carboxypeptidase and aminopeptidase. The amino acids are then absorbed and carried by the blood to the liver, where most reactions of amino acids occur.

The principal reaction of amino acids is transamination, a reversible reaction that can be used for both the synthesis and breakdown of amino acids.

$$\text{L-glutamate} + \alpha\text{-keto acid} \rightleftharpoons \alpha\text{-ketoglutarate} + \text{amino acid}$$

transamination

This reaction, and most others of amino acids, requires pyridoxal phosphate as a coenzyme.

L-Glutamate can also be deaminated to form α-ketoglutarate. By coupling transamination and deamination, nitrogen can be removed from amino acids. This nitrogen (ammonia) must be excreted to maintain a nitrogen balance. Because ammonia is very toxic, it is converted into urea, a water-soluble, nontoxic compound that is excreted in urine.

Each nonessential amino acid has a pathway of synthesis (anabolism), and all amino acids have a pathway (or pathways) of breakdown (catabolism). There are, therefore, a large number of individual pathways. However, some general aspects of amino acid metabolism can be described. Amino acids can be classified as essential or nonessential, depending upon their anabolism, and as glycogenic or ketogenic, depending on their catabolism. Essential amino acids are those for which there is no pathway (or an inadequate pathway) for their synthesis in the human body and must, therefore, be supplied in the diet. Nonessential amino acids are those that can be made in the human body. Glycogenic amino acids can be used to form carbohydrate (D-glucose). Ketogenic amino acids form ketone bodies.

Amino acids can also be decarboxylated to form amines. Many of these amines formed by decarboxylation of amino acids have potent and important physiological activities.

ADDITIONAL PROBLEMS

14. Complete the following table.

Vitamin	Coenzyme	Reaction
Thiamine		
	NAD	
		Carboxylations
Pyridoxal		
Lipoic acid		
	Coenzyme A	
	FAD	

15. The hormone insulin is a polypeptide. Explain why it cannot be administered orally to diabetics but must be given by injection.

16. Pepsinogen belongs to what class of molecules?

17. What is the function of enterokinase in digestion?

18. Complete the following equation with structural formulas.

$$\text{L-glutamate} + \text{pyruvate} \rightleftharpoons$$

19. Explain what is meant by the phrase *dynamic state of tissue protein.*

20. What is an incomplete protein?

APPENDIX A

SCIENTIFIC NOTATION AND SIGNIFICANT FIGURES

A-1 SCIENTIFIC NOTATION

Chemists often use very large and very small numbers. For example, the number of water molecules in a 12-ounce can of a soft drink is approximately 10,000,000,000,000,000,000,000,000 (10 septillion) and the diameter of a hydrogen atom is approximately 0.00000001 (1 one-hundred millionth) cm. Reading these numbers or doing calculations with them as written here is a very difficult task. To alleviate the problem scientists write such numbers in what is known as *scientific notation.*

To understand scientific notation we must understand how numbers can be written in terms of powers of other numbers. For example, the number 16 can be written as $4 \times 4 = 4^2$ and 1000 as $10 \times 10 \times 10 = 10^3$. For these numbers the 2 and the 3 are the powers (or exponents) to which the base numbers 4 and 10 are raised to obtain the desired numbers 16 and 1000, respectively. The numbers 4^2 and 10^3 are referred to as exponential numbers.

Any base can be used in scientific notation; however, a base of 10 is usually preferred. When a base of 10 is used, the number system is called a decimal system. Some simple examples of the decimal system follow.

$$1,000,000 = 10 \times 10 \times 10 \times 10 \times 10 \times 10 = 10^6$$
$$1000 = 10 \times 10 \times 10 = 10^3$$
$$100 = 10 \times 10 = 10^2$$
$$10 = 10 = 10^1$$
$$1 = (10)^0 = 10^0$$

In the above examples, it is important to note that any number (in this case 10) raised to a power of zero is mathematically defined as 1, as shown by the last equality. The number of molecules of water in a 12-ounce can is expresed as 10^{25}, using scientific notation.

A similar procedure is used for numbers less than 1, with one modification. This is illustrated by the number one one-hundredth (0.01), which can be written

$$0.01 = \frac{1}{100} = \frac{1}{10 \times 10} = \frac{1}{10^2}$$

To represent numbers such as this, we follow the convention that an exponential number appearing in the denominator (on the bottom) of a fraction is equivalent to the same number written in the numerator (on top) with a negative sign in front of the exponent. Thus we have

$$0.01 = \frac{1}{10^2} = 10^{-2}$$

Using this procedure, we can write the following equalities:

$$0.000001 = \frac{1}{10^6} = 10^{-6}$$

$$0.001 = \frac{1}{10^3} = 10^{-3}$$

$$0.01 = \frac{1}{10^2} = 10^{-2}$$

$$0.1 = \frac{1}{10^1} = 10^{-1}$$

Using scientific notation, we say that the diameter of a hydrogen atom (0.00000001 cm) is equal to 10^{-8} cm.

For more complex numbers such as 5000 and 0.000000005 we follow a similar procedure, but the number written in scientific notation consists of both an exponential part, as before, and a coefficient portion that precedes the exponential part. This is illustrated for the number 5000 by $5000 = 5 \times 1000 = 5 \times 10^3$ and for 0.000000005 by $0.000000005 = 5 \times 0.000000001 = 5 \times 10^{-9}$.

Rules summarizing the procedure to be followed in writing any number in scientific notation follow.

Numbers Greater Than 1

As an example of a number greater than 1, we will consider an important number in chemistry known as *Avogadro's number*. Avogadro's number is 602,000,000,000,000,000,000,000. (More is said about this number in Chapter 4 where the mole concept is introduced.)

1. *Move the decimal to the right of the first digit that appears in the number as read from left to right.* For Avogadro's number, we have

$$\underline{6020000000000000000000000.}$$

In general, numbers in which there is no decimal may be assumed to have a decimal at the end of the number, as read from left to right. For example, 5280 can be considered to be 5280., where a decimal has been added.

2. *Count the number of digits between the original and final positions of the decimal (n).* For Avogadro's number, $n = 23$.

3. *Drop the unimportant zeros.* The meaning of this will be clear when significant figures are introduced in Section A-2. In simple terms, this means that we drop zeros that only tell us the magnitude, or size, of the number and are not digits that are actually known. For example, in Avogadro's number the zero between the 6 and the 2 is actually known and is retained. The remaining zeros simply tell us how large the number is and are dropped. Thus, only the 6.02 is retained.

4. *Multiply the number obtained in step 3 by 10^n.* Using the result obtained in step 2, we have 6.02×10^{23}. When a number appears in this form, with the coefficient having a value between 1 and 10, the number is considered to be in standard, or proper, form. Occasionally the decimal appears elsewhere in numbers written in scientific notation. For example, Avogadro's number may be written 60.2×10^{22}. However, scientists usually report numbers in standard form unless there is some special reason for writing them in another form.

Numbers Between Zero and One

To illustrate the procedure to be followed for numbers between zero and one, we consider the mass of a single hydrogen atom, 0.000000000000000000000000167 g.

1. *Move the decimal to the right of the first nonzero digit that appears in the number when read from left to right.* For the above number we have

$$0.000000000000000000000000167$$

2. *Count the number of digits between the original and final positions (m).* For the number above, $m = 24$.

3. *Drop all zeros to the left of the nonzero digits.* For the number considered, this leaves 1.67. All unimportant zeros have been dropped as in the case of Avogadro's number. The zero to the left of the decimal in the original number is included to emphasize the fact that the number is less than 1. For the number considered, this is not really necessary, but for numbers such as 0.123, the appearance of the zero immediately alerts us to the fact that the number is less than 1, not 123. In most cases, scientists include this zero regardless of the size of the number as long as the number is less than 1.

4. *Multiply the final number by 10^{-m}.* Thus we have 1.67×10^{-24}.

A third type of mathematical manipulation involving scientific notation deals with *moving the decimal within the coefficient portion of a scientific number.* Two rules apply to such changes:

1. *For each position the decimal is moved to the right in the coefficient, the exponent is decreased by 1.* This is the procedure that was followed above when 6.02×10^{23} was changed to 60.2×10^{22}.

2. *For each position the decimal is moved to the left in the coefficient, the exponent is increased by 1.* Using this rule, we say

$$1.67 \times 10^{-24} = 0.0167 \times 10^{-24+2} = 0.0167 \times 10^{-22}$$

A-2 SIGNIFICANT FIGURES

To illustrate the meaning of the term *significant figures,* let us consider the reporting of the attendance at a sports event. When a figure such as 50,000 people is reported, most of us realize that there are not exactly 50,000 individuals there but that there are approximately 50,000 people. In reality, there may be 47,231, 53,918, or some other number near 50,000 in attendance.

Such approximations are necessary not only in sports but in all measurements, including scientific measurements. The number of molecules of water in a 12-ounce can and the diameter of a hydrogen atom, given in Section A-1, really are only approximate numbers.

When dealing with numbers such as the attendance at a sports event or the value of the national debt, it is not particularly important to most of us that we have an exact value or that we know how accurately the number is known. In science, however, how accurately the number is known is frequently important. The accuracy is indicated by the number of significant figures. Thus, *significant figures are the digits in the numerical value of a quantity that are precisely, or accurately, known.* They are all the digits that appear in a number except zeros that serve only to indicate the position of the decimal point.

As an example of this, let us consider that a friend has just determined his or her height and indicates that it is 69 inches (5 feet 9 inches). Measuring to the nearest inch is fairly easy to do and so when our friend says 69 inches, we can assume that he or she is closer to 69 than to 68 or 70 inches tall. In this case both the 6 and the 9 are accurately known; therefore, there are two significant figures in the number.

A second example of significant figures is that the number of feet in a mile is 5280. In a number such as this, all four digits are significant because by the definitions of *foot* and *mile,* there are exactly 5280 feet per mile. In this case we would say there are four significant figures. Actually there are an infinite number of significant figures in this case because the relationship 1 mile = 5280 feet is exact. That is, we can really write 1 mile = 5280.000 . . . feet. In general, however, only the four digits shown are considered.

As already stated, *significant figures are all the digits that appear in a number except zeros that serve only to indicate the position of the decimal point.* Although this rule is probably adequate for determining the number of significant figures, a few simple additional rules may be helpful. These rules, like the above rule, apply only when a number appears with the correct number of significant figures.

1. *All nonzero digits are significant regardless of the magnitude, or size, of the number.* There are three significant figures in the number 145,000 and four significant figures in 0.0009215.
2. *All zeros appearing between nonzero digits are significant.* An example of this is that the zero in 6.02 is significant and all three zeros in 1010101 are considered to be significant.
3. *Zeros to the right of the decimal point following nonzero digits are significant.* Using this rule, we know that there are four significant figures in 123.0 and in 1.230.
4. *Zeros preceding nonzero digits in numbers less than 1 are not significant.* Using this rule, we know that there are only two significant digits in 0.0014 and three in 0.000521.

5. *Zeros following nonzero digits in numbers greater than 1 are not significant unless a decimal point appears in the number.* An example of this is that the number 546,000. has six significant digits, while 546,000 is considered to have only three.

Determining the number of significant figures in numbers of the type described in rule 5 above frequently is difficult because people may not properly use decimal points to specify the number of significant digits. Most of us would probably realize that the number 50,000, referred to earlier, does not contain five significant figures. However, we might wonder whether there are one, two, or three significant figures in the number. The accuracy of the number depends upon how it is determined. If someone simply looks at the crowd and makes an estimate, there is probably only one significant figure. However, if the number of seats in each portion of the stadium is known or there is a capacity crowd, there may be two, three, or even more significant figures in the number 50,000, referred to earlier.

A similar problem arises with respect to a figure such as 5280 feet. The number is frequently written without a decimal and unless we know that because of the way it is defined the number is exact (for example, 5280 feet = 1 mile), we may assume that there are only three significant figures.

Scientific notation provides a convenient way of indicating the significant figures in a number. For example, the one significant figure in 50,000 is quite easily indicated when it is written 5×10^4 and three significant figures are represented by 5.00×10^4, as discussed in rule 3. Likewise, the four significant figures in 5280 are obvious when the number is written as 5.280×10^3.

In this text, the authors have not asked the student to use significant figures. However, the answers to problems that are provided by the authors follow the rule that *answers to a calculation involving multiplication or division should not contain more significant figures than appear in the least accurate number involved in the calculation.* For example,

$$\frac{11.2 \times 54.11}{2.3 \times 19.724} = 13$$

Only two significant figures are retained in 13 because there are two significant figures in 2.3. Similarly, the authors have followed the rule that *answers to a calculation involving addition and subtraction should not contain digits that have been added or subtracted in a column where a digit is not accurately known in one or more of the numbers involved in the calculation.* For example, in the calculation

$$
\begin{array}{r}
11.2 \\
54.11 \\
2.3 \\
+\ 19.724 \\
\hline
87.3
\end{array}
$$

the answer is reported to only the nearest tenth because two of the numbers involved in the calculation are only known to this degree of accuracy.

APPENDIX B

VAPOR PRESSURE OF WATER AT DIFFERENT TEMPERATURES

Temperature, °C	Pressure		Temperature, °C	Pressure	
	mmHg	atm		mmHg	atm
0	4.6	0.0061	23	21.1	0.0278
1	4.9	0.0065	24	22.4	0.0295
2	5.3	0.0070	25	23.8	0.0313
3	5.7	0.0075	26	25.2	0.0332
4	6.1	0.0080	27	26.7	0.0351
5	6.5	0.0086	28	28.3	0.0372
6	7.0	0.0092	29	30.0	0.0395
7	7.5	0.0099	30	31.8	0.0418
8	8.0	0.0105	35	42.2	0.0555
9	8.6	0.0113	40	55.3	0.0728
10	9.2	0.0121	45	71.9	0.0946
11	9.8	0.0129	50	92.5	0.1217
12	10.5	0.0138	55	118.0	0.1553
13	11.2	0.0147	60	149.4	0.1966
14	12.0	0.0158	65	187.5	0.2467
15	12.8	0.0168	70	233.7	0.3075
16	13.6	0.0179	75	289.1	0.3804
17	14.5	0.0191	80	355.1	0.4672
18	15.5	0.0204	85	433.6	0.5705
19	16.5	0.0217	90	525.8	0.6918
20	17.5	0.0230	95	633.9	0.8341
21	18.7	0.0246	100	760.0	1.0000
22	19.8	0.0261	105	906.1	1.1192

INDEX

NOTE: The letters d, f, t, p, and n following page numbers indicate definition, figure, table, problem, and footnote, respectively.